Risk Management in Civil Infrastructure

Civil Infrastructure Health and Sustainability
Series Editors: Mohammed M.Ettouney and Sreenivas Alampalli

Risk Management in Civil Infrastructure
Mohammed M. Ettouney and Sreenivas Alampalli

Multihazard Considerations in Civil Infrastructure
Mohammed M. Ettouney and Sreenivas Alampalli

Risk Management in Civil Infrastructure

Mohammed M. Ettouney
Sreenivas Alampalli

CRC Press
Taylor & Francis Group
Boca Raton London New York

CRC Press is an imprint of the
Taylor & Francis Group, an **informa** business

CRC Press
Taylor & Francis Group
6000 Broken Sound Parkway NW, Suite 300
Boca Raton, FL 33487-2742

ISBN-13: 978-1-4822-0844-3 (hbk)

Visit the Taylor & Francis Web site at
http://www.taylorandfrancis.com

and the CRC Press Web site at
http://www.crcpress.com

Dedication

To Cairo, Egypt: The mother of the whole universe.

Mohammed M. Ettouney

To my wife, Sharada, and son, Sandeep, for their love and patience.

Sreenivas Alampalli

Contents

Preface

A few years ago, during a discussion on probabilistic issues involving the propagation of uncertainties during progressive collapse events of buildings, an infrastructure owner raised a question: What am I going to do with this information? This question illustrates the essentiality of risk management in the infrastructure field. In practice, decisions are based on consequences that are not always considered during the design and analysis process. As the infrastructure ages, this disjointed process between design/analysis and consequences/costs keeps growing. This process can only result in a less-than-optimal performance of the system as well as increased costs to the owners and society at large in the long run. The paradigm of risk management, as defined in this book, aims at eliminating the disjoint throughout the system life cycle starting from the initial design phase. We also widen the reach of risk management to cover a single infrastructure asset as well as a network (or community) of assets and account for the multitudes of hazards* that might affect those individual assets or asset networks.

The concept of risk management in civil infrastructures is not new: it is practiced informally all the time. Faced with any uncertainty, humans usually establish, from their respective viewpoints, the most rewarding course of action using their past experiences or any additional tools available, such as expert opinions. The risks are observed and communicated until the uncertainty associated with these situations is reduced or resolved. These steps are a microcosm of the risk management process, albeit in a simple and subjective manner. The purpose of this book is to provide an objective dissertation of a risk management (in all of it components) process that can be applied to civil infrastructure.

Risk, at its foundation, is built to address threats, vulnerabilities, and consequences all at once in a balanced and objective manner. Chapter 1 introduces the concept of risk from several viewpoints. Initially, the chapter discusses the concept of risk and then offers a suitable definition of risk that serves us best to explore risk management concepts in civil infrastructure. We then discuss the three basic components of risk while emphasizing the universality of risk concept. The relation between risk and other popular concepts such as reliability and resilience is explored, followed by a discussion of five risk management concepts. We end the chapter with a discussion of essential attributes that need consideration for risk management and the benefits (potential return on investment) to different stakeholders for implementing risk management in an appropriate manner.

Several numerical processes, including probabilistic graph networks (GNs), will be used throughout the book. Chapter 2 provides a brief introduction to the main attributes of GNs. We start by a general offering of GNs and their main components. We then explore several subcategories of GNs that include Bayesian networks (BNs) and Markov networks (MNs) as well as chain graphs (CGs). CGs are GNs that combine some of the attributes of both BNs and MNs. We also briefly discuss an influence diagram that accommodates decisions under uncertainty and finally present a dynamic graph network (DGN) that accommodates time marching. For each GN type, we briefly discuss the theoretical basis and then offer simple examples from the field of civil infrastructure.

Chapter 3 concentrates on the first component of the risk management process, that is, risk assessment. We start the chapter by addressing several pertinent issues regarding risk assessment in civil engineering. Two solution methods that best suit civil infrastructure applications are weighted averages and GNs. We then offer several case studies to showcase the use of these methods for the risk assessment of single assets as well as a community of assets. We end the chapter by exploring resilience assessment, which we will argue to be a special manifestation of risk assessment.

* The subject of multihazards will not be covered in this book. See Ettouney and Alampalli (2016) for detailed coverage of the subject.

Risk acceptance is the next, and perhaps the most difficult, component of risk management. Chapter 4 begins with an overview of risk acceptance methods. Even though risk acceptance methods are subjective in nature, we differentiate between subjective and semisubjective methods. We then offer several case studies of interest to civil infrastructure stakeholders. These include the acceptance at different limit states: lower limit states, ductile limit states, or brittle limit states. We also offer case studies that use decision under uncertainty processes. The chapter ends by exploring the grandfathering practice in the field of civil infrastructure from a risk viewpoint.

Chapter 5 is devoted to risk treatment processes. This chapter first discusses phases of risk treatment and differences between risk treatment and risk mitigation. The remainder of the chapter covers different processes of risk treatment categories through specific and simple examples. Among the several case studies presented in this chapter, we compare weighted averages and GN methods for risk prioritization. Several case studies offer applications of decisions under uncertainties, which are the main attributes of any risk treatment effort.

Because all civil infrastructures are meant to be in service for a relatively long period, it is reasonable to assume that assessed risk, accepted risk thresholds, and risk treatment change with time. This explains the importance of the next risk management component, that is, risk monitoring. Chapter 6 covers some of the salient aspects of risk monitoring. We discuss briefly the monitoring attributes of risk components, with special attention given to consequence monitoring because it is often overlooked. We then offer practical case studies to illustrate various aspects of risk monitoring.

Risk communication often constitutes, in our opinion, the most important component, but it is neither well understood nor well practiced. Chapter 7 covers risk communication in the civil infrastructure field. The chapter first explores three important tenets of risk communications: its centrality in the risk management process, the different theories that aim at improving understanding the nature of risk communications, and the rich and varied objectives of risk communications. We follow that by delving in-depth using the Lasswell model of risk communications, especially as it applies to civil infrastructures. We then offer a communications model for the objective assessment of the effectiveness of the risk communications based on the Laswell model. The chapter ends with a discussion of the specifics of communications of risk (vulnerability, threat/hazard, consequences, reliability, and/or exposure) and risk management components. We argue that communicating each risk component might require different considerations. Similarly, we show that communicating different risk management components may require different communication strategies.

Chapter 8 explores the varied risk management applications in the field of civil infrastructures. This chapter covers applications in bridge security, critical findings practice, bridge network management, cascading effects, redundancy, community resilience, deteriorated structures, and functionally deficient structures. Through these applications and case studies, we show that risk management offers an objective solution to problems that have been generally approached subjectively.

Mohammed M. Ettouney
West New York, New Jersey

Sreenivas Alampalli
Loudonville, New York

Acknowledgments

This book was possible only with help and encouragement from many individuals. I am indebted to my friends, Dr. Amar Chaker, Dr. Elsayed A. Elsayed, Dr. Raymond (Ray) Daddazio, Mr. Michael (Mike) Udvardy, and Mr. Amr Aly, for all of their support during the preparations of this book. Special thanks are due to Mr. Eric Letvin, Esq., for the long and beneficial collaborations we have had over the years, with regard to risk management of civil infrastructure.

This book is also the result of the close collaboration with my friend, Dr. Sreenivas Alampalli, without whom it could not have been materialized. The help and guidance of Mr. Joe Clements in managing the whole project and the editorial staff of CRC Press/Taylor & Francis Group is deeply acknowledged and appreciated.

Finally, and most importantly, there are no words that can describe the immense gratitude, appreciation, and acknowledgments for Ms. Milagros (Mila) Kennett. Mila is the pioneer of risk management in the field of civil infrastructure. Those very pioneering efforts and continued innovation and friendship are definitely the very foundation upon which this book was founded.

Mohammed M. Ettouney

It was a pleasure working with Dr. Mohammed Ettouney in writing this book. Without his interest and push, this might not have happened. This book has benefited from the experience gained by working with many from government, industry, and academia over the years. I thank Joseph Clements and his colleagues from CRC Press/Taylor & Francis Group for publishing this book and the support they have provided during the publication process.

Given that research and publications are a hobby of mine, any time spent writing this book took time away from my family. I am deeply indebted to the support, encouragement, and patience from my wife, Sharada Alampalli, and son, Sandeep Alampalli, during this period.

Sreenivas Alampalli

Authors

Mohammed M. Ettouney, PhD, PE, MBA, F AEI, Dist M ASCE, received an honors award from the National Institute of Building Sciences (2015) for developing an Advanced Materials Database, Owners Performance Methods, and co-developing School Safety procedures. His developments of risk and resilience processes for projects related to the US Department of Homeland Security earned him the Anachin Innovation Award (2013). He is a distinguished member in the American Society of Civil Engineers (ASCE) since 2011). He was conferred the Innovators Award in 2008 by the New Jersey Inventors Hall of Fame after he was nominated to receive such a great honor by the ASCE. Dr. Ettouney also received the Homer Gage Balcom Lifetime Achievement Award from the Metropolitan Section of ASCE (2008). He won the Project of the Year Award, Platinum Award (2008) for the New Haven Coliseum Demolition Project (ACEC, New York). He is a fellow of the Architectural Engineering Institute (AEI) and a distinguished member in the ASCE. Among his other recent achievements are the pioneering work on "Theory of Multihazards of Infrastructures;" the "Theory of Progressive Collapse;" pioneering work in multihazards/multidisciplinary evaluation of risk and resilience of buildings, tunnels, bridges, and transit stations; and an innovative green design method for protecting utilities from demolition/blasting (City of New Haven, Connecticut). He has professional interest in diverse areas of engineering mechanics, structural engineering, architectural engineering, and asset management as demonstrated by the list of his publications, invited presentations, seminars, and sessions organized during national/international conferences, besides his membership in the different professional organizations.

Dr. Ettouney is a consulting engineer. He received his Doctor of Science in structural mechanics from the Massachusetts Institute of Technology, Cambridge, Massachusetts, in 1976. Since then, his interest in the field, both as a practitioner and a researcher, has been in multihazards safety of structures, progressive collapse of buildings, uncertainties in structural stability, blast mitigation of numerous buildings around the world, underwater acoustics, seismic analysis and design, structural health monitoring, risk and resilience management of civil infrastructures, and innovative concepts such as probabilistic boundary element method, scale-independent elements, and framework for evaluation of lunar-based structural concepts. He is a former president and former member of the board of governors of AEI, member of the board of directors of the Building Security Council, and member of several technical committees on building/infrastructure security, earthquake hazards, architectural engineering, nondestructive testing, and structural health monitoring. He chaired several symposia and national engineering-related conferences.

Dr. Ettouney has authored or coauthored more than 400 publications and reports and contributed to and authored several books. He introduced several new practical and theoretical methods in the fields of earthquake engineering, acoustics, structural health monitoring, progressive collapse, blast engineering, and underwater vibrations. He co-invented the "seismic blast" slotted connection. More recently, he introduced the "economic theory of inspection," "general and special theories of instrumentation," coined the concept of "resilience management," and numerous principles and techniques in the field of infrastructure health—all pioneering efforts that can help in developing durable infrastructures at reasonable costs.

Sreenivas Alampalli, PE, MBA, is director of the Structures Evaluation Services Bureau at the New York State Department of Transportation (NYSDOT). Before taking up the current responsibility in 2003, Dr. Alampalli was director of the Transportation Research and Development Bureau, where he worked for about 14 years in various positions. He also taught at Union College and Rensselaer Polytechnic Institute as an adjunct faculty.

Dr. Alampalli earned his PhD and MBA from Rensselaer Polytechnic Institute, Troy, New York, his MS from the Indian Institute of Technology (IIT), Kharagpur, India, and his BS from Sri Venkateswara University, Tirupati, India. His interests include infrastructure management, innovative materials for infrastructure applications, nondestructive testing (NDT), structural health monitoring, and long-term bridge performance. Dr. Alampalli is a fellow of the American Society of Civil Engineers (ASCE), American Society for Nondestructive Testing (ASNT), and International Society for Health Monitoring of Intelligent Infrastructure (ISHMII). He was the recipient of the Bridge NDT Lifetime Service Award in 2014 from ASNT for outstanding voluntary service to the bridge and highway NDT industry. In 2013, he also received ASCE's Henry L. Michel Award for Industry Advancement of Research. Other notable awards he has received include the ASCE Government Civil Engineer of the Year in 2014 and the prestigious Charles Pankow Award for Innovation from the Civil Engineering Research Foundation in 2000. He has authored or coauthored more than 250 technical publications, including two books on infrastructure health in civil engineering.

Dr. Alampalli is an active member of several technical committees in the Transportation Research Board (TRB), ASCE, and ASNT. He currently chairs the TRB Standing Technical Committee on Field Testing and Nondestructive Evaluation (NDE) of Transportation Structures. He has served as the TRB representative for NYSDOT and also as a member of the National Research Advisory Committee (RAC). He is a book review editor of the ASCE's *Journal of Bridge Engineering* and serves on the editorial boards of the journals *Structure and Infrastructure Engineering: Maintenance, Management, Life-Cycle Design and Performance* and *Bridge Structures: Assessment, Design and Construction.*

1 Introduction

1.1 OVERVIEW

1.1.1 Risk Management Landscape

Recent devastating man-made and natural hazards brought to focus the need for civil infrastructures to implement measures to reduce their harmful effects. Implementing many such measures, whether physical or operational, is very expensive and could pose a major problem to infrastructure stakeholders. Budget and personnel resource limitations in the current environment make this problem even more critical. To balance the need for adequate performance at optimal/reasonable costs, risk management can offer the best, perhaps only, management paradigm to face the challenges. This chapter introduces the major components of risk and risk management.

It should be recognized that only by finding an appropriate balance between expected performance levels and cost can an expected level of service from the infrastructure during its life span be realized. Risk, at its foundation, is built to address threats, vulnerabilities, and consequences all at once in a balanced and objective manner. Thus, risk management is the appropriate venue for addressing this issue. To show how risk can assist in overcoming this seemingly impossible task, the present chapter introduces the concept of risk from several viewpoints. Initially, the chapter discusses the concept of risk and then offers a suitable definition of risk that serves us best to explore risk management concepts in civil infrastructures. We then discuss the three basic components of risk and emphasize the universality of risk concept. The relation between risk and other popular concepts, such as reliability and resilience, is explored followed by a discussion of five risk management concepts. We end the chapter with a discussion of the essential attributes for risk management and the benefits (potential return on investment) to different stakeholders for implementing appropriate risk management. The layout of the chapter is shown in Figure 1.1.

1.1.2 What Is *Risk*?

We start by asking the simple and obvious question: *What is risk?* The answer is also simple, and perhaps is obvious. Risk, in its most general form, is *an estimate of a potential of a nondesirable outcome due to an event*. The many manifestations of risk include the following:

- *Hazards risk*: Hazardous events abound. These can be natural hazards such as floods or man-made hazards such as bomb blasts and hazardous material spills. Additionally, hazards can be related to health or result from climate change, and so forth. All these hazards have the potential of producing undesirable outcomes. The metric we normally use to measure these potential outcomes are hazard risks. They can be abnormal (such as earthquakes, floods, or bomb blasts) or mundane and stealthy (such as routine wear and tear, fatigue, corrosion, or freeze–thaw cycles).
- *Operational risk*: Because most decisions in all operations involve some uncertainty, some uncertainty is associated with all outcomes. Thus, some have the potential to yield less than desirable outcomes and can generically be defined as risk. We reach the conclusion that, irrespective of type of the operation, some measure of operational risk is associated with it, and the decision makers' responsibility is to reduce such an operational risk.
- *Financial risk*: This type of risk involves potential of monetary losses associated with a particular financial portfolio or project execution. Though usually associated with the finance community, it is becoming relevant to civil infrastructures because of innovative

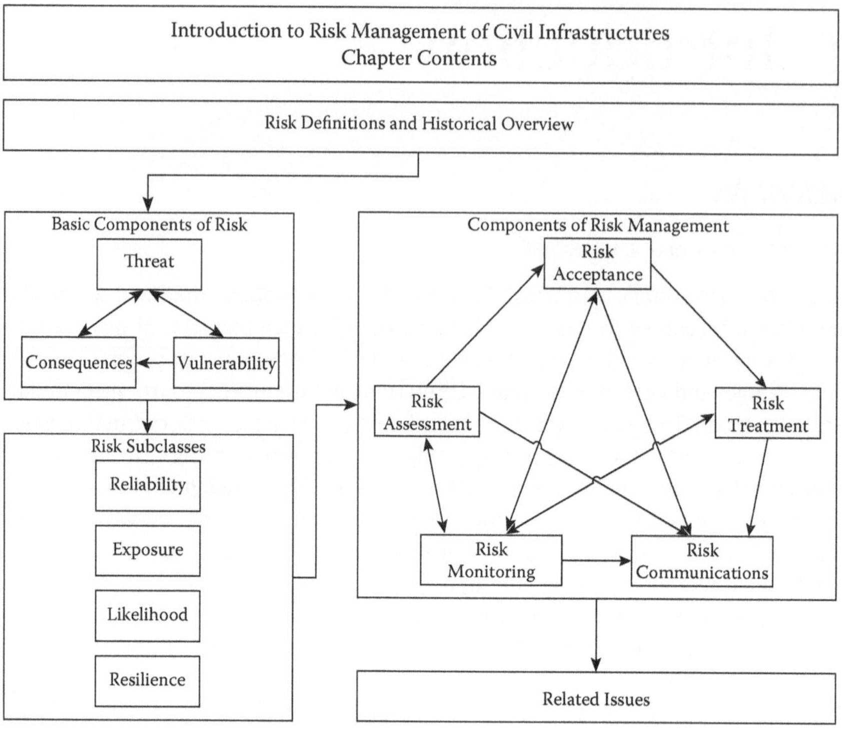

FIGURE 1.1 Chapter layout.

infrastructure project delivery/finance options, such as design–build and design–build–maintain with public–private partnerships.

- *Strategic risk*: This risk deals with the risk associated with strategic decisions and depends on the topological scale of interest (such as international, national, local, corporate, or networkwide civil infrastructure). The nature of strategic risk depends on the topology, the events, and the metric of strategic risk. For example, it can be a national security risk, international risk of health-related issues, or a risk of civil infrastructure aging in a specific network or private financing used for a public works project.

This book includes any type of the above-discussed risks that can afflict civil infrastructures. As such, risks associated with oil spills, epidemic health, or strategic international security are not of immediate interest; thus, we limit our context to risks that afflict single infrastructure assets (buildings, bridges, tunnels, etc.) and/or infrastructure networks.

1.1.3 History of Design Paradigms and Risk

Most design specifications started with working stress methodologies that did not consider risk explicitly. Load factor designs that accounted for uncertainty in loading followed. In recent years, load and resistance factor design (LRFD) accounts for uncertainties in both demand and capacities. LRFD methodologies consider reliability as one component of risk but do not explicitly account for consequences. As we are moving toward performance-based design, when reliability of a structure and consequences are accounted for, risk-based design concepts are evolving rapidly. As software and computing capabilities mature and data required for risk-based approaches are well understood, a risk paradigm in design will become more predominant. This section gives an overview of some work noted in the recent literature.

It has been fairly well recognized that possible structural responses are necessary for probabilistic performance assessment of structures and risk analysis. Incremental dynamic analysis (IDA) (Vamvatsikos and Cornell 2002) can assist in this aspect for more detailed seismic performance predictions of structures subjected to different seismic excitation levels. IDA can be used for fragility curves development and to predict failure probabilities. Tehrani and Mitchell (2014) note that this method has been successfully used for buildings and have applied this method to predict the seismic risk assessment of four-span bridges, designed using Canadian bridge design code with different column heights and diameters in Montreal. They studied the seismic risk associated with exceeding different damage states in the columns, including yielding, cover spalling, bar buckling, and structural collapse (i.e., dynamic instability). Based on the estimated mean annual probability of exceeding different damage states in the columns, they concluded that the probabilities of exceeding different damage states were reasonably low.

Uncertainties in flooding and flood-related hazards have been studied from the risk viewpoint. For example, Archetti and Lamberti (2003) examined failure, damage, and risk in the context of a debris-flow-prone basin and presented a risk based on the structural reliability methods (SRMs) of Thoft-Christiansen and Baker (1982) and Burcharth (1992). They carefully examined the risk from an individual point of view when a single person is choosing whether to carry out a single activity and the societal point of view when many people are exposed to hazardous activities (Vrijling et al. 1995; IUGS 1997; Hughes 2000). Failure, in the context of debris-flow events, was defined as any event that flows out of the channel or causes the destruction of structures on banks. Rationalizing the existing good practice, the risk acceptability criterion was set as "houses cannot be exposed to any appraisable hazard and roads can be exposed provided that the failure probability remains below a certain limit, depending on contact probability." They determined the failure probability functions and system fault trees, described all variables entering in the failure functions, and carried out a risk analysis. Based on the example of risk assessment presented, they showed the importance of some variable factors (precipitation, sediment/water concentration, basin capacity, bank elevation) and the minor importance of debris rheology. They illustrated this methodology using a case study.

A major application of risk management can be found in the civil infrastructure construction field. Mitropoulos and Namboodiri (2011) contend, based on accident statistics, that designing highly productive as well as safe operations remains a significant challenge. According to these researchers, though productivity analysis methods are available (Halpin and Woodhead 1976; Ioannou 1989; Oglesby et al. 1989; Howell et al. 1993; Martinez and Ioannou 1999) to improve productivity, ability to analyze or simulate construction activities regarding safety risks is limited; no methodology allows for the measurement of safety risk variations over the duration of an activity and evaluation of the effect of production variables on the safety risk of the operation. Hence, they introduced the task demand assessment methodology, grounded in a cognitive perspective, for measuring the accident potential of an activity based on production variables. The task demand reflects the difficulty to perform the activity safely and is based on the exposure to a hazard and the presence and level of observable task demand factors—that is, risk factors that can increase the potential for an accident. They applied the methodology to roofing and concrete paving operations to demonstrate its feasibility and applicability as well as limitations of the methodology.

Another major study of risk management in construction recognized that risk management is an important task for construction companies, in addition to being a legal requirement, because of uncertainties involved with tasks, contract duration, contract requirements, change orders, changes in legislation or regulations, design standards, bidding process assumptions, adoption of nonstandard building contracts and site conditions, different participants, construction methods, and the number of personnel involved with different skills and experience. Thus, proper risk management is required to deal with uncertainties, efficient use of resources, avoiding financial loss through appropriate mitigation measures depending on risk, and effective decision-making. Zou et al. (2010) contend that a risk management maturity model can help identify the strengths and weaknesses of an organization because it reflects the sophistication of an organization's understanding of its risk portfolio and how to manage those risks, as well as the internal business continuity systems needed to cope with and recover from their eventuality. Hence, they tried to develop a web-based risk management maturity model tool (RM3) to benefit construction organization.

The RM3 tool has five attributes, namely, management, risk culture, ability to identify risk, ability to analyze risk, and application of a standardized risk management process/system. These attributes were measured against four levels: initial, repeated, managed, and optimized. Authors chose a web-based model data collection method in development process due to cost savings, convenience, and time factors and also because it can be administered to a number of organizations located in different areas.

Construction organizations can use this model to assess and get a basic understanding of their risk management maturity and capability at the organization level or to compare risk management maturity and capability between different departments and project teams. Using the understanding gained and the determination of the current level, strategies can be developed to increase maturity level.

FEMA/US-DHS (FEMA 2005; DHS 2011a, 2011b, 2011c) developed a set of risk assessment methods. The risk model was based on a modified weighted averages approach, which accommodates interactions (context aware) between problem variables to change the weights according to the assessed variables. Several types of civil infrastructures, such as buildings, mass transit stations, and tunnels, were considered. In addition, the developers took into account the interactions of several man-made hazards (including bomb blast, chem-bio, and arson) and natural hazards (including earthquake, wind, and floods).

1.1.4 Convenient Definition for Risk

It is essential to offer a concise definition of risk at the onset of this book to guide us throughout this book. It is not an easy task, given the immense landscape that the subject of risk spans, as was made clear in our earlier discussion of the nature of risk in Section 1.1.2. It is exactly because of this extensive landscape that we are faced with a plethora of definitions of risk in the literature. Should we choose an existing definition or offer a definition that may suit us better with subsequent content in this book?

We immediately notice a risk definition in the literature does not exactly suit us: many use risk as the probability of an undesired occurrence or an event. This definition is not adequate to civil infrastructure applications in general. We need our risk definition to include uncertainties from the ground up, while being more cognizant to civil infrastructure practices. It should accommodate uncertainties, analytics, as well as be practical for civil infrastructure stakeholders, including management, and be based on historically accepted risk definitions. We also note that risk is often defined as the relationship between a particular hazard (or threat) that might degrade the performance of the infrastructure under consideration and the associated consequences (Gutteling and Wiegman 1996; FEMA 2005; NRC 2010). Such a definition seems to be useful for the civil infrastructure field. Thus, we modify it slightly to reflect that we are only dealing with civil infrastructures (whether single asset or a community/network) and define risk for the purpose of this book as: "Risk is a description of an uncertain objective or subjective *alphanumeric expression*,* which describes an *outcome* of an unfavorable,† uncertain *event*, which might degrade performance of a single (or community of) civil infrastructure *asset* (or *assets*)."

This definition is consistent with all popular risk definitions in the cited references as well as in most other communities that deal with the concept of risk, yet is also useful for our applications in the civil infrastructure field.

---

* Usually a single-valued expression.

† The definition can also be used for favorable events, for reward instead of risk.

1.2 RISK BASIC COMPONENTS

1.2.1 Overview

Risk as defined in Section 1.1.1 is obviously a complex and overreaching metric. As such, it is the foundation of a whole management paradigm, which, if followed properly, can lead to the desired performance at reasonable costs for the civil infrastructures' stakeholders. To understand and apply the risk concepts in a manageable fashion, risk can be generally divided into three basic components that will be examined in this section.

1.2.2 Threat/Hazard/Demand

Hazard indicates a situation that can afflict a particular system or component and has potential to degrade the performance of that system. Generally described in an objective manner, in some situations, it can also be described in a subjective manner. In most cases, different practitioners use the terms hazards, threats, or demands to denote specific situations. In some infrastructure security applications, target attractiveness (see DHS 2011a, 2011b, 2011c) might be used as an objective descriptor of security threats. The common denominator among all three terms is that they potentially have a degrading effect on the civil infrastructure exposed to them. Hence, for simplicity, throughout this book, we will be using the words hazards, threats, and demands interchangeably to denote a degrading event or process.

As a basic component of risk Ri, it is clear that threat, T, is directly proportional to Ri:

$$Ri \propto T \tag{1.1}$$

also

$$Ri \to 0 \tag{1.2}$$

as

$$T \to 0 \tag{1.3}$$

1.2.3 Vulnerability/Capacity

Vulnerability, V, is a popular metric that is used as one of the basic components in describing risk. It indicates the susceptibility of the system or one of its components to a particular degrading event or group of events. Vulnerability is usually expressed subjectively. As such, it can be useful to use when an analyst is considering highly subjective situations that cannot be described in an objective manner.

From a different, but related, viewpoint, the term capacity, Cap, is also a well-known descriptor in civil infrastructure design. It indicates the capability of a particular system or component in performing as it is exposed to an event. Generally described in an objective manner, it is sometimes described in a subjective manner.

Objectively, we might want to relate vulnerability to capacity. For example, the two statements:

- A particular system (or component) has a low vulnerability
- A particular system (or component) has a high capacity

are subjectively equivalent. The reverse is also true. We recognize that though capacity Cap is a term with a unique use in the community, vulnerability, V, has different definitions (Cadona 2012). Our current use of V in this book is consistent with the DHS use (DHS 2011a, 2011b, 2011c).

We prefer this as it aligns with *Cap* well, thus simplifying the utilization of both *V* and *Cap* in any risk-based management efforts. But stakeholders have to keep in mind that *V* and *Cap* have opposing meanings on the physical system such that

$$V \propto (Cap)^{-1} \tag{1.4}$$

Similar to *Threat*, risk is directly proportional to vulnerability, *V*, such that

$$Ri \propto V \tag{1.5}$$

also

$$Ri \rightarrow 0 \tag{1.6}$$

as

$$V \rightarrow 0 \tag{1.7}$$

or

$$Cap \rightarrow \infty \tag{1.8}$$

1.2.4 Consequences

Consequences, *C*, is the third basic component of risk. It describes the impact* or results of a particular event. The sources of consequences can be

- Directly related to the asset, such as costs of repair or retrofit
- Indirectly related to the asset, i.e., because of an event or particular decision, such as the social or economic impact of closing a particular bridge or building

In most situations, both types of consequences should be modeled for accurate risk management practice. Note that the direct and indirect consequences might be strongly related. However, stakeholders need to recognize the basic differences between the two consequences types and model them as well as their interactions, as pertinent.

As a basic component of risk *Ri*, consequences, *C*, is directly proportional to *Ri*, such that

$$Ri \propto C \tag{1.9}$$

and

$$Ri \rightarrow 0 \tag{1.10}$$

as

$$C \rightarrow 0 \tag{1.11}$$

Sometimes the importance (also referred to as *criticality*) of the asset under considerations is used in risk-based management. In this book, we consider importance (or criticality) as a subset

* Some authors use the term impact instead of consequences. However, we submit that both terms are/should be identical, and we use the term consequences in this book to follow the use of DHS (2011a, 2011b, 2011c).

of consequences. An example of such use can be found in (DHS 2011a, 2011b, 2011c). We also provide several examples of the use of importance as a subset of consequences in a case study (see, e.g., Case Studies 4.5 and 4.6). Note also that importance (or criticality) can also be a factor in assessing *T*. Infrastructure security is one of the instances where importance can contribute to both *T* and consequences.

Even though consequences of hazard events are one of the basic components of risk-based management of infrastructure, many do not give it proper attention. One situation where authors gave due attention to consequences can be found in Changnon (2009). This study investigated consequences (or losses) due to tornadoes in the United States based on insured tornado loss data that considers temporal changes, inflation, and insurance coverage. It considered the interaction between hazards (i.e., tornadoes, hailstorms, flooding, and hurricanes) to isolate loss only from tornadoes. It also investigated the frequency and number of tornado-related catastrophes as well as the temporal distribution, which is very important in estimating the vulnerability (and risk) of infrastructure in such events. It concluded, along with others, that the insurance loss data for tornadoes during 1949–2006 revealed much higher annual average losses than prior published estimates had indicated. Tornado-only catastrophes showed no upward trends with time and are not suggestive of increases potentially related to global climate change. The tornado, hail, and flood catastrophes had upward time trends, and this was due to increases in flooding resulting from upward trends in heavy rains during the 20th century. The number and losses of tornado and hurricane catastrophes trended upward with time, a result of upward trends in storm-producing conditions and enhanced societal vulnerability along the coasts. The study illustrates the value of a good analysis of data for proper and accurate risk assessment (vulnerability, as well as consequences) in multihazard conditions due to interactions between hazards.

1.2.5 Basic Risk Equation

We can now express risk formally as a function of its three basic components:

$$Ri = f(C, T, V) \tag{1.12}$$

The function $f(\,)$ is a complex function, which in a vast majority of situations cannot be expressed analytically. We also note that the three risk components, *C*, *T* and *V*, can also be codependent and can further complicate $f(\,)$. We devote a major part of this book in addressing the function $f(\,)$ or some of its derivatives. They are discussed in Section 1.3. Its role in the risk management landscape is summarized in Section 1.4, and its attributes are briefly mentioned in Section 1.5.

1.2.6 Risk Analogy Principle

We established that risk as a concept, metric, and management paradigm results from an event *T* (threat/hazard/demand) and affects a physical system with particular physical attributes *V* (vulnerability/capacity), which results in a set of consequences *C*. We identified the union of these three components as *risk*. Let us reflect further on this union in the context of Newton's second law of motion:

$$F = ma \tag{1.13}$$

where *F*, *m*, and *a* represents force, mass, and acceleration, respectively. In civil engineering, Equation 1.13 can be rewritten as the structural equilibrium equation:

$$P = KU \tag{1.14}$$

where P, K, and U represent force, displacement, and stiffness, respectively. In either Equation 1.13 or 1.14, we recognize an event P (or F) that is applied to a physical entity, which has an attribute of K (or m) with a resulting response measured by U (or a). There is an obvious analogy with the risk components T, V, and C as described earlier.

Given the analogy between risk components and the components of the universal equation of motion (1.13) or the structural equilibrium equation (Equation 1.14), we may want to take this analogy a bit further and postulate that risk (as represented by T, V, and C) might be an analogous to energy as represented by F, m, and a (for general physical systems) or P, K, and U (for specific mechanical/structural systems). Note that from Equation 1.13, system energy can be represented by:

$$E = f_1(F, m, a) \tag{1.15}$$

And for Equation 1.14, system energy can be represented by:

$$E = f_2(P, K, U) \tag{1.16}$$

Note how the three components of both risk situations and the physical or structural systems are analogous in how they flow from one component to the other as we observed above. In the risk equation (Equation 1.12), which is the symbolic relationship between risk and its components, risk has an affinity to its three components. As any of the three components increase, the risk will also increase. The reverse is also true. It can be easily shown that the system energy, E, has similar affinity with its three components. As any of its three components increase, the system energy increases. The reverse is also true.

Base on the above, we can state that:

1. There is an analogy between risk components and the three components of the universal law of motion, or the general structural equilibrium equation.
2. There is an analogy between risk and system energy.

We call this risk analogy principle, and it can be of value. Generically, given the primacy of energy in understanding the engineering phenomena, we can also pose risk in the same context, as it is instrumental in understanding and explaining many of the issues civil infrastructure stakeholders face. In many of the ensuing case studies, we show that risk management is routinely practiced subjectively based on experience and that it can be explained and dealt with in a quantitative manner using concepts shown later in this book.

1.3 RELIABILITY, EXPOSURE, LIKELIHOOD, RESILIENCE, AND SUSTAINABILITY

1.3.1 Overview

Even though the three basic components of risk are useful in many risk-based management applications, five risk subclasses have been in use or are emerging in the field of civil infrastructures. They are reliability, exposure, likelihood, resilience, and sustainability. We are interested in exploring their relationships with risk to enable us to make effective use of different risk management tools and processes to understand these emerging concepts. For example, we might use risk assessment concepts to assess resilience or sustainability. The reverse is also true. Several of the concepts, such as reliability, have been studied thoroughly in the literature; these methodologies can be used in risk management with appropriate changes. This section explores these interrelationships.

1.3.2 Reliability

The concept of reliability, *Rel*, in civil engineering has a long history (see Elishakoff 1983; Rausand 2004). In the most general sense, reliability concerns itself with a relationship between the capacity of the system of interest, *Capacity*, and the demands on that system, *Demand*. We express this general reliability relationship as

$$Rel = f_{RELIABILITY}(Capacity, Demand) \tag{1.17}$$

In order to ensure system safety, Equation 1.17 is always subjected to the condition

$$Capacity \geq Demand \tag{1.18}$$

The failure limit state of the system is reached when

$$Capacity = Demand \tag{1.19}$$

A more strict form of *Rel* used in a design problem can be expressed as

$$Rel = p\,(Capacity \geq Demand) \tag{1.20}$$

In this book, we use the more general definition of *Rel* as expressed by Equation 1.17 while subjected to condition (Equation 1.18) to highlight essential differences between reliability and risk without loss of the generality expressed by Equation 1.20. On a conceptual level, given the basic risk and reliability equations (Equations 1.12 and 1.17, respectively), and recalling that *Capacity*, in a qualitative sense, is the inverse of vulnerability and that *Demand* is another expression of Threat/Hazard, *T*, we can express *Rel* as a subclass of *Ri* such that

$$Ri = f\,(Rel, C) \tag{1.21}$$

The qualitative relationship is illustrated in Figure 1.2.

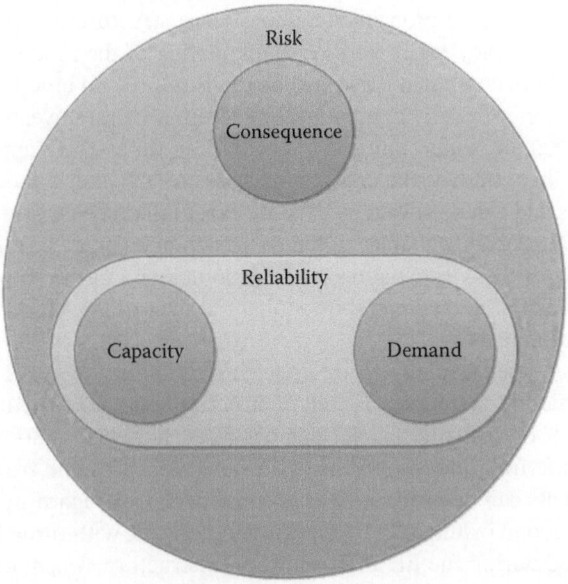

FIGURE 1.2 Relationship between risk and reliability.

When a structure (or system) is in service for a long time, changes are made to them as part of maintenance and renewal to adopt to new in-service conditions or regulations. Often these changes are made without a complete system analysis and, thus, can introduce unforeseen risks. Hence, it is required to do a system analysis periodically to identify unexpected risks that may need mitigation efforts. Vo et al. (1997) describe such an analysis that the Pacific Northwest Laboratory for the U.S. Army Corps of Engineers (USACE) conducted. They evaluated the risks associated with nonroutine closure of water flow through the turbines of powerhouses along the Columbia and Snake Rivers because the intake gates at some hydroelectric stations have been removed or raised to improve fish guidance.

Balas and Ergin (2002) described the first application of reliability-based project management practice for coastal projects in Turkey to estimate project parameters by considering the structural damage risk to quantify extreme project risks encountered in coastal construction investments. They estimated the project completion time by specifying the exceedance probability of the damage level at the design stage and obtaining the delay times of construction activities due to damage at the project-planning phase. For the network evaluation, they derived correlation factors for design parameters and project activities using Monte Carlo and Latin Hypercube sampling. They noted that the application of Monte Carlo simulations (Level III) has the advantage of robustness, provided the probability distributions of random variables and their correlation are determined from the data compiled over a sufficient number of years.

Straub and Kiureghian (2010a,b) combined Bayesian networks (BNs) and Structural Reliability Methods (SRMs) to create a new computational framework, an enhanced BN (eBN) to take advantage of both the methods for reliability and risk analysis of engineering structures and infrastructure. SRMs are used to solve the probability of an event, E, that is given in an integral form. SRMs are well documented in the literature (e.g., Ditlevsen and Madsen 1996; Rackwitz 2001; Der Kiureghian 2005; Ellingwood 2006). SRMs enable accurate assessment of probabilities of rare events represented by computationally demanding physically based models.

BNs are probabilistic models that facilitate efficient representation of the dependence structure among random variables by graphical means and have been used in many fields, such as statistical modeling, language processing, image recognition, machine learning, and risk analysis (Friis-Hansen 2000, 2004; Faber et al. 2002; Russell and Norvig 2003; Mahadevan and Rebba 2005; Grêt-Regamey and Straub 2006; Bensi et al. 2009; Nishijima et al. 2009). BNs are efficient in representing and evaluating complex probabilistic dependence structures, as present in infrastructure and structural systems, and they facilitate Bayesian updating of the model when new information becomes available. Authors combined these two methods to provide an eBN framework for a unified and powerful tool for efficiently computing probabilities of rare events in complex structural and infrastructure systems in which information evolves in time. Their approach employs a node elimination algorithm that removes the continuous nodes to arrive at a reduced BN (rBN), which only has discrete nodes and can be solved by existing exact inference algorithms. SRMs are used to compute the conditional probability tables of the rBN as component or system reliability problems. To evaluate the number of SRM calculations and the complexity of the resulting rBN, the concept of Markov envelopes was introduced and showed that the sizes of the Markov envelopes determine the computational requirements.

Authors applied their methodology to time-varying reliability analysis of a single structural system and an infrastructure system composed of structural and nonstructural elements. The first example (a basic model of a ductile frame) demonstrated how information obtained during the life cycle of the structure from measurements of element capacities and from observations of past and current hazard events can be used to make optimal decisions regarding repair or replacement of the structure. The second example (a transportation network with three bridges) demonstrated how information gained during the life of an entire infrastructure system can be used to probabilistically update our state of knowledge about the future performance of the infrastructure and its elements.

1.3.3 Exposure

In the context of risk, there are two specific and somewhat different definitions of exposure. Cardona et al. (2012) offered a definition for exposure as a function of a combination of hazards and their impact. Exposure is also defined in the finance industry, which relates exposure with the impact of financial transactions (Jorion 2007) to denote the potential of monetary losses.

We use the term *Exposure* in a different manner in this book because it is difficult to clearly delineate or quantify hazards and vulnerabilities, though it is easy to quantify their combined effects. For example, in the case of visual inspection of civil infrastructures, the condition data (a form of *exposure*) is a subjective estimate of the combined effects of the hazard (e.g., corrosion or flood) and the vulnerability of the component to that hazard. This is because it is very hard to separate the effect of individual hazards such as wear and tear, fatigue, and corrosion. Thus, exposure plays a major role in the civil infrastructure field, and we would like to build on this existing role.

Formally, we define exposure, E, as a relationship between hazards or threats, T, and vulnerability in the form

$$E = f_2(V, T) \tag{1.22}$$

In many situations, the use of E is beneficial.

Note that by following the definition of exposure in Equation 1.22 and by noting the general reliability relationship, we allow exposure to be the inverse of reliability. Thus, in the general sense, higher reliability implies lower exposure and lower reliability implies higher exposure. In general, exposure is usually expressed in a subjective manner whereas reliability and likelihood (Section 1.3.4) are usually expressed and computed in an objective manner.

1.3.3.1 Exposure Application: Bridge Inspection

The bridge inspection process can be recast as an exposure to the risk process as shown in Figure 1.3. The figure envisions the inspection process as a multistep process:

- *Causes (hazards) and effects (deterioration/exposure)*: Different hazards/demands degrade infrastructures components. For example, environmental conditions combined with deicing salts will cause steel members or reinforcing rebars to corrode. Similarly, traffic demands,

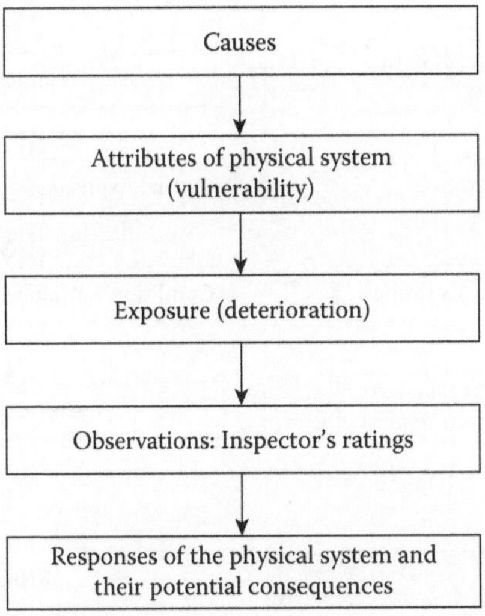

FIGURE 1.3 Inspection process.

freeze–thaw cycles, and other environmental demands will cause deterioration (exposure) of roadway pavements.

- *Observable effects*: Because of the causes, components are affected in observable ways. For example, concrete components display discoloration, delamination, spalling, or peeling. Soil erosion results from scouring processes. These observable effects are described in detail in various inspection manuals (FHWA 1970 and NYSDOT 2006.) and experienced bridge inspectors are very familiar with them.
- *Condition rating* (*CR*): This is the subjective numerical rating that an inspector assigns to the component based on observations. CR is usually assigned on a value scale (e.g., 0–9 scale), with higher values generally indicating good condition and lower values indicating deteriorated condition.
- *Responses/consequences*: This is the expected response/behavior of the observed/rated component in its observed condition due to all hazards that component experienced (includes traffic and other gravity loads, wind, flood/scour, collision/impact, etc.).

If we accept the process of Figure 1.3, we immediately observe that the process is analogous to a risk process. Here, we have the cause as a hazard and the attributes of the system to be affected as the vulnerability. The combined effect of hazards and vulnerability is the exposure. The inspector of the exposure estimates the observations (*CR*). The management response to the *CR* is the consequences. The analogies between the inspection process and risk process are shown in Figure 1.4. A risk-based inspection process is also shown in the figure. It requires that the inspection effort be governed by the resulting risk from the process. This immediately points to the necessity of including an accurate estimation of consequences in any risk-based inspection process.

1.3.4 Likelihood

Likelihood, L, is another form of risk subclasses. Many fields that handle risks and hazards (Vose 2009; Ayyub 2014) use it. It is a function of vulnerability, V, and hazard, H, in the form

$$L = f(V, H) \tag{1.23}$$

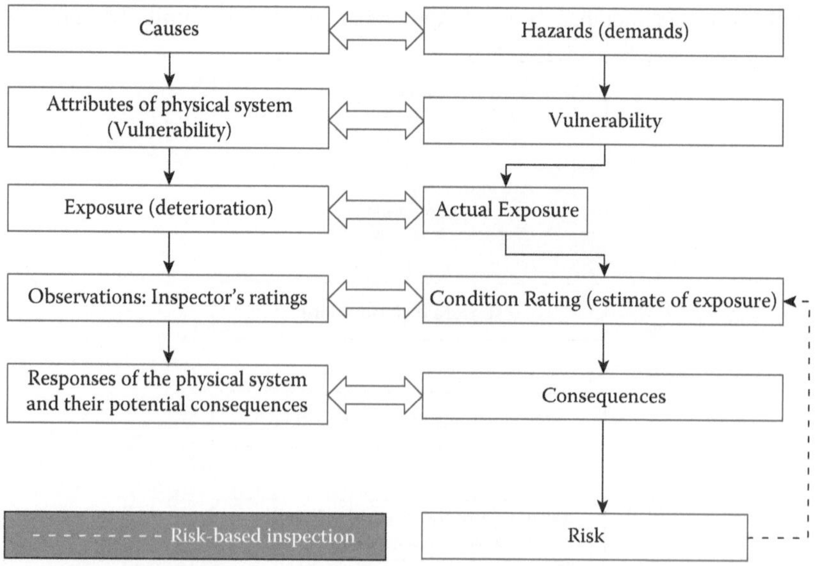

FIGURE 1.4 Analogies of inspection processes.

The function $f(\,)$ can be fairly simple, such that

$$L = V.H \tag{1.24}$$

or a bit more complex

$$L = \int p(H)\, V \tag{1.25}$$

with $p(H)$ expressing the probability of a particular hazard level. Several authors (Vose 2009, and FEMA 455) have used Equation 1.24, and Equation 1.25 is well known in the field of performance-based designs (see Hawk 2003; Ettouney and Alampalli 2012b).

In addition to being a risk subclass, which can aid in assessing risk, it has also been used as a risk treatment aid via the $L-C$ chart (Vose 2009; Ayyub 2014). We take a closer look at the usefulness and accuracy of the $L-C$ chart in Chapter 5.

1.3.5 Resilience

1.3.5.1 Overview

Natural and man-made hazardous events can impose a devastating cost to society. As Figure 1.5 shows, the costs of some of these disasters in the United States alone can be staggering. Stakeholders of civil infrastructure have a stake in reducing these costs by improving and maintaining operational and physical performance.

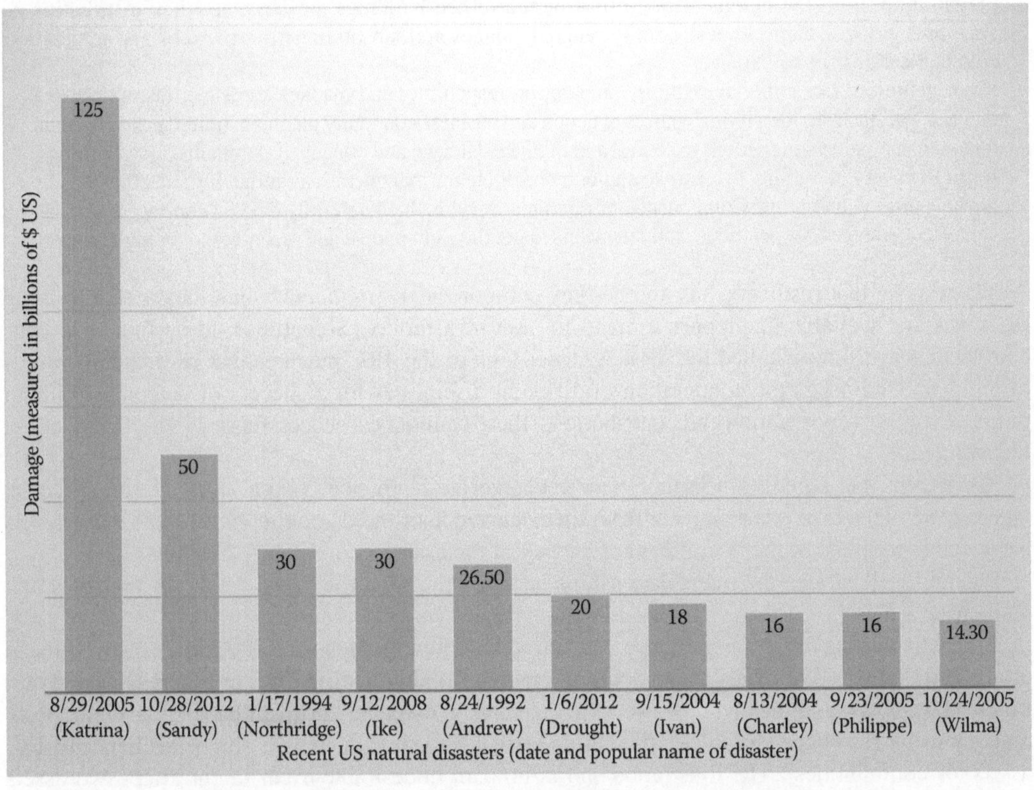

FIGURE 1.5 Damages from recent natural disasters in the US. (From EM-DAT, *EM-DAT: The OFDA/CRED International Disaster Database*, Université Catholique de Louvain, Brussels, Belgium, 2014. With permission.)

Throughout history, infrastructure resilience has been defined in numerous ways. The most widely used and most objective is that of the National Infrastructure Advisory Council (NIAC 2009), which states:

> Infrastructure resilience is the ability to reduce the magnitude and/or duration of disruptive events. The effectiveness of a resilient infrastructure or enterprise depends upon its ability to anticipate, absorb, adapt to, and/or rapidly recover from a potentially disruptive event.

With the emergence of resilience as a major civil infrastructure paradigm, it is only logical to wonder about its relationship with risk. Therefore, in this section, we discuss briefly the components of resilience. We then explore some of the basic issues related to resilience to lead us to a discussion about the interrelationships between resilience and risk. We end this section with an obvious conclusion (in view of the universality of risk, as proposed earlier)—that resilience is a subcomponent of risk. This relationship between resilience and risk is an important one. If we accept such a relationship, we then can apply processes and management activities that exist for risk management to resilience management, perhaps with some modifications. On the other hand, ignoring such a relationship between resilience and risk might cause inefficient resilience processing and wasted effort and resources resulting in less than optimal resilience. Because of this, we offer some case studies regarding resilience later in this book. Our handling of resilience is similar to our handling of risk, with appropriate changes, which account for the special needs of resilience.

1.3.5.2 Resilience and Its Components

NIAC (2009) specified three components of resilience:

> Robustness: the ability to maintain critical operations and functions in the face of crisis. This includes the building itself, the design of the infrastructure (office buildings, power generation, distribution structures, bridges, dams, levees), or in system redundancy and substitution (transportation, power grid, communications networks).
>
> Resourcefulness: the ability to skillfully prepare for, respond to, and manage a crisis or disruption as it unfolds. This includes identifying courses of action and business continuity planning; training; supply chain management; prioritizing actions to control and mitigate damage; and effectively communicating decisions.
>
> Rapid Recovery: the ability to return to and/or reconstitute normal operations as quickly and efficiently as possible after a disruption. Components [of rapid recovery] include carefully drafted contingency plans, competent emergency operations, and the means to get the right people and resources to the right places.

We propose that resilience has another key component *redundancy*, which means that back-up resources are available to support demand in case of a failure. Sometimes, these four resilience components are simply called the 4Rs. A closer look at the 4Rs indicates that resilience is multidisciplinary and needs the cooperation of different disciplines for a successful outcome. Without multidisciplinary cooperation and contributions, there cannot be a successful or an efficient resilient infrastructure.

A different, but equally important, way to categorize resilience is on a *Time–Function** scale, where *Time* is the time scale starting from the occurrence of the damaging hazard and *Function* is a subjective estimate of the operational efficiency of the asset or community of assets.

Figure 1.6 illustrates the interrelationships between the 4Rs, the *Time–Function* variables and resilience.

A beneficial illustration of resilience was introduced first by Hynes (2001) and then by Bruneau et al. (2003, 2007). We recast the concept in terms of operational quality in Figures 1.7 and 1.8. Figure 1.7 shows graphically how we can objectively estimate the resilience of a community asset by using resilience charts. We then use Figure 1.8 to show a comparison of resilience of two assets (or communities). The illustration shows how an undesirable event might affect two assets

* Sometimes called operational efficiency (OE), operational level, quality of operation, or operational quality.

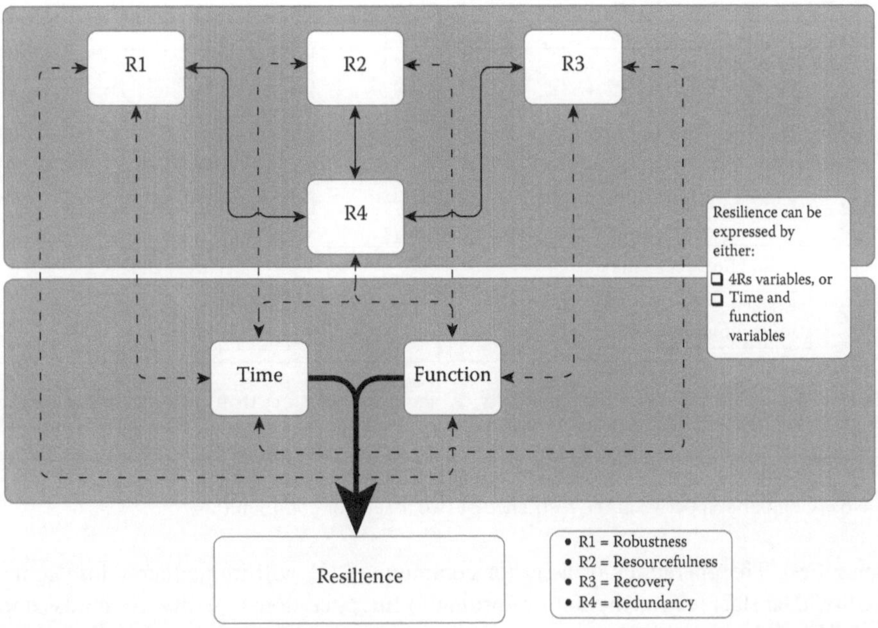

FIGURE 1.6 Resilience and its components.

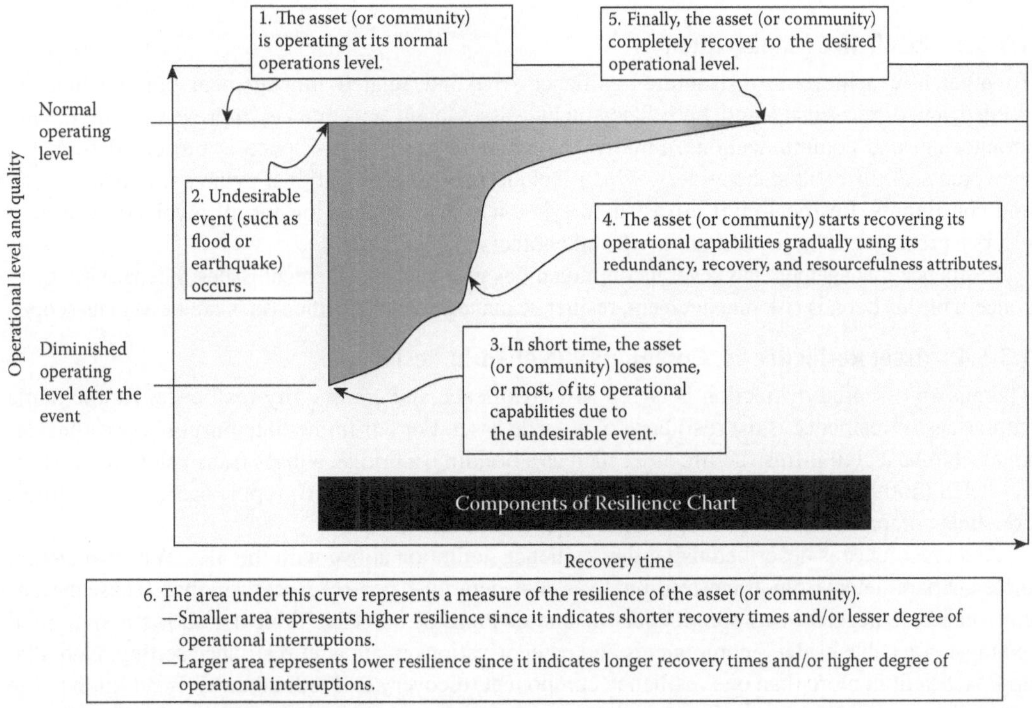

FIGURE 1.7 Resilience definition in the time-operational quality space.

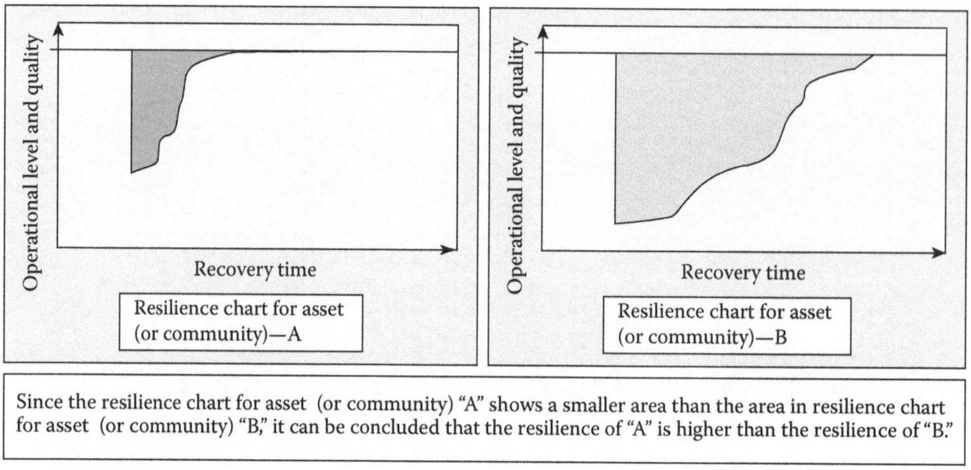

FIGURE 1.8 Comparison between the resilience of two assets (or communities).

(or communities). The operations of asset (or community) "A" will immediately lose some operational quality, then start recovering until returning to full operational quality. As for asset (or community) "B," it will lose much of its operational quality when subjected to the same undesirable event. It will recover but at a much slower pace than asset "A." Thus, "A" is more resilient than "B." The area under the curve can be used to describe the time-operational quality of the asset in a community to objectively assess the resilience of the community. This illustration shows that communities with resilient infrastructure return to normal operating levels much faster than those communities lacking resilient assets and provides further evidence that infrastructure investment translates to efficiency. We finally illustrate the relationships between the levels of the 4Rs and the *Time–Function* scale in Figure 1.9.

1.3.5.3 Resilience Management

To efficiently achieve infrastructure resilience, five interrelated management components are needed, and discussed at length here. They include assessment, acceptance, improvement/treatment, monitoring, and communications. Finally, the business case for resilience is presented from the viewpoints of different stakeholders, which include public organizations, owners and contractors, and consultants. Resilient civil infrastructure is a *win–win* proposition for all involved because it offers a great return on investment to all stakeholders.

Resilience management has several commonalities with risk management issues. Because the main concern of this book is risk management, resilience management–specific issues are beyond its scope.

1.3.5.4 Asset Resilience vs. Community (Network) Resilience

There is an essential distinction between asset resilience and community resilience. As the name implies, asset resilience is the resilience of a single asset. For our immediate purposes, we consider an asset to be a civil infrastructure asset such as a building, a bridge, a mass transit station, or a tunnel. DHS (2009) and the American Society of Civil Engineers (ASCE) report card, ASCE (2013), contain a comprehensive list of types of civil infrastructure assets.

Asset resilience is described using the resilience definition above with the 4Rs. Within an asset, different parameters (sometimes referred to as considerations) control asset resilience. These parameters can be categorized as components of one or more of the 4Rs. Table 1.1 shows a simplified example of a cable bridge's components and categorizations in an asset resilience setting. Note that approaches fit in more than one resilience component (recovery and redundancy). In addition to the categorizations of Table 1.1, a functional diagram, called a graph network, needs to be established. This network shows dependencies of different parameters. Arrows from the controlling parameter to

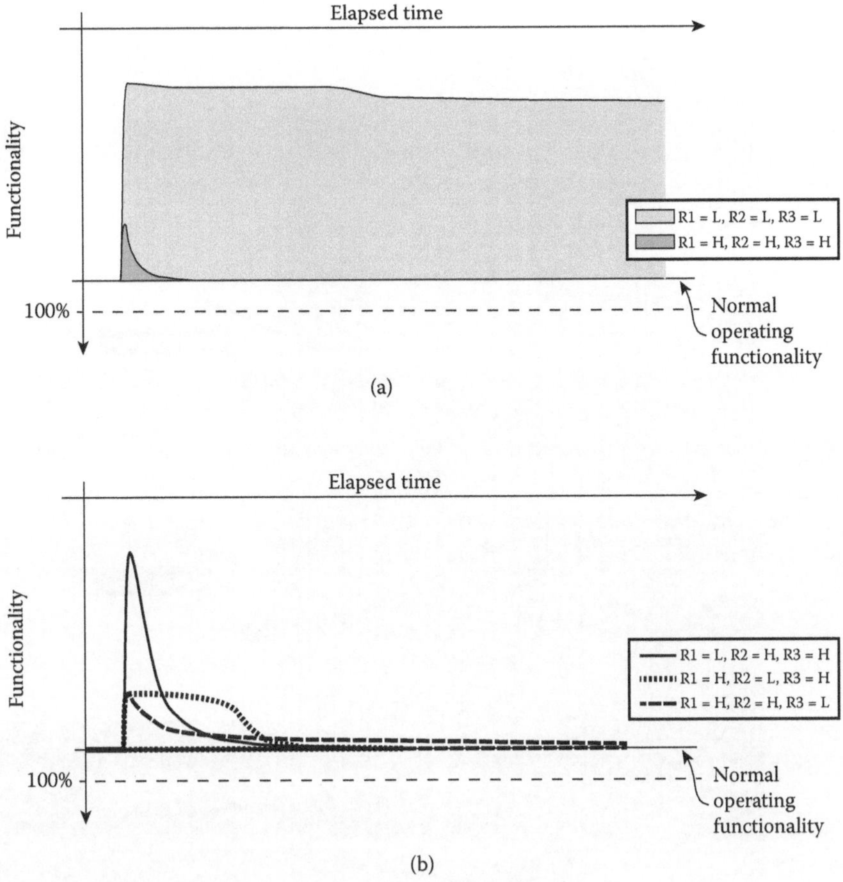

FIGURE 1.9 Levels of 4Rs as they can affect resilience performance in a *Time–Function* scale. (a) Extreme limits of the resilience components and (b) intermediate limits of the resilience components.

TABLE 1.1
Cable Bridge Asset Resilience Example

Resilience Component	Cable Bridge Parameter (Considerations)
R1: Robustness	• Suspension cables • Stiffening girders/trusses
R2: Resourcefulness	• Inspection (special) • Memorandum of understandings between different organizations
R3: Recovery	• Approaches (ramps) of roadway to bridge • Training
R4: Redundancy	• Approaches (ramps) of roadway to bridge • Suspenders

the dependent parameter express the dependencies. If there is no obvious dependence between two linked parameters, then a simple line connecting the two parameters is used. Similarly, Figure 1.10 shows a simple network for asset resilience of a medium-size building. Capturing important parameters (both operational and physical) as well as their interdependencies (as shown in Table 1.1 and Figure 1.10) is an essential first step in developing successful asset resilience management.

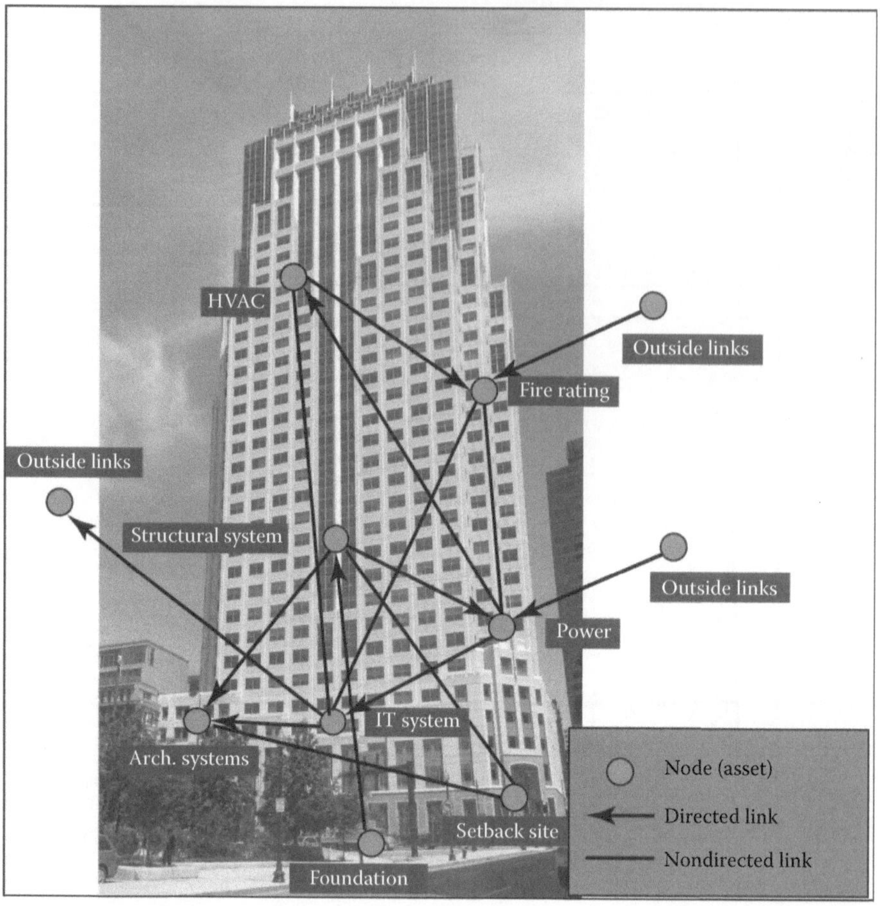

FIGURE 1.10 Asset resilience links for resilience management (simplified tall building example).

We turn our attention now to community resilience. As the name implies, a community is comprised of several assets (nodes) interconnected via links that may be assets themselves. The nodes and links constitute a community's network. Community resilience depends on the resilience of the network's individual asset components (both nodes and links). Thus, as an essential step in community resilience management, we need to have a greater understanding of the resilience of nodes and links. In addition, community resilience depends on the topology of the network and the linkages between different nodes.

The size of the community is completely subjective. A community could be a simple campus comprising a small number of buildings (such as a small hospital or college). A community could be a transportation network, a small town, a region, a whole county—or even a state. Each of the 4Rs of community resilience is a function of all of the 4Rs of its nodes and links as well as the topology of the network. Figure 1.11 shows a simple resilience network for a small community.

1.3.6 Risk Relationships

Following the definitions of risk, reliability, exposure, and likelihood, we can formally relate risk to these sub components as

$$Ri = f_3\,(E, C) \tag{1.26}$$

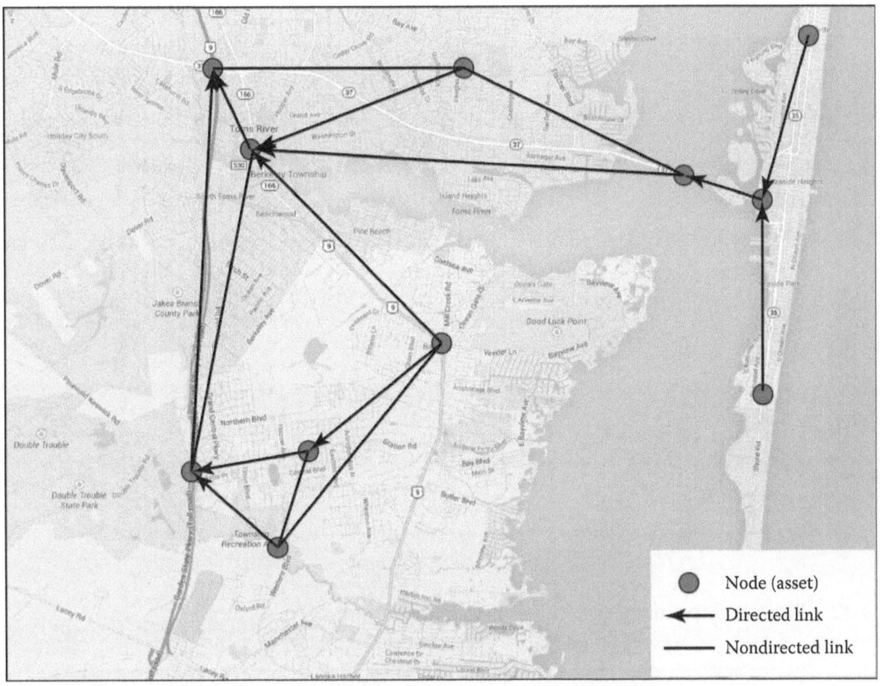

FIGURE 1.11 Community network for resilience management.

$$Ri = f_4\,(Rel,\ C) \tag{1.27}$$

$$Ri = f_5\,(L,\ C) \tag{1.28}$$

We use all these risk expressions throughout the book, as pertinent. As noted earlier, *E, C, Rel,* and *L* represent exposure, consequences, reliability, and likelihood, respectively. Figure 1.12 illustrates the interrelationships between risk, exposure, and reliability.

1.3.6.1 Risk Relationship with Resilience

One objective for this section is to clarify the distinctions and relationships between risk and resilience. We will also add to the mix the subject of sustainability; however, full treatment of sustainability is outside of the scope of this book. We start with risk, which is controlled generically by Equation 1.12 as a function of *T, V*, and *C*. We also have seen earlier that a reasonable resilience definition relates resilience to robustness, resourcefulness, recovery, and redundancy (the 4Rs). Upon further reflection, we can see that the 4Rs can be recast as a subset of *T, V*, and *C*. Ettouney and Alampalli (2012) proposed qualitative relationships similar to those in Table 1.2 and Figure 1.13.

Thus, resilience can be treated as a special form of risk. Though resilience focuses mainly on continuity of operations or the degree and length of operational interruptions, these issues can be recast into the three basic components of risk as illustrated in Figure 1.13. From a different viewpoint, Figure 1.14 shows the relationships between risk basic components and resilience's 4Rs.

Sustainability is a bit more difficult to define because of the immense number of definitions. Some sustainability objectives include operating within acceptable and renewable limits; allowing environmental, societal, economical, and humanitarian concerns govern actions; or having respect for available natural and universal resources. (See Sustainable Measures 2014 for more about sustainability issues.)

Reflecting on sustainability objectives, we can recast them again in terms of a risk's threats, vulnerability, and consequences. Sustainability, with all of its varying interpretations, is also a subset of risk. Unfortunately, because there is no universal definition of sustainability components similar to resilience components, we can't offer a table showing relationships between sustainability

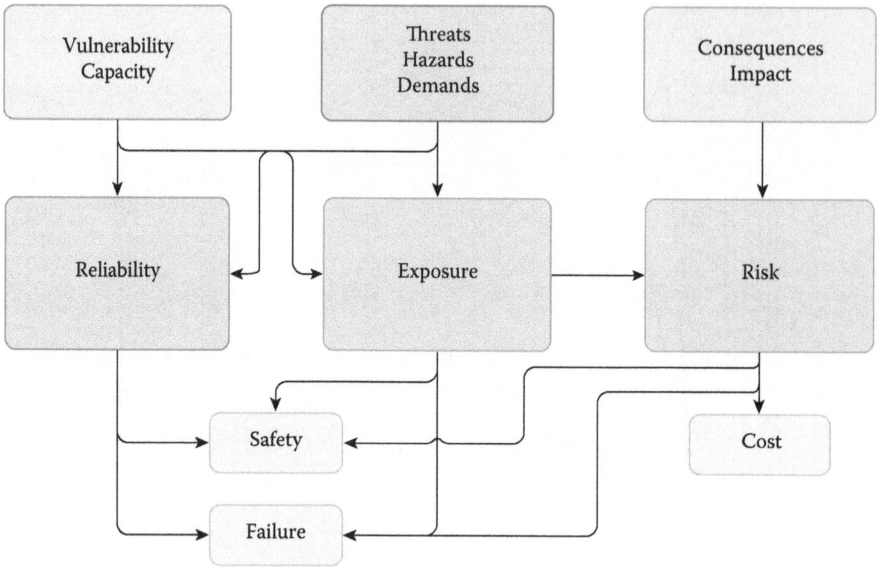

FIGURE 1.12 Relationship among risk, reliability, and exposure.

TABLE 1.2
Relationship between Risk and Resilience

Risk Components	Resilience Components: Robustness	Resourcefulness	Recovery	Redundancy
Consequences	Minor	Major	Major	Major
Threat	Major	Minor	Minor	Major
Vulnerability	Major	Minor	Minor	Major

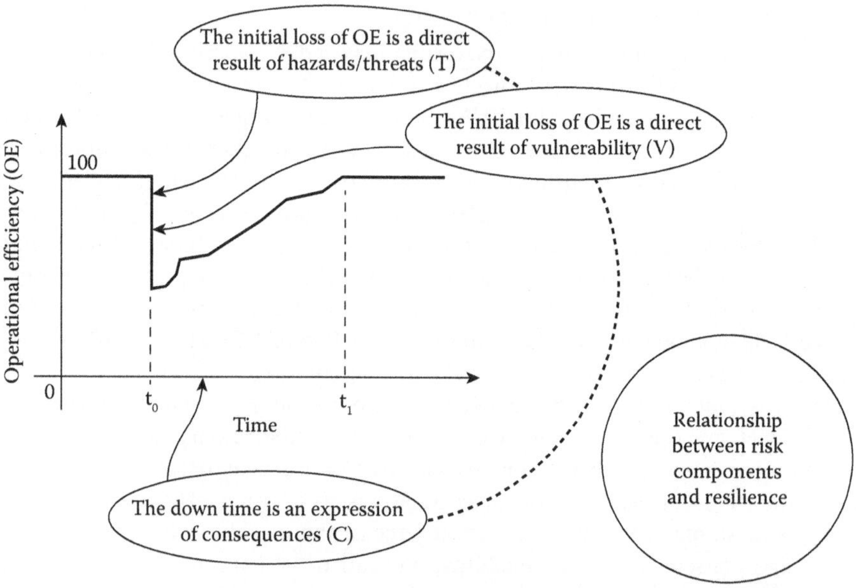

FIGURE 1.13 Simplified relationship between resilience *Time–Function* scales and risk basic components.

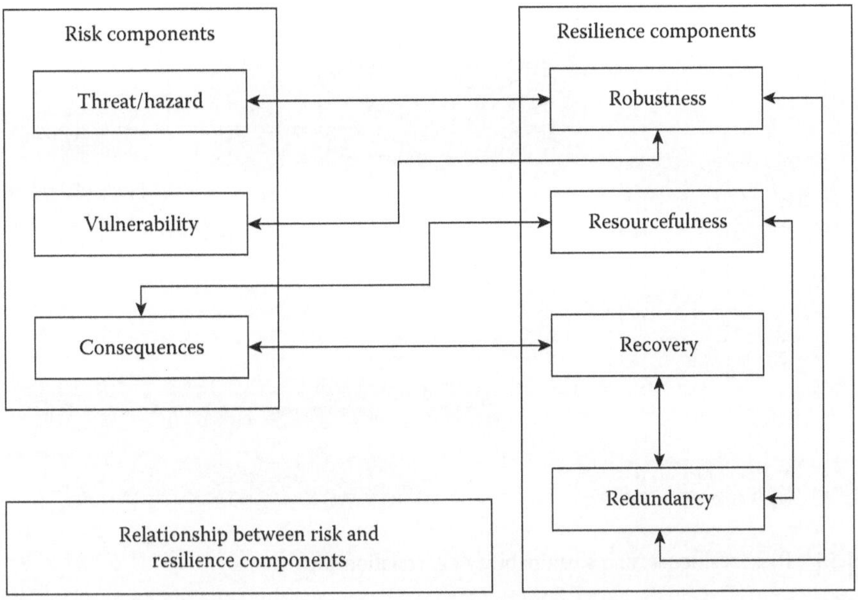

FIGURE 1.14 Simplified relationship between resilience's 4Rs and risk basic components.

TABLE 1.3
Risk, Resilience, and Sustainability

Item	Risk	Resilience	Sustainability
Components	• Vulnerability • Threat • Consequences	• Robustness • Resourcefulness • Recovery • Redundancy	Varies, but can be also traced to Vulnerabilities, Threats, and Consequences
Management components	• Assessment • Acceptance • Treatment/Improvement • Monitoring • Communications		
Metric properties	• Objective • Subjective • Relative • Absolute (can be monetary, time, or combinations of same)		
Main emphasis	Cost or cost benefits	Continuity of operations	Long-term consequences Conservation of resources

components and risk components as we have done in Table 1.2 for risk-resilience interrelationships. Suffice it to say, threats, vulnerabilities, and consequences in sustainability can also be considered as a subset of the risk application.

Table 1.3 shows comparisons of different management issues that concern risk, resilience, and sustainability. Figure 1.15 illustrates how risk is a superset of both resilience and sustainability. It also shows how risk can be considered a superset of reliability and safety as noted earlier in Section 1.3.2. To summarize, though risk resilience and sustainability have common components and many overlapping objectives and goals, risk acts as an umbrella that contains all of those paradigms as subsets. This is an important conclusion because it permits us to use, with modifications as the situations calls for, all objective and subjective risk tools, processes, and methodologies to address issues related to resilience and sustainability.

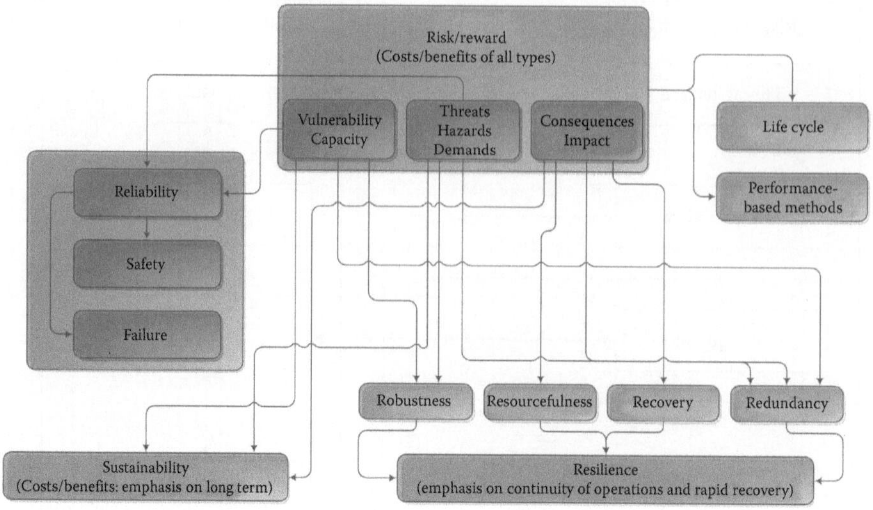

FIGURE 1.15 Risk, resilience, and sustainability interrelationships.

1.4 COMPONENTS OF RISK MANAGEMENT

1.4.1 Overview

We introduced risk and its components in the previous sections. We also addressed many of the parameters that affect risk. We now turn our attention to the subject of how stakeholders can manage different types of risks that can afflict their assets or communities. We note that the first step should be to *assess* the state of risk amid a natural or man-made hazard or threat. Second, decide whether to *accept* this assessment. If the assessed risk level is not acceptable, a risk *treatment/improvement* program should be initiated to bring the assessed risk down to an acceptable level. Unfortunately, a universal fact is that the passage of time has a tendency to reduce physical capacity (aging effects) and/or operational efficiency (e.g., budget cuts), while demands from infrastructure tend to increase (e.g., traffic, intensity of storms, climate change). These universal truths will have the inescapable result of risk increase as time goes on. This will necessitate the need for risk *monitoring* to ensure that risk remains within an acceptable level. Finally, all of those activities need to be *communicated* between all stakeholders, including legislators and the public, to maintain public confidence as well as the appropriate level of project funding.

All five components (assessment, acceptance, improvement, monitoring, and communications) constitute an essential framework for complete and successful risk management, as shown in Figure 1.16, which also reveals the complex interrelationships between the five risk management components. Additionally, the figure shows the need for communication while executing all other risk management components. The latter fact is illustrated in Figure 1.17.

1.4.2 Risk Assessment

The first and most widely practiced component of risk management is risk assessment. Some of the methods for assessing risk are based on subjective processes (Gutteling and Wiegman 1996), while others are based on objective processes (DHS 2011b, 2011c). There are detailed relative and absolute risk assessment methods available (Ettouney and Alampalli 2012a, 2012b) as well as probabilistic or deterministic risk assessment methods (Fenton and Neil 2013; Ettouney and Alampalli, 2016). Masada et al. (2007) describe a study that validated the Ohio Department of Transportation (ODOT)'s culvert inspection program instituted in 2003 (ODOT 2003) and proposed a risk assessment method to compute the overall structural health rating for any inspected culvert to recommend

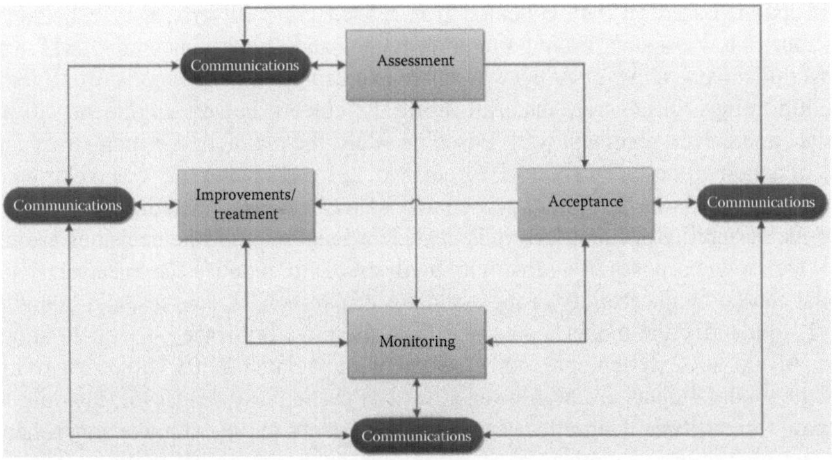

FIGURE 1.16 Risk management components and their interrelationships.

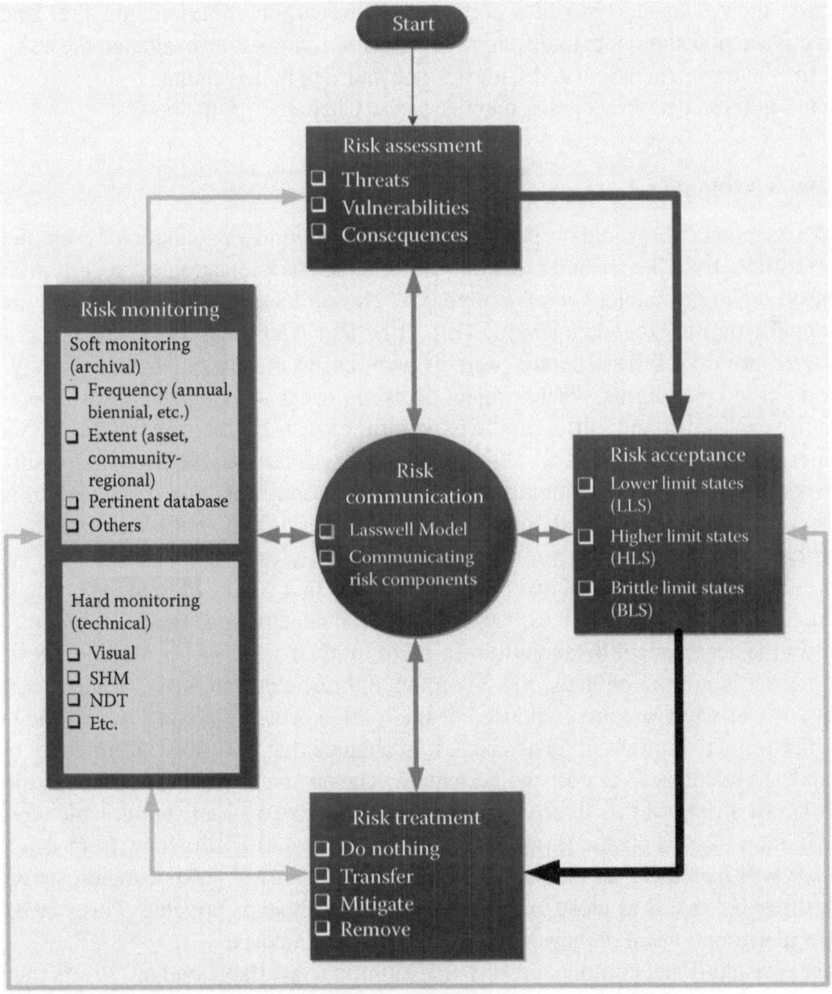

FIGURE 1.17 Centrality of risk communications in risk management practice.

a course of action. Based on data collected from 25 concrete culverts, they concluded that the inspection approach was sound. From a literature review and discussions with ODOT, a risk index (adjusted overall culvert rating, or AOR) was calculated using the ratio of soil cover thickness, rise, and inspection ratings for concrete material inside the culvert, culvert alignment, culvert seams/joint, top slab, and culvert abutment wall. Based on AOR, the maintenance immediacy and inspection course of action proposed by NCHRP Report No. 251 (NCHRP 1982) was recommended.

Sensitivity analysis is an important component of any analysis, including risk assessment, to ensure that the variability of input factors is considered and appropriate decisions are made. Tsai et al. (2011) state two purposes of a sensitivity analysis: (1) to quantify the uncertainty in the calibrated model caused by uncertainty in the estimates of parameters, stresses, and boundary conditions and (2) to identify the model inputs that have the most influence on model calibration and predictions (ASTM 2002; Gribb et al. 2002; Albrecht and Miquel 2010). If the sensitivity analysis shows that the model outputs are highly sensitive to a particular input factor, then the associated uncertainty or variability will significantly affect the model's ability to make consequential interpretations and predictions (Seuntjens et al. 2002). Through sensitivity analysis, it is possible to decompose the model output variation back to the input factors and to identify those that need to be measured or estimated accurately to achieve a precision improvement in the model output. With this understanding, authors conducted a sensitivity analysis to evaluate the effectiveness of input parameters on the calculated risks at a petroleum-hydrocarbon contaminated site. Under a risk-based corrective action approach, as developed by ASTM (2004), they evaluated the risks to human health and the environment associated with this site and risk management.

We discuss in more detail risk assessment issues in Chapter 3 of this book.

1.4.3 Risk Acceptance

Setting risk acceptance thresholds is perhaps the most difficult step in asset (or community) management. Traditionally, a prescribed acceptance threshold has been adopted in civil infrastructure projects based on a reasonable, yet subjective (i.e., chosen based on personal judgment or experience) probabilistic nonexceedance value (Ettouney and Alampalli 2016). Such a practice was reasonable because civil infrastructure projects were based mostly on ensuring safety. With the advent of risk-based paradigms, where project decisions are based on safety as well as reasonable costs (both life cycle costs and initial capital expenditures, *CapEx*), setting risk acceptance thresholds became more difficult to define. Objectively setting a reasonable acceptance threshold for different risks of the asset or community under consideration is as important as defining/assessing those risks themselves. For without setting an appropriate threshold, risk treatment projects might be either unnecessarily costly, or they might result in lesser performance, see Figure 1.18. Unfortunately, there is little known on objective methods that can be of help in setting reasonable risk acceptance thresholds, but one can still make some general suggestions. A simple approach is to try to set the acceptance threshold to minimize total costs as will be detailed next.

Even though it is not too obvious, risk acceptance thresholds can have immense cost implications. Total costs of any event are composed of the total expected damage costs of the event itself and costs of retrofits that might mitigate risks. It is obvious that as costs of retrofits increase, risks decrease, and as a result damage costs will decrease. The reverse is true. This relationship is shown in Figure 1.18. As shown in this figure, the total cost will have a *sweet spot* at which total costs is minimum. If the risk acceptance threshold moves away, in either direction from this threshold, the total costs will increase—an undesirable result. Thus, decision makers should strive for a risk acceptance threshold that is as close to this sweet spot threshold as possible. This can be achieved only by careful estimations of damage costs as well as retrofit costs.

There are two additional complications to this situation. Available budget, Ri_A, is the first issue as shown in Figure 1.18. If the available budget can only afford the decision maker a higher risk level than the sweet spot offer, then the resulting retrofits will be far from optimal. Similarly, if

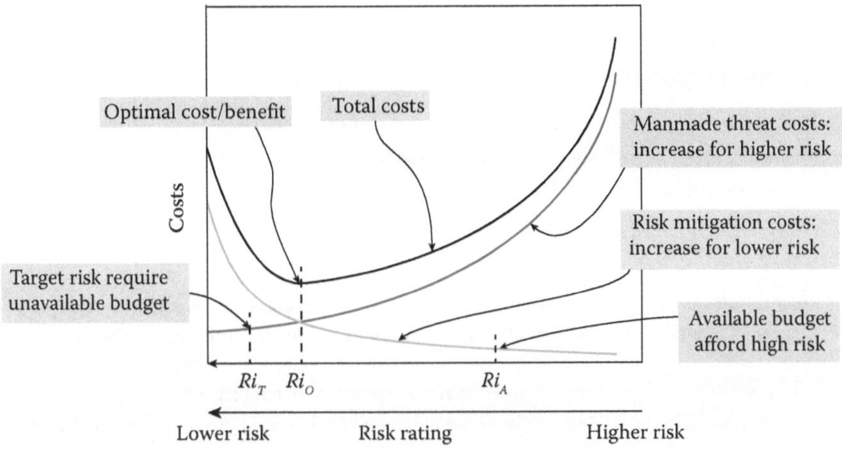

FIGURE 1.18 Practical vs. optimal risk acceptance conditions.

the target risk, Ri_T (desired risk rating) is much lower than either the sweet spot or the available budget, such a situation will result in a nonoptimal risk situation, and it will also result in potential budget-security mismatch. Additional studies might be needed to see how an acceptable risk vs. available budget compromise might be reached.

Figure 1.18 shows only one level of a risk-related damaging event. Similar illustrations can be made for all possible event levels. The risk optimal acceptance limit is expected to differ for different event levels. Also, it is advisable to include two additional factors while looking for this risk acceptance sweet spot: the life cycle costs as well as benefits and the effect of multihazards if the asset or communities are susceptible to more than one potentially damaging hazard.

From the discussion above, we conclude three important points:

1. Setting appropriate risk acceptance thresholds are of immense importance in any risk-based management effort,
2. The subject of risk acceptance is as complex as any in the field of risk management, and
3. There are also immense knowledge gaps in the subject of risk acceptance.

We try to address some of the issues relating to risk acceptance in Chapter 4 and offer some practical examples of the topic.

1.4.4 Risk Treatment/Improvement

Risk treatment activity is generally considered when assessed risk of the asset (or asset network) of interest is found to be higher than the risk acceptance threshold. The risk treatment component in a risk-based management program then includes three phases:

- Phase I: Baseline mitigation
- Phase II: Project plans and prioritizations
- Phase III: Project execution

In Phase I, risk treatment efforts try to decide upon the type of risk reduction that is most efficient. There are four possibilities as shown in Table 1.4 and Figure 1.19.

If the outcome of Phase I is to embark on a risk mitigation effort (see Figure 1.19), the decision maker then moves into Phase II that is preparing potential risk mitigation projects. Risk mitigation projects should account for potential improvements to some or all of the C, T, and/or V components.

TABLE 1.4
Decision Options for Phase I of Risk Treatment

Type of Action	Comments
Do nothing	Some of the reasons can be: • The margin between accepted threshold and actual risk ratings are minimal • Budget constraints • Other subjective circumstances
Remove potentially damaging events	If threats or hazards can be removed entirely, it is a good option. Man-made hazards are a prime example for this option. Natural hazards are difficult to remove, except if the whole facility can be moved away from the exposure to natural hazard.
Transfer responsibility	Reasons for potential interruption to operations might be transferred by opting for insurance. In case of many civil infrastructures such as bridges or tunnels, this option is normally not feasible.
Mitigate deficiencies	Mitigations include physical retrofits (see Figure 1.19) and/or operational improvements. (Note that risk mitigation is a subset of risk treatment in our considerations.)

FIGURE 1.19 Typical retrofit effort.

Often decisions are made to physically improve facilities that are usually accomplished by improving asset vulnerability, *V*. Sometimes, it might be more efficient to improve threat, *T*, or consequence ratings, *C*. Careful studies are needed to achieve a balance between the improvements of the *C*, *T*, and/or *V* components to achieve the optimal costs and performance. For example, in a bridge security risk improvement effort, it might be more efficient to improve access control or setback

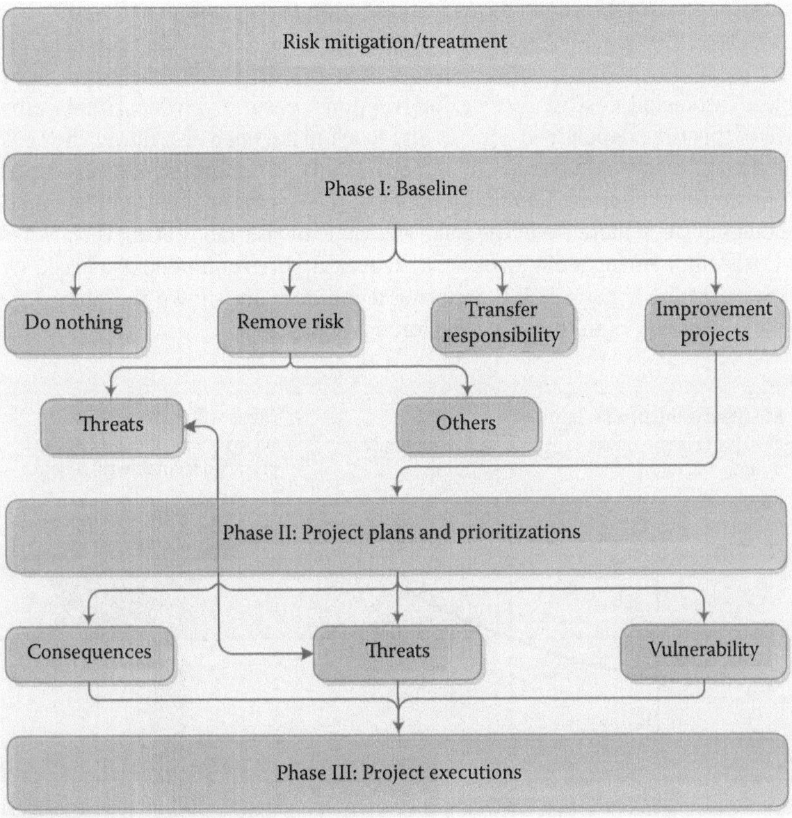

FIGURE 1.20 Three phases of bridge security risk mitigation/treatment.

than bridge hardening. It is also very important to consider potential undesirable/unintended effects due to possible interactions between hazards before embarking on a bridge security rehabilitation effort.

After identifying potential candidate improvement projects, a project prioritization effort is needed to select the most suitable and cost effective project to achieve the required risk threshold level. This should consider costs, benefits, project scale, and life cycle considerations (see Fenton and Neil [2013] for theoretical background; Ettouney and Alampalli [2012a, 2012b] for practical examples). We discuss the issue of risk treatment at length and offer several practical case studies in Chapter 5. After deciding on the most suitable project for risk improvements, Phase III entails project execution which is beyond the scope of this book. Figure 1.20 shows the three phases of risk improvement.

1.4.5 Risk Monitoring

It is given that assessed risks ratings, accepted risk thresholds, and risk treatment efforts of assets or asset networks will tend to change/degrade with the passage of time as a result of potential increases in one or more of the *C*, *T*, and/or *V* components. For example, physical deterioration (i.e., increased vulnerabilities) of bridges in general is well documented through deterioration rating studies (see Agrawal and Kawaguchi 2009). Mahmoud (2011) documented the deterioration issue of suspension cables as a function of time. The rate of physical degradation can be steeper than the forecasted values and accepted thresholds can be reached much faster than that predicted by initial risk studies. Of course, the vulnerability component of risk is not the only risk component that increases with time. For example, security threats tend to increase with time passage as target attractiveness or importance increases. Detailed discussions on target attractiveness can be found elsewhere (FEMA 2005; DHS 2011a, 2011b, 2011c). Finally, consequences do increase as time passes due to increased traffic volumes, associated cost of service interruptions, costs of physical replacements, and so forth. Because of this time dependency of risk and to avoid having a risk rating above an acceptable threshold, the state of risk should be monitored periodically as a component of a comprehensive risk management process and appropriate modifications to the original plans should be made as needed to avoid the potential steep increase of the risk. The value of risk monitoring is shown conceptually in Figure 1.21. Risk monitoring techniques should cover all three components of risk: *C*, *T*, and/or *V*, as pertinent. Some of the general risk monitoring techniques are shown in Tables 1.5 and 1.6 and Figure 1.22. Details of risk monitoring are explored in Chapter 6.

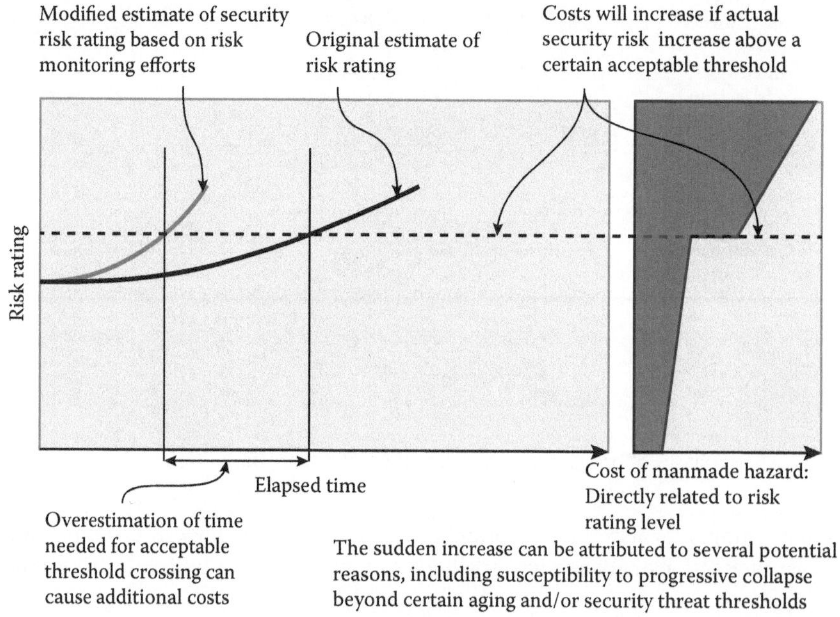

FIGURE 1.21 Value of risk monitoring.

TABLE 1.5
Examples of Risk Monitoring Techniques

Component	Monitoring Techniques
Consequences	• Average daily traffic (ADT) • Weigh in motion data for demands • Costs of service interruptions • Costs of physical retrofits and replacements • Changes in other potential consequences (see DHS 2011a, 2011b, 2011c)
Threats	• Different demand or threat monitoring techniques (see FEMA [2005] for infrastructure security, for example) • For man-made hazards: follow changes in target attractiveness issues (see DHS 2011a, 2011b, 2011c)
Vulnerabilities	• Structural health monitoring methods (see Ettouney and Alampalli 2012a, 2012b) • Pertinent inspection and maintenance processes (see Table 1.6)

TABLE 1.6
Types of Vulnerability Monitoring/Inspection Practices for Structures Such as Bridges

Type of Inspection/Monitoring	Description	Impact on Security Consideration
Routine	Visual inspection	Can also result in some observations that have risk retrofit implications
Special	In-depth testing	By detecting degradations in properties, as-is robustness of assets and their components to resist different hazards can be estimated rather than the design robustness
Structural health monitoring, see also Figure 1.22	Modal and parameter testing and global behavior of the bridge	Potential intersections/interactions between hazards can be estimated
Nondestructive testing	Local area or specific component testing	By detecting degradations in properties, as-is robustness of cables to different hazards, such as earthquakes or wind, can be estimated rather than the design robustness

FIGURE 1.22 Sensors for structural health monitoring of performance of box girder bridger.

1.4.6 Risk Communications

The final and perhaps most important component in risk management is risk communications. Communications is the bridge between the technical and professional community, decision makers, elected officials, funding sources, and the public at large. There are numerous objectives to risk communications, some of which are shown in Table 1.7. Obviously all of the above objectives are important. Unfortunately, risk communications usually receive the least degree of attention, even though due to its paramount importance, it should receive the most attention.

Lasswell (1948) and Gutteling and Wiegman (1996) subdivided risk communications into five essential modules: originator (source), content (message), recipient, medium (channel), and objectives (destination). These five modules are interrelated as shown in Figure 1.23. Any successful risk communication plan has to accommodate all the five modules in a careful and well-thought manner. Additional discussions of risk communications for civil infrastructures in general are in Chapter 7.

1.5 MISCELLANEOUS ISSUES RELATED TO RISK MANAGEMENT

1.5.1 Overview

Whenever we deal with a risk management effort, several additional issues need to be accommodated, as pertinent. In this section, we offer an overview of some important risk-related issues. Some of these topics are discussed in more detail in later chapters.

TABLE 1.7
Objectives of Risk Communications

Objectives	Comments
Education	This includes educating decision makers (message can be technical or nontechnical) and educating public (message would be nontechnical)
Security warnings	Communicating upcoming hazard or threat and susceptibility of community or asset
Behavior change	Communicating status of some of the risk components and aims at modifying the behavior of stakeholders to improve or comply with different needs to improve risk
Mitigation efforts	Informing stakeholders (public, managers, etc.) of upcoming or ongoing improvement efforts as needed
Funding	Relay needs for funding to appropriate personnel (e.g., owners, elected officials, federal, state, and local agencies)

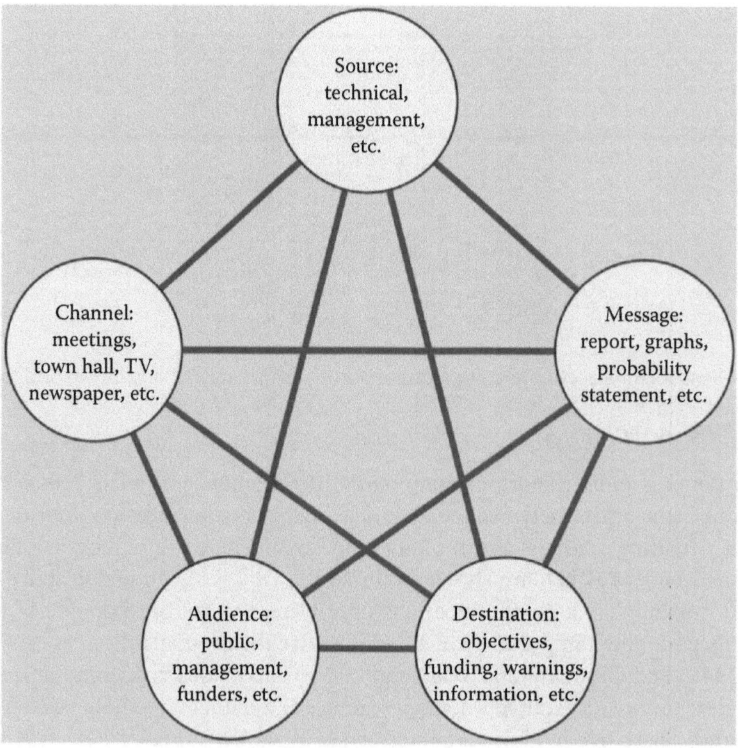

FIGURE 1.23 Risk communication modules as pertained to suspension bridges.

1.5.2 Uncertainty

Uncertainty considerations are so basic in risk management that many authors just used probability of undesirable events as simply the definition of risk itself. Even though, we mentioned such an approach is not acceptable, we still consider proper dealings with uncertainties to be a major component in risk management. Because of this, most of the processes we offer in this book consider uncertain variables, which result in describing the risk. In some risk methods, such as the weighted averages method, the uncertainties are included in the assumed weights. Fenton and Neil (2013) have discussed the relationships between weights and the probability distributions (or histograms).

1.5.3 Intersections/Links/Networks

Numerous variables control even the simplest risk models. In realistic problems, most of these variables are codependent, sometimes in fairly complex fashion. Because of this, these codependencies are accommodated in all risk management components, as expected. Obviously, ignoring these linkages or interactions when performing any of the risk management components could lead to erroneous decisions or computations, which might, in turn, lead to costly operations and the potential for unintended consequences.

1.5.4 Multihazards Considerations

If we define a hazard as any event that might degrade the performance of an asset or a community, as in Section 1.2.2, we can conclude that assets or communities are rarely subjected to just one type of hazard. Applying this general definition of a hazard, we believe that abnormal man-made events (such as bomb blasts) or natural events (such as earthquakes or super storms) are considered hazards. We must also consider lesser dramatic, but equally destructive, processes such as corrosion or normal wear and tear as hazards. Ettouney et al. (2005) and Ettouney and Alampalli (2016) introduced a theory of multihazards, postulating that decisions regarding a hazard in the civil infrastructure community will affect the behavior of the infrastructure or community due to any other hazard. This means that a multihazard risk-based strategy needs to be implemented as opposed to dealing with one hazard at a time. Failing to follow a multihazard risk-based strategy might result in costly unintended consequences for assets and/or communities.

1.5.5 Snapshot vs. Time Marching

Civil infrastructures are built to last for a long time. During this time of service, they are expected to offer the stakeholders an adequate performance at reasonable costs. These two intertwining and closely related goals are what makes risk-based management an essential paradigm for decision makers that no other paradigm can replace. This means that risk management processes need to accommodate current/immediate goals while offering potential for an optimal cost/performance in the future as time marches. Four out of the five risk management components might then be affected by time marching considerations: acceptance, treatment, monitoring, and communications. Some techniques that can aid future planning in a risk management landscape are:

- Processes such as life cycle analysis (see Hawk 2003)
- Dynamic graph networks*
- Decisions under uncertainties tools such as decision trees or influence diagrams†

1.5.6 Subjective vs. Objective Risk

Defining risk scales/metrics is an essential step in ensuring efficient risk management such that optimal decisions are made. These metrics can be subjective or objective. An example for subjective risk scales are *High*, *Medium*, and *Low*. Subjective risk scales are relative by definition. They offer simplicity, yet they can't be used in situations where objective risk metrics are needed.

* Presented in Chapter 2, with several case studies throughout this book.

† Presented in Chapter 2, with several case studies throughout this book.

Several authors have used this approach to describe risk in a subjective fashion (FEMA 2005, 2011a, 2011b, 2011c; Fenton and Neil 2013).

Objective risk scales can be subdivided into two categories: relative or absolute. Relative objective risk or any of its components can be described objectively on an analytical scale, say from 0 to 10 or 0 to 100. In most cases, lower risk scores indicate a better situation. A perfect risk condition would be at the bottom of the scale, say a risk of 0. The relative risk scales were also used by DHS (2011a, 2011b, 2011c). An absolute risk scale is generally measured with monetary units. An example would be: average value at risk of a particular situation is $50. The lower number indicates less value at risk or simply less risk. The absolute risk scales have been used by numerous researchers such as Jorion (2007) in the field of finance and Ettouney and Alampalli (2012a, 2012b) in the field of civil infrastructure. A comparison between advantages and disadvantages of risk scales is shown in Table 1.8.

1.5.7 Cascading Effects

Ettouney (2014) offered a definition of cascading effect (CE) process that stated:

> When an undesirable event affects a particular asset (or a component of an asset or community), and degrades its performance, such that the influences of this initial degradation propagate downstream through the asset (or community). During such propagation, the operations of other components become degraded too. The degradation process will continue throughout the asset (or community). The net result, when the degradation propagation finally stops, might be a severe deterioration of the operation of the asset or the community.

There are numerous examples of CE in the field of civil infrastructures. The progressive collapse of buildings is a classic example. Another recent example of CEs occurred during the 2012 super storm Sandy (see Figures 1.24 through 1.26) in the New York–New Jersey region. In short, we can find examples of CEs in the aftermath of any natural or man-made disaster.

Clearly, CE behavior plays an integral role in the risk-based management of an asset or community. A lower CE score translates to lower risk, and the reverse is true. This is another reason for the importance of links in Figures 1.24 (single asset) and 1.25 (community of assets). Without accurate representations of links in any risk management process, CE and the whole risk management process will be neither accurate nor efficient.

Given the obvious importance of considering CE within the context of risk management, we need a method to model it objectively. Unfortunately, there currently is no standard available for standardizing CE measures within an asset or community. Recently, two measures have been proposed: the cascading effects extent (CEE) and the cascading effects strength (CES) (Ettouney 2014).

TABLE 1.8
Advantages and Disadvantages of Risk Scales

Risk Scale		Advantages	Disadvantages
Subjective	Relative-discrete	Simple	Difficult to use for detailed risk management components other than assessment
Objective	Relative-continuous	Simple. Can be used for large-scale assessment and project prioritizations	Requires more computational resources
	Absolute	Most useful for risk scales	Requires most computational resources; can be complex at times

FIGURE 1.24 Cascading effects during super storm Sandy: high wind resulted in broken glass, which resulted in rain damage of the store. (Courtesy of Albert Di Bernardo.)

FIGURE 1.25 Cascading effects on private homes. Wind and flood damage led to fire damage. (Courtesy of Marco Coco.)

CEE measures the extent of the effects that a particular consideration propagates within the asset or community. CES is a metric describing the relative importance of parameters or considerations affected by the consideration of interest, their condition ratings, and the geometric extent of the propagation of the effects of such a consideration throughout the facility or community. CEE is a purely geometric metric, and CES is a metric that combines geometry, importance, and CR in

FIGURE 1.26 CE effects during super storm Sandy: storm surge led to the flooding of subway stations, which led to interruption of services. The resulting effects on numerous businesses throughout NYC were immense. (Courtesy of Albert Di Bernardo.)

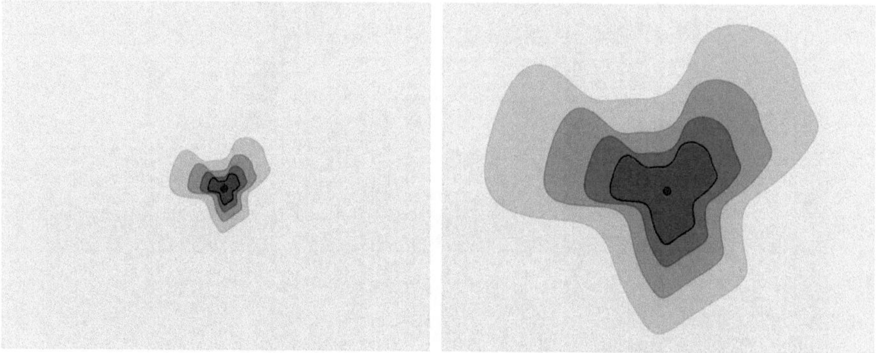

Limited CEE   Significant CEE

FIGURE 1.27 Examples of CEE.

one metric. For simplicity, CEE and CES are taken to be in the range of zero to 100. Because CEE is merely a geometric measure, its value is not an indication of higher or lower performance. If CES is low, it would indicate lower performance; higher performance is inferred for higher values. Figure 1.27 shows schematically different performances of CEE.

1.5.8 Unintended Consequences

One of the most desired goals of adequate risk management is to try to avoid unintended consequences. For example, when devising a risk improvement project for a particular asset of the community for flood mitigation, if the right precautions are not taken, the project could prove harmful later if/when an earthquake strikes the community (see Ettouney and Alampalli 2012b). Some of the concerns that can produce unintended consequences are:

1. Ignoring interactions within an asset or community
2. Cascading effects
3. Ignoring interactions between hazards

REFERENCES

Agrawal, A.K., and Kawaguchi, A. (2009). *Bridge Element Deterioration Rates,* Final report on project C-01-51, New York State Department of Transportation, Albany, NY.

Alampalli, S., and Ettouney, M. (2010). Resiliency of Bridges: A Decision Making Tool. *Bridge Struct.* 6(1), 65–72.

Albrecht, A., and Miquel, S. (2010). Extension of Sensitivity and Uncertainty Analysis for Long Term Dose Assessment of High Level Nuclear Waste Disposal Sites to Uncertainties in the Human Behavior. *J. Environ. Radioact.* 101(1), 55–67.

Archetti, R., and Lamberti, A. (2003). Assessment of Risk due to Debris Flow Events. *Nat. Hazards Rev.* 4(3), 115–125.

ASCE. (2013). *ASEC Report Card for America's Infrastructure*, American Society of Civil Engineers, Reston, VA.

ASTM. (2002). *Standard Guide for Risk-Based Corrective Action Applied at Petroleum Release Sites*, E1739-95, ASTM, West Conshohocken, PA.

ASTM. (2004). *Standard Guide for Risk-based Corrective Action*. E2081-00, West Conshohocken, PA.

Ayyub, B.M. (2014). *Risk Analysis in Engineering and Economics*, 2nd ed., CRC, Boca Raton, FL.

Balas, C.E., and Ergin, A. (2002). Reliability-Based Risk Assessment in Coastal Projects: Case Study in Turkey. *Journal of Waterway, Port, Coastal, and Ocean Engineering*, 128(2): 52–61.

Bensi, M. T., Straub, D., Friis-Hansen, P., and Der Kiureghian, A. (2009). Modeling infrastructure system performance using BN. *Proc. ICOSSAR'09*, CRC Press, Boca Raton, FL.

Bruneau, M., Chang, S., Eguchi, R., O'Rourke, T., Reinhorn, A., Shinozuka, M., Tierney, K., Wallace, W. and Winterfelt, D. (2003). A Framework to Qualitatively Assess and Enhance the Seismic Resilience of Communities. *Earthquake. Spectra. J.* 19(4), 733–752.

Bruneau, M., and Reinhorn, A. (2007). Exploring the Concept of Seismic Resilience for Acute Care Facilities. *Earthquake Spectra.* 23(1), 41–62.

Burcharth, H. (1992). Reliability Evaluation of a Structure at Sea. In *Proceedings of the Short Course on Design and Reliability of Coastal Structures*, Lamberti, (Ed.). Bologna, Italy, pp. 21.1–21.48.

Cardona, O.D., Van Aalst, M.K., Birkmann, J., Fordham, M., McGregor, G., Perez, R., Pulwarty, R.S., Schipper, E.L.F., and Sinh, B.T. (2012). Determinants of Risk: Exposure and Vulnerability. In: *Managing the Risk of Extreme Events and Disasters to Advance Climate Change Adaptation*, Field, C.B., Barros, V., Stocker, T.F., Qin, D., Dokken, D.J., Ebi, K.L., Mastrandrea, M.D., et al. (Eds.). A Special Report of Working Groups I and II of the Intergovernmental Panel on Climate Change (IPCC). Cambridge University Press, Cambridge, UK, pp. 65–108.

Changnon, S. (2009). Tornado losses in the United States. *Nat. Hazards Rev.* 10, 145–150.

Der Kiureghian, A. (2005). First- and Second-order Reliability Methods. *Engineering Design Reliability Handbook*, Chap. 14, E. Nikolaidis et al. (Eds.) CRC, Boca Raton, FL.

DHS. (2009). *National Infrastructure Protection Plan*, Department of Homeland Security, Washington, DC.

DHS. (2011a). *Integrated Rapid Visual of Screening of Buildings*, Building and Infrastructure Protection Series, DHS, Washington, DC.

DHS. (2011b). *Integrated Rapid Visual Screening of Mass Transit Stations*, Building and Infrastructure Protection Series, DHS, Washington, DC.

DHS. (2011c). *Integrated Rapid Visual Screening of Tunnels*, Building and Infrastructure Protection Series, DHS, Washington, DC.

Ditlevsen, O., and Madsen, H. O. (1996). *Structural Reliability Methods*, Wiley, New York.

Elishakoff, I. (1983). *Probabilistic Methods in the Theory of Structures*, Wiley, Hoboken, NJ.

Ellingwood, B. R. (2006). Structural safety special issue: General purpose software for structural reliability analysis. *Struct. Safety*, 28(1–2), 1–2.

EM-DAT. (2014). *EM-DAT: The OFDA/CRED International Disaster Database*, Université Catholique de Louvain, Brussels, Belgium.

Ettouney, M.M. (2014). Cascading Effects: A Needed Metric for Resilient Infrastructures. *J. Natl. Inst. Build. Sci.* 2(2).
Ettouney, M.M., and Alampalli, S. (2012a). *Infrastructure Health in Civil Engineering: Theory and Components*, CRC Press, Boca Raton, FL.
Ettouney, M.M., and Alampalli, S. (2012b). *Infrastructure Health in Civil Engineering: Applications and Management*, CRC Press, Boca Raton, FL.
Ettouney, M.M., and Alampalli, S. (2016). *Multihazard Considerations in Civil Infrastructure*, CRC Press, Boca Raton, FL.
Ettouney, M.M., Alampalli, S., and Agrawal, A. (2005). *Theory of Multihazards for Bridge Structures*, IO Press, New York.
Faber, M. H., Kroon, I. B., Kragh, E., Bayly, D., and Decosemaeker, P. (2002). Risk Assessment of Decommissioning Options Using Bayesian Networks. *J. Offshore Mech. Arct. Eng.*, 124(4), 231–238.
FEMA. (2005). *Risk Assessment: A How-To Guide to Mitigate Terrorist Attacks*, Risk Management Series, FEMA 452, Federal Emergency Management Agency, Washington, DC.
Fenton, N., and Neil, M. (2013). *Risk Assessment and Decision Analysis with Bayesian Networks*, CRC Press, Boca Raton, FL.
FHWA. (1970). Bridge Inspector's Reference Manual. Washington, D.C.: United States Department of Transportation, 2002, Revised 2006.
Friis-Hansen, A. (2000). Bayesian networks as a decision support tool in marine applications. Ph.D. thesis, DTU, Lyngby, Denmark.
Friis-Hansen, P. (2004). Structuring of Complex Systems Using Bayesian Network. *Proceedings Workshop on Reliability Analysis of Complex Systems*, Technical Univ. of Denmark, Lyngby, Denmark.
Grêt-Regamey, A., and Straub, D. (2006). Spatially Explicit Avalanche Risk Assessment Linking Bayesian Networks to a GIS. *Nat. Hazards Earth Syst. Sci.*, 6(6), 911–926.
Gribb, M.M., Bene, K.J., and Shrader, A. (2002). Sensitivity Analysis of a Soil Leachability Model for Petroleum Fate and Transport in the Vadose Zone. *Adv. Environ. Res.* 7(1), 59–72.
Gutteling, J., and Wiegman, O. (1996). *Exploring Risk Communications*, Kluwer Academic, Dordrecht, The Netherlands.
Halpin, D.W., and Woodhead, R.W. (1976). *Design of Construction and Process Operations*, Wiley, New York.
Hawk, H. (2003). *Bridge Life Cycle Cost Analysis*, National Cooperative Highway Research Program, NCHRP Report 483, National Research Council, Washington, DC.
Howell, G.A., Laufer, A., and Ballard, G. (1993). Interaction between Subcycles: One Key to Improved Methods. *J. Construct. Eng. Manage.* 119, 714.
Hughes, A. (2000). *Risk Management for UK Reservoirs*. CIRIA Rep. No. C542, London.
Hynes, M.E. (2001) Personal Communications.
Ioannou, P.G. (1989). *UM-CYCLONE Discrete Event Simulation System, Reference Manual*, Report No. UMCE-89-11, Department of Civil Engineering, University of Michigan, Ann Arbor, MI.
Jorion, P. (2007). *Value at Risk: The New Benchmark for Managing Financial Risk*, McGraw-Hill, New York.
Lasswell, H.D. (1948). The Structure and Function of Communication in Society. In: *The Communications of Ideas*, Bryson, L. (Ed.). Harper, New York, pp. 37–51.
Mahadevan, S., and Rebba, R. (2005). Validation of Reliability Computational Models Using Bayes Networks. *Reliab. Eng. Syst. Saf.* 87(2), 223–232.
Mahmoud, K.M. (2011). BTC Method for Evaluation of Remaining Strength and Service Life of Bridge Cables. Report C-07-11. Submitted to the New York State Department of Transportation, Albany, NY.
Martinez, J.C., and Ioannou, P.G. (1999). General-purpose Systems for Effective Construction Simulation. *J. Construct. Eng. Manage.* 125(4), 265–276.
Masada, T., Sargand, S.M., Tarawneh, B., Mitchell, G.F., and Gruver, D. (2007). Inspection and Risk Assessment of Concrete Culverts under Ohio's Highways. *J. Perform. Construct. Facil.* 21(3), 225–233.
Mitropoulos, P., and Namboodiri, M. (2011). New Method for Measuring the Safety Risk of Construction Activities: Task Demand Assessment. *J. Construct. Eng. Manage.* 137(1), 30–38.
National Cooperative Highway Research Program (NCHRP). (1982). Assessment of Deficiencies & Preservation of Bridge Substructures below the Waterline. Rep. No. 251, Transportation Research Board, National Research Council, Washington, DC.
Nishijima, K., Maes, M. A., Goyet, J., and Faber, M. H. (2009). Constrained Optimization of Component Reliabilities in Complex Systems. *Struct. Safety*, 31(2), 168–178.

NIAC (National Infrastructure Advisory Council). (2009). *Critical Infrastructure Resilience Final Report and Recommendations*, NAIC, Washington, DC.

NRC. (2010). *Review of the Department of Homeland Security's Approach to Risk Analysis*, National Academic Press, Washington, DC.

NYSDOT. (2006). *Bridge Inspection Manual* (1997 with updates through 2006). New York State Department of Transportation, Albany, NY.

Oglesby, C.H., Parker, H., and Howel, G. (1989). *Productivity Improvement in Construction*, McGraw-Hill College, NY.

Ohio Dept. of Transportation (ODOT). (2003). *Culvert Management Manual*, Columbus, OH. Available at www.dot.state.oh.us/se/hy/Culvert.htm

Rackwitz, R. (2001). Reliability Analysis—A Review and Some Perspectives. *Struct. Safety*, 23(4), 365–395.

Rausand, M. (2004). *System Reliability Theory: Models, Statistical Methods, and Applications*, John Wiley & Sons, Hoboken, NJ.

Russell, S. J., and Norvig, P. (2003). *Artificial Intelligence: A Modern Approach*, 2nd Ed., Prentice-Hall, Upper Saddle River, NJ.

Seuntjens, P., Mallants, D., Simunek, J., Patyn, J., and Jacques, D. (2002). Sensitivity Analysis of Physical and Chemical Properties Affecting Field Scale Caatdmium Transport in a Heterogeneous Soil Profile. *J. Hydrol.* 264(1–4), 185–200.

Straub, D., and Kiureghian, A.D. (2010) Bayesian Network Enhanced with Structural Reliability Methods: Methodology. *J. Eng. Mech.*, 136(10), 1248–1258.

Straub, D., and Kiureghian, A.D. (2010) Bayesian Network Enhanced with Structural Reliability Methods: Application. *J. Eng. Mech.*, 136(10), 1259–1270.

Sustainable Measures. (2014). Available at http://www.sustainablemeasures.com/node/35 (accessed April 7, 2014).

Tehrani, P., and Mitchell, D. (2013). Seismic Risk Assessment of Four-Span Bridges in Montreal Designed Using the Canadian Bridge Design Code. *J. Bridge. Eng* 10.1061/(ASCE)BE.1943-5592.0000499, A4014002.

Thoft-Christiansen, P., and Baker, M.J. (1982). *Structural Reliability Theory and Its Application*, Springer, Berlin.

Tsai, T.T., Kao, C.M., Surampalli, R.Y., Huang, W.Y., and Rao, J.P. (2011). Sensitivity Analysis of Risk Assessment at a Petroleum-Hydrocarbon Contaminated Site. *J. Hazard Toxic. Radioact. Waste*, 15(2), 89–98.

Union of Geological Sciences (IUGS), Working Group on Landslides, Committee on Risk Assessment. (1997). *Quantitative Risk Assessment for Slopes and Landslides—The State of the Art in Landslides Risk Assessment*. Rep., Cruden, D.M. and Fell, R. (Eds.). Balkema, Rotterdam, The Netherlands.

Vamvatsikos, D., and Cornell, C.A. (2002). Incremental Dynamic Analysis. *Earthquake Eng. Struct. Dynam.* 31(3), 491–514.

Vo, T.V., Blackburn, T.R., Casazza, L.O., Heasler, P.G., Mara, N.L., and Phan, H.K. (1997). Reliability Analysis in Support of Risk Assessment for Hydroelectric Generating Stations, *Journal of Water Resources Planning and Management*, 123(3): 179–188.

Vose, D. (2009). *Risk Analysis a Quantitative Guide*, Wiley, Hoboken, NJ.

Vrijling, J.K., van Hengel, W., and Houben, R.J. (1995). A Framework for Risk Evaluation. *J. Hazard. Mater.* 43, 245–261.

Zou, P.X.W., Chen, Y., and Chan, T.Y. (2010). Understanding and Improving Your Risk Management Capability: Assessment Model for Construction Organizations. *J. Construct. Eng. Manage.* 136(8), 854–863.

2 Graph Networks

2.1 INTRODUCTION

2.1.1 Overview

Detailed study and accurate analysis of topics such as risk, resilience, redundancy, exposure, consequences, and hazards associated with infrastructure require adherence to several attributes noted below. Improper consideration of these attributes may lead to incomplete and/or inaccurate results and will lead to (wrong) decisions with potential negative consequences. Attributes of interest include the following:

- Realistic and accurate modeling linkages/interactions among variables
- Accommodation of uncertainties associated with different variables
- Accurate modeling of observations at any given time (snapshot or a time period)
- Accommodation of objective and subjective variables and pertinent combinations
- Seamless accommodation of decision under uncertainty (DUU) while accounting for all of the above attributes
- Ease of simulating complex models
- Ease of changing, adding to, or trimming complex models

Closed form solutions (Vose 2009), decision tree (Neapolitan 2004), and other tree-related constructs and weighted averages models (Fenton and Neil 2013) are some of the possible analytical models that can be used to accommodate above attributes However, each of these methods can only accommodate some, but not all, attributes mentioned earlier. Probabilistic graph networks (GNs) can accommodate all of these attributes, and they have potential for use in the field of infrastructure management.

GN methods have been used extensively in the recent past in the fields of computer science and medicine, as well as several other fields (Jensen 1996; Koller and Friedman 2009). But they have not been used extensively in the field of civil infrastructure. In this book, we demonstrate applications of GNs with its different facets. Because the reader is probably not familiar with this method, it will be explained in this chapter. Numerous taxonomies and notations have been used in the literature, but no standard taxonomy exists for the application of GNs in the civil engineering field. Hence, whenever applicable, for this chapter and the rest of the book, the taxonomy of Kjaerulff and Madsen (2013) is used because of its simplicity and generality.

2.1.2 Contents of This Chapter

This chapter provides a brief introduction to the main attributes of GNs and their different subcategories, which can be of help while studying the main topics in this book. We start by a general offering of GNs and their main components. We then explore three subcategories of GN:

1. Static (or snapshot) GN: We illustrate the use of three different incarnations of Static GN, namely, Bayesian network (BN), Markov network (MN), and chain graph (CG) network (CGN).
2. Influence diagram (ID) that accommodates decisions under uncertainty.
3. Dynamic graph network (DGN) that accommodates time marching.

Figure 2.1 presents the layout of this chapter. For each GN type, we briefly discuss its theoretical basis and then offer simple examples to further explain it. Whenever possible, we offer simple,

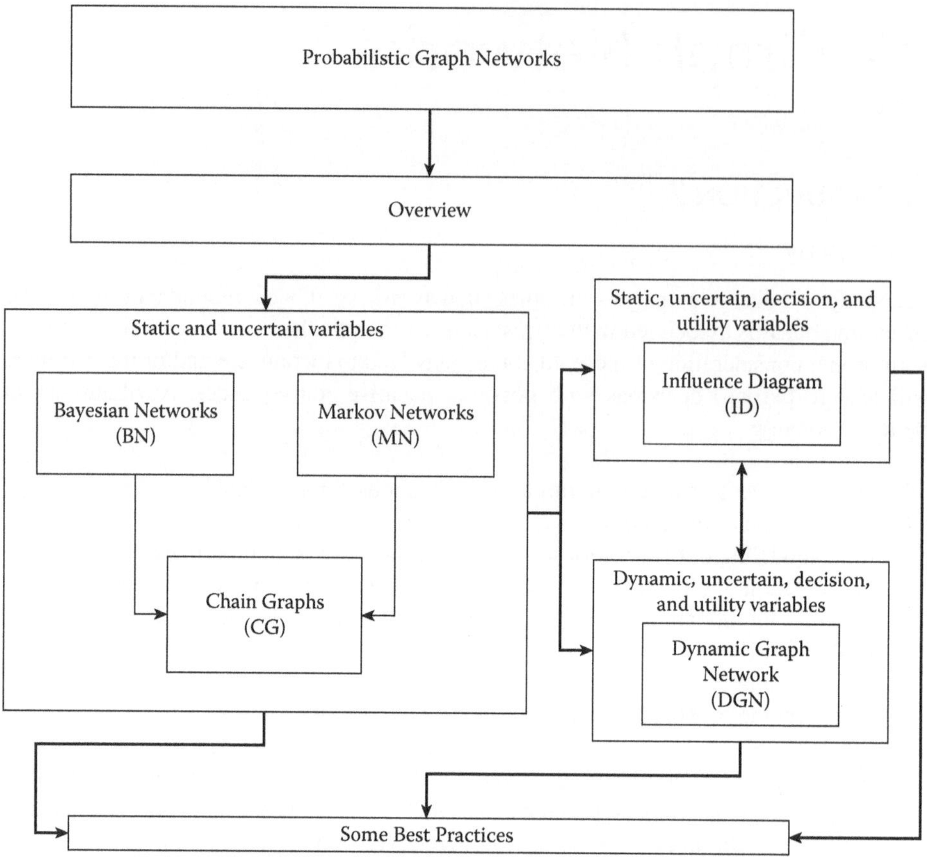

FIGURE 2.1 Chapter layout.

practical examples from the field of civil infrastructures while avoiding the theoretical complexities of the GN process.

Readers should note that this chapter only gives a brief introduction of the rich and demanding field of GNs. Readers are encouraged to research the references that are cited herein for additional discussions and presentations of this rapidly developing and beneficial field.

2.2 FORMAL MODELING OF GRAPH NETWORKS

2.2.1 Overview

A graph network, $\mathcal{G}$, can be defined as

$$\mathcal{G} = (\mathcal{X}, \mathcal{L}) \tag{2.1}$$

where $\mathcal{X}$ represents a finite set of nodes (sometimes referred to as vertices) and $\mathcal{L}$ represents a finite set of links (sometimes referred to as edges or connectors) between the nodes. In general, the nodes represent random variables $X_i \in \mathcal{X}$ (In addition to random variables, a node in $\mathcal{G}$ might contain other types of variables. These will be discussed later.). Within $\mathcal{G}$, an ordered pair of nodes, X_i and X_j, are connected by a link such that $(X_i, X_j) \in \mathcal{L}$. This linkage is generally denoted as $X_i \sim X_j$.

2.2.2 Components of Graph Networks

Throughout this book, we consider three types of nodes and two types of links.

2.2.2.1 Nodes

Three types of nodes are used in BN to represent variable parameters. These are nodes that represent random variables, nodes that represent decision variables, and nodes that represent utility variables. We discuss the details of each of the node types next.

Nodes representing random variables: Variables of this type are random, which can be described either by histograms with discrete states or probability density functions with continuous states. States are also referred to as *ranks*. The variables of all case studies in this book are described using histograms with a discrete number of states. Fenton and Neil (2013) described methodologies for solving BN with continuous state variables. We note that solutions of BN, or general GN, can be simplified greatly if all the random variables within the network are Gaussian (see Koller and Friedman 2009). We represent nodes containing random variables with oval shapes, as shown in Figure 2.2. The applicable type of GN, where random variables can be used, is shown in Table 2.1.

Nodes representing decision variables: Decision-making is seldom made under complete and certain conditions. Most decisions rather rely on uncertain or incomplete information that is usually linked together in a complex manner and can be effectively modeled using GNs. In such situations, the different potential choices (or decisions) available to decision makers can be modeled using decision nodes/variables. We represent nodes containing decision variables with rectangular shapes as shown in Figure 2.2. The applicable type of GN in which decision variables can be used in is shown in Table 2.1.

Nodes representing utility variables: Utility is a means to compute the consequences of various events or decisions. Utility can also be used as an expression of risk or reward both subjectively (in a relative manner) or objectively (in an absolute manner). When utility is expressed in an objective manner, it is generally expressed in a monetary fashion. Von Neumann and Morgenstern (1944) were first to address the concept of utility and several other authors address them extensively

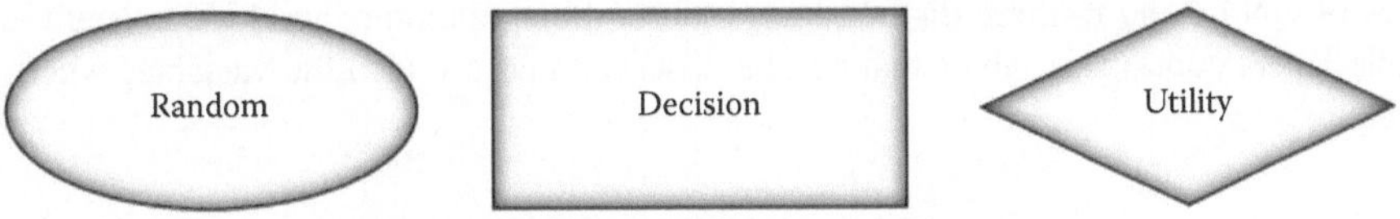

FIGURE 2.2 Representations of different types of nodes/variables in a GN.

TABLE 2.1
Relationship between Network Type and Nodes

Node (Variable)	GN
Random	BN, MN, CG, ID, DBN, or DGN
Decision	ID[a]
Utility	ID[a]

[a] There is no reason to prevent using decision nodes in a dynamic GN. However, for simplicity, we do not address this type of GN in this book.

(Trueman 1977; Neapolitan 2004). In a GN setting, utilities can be included as variables that depend on other random and/or decision variables. We represent nodes containing utility variables with diamond shapes as shown in Figure 2.2. The applicable type of GN in which utility variables can be used in is shown in Table 2.1.

2.2.2.2 Links

Two types of links are used to connect nodes/variables—directed and undirected. A directed link from X_i to X_j is denoted as $X_i \rightarrow X_j$. It indicates a conditional dependence of X_j on X_i. In such cases, X_i is referred to as the parent node and X_j is referred to as the child node. As an example, we know that the vulnerability of a bridge, *V*, depends on its condition (rating), *CR*. In a network setting, *CR* is said to be the parent node to the child node *V*, i.e., $CR \rightarrow V$. The directional links in a probabilistic network is expressed objectively by conditional probabilities tables (CPTs).

In many civil infrastructure applications, we encounter situations where the two nodes are linked, but there is no obvious directional dependency between them. In such cases, a link in a $\mathcal{G}$ will have no direction and is denoted as $X_i - X_j$. The nondirectional relationships in a probabilistic network are expressed objectively by potentials, ϕ. For example, average daily traffic of a route or a bridge, *ADT*, is related to the functional classification of the route (interstate, arterial, local, etc.), *Functional*. Functional class is usually high when *ADT* is high, and vice versa. Though an affinity exists between these two variables, there is no clear directionality between the two. Based on this, we can formally express the link between the two as *ADT–Functional*.

BN is a $\mathcal{G}$ with only directional links. Similarly, MN is a $\mathcal{G}$ with only nondirectional links (some authors refer to it also as a Markov random fields, or MRF). When a $\mathcal{G}$ has both directional and nondirectional links, it is called a CG. We discuss each of these three networks in subsequent sections. We model directed links using a line with a single arrowhead pointing from the parent variable toward the child variable (see Figure 2.3). A nondirected link is modeled as a line with no arrowheads as shown in Figure 2.3.

2.2.2.3 Layers

The nodes of GN belong to three distinct layers: observable, outcome, and hidden (see Figure 2.4). Observable layers contain variables that can be observed and are also the variables where the user

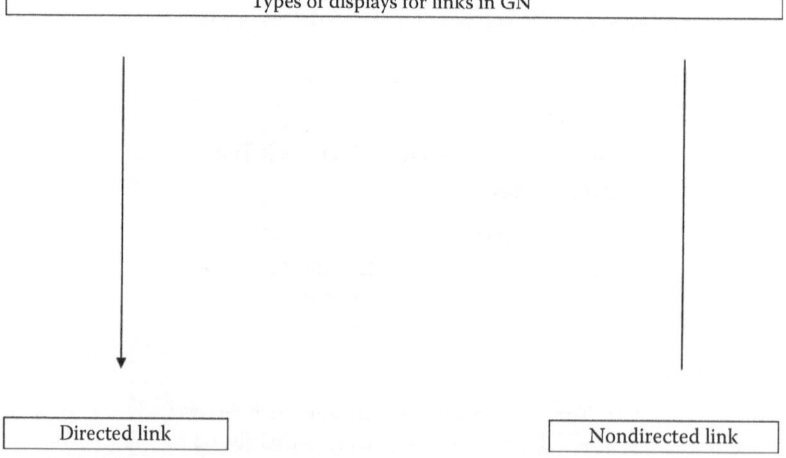

FIGURE 2.3 Representations of links in a GN.

can introduce evidence. For example, a variable named *V.I.* that contains visual inspection results of an infrastructure component lies within an observable layer. The outcome layer contains variables where the user might be interested in computing the values of the variable. For example, risk, reliability, or resilience generally resides in the outcome layer. Hidden layers contain variables that do not belong to observable or outcome layers. They are added to the network to make the model easier to build and more efficient to compute. Of course, this categorization does not need to be adhered to at all times. For example, a user might use the inference capability of GNs and enter a particular risk value as evidence (either as soft or hard evidence, see Section 2.2.5.1) to infer the value of a visual inspection, *V.I.*, state that produces the risk value. Figure 2.4 illustrates the concept of the three layers in a GN. This figure represents a simple risk model for structurally deficient bridges (FHWA 1995). The observable layer includes condition ratings for three components of the bridge: bridge deck (*CR_1*), bridge superstructure (*CR_2*), and bridge substructure (*CR_3*) as well as social and economic consequences (*C_O*). The outcome layer contains the risk variable, *Ri*. The remainder of the variables are all hidden layers.

2.2.3 Tree vs. Network

A connected $\mathcal{G}$ contains paths between any pair of nodes X_i and X_j; such a path is defined by the nodes along the path: $X_i \ldots X_j$. A tree is a special type of network where a path between any given pair of nodes is unique. This points to the limitation of trees in modeling even the simplest relationships for realistic risk/multihazard models. We offer a comparison between the uses of trees vs. GNs in Section 2.6.

2.2.4 Cyclic vs. Acyclic Networks

A cycle in a $\mathcal{G}$ exists along the path formed starting from a particular node along the directed links from one node to the other and ending at the starting node. Any $\mathcal{G}$ with at least one cycle is called a cyclic graph. A $\mathcal{G}$ is called acyclic if it does not contain any cycles. All directed graphs in this book are acyclic.

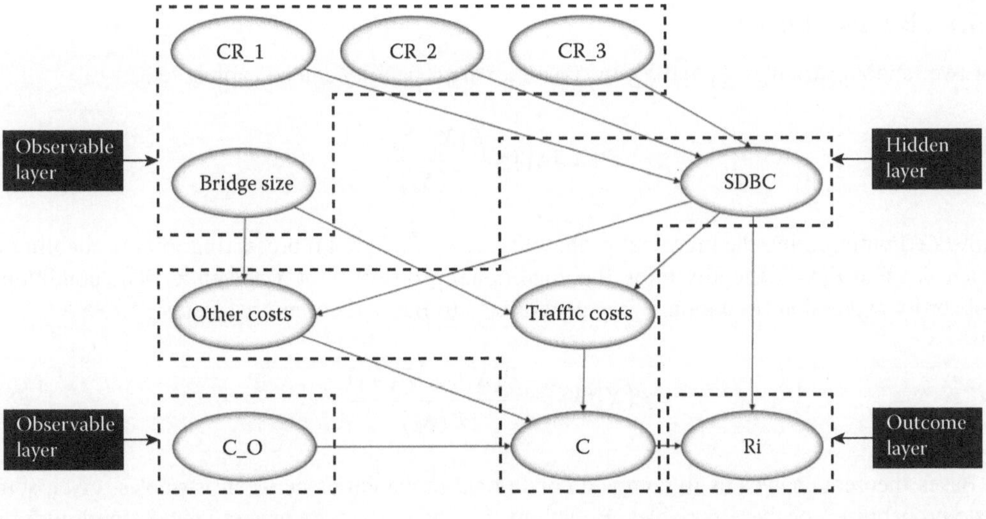

FIGURE 2.4 Layers in a GN.

2.2.5 Observations (Evidence)

One of the basic components in GNs is the link between different variables. These links as described objectively using one of two methods. Directional links are described by a CPT that contains the probabilities of the states of the child node given the states of the parent node. Nondirectional links are described by potentials that contain the affinities between the states of the different nodes within a group of nodes (called cliques) that are connected by nondirectional links. A clique may contain two or more nodes.

The CPTs and/or potentials within a chain graph can then be solved to produce the historical marginal probabilities of all random variables within the GN. We give specific examples of different types of GNs later in the chapter.

The power of GN extends far beyond obtaining historical (prior) probabilities. The true power of GN lies in its capability of inference, or introducing evidence/observation of one or more variables within the network and calculating the corresponding probabilities of other variables given the observations. (We use evidence and observations interchangeably throughout this book.) For example, a visual inspection, *V.I.*, variable within a GN might have a specific historical histogram. However, after an inspection process, the inspector might observe that the condition of a component to be in *adequate* state. Thus, this observation/evidence of *V.I.* = *adequate* can be introduced in the GN and corresponding probabilities of the remaining variables can now be updated.

2.2.5.1 Soft Evidence vs. Hard Evidence

Evidence can be introduced to a GN through hard or soft evidence. Evidence is regarded as hard evidence when the observation is introduced as a 100% probability of a certain state of the variable. For example, if the aforementioned variable, *V.I.*, has four states (*good, adequate, poor, fail*) and the observation introduced is *V.I.* = *adequate*, then it implies that the probability of *V.I.* = *adequate* is 100%. Hence, this also implies that probabilities of the other three states of *V.I.* are equal to zero.

In many situations, evidence is not that certain. For example, due to such uncertainty in observations, let us assume that the inspector assigns (0%, 90%, 10%, 0%) as probabilities associated with the four above mentioned *V.I.* states. Introducing this observation histogram to the network is called soft evidence. Soft evidence involves more computations than hard evidence because it involves a process of linear summations (see Huang 2015).

2.3 BAYESIAN NETWORKS

2.3.1 Bayesian Theory

For two random variables, X_1 and X_2, the conditional probability can be expressed as:

$$P(X_2|X_1) = \frac{P(X_1, X_2)}{P(X_1)} \tag{2.2}$$

with $P(X_1)$ representing the marginal probability of X_1, $P(X_1, X_2)$ representing joint probability of X_1 and X_2, and $P(X_2|X_1)$ representing the conditional probability of X_2 given X_1. The conditional probability expression (Equation 2.2) is generalized to Bayes theorem that states:

$$P(X_2|X_1) = \frac{P(X_1|X_2)P(X_2)}{P(X_1)} \tag{2.3}$$

Bayes theorem enables us to compute conditional probabilities between variables, given some basic probabilities of these variables. Equations 2.2 and 2.3 can be generalized for multitudes of variables (see Fenton and Neil 2013). As the number of random variables increases, it exponentially increases the complexity of modeling and solution. Hence, there is a need to both simplify the modeling efforts and increase the efficiency of solutions. BNs offer such simplification and increase the solution efficiency.

2.3.2 Definition of Bayesian Networks

BN $\mathcal{G}$ is used to compute joint probabilities for a set of variables $\mathcal{X}$ that are connected with a set of directional links $\mathcal{L}$. Formally, we can express such a joint probability as:

$$P(X_1, X_2, \ldots X_n) = \prod_{i=1}^{n} P\left(X_i \mid (Parents)_{X_i}\right) \tag{2.4}$$

In Equation 2.4, the total number of random variables in $\mathcal{G}$ is n. Note that the number of random variables is identical to the number of nodes in a BN. This is because of the absence of other types of nodes (decision nodes or utility nodes) in BN. The expression $X_i \mid (Parents)_{X_i}$ represents the conditional probabilities of X_i, given all of its parents. As noted earlier in this chapter, these conditional probabilities are the building blocks of BN and are usually depicted in a single table, CPTs, for each variable.

Equation 2.4 can be computationally demanding, even for a limited number of random variables. Therefore, several efficient computational techniques were developed to simplify its solutions and are based on the basic principles of probability theory. Detailed description of BN solvers is not within the scope of this book as the focus is on the use of BN for solving risk/multihazards problems in civil infrastructures. More details on BN solvers can be easily found in the literature (e.g., Koller and Friedman 2009; Fenton and Neil 2013; Kjaerulff and Madsen 2013).

2.3.3 Templates of BN Examples

Throughout this book, we use BN to solve infrastructure-related risk management/multihazard problems and case studies. Whenever possible, and for the sake of clarity, we use the following specific steps when presenting these problems and case studies.

Problem description/introduction: We usually start each BN example with a short introduction that describes the overall problem within the context of civil infrastructure. We then discuss the practical issues involved with the problem and present any pertinent special features associated with the example.

Model description: This section of the case study introduces the nodes, $\mathcal{X}$, the directional links, $\mathcal{L}$, the ranks and the definitions of the ranked variables, as well as their discrete states. We use ranked/discrete variables throughout this book. Whenever possible, same examples are repeated to illustrate other types of variables. The model section also includes one or more illustrations of the BN graph model.

CPTs: As explained in Section 2.3.2, each node will have a corresponding CPT. CPT can be generated in several ways. A CPT structure used in this book contains n_{ROW} rows that represent each of the states of the n-th node under consideration and n_{COL} columns that represent all potential states of all parent nodes. As such:

$$n_{ROW} = n_{STATE}\,|_n \tag{2.5}$$

$$n_{COL} = \prod_{i=1}^{i=n_{PARENT}} n_{STATE}\,|_i \tag{2.6}$$

with the number of discrete states of the histogram representing the kth node is $n_{STATE}\,|_k$.

The content of the j-th row represents the probability of the n-th node at the j-th state conditioned on all the different states of all the parents of the node. There are numerous sources regarding objective and/or subjective methods to develop CPTs. (A good discussion of the subject can be found in Fenton and Neil [2012]; Neapolitan [2004].) In most situations, the CPTs of the examples are presented in appendices at the end of each chapter.

Results: After executing the BN, the resulting histograms will be presented. Both historical (or marginal) histograms and histograms resulting from entering observations of one or more variables are found in most cases.

Concluding remarks: To fully benefit from each example, whenever pertinent, we present closing remarks summarizing the value the example offers to readers.

2.3.4 Case Study 2.1: Absolute Risk vs. Relative Risk

The first example is a 2-node example, and it is easy to follow the solution steps manually, if desired. Even though this case study is relatively simple, it addresses an important aspect of risk assessment: relative risk vs. absolute risk.

2.3.4.1 Overview

In many risk problems, the risk measures are relative; that is, risk is measured on a dimensionless scale, say from 0 to 100. Other relative risk measures can be subjective, that is, low, medium, or high. Relating a subjective risk metric to an absolute scale (such as monetary value at risk) is an important topic and might be of interest to many decision makers (see Jorion 2007). Developing such a relationship is not an easy task, and in most cases, to assign definitive absolute values to subjective criterion is almost impossible. Hence, such relationships essentially are probabilistic. We define such a probabilistic relationship between relative and absolute risk using CPTs in this BN example.

2.3.4.2 Model

The model shown in Figure 2.5 contains only two nodes, each representing a random variable simulating relative risk and absolute risk, *Ri_Rel* and *Ri_Abs*, respectively. For the purpose of this example, absolute risk conditionally depends on relative risk; as such, a single directional link points from *Ri_Rel* to *Ri_Abs*. Three states were assumed for relative risk, and five monetary states of absolute risk are shown in Table 2.2.

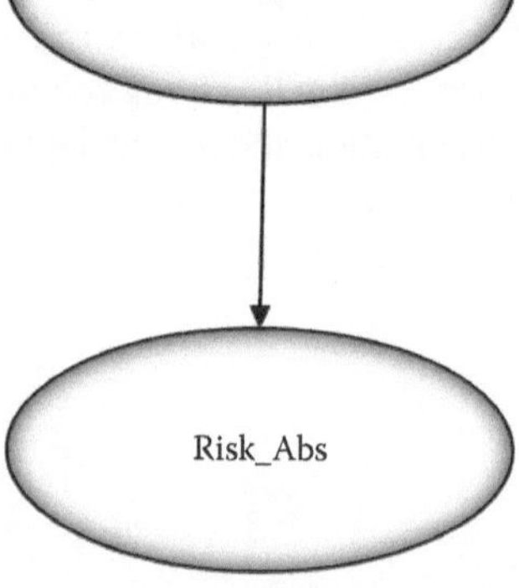

FIGURE 2.5 BN model for absolute risk vs. relative risk.

2.3.4.3 Conditional Probabilities Tables

As noted earlier, it is fairly reasonable to assume that the conditional relationship between *Ri_Abs* and *Ri_Rel* is probabilistic. Therefore, we can form a CPT to describe such a relationship based on decision makers' past experiences, personal judgment, or a closed form probabilistic relationship. A reasonable CPT of *Ri_Rel* is shown in Table 2.3, and the CPT of *Ri_Abs* is shown in Table 2.4 with their respective histograms.

TABLE 2.2
Nodal States for *Ri_Abs* and *Ri_Rel*

Variable	# of States	Ranks (States)
Ri_Rel	3	High, Medium, Low
Ri_Abs	5	> \$100.00, \$100.00, \$50.00, \$10.00, <\$1.00

TABLE 2.3
CPT of *Ri_Rel*

Risk_Rel	Low	0.6
	Medium	0.3
	High	0.1
Histograms for conditional probabilities		

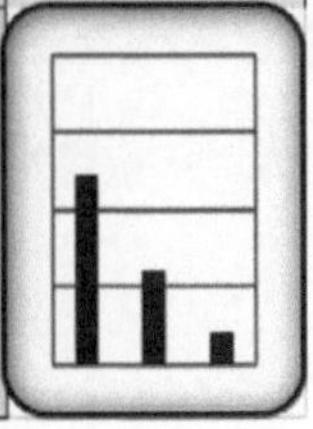

TABLE 2.4
CPT of *Ri_Abs*

Risk_Rel		Low	Medium	High
Risk_Abs	_< \$ 1	0.4811	0.1506	0.0016
	_\$ 10	0.3823	0.2573	0.0254
	_\$ 50	0.1208	0.2863	0.1615
	_\$ 100	0.0151	0.2077	0.4057
	_> \$ 100	0.0007	0.0981	0.4057
Histograms for conditional probabilities				

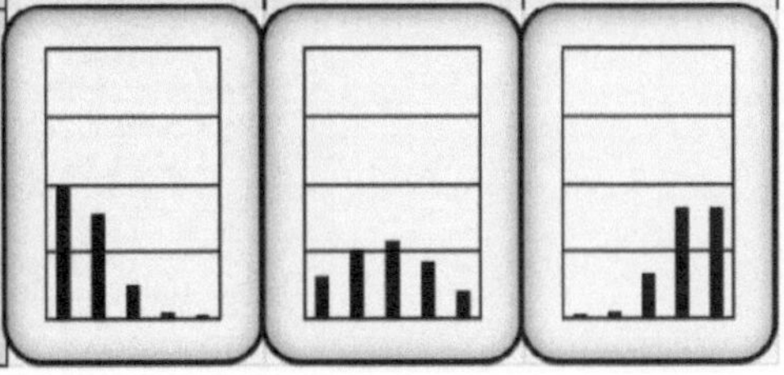

2.3.4.4 Results

With the BN completely defined, we can execute the model first to obtain the prior (historical) probabilities. The resulting histograms are shown in Figure 2.6. Of course, the power of BN is in its capability to assign observations to different variables and evaluate the resulting probabilities of other variables. For example, let us assume that through a process of risk assessment, it is observed that *Ri_Rel* is *Medium* (M). We can specify such an observation in the BN model and find how the probabilities of *Ri_Abs* change. The results are shown in Figure 2.7.

To get an idea of how *Ri_Abs* changes as the observation of *Ri_Rel* changes, we first compute the average historical *Ri_Abs* from the histogram of Table 2.4. We assign monetary values to each of the five bins as $1 M, $10 M, $50 M, $100 M, and $200 M. We can then compute the average historical risk at $ 37.374 M. When *Ri_Rel* of the facility is observed to be *Medium* (M), we can use the results of Figure 2.7 to compute the average absolute risk *Ri_Abs* to be $ 57.521 M. This and similar results can help decision makers in making appropriate risk acceptance, treatment, and communications.

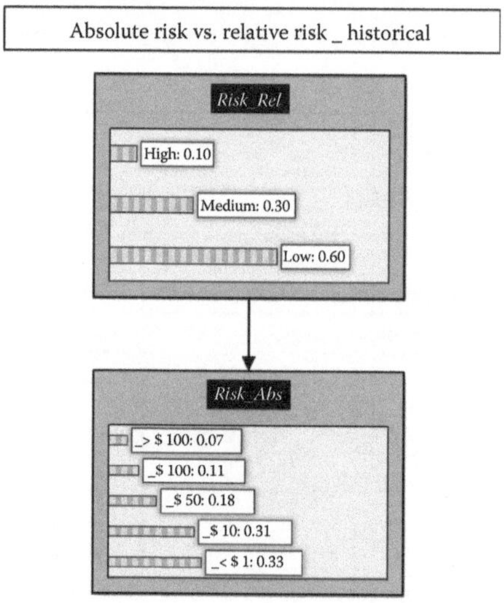

FIGURE 2.6 Historical (prior) probabilities of an *Ri_Rel* and *Ri_Abs* network.

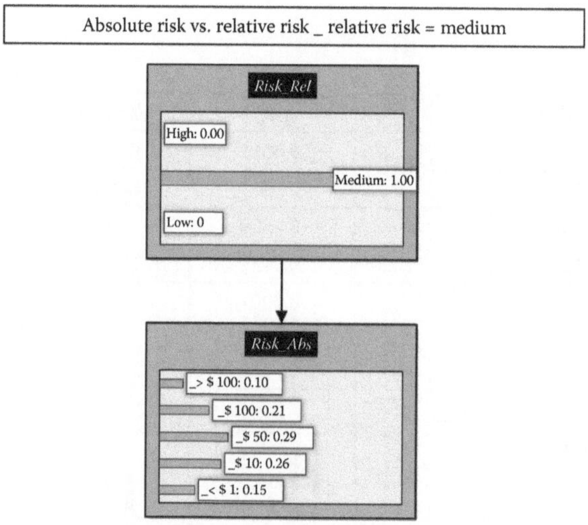

FIGURE 2.7 Histograms of network when *Ri_Rel* = M.

2.3.4.5 Concluding Remarks

Most decision makers would like to see risk values in monetary or other similar absolute fashion. Providing such data associated with the probabilities of their occurrence can help decision makers in developing appropriate risk acceptance, treatment, and communication methods. These values can also be used in project evaluation, projects prioritization, and justification of funding requests.

2.3.5 Case Study 2.2: Infrastructure Security

This case study offers an example of BN application in the infrastructure security field. Security vulnerability assessment of infrastructure has been an important aspect for infrastructure owners. Several factors influence security vulnerability of an infrastructure asset. Of these factors, setback distance and hardening play an important role. Higher setback distance lowers the chances of the threat, while hardening of the structure reduces the effects of the blast. This case study reviews the use of BN in evaluating the influence of setback distance and hardening in security vulnerability assessment.

2.3.5.1 Overview

Among other parameters, the interrelationship between setback distance, *Setback*, and structural hardening, *Hardening*, affects the overall vulnerability of a structure, *Vulnerability*. BN is a natural venue to express this relationship objectively, which is the subject of this example.

2.3.5.2 Model

Let us consider the BN model of Figure 2.8. It contains three variables: *Setback*, *Hardening*, and *Vulnerability*. Since *Vulnerability* of a structure depends on *Setback* and *Hardening*, we model it as a child of both variables. Also, note that *Setback* is independent of *Hardening*, and hence, there is no link between them. We choose three discrete states (ranks) for each of the variables as shown in Table 2.5.

2.3.5.3 Conditional Probabilities Tables

The CPTs of the three variables are shown in Tables 2.6 through 2.8. Tables 2.6 and 2.7 show the historical probabilities of *Setback* and *Hardening*; such probabilities can be estimated from historical frequencies or personal judgment. Table 2.8 shows the conditional probabilities of *Vulnerability* as they relate to the two parent variables. Those probabilities can be obtained either

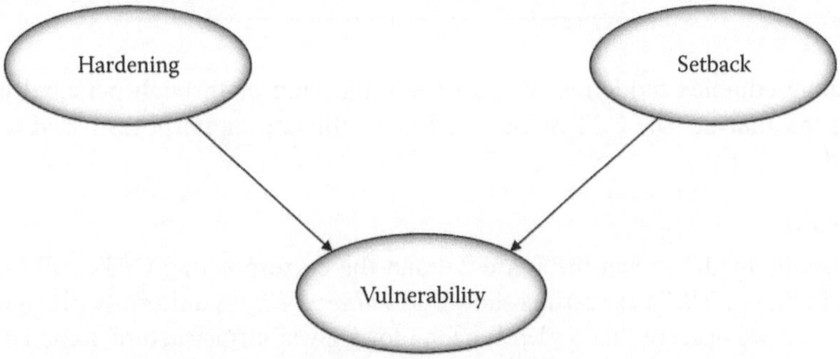

FIGURE 2.8 BN model for setback vs. hardening in a building vulnerability problem.

TABLE 2.5
Discrete States (Ranks) for Model Variable

Variable	# of States	Ranks (States)
Setback	3	Small, Medium, Long
Hardening	3	Limited, Medium, Hardened
Vulnerability	3	High, Medium, Low

TABLE 2.6
CPT of *Setback*

Setback	Long	0.2
	Medium	0.5
	Small	0.3
Histograms for conditional probabilities		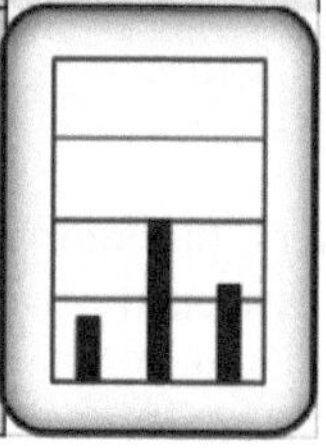

TABLE 2.7
CPT of *Hardening*

Hardening	Hardened	0.15
	Medium	0.35
	Limited	0.5
Histograms for conditional probabilities		

objectively using complex modeling of the system on hand or through personal judgment and experience of the analyst. The CPT in Table 2.8 is for illustration purposes based on the authors' best judgment.

2.3.5.4 Results

Solving BN for the model shown in Figure 2.8 and the corresponding CPTs will yield the histograms shown in Figure 2.9. These results show that *Vulnerability* is uniformly distributed across its states. However, if we observe that *Setback* is long for a given infrastructure, as noted during a risk assessment effort, the resulting histograms shown in Figure 2.10 show an improved *Vulnerability* rating. Similarly, if we observe that the infrastructure is hardened to the hazard under consideration,

TABLE 2.8
CPT of *Vulnerability*

Hardening		Hardened			Medium			Limited		
Setback		Long	Medium	Small	Long	Medium	Small	Long	Medium	Small
Vulnerability	Low	0.8240	0.7677	0.6999	0.7652	0.3553	0.1886	0.6811	0.0355	0.0000
	Medium	0.1726	0.2260	0.2885	0.2257	0.5527	0.5121	0.2927	0.4292	0.0117
	High	0.0033	0.0063	0.0116	0.0091	0.0921	0.2994	0.0263	0.5353	0.9883
Histograms for conditional probabilities										

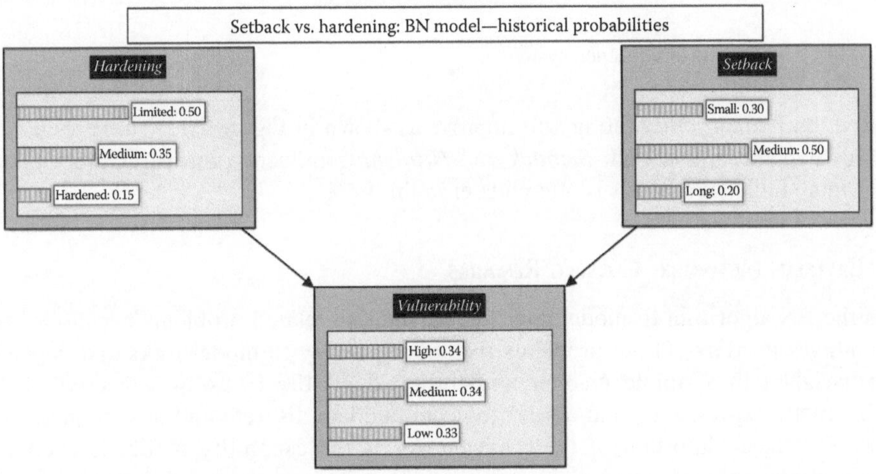

FIGURE 2.9 Historical (prior) probabilities for *Setback*, *Hardening*, and *Vulnerability*.

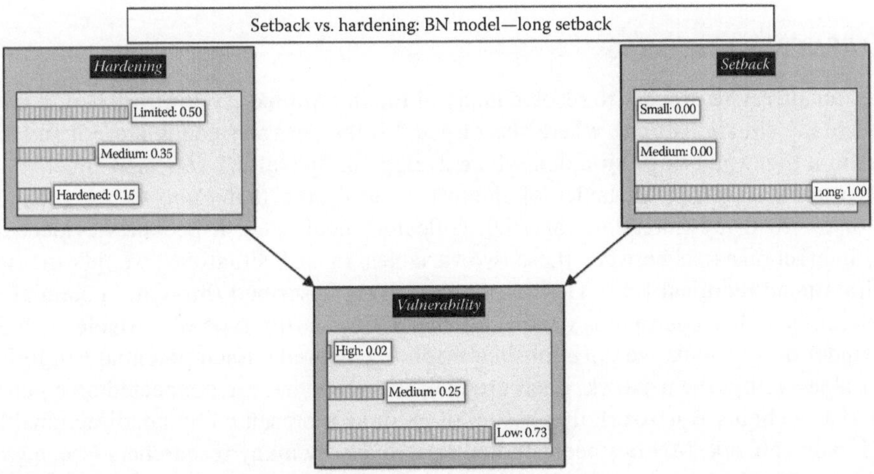

FIGURE 2.10 Histograms when *Setback* is long.

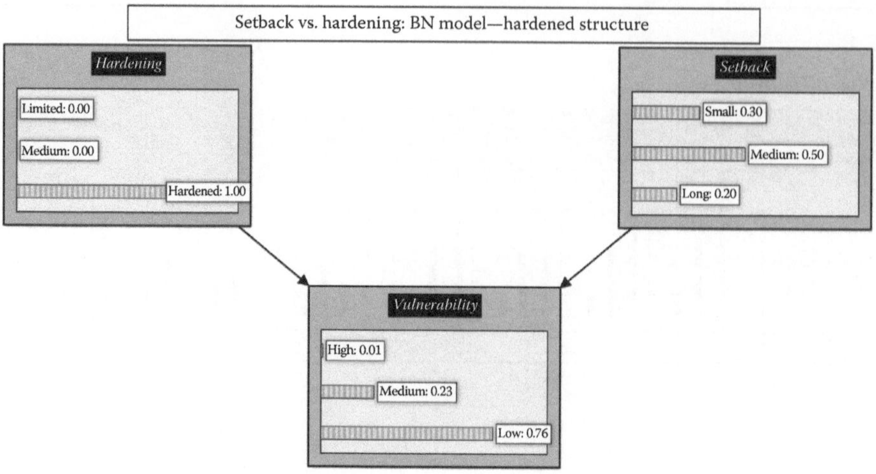

FIGURE 2.11 Histograms of hardened system.

as expected, the *Vulnerability* rating will improve as shown in Figure 2.11. These results indicate the obvious hypothesis that both *Setback* and *Hardening* influence infrastructure *Vulnerability* greatly. A more in-depth analysis is offered later in this book.

2.3.6 Bayesian Network: Closing Remarks

We used the BN algorithm to model and execute the risk-related problems because of the several advantages it offers. These problems include capability to model links and dependencies between variables in a simple manner; scalability; flexibility in using subjective and objective probabilistic expressions; and ability to accommodate discrete and/or continuous random variables. The most important of these advantages is the capability of BN to accommodate linkages/interrelationships between random variables. This is a crucial factor as its absence might render the accuracy of risk assessment in question. The main disadvantage of BN is the potential difficulties in obtaining CPTs between dependent variables. There are many methods to generate CPTs. Some of the methods that can be used in generating CPTs are described in Section 2.8.

2.4 MARKOV NETWORKS

BN offers an attractive process to resolve many of the interlinked civil infrastructure problems. However, many situations exist where the directed links between nodes are not suitable. For example, in a highway/bridge situation where average daily traffic, *ADT*, and functional class, *Func_Cl*, representing two variables of interest. Considering that *Func_Cl* represents classification of the roadway (interstate, arterial, collector, local, etc.), it becomes evident that the two-way interaction exists between these two variables. In such situations, we link the two variables with a nondirectional link. The link is objectively described through a potential, ϕ. The potential can be either continuous or discrete depending on the type of variables it links. For the remainder of this book, we only consider discrete potentials. Each potential can link two or more variables within the network. Each group of variables that are connected by a potential is referred to as a clique. A network that has all of its links represented by nondirectional links is called Markov network. MN has been studied extensively by many researchers (see, e.g., Koller and Friedman 2009) who offer a detailed description of MN as well as some of its solution techniques and processes.

In general, MN concerns itself with the joint probability of variables within the network and can be expressed as:

$$P(\mathcal{X})=\frac{1}{Z}\prod_{i=1}^{m}\phi_i(\mathcal{X}_i) \tag{2.7}$$

where $X_1, X_2, \ldots, X_n \in \mathcal{X}$ represent all the network variables and the total number of cliques in a network is m. The subset $X_i \subseteq \mathcal{X}$ represents the variables interlinked by the potential $\phi_{i.}$. Generally speaking, the affinities within ϕ_i are non negative, are described by a table, and don't have to be <1.0 as they are not necessarily probabilities. The function Z is a normalization function that ensures that the sum of the marginal probabilities of a variable in the network is unity. It can be expressed as:

$$Z=\sum_{\mathcal{X}}\prod_{i=1}^{m}\phi_i(\mathcal{X}_i) \tag{2.8}$$

Equations 2.7 and 2.8 define the complete process for computing joint probabilities for variables within MN. Two examples are shown below on the use of these equations for solving MN models.

2.4.1 Case Study 2.3: Traffic and Functional Class

2.4.1.1 Overview

Average daily traffic of a route and its functional class, *ADT* and *Func_Cl*, respectively are mutually dependent or interlinked, and therefore, their relationship can only be represented by a nondirectional link in an MN. The 2-node MN that is composed of *ADT* and *Func_Cl* is shown in Figure 2.12. Let us assume three states for *ADT* and four states for *Func_Cl* as shown in Table 2.9. A reasonable potential for a clique that comprise the two variables (actually, it is the only possible clique in this 2-noded network) is shown in Table 2.10. Using the available information, marginal historical (prior) probabilities of the two variables can be obtained using Equations 2.7 and 2.8. Observations (evidence) can also be introduced in an MN. The process is detailed by many authors (see, e.g., Koller and Friedman 2009).

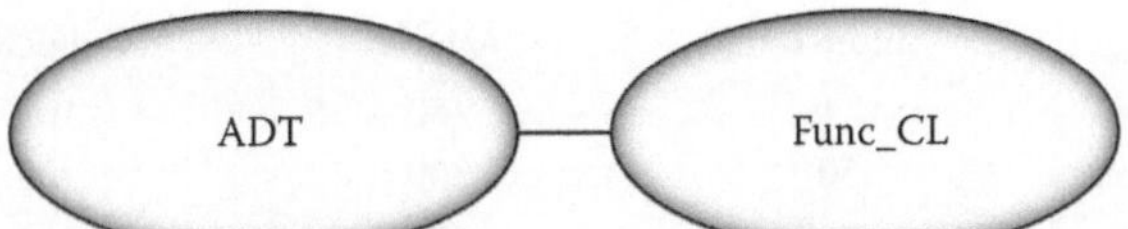

FIGURE 2.12 MN of a 2-nodes model of *ADT* and *Func_Cl*.

TABLE 2.9
Nodal States for *ADT* and *Func_Cl*

Variable	# of States	Ranks (States)
ADT	3	High, Medium, Low
Func_Cl	4	Local, Collector, Arterial, Interstates

2.4.1.2 Potentials

The potential of a set of nodal variables is a multidimensional relationship that describes objectively the affinities between each state of all the nodal variables in the set. In this example, the potential expresses the affinity between two variables, *ADT*, and *Func_CL*. Because the number of states in the two variables are 3 and 4, respectively, the potential of interest would be a 2D matrix as shown in Table 2.10. Note that the components of the matrix are all positive and none are zero.

2.4.1.3 Results

From Equations 2.7 and 2.8, we obtain historical probabilities that are compatible with the potentials given in Table 2.10 and that are shown in Figure 2.13. Figure 2.13 shows that historically, for this case study, though the distribution of the *Func_Cl* is almost even between all bridges under consideration, the *ADT* is mostly *high* (53%). This is beneficial information to the decision maker; however, sometimes it is desired to compute probabilities based on specific observations or introducing evidence of some of the variables. We discuss this using an MN in the next section.

2.4.2 Case Study 2.4: Architectural Vulnerability

2.4.2.1 Overview

In this section, we examine the process of introducing observations (evidence) in an MN. We accomplish this using a simple example that is usually encountered in building security risk assessment. We first build the network and specify the controlling potentials. We then solve for historical probabilities. Finally, we introduce building specific observations during the risk assessment process.

2.4.2.2 Model

The BN model we used is shown in Figure 2.14. It is a 3-node network with all nodes linked using nondirectional links. Each variable describes a particular attribute of the building under consideration. *Accessibility* is a subjective measure that describes the degree and ease of the building access, *Site* is a subjective description of the overall vulnerability of the site, and *Architecture* is a subjective description of the overall vulnerability of the building's architectural features. More details regarding

TABLE 2.10
Potential ϕ (*ADT, Func_CL*) Showing Affinities between *ADT* and *Func_Cl*

Func_CL		Interstate	Arterial	Collector	Local
ADT	Low	1	50	100	400
	Medium	50	300	300	100
	High	500	400	300	250

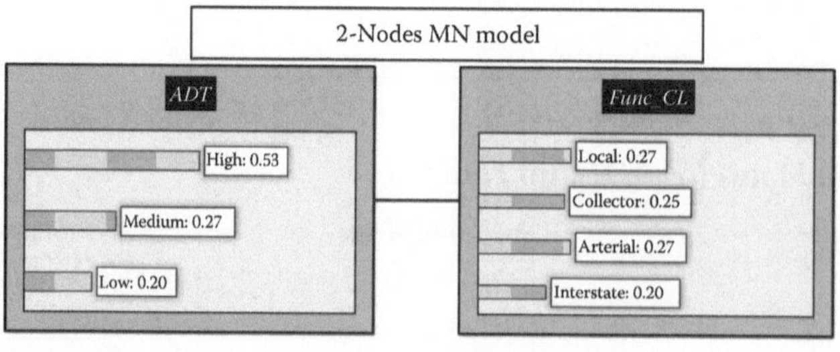

FIGURE 2.13 Historical (prior) probabilities of *ADT* and *Func_Cl*.

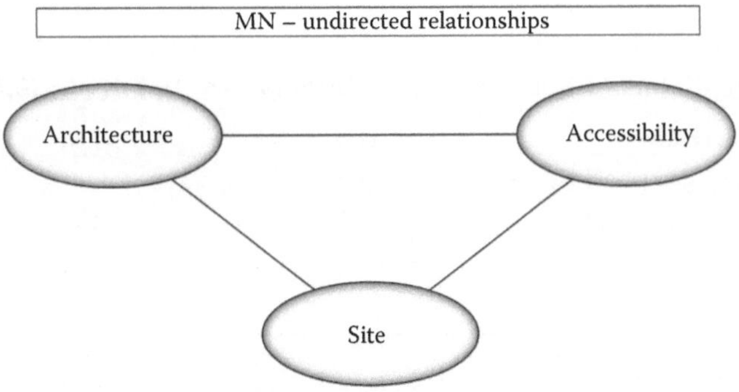

FIGURE 2.14 MN 3-nodes model building security.

TABLE 2.11
Nodal States for *Accessibility, Architecture,* and *Site*

Variable	# of States	Ranks (States)
Accessibility	2	High, Low
Architecture	2	High, Low
Site	2	High, Low

TABLE 2.12
Potential of Affinities between *Accessibility* and *Site*

Site		Low	High
Accessibility	Low	200	150
	High	50	100

these subjective variables can be found in FEMA (2005) and DHS (2011c). All the variables are interlinked in a bidirectional manner as vulnerability of the *Site* can affect the overall vulnerability of *Architecture* and/or *Accessibility*; and the reverse is also true. Similar statements can be made about the relationships among the three variables. This makes the network model an MN. The states of nodal variables of this network are shown in Table 2.11.

2.4.2.3 Conditional Probabilities Tables

We now model the affinities among the three variables using potentials that are given in Tables 2.12 through 2.14. Note that each of the three potentials describes bilateral affinities between the variables. There are many other ways to describe these links including use of a single potential that describes trilateral affinities between the variables. But the use of single potential would require more information than the three potential approach used here.

2.4.2.4 Results

Historical joint probabilities can now be evaluated using Equations 2.7 and 2.8 by using the data from Tables 2.12 through 2.14. The calculation procedure is shown in Table 2.15. The resulting joint probabilities for each combination of the states of the three variables before and after the

TABLE 2.13
Potential of Affinities between *Architecture* and *Accessibility*

Accessibility		Low	High
Architecture	Low	100	75
	High	25	50

TABLE 2.14
Potential of Affinities between *Site* and *Architecture*

Architecture		Low	High
Site	Low	200	50
	High	100	300

TABLE 2.15
Evaluation of Joint Probabilities in a 3-Node MN

Variable, X_{ij}			ϕ_1 (*Architecture, Site*)	ϕ_2 (*Accessibility, Architecture*)	ϕ_3 (*Site, Accessibility*)	$\phi_1\phi_2\phi_3$ Before Normalizing	Normalized
X_{11}	X_{21}	X_{31}	200	100	200	4000000	0.4
X_{11}	X_{21}	X_{32}	200	75	50	750000	0.075
X_{11}	X_{22}	X_{31}	50	25	200	250000	0.025
X_{11}	X_{22}	X_{32}	50	50	50	125000	0.0125
X_{12}	X_{21}	X_{31}	100	100	150	1500000	0.15
X_{12}	X_{21}	X_{32}	100	75	100	750000	0.075
X_{12}	X_{22}	X_{31}	300	25	150	1125000	0.1125
X_{12}	X_{22}	X_{32}	300	50	100	1500000	0.15

Note: i indicates variable number: 1 for *Site*, 2 for *Architecture*, and 3 for *Accessibility*; j refers to state: 1 for *Low* state and 2 for *High* state.

normalization are shown in the last two columns. The normalization factor used in this example is $Z = 1.00E + 07$. From Table 2.15, note that the joint probability of *Site* = *Low*, *Architecture* = *High*, and *Accessibility* = *High* is 0.0125. Similarly, the joint probability of *Site* = *High*, *Architecture* = *Low*, and *Accessibility* = *High* is 0.075.

Using the well-known relation between joint probability $P(X_1, X_2, \ldots, X_n)$ and marginal probability $P(X_j)$ (see Ayyub and McCuen 2011) shown below, we can compute the marginal probabilities for all nodes using Table 2.15.

$$P(X_j) = \sum_{X_j} P(X_1, X_2, \cdots X_n) \tag{2.9}$$

The corresponding results are shown in Figure 2.15. These probabilities are based on historical (prior) information.

We can still use Equations 2.7 and 2.8 to evaluate conditional probabilities based on any introduced observation (evidence). For example, upon a field visit, if an assessor observes that the *Accessibility* to the site is in a *Low* state, we can compute the resulting joint probabilities using

Equations 2.7 and 2.8 by including only the terms that correspond to *Accessibility* = *Low*. This is a subset of Table 2.15 and the results are shown in Table 2.16. The normalization factor $Z = 6.88E + 0.6$. Exercising Equation 2.9, we can evaluate the conditional probabilities of all the nodes resulting from such observation, as shown in Figure 2.16. Similarly, we can compute the marginal probabilities if the assessor observed that the *Accessibility* is in *High* state; corresponding joint and marginal probabilities are shown in Table 2.17 (the normalization factor now is $Z = 3.13E + 0.6$.) and Figure 2.17, respectively. Obviously, as the state of *Accessibility* changes, it promotes corresponding changes in the states of *Architecture* and *Site*. The potentials in Tables 2.12 through 2.14 cause these changes. Because the links in this model are not directional, it is expected that any observations in any other nodes will similarly promote changes in the rest of the model.

2.4.3 Case Study 2.3 Revisited: Traffic and Functional Class

Let us revisit the case study described in Section 2.4.1. If we are interested in the histogram of *ADT* if the *Func_Cl* of the route of interest is *interstate*, we can apply the observation (evidence). The resulting histograms of the network are shown in Figure 2.18. The resulting marginal probabilities of *ADT* is as expected: *High* (91%) probability of *ADT*, fairly negligible for *Medium ADT*, and nonexistent (~0%) for *Low ADT*.

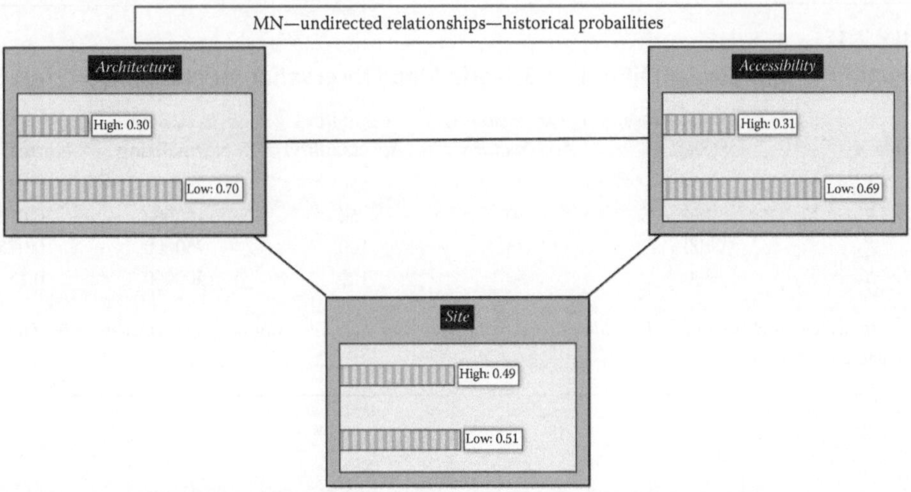

FIGURE 2.15 Historical probabilities of building security.

TABLE 2.16
Evaluation of Joint Probabilities in a 3-Node MN: Observation *Accessibility* = Low

Variable X_{ij}			ϕ_1 (*Architecture, Site*)	ϕ_1(*Accessibility, Architecture*)	ϕ_1 (*Site, Accessibility*)	$\phi_1\phi_2\phi_3$ Before Normalizing	Normalized
X_{11}	X_{21}	X_{31}	200	100	200	4000000	0.4
X_{11}	X_{22}	X_{31}	50	25	200	250000	0.025
X_{12}	X_{21}	X_{31}	100	100	150	1500000	0.15
X_{12}	X_{22}	X_{31}	300	25	150	1125000	0.1125

Note: *i* indicates variable number: 1 for *Site*, 2 for *Architecture*, and 3 for *Accessibility*; *j* refers to state: 1 for *Low* state and 2 for *High* state.

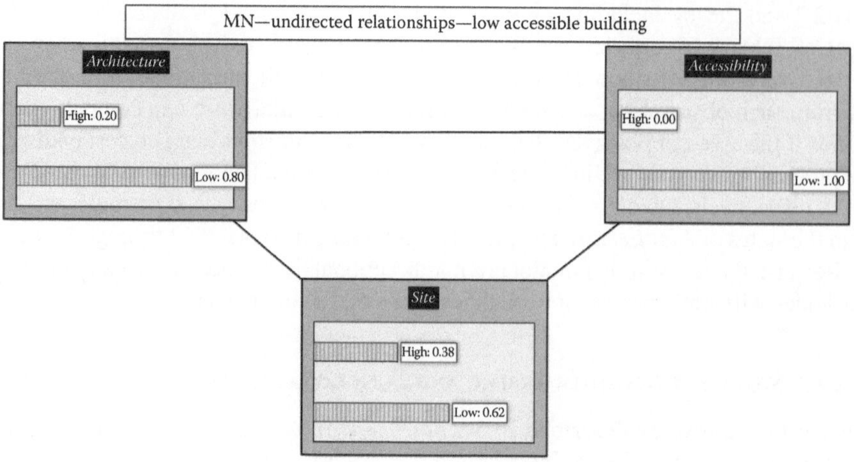

FIGURE 2.16 Histogram of building security when *Site* is difficult to access (Low).

TABLE 2.17
Evaluation of Joint Probabilities in a 3-Node MN: Observation *Accessibility = High*

Variable X_{ij}			ϕ_1 (*Architecture, Site*)	ϕ_2 (*Accessibility, Architecture*)	ϕ_3 (*Site, Accessibility*)	$\phi_1\,\phi_2\,\phi_3$ Before Normalizing	Normalized
X_{11}	X_{21}	X_{32}	200	75	50	750000	0.075
X_{11}	X_{22}	X_{32}	50	50	50	125000	0.0125
X_{12}	X_{21}	X_{32}	100	75	100	750000	0.075
X_{12}	X_{22}	X_{32}	300	50	100	1500000	0.15

Note: i indicates variable number: 1 for *Site*, 2 for *Architecture*, and 3 for *Accessibility*; j refers to state: 1 for *Low* state and 2 for *High* state.

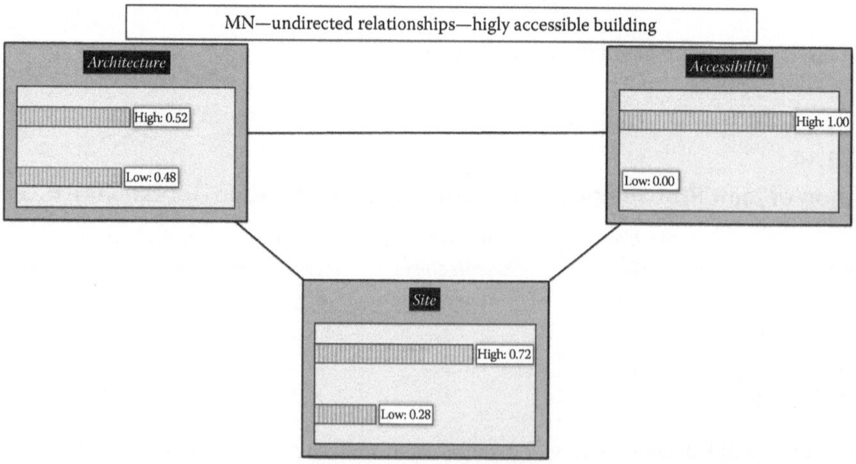

FIGURE 2.17 Histogram of building security when *Site* is difficult to access (High).

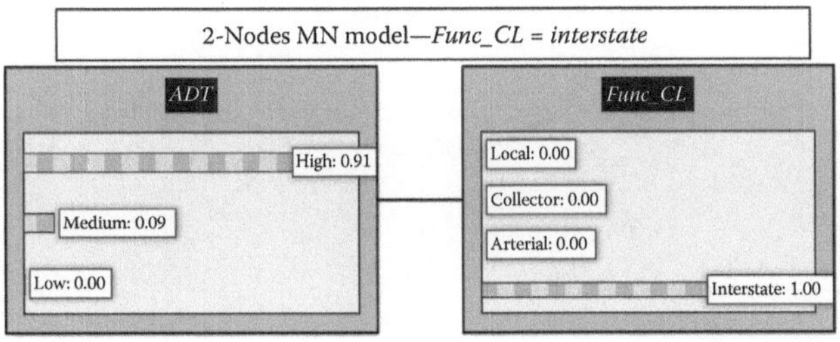

FIGURE 2.18 Probabilities of *ADT* for *Func_Cl* = interstate.

2.4.4 Concluding Remarks

We presented simple case studies that illustrate the main concepts of MN. In practice, the relation between the variables in a network are always not uniformly directional as in a BN or nondirectional in an MN. A network with mixed types of links is known as *chain graph* and is discussed in the next section.

2.5 CHAIN GRAPHS: COMBINING BAYESIAN AND MARKOV NETWORKS

2.5.1 Overview

Many problems in the field of civil infrastructures will not follow strictly either BN or MN models; that is, links between its nodes might not all be directed or nondirected links. Those types of networks are referred to as CGNs. Koller and Friedman (2009) discussed the mathematical features of CGN and methods of analysis at length. We present several examples featuring CGN in this book. For now, we offer a simple example to illustrate the concept.

2.5.2 Case Study 2.5: Risk and Its Components

2.5.2.1 Overview

A fairly simple, yet important CG example would be the risk, *Ri*, interrelationship with its three components: Vulnerability, Threat, and Consequences, or *V*, *T*, and *C*, respectively. *Ri* is a function of *V*, *T*, and *C*, and hence, directional links from the components (as parents) toward *Ri* (as a child) are used. However, in many situations, these three components influence each other, and thus, can be connected using nondirectional links. This section explores a simple CG network.

2.5.2.2 Model

The CG model is shown in Figure 2.19 with corresponding states of nodal variables shown in Table 2.18. As seen from Figure 2.19, both directional and nondirectional links between the variables are used. Therefore, this network is a CGN. A potential explanation of why the links between *V*, *T*, and *C* are two ways (or nondirectional) is shown in Table 2.19.

2.5.2.3 CPTs and Potentials

By definition, a CG model includes a mix of potentials (for describing the nondirectional links) and CPTs (for describing the directional links between child and parent variables). Thus, this example includes three potentials and a single CPT as shown in Tables 2.20 through 2.23. These tables give us a vivid contrast between the properties of potentials and CPTs. All CPTs include probabilities, so all the entries are ≥0.0 and ≤1.0, and the only requirement for entries of potentials is that they are positive.

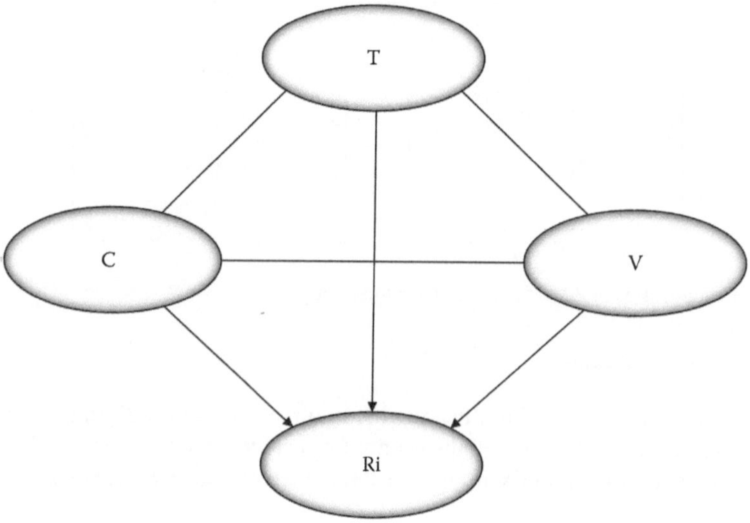

FIGURE 2.19 Chain graph model for risk.

TABLE 2.18
States of Nodal Variables

Variable	# of States	Ranks (States)
V	2	High, Low
T	2	High, Low
C	2	High, Low
Ri	2	High, Low

TABLE 2.19
Reasons for Nondirectional Links between Risk Components

	Affected Risk Component		
Risk Component	***V***	***T***	***C***
V	N.A.	A structure on soft soil can result in a higher earthquake threat	A vulnerable structure will cause higher undesirable consequences than a more robust structure
T	A higher threat will lead to nonlinear structural behavior, which is akin to an increased vulnerability	N.A.	A higher flood demand, e.g., will lead to higher direct and indirect costs overall
C	Higher consequences/costs may affect decisions regarding system retrofits	Higher consequences may be a cause for higher threat in certain situations such as man-made threats	N.A.

TABLE 2.20
Potential of Affinities between *T* and *V*

V		**Low**	**High**
T	Low	100	300
	High	75	500

TABLE 2.21
Potential of Affinities between *V* and *C*

C		**Low**	**High**
V	Low	150	75
	High	50	300

TABLE 2.22
Potential of Affinities between *T* and *C*

T		**Low**	**High**
C	Low	250	200
	High	150	300

TABLE 2.23
CPT for Relating *V*, *T*, *C* to R_i

C		Low				High			
V		Low		High		Low		High	
T		Low	High	Low	High	Low	High	Low	High
Ri	Low	0.9482	0.7644	0.9067	0.1180	0.8413	0.2356	0.1698	0.5375
	High	0.0518	0.2356	0.0933	0.8820	0.1587	0.7644	0.8302	0.4625
Histograms for conditional probabilities		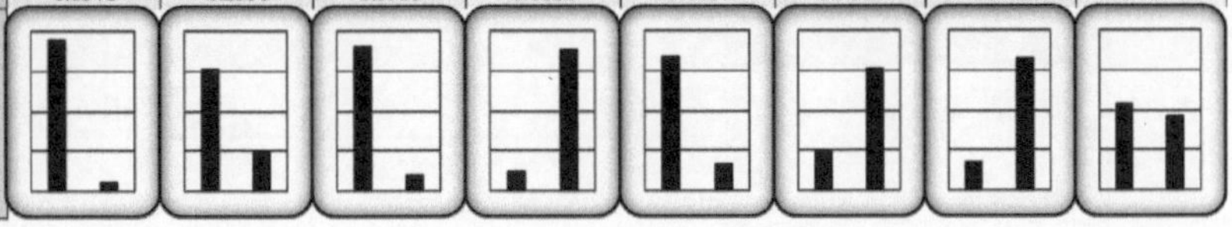							

2.5.2.4 Results

Exercising the methods of BN and MN as described earlier, the historical probabilities of this model are shown in Figure 2.20. If we observe that the vulnerability of the system is *Low*, then by entering this observation as evidence to the network, we obtain the probabilities shown in Figure 2.21. Note that the observation not only affected the resulting risk but also affected the threat and consequences, via the nondirectional links, which in turn have an even stronger effect on the resulting risk. A direct conclusion out of this simple analysis is that we need to account for all potential interactions between the components; otherwise the resulting risk computations might be erroneous.

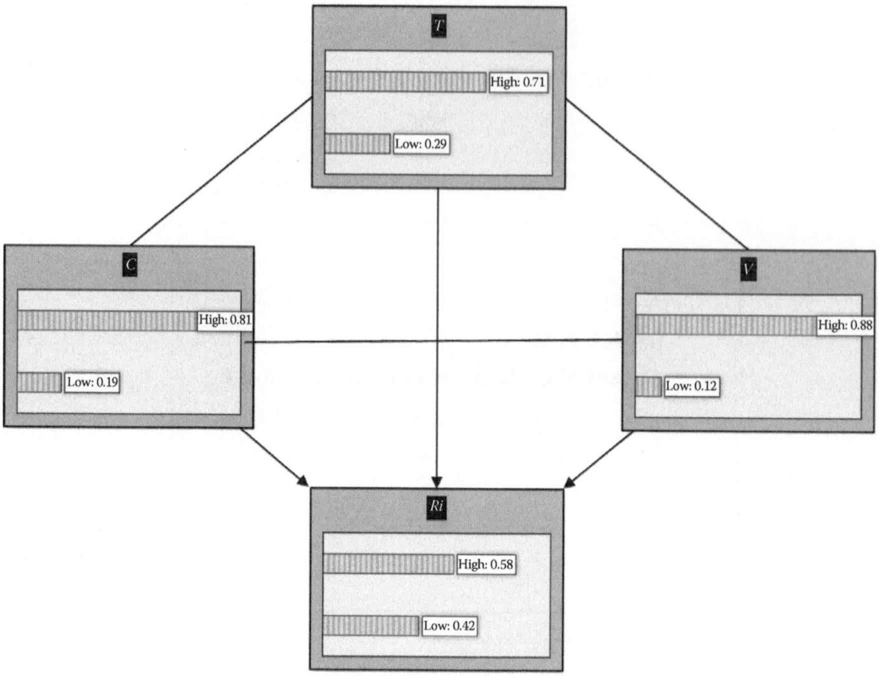

FIGURE 2.20 Historical probabilities of the chain graph risk model.

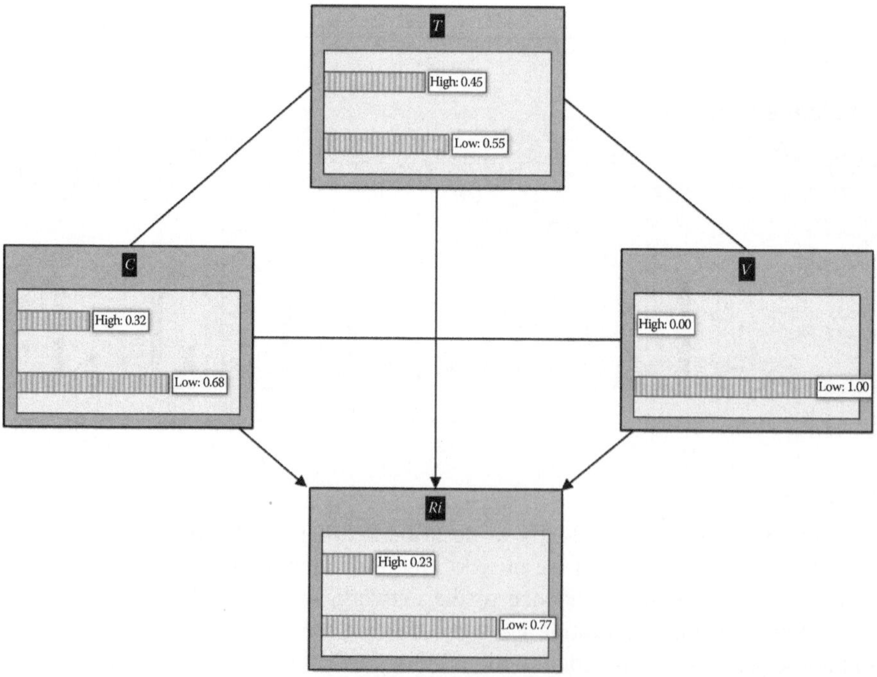

FIGURE 2.21 Probabilities of the chain graph risk model for V = Low.

2.6 DECISION UNDER UNCERTAINTY: INFLUENCE DIAGRAMS

2.6.1 Overview

Most, if not all, civil infrastructure decisions are made under uncertain conditions. These uncertainties need to be considered while making such decisions; otherwise the resulting outcomes might produce nonoptimal performance or unnecessary additional costs. Risk management components, especially risk acceptance, risk treatment, and risk monitoring, depend very much on the concept of Decision Under Uncertainty (DUU) and, thus, we have to include DUU when considering risk management from the ground up. This section includes a quick overview of the traditional methods of DUU, notes some of their shortcomings, and expresses the need for a more robust approach for DUU. We then discuss the concept of Influence Diagrams (IDs), which combine DUU traditional components with graphical network concepts. The resulting ID methodology will prove to be very useful and simple in risk management and multihazard (MH) applications that are discussed later. Readers should note that the entire MH decision theory is based on hazard interaction through decisions and will be a topic of another book.

2.6.2 Decision Trees and Utility

2.6.2.1 Overview

The Decision Tree Method is a well-known graphical technique for solving DUU situations (see Benjamin and Cornell 1970; Neapolitan 2004; Ettouney and Alampalli 2012a, 2012b) and was demonstrated to be useful for decision-making techniques such as posterior, preposterior, or sequential analysis (Benjamin and Cornell 1970). Similar to GNs, it can present the global overview of the entire problem graphically, in contrast to methods such as Basic Utility and Decision Table (Benjamin and Cornell 1970). Benjamin and Cornell (1970) also explored the use of decision trees in civil infrastructures.

This method is based on subdividing the decision problem into three types of sets: random variables, $\mathcal{V}$, decision process, $\mathcal{D}$, and utility (Cost), $\mathcal{U}$. The decision tree is constructed by branching at any of these three types of nodes, as pertinent. For each of the possible decisions $d_j \in \mathcal{D}$, there will be an expected utility (cost), $EU(d_j)$, that is computed as:

$$EU\left(d_j\right)=\sum_i U\left(d_j,h_i\right)P\left(h_i,\varepsilon\right) \tag{2.10}$$

The state set, $\mathcal{H}$, contains all possible states, h_i, that might result from different decisions d_j such that $h_i \in \mathcal{H}$. The probability of a particular state, h_i, given observation ε is $P(h_i, \varepsilon)$. The decision tree allows the decision maker to compute $EU(d_j)$ in a formal manner. The optimal decision, $d|_{OPTIMAL}$, is the one which produces optimal $EU(d_j)$. If cost is used as a utility metric, we have:

$$d|_{OPTIMAL}=\underset{d_j \in D}{MIN}\left(EU\left(d_j\right)\right) \tag{2.11}$$

2.6.2.2 Utility vs. Risk

The use of utility at a metric in DUU is popular in many communities (see Kjaerulff and Madsen 2013), and we explore it later in this book. Upon further reflection on the utility metric (U) in the field of civil infrastructure, we note that it is an outcome of exposure (E) and any pertinent consequences (C) for making decisions (or lack of them), such that

$$U = U\,(E, C) \tag{2.12}$$

Since we defined E to be a result of vulnerability (V) and threats (T), it can be written as:

$$U = U(C, T, V) \tag{2.13}$$

Upon further reflection, we conclude that utility is another expression of risk as defined in this book. Moreover, given risk can be expressed in a relative or absolute manner, we further conclude that if the utility metric is relative utility (e.g., a scale from 0–100), it is akin to relative risk; if utility metric is absolute utility (e.g., monetary cost is used to represent utility), it is akin to absolute risk.

An example of decision tree for DUU is shown in Figure 2.22. It shows a simplified bridge progressive collapse problem. Even though it contains only five tree-branching nodes and binary branches, the number of resulting potential instances is 24. This illustrates one of the limitations of the decision tree method that we discuss in a later section.

When we use optimal utility, as in decision trees or IDs in the following section, we are referring to optimal risk. For the remainder of this book, we use the term utility in DUU problems to be consistent with the DUU community. However, we may interchange the usage of the expressions risk with utility during the discussion of different DUU examples.

2.6.2.3 Case Study 2.6: Structure Inspection and Monitoring Methods Using Decision Tree

We reiterate the decision tree problem presented by Ettouney and Alampalli (2012). When an owner decides that a structure requires inspection, the owner has several options, such as visual inspection and sensor-based structural health monitoring (SHM). We model these variables as decision variables that a decision maker has an option to choose from:

1. d_1: Conduct a manual visual inspection (V.I.)
2. d_2: Utilize a SHM techniques (SHM)
3. d_3: Use a combination of the above two methods (V.I. + SHM)

Two possible uncertain variables can control this situation: needed repairs (*N.R.*) and observability of the chosen method or inspection results (*I.R*). The decision maker estimated that after the monitoring/inspection process is finished, there can be three outcomes for needed repairs, *N.R.*:

1. There is a need for immediate repair.
2. Some repair is needed within a year or two years.
3. There no obvious need for repair.

Also, there can be two outcomes for inspection results, *I.R.*:

1. The damage of interest can be observed by the process.
2. The damage can't be observed by the process.

A summary of all of the variable and decision states is shown in Table 2.24.

Based on previous experiences, the decision maker assigned different probabilities to possible potential outcomes. Therefore, the probabilities of damage occurrences are estimated as a CPT in Table 2.25 and the potential of observing damage from different methods are estimated as a CPT in Table 2.26. Potential monetary costs are estimated in Tables 2.27 and 2.28, respectively, for direct costs of each monitoring method and the different consequences of inspection as related to the need for repairs. In this example, visual inspection has the least direct costs compared to the other two methods available. Finally, Table 2.29 shows the total utilities (costs) for different scenarios

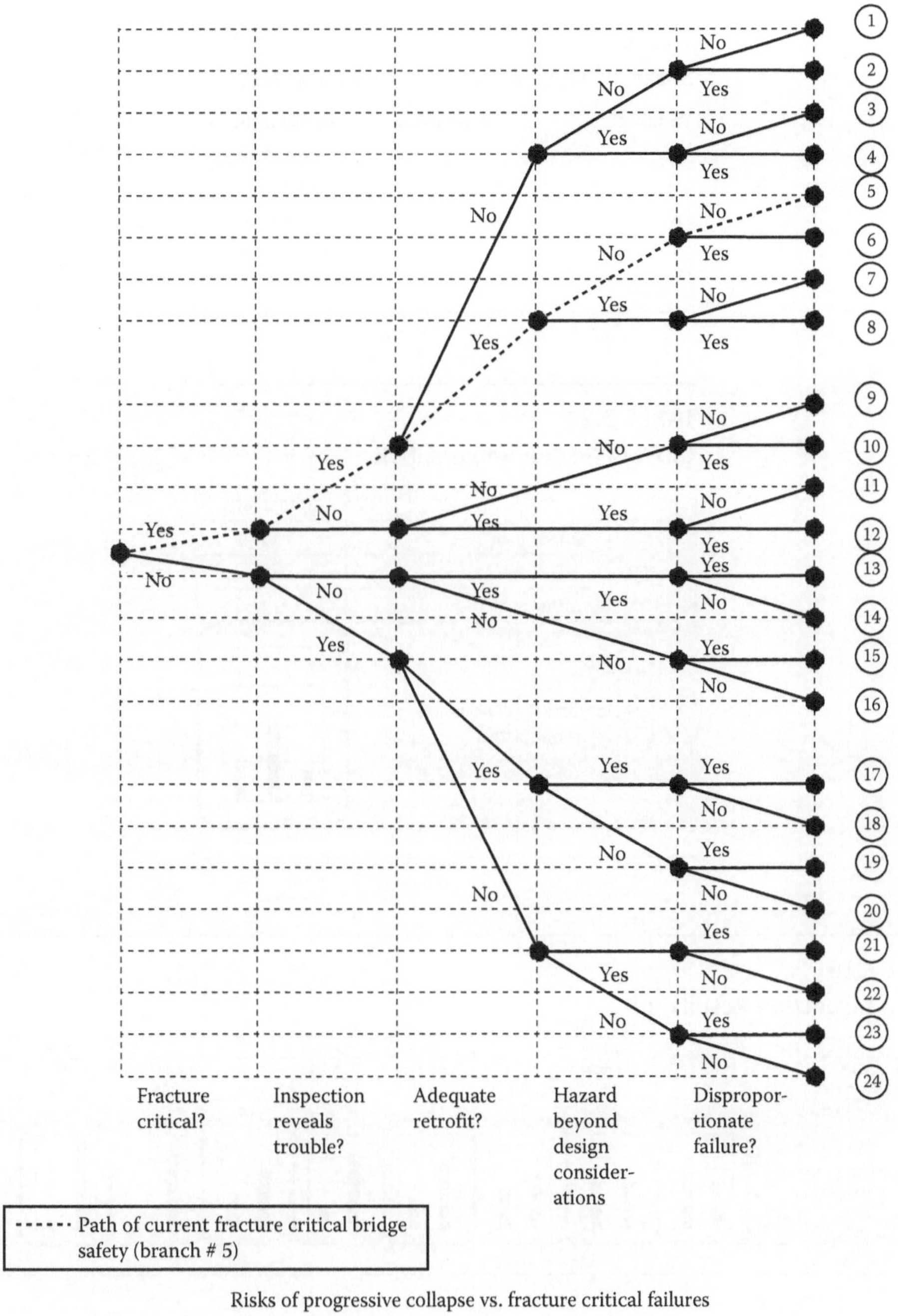

FIGURE 2.22 Decision tree of simplified bridge progressive collapse problem.

of this problem. Using Equation 2.10, the total cost of all three inspection methods can be computed by multiplying each probability along the branch with the total cost of each branch and then adding all branches within each method. The process is shown in Figure 2.23. The total expected costs for the methods are \$2.11 M, \$2.17 M, and \$2.15 M, for visual inspection, SHM only, and combined method. In this example, the visual inspection method produces the lowest risk, and using the SHM method produces the highest risk when measured in monetary terms.

TABLE 2.24
States of Decision and Uncertain Variables

Variable	# of States	Ranks (States)
N.R.	3	None, 1–2 years, immediate
I.R.	2	Not observed, observed
D	3	d_1, d_2, d_3

TABLE 2.25
CPT of Needed Repairs, N.R.

N.R.		Immediate
N.R.	Immediate	0.2
	1–2 Years	0.5
	None	0.3
Histograms for conditional probabilities		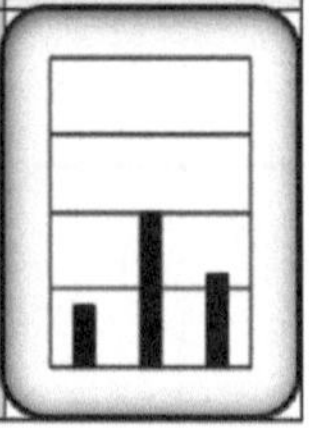

TABLE 2.26
CPT of Inspection Results, I.R.

D		*d1*			*d2*			*d3*		
N.R.		Immediate	1–2 Years	None	Immediate	1–2 Years	None	Immediate	1–2 Years	None
I.R.	Observed	0.4	0.5	0.6	0.7	0.8	0.6	0.85	0.95	0.9
	Not Observed	0.6	0.5	0.4	0.3	0.2	0.4	0.15	0.05	0.1
Histograms for conditional probabilities		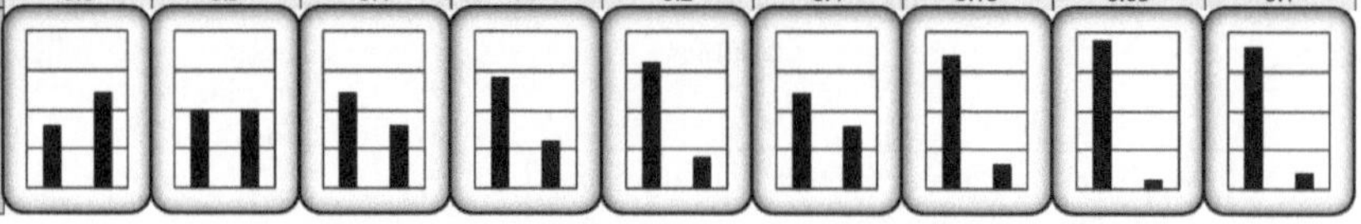								

TABLE 2.27
Direct Costs

Decision	Cost in $M
d_1 (visual inspection, V.I.)	0.50
d_2 (SHM)	1.00
d_3 (V.I. + SHM)	1.20

TABLE 2.28
General Costs of Different Outcomes

Needed Repairs (N.R.)	Costs in $M					
	Immediate		1–2 Years		None	
Inspection Results (I.R.)	Observed	Not Observed	Observed	Not Observed	Observed	Not Observed
General costs	1.50	5.50	1.00	2.30	0.00	0.00

Note: Cost of inspection or monitoring includes both direct and indirect costs in current dollars. General costs include estimated rehabilitation, social, and long-term costs.

TABLE 2.29
Utility Table

Needed Repairs (N.R.)	Utilities in $M					
	Immediate		1–2 Years		None	
Inspection Results (I.R.)	Observed	Not Observed	Observed	Not Observed	Observed	Not Observed
d_1 (Visual inspection, V.I.)	2.00	6.00	1.50	2.80	0.50	0.50
d_2 (SHM)	2.50	6.50	2.00	3.30	1.00	1.00
d_3 (V.I. + SHM)	2.70	6.70	2.20	3.50	1.20	1.20

2.6.3 Influence Diagrams

2.6.3.1 Overview

The process of computing probabilities for each possible decision, $P(h_i, \varepsilon)$, in Equation 2.10 involves computing a joint probability for all variables and decisions that control the utility under consideration. This results in increasing the size of decision trees geometrically as the number of nodes increase. Another limitation of decision trees is the directional dependence between nodes, that is, a BN type CPT relationship. At present, there is no information available in the literature on the use of a decision tree with a nondirectional relationship between the nodes. Many authors (e.g., Neapolitan 2004; Kjaerulff and Madsen 2013) discuss IDs and its development and solutions.

2.6.3.2 Case Study 2.7: Structure Inspection and Monitoring Methods Using ID

2.6.3.2.1 Overview

Let us revisit the case study on Structure Inspection and Monitoring Methods and explore it using an ID model. Until now, in our considerations of BN, MN, and CG, we only used random variable nodes. But an ID model includes random variable nodes as well as decision and utility nodes. A useful way to look at ID is that it has a base of BN, MN, or CG, with added topology for logic of decisions and utilities. There can be several decision variables or several utility variables, in addition to the random variables in a single ID model. The solution procedure of an ID model is similar to the solution procedures of BN, MN, or CG. Some additional steps are described below:

1. The number, type, links, and hierarchy of decisions that are introduced in the model as decision variables need to be well thought out and logically intertwined with the base BN, MN, or CG.

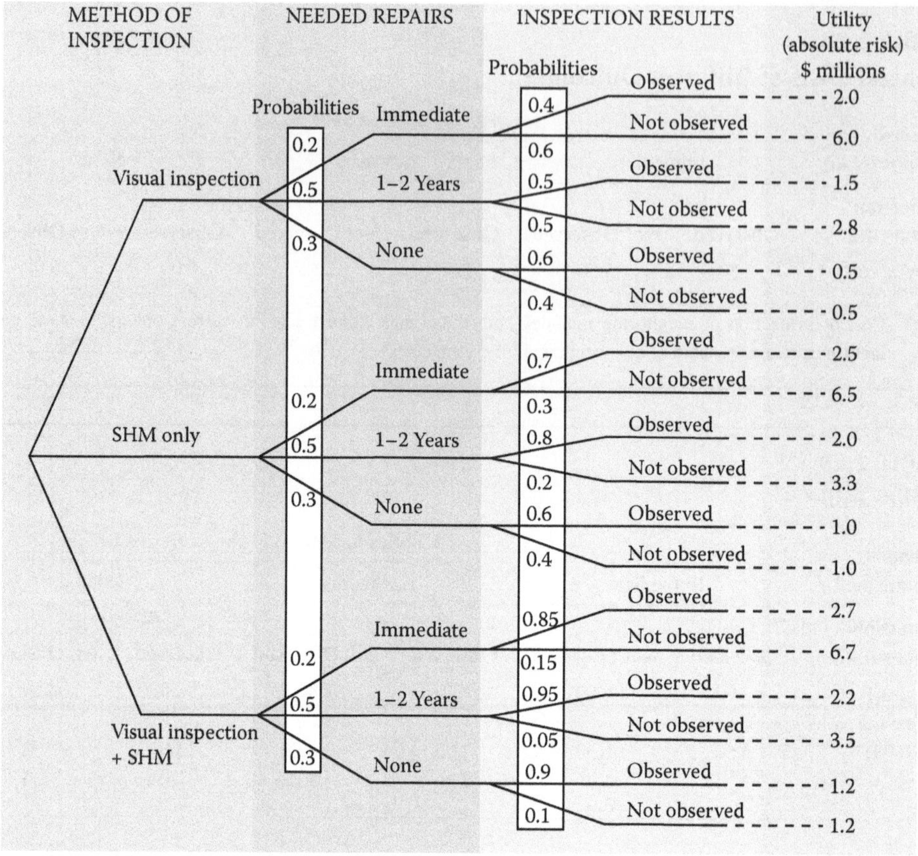

FIGURE 2.23 Decision tree graph.

2. The number, links, and hierarchy of utilities that are introduced to the model as utility variables need to be well thought out and logically intertwined with the base BN, MN, or CG as well as with the decision variables.
3. A utility or utilities table that contains costs as related to different states of variables (both random and decision) has to be constructed.
4. The resulting joint probabilities need computing, in addition to the marginal probabilities.
5. The utilities of different decisions need to be computed. This is done by using the computed joint probabilities of #4 and the utilities of #3.
6. The optimal decision is then computed from #5.

Next, we show a simple ID example; more realistic examples are offered later in this book.

2.6.3.2.2 Model

Let us reconsider the Case Study on Structure Inspection and Monitoring Methods where a structure may or may not need repairs.

Let us assume that the inspection of the structure will result in either observing a need for repair (N.R.) or not (D) and that the decision to repair or not is based solely on the inspection results (I.R.). Of course, a cost of different permutations is included in the utilities table. Hence, this example uses two random variables: *N.R.* and *I.R.* It includes one decision variable, *D*, and one utility variable, *U*. The resulting network is shown in Figure 2.24. The variables *N.R.*, *I.R.*, and *D* all affect *U*. As such, the utility table presented in Table 2.29 still applies.

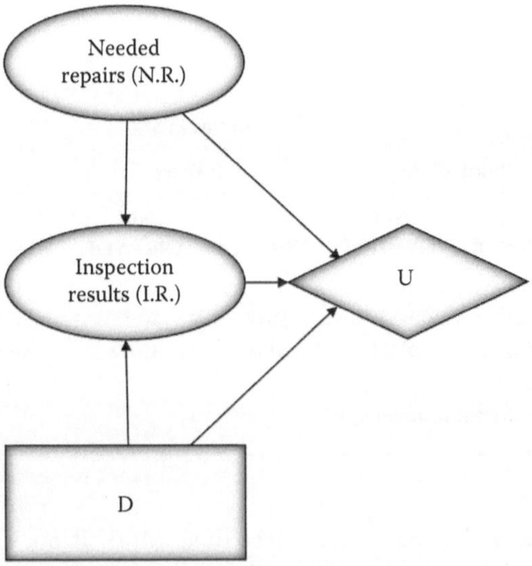

FIGURE 2.24 ID model for inspection type.

TABLE 2.30
Joint Probabilities

Needed Repairs (N.R.)	Utilities in $M					
	Immediate		1–2 Years		None	
Inspection Results (I.R.)	Observed	Not Observed	Observed	Not Observed	Observed	Not Observed
d_1 (V.I.)	0.0800	0.1200	0.2500	0.2500	0.1800	0.1200
d_2 (SHM)	0.1400	0.0600	0.4000	0.1000	0.1800	0.1200
d_3 (V.I. + SHM)	0.1700	0.0300	0.4750	0.0250	0.2700	0.0300

2.6.3.2.3 Conditional Probabilities Tables

CPTs offered in Tables 2.25 and 2.26 can be used to describe this problem.

2.6.3.2.4 Results

Using Tables 2.25 and 2.26, we can execute the underlying BN problem to generate the required joint probabilities shown in Table 2.30. Combining those joint probabilities with the utilities shown in Table 2.29, we can calculate the expected utility results shown in Table 2.31. This table can be used to find the optimal decision. The results are identical to the decision tree results, as expected. However, the ID model of Figure 2.24 is more efficient and simpler than the decision tree model of Figure 2.23. This advantage of ID can only be increased for models with a larger number and type of nodes and more complex interrelationships between the variables as we will invariably need in more realistic problems. We cover much more complex ID models later in this book.

2.7 DYNAMIC GRAPH NETWORKS

2.7.1 Overview

The variables in GN applications in the field of civil infrastructures are rarely time independent. Most, if not all, variables change with time. For example, deterioration of bridges, tunnels, or even

TABLE 2.31
Expected Utilities

Needed Repairs (N.R.)	Utilities in $M: Immediate		1–2 Years		None		
Inspection Results (I.R.)	Observed	Not Observed	Observed	Not Observed	Observed	Not Observed	E(U) in $M
d_1 (V.I.)	$0.16	$0.72	$0.38	$0.70	$0.09	$0.06	$2.11
d_2 (SHM)	$0.35	$0.39	$0.80	$0.33	$0.18	$0.12	$2.17
d_3 (V.I. + SHM)	$0.46	$0.20	$1.05	$0.09	$0.32	$0.04	$2.15
Optimal E(U) in $M	$2.11						
EU(D) =	d_1 (Visual Inspection, V.I.)						

buildings can affect vulnerability parameters over time. Most, if not all, consequences are also time dependent. Similarly, both man-made hazards that affect infrastructure security and natural hazards, such as frequency and magnitude of high winds or floods, are all time dependent. Thus, all the stakeholders should account for these changes.

One way to accommodate these time-dependent changes in BN properties is using a process known as dynamic Bayesian network (DBN) if the underlying network of interest is BN. Such a procedure was discussed by several authors including Neapolitan (2004) and Kjaerulff and Madsen (2013). We generalize such a concept for any type of GN using DGN.

In general, a typical DGN consists of the following:

1. *Time Slice*: Time slices are discrete times, at which the values, generally probabilities, of the vertices (nodal variables) have to be computed. These time slices can be in the past, present, or future. For example, if we need to forecast histograms of condition ratings of structural elements or risk in the next 20 years at two-year intervals starting from present, then we will have *11* time slices at which network evaluation is required.
2. *Template*: Template is the GN construct (e.g., BN, MN, or CG) that is repeated at each time slice. There is no reason for templates to be constant at each time slice. However, in this book, we only consider constant templates.
3. *Temporal Links*: Temporal links are used to move from one time slice to another. The temporal links will relate a variable random X_i at time t_i to the same random variable X_j at an adjacent time slice t_j. The time marching can be forward ($t_j > t_i$) or backward ($t_j < t_i$). The relationship is formalized as:

$$X_j = [T]X_i \tag{2.14}$$

   with $[T]$ representing a transformation matrix. This matrix, obviously, will have the same size as the number of the states of the discrete random variable X. There can be several temporal links in any DGN depending on the need, but there is no need for $[T]$ to be constant during the time marching process.
4. *Interface*: Interfaces are the vertices/nodes within a template that are connected by temporal links. These interface variables will be computed first in the new time slice using the relationship shown in Equation 2.14. The computed histograms of the interface variables will then be used as soft evidence (observations) in the new time slice to compute the probabilities of the rest of the network. Figure 2.25 shows the components of a typical DGN.

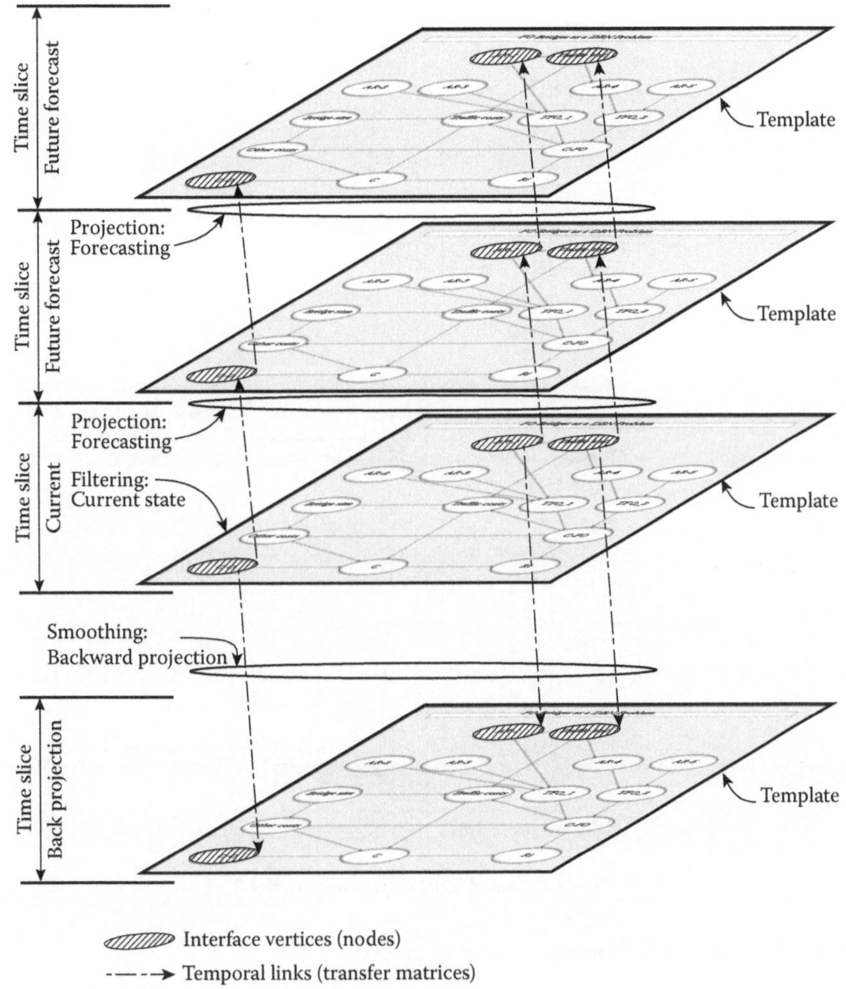

FIGURE 2.25 Components of DGN.

Depending on the type of time marching, there can be three types of DGN:

1. *Smoothing*: Smoothing is for backward projection.
2. *Filtering*: Filtering is used for current state.
3. *Prediction*: Prediction is used for forecasting of the variables and perhaps is the most important DGN process in the field of civil infrastructures.

A special case of DGN, when the number of nodes is reduced to a single node, is shown in Figure 2.26. Careful examination of this limit state reveals that this is none other than the popular Markov chain (Ettouney and Alampalli 2012a, 2012b). Thus, we reach two important conclusions, namely:

1. DGN is a superset of Markov chains.
2. In most complex civil infrastructure applications, such as the deterioration modeling of structures, use of a single variable in a Markov chain process can lead to inaccurate results. This is because there is seldom a circumstance where such a variable as deterioration is independent of other variables. A network approach such as DGN is recommended to accommodate the interaction between variables with time.

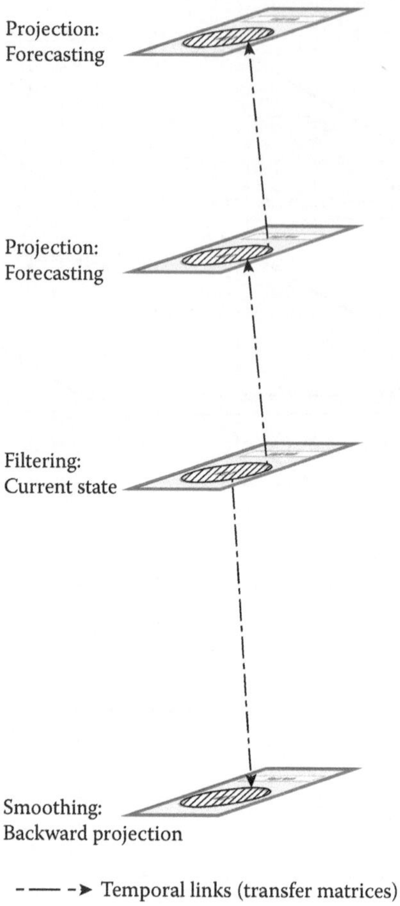

FIGURE 2.26 Single node DGN (defaults to Markov process).

2.7.2 Time Marching in DGN

We look to time marching in DGN in a similar manner as time marching in numerical integration of Newton's second law. The steps are as follows (see Figure 2.27):

1. Initialize variables by executing a prior computation for the GN of the template.
2. Apply observations in initial time slice.
3. March to next time slice by applying transfer matrices to interface variables.
4. Apply soft evidence using the newly generated probabilities at temporal links to generate a new marginal at the remainder of nodes in the BN template.
5. If there is any other evidence (soft or hard) at this time slice, apply it and regenerate marginal, if needed.
6. Update/change transfer matrices and temporal links, if needed.
7. Repeat steps 3 through 6 to move to new time slice.

2.7.3 Case Study 2.8: Observed vs. Actual Infrastructure Condition Rating

The condition rating, *CR*, of a structure is a metric routinely used to describe, subjectively, the state of deterioration of a structure and is generally obtained through visual inspection, *V.I.*, observations. It has been proven effective and useful for managing bridges for decades. A recent

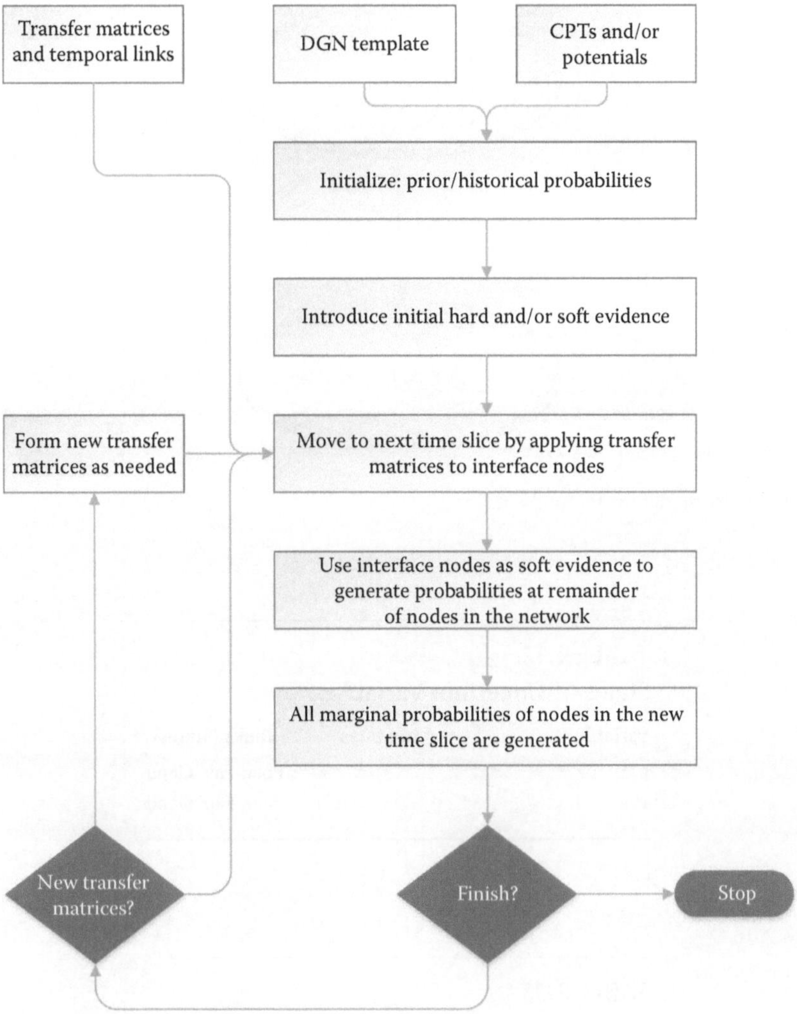

FIGURE 2.27 Steps of DGN.

study, Agrawal et al. (2013), showed that *V.I.* estimates the actual structure condition (i.e., CR) in a consistent manner. However, those same results also showed that some discrepancies, albeit small, exist between the *V.I.* and the actual bridge condition rating *CR*.

Forecasting bridge condition is usually needed during different bridge management activities; therefore, *V.I.* has been used in such forecasts. Time marching is sometimes employed through the use of Markov processes via transfer matrices (Agrawal et al. 2009). However, what is forecasted is the *V.I.*, not the actual condition rating *CR*. This dilemma results from implicitly assuming that the two time-varying variables, condition observed through *V.I.* and *CR*, are one and the same, and this is not true. The Markov process, as employed currently, is not adequate for handling a two-random variable problem. However, DBN is well suited for such time marching modeling.

We use a 2-node BN as a template as shown in Figure 2.28. The two nodes represent *CR* and *V.I.* Obviously, *CR* is the parent of *V.I.*, because the true state of the structure is represented by CR We assume three states for both variables as shown in Table 2.32. The governing CPTs are shown in Tables 2.33 and 2.34. The time marching scheme of this DBN is shown in Figure 2.29.

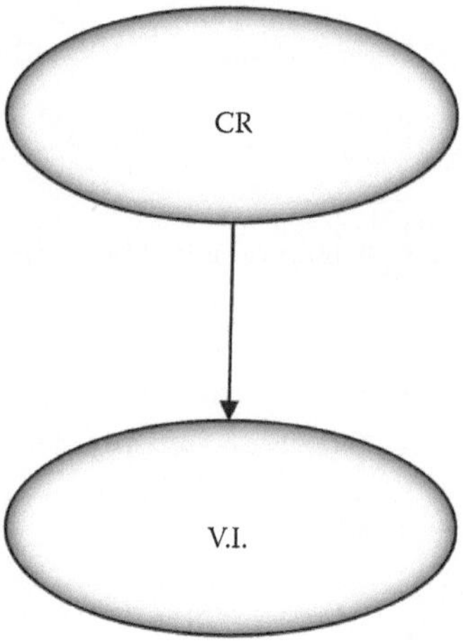

FIGURE 2.28 DBN model for *V.I.* and *CR*.

TABLE 2.32
States of Uncertain Variables

Variable	# of States	Ranks (States)
CR	3	Poor, Fair, Good
V.I.	3	Poor, Fair, Good

TABLE 2.33
CPT for *CR*

CR	Good	0.4448
	Fair	0.4967
	Poor	0.0585
Histograms for conditional probabilities		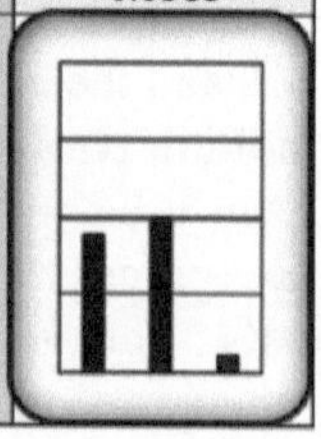

A transformation matrix $[T]$ is assumed, based on the decision maker's experience, to relate $\{p(V.I.)\}_{t}$, denoting the probabilities of different states of *V.I.* as they change between two time periods, t_i and t_{i+1} as:

$$\{p(V.I.)\}_{t_{i+1}} = [T]\{p(V.I.)\}_{t_i} \tag{2.15}$$

TABLE 2.34
CPT for *V.I.*

CR		Good	Fair	Poor
V.I.	Good	0.9883	0.2717	0.0001
	Fair	0.0117	0.5898	0.0800
	Poor	0.0000	0.1385	0.9199
Histograms for conditional probabilities		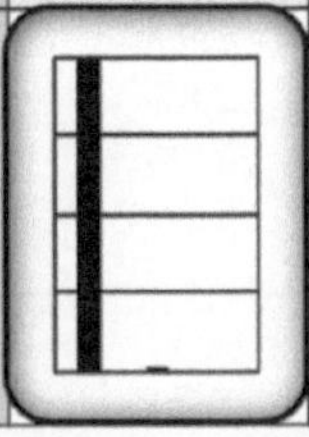	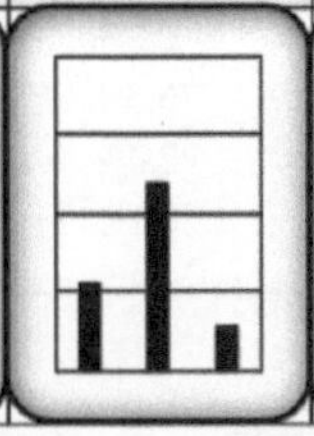	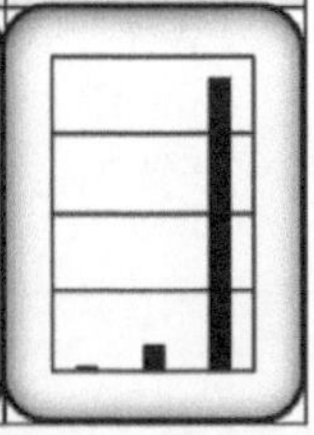

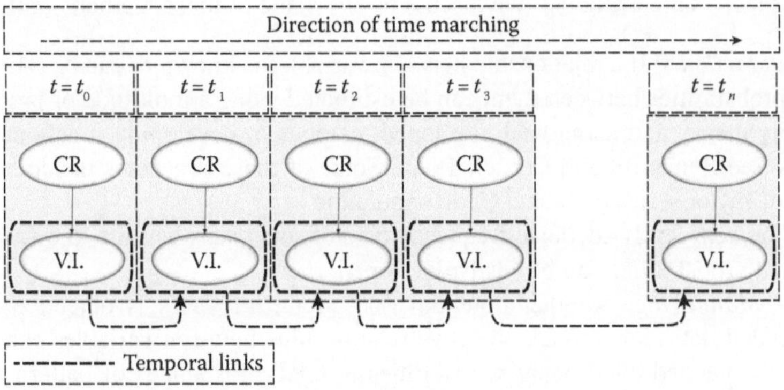

FIGURE 2.29 Time marching scheme for the 2-nodes DBN template.

For our example, we assume:

$$[T] = \begin{bmatrix} 0.95 & 0.00 & 0.00 \\ 0.080 & 0.80 & 0.00 \\ 0.02 & 0.20 & 1.00 \end{bmatrix} \tag{2.16}$$

We also assume an initial observation at t_0 of the states of *V.I.* to be:

$$\{p(V.I.)\}_{t0} = \begin{Bmatrix} 1 \\ 0 \\ 0 \end{Bmatrix} \tag{2.17}$$

We now have all the needed information to solve the needed DBN problem. Table 2.35 shows the state of both *V.I.* and *CR* at different time periods. Obviously, the differences between the expectations of the two variables, even though start fairly similar, diverge greatly in the forecasted periods. This can have major implications of the plans, budgeting, and expenditure of bridge managers.

TABLE 2.35
Initial and Forecast Estimates of *V.I.* and *CR*

Bridge Component		Probabilities at Start of Time Periods					
		$t = t_0$	$t = t_1$	$t = t_2$	$t = t_3$	$t = t_4$	$t = t_5$
V.I.	Good	100.0%	95.0%	90.3%	85.7%	81.5%	77.4%
	Fair	0.0%	8.0%	14.0%	18.4%	21.6%	23.8%
	Poor	0.0%	2.0%	5.5%	10.1%	15.5%	21.5%
CR	Good	76.5%	72.8%	69.3%	65.9%	62.7%	59.6%
	Fair	23.4%	31.1%	37.7%	43.5%	48.6%	53.1%
	Poor	0.1%	1.1%	2.7%	4.8%	7.2%	9.9%

2.8 SOURCES OF CPTs

One of the basics of GN is the estimation on CPTs. Several objective and subjective methods for developing those CPTs are available. We discuss briefly some of those methods in this section.

- *Analytical Methods*: If a relationship between variables is known or can be estimated, conditional probabilities between them can be estimated using a multitude of processes from probability theory and using well-developed methods of developing functions of random variables (see Benjamin and Cornell 1970). Some of these processes include, but are not limited to, Taylor series or Monte Carlo simulation.
- *Known Frequencies*: If adequate frequency data of some variables are known, the needed conditional probabilities can be easily developed.
- *Objective Monitoring*: Another potential rich source of CPTs is objective monitoring of the needed data. Sometimes objective relationships between variables can be used to generate the needed conditional probabilities in CPTs. Principles of pattern recognition (Theodoridis and Koutroumbas 2009) can also be used when there is a reasonable body of data about particular variables of interest to generate the CPTs relating these variables.
- *Subjective (Expert) Estimation*: The CPTs can also be generated subjectively using past experience and/or expert opinions.
- *Sensitivity Analysis*: Koller and Friedman (2009) suggested a sensitivity analysis process for developing CPTs. This process entails:
  - Assign specific values, from whatever source, including the user's own subjective expert estimate, for the CPT parameters.
  - Execute GN.
  - Observe the results.
  - Make reasonable changes to the CPTs that are under investigation. Repeat process until reasonable results are obtained.

Of course, a combination of any of the above methods can also be used when modeling GNs. Some examples of potential sources of CPTs for typical situations are shown in Table 2.36.

2.9 CONCLUDING REMARKS

In all, our aim is to explore as many practical risk management/multihazards situations as possible in the civil infrastructure field. We recognize that all the examples presented can be made more realistic and more involved than presented. However, for simplicity and limited space, we opted for examples that are simple to follow and easy to develop. Readers are encouraged to build on these to suit their applications, as needed.

TABLE 2.36
Potential Sources of CPTs for Some Typical Situations

Encountered situation	Possible source of information (CPT).
Measurements	Sensor measurements. Sensors can be strains, accelerations, temperature, tilt, etc. Also, general nondestructive testing (NDT) methods can be used to generate useful and accurate information.
Condition rating of a structure	Routine or special inspection reports form the basis of information of this node.
Live load	Weigh in motion (WIM) archives constitute the basis of information of this node. In some cases, WIM might be used in the real time. In such a situation, special attention to potential practical problems that might arise from the large amount of data available is needed.
Social and economic consequences	Social and economic information needs to be vetted carefully. Even though there is a large body of data available, useful and pertinent information that can be used in this particular type of problem might be difficult to attain. Discussions of this situation can be found in Ettouney and Alampalli (2012a and 2012b) and DHS (2011a, 2011b, and 2011c).
Exposure	As of the writing of this book, only subjective methods that can develop this CPT are available. Historical data and statistics can be used in the future to objectively compliment these subjective CPTs.
General consequences	Consequences CPTs can be evaluated subjectively using the process documented in Ettouney and Alampalli (2012a, 2012b) and DHS (2011a, 2011b, 2011c). Again, historical data and statistics can be used in the future to objectively compliment these subjective CPTs.
Safety/reliability	An immense body of objective and subjective methods relate reliability (as represented by safety) to exposure (as represented by visual observations). See, e.g., NYSDOT (2014) and FHWA (1995).
Risk	Relating risk to exposure or consequences can be done either objectively or subjectively. See Ettouney and Alampalli (2012a and 2012b) and DHS (2011a, 2011b, and 2011c).

REFERENCES

Agrawal, A., Kawaguchi, A., and Qian, G. (2009). "Bridge Element Deterioration Rates: Phase I Report," *New York State Department of Transportation*, Project # C-01-51, Albany, NY.

Agrawal, A., Washer, G., and Xu, G. (2013). "Consistency of the New York State Bridge Inspection Program," *New York State Department of Transportation*, Project # C-07-17, Albany, NY.

Ayyub, B., and McCuen, R. (2011). *Probability, Statistics, and Reliability for Engineers and Scientists*, Third Edition, CRC Press, Boca Raton, FL.

Benjamin, J., and Cornell, A. (1970). *Probability, Statistics, and Decisions for Civil Engineers*, McGraw-Hill, NY.

DHS. (2011). *Integrated Rapid Visual of Screening of Buildings*, Building and Infrastructure Protection Series. Washington, DC.

DHS. (2011a). *Integrated Rapid Visual Screening of Mass Transit Stations*, Building and Infrastructure Protection Series, DHS, Washington, DC.

DHS. (2011b). *Integrated Rapid Visual Screening of Tunnels*, Building and Infrastructure Protection Series, DHS, Washington, DC.

DHS. (2011c). *Integrated Rapid Visual of Screening of Buildings*, Building and Infrastructure Protection Series, DHS, Washington, DC.

Ettouney, M.M., and Alampalli, S. (2012a). *Infrastructure Health in Civil Engineering: Theory and Components*, CRC Press, Boca Raton, FL.

Ettouney, M.M., and Alampalli, S. (2012b). *Infrastructure Health in Civil Engineering: Applications and Management*, CRC Press, Boca Raton, FL.

FEMA. (2005). *Risk Assessment: A How-To Guide to Mitigate Terrorist Attacks*, Risk Management Series, FEMA 452, Federal Emergency Management Agency, Washington, DC.

Fenton, N., and Neil, M. (2013). *Risk Assessment and Decision Analysis with Bayesian Networks*, CRC Press, Boca Raton, FL.

FHWA. (1995). *Recording and Coding Guide for the Structure Inventory and Appraisal of the Nation's Bridges,* Report no. FHWA-PD-96-001, Federal Highway Administration, Washington, DC, December 1995.

Huang, J. (2015). "Reasoning with Bayesian Networks." Available at http://users.cecs.anu.edu.au/~jinbo/pr/lecture01.pdf [accessed September 9, 2015].

Jensen, F. (1996). *An Introduction to Bayesian Networks*, Springer, New York, NY.

Jorion, P. (2007). *Value at Risk*, McGraw-Hill, New York.

Kjaerulff, U., and Madsen, A. (2013). *Bayesian Networks and Influence Diagrams: A Guide to Construction and Analysis*, Springer Science, New York, NY.

Koller, D., and Friedman, N. (2009). *Probabilistic Graphical Models: Principles and Techniques*, the MIT Press, Cambridge, MA.

Neapolitan, R.E. (2004). *Learning Bayesian Networks,* Prentice-Hall, Upper Saddle River, NJ.

NYSDOT. (2014). *Bridge Inspection Manual,* New York State Department of Transportation, Albany, NY.

Theodoridis, S., and Koutroumbas, K. (2009). *Pattern Recognition*, Academic Press, London, UK.

Trueman, R.E. (1977). *An Introduction to Quantitative Methods for Decision Making*, Harcourt School.

Von Neumann, J., and Morgenstern, O. (1944). *Theory of Games and Economics*, Princeton University Press, Princeton, NJ.

Vose, D. (2009). *Risk Analysis a Quantitative Guide*, John Wiley & Sons, Hoboken, NJ.

3 Risk Assessment

3.1 INTRODUCTION

3.1.1 Overview

Risk assessment perhaps is the most practiced and studied component of risk management and can be considered the fulcrum of the risk management efforts. For without assessment, the other components cannot be processed. This chapter addresses risk assessment from a civil infrastructure perspective. We start the chapter by exploring the rich body of knowledge regarding risk assessment in civil engineering followed by an overview of the remainder of the chapter.

3.1.2 Risk Assessment in Civil Infrastructures

A large body of knowledge regarding risk assessment exists, and some recent studies pertaining to infrastructure management are summarized in this section.

3.1.2.1 Natural Hazards

Natural hazards effects on civil infrastructures witnessed many applications and use of risk-based practices. For example, Wolshon et al. (2005a, 2005b) reviewed hurricane evacuation practices and research from a transportation engineering perspective based on a survey of state evacuation plans and practices performed at Louisiana State University and a study of literature reviews and ongoing research projects. The focus was on evacuation policies and practices by states use in the areas of transportation system planning, preparedness, and response and implementation and management of new and innovative evacuation techniques and systems, including the use of contraflow traffic operations and the application of intelligent transportation systems related technologies and techniques.

Key findings addressing policies and practices for transportation system planning, preparedness, and response include: (1) Different states use several types of evacuation orders, and the three main types are mandatory, recommended, and voluntary evacuations. Governors generally maintain the authority to issue evacuation orders, but in most states, that authority has been delegated to local officials. In general, evacuation planning and decision-making occurs at the local level. (2) Most states use a combination of several different public awareness campaigns and media types throughout the year to raise the general level of hurricane and evacuation preparedness. (3) Encouraging signs of the improvement of hurricane evacuations in the United States and increased involvement of transportation planners and engineers has brought expertise and experience in dealing with transportation-related issues, including forecasting evacuation travel demand, evacuation traffic analysis and modeling, and the application of intelligent transportation systems (ITS) technologies. (4) Further education and a greater exchange of information is still needed, and many federal agencies and national organizations are actively trying to fill this gap. (5) Standards development and best practices guidelines in this area are needed to benefit state and local-level transportation department and emergency management officials and to ensure a basic level of practice uniformity from location to location.

The key findings addressing policies and practices from the perspective of evacuation traffic operations, management, and control include: (1) Planned use of contraflow freeway operations has been the most significant innovation because the capacity of existing evacuation routes won't accommodate the enormous demand associated with the evacuation of many medium-to-large population centers. But it has many inherent difficulties and poses challenges requiring effective communication and close cooperation between all stakeholders. (2) There is a vital need for emergency management agencies to have reliable and current information on traffic and weather conditions to coordinate evacuations on a strategic level. As data volumes increase with the latest technologies, tools for data management for effective evacuation decision support are necessary. (3) Another critical need for emergency management agencies is the capability to disseminate timely, accurate, and useful travel information out to evacuees. (4) Because of the impact construction activities can have on effective evacuation, transportation agencies should review appropriate risk mitigation efforts to retain full capacity should an evacuation be necessary and also have procedures in place to inform emergency management agencies of construction plans and schedules. (5) New research and development projects are ongoing to solve some of the evacuation related issues. These include assessments of temporal-spatial guidelines for beginning and ending contraflow, simulations of traffic operations along contraflow evacuation segments, and measurement and analysis of reverse flow traffic streams, and so forth.

Risk-based designs were also used in earthquake engineering. Many highway infrastructure owners commonly use the *Seismic Retrofit Manual for Highway Bridges* (Buckle and Friedland 1995). This manual stipulates that for bridges near unstable slopes, detailed geotechnical investigations should be carried out to assess the potential for slope instability under seismic excitations. This task requires detailed investigations and considerable resources to implement. According to Zatar et al. (2008), there is current interest in a careful assessment of the *most critical* embankments along priority routes, and no such complete methodology exists to identify and prioritize highway embankments that are susceptible to seismic failure. Hence, they developed a methodology based on seismic slope stability capacity to demand ratio, anticipated embankment displacement, and liquefaction potential obtained for each of the embankment sites. Categorization (A, B, or C) to identify failure risk of the embankment behavior during a specified seismic event is based on the significance of these three parameters.

Zatar and Harik (2008) applied this methodology to 127 bridge embankments on a priority route in western Kentucky and investigated the resulting seismic vulnerabilities during projected 50-, 250-, and 500-year seismic events.

Many researchers in seismic vulnerability and risk assessment studies of infrastructure have used fragility analysis. Fragility describes the probability of failure as a function of demand. Celik and Ellingwood (2009) studied the performance of reinforced concrete frames in the Central and Eastern United States (CEUS) when they are subjected to earthquakes that are believed to be plausible in the New Madrid Seismic Zone. They derived seismic fragilities for low-, mid-, and high-rise gravity load designed reinforced concrete frames using the simulation-based reliability analysis and used this data to evaluate the seismic vulnerability of the reinforced concrete frame inventory in Memphis, Tennessee, based on the performance-based design objectives in Federal Emergency Management Agency (FEMA) 450 (2003). Using these results, they concluded that it is unlikely that the majority of existing frames designed before 1990 design practices would meet the life safety and collapse prevention performance objectives advocated by prevalent building codes and guidelines for performance-based earthquake engineering.

3.1.2.2 Man-made Hazards

Risk assessment of man-made hazards has witnessed a large body of work, especially in the past few decades. The risk management series, which FEMA and US-DHS produced, contain several risk assessment documents for assessing man-made and natural hazards risks (see FEMA 2005, 2010; DHS 2011a, 2011b, 2011c). We use the methodologies of DHS (2011a, 2011b) as the basis of Case Studies 3.1 and 3.2.

3.1.2.3 Flood and Scour Hazards

Niezgoda and Johnson (2007) developed a two-step method of incorporating uncertainty and risk in stream restoration design as a combination of design failure modes, effects analysis (DFMEA), and risk quantification. They applied their procedure to a stream relocation project in Pennsylvania. DFMEA is a methodology intended to ensure that all design deficiencies have been identified and addressed through analysis of all components of a design in terms of the consequence of failure, the likelihood of a component failure, and the level of difficulty required to detect failure before the project is initiated. This information can be applied to cost terminology by adding the initial project cost (including assessment, design, and construction) and the summation of product of probability of failure of each component and consequence of failure of that component (in terms of costs of repair, replacement, and damage).

The role of subjectivity in the application of DFMEA was addressed by evaluating not only relative risk but also the potential variation in results when a group of knowledgeable stakeholders complete the DFMEA.

Gonzalo et al. (2012) note that because high flows occur relatively rarely, low flows will have an excessive weight in the parameter estimation, which leads to model parameters that result in models performing well under low-flow conditions but not under high flows. A possible solution to this problem is to use two different simulations, one calibrated for high flows and the other calibrated for low flows (Oudin et al. 2006). Hence, they reviewed the lumped hydrologic model currently used for flood warning systems in Costa Rica and then proposed an ad hoc hydrologic model to predict the hydrological response of midsize mountain watersheds with short lag times, thus providing more accurate predictions of the magnitude, arrival, and duration of floods. The modeling approach was expected to meet reasonable time accuracy below the hourly level to be suitable to analyze flash floods occurring in watersheds with short lag times.

Preservation of structures with historic or heritage value is an important subject, and often many risk assessment methodologies do not cover historic and heritage issues. Holický and Sýkora (2010) note that the risk assessment guidelines issued by the United Nations Disaster Relief Organization related to natural disasters (UNESCO-UNDRO 1979) and the European Directive 2007/60/EC (2007) do not include cultural heritage issues (Drdácký et al. 2007). Because assessing cultural heritage value in monetary terms is difficult (Canuti et al. 2000), authors state that risk assessments of such assets are normally based on a combination of quantitative and qualitative criteria where likelihoods of unfavorable events and expected consequences are evaluated in risk-rating matrices. Noting the damage to cultural heritage sites during fluvial flooding in August 2002 in the Czech Republic, researchers proposed a general framework of risk assessment using probabilistic methods and a simplified Bayesian network (BN), which requires a theoretical model suitable for predicting flows and the extent of future floods. Authors statistically analyzed hydrologic data for annual maximum flows of the Vltava River in Prague dating back to 1827 and estimated extreme flows, for different return periods (10 to 1000 years), needed to assess the flood risk to endangered sites and to make decisions on protective measures.

3.1.2.4 Geotechnical Engineering

Geotechnical risk-based applications also witnessed numerous uses of risk-related practices. Baker and Faber (2008) contend that liquefaction triggering assessments performed for individual locations provide little information on expected spatial extent of liquefaction events and proposed a method to quantify the potential extent of liquefaction by accounting for spatial dependence of soil properties and potential future earthquake shaking. To produce informative assessments of potential liquefaction extent, they combined available geotechnical tools for modeling liquefaction occurrence, geostatistical tools to model uncertain soil properties, and probabilistic seismic hazard models to compute the distribution of intensity of future ground motion shaking. Through an example, they illustrated their approach and demonstrated its use. Using their methodology, annual rates of liquefaction can be computed such that projects can be designed with uniform levels of risk and the extent of liquefied area under a building can be characterized by considering spatial dependence of soil properties. According to authors, estimating the needed probability distributions for soil properties at a given site probably is the greatest challenge for implementing this approach.

Li et al. (2010) present a quantitative risk assessment methodology to assess the risk for cut-slope projects under construction, applied the methodology to an actual situation, and then compared the results to make relevant conclusions to such projects. The proposed methodology involved hazard identification, probabilistic analysis, consequences analysis, and risk calculation. Realizing that the reliability of cut-slope is usually greater near project completion than during the initial phases of a project, landslide occurrence probability and corresponding consequences at every step in construction were analyzed to identify a critical scenario. Uncertainties of soil and rock strength parameters were accounted for using the Monte Carlo simulation method in this study. Landslide probability was estimated at the most dangerous construction step. Consequence analysis involved identifying the landslide magnitude, estimating the potential impacted area, identifying the exposed elements in the impacted area, and estimating the extra time due to landslide. Risk was calculated by integrating probability and consequence analysis using occurrence probability of the landslide event, temporal probability that landslide occurs in different time intervals, probability of spatial impact, and individual vulnerability given the spatial impact.

This methodology was applied on a large cut-slope project in which a landslide did occur and resulted in two deaths, four missing, several injuries, and complete destruction of protecting structures. According to the authors, this method proved effective and reasonably accurate, but the predicted time extension and the economic loss for time-dependent cost were significantly less than the actual values. They also concluded that the time overrun was difficult to estimate because both landslide magnitude and human factors influence it.

Reuse instead of disposal of materials is increasingly being considered in managing contaminated sediments dredged from harbors. Ndiba and Axe (2010) examined the risk of metal leaching into groundwater in reuse of phosphate and thermal-treated sediments. Their risk assessment showed that phosphate and thermally treated sediments do have potential for reuse in construction, based on the metals studied.

3.1.2.5 Offshore Platforms and Coastal Engineering

Offshore platforms, similar to other infrastructure, are also aging and require application of risk management by owners and regulators. Staneff et al. (1995) presented the California Coastal Platform Information Management System (CA IMS), a software implementation of the first level of a screening system for the reassessment and requalification of offshore platforms, such as those Bea and Craig (1993) and Aggarwal (1991) proposed. The software incorporates simplified structural-integrity, failure-consequence, and risk assessment routines, as well as platform data management and an advanced probabilistic environmental data mechanism. The structural-reliability-assessment procedure is employed by introducing a qualitative scoring factor model known as *reserve strength ratio* (RSR), obtained by dividing the structure's ultimate lateral-load capacity by its design or *reference* lateral loading, and then it is related to the probability of failure. A qualitative assessment of consequences is obtained and then converted into a numeric value and integrated into one combined consequence measure determined through utility functions that express user's risk aversion. Probability of failure and consequences are related and acceptable and marginal acceptable lines plotted in a graphical format. This is then used to decide if an offshore platform is considered acceptable or not (i.e., fit for purpose or not).

Research studies on structural reliability estimations were starting to gain popularity in the early 1990s. Tromans and Graaf (1994) outlined a structural reliability model/application based on Bea (1974) and Anderson et al. (1982) and applied this methodology to a steel jacket structure that experienced a severe hurricane loading without structural damage to calculate the risk of failure during that hurricane and also long-term reliability.

The analysis began with the hindcast database of meteorological-oceanographic conditions for a representative combination of extreme environmental conditions to generate a base caseload set. In the pushover analysis, this load set is progressively incremented to obtain the collapse sequence of the structure and its ultimate capacity. The collapse analysis is repeated for several directions to produce a failure surface for the structure, including the uncertainty in collapse strength. Using the hindcast, extreme response of a generic structure is studied to obtain the probability of failure by comparing the extreme load distribution with the strength distribution as defined by the failure surface.

3.1.2.6 Bridge Engineering

Safety and stability assessment of bridges, with unknown foundations over water, to assure safety and uninterrupted mobility after a large hydraulic event, such as flooding or a seismic event, is not easy, requiring considerable engineering judgment supported by previous experience. To assist engineers in such situations, Zayed et al. (2007) developed a model to assess risk factors of bridges with unknown foundations and to evaluate and prioritize various courses of action for managing these bridges. These researchers proposed a risk index (R) based on factors affecting physical risk. These factors were divided into three major risk parameters—remaining life, scour (see Figure 3.1), and seismic vulnerability, with each of these comprising several factors. The risk index was developed primarily to transfer the subjectivity of risk parameters into quantitatively determined values using the analytical hierarchy process (AHP) of Saaty (1980). Therefore, the risk index was a summation of the product of the weight of each factor along with its worth score (Zayed and Chang 2002; Zayed and Halpin 2004). Weights of risk parameters were determined using AHP, and the worth score was assessed using the utility function approach. If a bridge has a low R (0–0.2), routine or minor investigations were recommended, advanced investigation or major rehabilitation and/or replacement options for high risk ($R > 0.6$), and intermediate courses of action for intermediate value of R (0.3–0.6).

R-values were estimated, as part of a case study, for ten bridges in Florida, Indiana, and New York using both an abstract model and a detailed model to decide appropriate courses of action. The model was validated using a holistic evaluation method. The bridge engineers holistically evaluated each bridge in the ten-bridge case study on a scale from 0.00 to 1.00, using their experience, knowledge of the individual bridges, and input from inspection teams to develop their own diagnoses and courses of action. These decisions were compared to the recommendations rendered by the model and found to be in good agreement.

3.1.2.7 Construction Engineering

Due to trade agreements, such as the North American Free Trade Agreement, developments in Europe and the Soviet Union, and technological advancements such as the Internet have revolutionized the global economy and ways to do business. This also means that opportunities for companies,

FIGURE 3.1 Damaging effects of scour hazard.

including A/E/C firms (architectural, engineering, and construction firms), to compete globally. This makes risk assessment very important at several levels; that is, first the company has to decide if there is potential for expansion and doing short- as well as long-term business in other countries, analyze appropriate markets, and make sure they can deliver projects. Hastak and Shaked (2000) claim that most current models predominantly focus on political risk or instability, are not comprehensive, and do not address project level risks. Hence, they present a risk assessment model, the International Construction Risk Assessment Model (ICRAM-1), to assist the companies evaluate the potential risk involved in expanding operations in an international market by analyzing risk at the macro (or country environment), market, and project levels. They define macro level as the general risk to an international investor when expanding operations in a given country, market level as the risk associated with a specific international construction market, and project level as the risk associated with a project in that specific international construction market. To consider the interaction among these three levels, market level also includes the impact of the macro level on the construction market and the project level includes the impact of the macro and market levels on the project.

The ICRAM-1 analyzes 73 risk indicators at the above three interrelated levels, contains a risk assessment module for each level, and uses the AHP to analyze the hierarchy of risk indicators within each level and to determine the relative importance of the risk indicators by establishing priority among the criteria, subcriteria, and indicators. It also takes into account the transfer of risk between these levels. High-risk indicators (based on threshold-weighted risk for each of the three levels), the impact of country environment on a specific project, the impact of market environment on a specific project, and the overall project risk can be obtained from the model.

Authors also note that ICRAM-1 can also be used for analyzing different countries with respect to a specific project and in comparing different types of projects in a specific country. Using the model several times over a defined period can also facilitate monitoring of risk associated with a specific country with respect to the macro, market, and project levels. It should be noted that input to ICRAM-1 is very subjective and confidence in the results depends on the information available and the reliability of the sources.

For underground construction, Choi et al. (2004) developed risk assessment software that uses an assessment procedure for identifying, analyzing, evaluating, and managing the risks inherent in underground construction projects. Building upon previous studies (Zadeh 1965; Hadipriono et al. 1986; Fujino 1994; Huang et al. 2001) on the fuzzy even tree analysis technique, they integrated three possible approaches: (1) When historical data are sufficient and available, the probability of each risk on potential failure paths can be evaluated by using simple frequency analysis. (2) If the data are insufficient, probability theories such as Monte Carlo Simulation or the Bayesian approach are used as assessment methods; that is, insufficient data are complemented by simulated or updated data. (3) If the data are not available, the occurrence probability of each path may have to be assessed by subjective judgments based on experts' experiences and knowledge. The uncertainty model was also designed to consider the uncertainties involved in both probabilistic parameter estimates and subjective judgments. Authors applied the proposed methodology and software to the sixth subway line construction project in Seoul that suffered many accidents during construction, and they concluded that their methodology provides rational and practical solutions to the insurance companies and the contractors because of its flexibility, easy-to-follow procedure and tools, and robust uncertainty modeling capability.

3.1.2.8 Environmental Risk

Applications of risk-based techniques in environmental engineering abound. For example, the geographical information system (GIS) are commonly used for the representation and analysis of spatial data. These tools were relatively new in the beginning of this century. At that time, it was expected that risk managers would demand access to decision-support tools based on GIS that allowed them to manage and understand the complicated nature of a disaster (Zerger and Smith 2003). This is a reality now and GIS is widely used for risk assessment and mitigation for disaster management. Samadder and Subbarao (2007) analyzed health effects of arsenic pollution of drinking water in six administrative blocks of the central part of the Murshidabad district (West Bengal, India) using GIS, and reported the regional impact of arsenic contamination with an intent to assist in planning and implementing priority-based arsenic mitigation options.

3.1.2.9 Urban Risks

Water mains are an integral part of infrastructure in maintaining quality of life. Grigg (2013) notes that replacement to sustain the integrity of water mains in the United States is on a cycle of approximately once in 200 years, a slow rate that contributes the current grade of D– on the Infrastructure Report Card (ASCE 2013). The main indicator of physical integrity is frequency of water main breaks, but reducing these is difficult because of the complexity and scale of the problem and the lack of incentives for pipeline replacement. Grigg (2013) reviewed current research to explain the knowledge base about main breaks, the extent to which they are a serious problem deserving more attention, the risk formulation of the decision process for pipeline replacement, the current rates of renewal, and the reasons why utilities renew pipes at low rates.

According to Grigg, based on surveys of larger utilities and conservative estimates, 250,000–300,000 breaks per year occur in the United States because of failures such as structural breaks, joint failures, punctures, or corrosive failure of walls, with an estimated average expected cost of $500,000 due to a large pipe break. The author proposed a general framework for risk assessment considering failure probability and consequences and considering how pipes are exposed to threats that may cause a break that will cause flooding and may lead to other damage.

They also identify that the utilities replacement program is not solely based on risk assessment but on several other reasons such as gaining efficiencies or coordinating replacement to take advantage of other infrastructure replacements. Given the resource issues, they note that to reduce the failure rate in the long run, solutions to physical, financial, and institutional problems of water mains will continue to require enlightened and committed professional and political leadership.

Water distribution systems (WDSs) are designed to deliver sufficient water to regular users as well as for emergency conditions such as urban fire events with required pressure, volume, and quality. Given fire can be associated with an earthquake, deterioration, or man-made hazard, considering this interaction is very important in risk assessment of WDSs to plan for appropriate mitigation efforts, if necessary. Kanta and Brumbelow (2013) developed a methodology to examine the vulnerability and risk to WDSs during urban fire events considering the hazard interaction. They considered three failure modes in examining vulnerability and risk for a water distribution system during fire events: (1) accidental failure of pipes due to soil-pipe interaction, (2) accidental failure of pipes due to a seismic event, and (3) malevolent action resulting in pipe failure. This was considered an optimization problem, which maximizes the system risk during an urban fire event concurrent with failure or disruption of specific system components. An optimal solution to the problem is achieved by interfacing a dynamic programming (DP) optimization algorithm with a hydraulic simulation model, EPANET 2.0. The hydraulic simulation model is used to solve WDS hydraulics to evaluate the objective function at each iteration. The method was applied to a case study to show the viability of the method in integrating vulnerability and risk and to demonstrate and assess existing infrastructure and potential mitigation alternatives. The researchers, based on a case study, conclude that infrastructure managers must evaluate multiple sources of risk being used to determine the relative importance of each risk source. The relative risks from the methodology can provide justification for focusing on accidental failure modes and appropriate mitigation methods. Kanta and Brumbelow (2013) have also shown that topological remoteness from regions with pronounced exposure to particular failure modes does grant some measure of reduced sensitivity to those modes; the methodology allows for quantitative determination of this remoteness-insensitivity correlation.

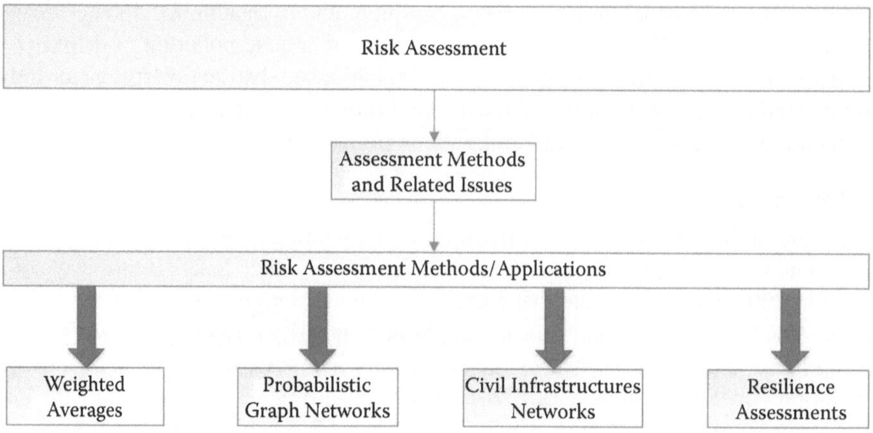

FIGURE 3.2 Contents of this chapter.

Wang et al. (2013) present the pipe burst risk assessment methods and procedures along with a sample mode applied to a Chinese City pipe network system for pipe burst risk classification examples. For risk assessment, identification of primary and secondary aspects is important for optimal service as well as durability. Keeping this in perspective, the researchers identify both hydraulic and nonhydraulic factors for pipe bursts. Pipe bursts can destroy a city's pipe network, leading to the water supply interruption, loss of life and property, as well as water quality problems. Hence, a mitigation measure was advocated through a pipe burst risk state assessment and a classification technology to estimate the approximate time when a pipe went out of service to formulate a pipe network maintenance plan in advance to avoid the occurrence of burst pipes. Based on hydraulic transient flow analysis, four risk factors were predicted and a four-layer classification map was created. A network classification map was created to provide technical support for the design, operation, and management of water supply networks. The technique proposed is limited in that it only works for pipes burst due to hydraulic transients.

3.1.3 This Chapter

The remainder of this chapter begins by addressing several pertinent issues regarding risk assessment in civil engineering. Though numerous analytical methods of risk assessment exist, weighted averages and graph networks (GNs) are the most suitable for civil infrastructure applications. We then proceed to offer several case studies to highlight the use of these methods that involve a risk assessment of both single assets and a community of assets. We end the chapter by exploring resilience assessment, which we argued earlier to be a special manifestation of risk assessment. The contents of the chapter are shown in Figure 3.2.

3.2 RISK ASSESSMENT METHODS AND RELATED ISSUES

3.2.1 Essential Ingredient of Risk Assessment

Risk assessment serves many objectives within any risk management effort. It is the foundation of all other components of the management effort. As such, the accuracy of an assessment endeavor needs to be as detailed and precise as possible to represent real situations on which to base objective decision-making. To achieve such accuracy, the nature of risk and the components contributing to it should be well identified and well understood (see Cordona et al. 2012). A system under consideration should be clearly identified: for example, is it a component, a whole asset, or a network of assets? The metric of the risk variable needs to be understood: is it subjective or objective; is it a snap shot in time or a life cycle? Another important consideration is the necessity, from ground up, to identify the need and objective of the whole risk assessment effort.

Risk assessors should identify all the uncertainties, both explicit and implicit, in advance. In many cases and for the sake of simplicity, the rigors of uncertainty computations might be hidden. However, this uncertainty underpinning the risk assessment process needs to be handled accordingly. Irrespective of the risk metrics on hand being subjective or objective, it should be recognized that the variables in any risk model can be a mix between subjective and objective variables. Variables in risk models, especially those in the civil infrastructure community, are interlinked in many cases. Modeling these links in the assessment model properly can mark the difference between useful versus erroneous results. Risk implies negative outcomes, sometimes expressed in monetary terms as costs. In some situations where models can be formed to assess rewards, as opposed to risk, positive outcomes will be produced. In other models, risk and rewards can be or need to be included in the same model.

3.2.2 Risk Assessment Methods (Risk as a Function of Random Variables)

An immense number of methods and processes are available in the literature, as noted in the previous section, that were developed over the years to assess risks associated with infrastructure management. Hence, in this section, we will describe some of the genres of risk assessment methods instead of going into the details of each method. We can formally define risk in its most general form as a function of three random variable functions.

$$Ri = f(X_1, X_2, X_3). \tag{3.1}$$

The three functions, X_i, which are also random variables, can be described as the consequences, C, threats, T, and vulnerability, V, of the physical and/or operational system under consideration. In other words

$$Ri = f(C, T, V). \tag{3.2}$$

Furthermore, we can, without any loss of generality, assume that each of the three basic components, C, T, and, V, are a function of several variables needed to completely describe Ri in as much detail as desired. These variables can be random or deterministic, as pertinent. The risk assessment process can now be summarized in three steps:

1. Identifying the basic variables of Equation 3.2, as pertinent
2. Defining the function $f(\,)$
3. Solving Equation 3.2 using appropriate process

We now look at genres of risk assessment processes that follow the above basic three steps. Essentially, this method of risk assessment aims at identifying the function $f(\,)$ of Equation 3.2. Sometimes, in simple situations, it is fairly straightforward to define the functional relationships underlying the risk, Ri, with its underlying variables. Several authors described this process in full detail (see Jorion 2007; Vose 2009). In situations where formalizing the function is still possible, but its analytical solution is not attainable, the Monte Carlo process can be used (see Benjamin and Cornell 1970; Jorion 2007; Vose 2009).

In the field of civil infrastructures, even for the simplest application, the function $f(\,)$ can be complicated. The numbers of variables as well as the description of the functional relationships between the different variables are complex and formal. Therefore, computational processes and efforts can increase quickly. We realize the theoretical and educational importance of this process; however, we decided to forgo additional discussions and use of this approach in this book to keep our presentations and discussions of practical value to the readers.

3.2.2.1 Weighted Averages

Due to its simplicity, weighted averages methods are used in many risk assessment fields, including civil infrastructures (see Section 3.1.2). The methods can be traced back to the function of random variables method (Vose 2009) as well as the GN method (Fenton and Neil 2013). DHS (2011a, 2011b, 2011c) use a combined version of GN and weighted averages models for assessing risk and resilience. We showcase the method in detail while assessing risk of mass transit stations and tunnels in Case Studies 3.1 and 3.2, respectively.

3.2.2.2 Probabilistic Tree Construct Methods: FTA and ETA

Tree constructs, such as Fault Tree Analysis (FTA) and/or Event Tree Analysis (ETA), can be shown to be subsets of GN. Fenton and Neil (2013) provided a comprehensive reasoning as to why the use of the more general GN is preferred over FTA or ETA. Because of this, we will not be using them in the case studies in this book.

3.2.2.3 Graph Networks

We discussed GNs (also referred to as probabilistic graph networks) in Chapter 2 extensively. GN is a numeric process that can accurately accommodate complex risk models with a large number of variables, while maintaining a complex set of relationships between these variables. Because GN is probabilistic from ground up, it can model the uncertainties of those variables better, and hence, we rely on GN modeling for risk as well as other metrics throughout this book.

Risk assessment of civil infrastructure needs to accommodate single as well as asset networks. When the stakeholder is interested in risks in relation to multitudes of civil assets, which form an asset network (or community), those asset networks* can be linked together by a complex or simple set of links or relationships. The existence of links in either of the two topologies also serves as a good reason to use GN as the assessment method because accommodating linkages is integrated in GN modeling and solution processes.

3.2.3 Combination of Risks

3.2.3.1 Overview

Civil infrastructures are interlinked systems on many scales. Perhaps the most recognized scale is what we can call the single asset scale. On this scale, we can identify single infrastructure assets such as buildings, bridges, roadways, tunnels, and so on. Each of these assets include, on a smaller scale, different components that are connected together to form the single asset. These components can be structural (such as beams, columns, decks, or tunnel liners) or functional (such as information technology, security systems, firefighting systems or heating, ventilating, and air conditioning HVAC). In addition, operational components such as training and emergency operations/readiness also should be included. Single assets can also be combined to form an asset network (or community of assets). Our immediate interest is to note that each one of these levels can have its own risk value. For example, we can evaluate security risk, risk of failure of a structural beam, or risk of HVAC malfunction. We can also combine the risks of different single asset subcomponents to define a total risk of that single asset by combining all the risks of the pertinent subcomponents. Conversely, if we know the risks of different single assets in an asset community, we can combine all of those risks to find the total risk of the asset community.

Several methods are used to combine risk components of a system into total system risk. For example, the financial community has long been computing investment portfolio risk (total risk)

* Should not be confused with graph networks.

from the individual risks of the portfolio's component investment (see Jorion 2007). Two methods are available for computing total risk of a system from the risks of the system's subcomponents. We explore the two methods next.

3.2.3.2 Analytical Methods

The formal problem is that we have a system with N subcomponents. We know the risk of each of these subcomponents, $Ri|_i$ with $i = 1, 2, \ldots, N$. We need to evaluate the total risk of the system, $Ri|_{TOTAL}$. Several authors provided analytical solutions to this problem (see, e.g., Jorion 2007; FEMA455 2009; Vose 2009; DHS 2011a, 2011b, 2011c; Ayyub 2014). The solutions vary from simple weighted averages process such that

$$Ri|_{TOTAL} = \sum_{i=1}^{i=N} \left(w_i \left(Ri|_i \right) \right) \tag{3.3}$$

where the assessor defines a weight, w_i, for the i-th risk component. The weight is usually subjective and depends on the experience of the assessor and the nature of the problem at hand.

Another approach combines the risks using exponent, n, in the form

$$Ri|_{TOTAL} = \left(\frac{\sum_{i=1}^{i=N} \left(\left(Ri|_i \right)^n \right)}{N} \right)^{1/n} \tag{3.4}$$

It should be noted that Equations 3.3 and 3.4 give identical results if $w_i = 1$ and $n = 1$. Jorion (2007) used Equation 3.3 while Vose (2009), Jorion (2007), FEMA455 (2009), and DHS (2011a, 2011b, 2011c) all used Equation 3.4. There seems to be no obvious advantage of either of the two equations; both offer the advantage of being simple. Only a system of validation, on a case-by-case basis, may justify the use of one of the equations over the other. For example, FEMA455 (2009) and DHS (2011a, 2011b, 2011c) have reported such a validation effort that supported their use of $n = 10$ in their reports.

3.2.3.3 GN-Based Methods

Fenton and Neil (2013) reported a slightly more complex technique in combining risks to obtain a total risk. Their approach is based on the use of GN, where component risks, $Ri|_i$, and total risk, $Ri|_{TOTAL}$, were connected together in a BN in form of a child-parent directed link model. The relationships between the children risks (component risks) and the parent risk (total risk) can be formalized using conditional probabilities tables (CPTs). This approach offers more flexibility than the analytical methods of Section 3.2.3.2 because the CPTs can be assigned to simulate the preference of the users or to simulate any available historical observations or data. This is the model we use in most of the case studies presented in this book.

3.2.4 Limit States and Risk Assessment

Management of civil infrastructures relies heavily on different types of limit states that provide conditions separating satisfactory and unsatisfactory conditions. Because of this, any risk management process has to be linked to these limit states. This of course starts with risk assessment. We now explore the relationships between limit states of different types and their risk.

3.2.4.1 Objective Limit States

The most recognized limit states in the civil field are limit states that describe structural (or material) behavior. The obvious one is the elastic (sometimes referred to as *linear*) limit state. Ductile limit states are higher structural limit states. Higher limit states are important in some design situations for blast and earthquake conditions. We offer some case studies relating to the higher limit states in Chapter 4. We can describe some of these limit states as objective because a large amount of objective research involves them (see ASCE/SEI 2011).

3.2.4.2 Subjective Limit States

We turn our attention now to subjective limit states. At first look, subjective limit states don't seem to be possible. Yet, upon further reflection, it becomes clear that subjective limit states are perhaps as prevalent and a well-used concept in civil infrastructures as their objective counterparts. These limit states are generally related to subjective measures, such as slow deterioration observed in infrastructure components, vibration experienced by structural users, deflection of bridges, and so forth, that are not directly related to the integrity of structures. Essentially, most serviceability limit states used in the infrastructure arena can be classified as subjective limit states. This includes excessive vibration or deflection limits used in structural design to make sure users are comfortable and allowable cracking that does not affect structural performance. The deflection limit (span/800) used in bridge design is an example of the subjective limit. The limit states can be described in a subjective fashion, for example, using high, medium, and low, or can have limits based on a combination of a subjective scale similar to condition ratings with use type, for example, a combination of condition ratings and functional classification of a bridge.

What makes the consideration of subjective limit states of extreme interest is that they avoid one of the major limitations of objective limit states—objective limit states can only be considered from a mechanical viewpoint, that is, a relationship between objective demand (force, temperature, etc.) and objective response (stresses, displacements, etc.). In general, this results in the lowest objective limit state being the elastic limit of components and limited use of lower limit states for situations where designs or decisions are based on limit states that are below the elastic limit. This is important as designs and demands on civil infrastructures are usually well below the elastic limit. For example: corrosion (see Figure 3.3), paint, spalling, scour, and fatigue demands and the decisions for mitigating them should all be well below the elastic limit of the corresponding components. On the other hand, bridge ratings, if considered limit states, can easily and accurately describe well-understood states, where there exists plethora of specific decisions and mitigating processes. Because of this, we base our risk-based objective limit states on inspection ratings.

3.2.4.3 Exposure Limit States

As time progresses, the condition rating, *CR*, of a specific infrastructure component decreases, assuming that no retrofit or maintenance actions are undertaken. Assuming a scale ranging from 0 to 9,

FIGURE 3.3 Corrosion as a result of a collision with traffic.

similar to the one recommended for bridge inspections by the Federal Highway Administration (FHWA 1995), where a *CR* of 9 indicates pristine condition and a *CR* of 0 indicates failed condition, the resulting rating at a given time *t* after the structure/component is newly built or replaced, is such that $0 \le CR \le 9$. We note that *CR* is discrete, reflecting the inspection process and the inspector's subjective estimation based on a relatively objective criterion. Further, we note that *CR* is the resultant of:

1. Cumulative demands on the components as of time t. These demands can be corrosion, freeze/thaw, traffic (see Figures 3.4 and 3.5), fatigue, etc.
2. Capacity/response of the components due to the above cumulative demands. These effects can also be considered as the inverse of the vulnerability of the components to the above demands (higher vulnerability for lower capacity and lower vulnerability for high capacity).

FIGURE 3.4 Traffic congestion.

FIGURE 3.5 Cumulative traffic and construction demands.

Based on the above and information from Section 1.3.3 on exposure, we immediately recognize that inspection *CR* is an inverse of the exposure. We can now define exposure, *E*, of the component as a function of *CR* as

$$E = f\left(\frac{1}{CR}\right) \tag{3.5}$$

In keeping up with the simplicity of the rating system, we choose the function *f*( ) to produce an exposure of 10 for a *CR* of 0 (failed condition) and an exposure of 1 for a *CR* of 9 (pristine condition). The change of scale of exposure to a 1–10 scale is needed to be consistent with the relative risk scale used throughout this book. Indeed, we now have quantified, albeit in a subjective manner, exposure limit states for structure components. We need not worry about the subjective nature of such quantification because researchers showed those subjective estimates to be accurate (Agrawal et al. 2013).

3.2.4.4 Consequence Limit States

For our purpose in this section, we define system related consequences (there are other consequences which are independent from the physical state of the system, such as social and/or economic consequences) as the costs that might be incurred when a particular component reaches a particular limit state (e.g., say a particular *CR* level). As such, these consequences, *C*, are expressed as

$$C = g(CR) \tag{3.6}$$

It is reasonable to expect that as exposure increases, costs (consequences) increase.

One advantage of our current approach is that there should be a well-documented body of knowledge of these costs (consequences) for different components for most civil infrastructure types. We have reduced the estimation of consequences, which is usually the most difficult step in the risk assessment process, in this particular subject to a simple effort.

3.2.4.5 Risk Limit States

Recall that in Chapter 1 we established risk as a function of exposure and consequences as

$$Ri = Ri\ (E, C) \tag{3.7}$$

Ignoring, for now, consequences other than those which are affected directly by and from Equations 3.5 and 3.6, we can write Equation 3.7 as

$$Ri = Ri\ (CR) \tag{3.8}$$

that indicates that risk limit states can be established as a function of *CR*. We use this conclusion several times in the remainder of this chapter as well as in the following chapters.

3.3 WEIGHTED AVERAGES

3.3.1 Overview

We note that in many cases, a practical simplifying assumption would be the independence of threats, vulnerability, and consequences. In such situations, we might be able to express risks for *NH* hazards as

$$Ri|_i = \sqrt[3]{C_i \cdot T_i \cdot V_i} \tag{3.9}$$

where C_i, T_i, and V_i represent consequences, threats, and vulnerability ratings for the system due to the i-th hazard, with $i = 1, 2, \ldots, NH$. Using a power of one-third in Equation 3.9 would produce a more reasonable risk value. See FEMA (2009) for a more detailed discussion on the choice of the power in Equation 3.9.

In its simplest form, the vulnerability rating can be expressed as a weighted average of several parameters (DHS 2011a, 2011b, 2011c; Fenton and Neil 2013), such that

$$\{V\} = [WT_V]\,\{VR\} \tag{3.10}$$

Similarly, the hazard (threat) and consequences ratings can be expressed as

$$\{H\} = [WT_H]\,\{HR\} \tag{3.11}$$

and

$$\{C\} = [WT_C]\,\{CR\} \tag{3.12}$$

respectively.

Matrices $[WT_V]$, $[WT_H]$, and $[WT_C]$ include a set of weights related to the system and a set of hazards under consideration. The order of $\{VR\}$ is the number of issues/considerations that affect different hazard vulnerability ratings, nv. Similarly, the orders of $\{HR\}$ and $\{CR\}$ are the number of issues/considerations that affect different hazards/threat ratings nh and the number of issues/considerations that affect different consequences ratings, nc. The main efforts in weighted averages process are to identify appropriate issues which control C, T, and V and to develop the weights in matrices, $[WT_V]$, $[WT_H]$, and $[WT_C]$. We offer two case studies that aim at computing risks for different hazards as they pertain to mass transit stations and tunnels.*

3.3.2 Case Study 3.1: Mass Transit Stations

This case study involves computing hazards' risks for a mass transit station. The hazards of interest are man-made hazards, earthquakes, fire, floods, and wind. The different issues involving vulnerability, threats, and consequences as well as the weights relating to these issues and the different hazards are shown in Tables 3.1 through 3.3, respectively. Note that the weights are relative and have been estimated by the analysts. When an inspector assesses a particular mass transit station, the different vectors $\{VR\}$, $\{HR\}$, and $\{CR\}$ are estimated as shown in Tables 3.4 through 3.6, respectively. We can evaluate vectors for hazards vulnerability, threat, and consequence, $\{V\}$, $\{H\}$ and $\{C\}$ from Equations 3.10 through 3.12. Finally, by applying Equation 3.9, we generate different hazards risks, $\{Ri\}$. The relative scores for $\{V\}$ $\{H\}$, $\{C\}$, and $\{Ri\}$ are shown in Table 3.7. To get the total risk from the individual hazards risks, we apply Equation 3.4 with $N = 4$ (4 hazards risk will contribute to the total risk). Table 3.8 shows the computed total risk for different values of n. Note that both Vose (2009) and DHS (2011a, 2011b, 2011c) have used a value of $n = 10$ in computing the total risk.

3.3.3 Case Study 3.2: Tunnels

We repeat the process shown in the previous section for tunnels. The hazards of interest are same as before: man-made hazard, earthquakes, fire, floods, and wind. The different issues involving vulnerability, threats, and consequences as well as the weights relating to these issues and the different hazards are shown in Tables 3.9 through 3.11, respectively. Note that the weights are relative

* These are simplified versions of the processes in DHS, 2011a and 2011b The reader may want to refer to them for more detailed discussions.

TABLE 3.1
Vulnerability Weights Matrix $[WT_V]$

Vulnerability Issues	Site	Architecture	Structure	Ventilation	Fire Protection	Operational Systems	Nonstructural Systems	Physical Security	IT Issues
Man-made	13	4	8	12	8	12	2	38	4
Earthquakes	6	11	39	4	12	7	9	3	9
Fire	5	10	20	9	33	10	7	2	6
Floods	44	3	4	9	4	18	5	5	8
Wind	50	3	10	4	5	20	5	2	2

TABLE 3.2
Threat Weights Matrix $[WT_H]$

Threat/Hazards	Traffic Issues	Redundancy	Attractiveness	Earthquake	Fire	Floods	Wind
Man-made	14	11	43	1	21	4	4
Earthquakes	0	0	0	100	0	0	0
Fire	9	13	13	15	19	15	15
Floods	4	0	0	0	0	74	22
Wind	0	0	0	0	0	23	77

TABLE 3.3
Consequences Weights Matrix [WT_C]

Consequences issues	Number of Tracks	Number of Levels	Station Elevation	Peak Number of Riders per Day	Station Locality	Nearby Facilities	High Target Values	Impact of Physical Loss	Social Impact	Replacement Value	Operational Redundancy	Downtime
Man-made	5	4	8	10	3	8	10	13	13	13	8	8
Earthquakes	7	5	9	11	1	3	3	15	15	15	9	9
Fire	6	5	9	11	1	2	2	15	15	15	9	9
Floods	7	5	9	11	1	3	3	15	15	15	9	9
Winds	7	5	9	11	1	3	3	15	15	15	9	9

TABLE 3.4
Vulnerability Assessment Vector {VR}

Site	3
Architecture	9
Structure	8
Ventilation	8
Fire protection	7
Operational systems	3
Nonstructural systems	2
Physical security	2
IT issues	5

TABLE 3.5
Threat Assessment Vector {*TR*}

Traffic issues	4
Redundancy	8
Attractiveness	8
Earthquake	7
Fire	5
Floods	3
Wind	7

TABLE 3.6
Consequences Assessment Vector {*CR*}

	CR
Number of tracks	6
Number of levels	2
Station elevation	9
Peak number of riders per day	7
Station locality	8
Nearby facilities	3
High target values	8
Impact of physical loss	7
Social impact	8
Replacement value	8
Operational redundancy	7
Downtime	8

TABLE 3.7
Vectors for Vulnerability, Threat, Consequences, and Risk for Different Hazards

Hazard	Vulnerability	Threat	Consequences	Risk
Man-made	42%	65%	71%	58%
Earthquakes	63%	70%	72%	68%
Fire	64%	60%	73%	65%
Floods	41%	39%	72%	49%

TABLE 3.8
Total Risk from Individual Hazards Risks

n	1	2	5	10
{*Ri*}	60%	60%	62%	63%

TABLE 3.9
Vulnerability Weights Matrix [WT_V]

Structural Vulnerability Issues	Method of Construction	Liner Cross-Section	Liner Relative Thickness	Construction Material	Known Retrofits	Longest Span	Controlling Height	Type of Framing	Seismic Design	Overall Structural Conditions
Man-made	13	4	7	11	7	11	2	37	4	4
Earthquakes	6	10	38	4	12	7	9	3	9	2
Fire	4	10	19	9	32	10	7	2	6	3
Floods	41	2	4	8	4	16	5	5	7	7

TABLE 3.10
Threat Weights Matrix [WT_H]

Threat/Hazards	Locality	Peak # of Users	Attractiveness	Significance	Function within the Region	Earthquake	Fire	Floods
Man-made	13	11	40	5	6	1	20	4
Earthquakes	0	0	0	12	15	73	0	0
Fire	9	13	13	7	8	15	19	15
Floods	3	0	0	21	18	0	0	59

TABLE 3.11
Consequences Weights Matrix [WT_C]

Consequences Issues	Number of Tracks	Number of Levels	Underwater Tunnels	Peak number of Riders per Day	Tunnel Locality	Nearby Facilities	High Target Values	Impact of Physical Loss	Social Impact	Replacement Value	Operational Redundancy	Downtime
Man-made	5	4	8	10	3	8	10	13	13	13	8	8
Earthquakes	7	5	9	11	1	3	3	15	15	15	9	9
Fire	7	5	9	11	1	3	3	15	15	15	9	9
Floods	7	5	9	11	1	3	3	15	15	15	9	9

TABLE 3.12
Vulnerability Assessment Vector {*VR*}

	VR
Method of construction	3
Liner cross-section	9
Liner relative thickness	8
Construction material	8
Known retrofits	7
Longest span	3
Controlling height	2
Type of framing	3
Seismic design	2
Overall structural conditions	5

TABLE 3.13
Threat Assessment Vector {*TR*}

	HR
Locality	9
Peak # of users	8
Attractiveness	8
Significance	7
Function within the region	7
Earthquake	2
Fire	3
Floods	2

and have been estimated by the analysts. When the particular tunnel is assessed by an inspector, the different vectors {*VR*}, {*HR*}, and {CR} are estimated as shown in Tables 3.12 through 3.14, respectively. As before, we evaluate vectors for hazards vulnerability, threat, and consequence, {*V*}, {*H*} and {*C*}, from Equations 3.10 through 3.12. Finally, we obtain the risk vector, {*Ri*}, by applying Equation 3.9. The resulting relative scores for {*V*}, {*H*}, {*C*}, and {*Ri*} are shown in Table 3.15. To get the total risk from the individual hazards risks, we apply Equation 3.4 with $N = 4$ (4 hazards risk will contribute to the total risk). Table 3.16 shows the computed total risk for different values of *n*.*

3.3.4 Limitations

The above case studies, which use the weighted averages method for computing subjective risk, have some limitations.

1. Interactions between parameters in matrices are not accommodated.
2. A limited set of issues control the risk value and its components.

The above limitations can be overcome by increasing the number of issues that control risk and its components. In addition, interactions between issues can be introduced using an algorithmic coding.

* Vose (2009) and DHS (2011a, 2011b, and 2011c) have used a value of $n = 10$ in computing the total risk.

TABLE 3.14
Consequences Assessment Vector {*CR*}

	CR
Number of tracks	6
Number of levels	2
Underwater tunnels	9
Peak number of riders per day	7
Tunnel locality	8
Nearby facilities	3
High target values	8
Impact of physical loss	7
Social impact	8
Replacement value	8
Operational redundancy	7
Downtime	8

TABLE 3.15
Vectors for Vulnerability, Threat, Consequences, and Risk for Different Hazards

Hazard	Vulnerability	Threat	Consequences	Risk
Man-made	45%	67%	71%	60%
Earthquakes	61%	33%	72%	53%
Fire	62%	52%	72%	61%
Floods	40%	41%	72%	49%

TABLE 3.16
Total Risk from Individual Hazards Risks

n	1	2	5	10
{*Ri*}	56%	56%	56%	57%

These more accurate considerations have been accommodated and described in (DHS 2011a, 2011b, 2011c). Another method, which can accommodate the above limitations, is the GN method that we discuss next.

3.4 PROBABILISTIC GN

3.4.1 Overview

We presented GN in Chapter 2 as a comprehensive method for using four of the five components of risk management. In the remainder of this chapter, we explain different uses of GN in risk assessment. First, we present some practical applications regarding risk assessment of single assets, and then we look at the use of GN for risk assessment of asset networks. Finally, we explore the use of GN for resilience assessment.

3.4.2 Case Study 3.3: Earthquake Risk

3.4.2.1 Overview

Many authors have studied assessment of earthquake risk (see, e.g., ASCE/SEI 2011 for a comprehensive review of the subject). In this section, we provide an application of a BN model for computing earthquake risk as applied to a single asset (building) using the process illustrated earlier. Note that for simplicity, the model considers only a few parameters. It can, however, be expanded into a more detailed model, as needed.

3.4.2.2 Model Description

Figure 3.6 shows the model we use to assess building earthquake risk. Note that the variables can be categorized into three subcategories: vulnerability, threat, and consequences. Also, note that vulnerability and threat are both components of a variable exposure, *E*. Introducing a separate variable for exposure has two advantages: (1) It can be of help in understanding the behavior of the network, and (2) it reduces computing demands of the network. Finally, exposure, *E*, and consequences, *C*, form the two components we define as the risk, *Ri*. This categorization is illustrated in Figure 3.7. The states of nodal variables are shown in Table 3.17. The definitions of these variables are shown in Appendix 3.I (Table 3.34).

3.4.2.3 Conditional Probabilities Tables

The different CPTs of each variable of the network shown in Figure 3.6 are shown in Appendix 3.II. We used our best judgment in filling the different tables in a realistic manner for illustration purposes. For more on different methods of developing CPTs, see Chapter 2.

3.4.2.4 Results

We can now execute the BN model of Figure 3.6 to find the historical (prior) probabilities of all the variables, including the historical risk, *Ri*, which are shown in Figure 3.8. If we are interested in investigating the effects of improvements of various vulnerability parameters, such as improving structural properties or site conditions, on the assessed risk, we can easily perform such vulnerability studies using the BN model. The improved risk models are illustrated in Figures 3.9 and 3.10 for structural improvement and site improvement, respectively.

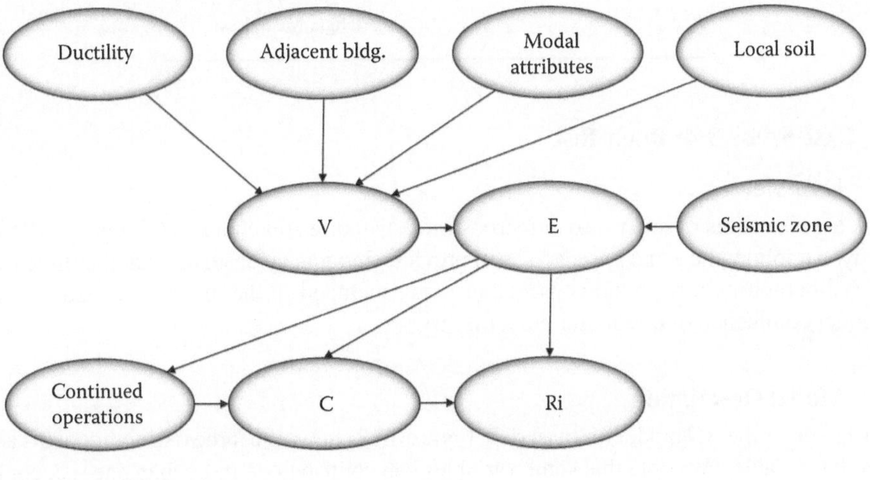

FIGURE 3.6 BN model for seismic risk.

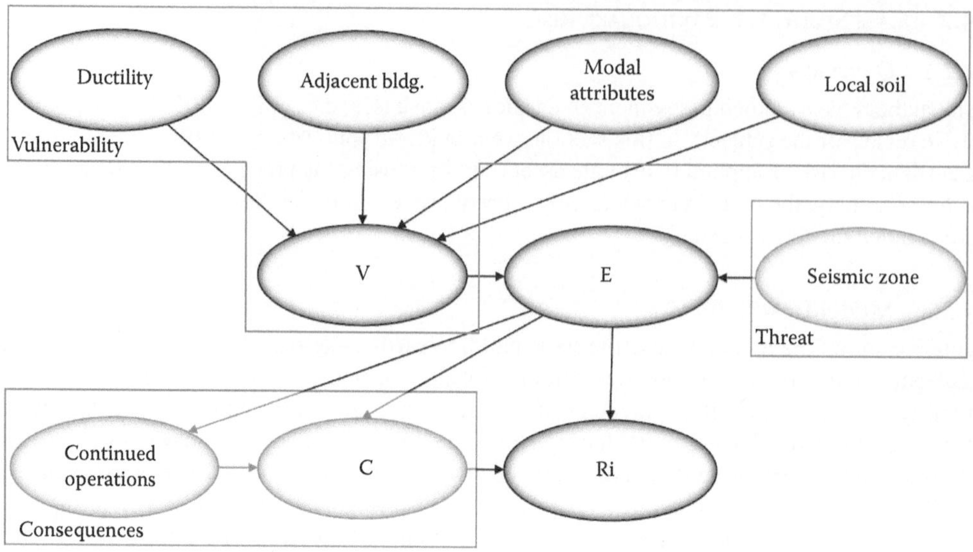

FIGURE 3.7 BN model for seismic risk showing risk components.

TABLE 3.17
States (Ranks) of Nodal Variables

Variable	# of States	States
Ductility	3	Low, Medium, High
Adjacent_Buildings	2	Close, Far
Modal_Attributes	2	Not Effective, Effective
Local_Soil	3	Weak, Strong, Rock
Seismic_Zone	3	High, Med., Low
Continued_Operations	3	Long, Med., Short
V	3	High, Med., Low
E	3	High, Med., Low
C	3	High, Med., Low
Ri	3	High, Med., Low

3.4.3 Case Study 3.4: Blast Risk

3.4.3.1 Overview

Blast risk assessment is of interest to numerous infrastructure stakeholders (Ettouney 2015). In this case study, we follow the same approach as in previous models by ensuring that the three risk components (vulnerability, threat, and consequences) are included in the model. We also use exposure (*E*) to model confluence of threat and vulnerability.

3.4.3.2 Model Description

Figure 3.11 shows the BN model for blast risk. Figure 3.12 shows the three risk components as related to the model variables. We note that some variables can contribute to more than one risk component.

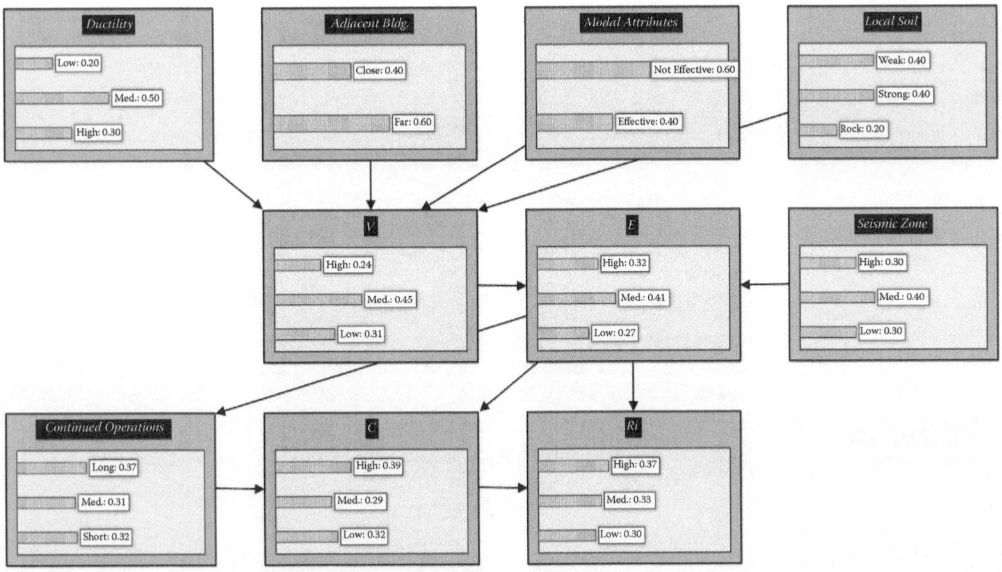

FIGURE 3.8 Seismic risk: historical probabilities.

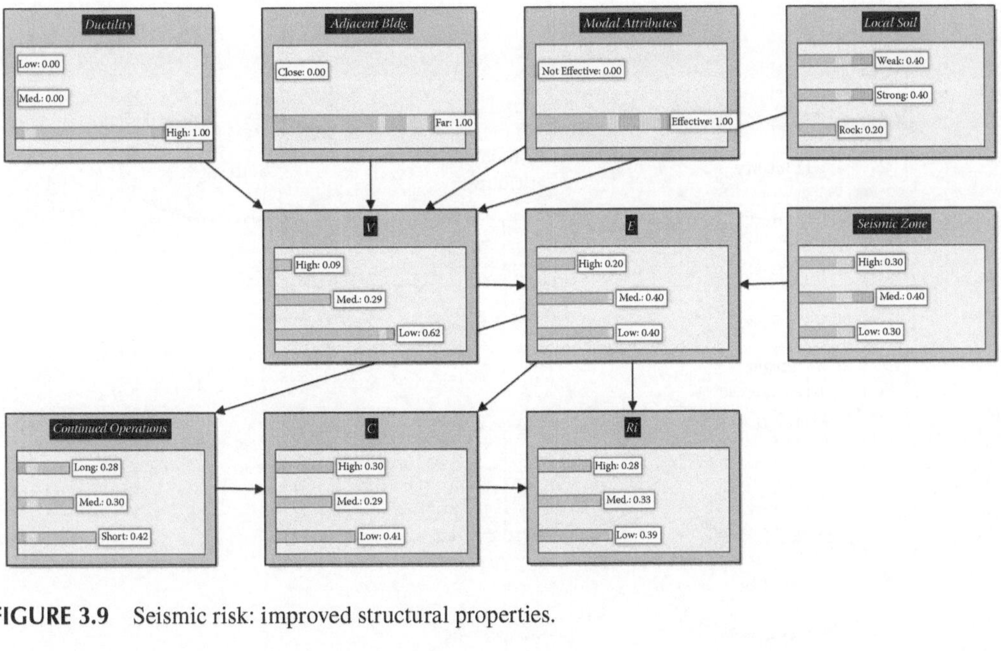

FIGURE 3.9 Seismic risk: improved structural properties.

For example, *Setback* contributes to both vulnerability and threat, while *Locality* contributes to vulnerability, threat, and consequences. This mix of contribution of variables is another reason why a graph network approach is needed in risk assessment methods. It is simply much more difficult, if not impossible, to accommodate such a mix by using any other available risk assessment method. To complete the model definition, we offer Table 3.18, which shows states of nodal variables, and Appendix 3.II for the definition of those variables.

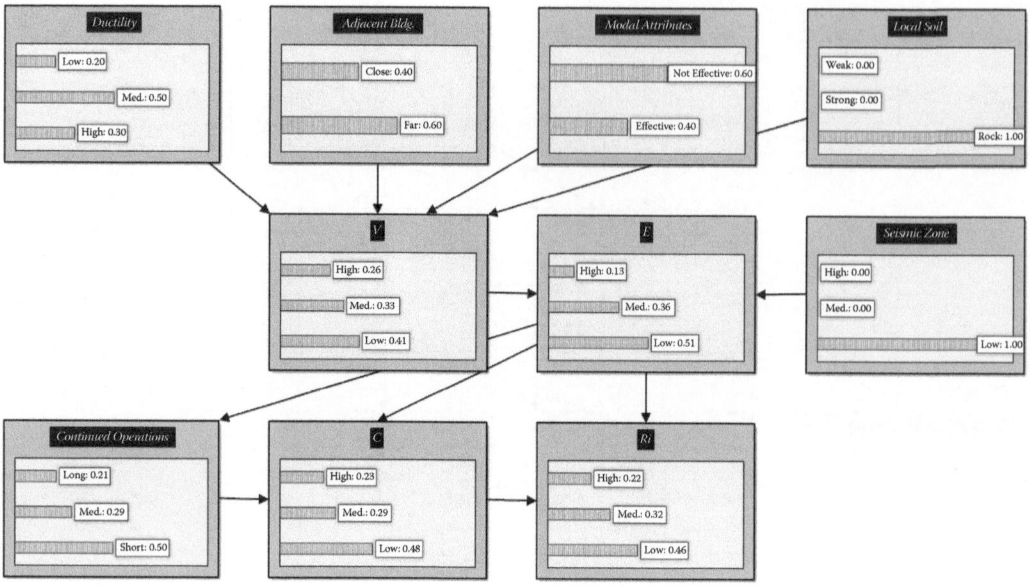

FIGURE 3.10 Seismic risk: improved building site and locality.

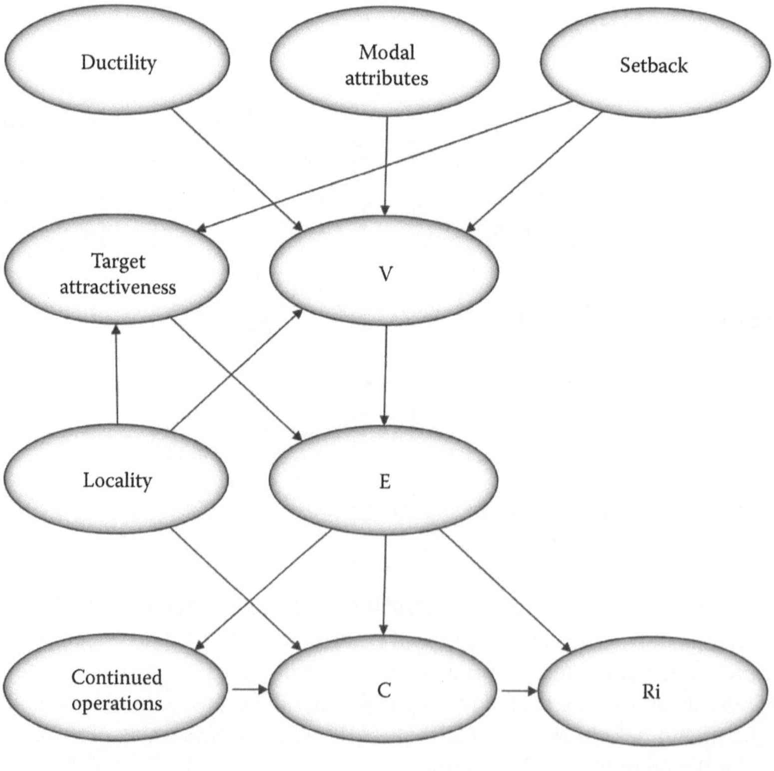

FIGURE 3.11 BN model for blast risk.

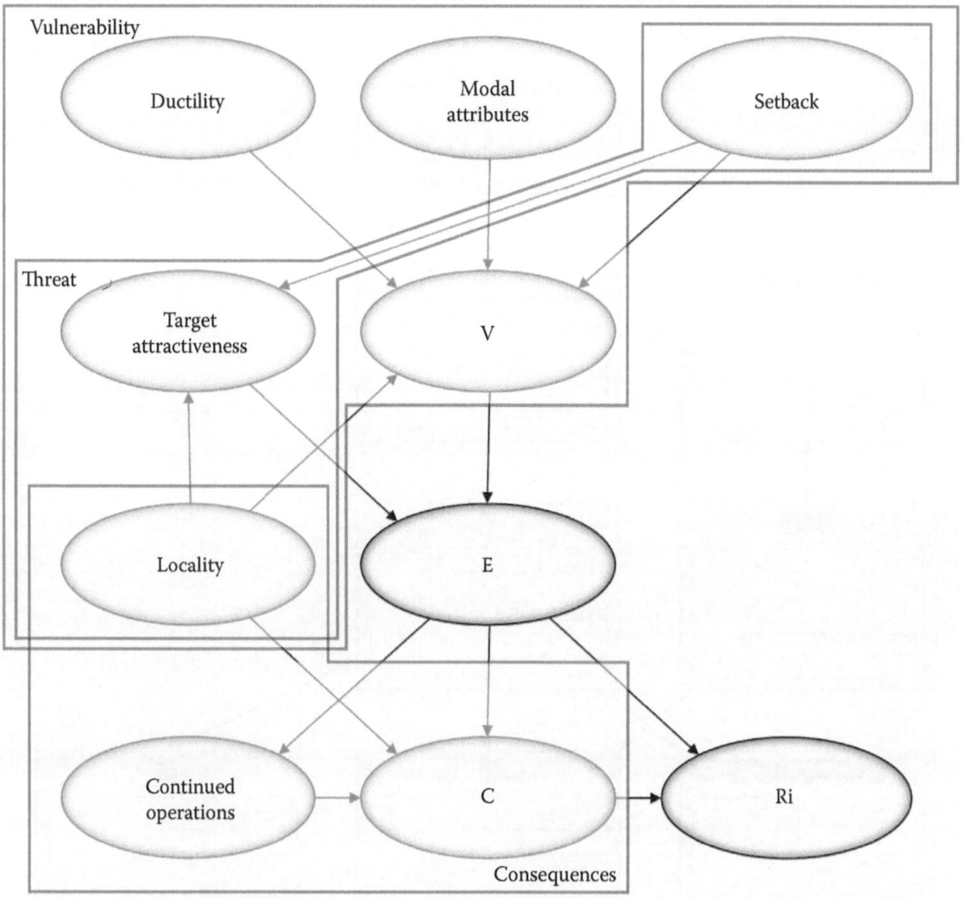

FIGURE 3.12 BN model for blast risk showing risk components.

TABLE 3.18
States (Ranks) of Nodal Variables

Variable	# of States	States
Ductility	3	Low, Med., High
Setback	2	Short, Long
Modal_Attributes	2	Not Effective, Effective
Target_Attractiveness	3	High, Med., Low
V	3	High, Med., Low
Locality	2	Attractive, Normal
Continued Operations	3	Long, Med., Short
E	3	High, Med., Low
C	3	High, Med., Low
Ri	3	High, Med., Low

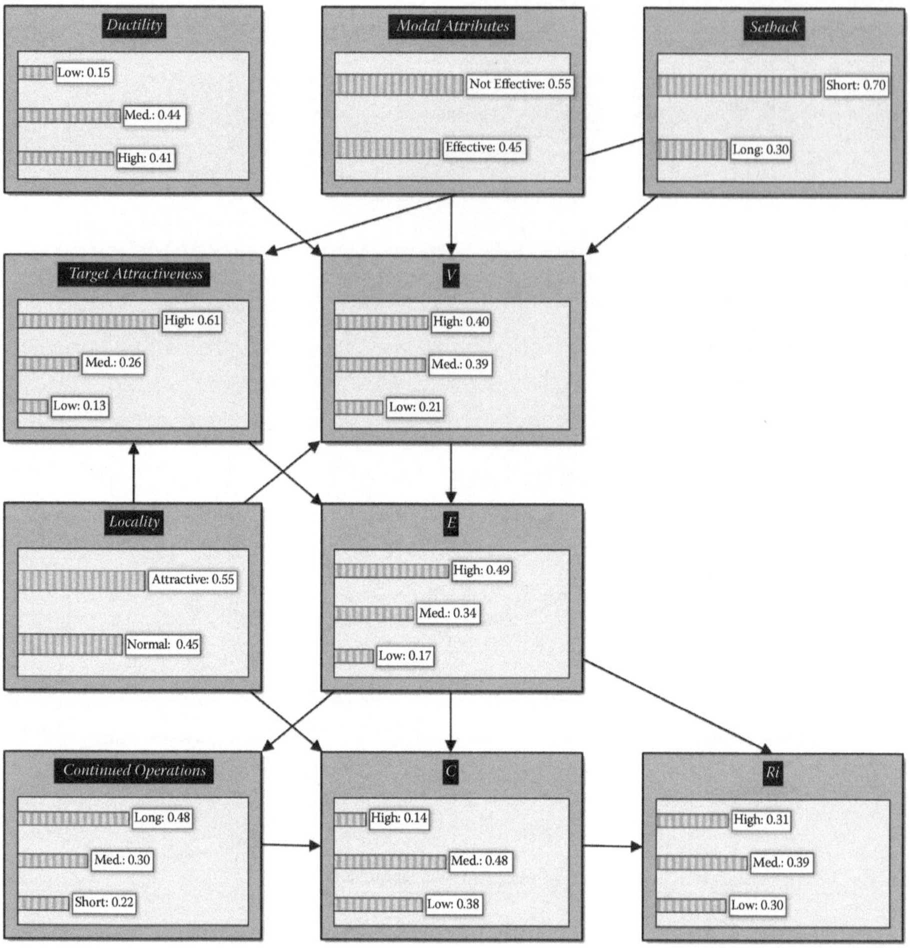

FIGURE 3.13 Historical probabilities for blast risk.

3.4.3.3 Conditional Probabilities Tables

The different CPTs of each variable of the network of Figure 3.11 are shown in Appendix 3.III. As usual, we used our best judgment in filling these tables in a realistic manner for illustration purposes. For more on different methods of developing CPTs, see Chapter 2.

3.4.3.4 Results

The historical probabilities of the risk, *Ri*, of the building blast hazard are shown in Figure 3.13. The chosen model and its CPTs seem to result in a medium historical risk, *Ri*. Let us assume that, even at such moderate risk values, the stakeholder considers improving the site condition as represented by *Locality*, *Attractiveness*, and *Setback*. The results of such improvements are shown in Figure 3.14. The changes result in a dramatic improvement in blast risk. On the other hand, if some structural modifications (such as retrofitting for another hazard or some wear and tear that resulted from structural aging conditions) resulted in an increase of blast vulnerability, and assuming that such changes affected *Ductility* (by reducing it from the historical norms) and/or *Modal_Attributes* (by shifting the modal characteristics of the structure into an unfavorable zone to blast behavior). These changes result in an increase of blast risk as shown in Figure 3.15.

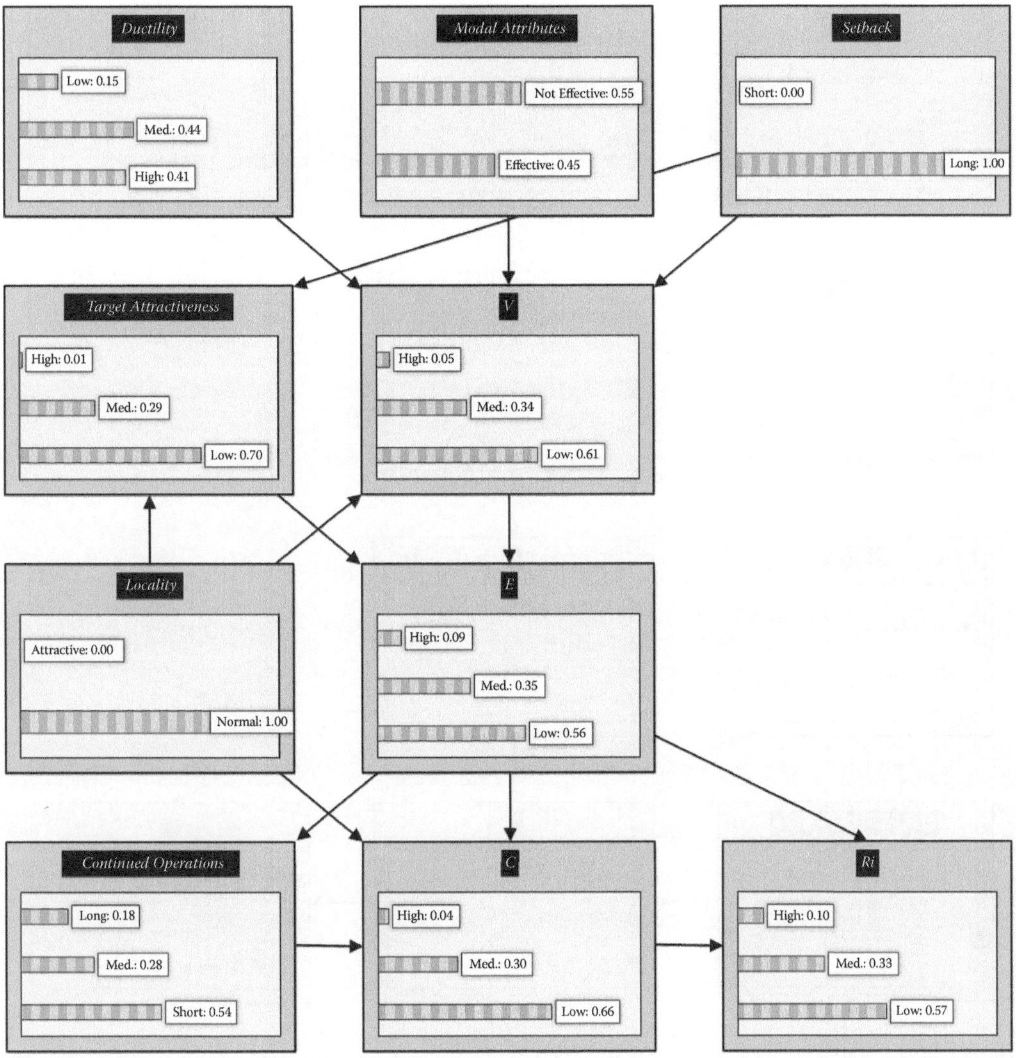

FIGURE 3.14 Blast risk for low site vulnerabilities and attractiveness.

We close this case study by noting that this blast risk assessment model is fairly simplified. For a more realistic assessment, the actual model needs to include many more variables than the ones we just offered here for illustration purposes.

3.4.4 Case Study 3.5: Building Security

3.4.4.1 Overview

FEMA 455 (2009) and later DHS (2011a, 2011b, 2011c) showed a simplified methodology for computing building security risk due to man-made bomb blasts. We note that the FEMA 455 methodology is based on weighted variables to compute risk (see Section 3.3 for more discussion of the weighted variables approach). This case study explores the use of a chain graph (CG) model in computing building security risk and uses the same simple variables of FEMA 455 (2009). The CG approach, as per Chapter 2, contains the two types of graph network links: directed and undirected. The model is intended to show the rich complexity and interlinks in a building security risk

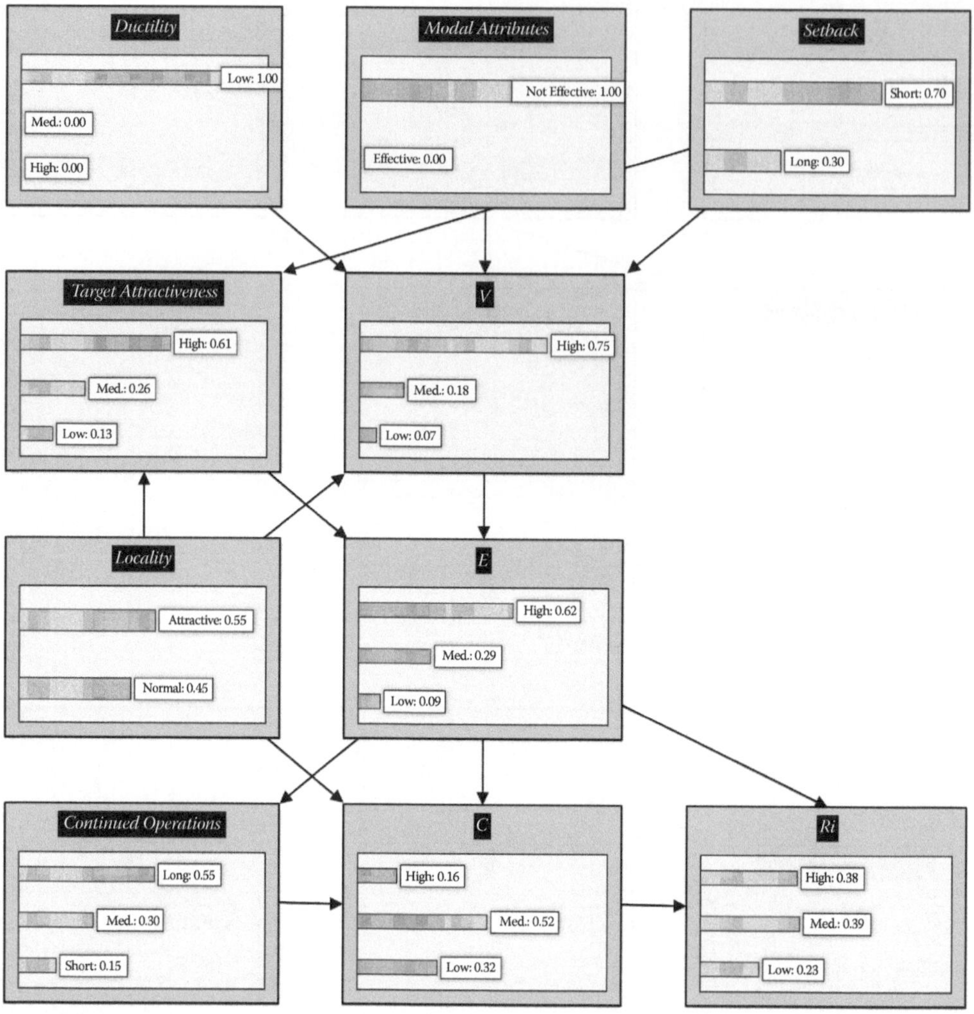

FIGURE 3.15 Blast risk for high structural vulnerabilities.

assessment process. We offer three assessment results: the historical (prior) risk, the low risk, and the high-risk results.

3.4.4.2 Model Description

The total building security risk model is shown in Figure 3.16. Note that the model shows the variables that constitute vulnerabilities, threats, and consequences in their respective sections. Some of the variables are common for more than one component. For example, *# of Occupants* is a common variable for both consequences, *C*, and threat, *T*. In addition, there are several nondirectional links that span between different components. This result in a CG graph model (see Chapter 2). The variable definitions and nodal variable states are shown in Appendix 3.II and Table 3.19, respectively.

3.4.4.3 Conditional Probabilities Tables

The CPTs and potentials of each variable of the network of Figure 3.16 are shown in Appendix 3.IV. Because this is a CG model, it includes a mix of CPTs (Tables 3.57 through 3.84) and potentials (Tables 3.55 through 3.56). As usual, we used our best judgment in filling these tables in a realistic manner for illustration purposes. For more on methods of developing CPTs, see Chapter 2.

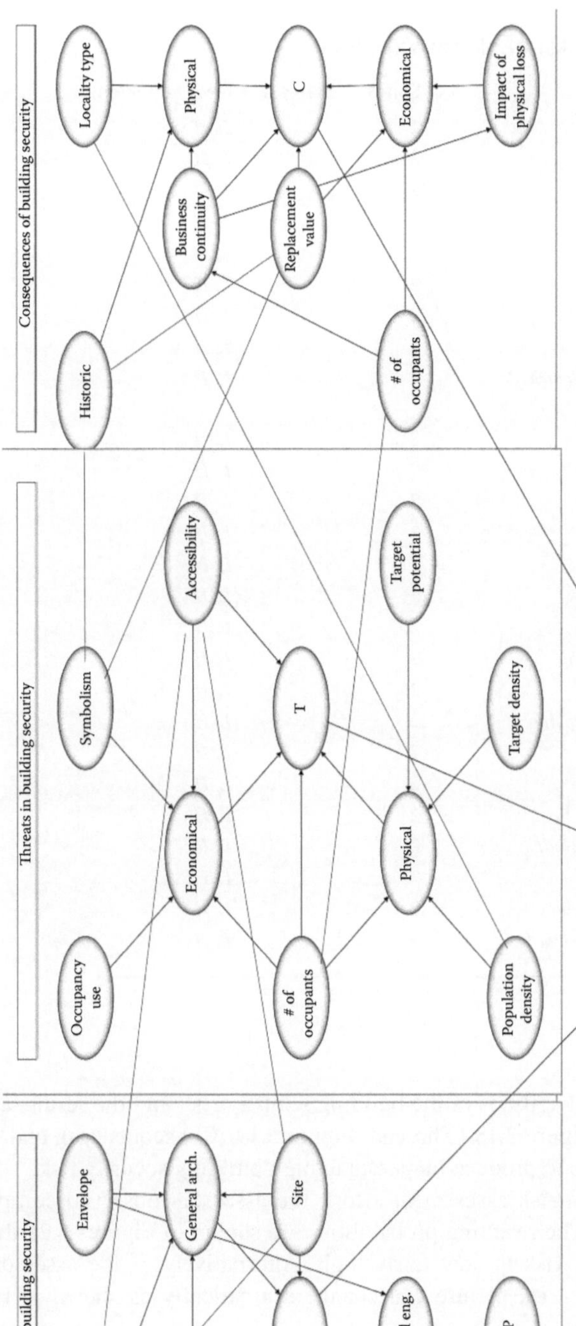

FIGURE 3.16 Chain graph model for building security risk.

TABLE 3.19
States (Ranks) of Nodal Variables

Variable	# of States	States (*L* = Low; *H* = High)
Architecture	2	*L*, *H*
Envelope	2	*L*, *H*
Security	2	*L*, *H*
General Arch.	2	*L*, *H*
V	2	*L*, *H*
Site	2	*L*, *H*
Structure	2	*L*, *H*
General Eng.	2	*L*, *H*
Mechanical/Electrical/ Plumbing (MEP)	2	*L*, *H*
Occupancy Use	2	*L*, *H*
Symbolism	2	*L*, *H*
Economical	2	*L*, *H*
Accessibility	2	*L*, *H*
T	2	*L*, *H*
# of Occupants	2	*L*, *H*
Physical_Sec	2	*L*, *H*
Target Potential	2	*L*, *H*
Target Density	2	*L*, *H*
Population Density	2	*L*, *H*
Risk, Ri	2	*L*, *H*
Historic	2	*L*, *H*
Locality Type	2	*L*, *H*
Business Continuity	2	*L*, *H*
C	2	*L*, *H*
Replacement Value	2	*L*, *H*
Impact of Physical Loss	2	*L*, *H*

3.4.4.4 Results

We developed the historical probabilities of the building security risk and the results are shown in Figure 3.17 (Vulnerabilities), Figure 3.18 (Threats), Figure 3.19 (Consequences), and Figure 3.20 (Risk). The chosen model seems to produce moderate to high building security risk.

Let us now assume that in a risk assessment effort, the assessors observed certain contributions from different variables. The resulting probabilities are shown in Figures 3.21 through 3.24. The resulting building security risk is now fairly high. Alternatively, if the assessors observed a low variable contribution, the risk picture can change dramatically as shown in Figures 3.25 through 3.28.

3.4.4.5 Concluding Remarks

We conclude this case study by noting that building security risk assessment, as we exhibited here, is a complex problem. It involves historical observations, building-specific observations, objective and subjective probabilistic CPTs and potentials as well as a complex network topology. In general, risk assessment efforts that lack one or more of these essential problem-solving attributes may result in inaccurate risk results, which may produce unnecessarily additional costs, or even worse, unsafe building situations.

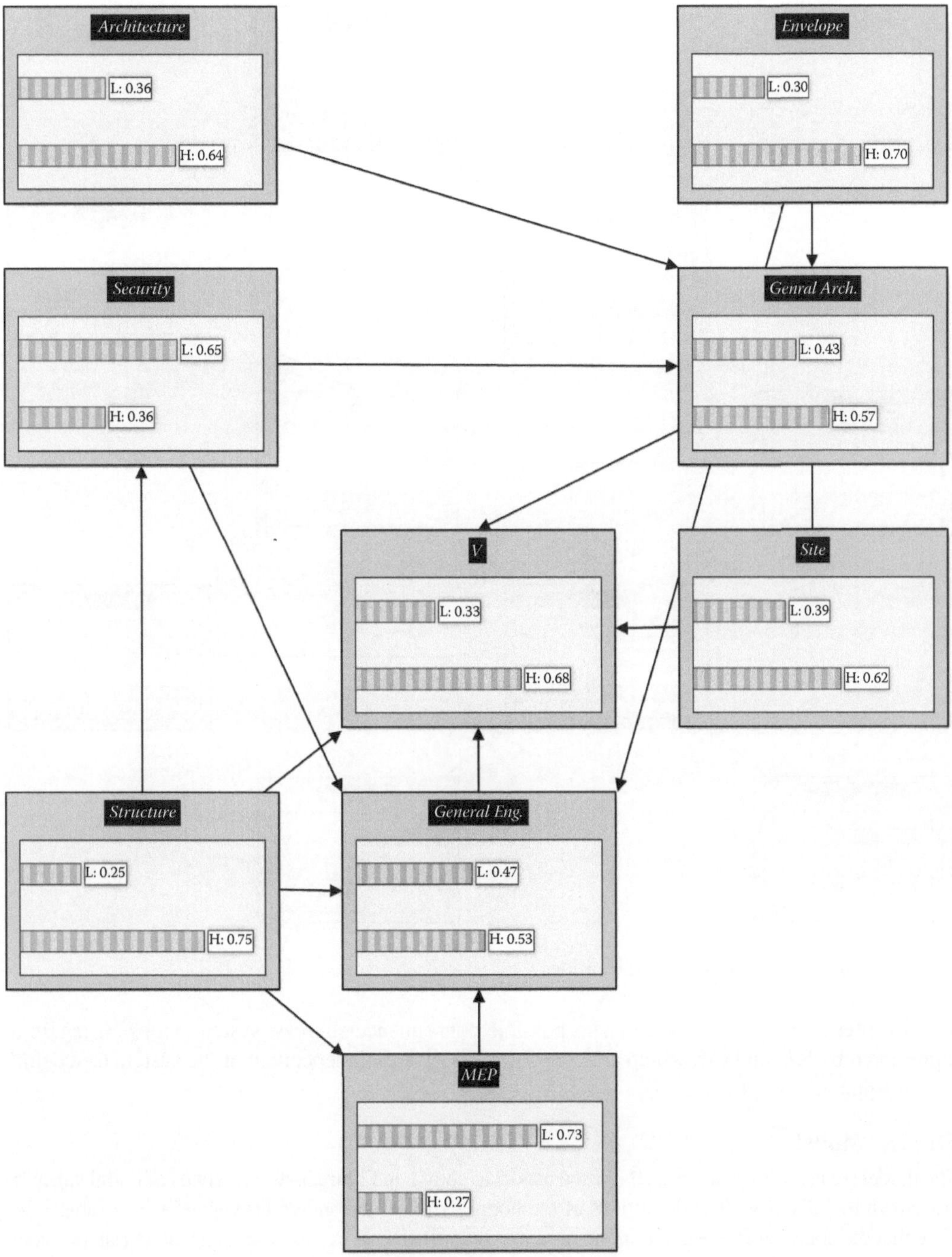

FIGURE 3.17 Vulnerabilities: historical probabilities.

3.4.5 Case Study 3.6: Absolute Risk

3.4.5.1 Overview

We already encountered absolute risk in a simple introductory case in Chapter 2. We further investigate absolute risk in this case study, where we define risk states in monetary terms and relate them to visual inspection, *V.I.* (as a subjective representative of exposure) and consequences, *C.*

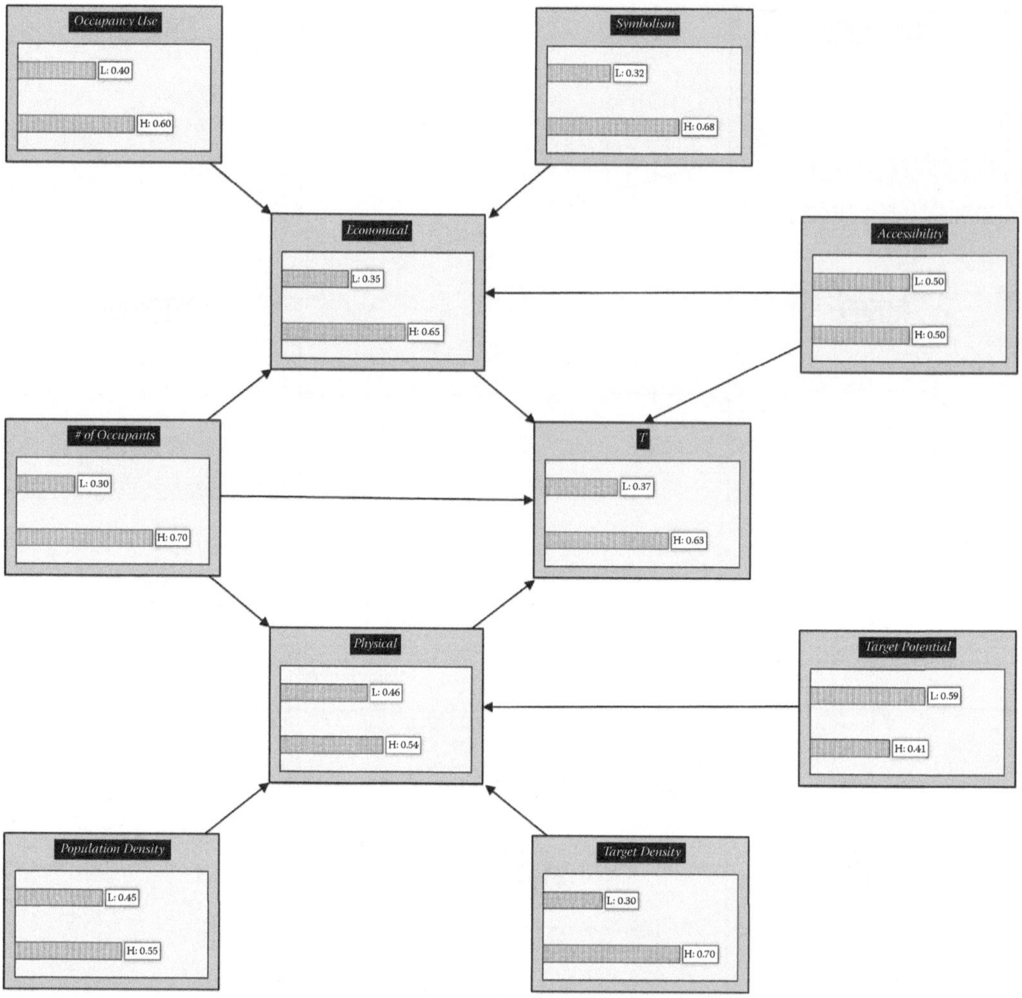

FIGURE 3.18 Threats: historical probabilities.

We also present *C* as a function of two possible consequence sources: system exposure (again as represented by *V.I.*) and *CO*, which are consequences that are independent of the system (see definition in Appendix 3.I [Table 3.34]).

3.4.5.2 Model

The model we use in this case is a BN-based model as shown in Figure 3.29. The states of nodal variables are shown in Table 3.20. The definitions of the nodal variables are shown in Appendix 3.I (Table 3.34). Note that the states on *Ri* are given in monetary terms, which makes this case an absolute risk problem.

3.4.5.3 Conditional Probabilities Tables

The different CPTs of each variable of the network of Figure 3.29 are shown in Appendix 3.V. As before, we used our best judgment in filling the different tables in a realistic manner for illustration purposes. For more on different methods of developing CPTs, see Chapter 2.

3.4.5.4 Results

We analyze the BN model for this case to obtain the historical probabilities first. The results are given in Figure 3.30. The historical risk in monetary value is $16,318.00 as computed in Table 3.21.

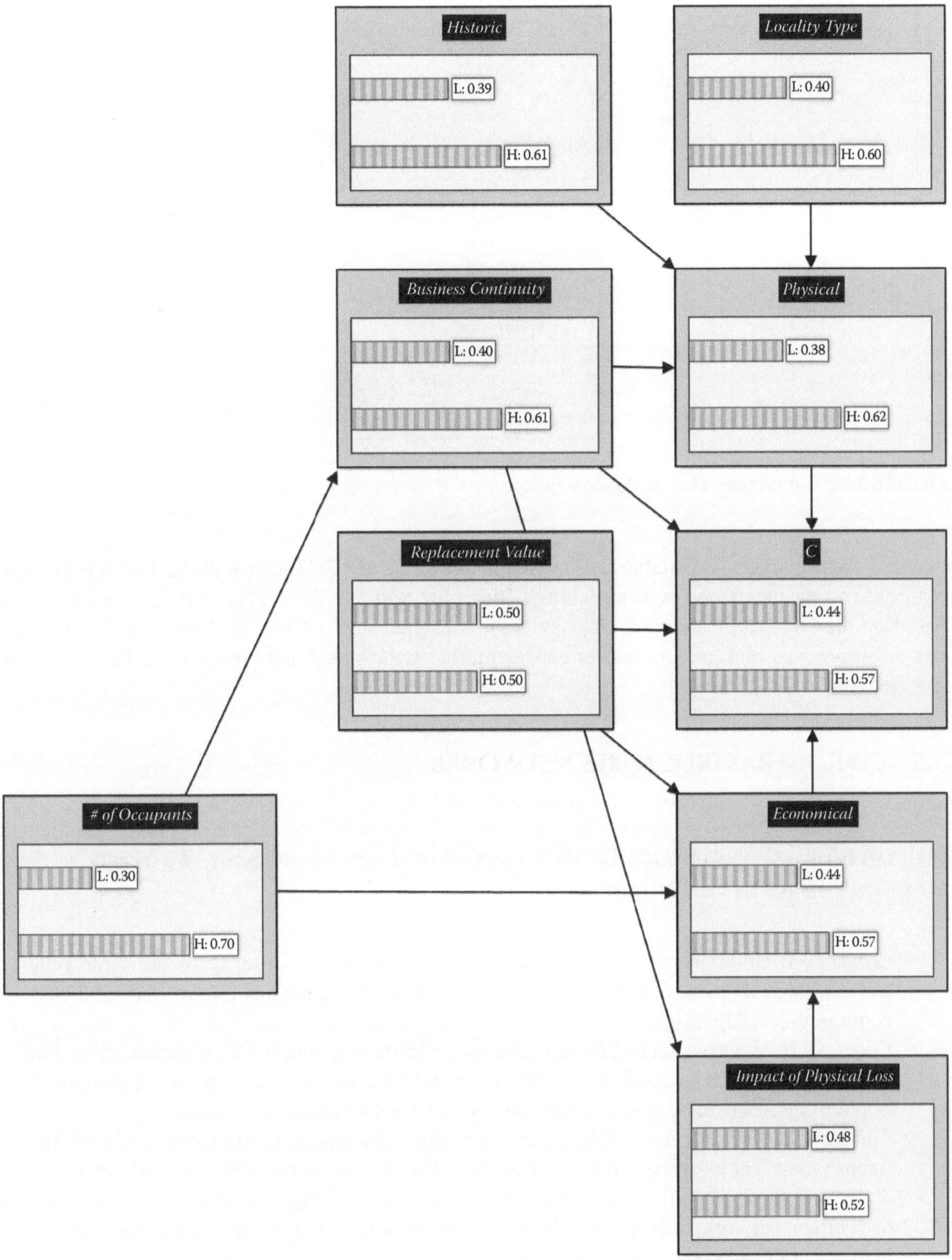

FIGURE 3.19 Consequences: historical probabilities.

If a new visual inspection effort is performed that showed a state of *Service (fairly poor)*, this can be introduced as an evidence (*V.I.*) observation. The resultant new risk is shown in Figure 3.31. Similarly, if other consequences, *CO*, away from the bridge system, such as changes in economic or societal conditions is observed, new evidence describing the change of the state of the variable can be introduced with results as shown in Figure 3.32. The respective resulting new absolute risk

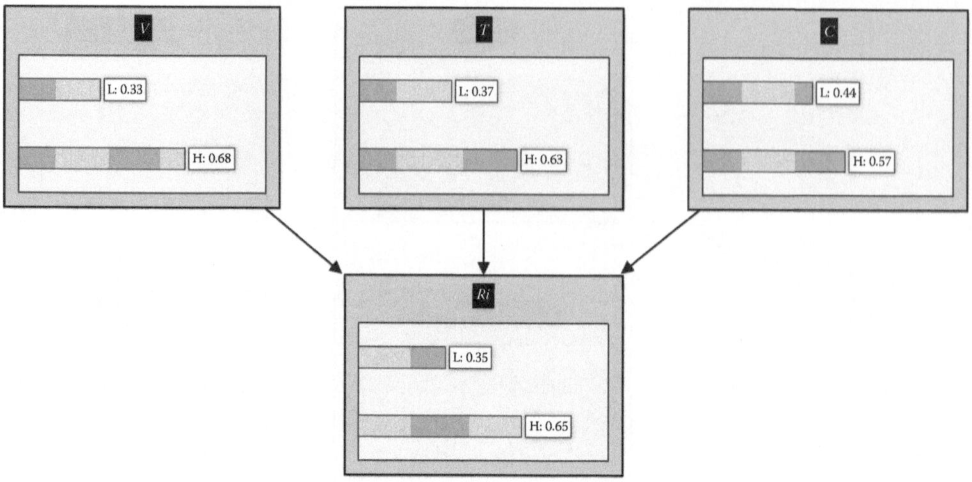

FIGURE 3.20 Total risk: historical probabilities.

based on these changes in variables is shown in Tables 3.22 and 3.23, respectively. The new average risk based on the newly observed visual inspection observation is $87,722, and the new average risk based on the changes in the new *High* external consequences is $18,079. Either of the two new risk environments might require closer consideration by the stakeholder because of the increase in risk from the initial condition.

3.5 CIVIL INFRASTRUCTURES NETWORKS

3.5.1 Overview

Civil infrastructure networks (CINs)* are composed of several single assets. We categorize these networks into three types as follows:

- *Category I*: Homogeneous/physical; the assets in the network are all of the same type and they are all physical assets. For example, a particular Category I network can be only composed of bridges.
- *Category II*: Heterogeneous/physical; the assets in the network are of different types, but they are all physical assets. For example, a particular Category II network can be composed of buildings, roadways, bridges, railways, mass transit stations, and tunnels.
- *Category III*: Heterogeneous/physical/nonphysical; the assets in the network are of different types. They can be physical or nonphysical. For example, this particular category of networks can contain tunnels, mass transit stations, building, and bridges as well as an operational network, operational information technology (IT) systems, and power grid that links all of the assets within the network.

We are interested in studying the risk (or resilience) of CINs in this book. In this section, we look at risk assessment examples of CINs.

* The reader should note the distinction between graph network (GN), which is a mathematical process, and civil infrastructure network (CIN), which is a topological entity.

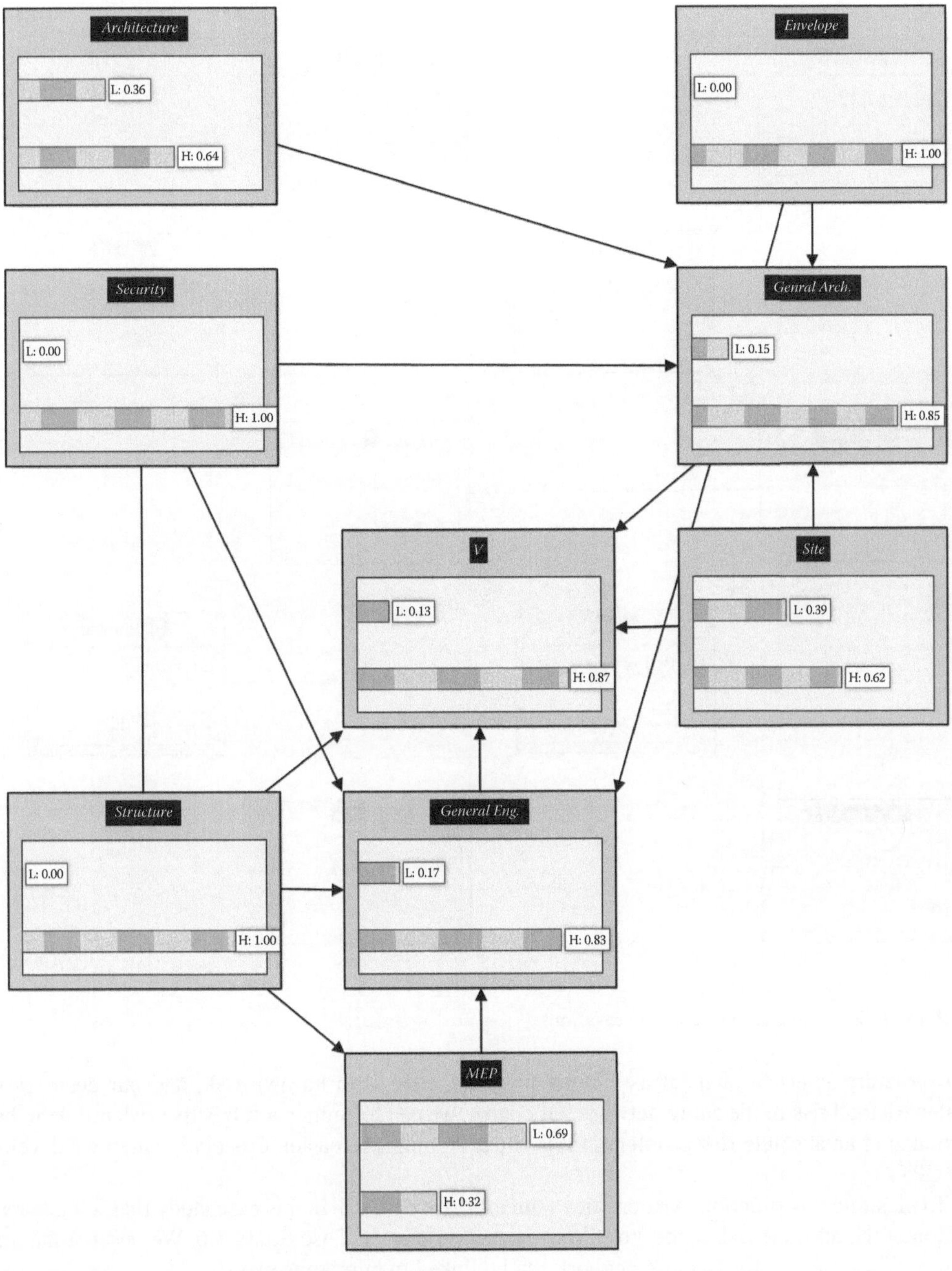

FIGURE 3.21 Vulnerabilities: high score observations.

3.5.2 Case Study 3.7: BN Risk Assessment of Bridge Network

3.5.2.1 Overview

One of the advantages of using absolute risk as a metric in a risk assessment is that we can combine individual risks from separate sources. A potential application of absolute risk is when we have several hazards, we can compute individual risks for each hazard and our desire is to compute a total combined risk from the individual hazards risks. Similarly, it is useful when we have discrete

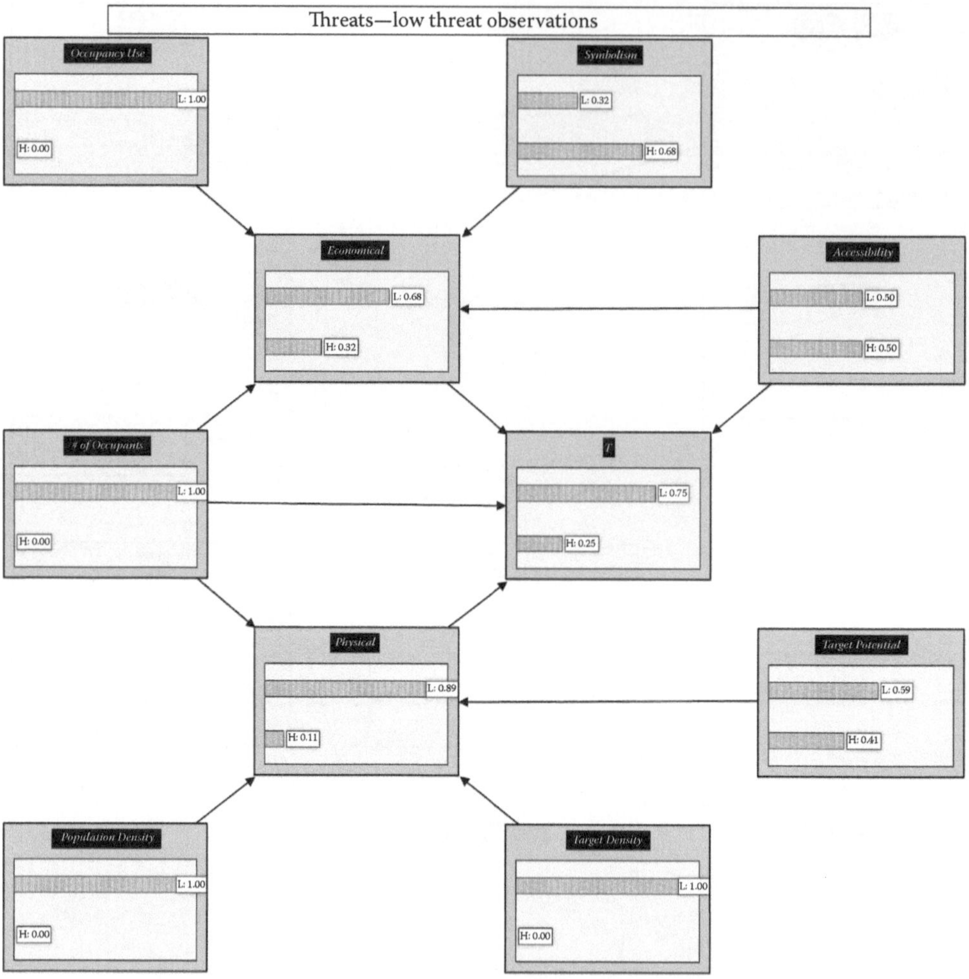

FIGURE 3.22 Threats: low score observations.

infrastructure assets (a civil infrastructure network), each with its own risk, and our desire is to evaluate a total risk of the entire network. Of course, we can formulate a subjective risk problem, but formulating an absolute risk problem in this situation might be easier, especially when we develop the CPTs.

To illustrate this situation, we consider a three-bridge network in this case study that is a generalization of the absolute risk model for a single asset (bridge) in Case Study 3.6. We observe that the assets within a civil infrastructure network can be linked in numerous ways:.

- *Consequences*: where single consequences can affect more than one asset (social, location, traffic, detours, symbolism, cost of replacement, etc.).
- *Hazards*: A single hazard can have effects on several assets in the network (earthquakes, floods, temperature, climate change, wind storms, etc.).
- *Vulnerabilities*: Vulnerabilities of assets within a network can be linked in a model if these assets are physically (bridge passing over a roadway) or virtually (a sewer or water system serving a building unit) linked. In some cases, the failure of these links can lead to a cascading failure effect.

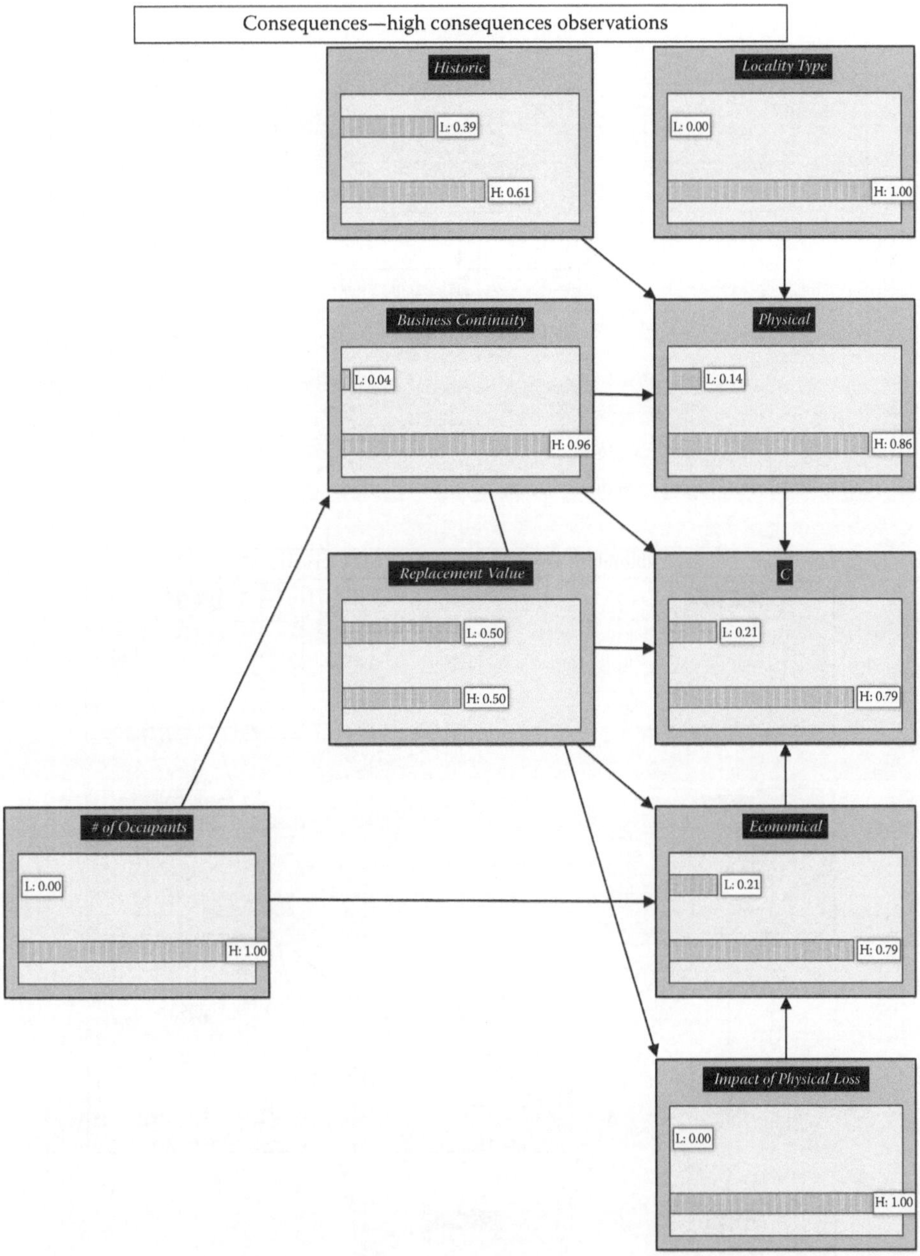

FIGURE 3.23 Consequences: high score observations.

In this example, we explore bridge network where the asset consequences in the network are linked; that is, a single factor can have an effect on the consequences of different assets in the network. This can represent, for example, the fact that some assets within the network are socially/economically more important than other assets. Of course, the links of consequences in this case are assumed directional, making this model a BN model. In case 07.08, we explore situations when links between assets are nondirectional, thus making that model a CG model.

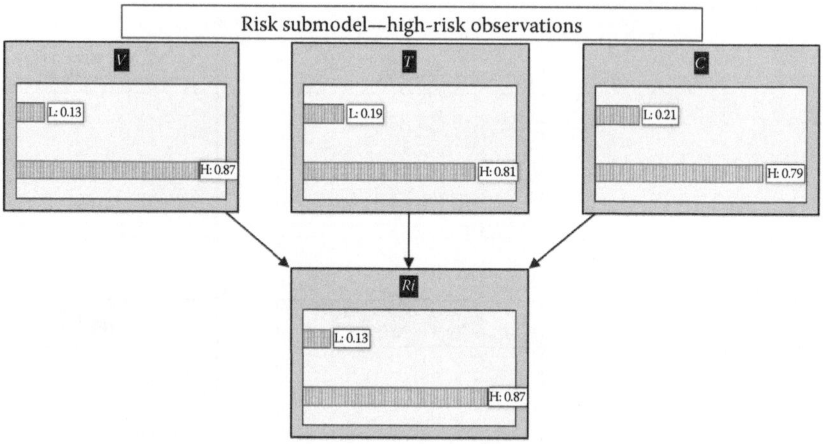

FIGURE 3.24 Total risk: high score observations.

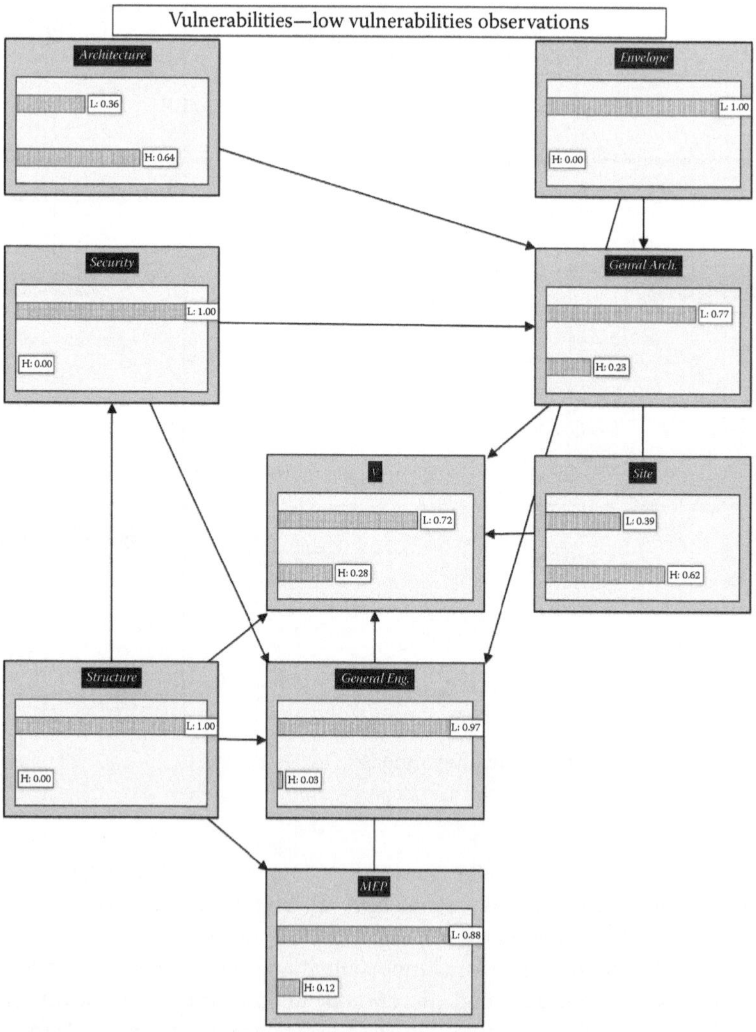

FIGURE 3.25 Vulnerabilities: low score observations.

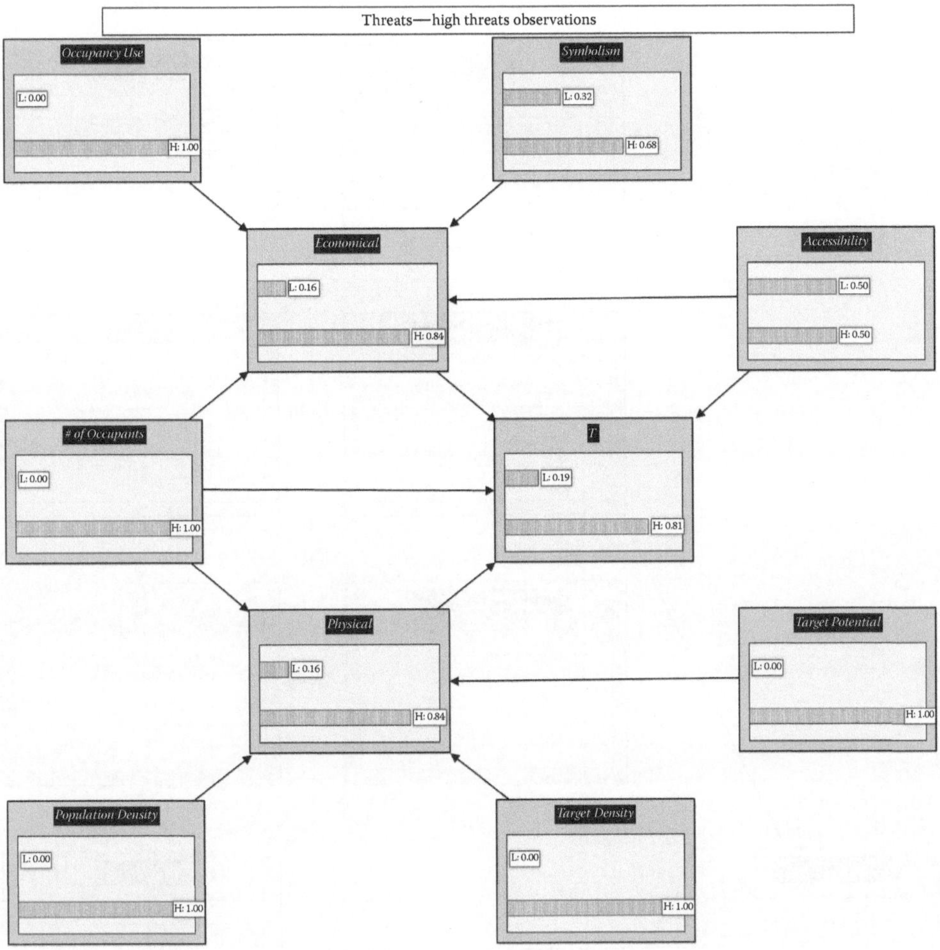

FIGURE 3.26 Threats: low score observations.

3.5.2.2 Model

In this BN model, we link the consequences of bridges 1 and 2, *C_1*, and *C_2*, respectively, through a single consequence variable, *C*. For example, if two bridges are in close proximity, where the traffic conditions might affect each of them, albeit to different degrees. We assume that the consequences of bridge 3, *C_3*, is independent of the consequences of the other two bridges, perhaps because of its location in a different region. Also, we assume that conditions observed from visual inspections, *V_1*, *V_2*, and *V_3*, of the three bridges are independent. The total risk of the network is also a link between the three bridges. Later in this chapter (Case Study 3.8), we examine another form of modeling interaction between consequences within a bridge network. Figure 3.33 shows the BN model used for the study, and Table 3.24 shows the states of nodal variables of the study. The definitions of the variables are shown in Table 3.34.

3.5.2.3 Conditional Probabilities Tables

The different CPTs of the variables in the network of Figure 3.33 are shown in Appendix 3.VI. We used our best judgment in filling the different tables in a realistic manner for illustration purposes. For more on different methods of developing CPTs, see Chapter 2.

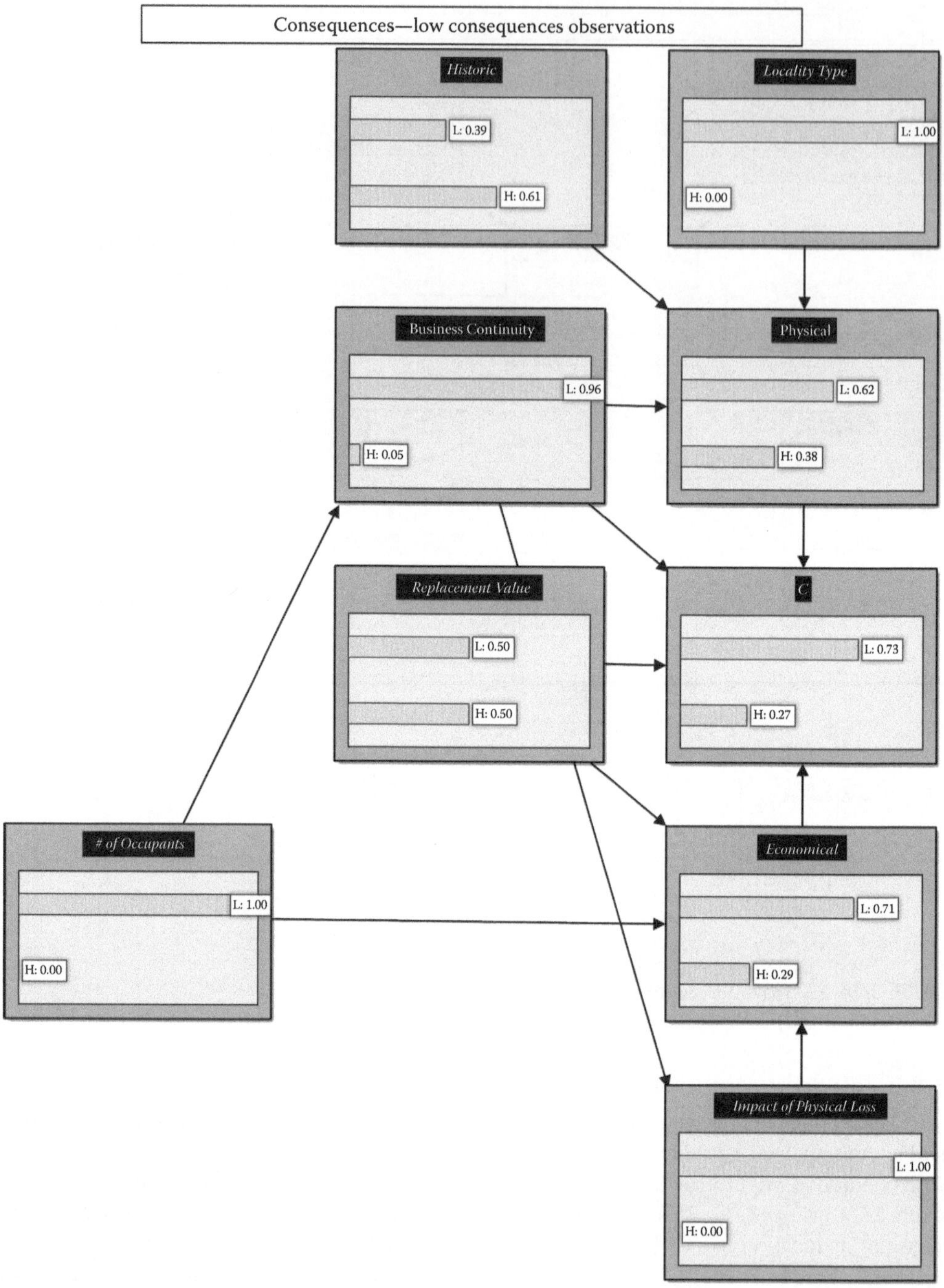

FIGURE 3.27 Consequences: low score observations.

3.5.2.4 Results

Historical risks of the individual assets and the whole network can be computed as shown in Figure 3.34. As computed, the average risk for the whole network can be computed using Table 3.25 as $68,409. Now, let us assume there is a recent bridge inspection cycle that generated inspection ratings of *Fair*, *Poor*, and *Poor*, for bridges 1, 2, and 3, respectively. Upon enforcing these observations in the network model, we get the updated probabilities of Figure 3.35. The new network risk

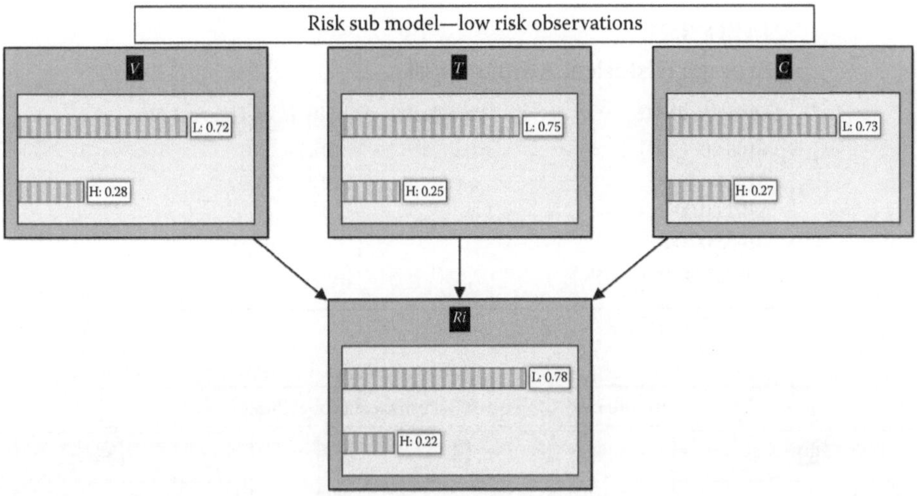

FIGURE 3.28 Total risk: low score observations.

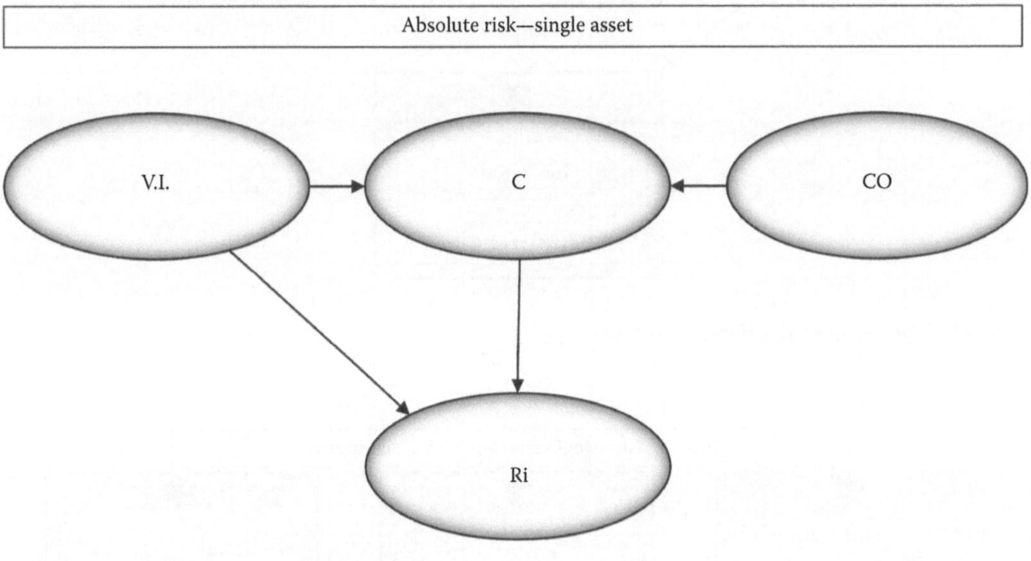

FIGURE 3.29 BN model for absolute risk.

TABLE 3.20
States of Nodal Variables

Variable	# of States	States
V.I.	4	Service (fairly poor), Poor, Fair, Good
CO	3	High, Medium, Low
C	3	High, Medium, Low
Ri	4	\$100,000, \$10,000,\$1000, \$100

TABLE 3.21
Average Historical Absolute Risk

Absolute Risk $	Historical Probability from Figure 3.30
$100.00	0.27
$1,000.00	0.341
$10,000.00	0.255
$100,000.00	0.134
Average historical risk $	$16,318

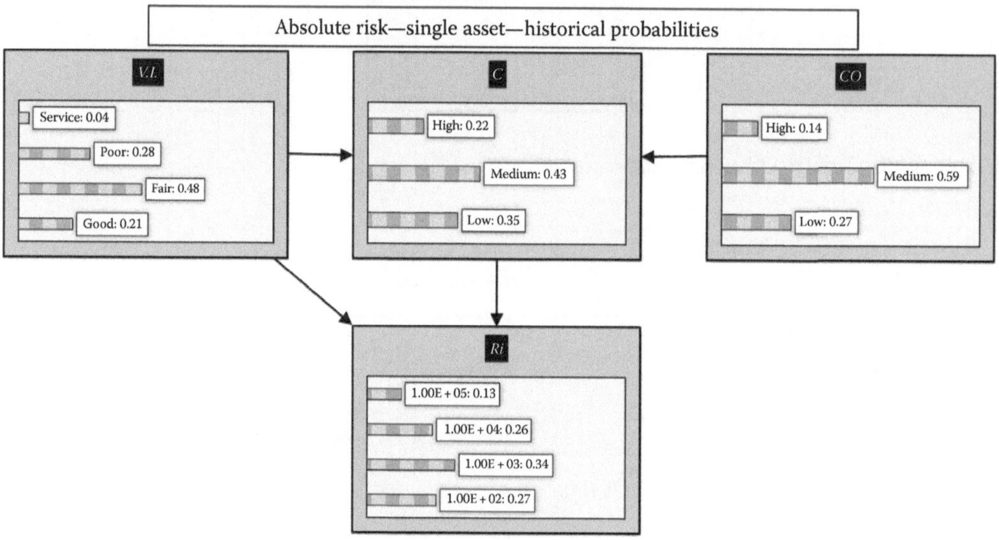

FIGURE 3.30 Historical (prior) probabilities.

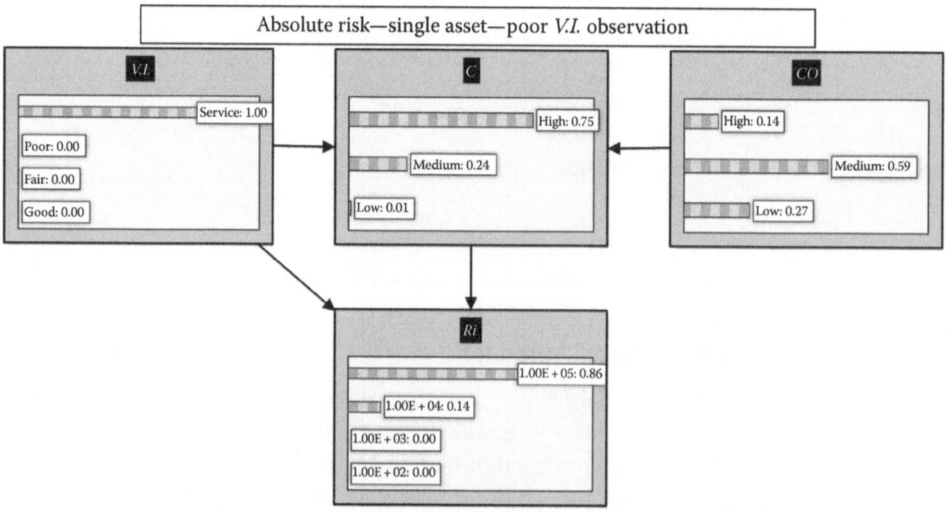

FIGURE 3.31 Risk after a new (fairly poor) *V.I.* observation.

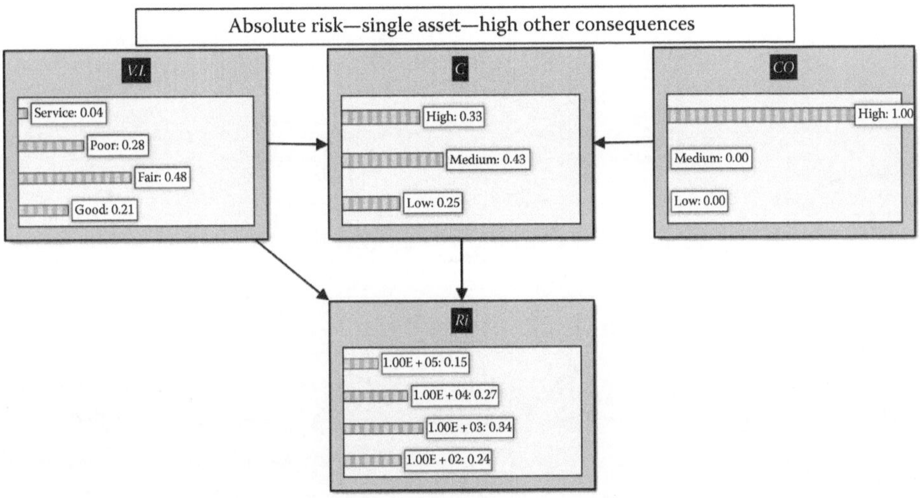

FIGURE 3.32 Risk after a change (higher) in consequences.

TABLE 3.22
Absolute Risk after New Visual Inspection Were Observed

Absolute Risk $	New Risk Probabilities from Figure 3.31
$100.00	0
$1,000.00	0
$10,000.00	0.13642
$100,000.00	0.86358
Average historical risk $	$87,722

TABLE 3.23
Absolute Risk after Higher Consequences Were Observed

Absolute Risk $	New Risk Probabilities from Figure 3.32
$100.00	0.243
$1,000.00	0.335
$10,000.00	0.272
$100,000.00	0.15
Average historical risk $	$18,079

is now $70,825 as computed from Table 3.26. Results such as these, as well as processes, are of importance to the stakeholders responsible for that bridge network.

3.5.3 Case Study 3.8: Chain Graph Risk Assessment of Bridge Network

3.5.3.1 Overview

Relationships (links) of a bridge or any civil asset network are versatile. They can be directional or nondirectional. For example, consequences for each asset of the network can be directional

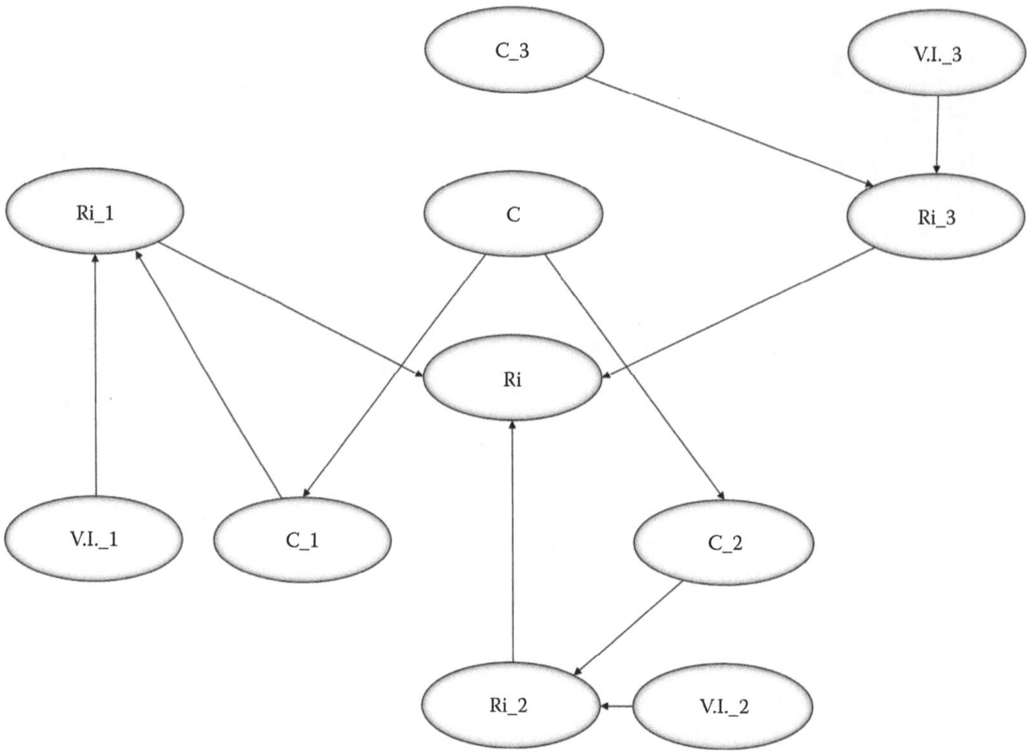

FIGURE 3.33 BN model for risk assessment of bridge network.

TABLE 3.24
States of Nodal Variables

Variable	# of States	States
C_1, C_2, C_3	3	High, Medium, Low
V.I._1, V.I._2, V.I._3	4	Service, Poor, Fair, Good
Ri_1, Ri_2, Ri_3	4	$100,000, $10,000, $1,000, $100
C	3	High, Medium, Low
Ri	4	$300,000, $30,000, $3,000, $300

(bridge detours can have directional effects because the closure of one bridge can have an effect on other bridges) or nondirectional (if 2 bridges are linked via a roadway, then 1 bridge's importance will have an affinity on the other bridge's importance). Case Study 3.7 explored a bridge network where all links are directional (BN). We now explore a situation where the links in the network are both directional and nondirectional (chain graph network). We also slightly modify how different consequences are linked in the model. We assume that even though the various assets in the model have different consequences, those consequences are correlated to each other through a 3-noded potential.

3.5.3.2 Model

In this model, we represent the interactions between the consequences of the three bridges in a nondirectional form. To do so, we model these interactions by a 3D potential that shows affinities among

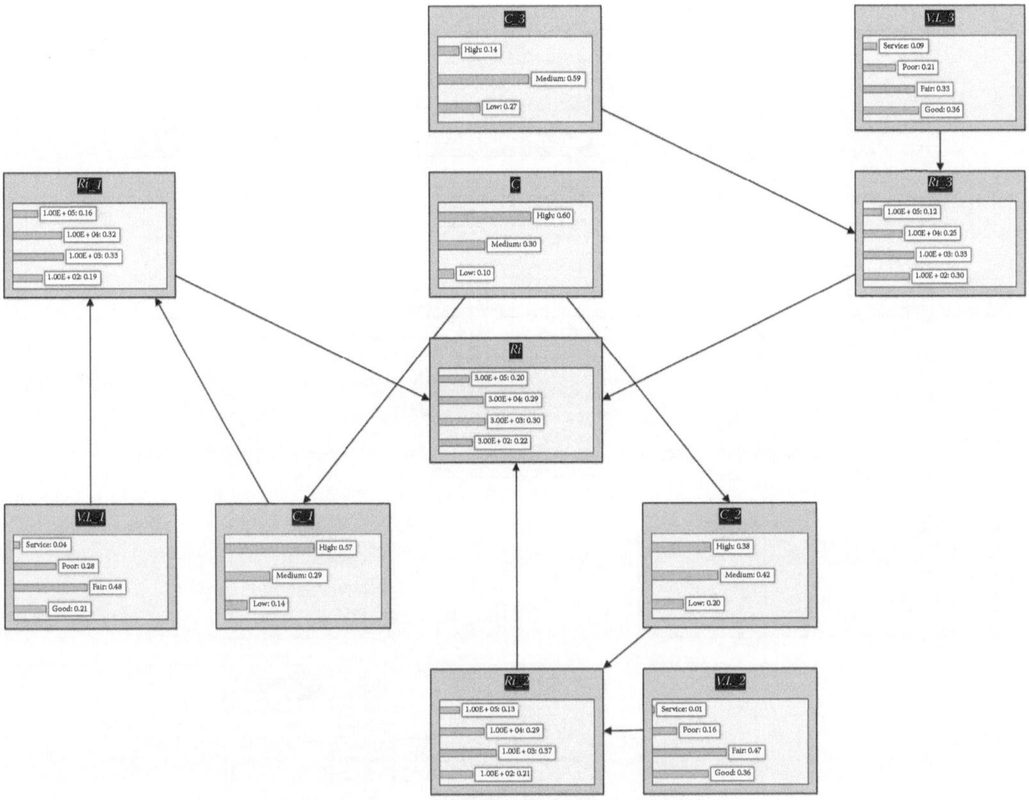

FIGURE 3.34 Network risk with BN modeling: historical probabilities.

TABLE 3.25
Absolute Historical Risk

Absolute Risk $	Historical Probability
$300.00	0.2164
$3,000.00	0.3005
$30,000.00	0.2871
$300,000.00	0.1961
Average historical risk	$68,409.42

these consequences. For more on nondirectional links and potential representations, see Chapter 2 and for a more comprehensive discussion, see Koller and Friedman (2009). The links in the rest of the model are directional and similar to the representations of Case Study 3.7. The differences between the directional and nondirectional links are subtle, but important. Incorporating a mix of directional and nondirectional links will produce a chain graph model as shown in Figure 3.36. The states of nodal variables are shown in Table 3.27. The definitions of those variables are presented in Appendix 3.I (Table 3.34).

3.5.3.3 Conditional Probabilities Tables

The different CPTs and potentials of each variable of the network of Figure 3.36 are shown in Appendix 3.VII. Because this is a CG model, it includes a mix of CPTs (Tables 3.101 through 3.110)

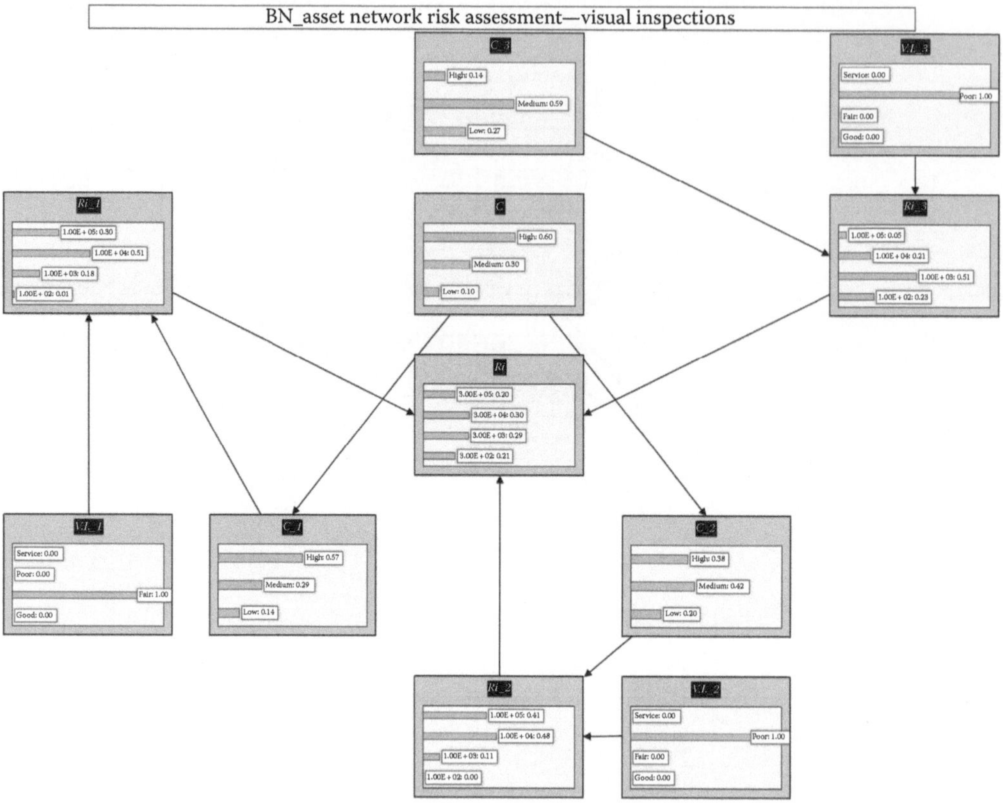

FIGURE 3.35 Network risk with BN modeling—with visual inspection (observations) for network assets.

TABLE 3.26
Absolute Risk after Observations

Absolute Risk $	Probability after Observation
$300.00	0.2056
$3,000.00	0.2928
$30,000.00	0.2985
$300,000.00	0.2031
Average historical risk	$70,825.08

and the 3-noded potential in Table 3.100. As usual, we used our best judgment in filling the different tables in a realistic manner for illustration purposes. For more on different methods of developing CPTs, see Chapter 2.

3.5.3.4 Results

We first execute the CG model to obtain historical probabilities. The result is shown in Figure 3.37. The historical network risk is also shown in Table 3.28, which is used to compute the monetary average risk of $67,148. The result is slightly different from that of Case Study 3.7 due to the slight changes in modeling consequences. This points to an important conclusion; that is, care needs to be exercised while modeling consequences and how they relate to the rest of the model. If new observations from different visual inspections of the three bridges become available, we now enforce those

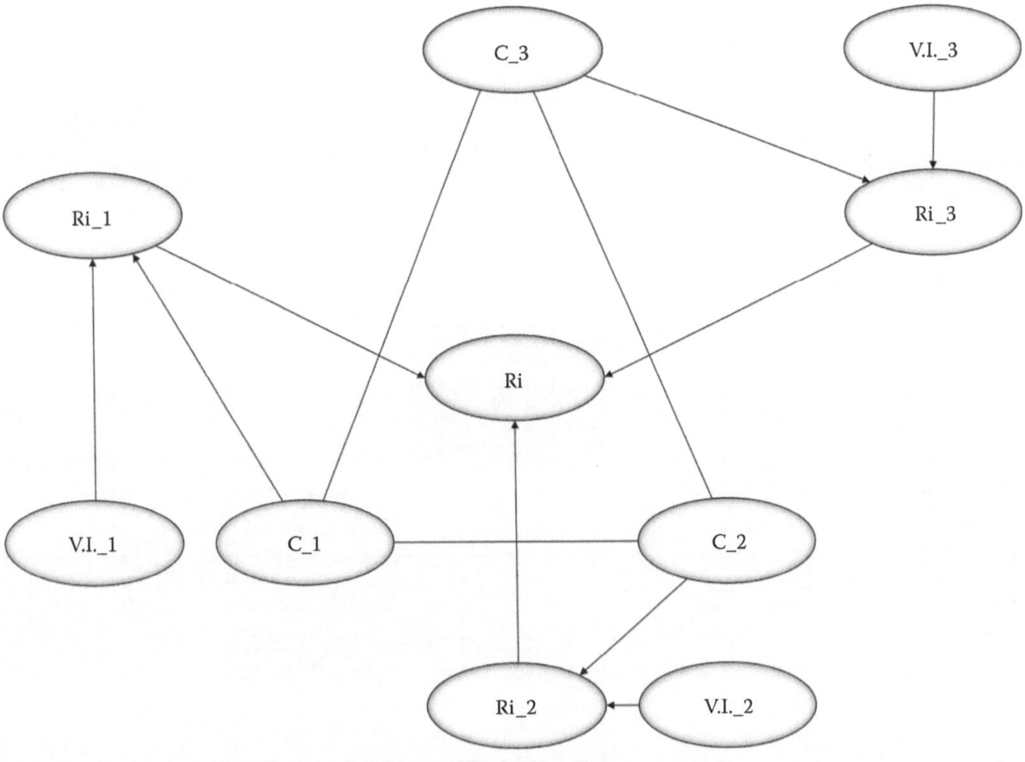

FIGURE 3.36 Chain graph (CG) model for risk assessment of bridge network.

TABLE 3.27
States of Nodal Variables

Variable	# of States	States
V.I._1, V.I._2, V.I._3	4	Service, Poor, Fair, Good
Ri_1, Ri_2, Ri_3	4	$100,000, $10,000, $1000, $100
Ri	4	$300,000, $30,000, $3,000, $300
C_1, C_2, C_3	3	High, Medium, Low

new evidences and solve the problem as shown in Figure 3.38. The new average network risk is now $71,677 as per Table 3.29. Again the slight difference from Case Study 3.7 is due to the changes in visual observations and consequence modeling.

3.6 RESILIENCE ASSESSMENT

3.6.1 Introduction

In Chapter 1, we noted that resilience is a special case of risk, and thus, we can employ risk assessment techniques and processes in resilience management/analysis. We use GN to illustrate practical cases of resilience assessment in this section. We also noted in Chapter 2 that objective resilience can be modeled using 4Rs or as a function of time and continuity of operations. We study these two approaches in the next two examples presented in this chapter.

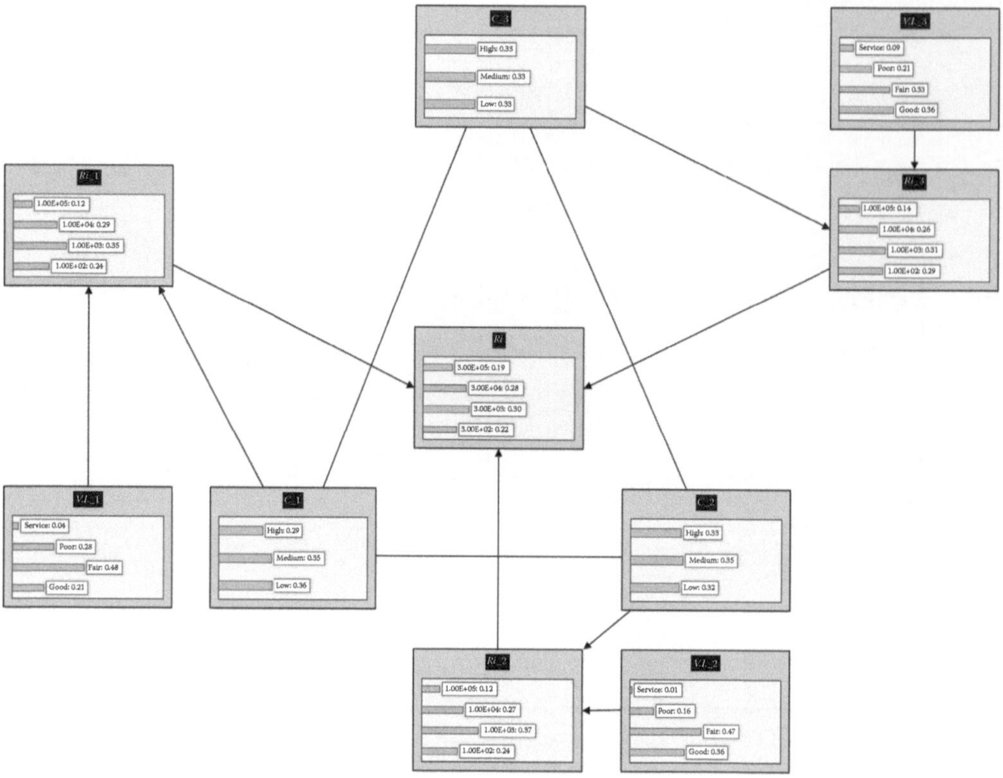

FIGURE 3.37 Network risk with CG modeling: historical probabilities.

TABLE 3.28
Historical Network Risk

Absolute Risk $	Historical Probability
$300.00	0.2214
$3,000.00	0.3026
$30,000.00	0.2838
$300,000.00	0.1922
Average historical risk	$67,148

3.6.2 Case Study 3.9: Flood (Scour) Resilience of Bridges, A 4-R Model

3.6.2.1 Overview

Resilience of bridges after a flood event is of interest to all bridge stakeholders. We offer a resilience assessment BN model that aims at assessing bridge flood resilience. This model will use the 4Rs in assessing resilience. Perhaps a good start is to isolate different issues that control the behavior of the first 3Rs: robustness, resourcefulness, and recovery. These factors are illustrated in Figure 3.39.

We note that some parameters in Figure 3.39 include redundancies, such as emergency plans in R_2, roadways in R_3, and so forth. Because redundancy is an essential metric in infrastructure resilience considerations, we would like to separate redundancy as a major resilience metric on its own

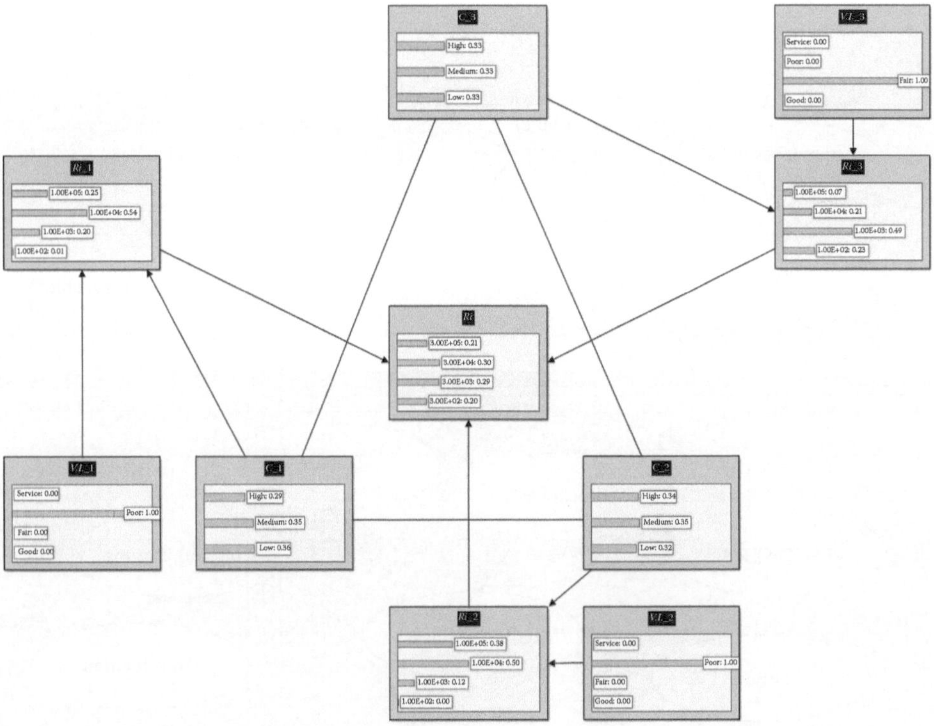

FIGURE 3.38 Network risk with CG modeling—with visual inspection (observations) for network assets.

TABLE 3.29
Network Risk after Observations

Absolute Risk $	Probability after Observation
$300.00	0.204
$3,000.00	0.292
$30,000.00	0.298
$300,000.00	0.206
Average historical risk	$71,677

and define it as R_4 We then define R_4i as a measure of redundancy within R_i with $i = 1, 2, 3$. We now recast the resilience *Re* as a function of the 4 Rs as

$$Re = Re\ (R_1, R_2, R_3, R_4) \tag{3.13}$$

The new 4Rs flood models are shown in Figure 3.40. This bridge resilience flood model can get to be even more complex. For example, several subissues control many of the issues in Figure 3.40. For example, if we look at the variable *Channel rating* we can identify several subissues, which can control its behavior as illustrated in Figure 3.41. For simplicity, the model we will offer herein uses a much smaller number of variables to illustrate the process. Readers are encouraged to solve the more complex models of Figures 3.40 and 3.41 as an exercise.

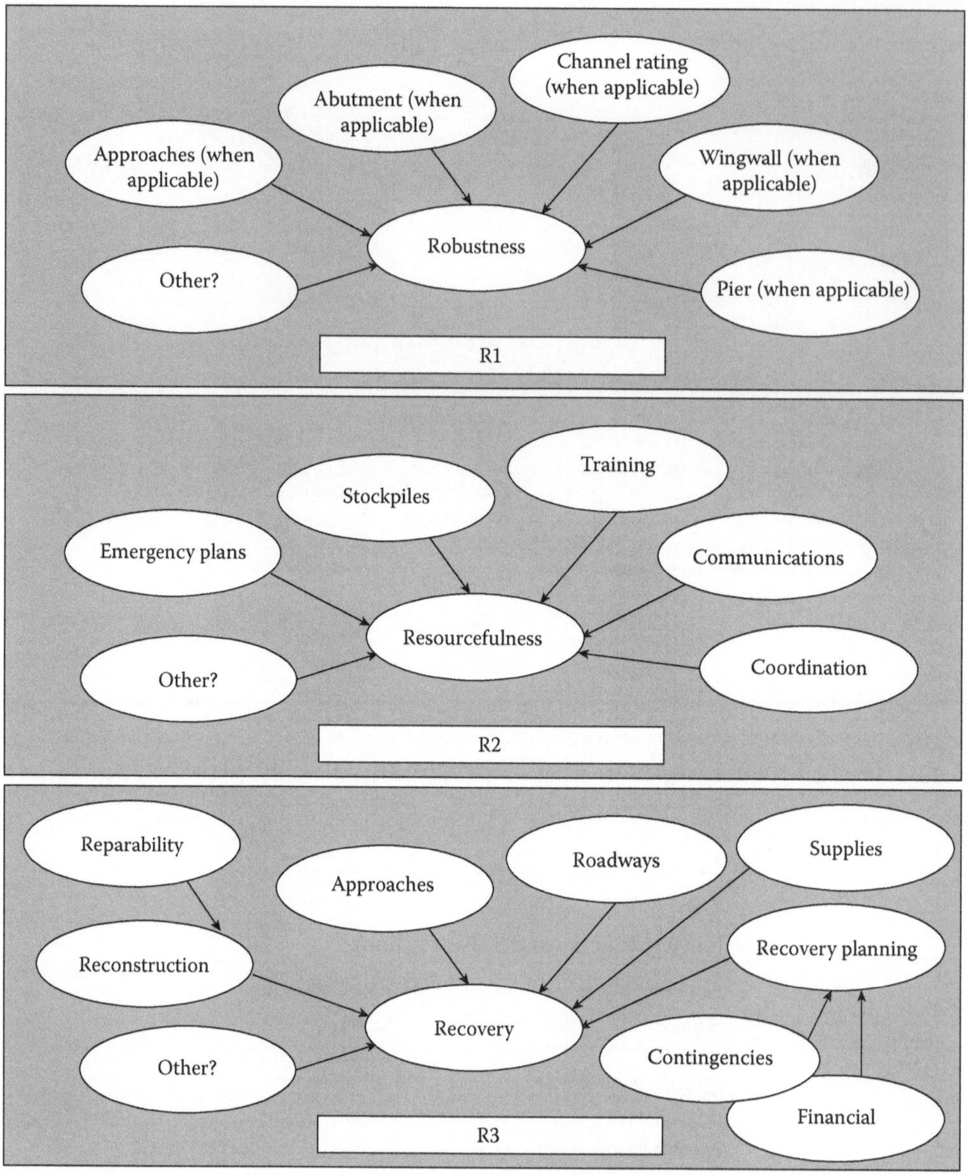

FIGURE 3.39 Parameters affecting the 3Rs.

3.6.2.2 Model Description

For now, we strive to complete building the 4Rs resilience model that will help us in objectively computing different ratings and probabilistic estimates for flood resilience. Recall that M.E. Hynes (2001, personal communications) and Bruneau and Reinhorn (2007) introduced a concept for resilience that was used later by Ettouney and Alampalli (2012a, 2012b). That concept relates resilience to the temporal behavior of continued operations (*CO*) such that

$$Re = \int CO(t)\,dt \tag{3.14}$$

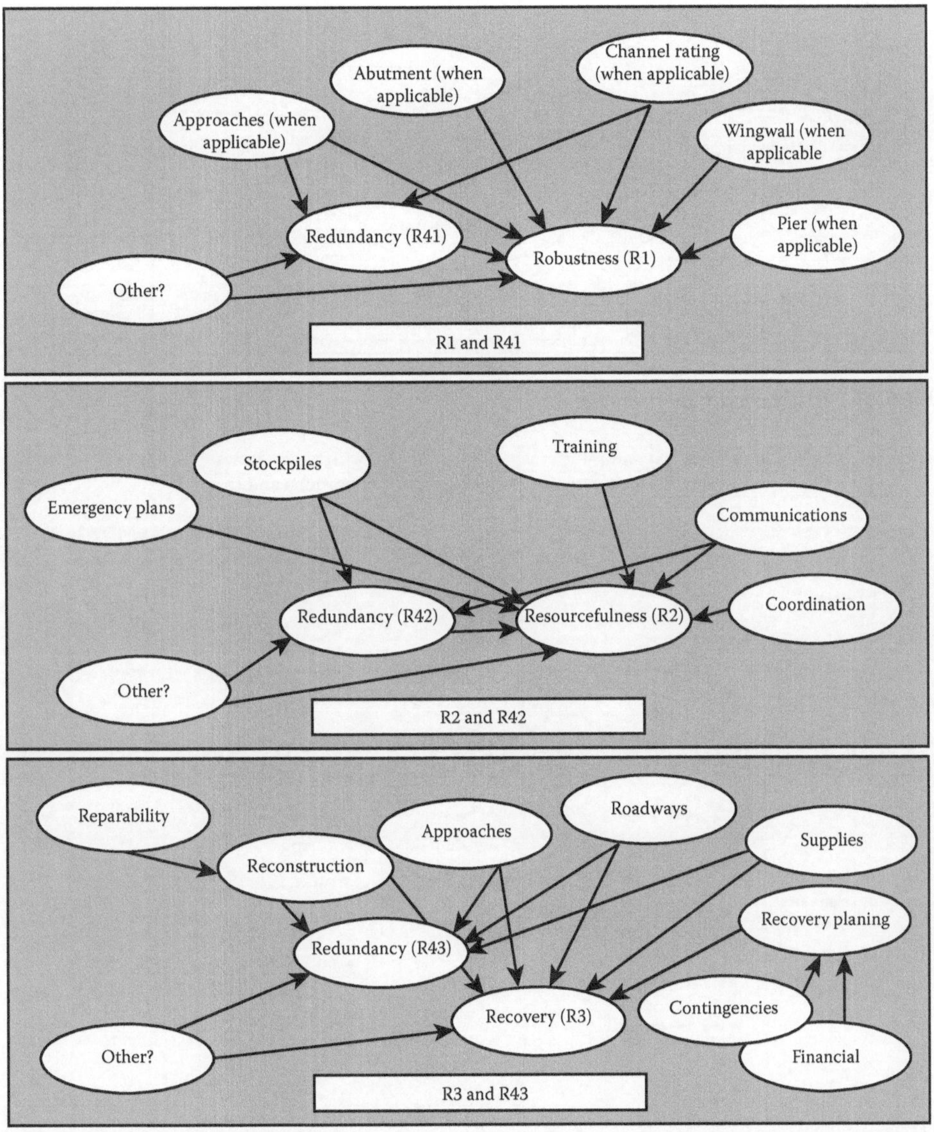

FIGURE 3.40 Flood resilience model with 4Rs.

From Equations 3.13 and 3.14 as well as the relationships of Figure 3.40, we now propose a simplified BN objective resilience model as shown in Figure 3.42. The states of nodal variables are given in Table 3.30. The definitions of the variables can be found in Appendix 3.I (Table 3.34).

There are some notes which need to be made regarding the model. First, the reader would observe that we used an abstract hazard, instead of specific hazard, such as a flood. We intend for this model to be applicable to any type of hazard, which prompted us to use the abstract name. Also, the hazard we chose affects only *R1* (robustness) for simplicity. A more general model would have *H* a parent of *R1, R2,* and *R3*. Also, note that even though there is no direct dependence from *R4* to *CO* or *T*, the redundancy (*R4*) still affects *CO* and *T* through *R14, R24*, and *R34*.

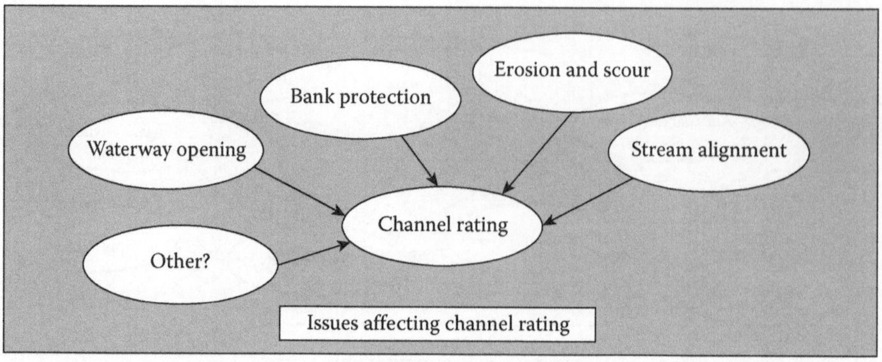

FIGURE 3.41 Issues controlling *Channel Rating.*

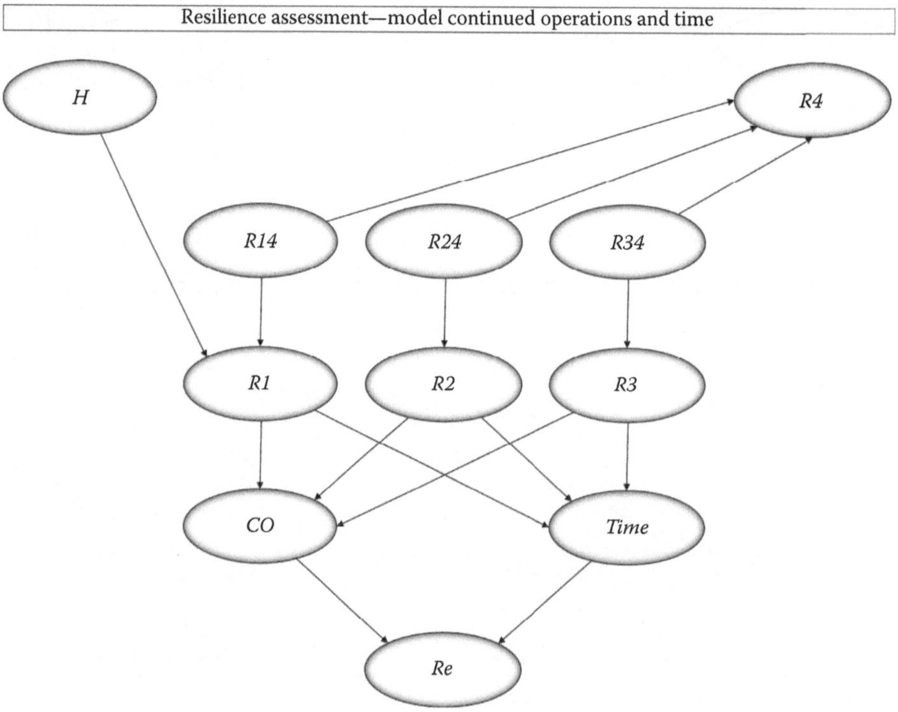

FIGURE 3.42 General BN resilience model.

3.6.2.3 Conditional Probabilities Tables

The different CPTs of each variable in the network of Figure 3.42 are shown in Appendix 3.VIII. As usual, we used our best judgment in filling the different tables in a realistic manner for illustration purposes. For more on different methods of developing CPTs, see Chapter 2.

3.6.2.4 Marginal Probabilities

We analyze the BN model for this case to obtain the historical resilience probabilities first. The results are given in Figure 3.43. The historical resilience average can be computed as 52%, using the process for three-bins histogram. If the stakeholder, in a resilience assessment effort, observed that the states of robustness and recovery, *R2* and *R3*, are *Low*, the resulting resilience can be computed

TABLE 3.30
States of Nodal Variables

Variable	# of States	States
R1	2	Low, High
R2	2	Low, High
R3	2	Low, High
R4	2	Low, High
R14, R24, and R34	2	Low, High
H	3	High, Medium, Low
CO	3	Low, Medium, High
Time	3	Low, Medium, High
Re	3	Low, Medium, High

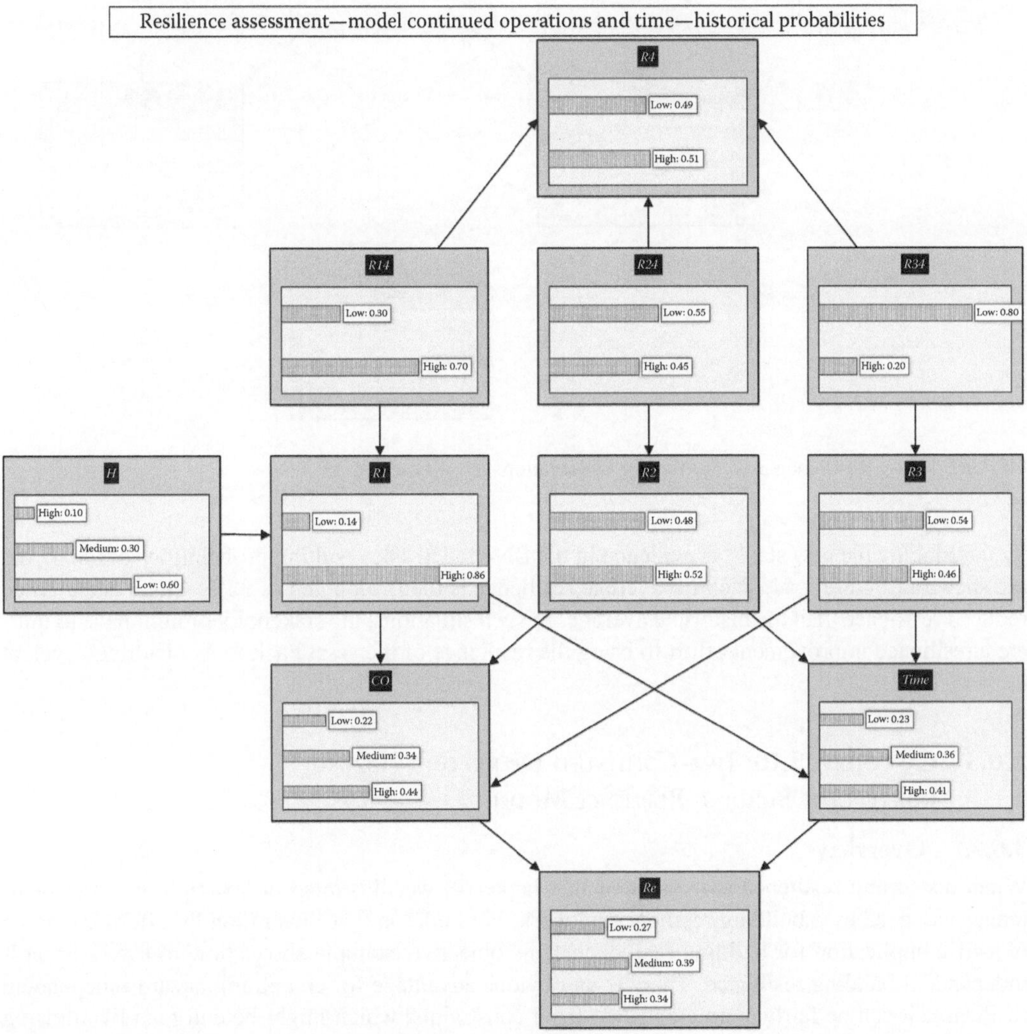

FIGURE 3.43 Historical resilience probabilities.

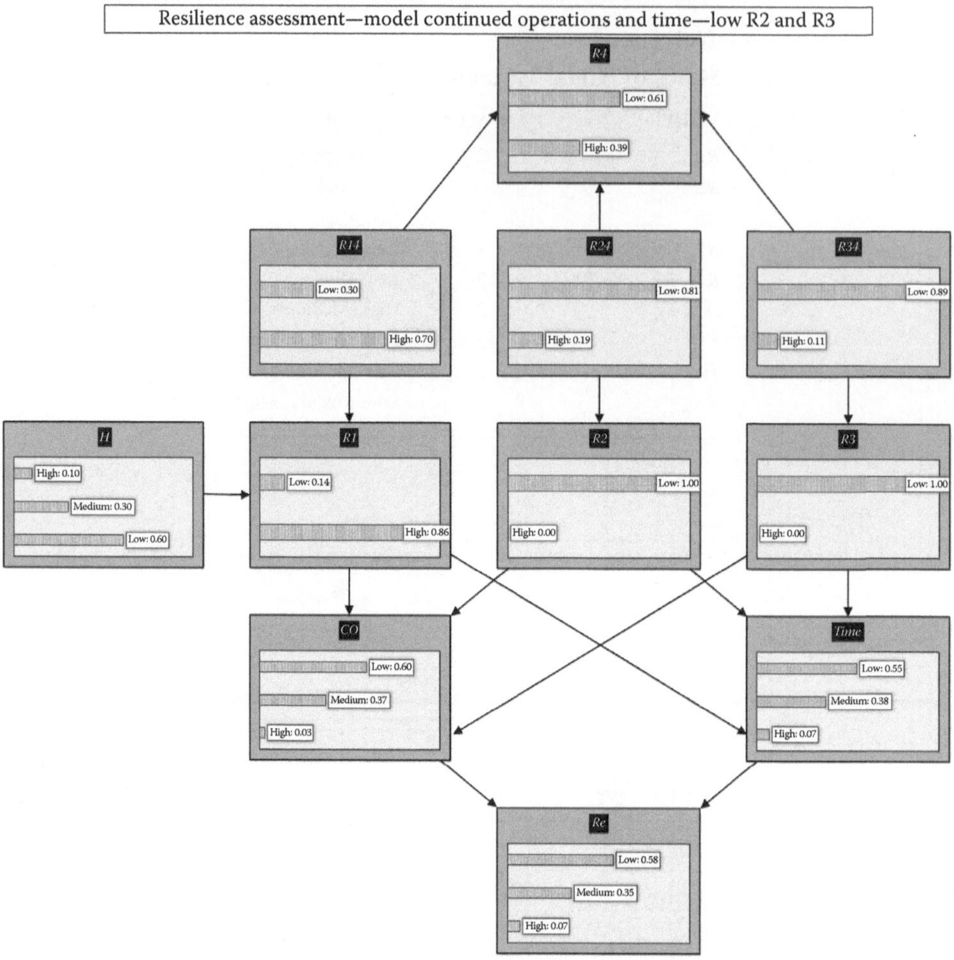

FIGURE 3.44 Resilience probabilities for observations *R2* = *Low* and *R3* = *Low*.

by introducing the new states as evidence in the BN model. The resulting probabilities of the model are shown in Figure 3.44. The new average resilience is then computed as 33%, which is a marked reduced resilience than the historical average. In such situations, the stakeholder might have to initiate a resilience improvement effort to bring the resilience of the asset back to its historical level, at the very least.

3.6.3 Case Study 3.10: Time-Continued Operations: Hazard Independent Building Resilience Model

3.6.3.1 Overview

When discussing resilience management in Chapter 2, we illustrated an example of asset resilience evidenced by a building resilience network. We build on that illustration by offering here an objective application for building resilience. This objective example shows how to assess hazard-independent building resilience. There is an obvious advantage for computing hazard-independent resilience: it can be fairly simple and free from constraints, which might be enforced by allowing for specific hazard or hazards in the model, thus producing a generic computational template, which can be reused often with minimal changes. Similarly, the obvious disadvantage is that it might be

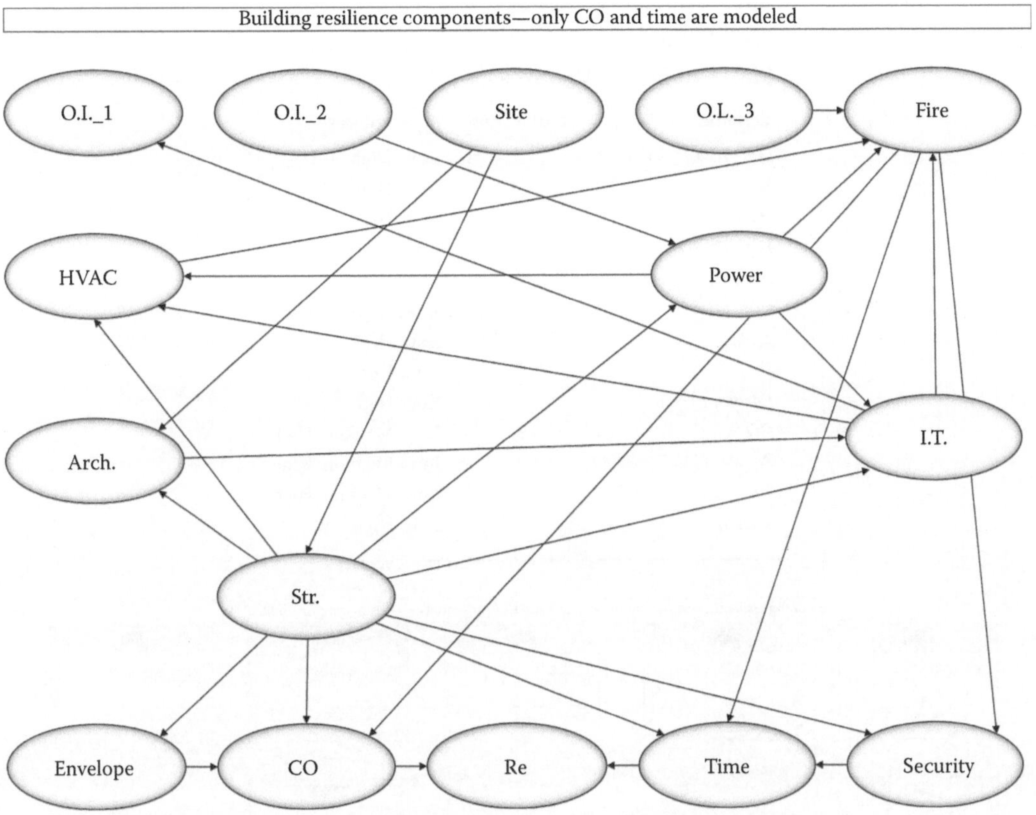

FIGURE 3.45 Simple BN building resilience model with nine building functions.

of limited value when a specific hazard (or hazards) is prevalent that is of importance to the stakeholders. Thus, we offer this case study as an example of resilience assessment of buildings. The readers can modify it by introducing a specific hazard or hazards or by adding more variables that are deemed of importance to the model. Note that Ettouney and Alampalli (2016) used a modified version of this model in a multihazards setting.

3.6.3.2 Model Description

Figure 3.45 shows the BN model for building resilience assessment. Note that we used the time, *Time*, and continued operations, *CO*, as the controlling parameter for evaluating resilience, *Re*, see Equation 3.14. Also, note that we used the variables *O.I.* to indicate outside influences, which can be a parent link (indicating that the building is a dependent to the outside link, e.g., *O.I._2*). In this case study, the building is dependent on the outside power. A child link (indicating that the building controls the outside, e.g., *O.I._1*) is also introduced. In this case study, the outside of the building is dependent on the *I.T.* operations of the building. Table 3.31 show the states of nodal variables of the model, and Appendix 3.I (Table 3.34) shows the definitions of these nodal variables.

3.6.3.3 Conditional Probabilities Tables

The different CPTs of the variables in the network of Figure 3.45 are shown in Appendix 3.IX. As usual, we used our best judgment in filling the different tables in a realistic manner for illustration purposes. For more on different methods of developing CPTs, see Chapter 2.

TABLE 3.31
States of Nodal Variables

Variable	# of States	States
O.I._1, O.I._2,O.I._2	2	Low, High
Site	2	Low, High
HVAC	2	Low, High
Fire_Protection	2	Low, High
Power	2	Low, High
I.T.	2	Low, High
Arch.	2	Low, High
Str.	2	Low, High
Envelope	2	Low, High
CO	3	Low, Medium, High
Re	3	Low, Medium, High
Time	3	Low, Medium, High
Security	2	Low, High

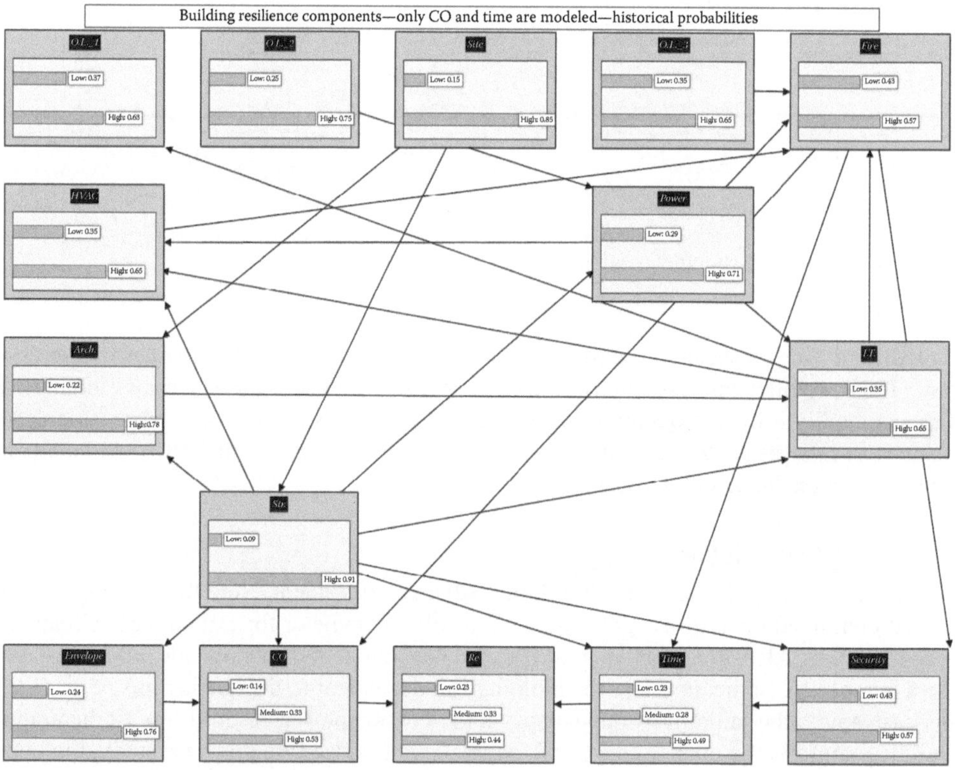

FIGURE 3.46 Historical building resilience probabilities.

3.6.3.4 Results

We start assessing building resilience, as usual, with assessing historical probabilities. The results are shown in Figure 3.46. The average historical resilience of this particular building is computed, using the process of Appendix 3.IV for histograms of three states and Table 3.32 to be 57%. To evaluate how resilience of this building varies, as say, two of its underlying building functions vary, we chose structural system (*Str.*) and fire system (*Fire*) to investigate. By changing the state of each of the two variables to *Low* and *High*, we obtain the four network solutions of Figures 3.47 through 3.50. We can use these results to compute the average resilience for each of the four situations, as shown in Table 3.33. A combination of a *High* rating of the structural system and a *High* rating of the fire system will improve the resilience of the building to 68%. Whereas a combination of a *Low* rating of the structural system and a *Low* rating of the fire system reduces the resilience of the building to 33%. Similar studies can highlight the most efficient combination of building functions ratings and can be used to provide for an optimal building resilience.

TABLE 3.32
Historical Building Resilience

Resilience State	Historical Resilience Probabilities
High	44%
Medium	33%
Low	23%
Average historical resilience (*Re*)	57%

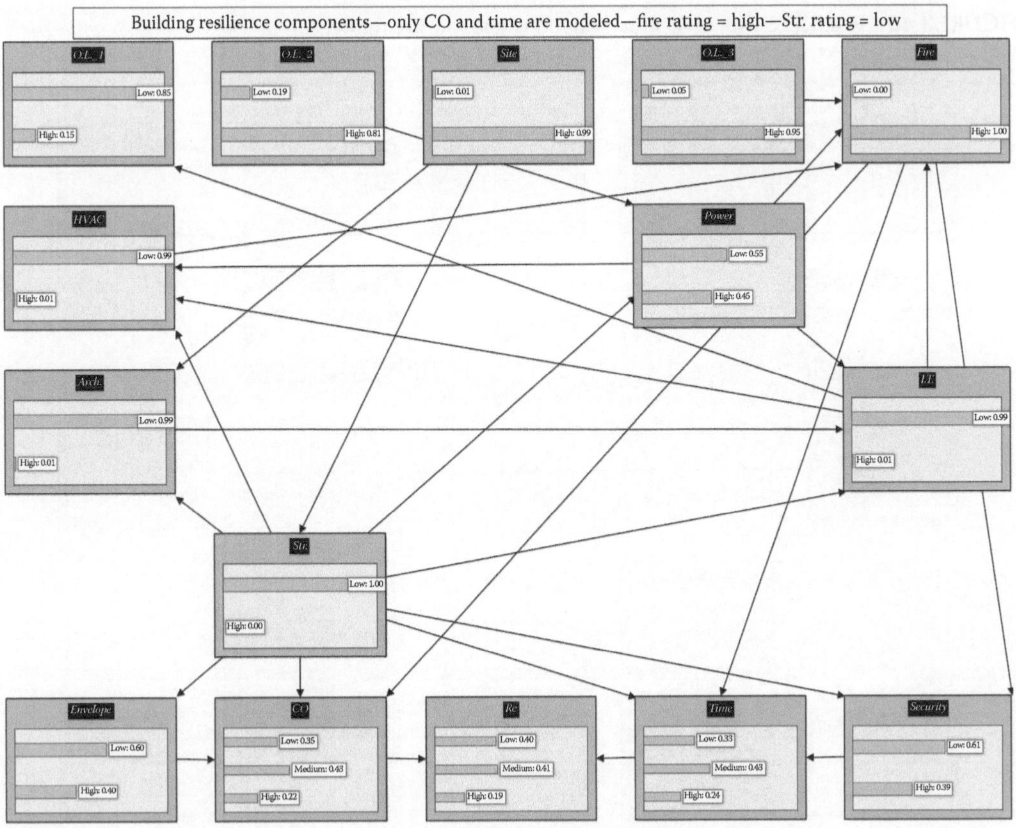

FIGURE 3.47 Resilience with low structural rating and high fire rating.

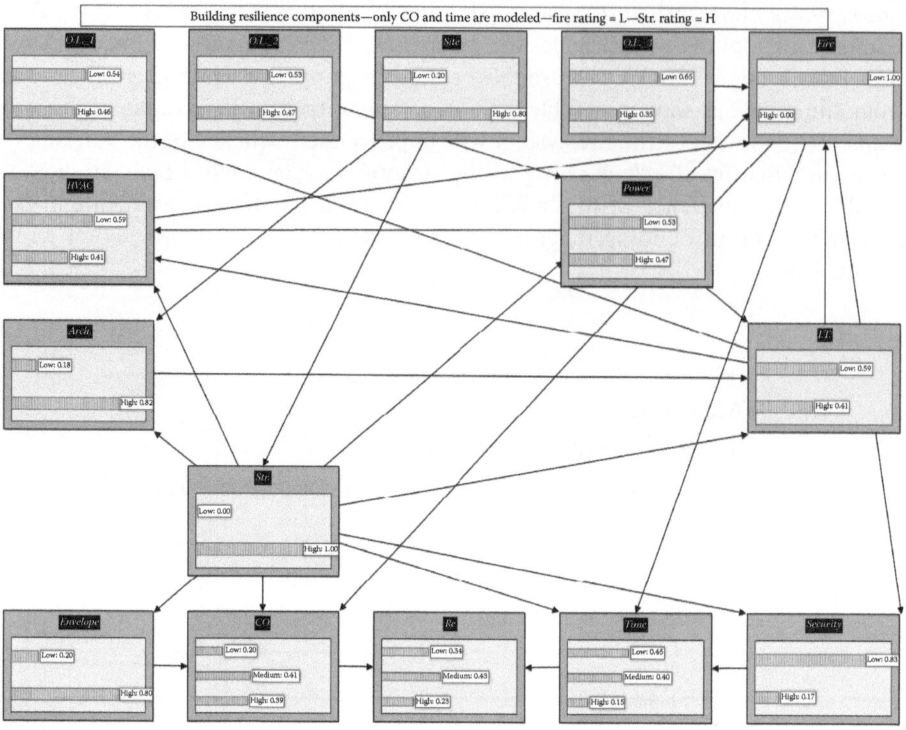

FIGURE 3.48 Resilience with high structural rating and low fire rating.

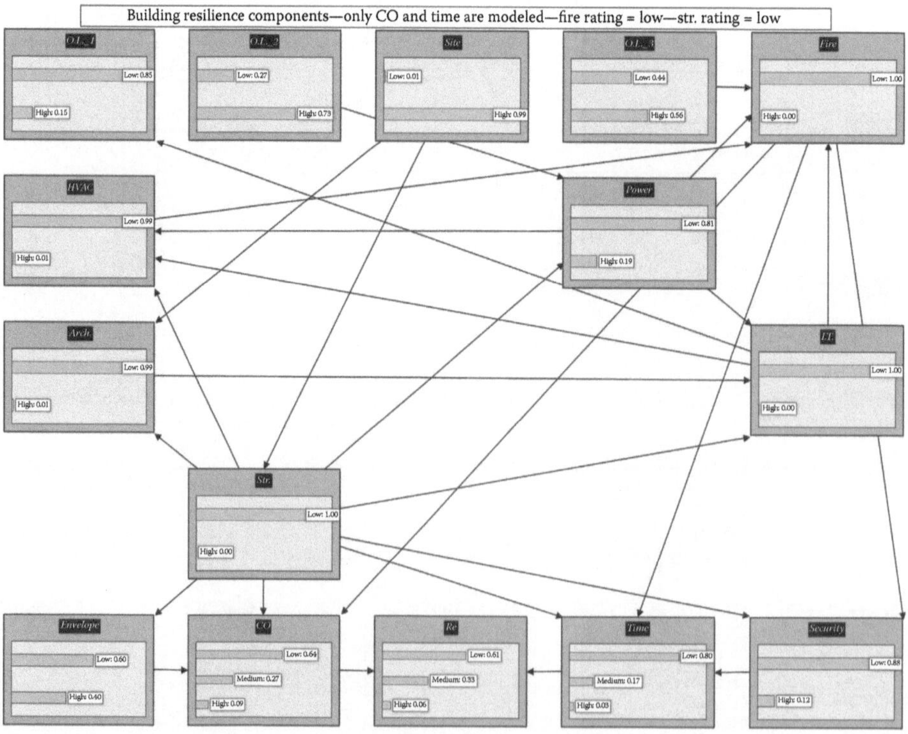

FIGURE 3.49 Resilience with low structural rating and low fire rating.

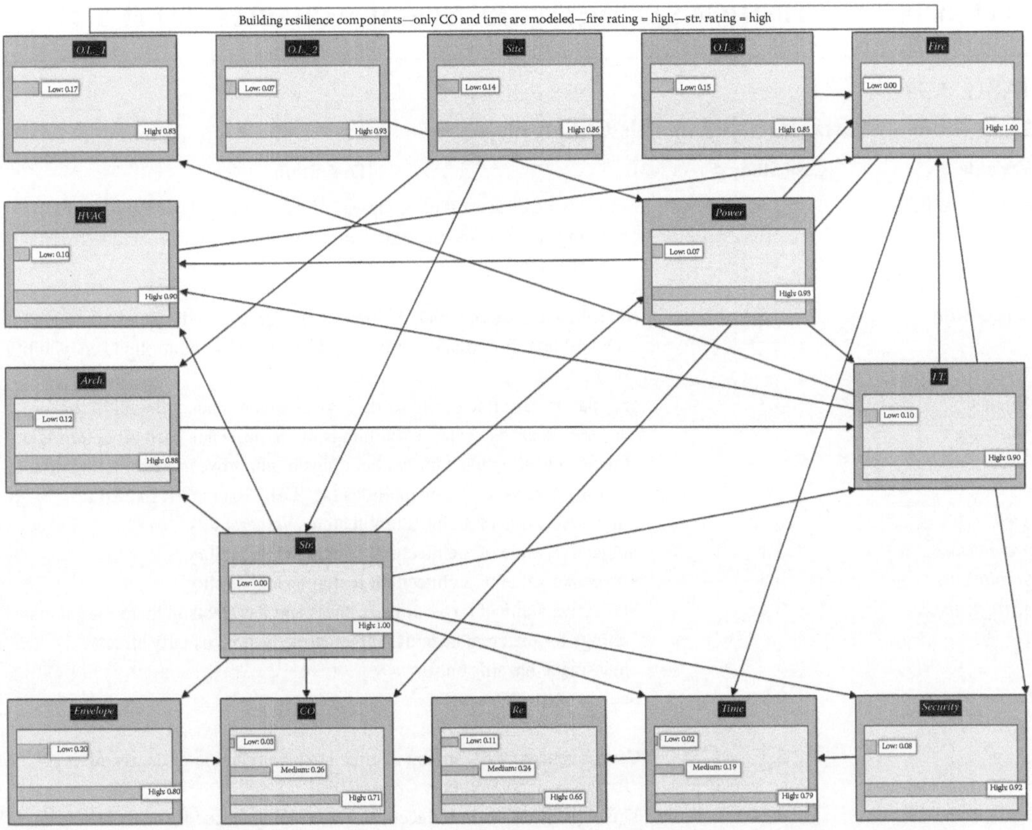

FIGURE 3.50 Resilience with high structural rating and high fire rating.

TABLE 3.33
Effects of *Fire* and *Str.* States on Building Resilience

	Probabilities after Introducing Observations			
	Str. = *Low*		*Str.* = *High*	
Resilience state	*Fire* = *Low*	*Fire* = *High*	*Fire* = *Low*	*Fire* = *High*
High	6%	19%	23%	65%
Medium	33%	41%	43%	24%
Low	61%	40%	34%	11%
Average resilience (*Re*)	32%	43%	46%	68%

APPENDIX 3.I: DEFINITIONS OF NETWORK NODAL VARIABLES (TABLE 3.34)

TABLE 3.34
Definitions of Variables Used in This Chapter

Variable	Case Study #	Definition
# of occupants	3.5	Subjective metric for # of occupants of the building. Densely occupied buildings are considered highly attractive for man-made threats (from security viewpoint).
Accessibility	3.5	Subjective measure, which describes the degree of and ease of accessing the building. For high security, it is desired to have controlled accessibility to the building.
Adjacent_ Buildings	3.3	This parameter affects both earthquake and man-made vulnerabilities for different reasons. Adjacent buildings might influence each other during a seismic event, while adjacent buildings might prove to affect both building security threats and vulnerabilities (see DHS 2011a, 2011b, 2011c).
Arch.	3.10	Subjective value of architectural features vulnerability
Architecture	3.1	Subjective value of architectural features vulnerability.
Architecture	3.5	Subjective value of architectural features vulnerability
Attractiveness	3.1	Subjective value of target attractiveness and its effects in increasing threat ratings for different hazards. Target attractiveness usually affects man-made hazards only.
Business Continuity	3.5	See *Continued Operations*.
C	3.3	Consequences of hazard are assumed to be a function of degree of *Exposure* and *Continued Operations*.
C	3.7	This variable, *C*, is introduced in this model to force identical consequences for two of the bridges in the network.
C	3.6	Objective estimate of total consequences. It is a function of two variables in this example: *V.I.* (*which is an approximation of system* exposure) and their consequences, *C.O*. Thus, we establish the presumption that total consequences rely on both the system itself via its exposure and other factors beyond the system itself, *C.O*.
C	3.4	Consequences
C	3.5	Consequences
C_1, C_2, C_3	3.7	*C_i* is the consequences for the i-th bridge in the network.
C_1, C_2, C_3	3.6	Same as above.
CO	3.9	Subjective estimate of continued operations. *CO* is one of the two parameters that define *Re*.
CO	3.6	Estimate of consequences beyond the system itself, these consequences can be, but are not limited to, cost of replacements, cost of traffic delays, symbolic nature of bridge, locality, and asset age. In these examples, *C.O*. is hazard independent.
CO	3.10	Same as above
Construction_ Material	3.2	Older construction material, such as nonreinforced masonry, is more vulnerable to many hazards such as earthquakes or bomb blast. Older wood construction might be more vulnerable to fire hazard.
Construction_ Method	3.2	Depending on the hazard, tunnel construction method can have major effects on tunnel vulnerability.
Continued Operations	3.4	Subjective variable that describes the length of continued operations interruption. It is desired to have as low *Continued Operations* as possible.

(Continued)

TABLE 3.34 ***(Continued)***
Definitions of Variables Used in This Chapter

Variable	Case Study #	Definition
Continued_ Operations	3.3	Subjective variable that describes the length of continued operations interruption. It is desired to have as low *Continued Operations* as possible.
Downtime	3.1	Expected long downtime after an occurrence of hazard will increase harmful consequences.
Ductility	3.3	Objective measure of the effective ductile behavior of the system, e.g., if the system is designed to experience high or low ductile behavior. Obviously, for high demand hazards such as earthquakes or blast, good ductile design is warranted.
Ductility	3.4	Same as above
E	3.3	Exposure is a result of both the hazard as well as the vulnerability of the system.
E	3.4	Same as above
Earthquake	3.1	Subjective value of earthquake threat.
Economical	3.5	Metric describing direct economic consequences.
Economical	3.5	Metric describing direct economic factors that might increase (or decrease) target attractiveness (bomb threats).
Elevation	3.1	Lower level (e.g., underwater) would have more consequences than grade- and dry-level stations.
Envelope	3.10	Objective measure, which describes vulnerability of the building envelope.
Envelope	3.5	Same as above
Fire_Protection	3.1	Subjective value of fire protection system vulnerability.
Fire_Protection	3.1	Subjective value of potential of fire hazard occurrence.
Fire_Protection	3.10	Same as above
Floods	3.1	Subjective value of flood threat.
Framing	3.2	Type of structural framing (systems) can have an effect on tunnel vulnerability. For example, ribbed cylinders offer more efficient structural performance than beam-column framing.
Function_ Within_Region	3.2	See *Impact_of_Loss*.
General Arch.	3.5	Objective variable, which describes total architectural vulnerability. It is a function of different architectural attributes of the building.
General Eng.	3.5	Objective variable, which describes total engineering vulnerability. It is a function of different engineering disciplines of the building, such as structural, envelope, Mechanical/Electrical/Plumbing (MEP), etc.
H	3.9	Subjective estimate of a hazard that might affect the system. A hazard can be any event that might degrade system performance.
Height	3.2	Longer controlling unsupported heights (columns/walls) attracts more internal forces within the structure, thus increases the potential of higher vulnerability.
High_Target_ Value	3.1	As value of station increases, potential harmful consequences also increase.
Historic	3.5	Objective variable that describes historical nature of the building. Historic buildings might prove to have more target attractiveness than nonhistoric buildings.
HVAC	3.10	Objective measure that describes vulnerability of building HVAC.
I.T.	3.1	Subjective value of adequacy and operational efficiency and resilience of Information technology (IT) systems.
I.T.	3.10	Same as above
Impact of Physical Loss	3.5	See *Impact_of_Loss*.

(Continued)

TABLE 3.34 ***(Continued)***
Definitions of Variables Used in This Chapter

Variable	Case Study #	Definition
Impact_of_Loss	3.1	Harmful consequences are directly related to impact of physical loss of the station.
Liner	3.2	Cross-section of liner will have a direct effect on the vulnerability of the tunnel.
Liner_Thickness	3.2	Liner thickness of tunnel will have direct effect on the vulnerability of the tunnel.
Local_Soil	3.3	Subjective estimate of soil condition.
Locality	3.4	Objective variable that describes locality from target attractiveness viewpoint. Some localities are more susceptible to man-made hazards than others are.
Locality Type	3.5	See *Locality*.
MEP	3.5	Objective measure that describes vulnerability of building MEP functions.
Modal_ Attributes	3.3	Subjective estimate on how vulnerability is affected by modal attributes of the system. For example, blast or impact will excite higher modes and wider frequency ranges than earthquakes.
Modal_ Attributes	3.4	Same as above
Nearby_ Facilities	3.1	Harmful consequences after a hazard occurrence can be magnified if the nearby facilities are damaged. The reverse is also true. See also *Adjacent_Buildings*.
No_of_Levels	3.1	Subjective value of number of levels in the station and their effect on overall consequences of the occurrence of different hazards.
No_of_Riders	3.1	Peak number of riders per day in the stations of interest: the larger the ridership, the more consequences.
No_of_Tracks	3.1	Subjective value of number for tracks and their effect on overall consequences of occurrence difference hazards.
No_of_Users	3.2	Number of users per day of the tunnel of interest: the larger the number of users, the more consequences.
None_Structural	3.1	Subjective value of nonstructural systems vulnerability.
O.L._1, O.L._2,O.L._2	3.10	*O.L._i* represents the operational efficiency of the i-th outside link that is related to one or more of building functions. For example, *O.L._2* simulates the external power source that controls the power operations of the building. Conversely, *O.L._1* simulates a satellite office, which uses the services of the *I.T.* operations within the building of interest.
Occupancy Use	3.5	
Op_Redundancy	3.1	Higher operational redundancy will minimize harmful consequences after a hazard occurrence.
Operational_ Systems	3.1	Subjective value of operational systems vulnerability.
Physical	3.5	An objective variable that is a function of some physical threat parameters.
Physical_Sec	3.1	Subjective value of adequacy of physical security systems and operations.
Population Density	3.5	This objective variable describes the density of population in the area/neighborhood in which the building is located.
Power	3.10	An objective measure of operational efficiency of power plant in the building of interest.
R1	3.9	Objective estimate of robustness, which is one of the four components of resilience.
R14, R24, and R34	3.9	Subjective estimate of redundancy, which might be due to robustness, robustness, and recovery.
R2	3.9	Objective estimate of resourcefulness, which is one of the four components of resilience.

(Continued)

TABLE 3.34 ***(Continued)***
Definitions of Variables Used in This Chapter

Variable	Case Study #	Definition
R3	3.9	Objective estimate of recovery, which is one of the four components of resilience.
R4	3.9	Objective estimate of redundancy, which is one of the four components of resilience.
Re	3.9	Objective estimate of resilience.
Re	3.10	Same as above
Redundancy	3.1	Subjective value of different redundancies (e.g., structural or operational redundancies) and its effects in increasing threat ratings for different hazards.
Replacement Value	3.5	
Replacement_ Value	3.1	Replacement value of the station is directly related to expected harmful consequences.
Retrofits	3.2	Retrofitted tunnels usually offer less vulnerability than nonretrofitted tunnels.
Ri	3.3	Objective estimate of risk, which is a function of *Exposure* and *C*.
Ri	3.7	Objective total network risk, which relates to the individual risk measures of all bridges in the network.
Ri	3.6	Objective estimate of risk, which is a function of Exposure (approximated in this case study by *V.I.* results) and *C*.
Ri	3.6	Same as Case Study 3.4.
Ri	3.4	In this example, risk, Ri, is a function of consequences and exposure. Numerous algorithms objectively relate risk to consequences and exposure.
Ri_1, Ri_2, Ri_3	3.7	*Ri_i* is the objective risk for the *i*-th bridge in the network.
Ri_1, Ri_2, Ri_3	3.6	Same as Case Study 3.4.
Risk	3.5	In this example, risk, Ri, is a function of vulnerability, threat, and consequences. Numerous algorithms objectively relate risk to vulnerability, threat, and consequences.
Security	3.10	Objective variable that describes the adequacy of security operations in the building.
Security	3.5	Same as above
Seismic_Design	3.2	Tunnels that are designed for seismic resistance are less vulnerable than those that are not seismically worthy.
Seismic_Zone	3.3, 3.4	Expected magnitude of the seismic hazard.
Significance	3.2	Significance of tunnel (historical, social, visibility) can have an important effect on harmful consequences after hazard occurrence.
Site	3.1, 3.5, 3.10	Subjective value of site vulnerability.
Social_Impact	3.1	Harmful consequences are directly related to the social importance and use of the station.
Span	3.2	Longer controlling spans attract more internal forces within the structure, thus increasing the potential of higher vulnerability.
Sta_Locality	3.1	Location of the station has a direct effect on the consequences of hazard occurrences.
Str.	3.10	See *Structure*.
Structural_ Condition	3.2	Tunnels with better overall condition ratings (no section losses due to corrosion, for example) are less vulnerable than tunnels with lower condition ratings.
Structure	3.5	Subjective value of structural system vulnerability.
Structure	3.1	Subjective value of structural system vulnerability.
Symbolism	3.5	See *Historical*.

(Continued)

TABLE 3.34 ***(Continued)***
Definitions of Variables Used in This Chapter

Variable	Case Study #	Definition
T	3.5	Objective variable describing overall man-made threat in the building of interest.
Target Density	3.5	Indicates whether there have been any signs of man-made hazards in the past.
Target Potential	3.5	See *Target_Attractiveness.*
Target_ Attractiveness	3.4	Objective measure indicating the propensity of the building to attract man-made hazards.
Time	3.9	Subjective estimate of elapsed time from the onset of hazard until the system achieves the desired level of efficient operations. *Time* is one of the two parameters that define *Re*.
Time	3.10	Same as above
Traffic	3.1	Subjective value of traffic related issues and its effects in increasing threat ratings for different hazards.
Tunnel_ Locality	3.2	Location of the tunnel will have direct effect on consequences of hazard occurrences.
Underwater_ Tunnel	3.2	Consequences of hazard occurrences might result in higher consequences than tunnels on dry sites.
V	3.3	Objective estimate of level of vulnerability of the system. This variable (in this example) is the combined effects of all vulnerabilities due to the hazard of interest. In this example, we assume that vulnerability is independent of hazard, which is not always a correct assumption.
V	3.4	Same as above
V	3.5	Same as above
V.I.	3.6	Visual inspection results are simulated by this objective variable (V.I.). It is a subjective measure of condition rating of the system of interest. Its estimates are a function of both hazards and vulnerability. Thus, it is related directly to actual exposure of the system. The number of states has to be consistent with the number of condition ratings the owner uses for visual inspection. Four condition ratings used in this example are only for simplicity.
V.I._1, V.I._2, V.I._3	3.7	*V.I._i* is the visual inspection rating for the i-th bridge in the network.
V.I._1, V.I._2, V.I._3	3.6	Same as Case Study 3.7.
Ventilation	3.1	Subjective value of ventilation vulnerability.
Wind	3.1	Subjective value of high wind threat.

APPENDIX 3.II: CPTs FOR CASE STUDY 3.3 (TABLES 3.35–3.44)

TABLE 3.35
CPT of *Adjacent Buildings*

Adjacent Bldg.	Far	0.6
	Close	0.4
Histograms for conditional probabilities		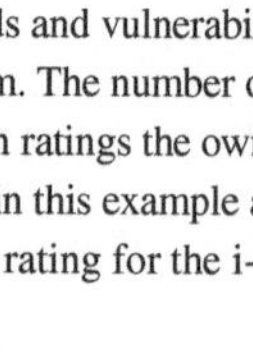

TABLE 3.36
CPT of *C*

E		Low			Medium			High		
Continued Operations		Short	Med.	Long	Short	Med.	Long	Short	Med.	Long
C	Low	0.8240	0.4637	0.2397	0.3629	0.2668	0.1867	0.2242	0.0355	0.0001
	Med.	0.1726	0.4148	0.4050	0.4931	0.4085	0.4152	0.4626	0.4292	0.0800
	High	0.0033	0.1215	0.3553	0.1439	0.3248	0.3981	0.3133	0.5353	0.9199
Histograms for conditional probabilities		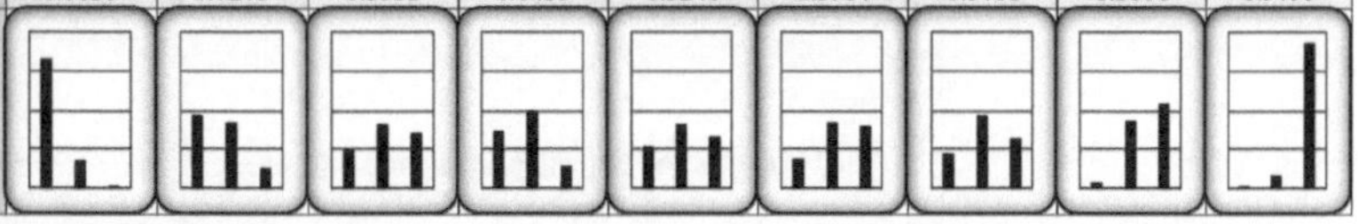								

TABLE 3.37
CPT of *Continued Operation*

E		Low	Med.	High
Continued Operations	Short	0.8240	0.2242	0.0063
	Med.	0.1726	0.4626	0.2260
	Long	0.0033	0.3133	0.7677
Histograms for conditional probabilities		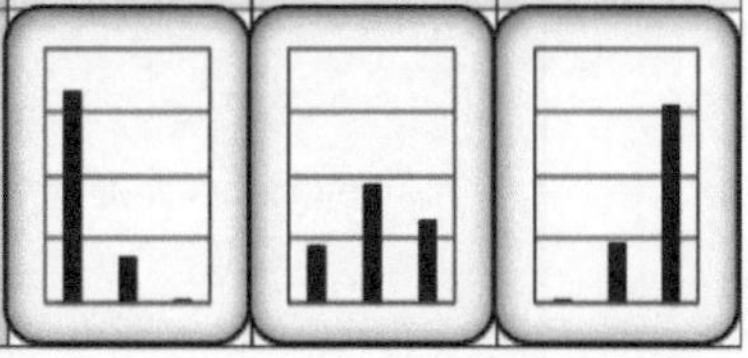		

TABLE 3.38
CPT of *Ductility*

Ductility	High	0.3
	Med.	0.5
	Low	0.2
Histograms for conditional probabilities		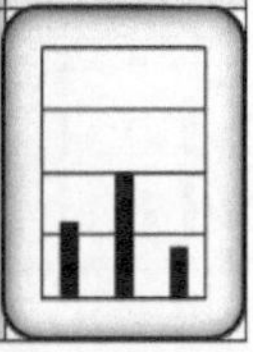

TABLE 3.39
CPT of *E*

Seismic Zone		Low			Medium			High		
V		Low	Med.	High	Low	Med.	High	Low	Med.	High
E	Low	0.8683	0.3553	0.1439	0.4448	0.1854	0.0288	0.3133	0.0778	0.0033
	Med.	0.1314	0.5527	0.4931	0.4967	0.4525	0.4007	0.4626	0.4258	0.1726
	High	0.0003	0.0921	0.3629	0.0585	0.3621	0.5705	0.2242	0.4963	0.8240
Histograms for conditional probabilities		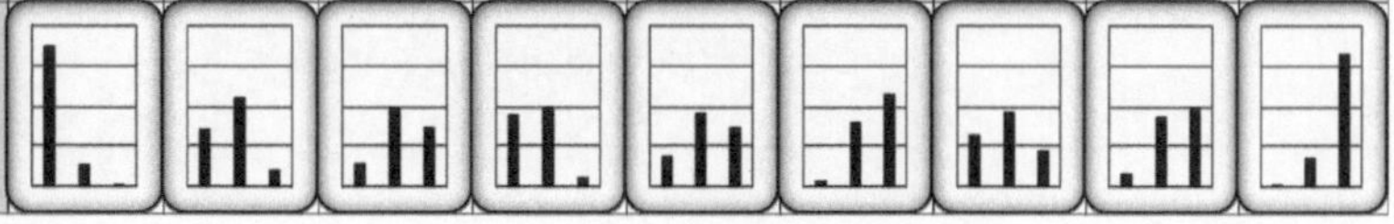								

TABLE 3.40
CPT of *Local Soil*

Local Soil	Rock	0.2
	Strong	0.4
	Weak	0.4
Histograms for conditional probabilities		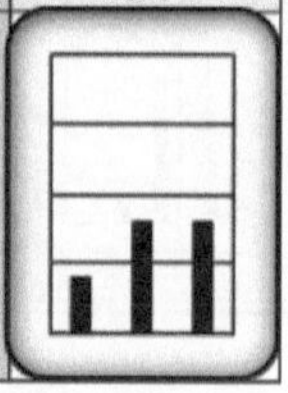

TABLE 3.41
CPT of *Modal Attributes*

Modal Attributes	Effective	0.4
	Not Effective	0.6
Histograms for conditional probabilities		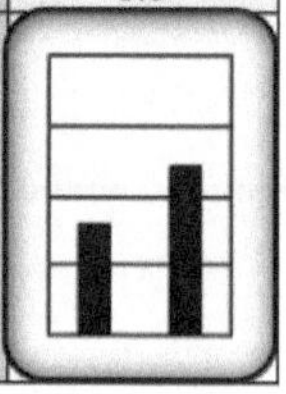

TABLE 3.42
CPT of *Ri*

E		Low			Medium			High		
C		Low	Med.	High	Low	Med.	High	Low	Med.	High
Ri	Low	0.8240	0.4630	0.0921	0.6216	0.1729	0.0207	0.3553	0.0355	0.0033
	Med.	0.1726	0.4839	0.5527	0.3577	0.6007	0.3577	0.5527	0.4292	0.1726
	High	0.0033	0.0531	0.3553	0.0207	0.2264	0.6216	0.0921	0.5353	0.8240
Histograms for conditional probabilities										

TABLE 3.43
CPT of *Seismic Zone*

Seismic Zone	Low	0.3
	Med.	0.4
	High	0.3
Histograms for conditional probabilities		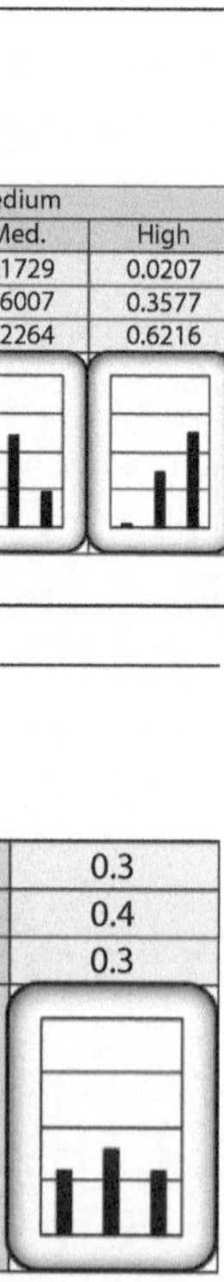

TABLE 3.44
CPT of *V*

Local Soil		Rock											
Modal Attributes		Effective						Not Effective					
Adjacent Bldg.		Far			Close			Far			Close		
Ductility		High	Med.	Low	High	Med.	Low	High	Med.	Low	High	Med.	Low
V	Low	0.9199	0.5961	0.3621	0.8683	0.5191	0.3133	0.5783	0.2851	0.1854	0.4504	0.2179	0.1511
	Med.	0.0800	0.3709	0.4525	0.1314	0.4294	0.4626	0.4142	0.5545	0.4525	0.5321	0.5643	0.4362
	High	0.0001	0.0330	0.1854	0.0003	0.0515	0.2242	0.0074	0.1604	0.3621	0.0175	0.2179	0.4126

Histograms for conditional probabilities

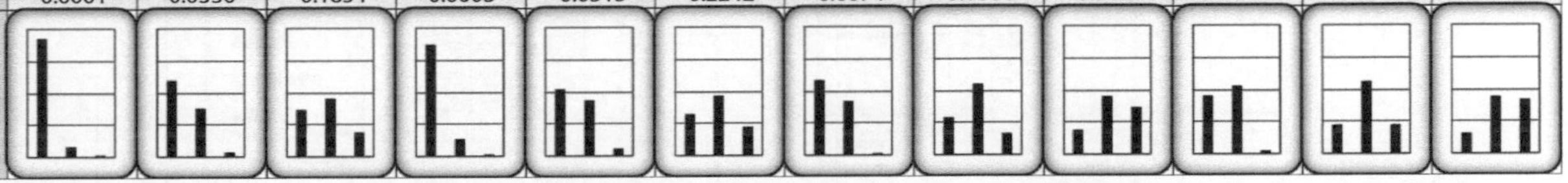

Local Soil		Strong											
Modal Attributes		Effective						Not Effective					
Adjacent Bldg.		Far			Close			Far			Close		
Ductility		High	Med.	Low	High	Med.	Low	High	Med.	Low	High	Med.	Low
V	Low	0.5961	0.3600	0.1604	0.5191	0.2851	0.1138	0.2851	0.1138	0.0330	0.2179	0.0779	0.0205
	Med.	0.3709	0.5262	0.5545	0.4294	0.5545	0.5262	0.5545	0.5262	0.3709	0.5643	0.4829	0.3120
	High	0.0330	0.1138	0.2851	0.0515	0.1604	0.3600	0.1604	0.3600	0.5961	0.2179	0.4392	0.6674

Histograms for conditional probabilities

Local Soil		Weak											
Modal Attributes		Effective						Not Effective					
Adjacent Bldg.		Far			Close			Far			Close		
Ductility		High	Med.	Low	High	Med.	Low	High	Med.	Low	High	Med.	Low
V	Low	0.3621	0.1604	0.0074	0.3133	0.1138	0.0029	0.1854	0.0330	0.0001	0.1511	0.0205	0.0000
	Med.	0.4525	0.5545	0.4142	0.4626	0.5262	0.3009	0.4525	0.3709	0.0800	0.4362	0.3120	0.0467
	High	0.1854	0.2851	0.5783	0.2242	0.3600	0.6962	0.3621	0.5961	0.9199	0.4126	0.6674	0.9533

Histograms for conditional probabilities

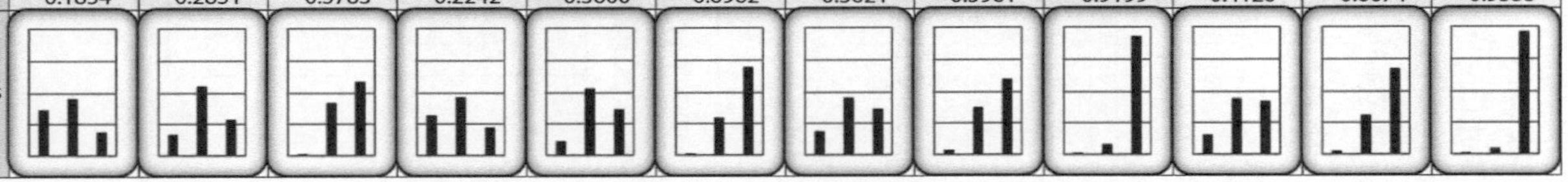

APPENDIX 3.III: CPTs FOR CASE STUDY 3.4 (TABLES 3.45–3.54)

TABLE 3.45
CPT of *C*

Locality		Normal								
E		Low			Med.			High		
Continued Operations		Short	Med.	Long	Short	Med.	Long	Short	Med.	Long
C	Low	0.8240	0.5353	0.1986	0.7677	0.4448	0.1385	0.6999	0.3553	0.0921
	Med.	0.1726	0.4292	0.6028	0.2260	0.4967	0.5898	0.2885	0.5527	0.5527
	High	0.0033	0.0355	0.1986	0.0063	0.0585	0.2717	0.0116	0.0921	0.3553
Histograms for conditional probabilities		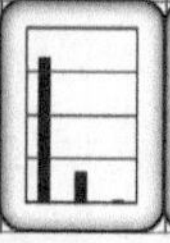	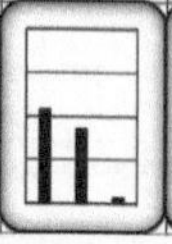	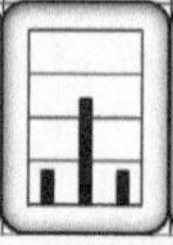			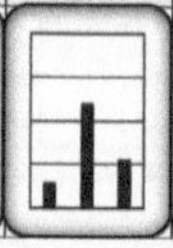	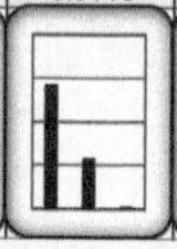	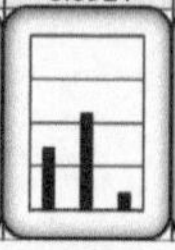	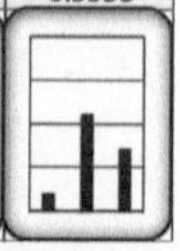

Locality		Attractive								
E		Low			Med.			High		
Continued Operations		Short	Med.	Long	Short	Med.	Long	Short	Med.	Long
C	Low	0.3629	0.1439	0.0385	0.2994	0.1071	0.0263	0.2407	0.0778	0.0176
	Med.	0.4931	0.4931	0.3377	0.5121	0.4635	0.2927	0.5186	0.4258	0.2497
	High	0.1439	0.3629	0.6238	0.1886	0.4293	0.6811	0.2407	0.4963	0.7327
Histograms for conditional probabilities		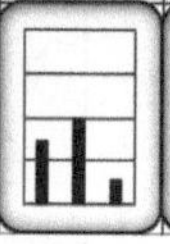	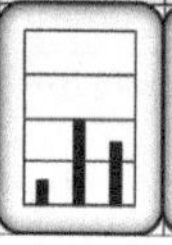	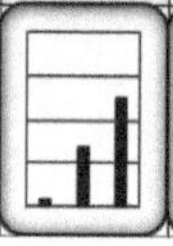	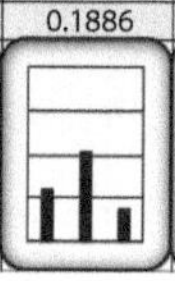	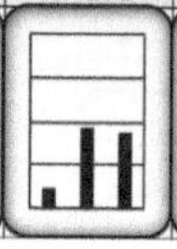	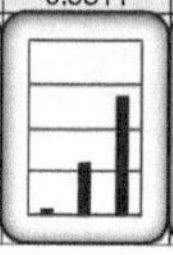	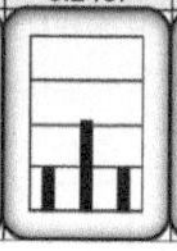	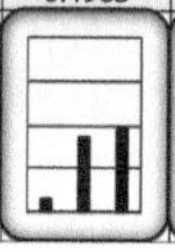	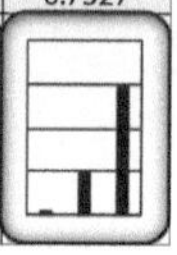

TABLE 3.46
CPT of *Continued Operations*

E		Low	Med.	High
Continued Operations	Short	0.8240	0.2242	0.0063
	Med.	0.1726	0.4626	0.2260
	Long	0.0033	0.3133	0.7677
Histograms for conditional probabilities		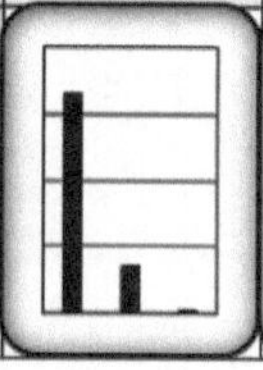	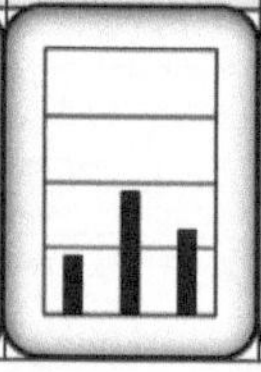	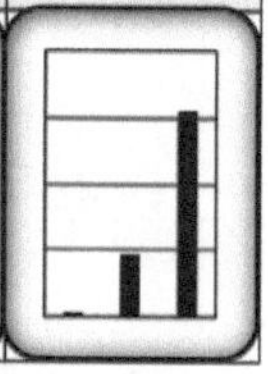

TABLE 3.47
CPT of *Ductility*

Ductility	High	0.4126
	Med.	0.4362
	Low	0.1511
Histograms for conditional probabilities		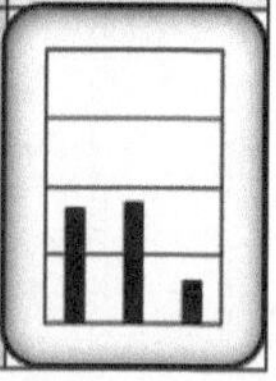

TABLE 3.48
CPT of *E*

Target Attractiveness		Low			Med.			High		
V		Low	Med.	High	Low	Med.	High	Low	Med.	High
E	Low	0.8683	0.3553	0.1439	0.4448	0.1854	0.0288	0.3133	0.0778	0.0033
	Med.	0.1314	0.5527	0.4931	0.4967	0.4525	0.4007	0.4626	0.4258	0.1726
	High	0.0003	0.0921	0.3629	0.0585	0.3621	0.5705	0.2242	0.4963	0.8240
Histograms for conditional probabilities		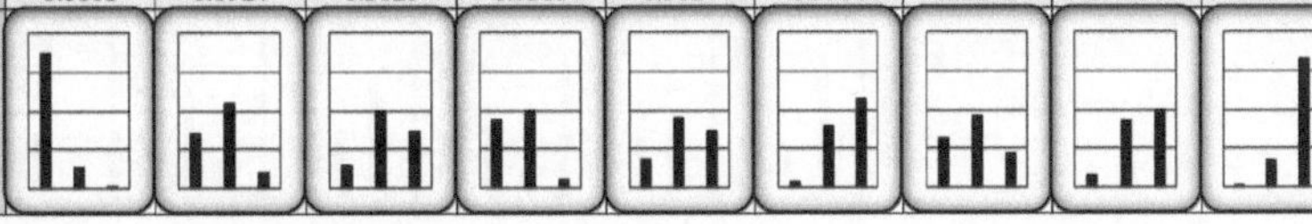								

TABLE 3.49
CPT of *Locality*

Locality	Normal	0.45
	Attractive	0.55
Histograms for conditional probabilities		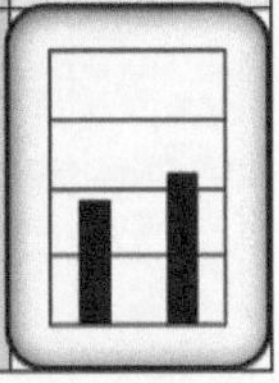

TABLE 3.50
CPT of *Modal Attributes*

Modal Attributes	Effective	0.45
	Not Effective	0.55
Histograms for conditional probabilities		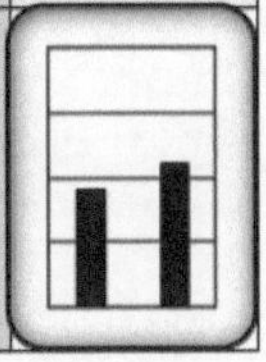

TABLE 3.51
CPT of *Ri*

E		Low			Med.			High		
C		Low	Med.	High	Low	Med.	High	Low	Med.	High
Ri	Low	0.8240	0.4630	0.0921	0.6216	0.1729	0.0207	0.3553	0.0355	0.0033
	Med.	0.1726	0.4839	0.5527	0.3577	0.6007	0.3577	0.5527	0.4292	0.1726
	High	0.0033	0.0531	0.3553	0.0207	0.2264	0.6216	0.0921	0.5353	0.8240
Histograms for conditional probabilities										

TABLE 3.52
CPT of *Setback*

Setback	Long	0.3
	Short	0.7
Histograms for conditional probabilities		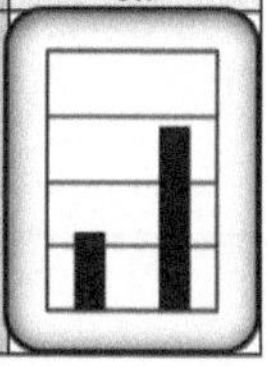

TABLE 3.53
CPT of *Target Attractiveness*

Setback		Long		Short	
Locality		Normal	Attractive	Normal	Attractive
Target Attractiveness	Low	0.6999	0.1511	0.0207	0.0000
	Med.	0.2885	0.4362	0.3577	0.0933
	High	0.0116	0.4126	0.6216	0.9067
Histograms for conditional probabilities		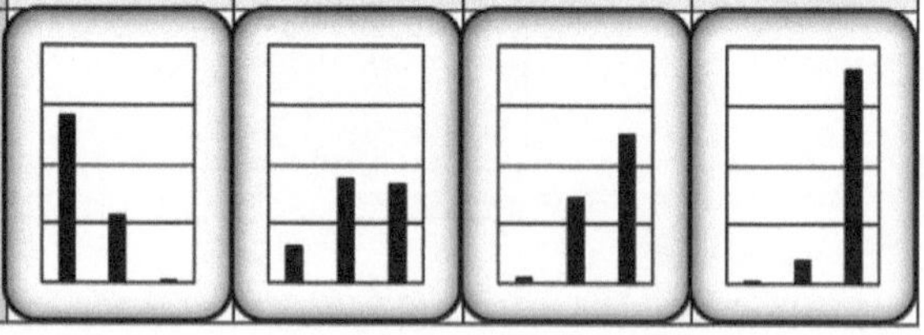			

TABLE 3.54
CPT of *V*

Setback		Long											
Modal Attributes		Effective						Not Effective					
Locality		Normal			Attractive			Normal			Attractive		
Ductility		High	Med.	Low	High	Med.	Low	High	Med.	Low	High	Med.	Low
V	Low	0.9199	0.6216	0.4963	0.5353	0.4072	0.3629	0.6216	0.4696	0.3133	0.3621	0.2872	0.1854
	Med.	0.0800	0.3577	0.4258	0.4292	0.5017	0.4931	0.3577	0.4417	0.4626	0.4525	0.5144	0.4525
	High	0.0001	0.0207	0.0778	0.0355	0.0911	0.1439	0.0207	0.0887	0.2242	0.1854	0.1984	0.3621
Histograms for conditional probabilities		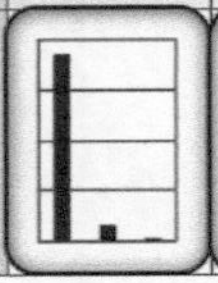	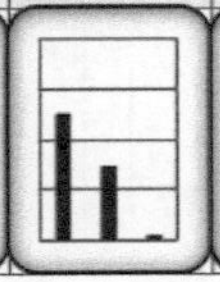	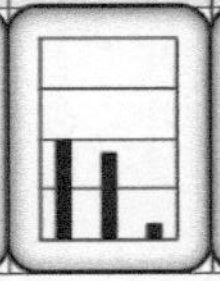	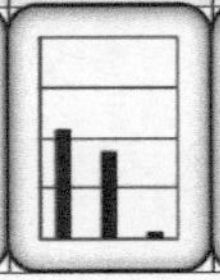	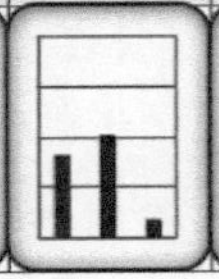	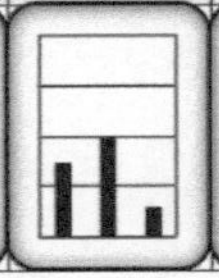	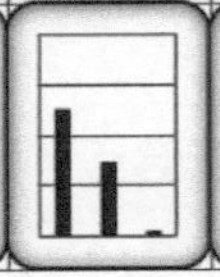	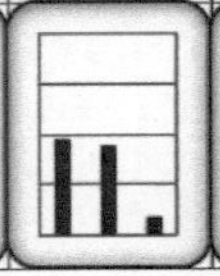	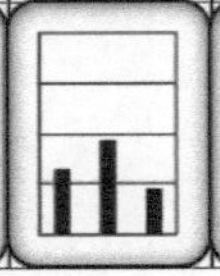	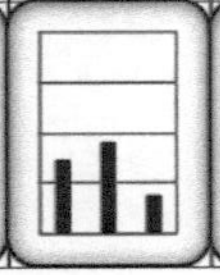	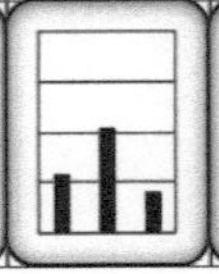	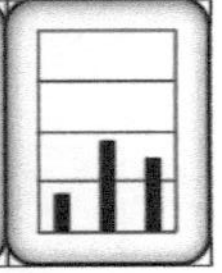

Setback		Short											
Modal Attributes		Effective						Not Effective					
Locality		Normal			Attractive			Normal			Attractive		
Ductility		High	Med.	Low	High	Med.	Low	High	Med.	Low	High	Med.	Low
V	Low	0.4963	0.2242	0.0918	0.1385	0.0772	0.0778	0.0585	0.0147	0.0003	0.0116	0.0056	0.0000
	Med.	0.4258	0.4626	0.3839	0.5898	0.5322	0.4258	0.4967	0.3155	0.1314	0.2885	0.2145	0.0117
	High	0.0778	0.3133	0.5243	0.2717	0.3906	0.4963	0.4448	0.6698	0.8683	0.6999	0.7799	0.9883
Histograms for conditional probabilities		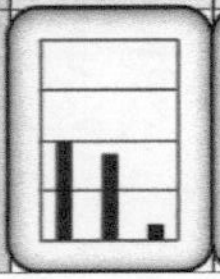	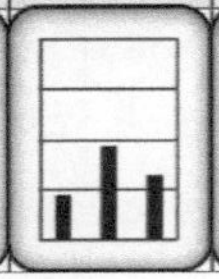	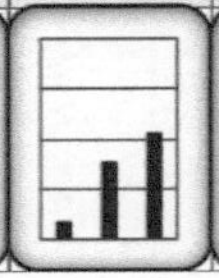	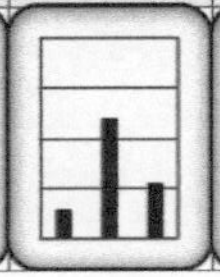	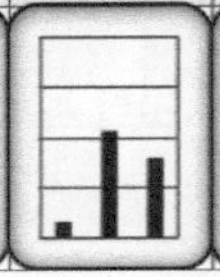	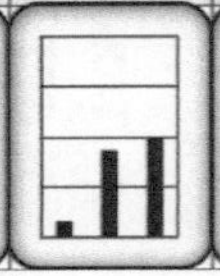	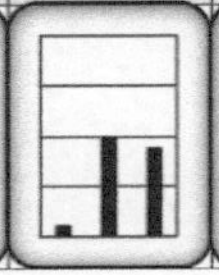	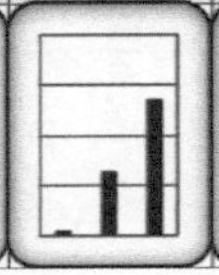	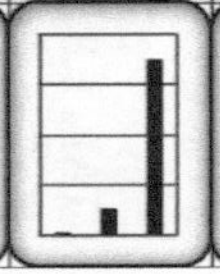	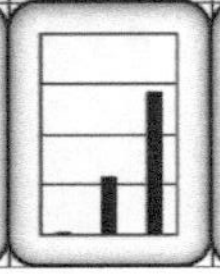	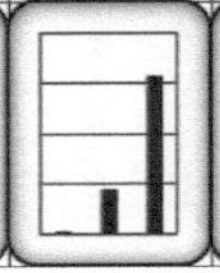	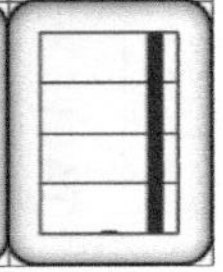

APPENDIX 3.IV: CPTs AND POTENTIALS FOR CASE STUDY 3.5 (TABLES 3.55–3.88)

TABLE 3.55
Potential ϕ (*Replacement Value, Historic, Symbolism*)

Replacement Value		**H**		**L**	
Historic		**H**	**L**	**H**	**L**
Symbolism	H	300	50	180	200
	L	130	90	100	400

TABLE 3.56
Potential ϕ (*Accessibility, Architecture, Site*)

Accessibility		**H**		**L**	
Architecture		**H**	**L**	**H**	**L**
Site	H	300	50	180	200
	L	200	90	100	500

TABLE 3.57
CPT of *# of Occupants*

# of Occupants	H	0.7000
	L	0.3000
Histograms for conditional probabilities		

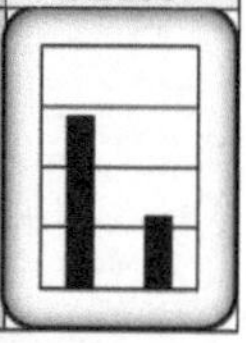

Note: This is same variable in both Consequences and Threat modules.

TABLE 3.58
CPT of *Accessibility*

Accessibility	H	0.50
	L	0.50
Histograms for conditional probabilities		

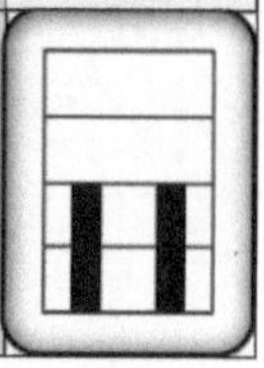

Note: These initial probabilities are obtained from a simple Markov network that is generated by the potential ϕ (*Accessibility, Architecture, Site*) as shown in Table 3.56, and the simple MN is a subset of the CG of Figure 3.16.

TABLE 3.59
CPT of *Architecture*

Architecture	H	0.641
	L	0.359
Histograms for conditional probabilities		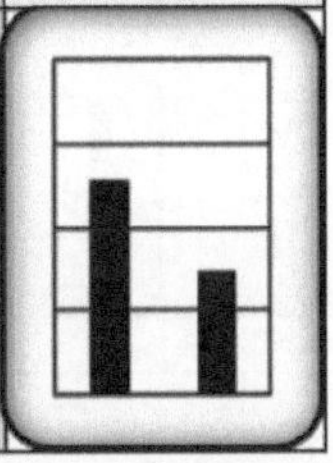

Note: These initial probabilities are obtained from a simple Markov network that is generated by the potential ϕ (*Accessibility, Architecture, Site*) as shown in Table 3.56, and the simple MN is a subset of the CG of Figure 3.16.

TABLE 3.60
CPT of *Business Continuity*

# of Occupants		H	L
Business Continuity	H	0.7918	0.1698
	L	0.2082	0.8302
Histograms for conditional probabilities		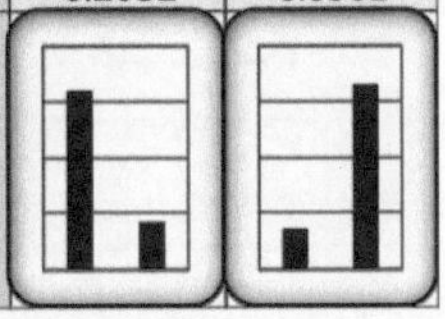	

TABLE 3.61
CPT of *C*

Business continuity		H							
Replacement value		H				L			
Economical		H		L		H		L	
Physical		H	L	H	L	H	L	H	L
C	H	0.9883	0.8820	0.7052	0.5550	0.8302	0.4450	0.4614	0.2948
	L	0.0117	0.1180	0.2948	0.4450	0.1698	0.5550	0.5386	0.7052
Histograms for conditional probabilities									

Business continuity		L							
Replacement value		H				L			
Economical		H		L		H		L	
Physical		H	L	H	L	H	L	H	L
C	H	0.7184	0.5644	0.3085	0.2918	0.3985	0.1873	0.0794	0.0000
	L	0.2816	0.4356	0.6915	0.7082	0.6015	0.8127	0.9206	1.0000
Histograms for conditional probabilities									

TABLE 3.62
CPT of *Economical* (Consequences)

Impact of physical loss		H				L			
Replacement value		H		L		H		L	
# of occupants		H	L	H	L	H	L	H	L
Economical	H	0.9883	0.6852	0.5955	0.1873	0.3399	0.5391	0.5500	0.0329
	L	0.0117	0.3148	0.4045	0.8127	0.6601	0.4609	0.4500	0.9671
Histograms for conditional probabilities									

TABLE 3.63
CPT of *Economical* (Threat)

# of Occupants		H							
Accessibility		H				L			
Symbolism		H		L		H		L	
Occupancy use		H	L	H	L	H	L	H	L
Economical	H	0.9976	0.9292	0.7589	0.5910	0.8872	0.4694	0.5109	0.3069
	L	0.0024	0.0708	0.2411	0.4090	0.1128	0.5306	0.4891	0.6931
Histograms for conditional probabilities									
# of Occupants		L							
Accessibility		H				L			
Symbolism		H		L		H		L	
Occupancy use		H	L	H	L	H	L	H	L
Economical	H	0.7847	0.6069	0.3773	0.2969	0.4074	0.1847	0.0708	0.0000
	L	0.2153	0.3931	0.6227	0.7031	0.5926	0.8153	0.9292	1.0000
Histograms for conditional probabilities									

TABLE 3.64
CPT of *Envelope*

Envelope	H	0.7
	L	0.3
Histograms for conditional probabilities		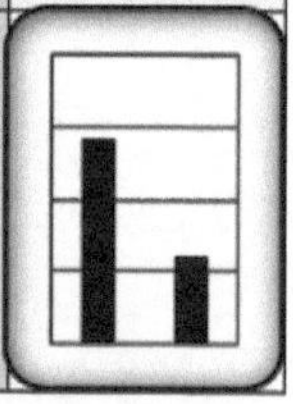

TABLE 3.65
CPT of *General Arch*

Security		H							
Envelope		H				L			
Architecture		H		L		H		L	
Site		H	L	H	L	H	L	H	L
General Arch.	H	0.9905	0.8932	0.7209	0.5691	0.8450	0.4592	0.4865	0.3109
	L	0.0095	0.1068	0.2791	0.4309	0.1550	0.5408	0.5135	0.6891
Histograms for conditional probabilities		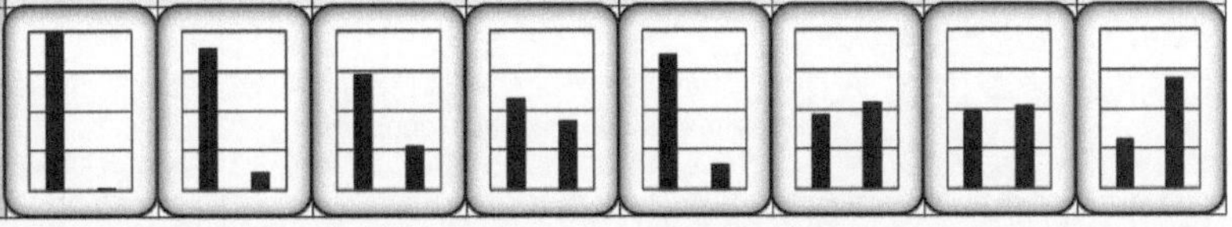							

Security		L							
Envelope		H				L			
Architecture		H		L		H		L	
Site		H	L	H	L	H	L	H	L
General Arch.	H	0.7389	0.5809	0.3557	0.3040	0.4071	0.1995	0.0883	0.0001
	L	0.2611	0.4191	0.6443	0.6960	0.5929	0.8005	0.9117	0.9999
Histograms for conditional probabilities									

TABLE 3.66
CPT of *General Eng.*

Security		H							
Envelope		H				L			
MEP		H		L		H		L	
Structure		H	L	H	L	H	L	H	L
General Eng.	H	0.9944	0.9173	0.7574	0.6034	0.8774	0.4945	0.5481	0.3523
	L	0.0056	0.0827	0.2426	0.3966	0.1226	0.5055	0.4519	0.6477
Histograms for conditional probabilities		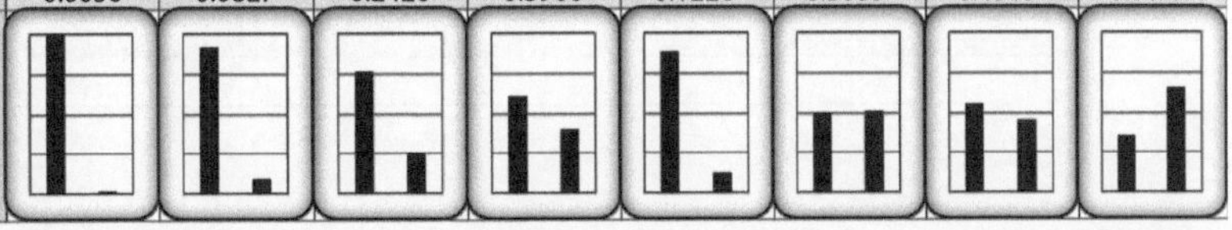							

Security		L							
Envelope		H				L			
MEP		H		L		H		L	
Structure		H	L	H	L	H	L	H	L
General Eng.	H	0.7856	0.6207	0.4801	0.3349	0.4285	0.2319	0.1136	0.0002
	L	0.2144	0.3793	0.5199	0.6651	0.5715	0.7681	0.8864	0.9998
Histograms for conditional probabilities									

TABLE 3.67
CPT of *Historic*

Historic	H	0.606
	L	0.394
Histograms for conditional probabilities		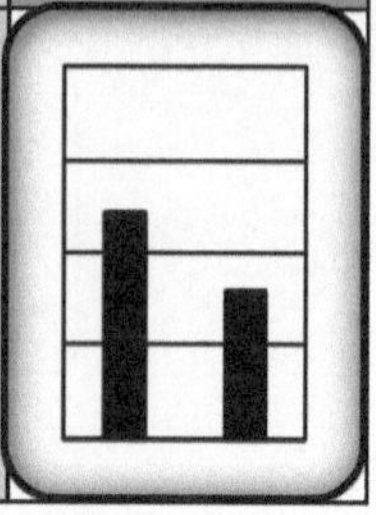

Note: These initial probabilities are obtained from a simple Markov network that is generated by the potential ϕ (*Replacement Value, Historic, Symbolism*) as shown in Table 3.55, and the simple MN is a subset of the CG of Figure 3.16.

TABLE 3.68
CPT of *Impact of Physical Loss*

Business Continuity		H	L
Impact of Physical Loss	H	0.7918	0.1109
	L	0.2082	0.8891
Histograms for conditional probabilities		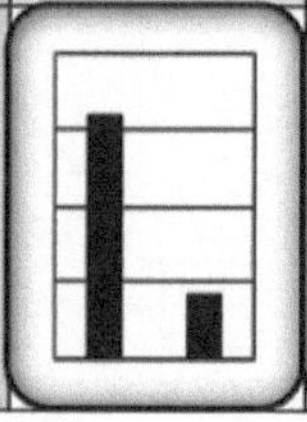	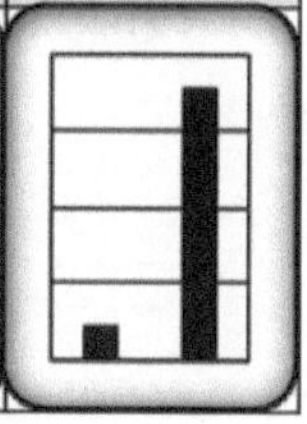

TABLE 3.69
CPT of *Locality Type*

Locality Type	H	0.6
	L	0.4
Histograms for conditional probabilities		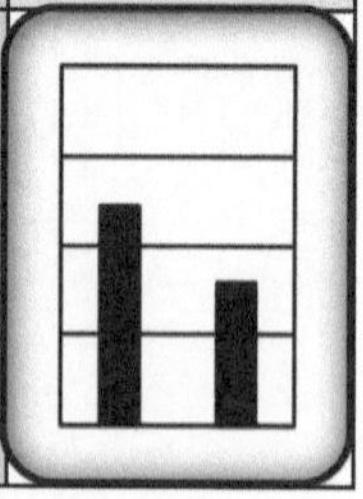

TABLE 3.70
CPT of *MEP*

Structure		H	L
MEP	H	0.3148	0.1180
	L	0.6852	0.8820
Histograms for conditional probabilities		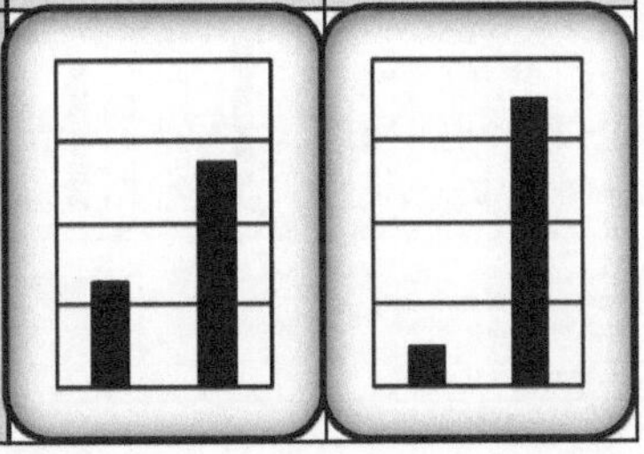	

TABLE 3.71
CPT of *Occupancy Use*

Occupancy use	H	0.6
	L	0.4
Histograms for conditional probabilities		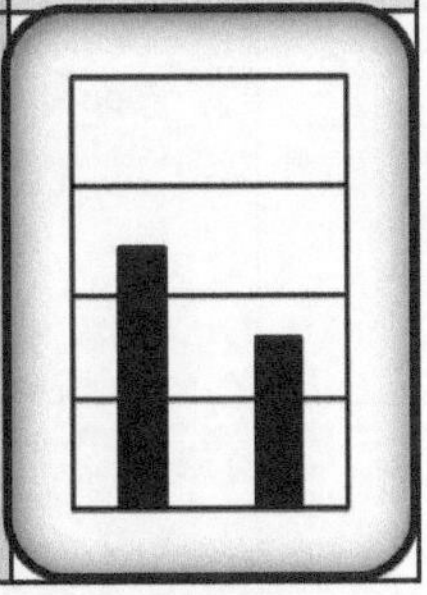

TABLE 3.72
CPT of *Physical* (Consequences)

Business continuity		H				L			
Historic		H		L		H		L	
Locality type		H	L	H	L	H	L	H	L
Physical	H	0.9949	0.7644	0.6852	0.2372	0.3913	0.5778	0.5550	0.0518
	L	0.0051	0.2356	0.3148	0.7628	0.6087	0.4222	0.4450	0.9482
Histograms for conditional probabilities									

TABLE 3.73
CPT of *Physical* (Threat)

Target potential		H							
Target density		H				L			
Population density		H		L		H		L	
# of occupants		H	L	H	L	H	L	H	L
Physical	H	0.9953	0.9042	0.7178	0.5526	0.8517	0.4305	0.4413	0.2638
	L	0.0047	0.0958	0.2822	0.4474	0.1483	0.5695	0.5587	0.7362
Histograms for conditional probabilities		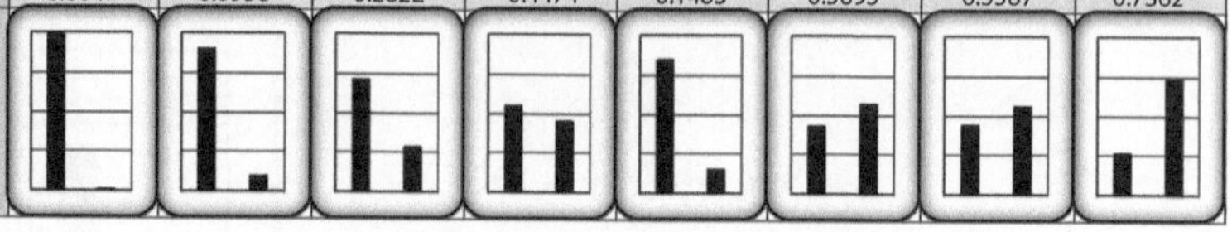							

Target potential		L							
Target density		H				L			
Population density		H		L		H		L	
# of occupants		H	L	H	L	H	L	H	L
Physical	H	0.7309	0.5619	0.2381	0.2647	0.3844	0.1539	0.0513	0.0000
	L	0.2691	0.4381	0.7619	0.7353	0.6156	0.8461	0.9487	1.0000
Histograms for conditional probabilities									

TABLE 3.74
CPT of *Population Density*

Population Density	H	0.55
	L	0.45
Histograms for conditional probabilities		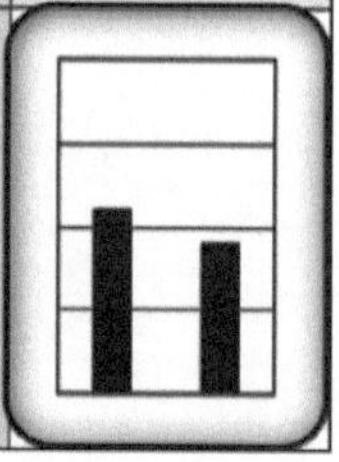

TABLE 3.75
CPT of *Replacement Value*

Replacement Value	H	0.500
	L	0.500
Histograms for conditional probabilities		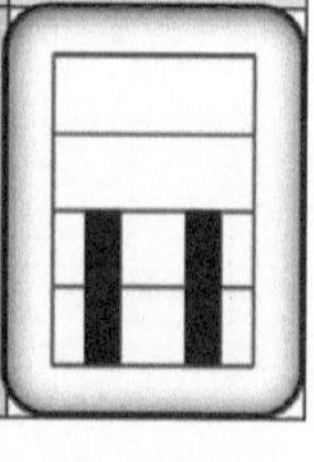

TABLE 3.76
CPT of *Ri*

V		H				L			
T		H		L		H		L	
C		H	L	H	L	H	L	H	L
Ri	H	1.0000	0.8302	0.8302	0.1698	0.8302	0.1698	0.1698	0.0000
	L	0.0000	0.1698	0.1698	0.8302	0.1698	0.8302	0.8302	1.0000
Histograms for conditional probabilities		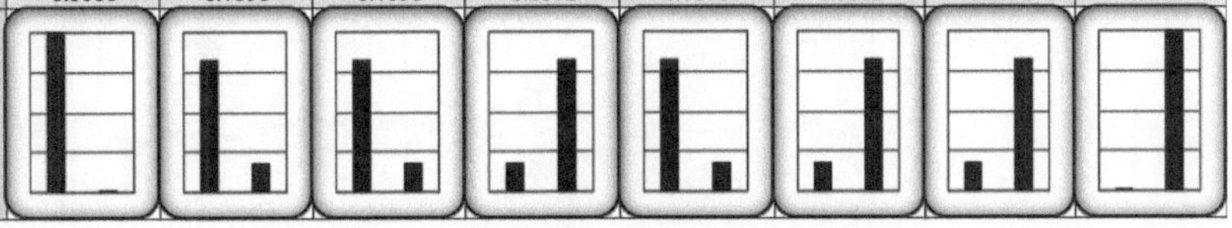							

TABLE 3.77
CPT of *Security*

Structure		H	L
Security	H	0.3913	0.2478
	L	0.6087	0.7522
Histograms for conditional probabilities		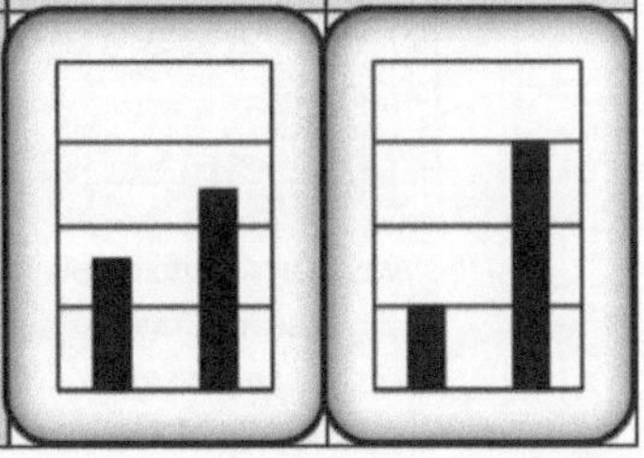	

TABLE 3.78
CPT of *Site*

Site	H	0.615
	L	0.385
Histograms for conditional probabilities		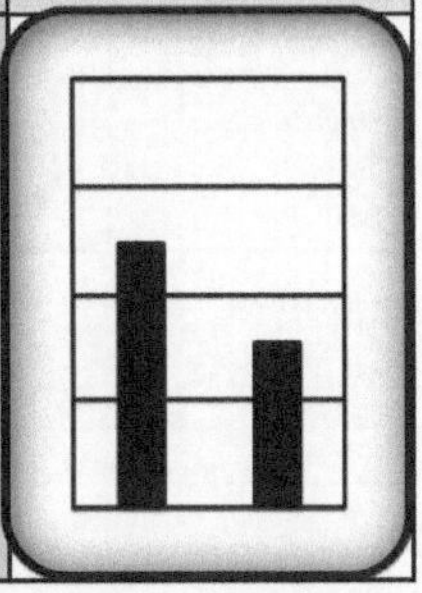

Note: These initial probabilities are obtained from a simple Markov network that is generated by the potential ϕ (*Accessibility, Architecture, Site*) as shown in Table 3.56, and the simple MN is a subset of the CG of Figure 3.16.

TABLE 3.79
CPT of *Structure*

Structure	H	0.75
	L	0.25
Histograms for conditional probabilities		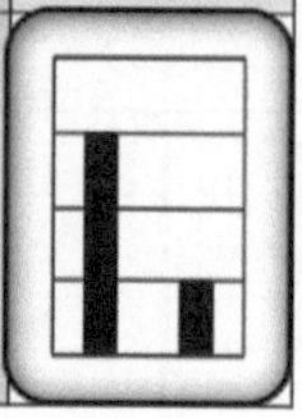

TABLE 3.80
CPT of *Symbolism*

Symbolism	H	0.6761
	L	0.3239
Histograms for conditional probabilities		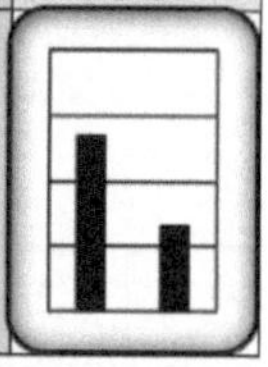

Note: These initial probabilities are obtained from a simple Markov network that is generated by the potential ϕ (*Replacement Value, Historic, Symbolism*) as shown in Table 3.55, and the simple MN is a subset of the CG of Figure 3.16.

TABLE 3.81
CPT of *T*

Economical		H							
Physical		H				L			
Accessibility		H		L		H		L	
# of occupants		H	L	H	L	H	L	H	L
T	H	0.9949	0.9206	0.7628	0.6087	0.8820	0.5000	0.5577	0.3590
	L	0.0051	0.0794	0.2372	0.3913	0.1180	0.5000	0.4423	0.6410
Histograms for conditional probabilities									

Economical		L							
Physical		H				L			
Accessibility		H		L		H		L	
# of occupants		H	L	H	L	H	L	H	L
T	H	0.7924	0.6268	0.5000	0.3399	0.4319	0.2372	0.1180	0.0002
	L	0.2076	0.3732	0.5000	0.6601	0.5681	0.7628	0.8820	0.9998
Histograms for conditional probabilities		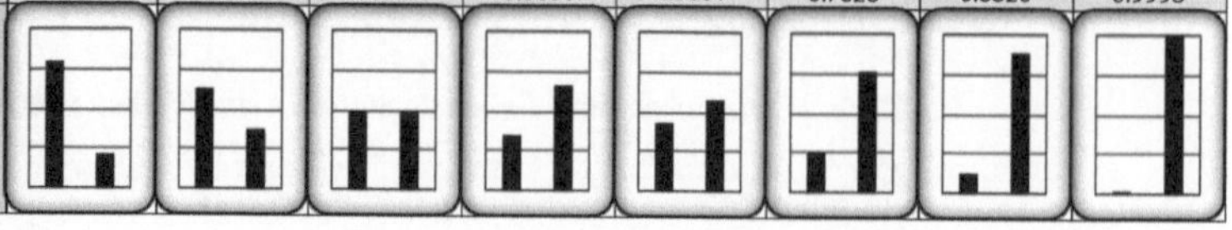							

TABLE 3.82
CPT of *Target Density*

Target density	H	0.7
	L	0.3
Histograms for conditional probabilities		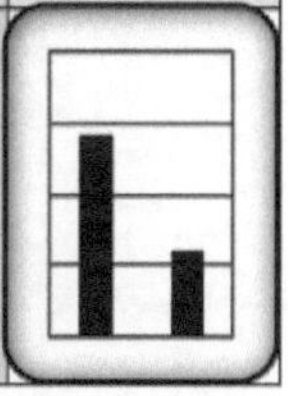

TABLE 3.83
CPT of *Target Potential*

Target potential	H	0.45
	L	0.65
Histograms for conditional probabilities		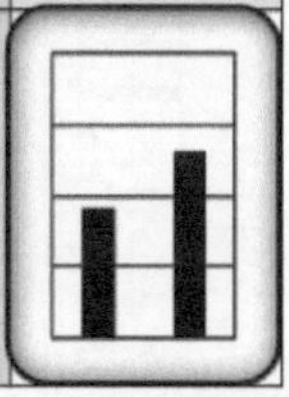

TABLE 3.84
CPT of *V*

Structure	H							
Site	H				L			
General eng.	H		L		H		L	
General arch.	H	L	H	L	H	L	H	L
V H	0.9901	0.9082	0.7592	0.6175	0.8713	0.5199	0.5851	0.3978
V L	0.0099	0.0918	0.2408	0.3825	0.1287	0.4801	0.4149	0.6022
Histograms for conditional probabilities	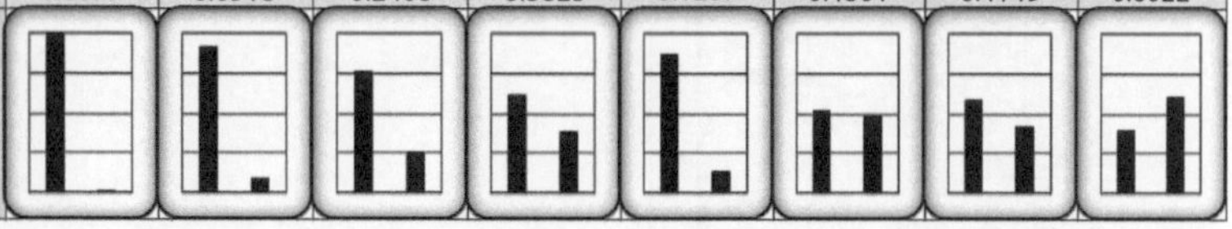							

Structure	L							
Site	H				L			
General eng.	H		L		H		L	
General arch.	H	L	H	L	H	L	H	L
V H	0.7901	0.6363	0.5662	0.3731	0.4498	0.2815	0.1657	0.0032
V L	0.2099	0.3637	0.4338	0.6269	0.5502	0.7185	0.8343	0.9968
Histograms for conditional probabilities	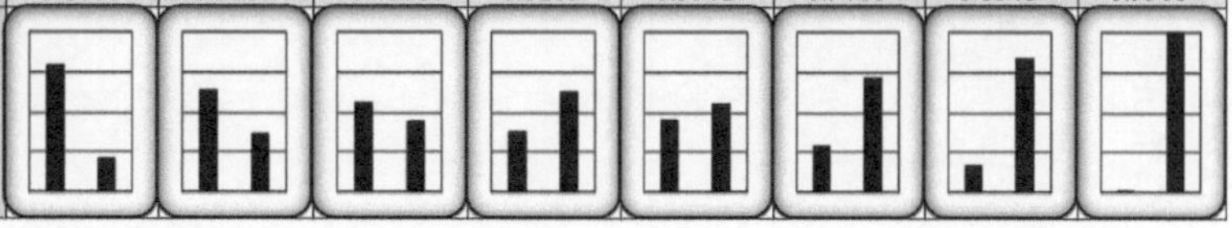							

APPENDIX 3.V: CPTs FOR CASE STUDY 3.6 (TABLES 3.85–3.88)

TABLE 3.85
CPT of C

	CO	Low				Medium				High			
	V.I.	Good	Fair	Poor	Service	Good	Fair	Poor	Service	Good	Fair	Poor	Service
C	Low	0.8240	0.5353	0.0921	0.0355	0.6238	0.3629	0.0828	0.0010	0.4637	0.2670	0.0778	0.0000
	Medium	0.1726	0.4292	0.5527	0.4292	0.3377	0.4931	0.4795	0.2048	0.4148	0.4660	0.4258	0.0117
	High	0.0033	0.0355	0.3553	0.5353	0.0385	0.1439	0.4376	0.7942	0.1215	0.2670	0.4963	0.9883

Histograms for conditional probabilities

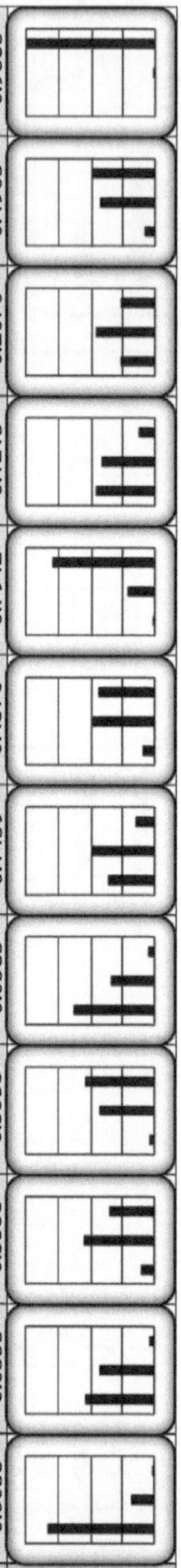

TABLE 3.86
CPT of CO

CO	Low	0.2717
	Medium	0.5898
	High	0.1385

Histograms for conditional probabilities

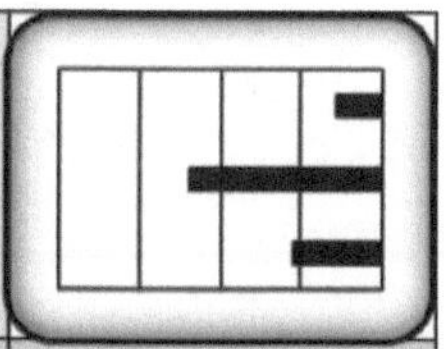

TABLE 3.87
CPT of *Ri*

C		Low				Medium				High			
V.I.		Good	Fair	Poor	Service	Good	Fair	Poor	Service	Good	Fair	Poor	Service
Ri	_1.00E+02	0.7878	0.3548	0.0098	0.0129	0.5231	0.2089	0.0113	0.0001	0.3464	0.1424	0.0129	2.9504E-13
	_1.00E+03	0.2071	0.5519	0.2436	0.1570	0.3976	0.4763	0.1889	0.0210	0.4164	0.3576	0.1570	5.0834E-06
	_1.00E+04	0.0051	0.0920	0.5911	0.4717	0.0759	0.2751	0.5326	0.3775	0.1994	0.3576	0.4717	0.0551
	_1.00E+05	9.8226E-06	0.0014	0.1554	0.3585	0.0035	0.0397	0.2672	0.6014	0.0378	0.1424	0.3585	0.9449
Histograms for conditional probabilities		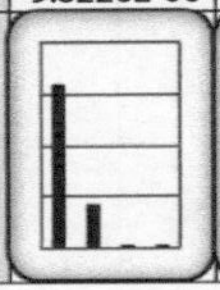	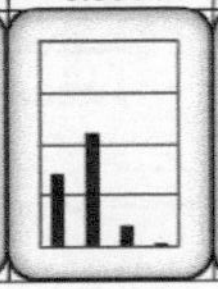	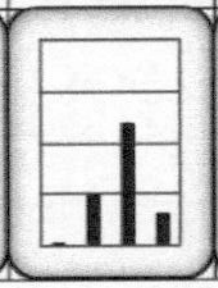	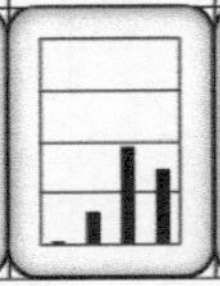	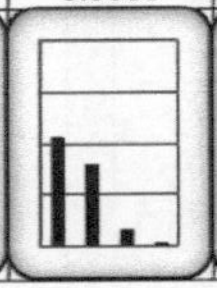	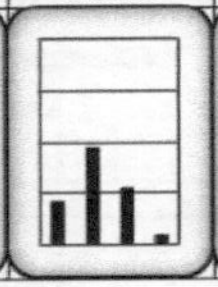	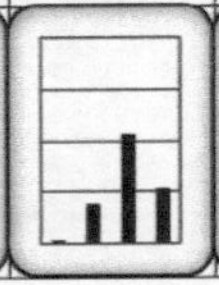	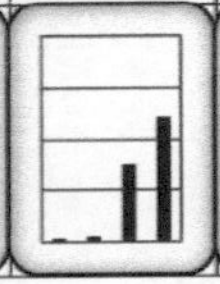	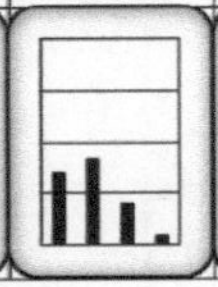	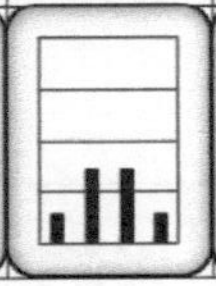	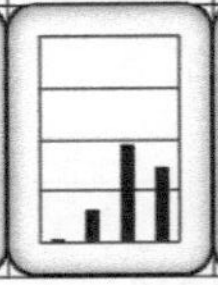	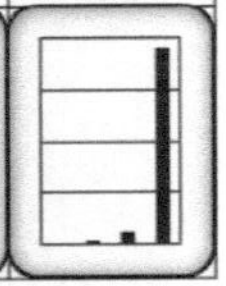

TABLE 3.88
CPT of *V.I.*

V.I.	Good	0.2089
	Fair	0.4763
	Poor	0.2751
	Service	0.0397
Histograms for conditional probabilities		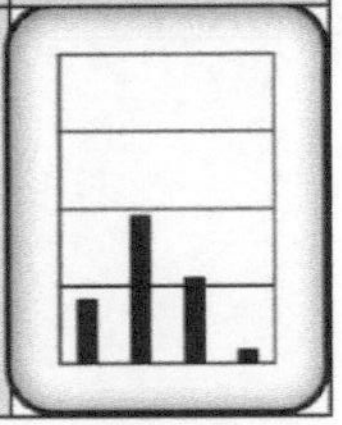

APPENDIX 3.VI: CPTs FOR CASE STUDY 3.7 (TABLES 3.89–3.99)

TABLE 3.89
CPT of *C*

C	Low	0.1
	Medium	0.3
	High	0.6
Histograms for conditional probabilities		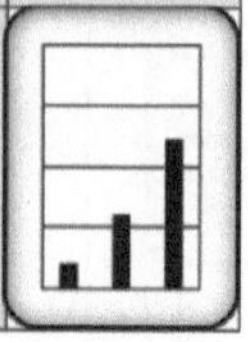

TABLE 3.90
CPT of *C_1*

C		Low	Medium	High
C_1	Low	0.8240	0.1854	0.0063
	Medium	0.1726	0.4525	0.2260
	High	0.0033	0.3621	0.7677
Histograms for conditional probabilities		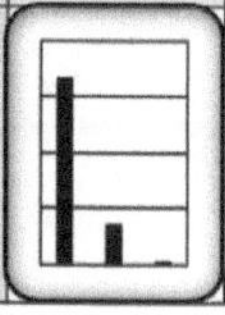	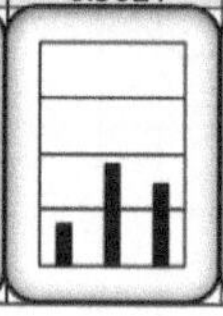	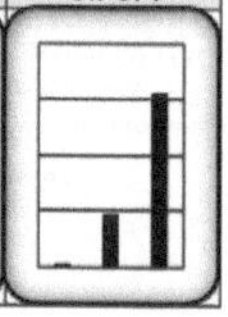

TABLE 3.91
CPT of *C_2*

C		Low	Medium	High
C_2	Low	0.6999	0.3621	0.0355
	Medium	0.2885	0.4525	0.4292
	High	0.0116	0.1854	0.5353
Histograms for conditional probabilities		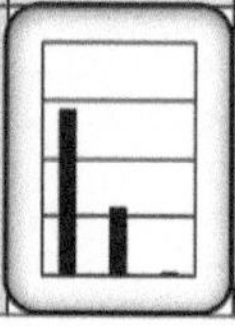	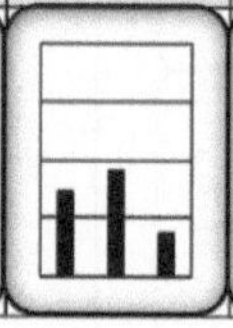	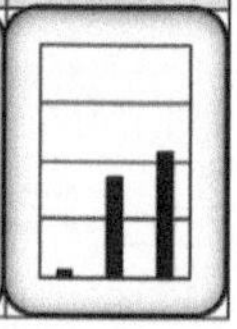

TABLE 3.92
CPT of *C_3*

C_3	Low	0.2717
	Medium	0.5898
	High	0.1385
Histograms for conditional probabilities		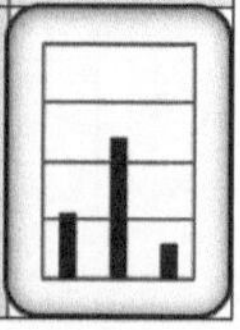

TABLE 3.93
CPT of *Ri*

Ri_3		_1.00E+02							
Ri_2		_1.00E+02				_1.00E+03			
Ri_1		_1.00E+02	_1.00E+03	_1.00E+04	_1.00E+05	_1.00E+02	_1.00E+03	_1.00E+04	_1.00E+05
Ri	_3.00E+02	0.8819	0.4202	0.2453	0.1317	0.4202	0.2453	0.1317	0.1774
	_3.00E+03	0.1180	0.4501	0.3634	0.2505	0.4501	0.3634	0.2505	0.2827
	_3.00E+04	0.0001	0.1217	0.2796	0.3264	0.1217	0.2796	0.3264	0.3088
	_3.00E+05	0.0000	0.0080	0.1116	0.2914	0.0080	0.1116	0.2914	0.2311
Histograms for conditional probabilities		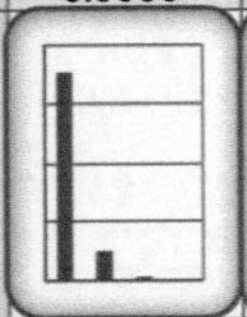	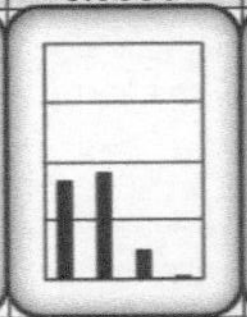	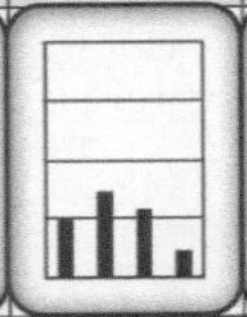	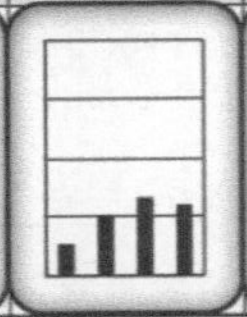	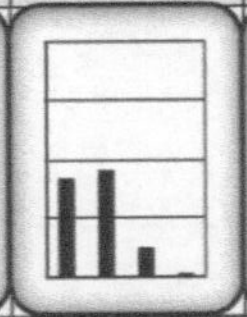		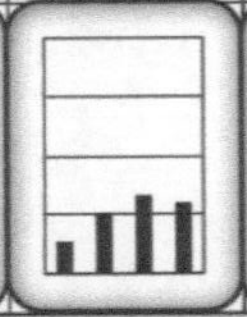	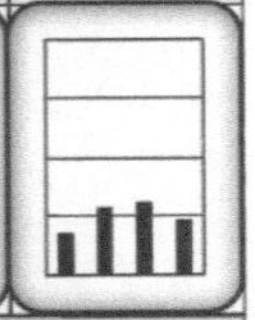
Ri_3		_1.00E+02							
Ri_2		_1.00E+04				_1.00E+05			
Ri_1		_1.00E+02	_1.00E+03	_1.00E+04	_1.00E+05	_1.00E+02	_1.00E+03	_1.00E+04	_1.00E+05
Ri	_3.00E+02	0.2453	0.1317	0.1774	0.2311	0.1317	0.1774	0.2311	0.2914
	_3.00E+03	0.3634	0.2505	0.2827	0.3088	0.2505	0.2827	0.3088	0.3264
	_3.00E+04	0.2796	0.3264	0.3088	0.2827	0.3264	0.3088	0.2827	0.2505
	_3.00E+05	0.1116	0.2914	0.2311	0.1774	0.2914	0.2311	0.1774	0.1317
Histograms for conditional probabilities		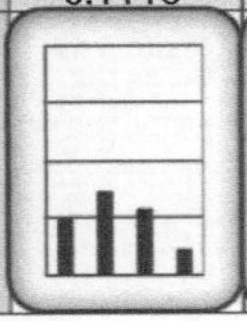	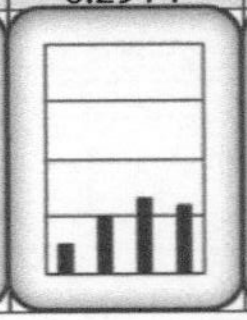	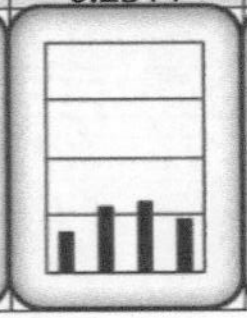	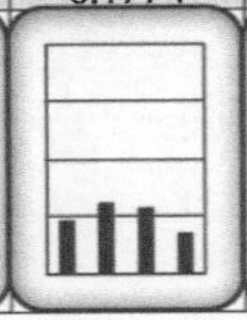	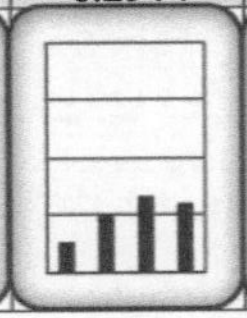	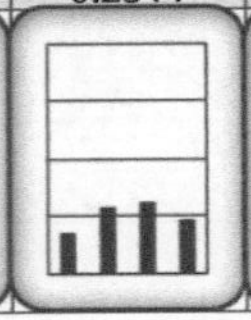		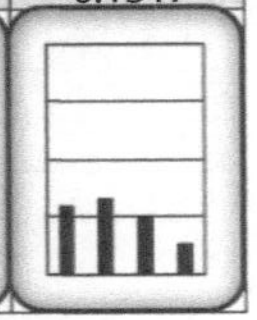

(Continued)

TABLE 3.93 ***(Continued)***
CPT of *Ri*

Ri_3		_1.00E+03							
Ri_2		_1.00E+02				_1.00E+03			
Ri_1		_1.00E+02	_1.00E+03	_1.00E+04	_1.00E+05	_1.00E+02	_1.00E+03	_1.00E+04	_1.00E+05
Ri	_3.00E+02	0.4202	0.2453	0.1317	0.1774	0.2453	0.1317	0.1774	0.2311
	_3.00E+03	0.4501	0.3634	0.2505	0.2827	0.3634	0.2505	0.2827	0.3088
	_3.00E+04	0.1217	0.2796	0.3264	0.3088	0.2796	0.3264	0.3088	0.2827
	_3.00E+05	0.0080	0.1116	0.2914	0.2311	0.1116	0.2914	0.2311	0.1774
Histograms for conditional probabilities		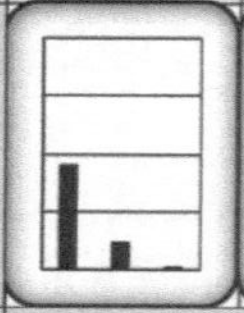	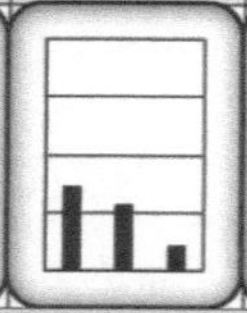	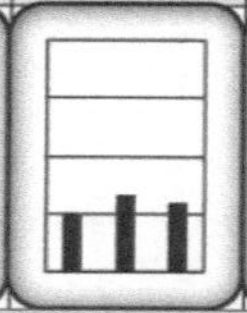	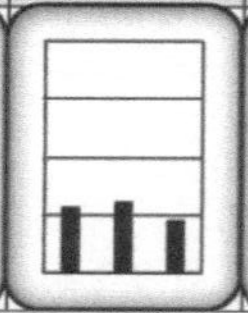	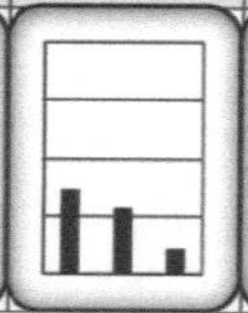		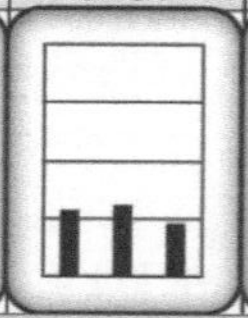	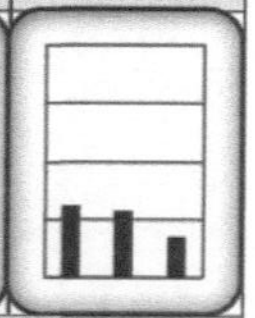
Ri_3		_1.00E+03							
Ri_2		_1.00E+04				_1.00E+05			
Ri_1		_1.00E+02	_1.00E+03	_1.00E+04	_1.00E+05	_1.00E+02	_1.00E+03	_1.00E+04	_1.00E+05
Ri	_3.00E+02	0.1317	0.1774	0.2311	0.2914	0.1774	0.2311	0.2914	0.1116
	_3.00E+03	0.2505	0.2827	0.3088	0.3264	0.2827	0.3088	0.3264	0.2796
	_3.00E+04	0.3264	0.3088	0.2827	0.2505	0.3088	0.2827	0.2505	0.3634
	_3.00E+05	0.2914	0.2311	0.1774	0.1317	0.2311	0.1774	0.1317	0.2453
Histograms for conditional probabilities		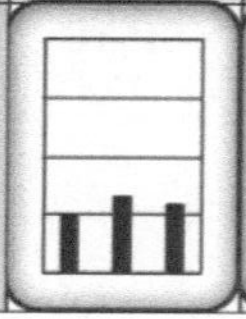	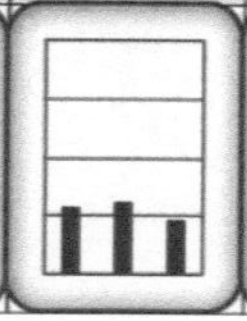	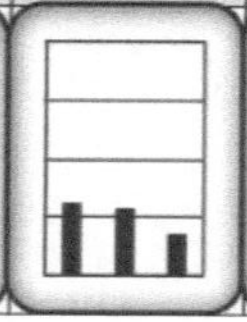	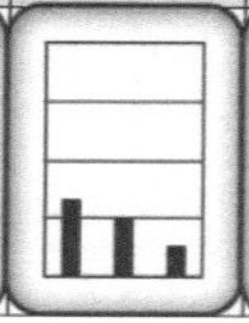	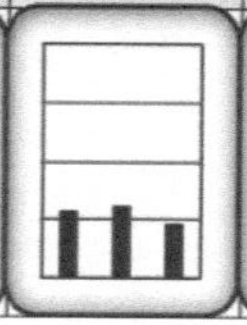	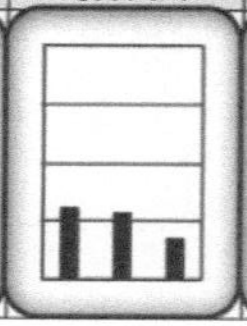	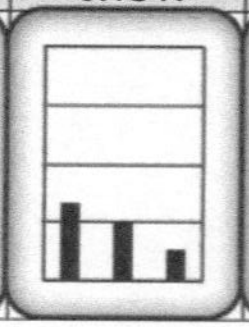	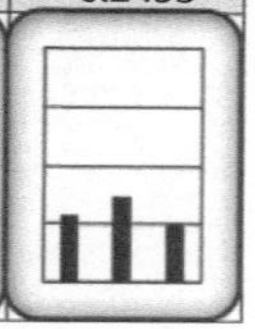

(Continued)

TABLE 3.93 ***(Continued)***
CPT of *Ri*

Ri_3		_1.00E+04							
Ri_2		_1.00E+02				_1.00E+03			
Ri_1		_1.00E+02	_1.00E+03	_1.00E+04	_1.00E+05	_1.00E+02	_1.00E+03	_1.00E+04	_1.00E+05
Ri	_3.00E+02	0.2453	0.1317	0.1774	0.2311	0.1317	0.1774	0.2311	0.2914
	_3.00E+03	0.3634	0.2505	0.2827	0.3088	0.2505	0.2827	0.3088	0.3264
	_3.00E+04	0.2796	0.3264	0.3088	0.2827	0.3264	0.3088	0.2827	0.2505
	_3.00E+05	0.1116	0.2914	0.2311	0.1774	0.2914	0.2311	0.1774	0.1317
Histograms for conditional probabilities		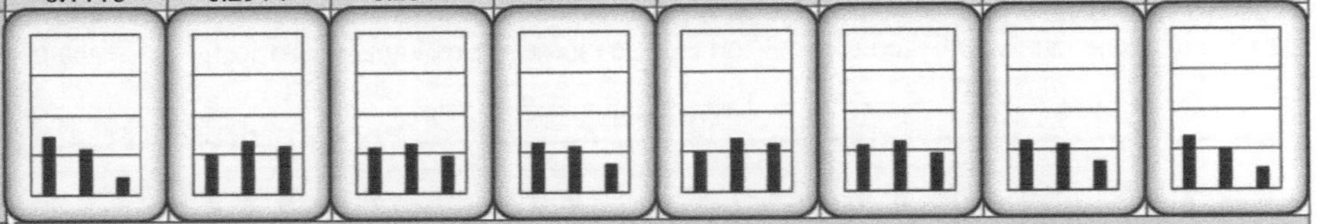							
Ri_3		_1.00E+04							
Ri_2		_1.00E+04				_1.00E+05			
Ri_1		_1.00E+02	_1.00E+03	_1.00E+04	_1.00E+05	_1.00E+02	_1.00E+03	_1.00E+04	_1.00E+05
Ri	_3.00E+02	0.1774	0.2311	0.2914	0.1116	0.2311	0.2914	0.1116	0.0080
	_3.00E+03	0.2827	0.3088	0.3264	0.2796	0.3088	0.3264	0.2796	0.1217
	_3.00E+04	0.3088	0.2827	0.2505	0.3634	0.2827	0.2505	0.3634	0.4501
	_3.00E+05	0.2311	0.1774	0.1317	0.2453	0.1774	0.1317	0.2453	0.4202
Histograms for conditional probabilities									

(Continued)

TABLE 3.93 *(Continued)*
CPT of *Ri*

Ri_3		_1.00E+05							
Ri_2		_1.00E+02				_1.00E+03			
Ri_1		_1.00E+02	_1.00E+03	_1.00E+04	_1.00E+05	_1.00E+02	_1.00E+03	_1.00E+04	_1.00E+05
Ri	_3.00E+02	0.1317	0.1774	0.2311	0.2914	0.1774	0.2311	0.2914	0.1116
	_3.00E+03	0.2505	0.2827	0.3088	0.3264	0.2827	0.3088	0.3264	0.2796
	_3.00E+04	0.3264	0.3088	0.2827	0.2505	0.3088	0.2827	0.2505	0.3634
	_3.00E+05	0.2914	0.2311	0.1774	0.1317	0.2311	0.1774	0.1317	0.2453
Histograms for conditional probabilities		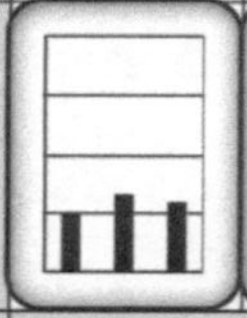	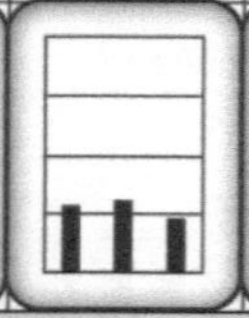	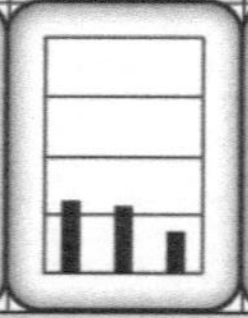		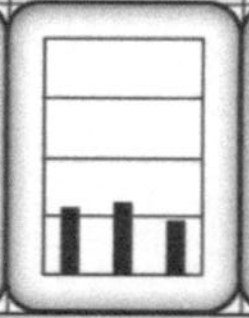	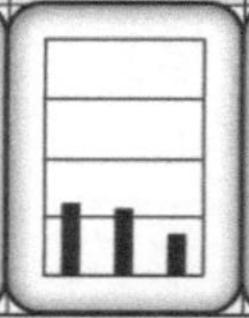	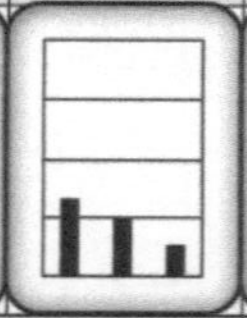	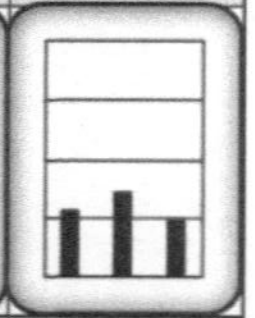
Ri_3		_1.00E+05							
Ri_2		_1.00E+04				_1.00E+05			
Ri_1		_1.00E+02	_1.00E+03	_1.00E+04	_1.00E+05	_1.00E+02	_1.00E+03	_1.00E+04	_1.00E+05
Ri	_3.00E+02	0.2311	0.2914	0.1116	0.0080	0.2914	0.1116	0.0080	0.0000
	_3.00E+03	0.3088	0.3264	0.2796	0.1217	0.3264	0.2796	0.1217	0.0001
	_3.00E+04	0.2827	0.2505	0.3634	0.4501	0.2505	0.3634	0.4501	0.1180
	_3.00E+05	0.1774	0.1317	0.2453	0.4202	0.1317	0.2453	0.4202	0.8819
Histograms for conditional probabilities		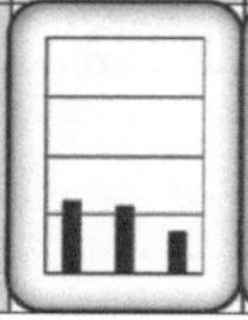	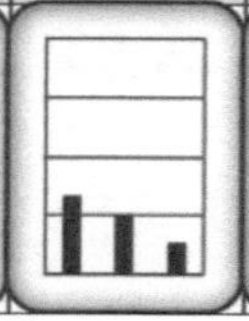	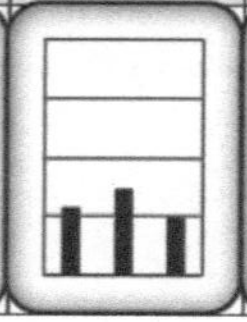		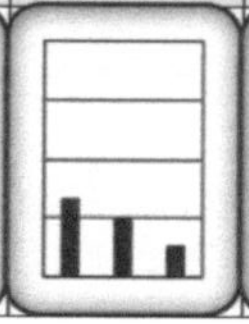	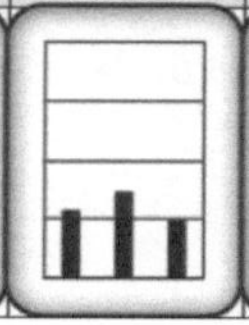	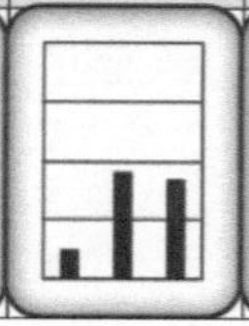	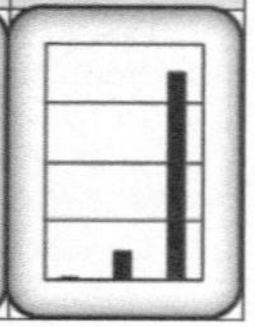

TABLE 3.94
CPT of *Ri_1*

C_1		Low				Medium				High			
V.I._1		Good	Fair	Poor	Service	Good	Fair	Poor	Service	Good	Fair	Poor	Service
Ri_1	_1.00E+02	0.7878	0.3548	0.0098	0.0129	0.5231	0.2089	0.0113	0.0001	0.3464	0.1424	0.0129	0.0000
	_1.00E+03	0.2071	0.5519	0.2436	0.1570	0.3976	0.4763	0.1889	0.0210	0.4164	0.3576	0.1570	0.0000
	_1.00E+04	0.0051	0.0920	0.5911	0.4717	0.0759	0.2751	0.5326	0.3775	0.1994	0.3576	0.4717	0.0551
	_1.00E+05	0.0000	0.0014	0.1554	0.3585	0.0035	0.0397	0.2672	0.6014	0.0378	0.1424	0.3585	0.9449
Histograms for conditional probabilities													

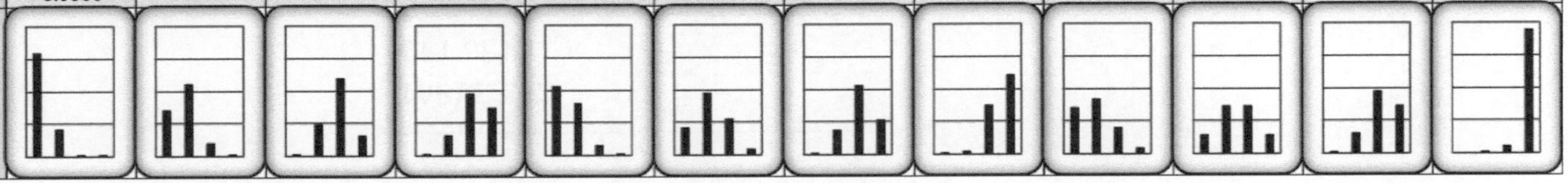

TABLE 3.95
CPT of *Ri_2*

C_2		Low				Medium				High			
V.I._2		Good	Fair	Poor	Service	Good	Fair	Poor	Service	Good	Fair	Poor	Service
Ri_2	_1.00E+02	0.6999	0.2451	0.0039	0.0068	0.3585	0.1487	0.0023	0.0000	0.2662	0.1065	0.0068	0.0000
	_1.00E+03	0.2884	0.5947	0.1564	0.1112	0.4717	0.4468	0.0806	0.0093	0.4140	0.3217	0.1112	0.0000
	_1.00E+04	0.0116	0.1564	0.5947	0.4410	0.1570	0.3397	0.4585	0.2793	0.2566	0.3867	0.4410	0.0362
	_1.00E+05	0.0000	0.0039	0.2451	0.4410	0.0129	0.0648	0.4585	0.7114	0.0631	0.1852	0.4410	0.9638
Histograms for conditional probabilities													

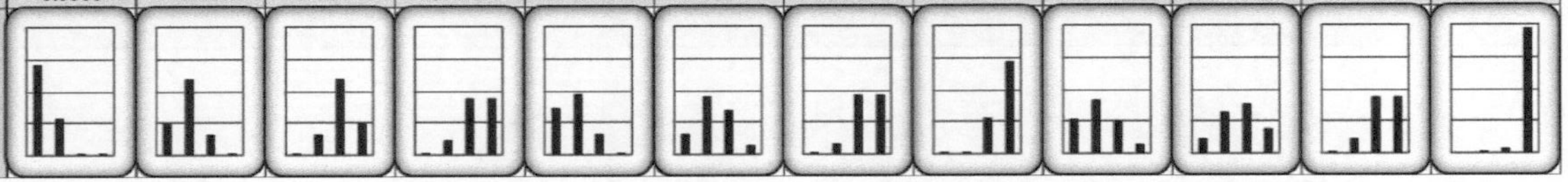

TABLE 3.96
CPT of *Ri_3*

C_3		Low				Medium				High			
V.I._3		Good	Fair	Poor	Service	Good	Fair	Poor	Service	Good	Fair	Poor	Service
Ri_3	_1.00E+02	0.9206	0.3076	0.0002	0.0014	0.5231	0.2089	0.0231	0.0001	0.3309	0.1709	0.0378	0.0000
	_1.00E+03	0.0794	0.6696	0.1584	0.0920	0.3976	0.4763	0.2125	0.0249	0.3772	0.3291	0.1994	0.0005
	_1.00E+04	0.0000	0.0228	0.7746	0.5519	0.0759	0.2751	0.4845	0.3815	0.2233	0.3291	0.4164	0.1319
	_1.00E+05	0.0000	0.0000	0.0668	0.3548	0.0035	0.0397	0.2799	0.5935	0.0685	0.1709	0.3464	0.8676
Histograms for conditional probabilities		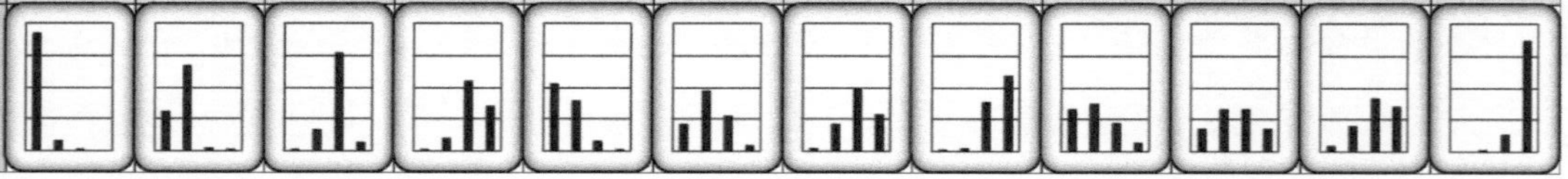											

TABLE 3.97
CPT of *V.I._1*

V.I._1	Good	0.2089
	Fair	0.4763
	Poor	0.2751
	Service	0.0397
Histograms for conditional probabilities		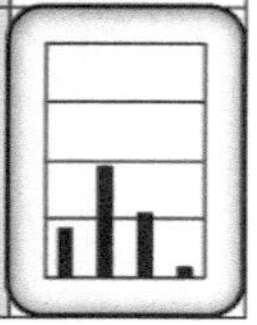

TABLE 3.98
CPT of *V.I._2*

V.I._2	Good	0.3585
	Fair	0.4717
	Poor	0.1570
	Service	0.0129
Histograms for conditional probabilities		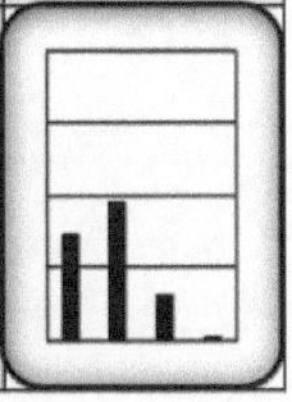

TABLE 3.99
CPT of *V.I._3*

V.I._3	Good	0.3607
	Fair	0.3344
	Poor	0.2124
	Service	0.0924
Histograms for conditional probabilities		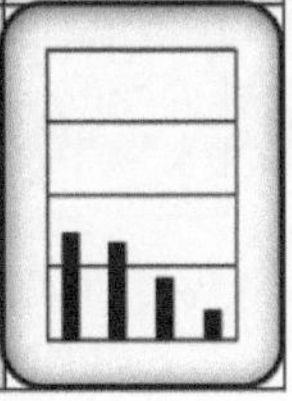

APPENDIX 3.VII: CPTs AND POTENTIALS FOR CASE STUDY 3.8 (TABLES 3.100–3.110)

TABLE 3.100
Potential ϕ (*C_1, C_2, C_3*)

C_3		Low			Medium			High		
C_2		Low	Medium	High	Low	Medium	High	Low	Medium	High
C_1	Low	220	190	140	210	150	130	170	120	100
	Medium	200	180	150	200	170	200	190	200	180
	High	120	150	180	170	200	250	200	240	300

TABLE 3.101
CPT of *C_1*

C_1	Low	0.357142869
	Medium	0.351190489
	High	0.291666677
Histograms for conditional probabilities		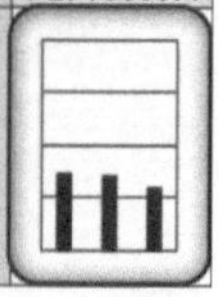

Note: These initial probabilities are obtained from a simple Markov network that is generated by the potential ϕ (*C_1*, *C_2*, *C_3*) as shown in Table 3.100, and the simple MN is a subset of the CG of Figure 3.36.

TABLE 3.102
CPT of *C_2*

C_2	Low	0.32142858
	Medium	0.345238108
	High	0.333333347
Histograms for conditional probabilities		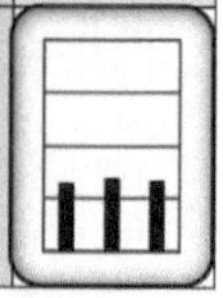

Note: These initial probabilities are obtained from a simple Markov network that is generated by the potential ϕ (*C_1*, *C_2*, *C_3*) as shown in Table 3.100, and the simple MN is a subset of the CG of Figure 3.36.

TABLE 3.103
CPT of *C_3*

C_3	Low	0.333333345
	Medium	0.333333345
	High	0.333333345
Histograms for conditional probabilities		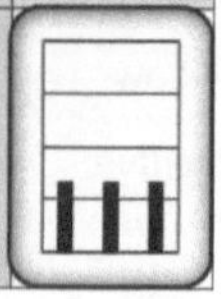

Note: These initial probabilities are obtained from a simple Markov network that is generated by the potential ϕ (*C_1*, *C_2*, *C_3*) as shown in Table 3.100, and the simple MN is a subset of the CG of Figure 3.36.

TABLE 3.104
CPT of *Ri*

Ri_3		_1.00E+02							
Ri_2		_1.00E+02				_1.00E+03			
Ri_1		_1.00E+02	_1.00E+03	_1.00E+04	_1.00E+05	_1.00E+02	_1.00E+03	_1.00E+04	_1.00E+05
Ri	_3.00E+02	0.881869797	0.420240782	0.2452786	0.131704486	0.420240782	0.2452786	0.131704486	0.177354386
	_3.00E+03	0.118031341	0.450063499	0.363439136	0.250498012	0.450063499	0.363439136	0.250498012	0.282734672
	_3.00E+04	9.88618E-05	0.121651878	0.279636601	0.326398478	0.121651878	0.279636601	0.326398478	0.308810682
	_3.00E+05	2.2947E-10	0.00804384	0.111645663	0.291399024	0.00804384	0.111645663	0.291399024	0.231100259
Histograms for conditional probabilities		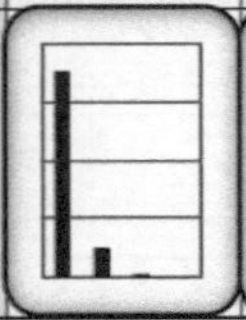	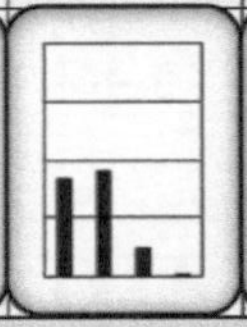	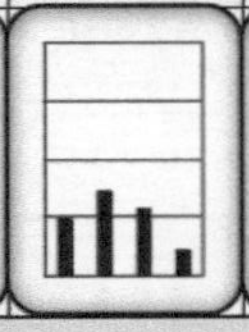	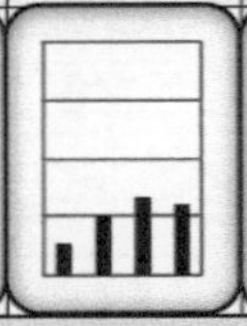	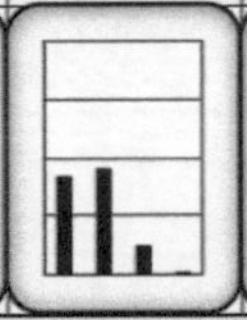	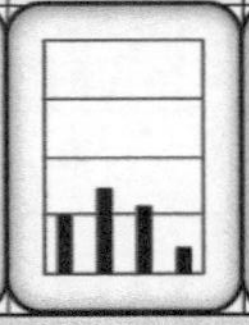	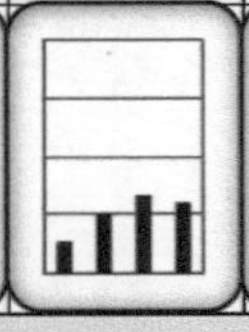	
Ri_3		_1.00E+02							
Ri_2		_1.00E+04				_1.00E+05			
Ri_1		_1.00E+02	_1.00E+03	_1.00E+04	_1.00E+05	_1.00E+02	_1.00E+03	_1.00E+04	_1.00E+05
Ri	_3.00E+02	0.2452786	0.131704486	0.177354386	0.231100259	0.131704486	0.177354386	0.231100259	0.291399024
	_3.00E+03	0.363439136	0.250498012	0.282734672	0.308810682	0.250498012	0.282734672	0.308810682	0.326398478
	_3.00E+04	0.279636601	0.326398478	0.308810682	0.282734672	0.326398478	0.308810682	0.282734672	0.250498012
	_3.00E+05	0.111645663	0.291399024	0.231100259	0.177354386	0.291399024	0.231100259	0.177354386	0.131704486
Histograms for conditional probabilities		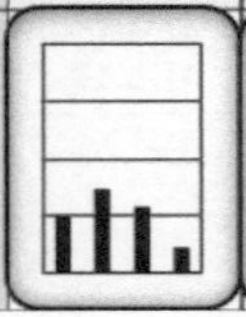		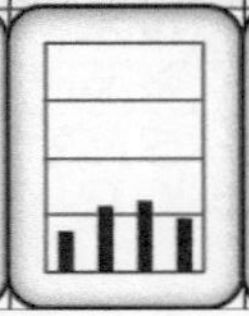	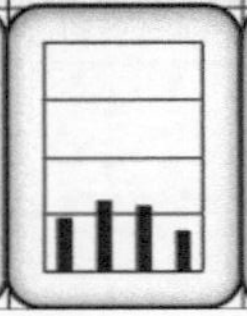		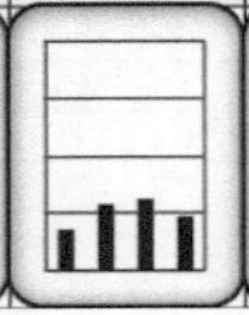	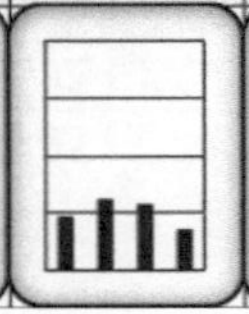	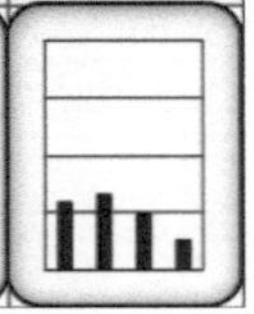

(Continued)

TABLE 3.104 ***(Continued)***
CPT of *Ri*

Ri_3		_1.00E+03							
Ri_2		_1.00E+02				_1.00E+03			
Ri_1		_1.00E+02	_1.00E+03	_1.00E+04	_1.00E+05	_1.00E+02	_1.00E+03	_1.00E+04	_1.00E+05
Ri	_3.00E+02	0.420240782	0.2452786	0.131704486	0.177354386	0.2452786	0.131704486	0.177354386	0.231100259
	_3.00E+03	0.450063499	0.363439136	0.250498012	0.282734672	0.363439136	0.250498012	0.282734672	0.308810682
	_3.00E+04	0.121651878	0.279636601	0.326398478	0.308810682	0.279636601	0.326398478	0.308810682	0.282734672
	_3.00E+05	0.00804384	0.111645663	0.291399024	0.231100259	0.111645663	0.291399024	0.231100259	0.177354386
Histograms for conditional probabilities			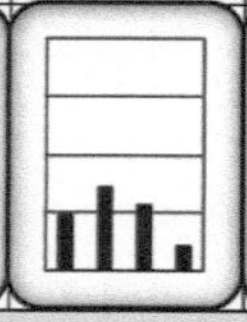	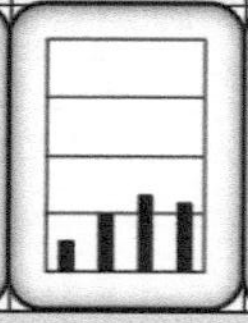	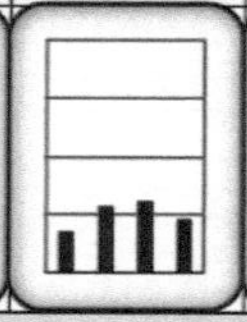	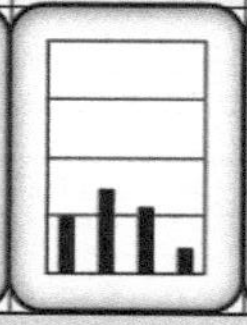		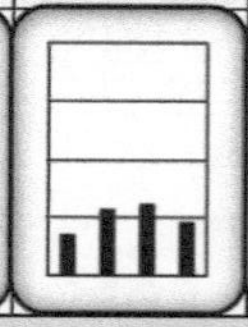	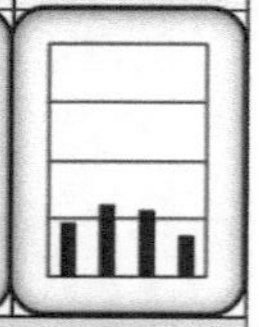
Ri_3		_1.00E+03							
Ri_2		_1.00E+04				_1.00E+05			
Ri_1		_1.00E+02	_1.00E+03	_1.00E+04	_1.00E+05	_1.00E+02	_1.00E+03	_1.00E+04	_1.00E+05
Ri	_3.00E+02	0.131704486	0.177354386	0.231100259	0.291399024	0.177354386	0.231100259	0.291399024	0.111645663
	_3.00E+03	0.250498012	0.282734672	0.308810682	0.326398478	0.282734672	0.308810682	0.326398478	0.279636601
	_3.00E+04	0.326398478	0.308810682	0.282734672	0.250498012	0.308810682	0.282734672	0.250498012	0.363439136
	_3.00E+05	0.291399024	0.231100259	0.177354386	0.131704486	0.231100259	0.177354386	0.131704486	0.2452786
Histograms for conditional probabilities		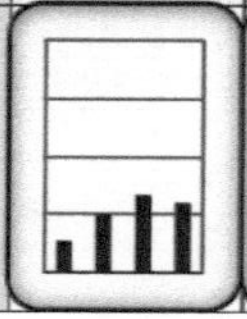	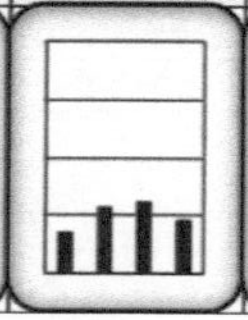	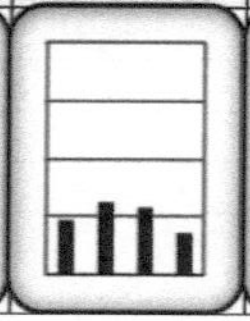	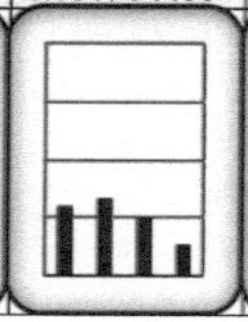	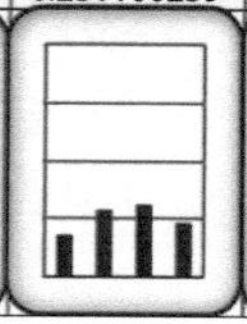	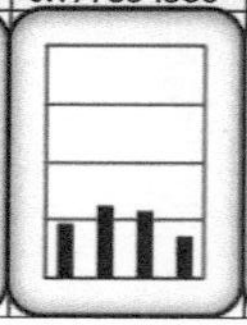	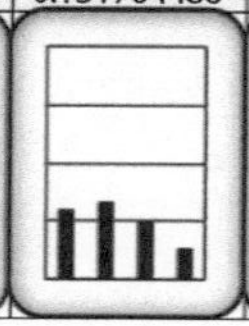	

(Continued)

TABLE 3.104 ***(Continued)***
CPT of *Ri*

Ri_3		_1.00E+04							
Ri_2		_1.00E+02				_1.00E+03			
Ri_1		_1.00E+02	_1.00E+03	_1.00E+04	_1.00E+05	_1.00E+02	_1.00E+03	_1.00E+04	_1.00E+05
Ri	_3.00E+02	0.2452786	0.131704486	0.177354386	0.231100259	0.131704486	0.177354386	0.231100259	0.291399024
	_3.00E+03	0.363439136	0.250498012	0.282734672	0.308810682	0.250498012	0.282734672	0.308810682	0.326398478
	_3.00E+04	0.279636601	0.326398478	0.308810682	0.282734672	0.326398478	0.308810682	0.282734672	0.250498012
	_3.00E+05	0.111645663	0.291399024	0.231100259	0.177354386	0.291399024	0.231100259	0.177354386	0.131704486
Histograms for conditional probabilities		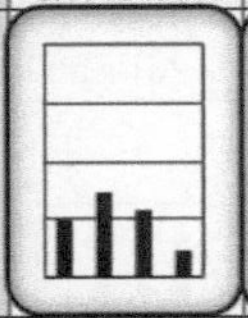	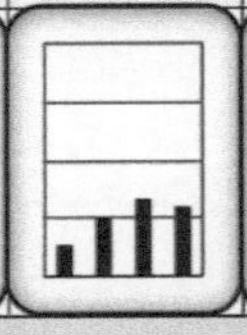	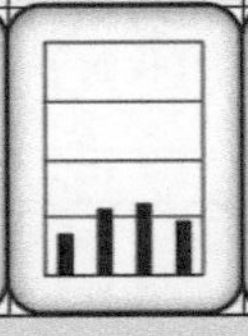	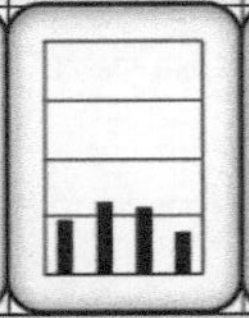	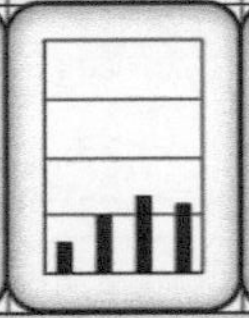	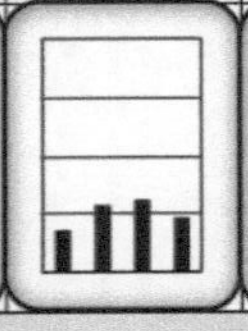	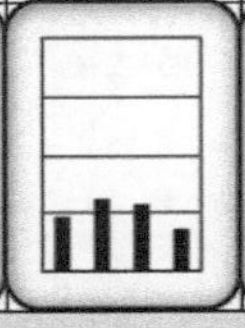	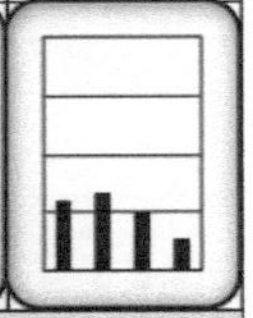
Ri_3		_1.00E+04							
Ri_2		_1.00E+04				_1.00E+05			
Ri_1		_1.00E+02	_1.00E+03	_1.00E+04	_1.00E+05	_1.00E+02	_1.00E+03	_1.00E+04	_1.00E+05
Ri	_3.00E+02	0.177354386	0.231100259	0.291399024	0.111645663	0.231100259	0.291399024	0.111645663	0.00804384
	_3.00E+03	0.282734672	0.308810682	0.326398478	0.279636601	0.308810682	0.326398478	0.279636601	0.121651878
	_3.00E+04	0.308810682	0.282734672	0.250498012	0.363439136	0.282734672	0.250498012	0.363439136	0.450063499
	_3.00E+05	0.231100259	0.177354386	0.131704486	0.2452786	0.177354386	0.131704486	0.2452786	0.420240782
Histograms for conditional probabilities		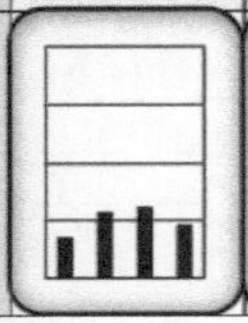	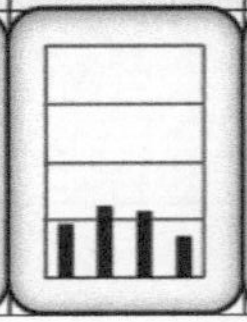	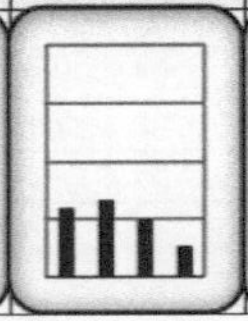	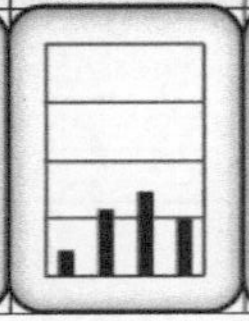	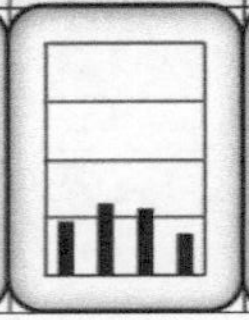	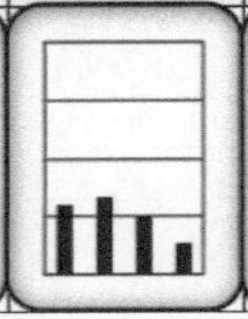	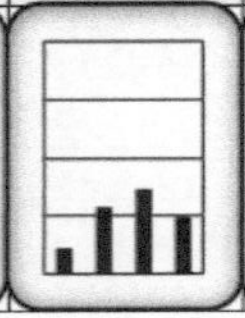	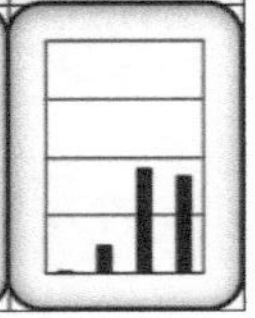

(*Continued*)

TABLE 3.104 ***(Continued)***
CPT of *Ri*

Ri_3		_1.00E+05							
Ri_2		_1.00E+02				_1.00E+03			
Ri_1		_1.00E+02	_1.00E+03	_1.00E+04	_1.00E+05	_1.00E+02	_1.00E+03	_1.00E+04	_1.00E+05
Ri	_3.00E+02	0.131704486	0.177354386	0.231100259	0.291399024	0.177354386	0.231100259	0.291399024	0.111645663
	_3.00E+03	0.250498012	0.282734672	0.308810682	0.326398478	0.282734672	0.308810682	0.326398478	0.279636601
	_3.00E+04	0.326398478	0.308810682	0.282734672	0.250498012	0.308810682	0.282734672	0.250498012	0.363439136
	_3.00E+05	0.291399024	0.231100259	0.177354386	0.131704486	0.231100259	0.177354386	0.131704486	0.2452786
Histograms for conditional probabilities		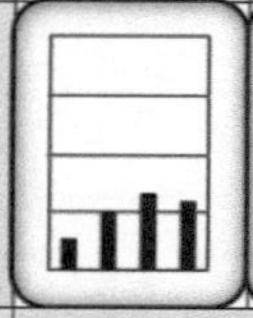	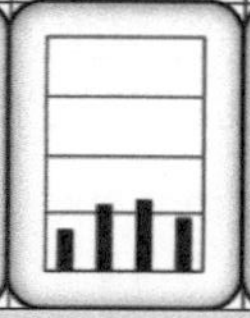	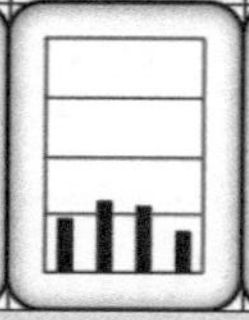	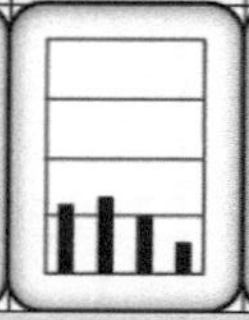	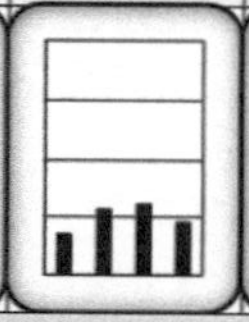	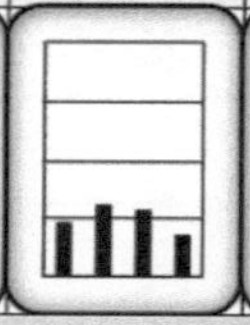	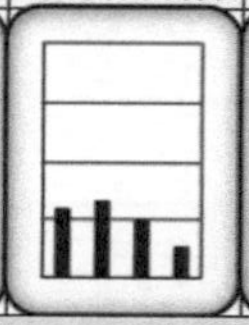	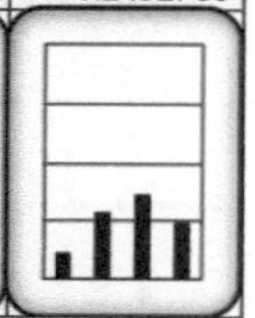
Ri_3		_1.00E+05							
Ri_2		_1.00E+04				_1.00E+05			
Ri_1		_1.00E+02	_1.00E+03	_1.00E+04	_1.00E+05	_1.00E+02	_1.00E+03	_1.00E+04	_1.00E+05
Ri	_3.00E+02	0.231100259	0.291399024	0.111645663	0.00804384	0.291399024	0.111645663	0.00804384	2.2947E-10
	_3.00E+03	0.308810682	0.326398478	0.279636601	0.121651878	0.326398478	0.279636601	0.121651878	9.88618E-05
	_3.00E+04	0.282734672	0.250498012	0.363439136	0.450063499	0.250498012	0.363439136	0.450063499	0.118031341
	_3.00E+05	0.177354386	0.131704486	0.2452786	0.420240782	0.131704486	0.2452786	0.420240782	0.881869797
Histograms for conditional probabilities		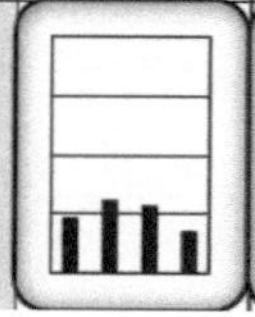	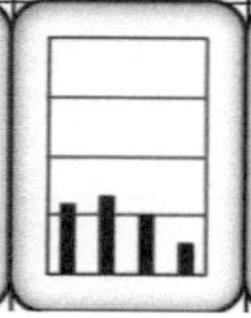			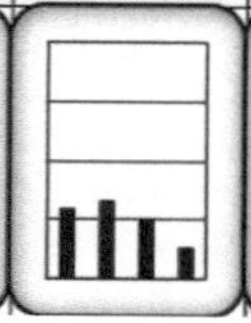	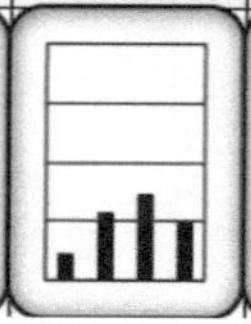	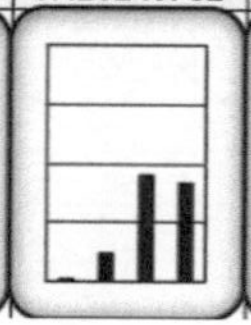	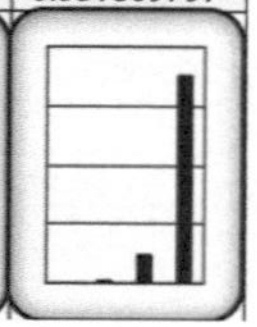

TABLE 3.105
CPT of *Ri_1*

C_1		Low				Medium				High			
V.I._1		Good	Fair	Poor	Service	Good	Fair	Poor	Service	Good	Fair	Poor	Service
Ri_1	_1.00E+02	0.7878	0.3548	0.0098	0.0129	0.5231	0.2089	0.0113	0.0001	0.3464	0.1424	0.0129	0.0000
	_1.00E+03	0.2071	0.5519	0.2436	0.1570	0.3976	0.4763	0.1889	0.0210	0.4164	0.3576	0.1570	0.0000
	_1.00E+04	0.0051	0.0920	0.5911	0.4717	0.0759	0.2751	0.5326	0.3775	0.1994	0.3576	0.4717	0.0551
	_1.00E+05	0.0000	0.0014	0.1554	0.3585	0.0035	0.0397	0.2672	0.6014	0.0378	0.1424	0.3585	0.9449
Histograms for conditional probabilities		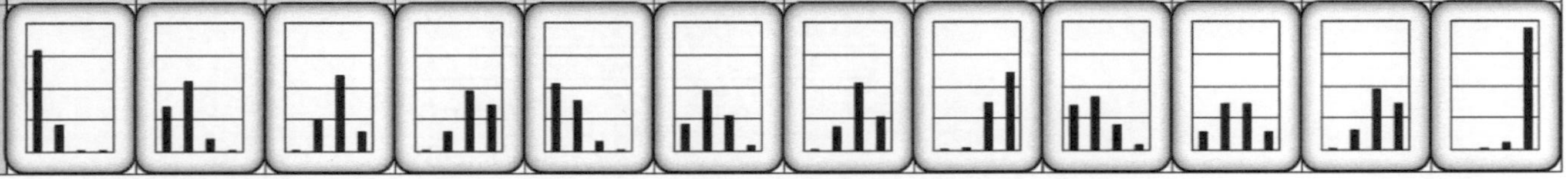											

TABLE 3.106
CPT of *Ri_2*

C_2		Low				Medium				High			
V.I._2		Good	Fair	Poor	Service	Good	Fair	Poor	Service	Good	Fair	Poor	Service
Ri_2	_1.00E+02	0.6999	0.2451	0.0039	0.0068	0.3585	0.1487	0.0023	0.0000	0.2662	0.1065	0.0068	0.0000
	_1.00E+03	0.2884	0.5947	0.1564	0.1112	0.4717	0.4468	0.0806	0.0093	0.4140	0.3217	0.1112	0.0000
	_1.00E+04	0.0116	0.1564	0.5947	0.4410	0.1570	0.3397	0.4585	0.2793	0.2566	0.3867	0.4410	0.0362
	_1.00E+05	0.0000	0.0039	0.2451	0.4410	0.0129	0.0648	0.4585	0.7114	0.0631	0.1852	0.4410	0.9638
Histograms for conditional probabilities		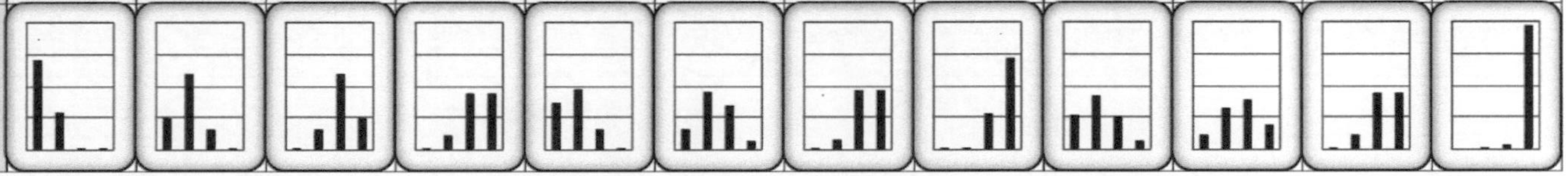											

TABLE 3.107
CPT of *Ri_3*

C_3		Low				Medium				High			
V.I._3		Good	Fair	Poor	Service	Good	Fair	Poor	Service	Good	Fair	Poor	Service
Ri_3	_1.00E+02	0.9206	0.3076	0.0002	0.0014	0.5231	0.2089	0.0231	0.0001	0.3309	0.1709	0.0378	0.0000
	_1.00E+03	0.0794	0.6696	0.1584	0.0920	0.3976	0.4763	0.2125	0.0249	0.3772	0.3291	0.1994	0.0005
	_1.00E+04	0.0000	0.0228	0.7746	0.5519	0.0759	0.2751	0.4845	0.3815	0.2233	0.3291	0.4164	0.1319
	_1.00E+05	0.0000	0.0000	0.0668	0.3548	0.0035	0.0397	0.2799	0.5935	0.0685	0.1709	0.3464	0.8676
Histograms for conditional probabilities		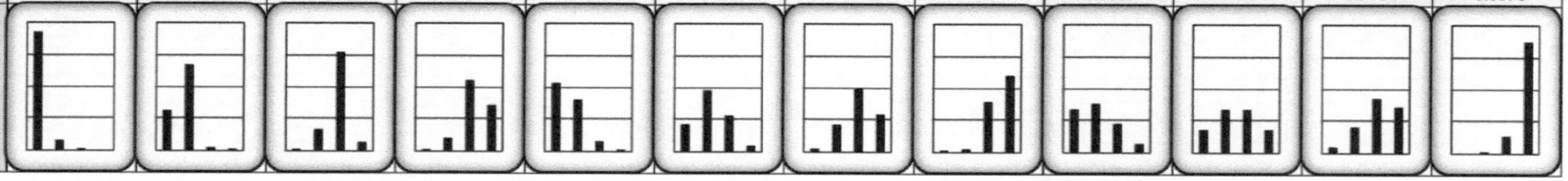											

TABLE 3.108
CPT of *V.I._1*

V.I._1	Good	0.2089
	Fair	0.4763
	Poor	0.2751
	Service	0.0397
Histograms for conditional probabilities		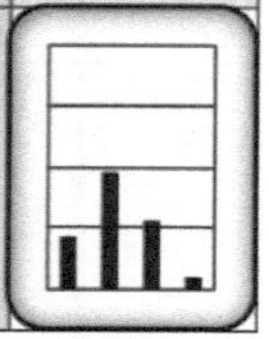

TABLE 3.109
CPT of *V.I._2*

V.I._2	Good	0.3585
	Fair	0.4717
	Poor	0.1570
	Service	0.0129
Histograms for conditional probabilities		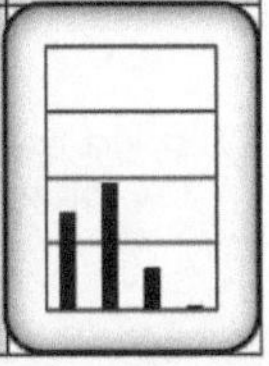

TABLE 3.110
CPT of *V.I._3*

V.I._3	Good	0.3607
	Fair	0.3344
	Poor	0.2124
	Service	0.0924
Histograms for conditional probabilities		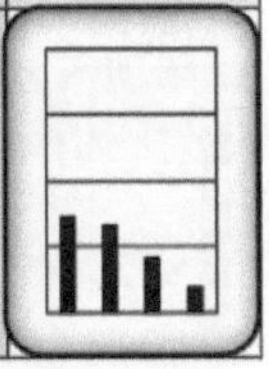

APPENDIX 3.VIII: CPTs FOR CASE STUDY 3.9 (TABLES 3.111–3.121)

TABLE 3.111
CPT of *CO*

R3		High				Low			
R2		High		Low		High		Low	
R1		High	Low	High	Low	High	Low	High	Low
CO	High	0.9199	0.4562	0.5353	0.2242	0.4637	0.0553	0.0355	0.0000
	Medium	0.0800	0.4493	0.4292	0.4626	0.4148	0.3829	0.4292	0.0325
	Low	0.0001	0.0945	0.0355	0.3133	0.1215	0.5618	0.5353	0.9675
Histograms for conditional probabilities		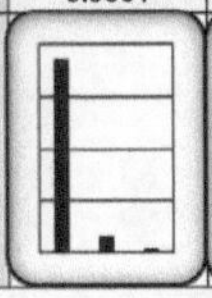	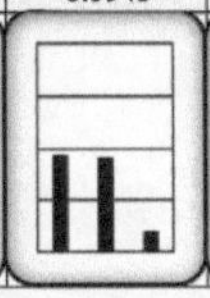	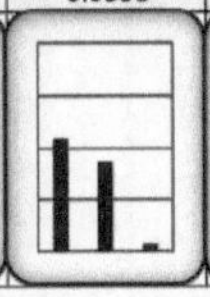	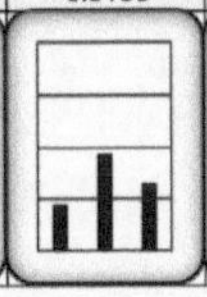	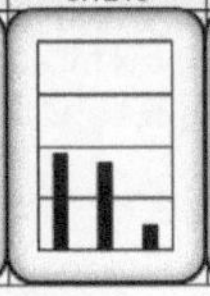	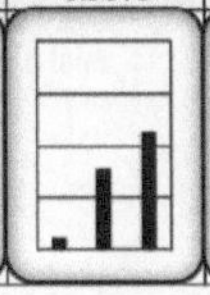	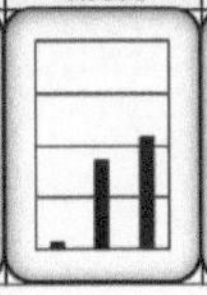	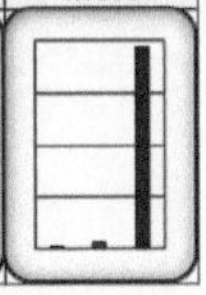

TABLE 3.112
CPT of *H*

H	Low	0.6
	Medium	0.3
	High	0.1
Histograms for conditional probabilities		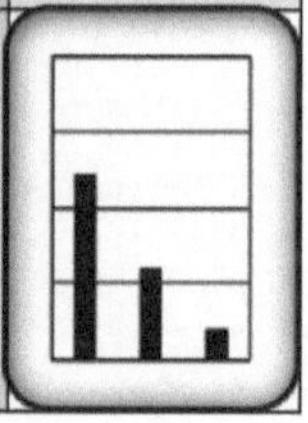

TABLE 3.113
CPT of *R1*

R14		High			Low		
H		Low	Medium	High	Low	Medium	High
R1	High	0.9206	0.7082	0.5681	0.9986	0.9067	0.6852
	Low	0.0794	0.2918	0.4319	0.0014	0.0933	0.3148
Histograms for conditional probabilities		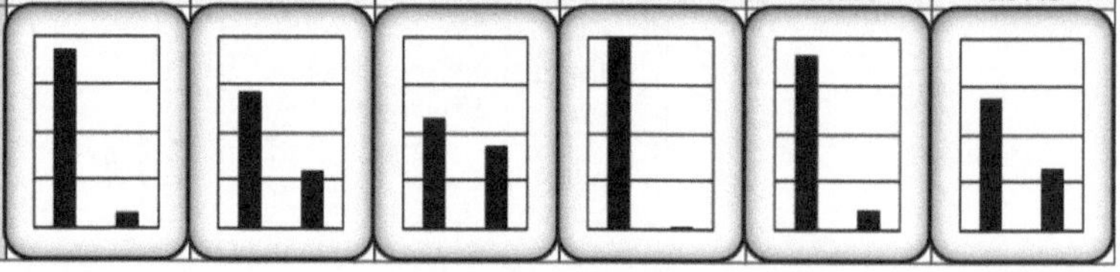					

TABLE 3.114
CPT of *R2*

R24		High	Low
R2	High	0.8	0.3
	Low	0.2	0.7
Histograms for conditional probabilities			

TABLE 3.115
CPT of *R3*

R34		High	Low
R3	High	0.7	0.4
	Low	0.3	0.6
Histograms for conditional probabilities		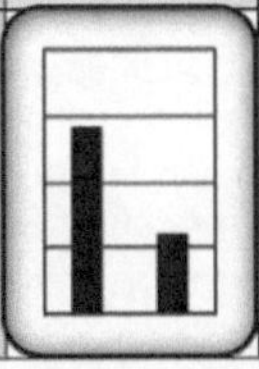	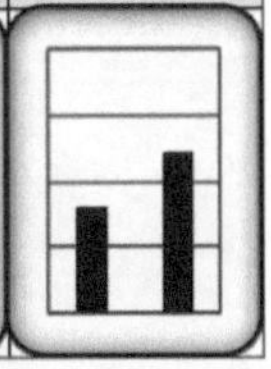

TABLE 3.116
CPT of *R4*

R34		High				Low			
R24		High		Low		High		Low	
R14		High	Low	High	Low	High	Low	High	Low
R4	High	0.9949	0.9239	0.6410	0.2918	0.8413	0.2082	0.4045	0.0129
	Low	0.0051	0.0761	0.3590	0.7082	0.1587	0.7918	0.5955	0.9871
Histograms for conditional probabilities		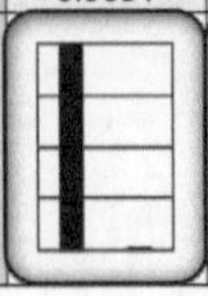	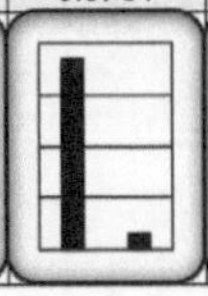	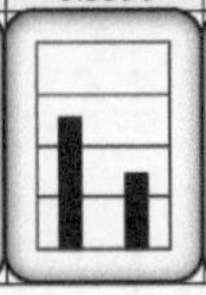	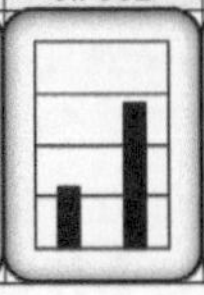	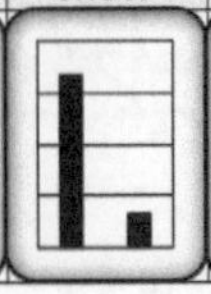	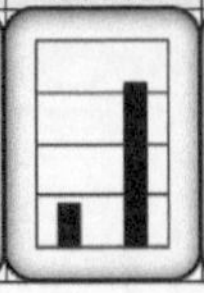	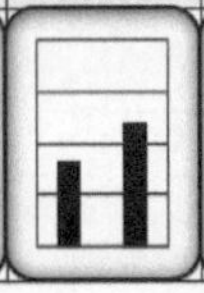	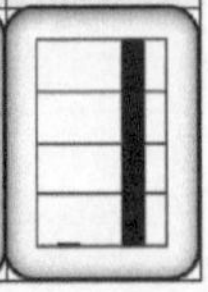

TABLE 3.117
CPT of *R14*

R14	High	0.7
	Low	0.3
Histograms for conditional probabilities		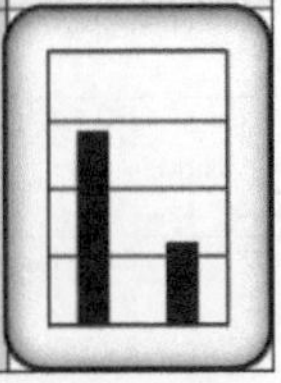

TABLE 3.118
CPT of *R24*

R24	High	0.45
	Low	0.55
Histograms for conditional probabilities		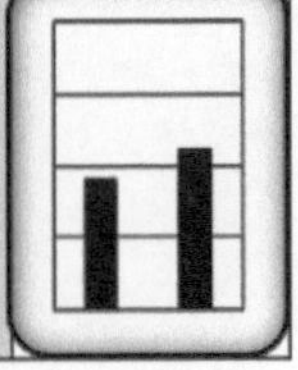

TABLE 3.119
CPT of *R34*

R34	High	0.2
	Low	0.8
Histograms for conditional probabilities		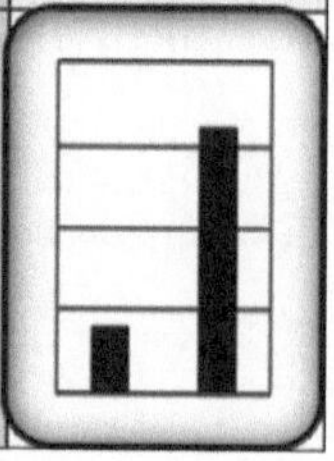

TABLE 3.120
CPT of *Re*

Time		High			Medium			Low		
CO		High	Medium	Low	High	Medium	Low	High	Medium	Low
Re	High	0.9199	0.2242	0.0585	0.5783	0.1854	0.0207	0.0743	0.1215	0.0056
	Medium	0.0800	0.4626	0.4967	0.4142	0.4525	0.3577	0.7084	0.4148	0.2145
	Low	0.0001	0.3133	0.4448	0.0074	0.3621	0.6216	0.2173	0.4637	0.7799
Histograms for conditional probabilities		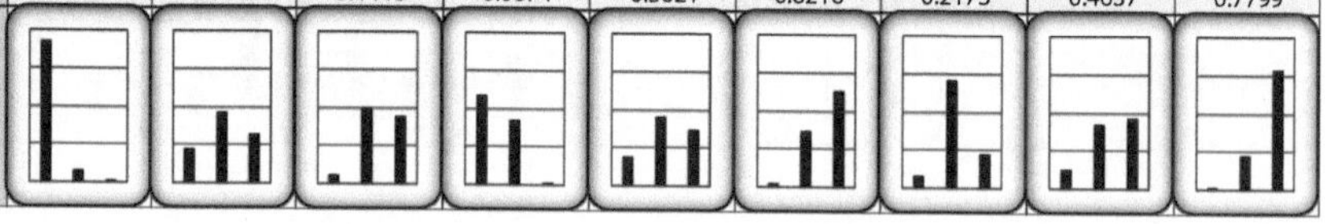								

TABLE 3.121
CPT of *Time*

R3		High				Low			
R2		High		Low		High		Low	
R1		High	Low	High	Low	High	Low	High	Low
Time	High	0.8240	0.4330	0.4963	0.2491	0.4379	0.0963	0.0778	0.0012
	Medium	0.1726	0.4282	0.4258	0.4303	0.4026	0.3894	0.4258	0.1359
	Low	0.0033	0.1387	0.0778	0.3206	0.1595	0.5143	0.4963	0.8629
Histograms for conditional probabilities		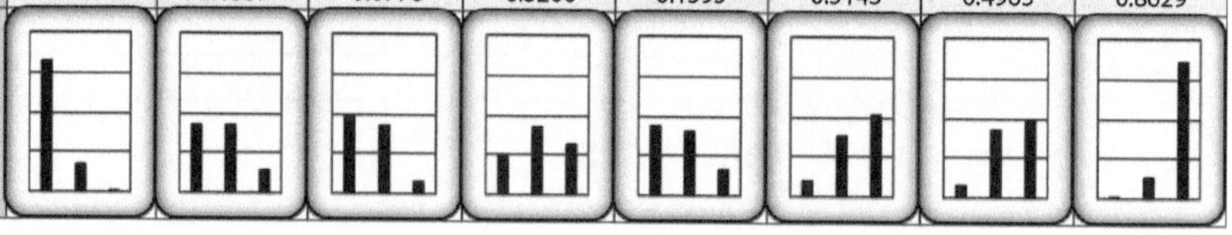							

APPENDIX 3.IX: CPTs FOR CASE STUDY 3.10 (TABLES 3.122–3.136)

TABLE 3.122
CPT of *Arch.*

Site		High		Low	
Str.		High	Low	High	Low
Arch.	High	1.0000	0.0117	0.1109	0.1180
	Low	0.0000	0.9883	0.8891	0.8820
Histograms for conditional probabilities		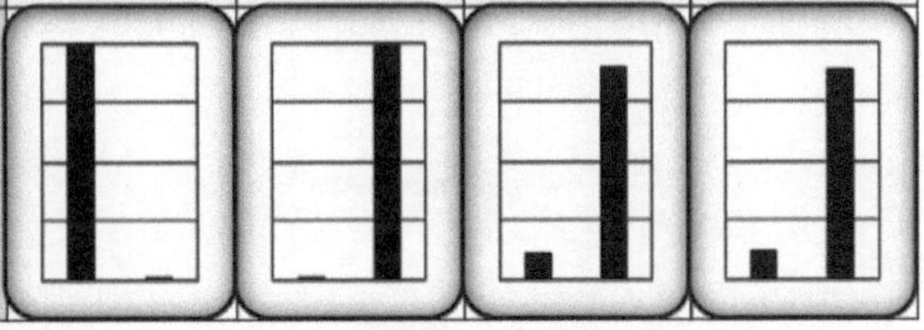			

TABLE 3.123
CPT of *CO*

Envelope		High				Low			
Str.		High		Low		High		Low	
Fire		High	Low	High	Low	High	Low	High	Low
CO	High	0.7677	0.4637	0.4562	0.2166	0.4618	0.0875	0.0681	0.0019
	Medium	0.2260	0.4148	0.4493	0.4246	0.3936	0.3783	0.4091	0.1672
	Low	0.0063	0.1215	0.0945	0.3588	0.1445	0.5343	0.5228	0.8309
Histograms for conditional probabilities		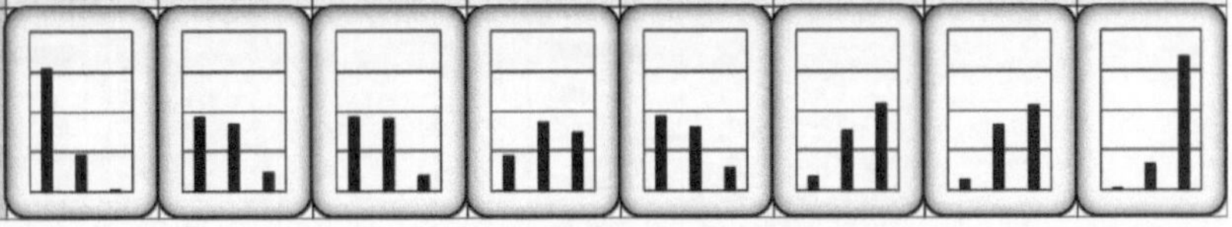							

TABLE 3.124
CPT of *Envelope*

Str.		High	Low
Envelope	High	0.8	0.4
	Low	0.2	0.6
Histograms for conditional probabilities		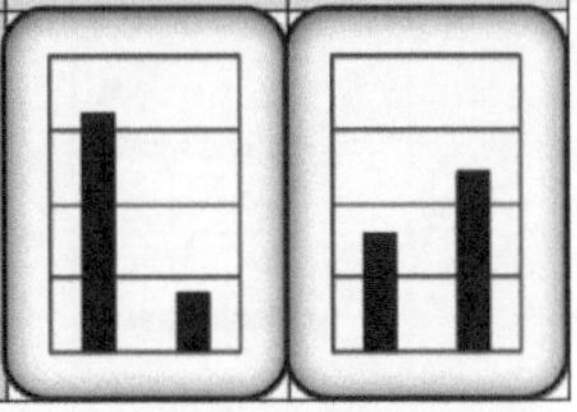	

TABLE 3.125
CPT of *Fire*

O.I._3		High								Low							
Power		High				Low				High				Low			
HVAC		High		Low		High		Low		High		Low		High		Low	
I.T.		High	Low	High	Low	High	Low	High	Low	High	Low	High	Low	High	Low	High	Low
Fire	High	0.9949	0.8820	0.7466	0.5955	0.9067	0.5955	0.2534	0.2356	0.3699	0.2356	0.0250	0.0518	0.0502	0.0518	0.0051	0.0204
	Low	0.0051	0.1180	0.2534	0.4045	0.0933	0.4045	0.7466	0.7644	0.6301	0.7644	0.9750	0.9482	0.9498	0.9482	0.9949	0.9796
Histograms for conditional probabilities																	

TABLE 3.126
CPT of *HVAC*

I.T.		High				Low			
Str.		High		Low		High		Low	
Power		High	Low	High	Low	High	Low	High	Low
HVAC	High	1.0000	0.0794	0.0518	0.0000	0.0933	0.0051	0.0117	0.0000
	Low	0.0000	0.9206	0.9482	1.0000	0.9067	0.9949	0.9883	1.0000
Histograms for conditional probabilities		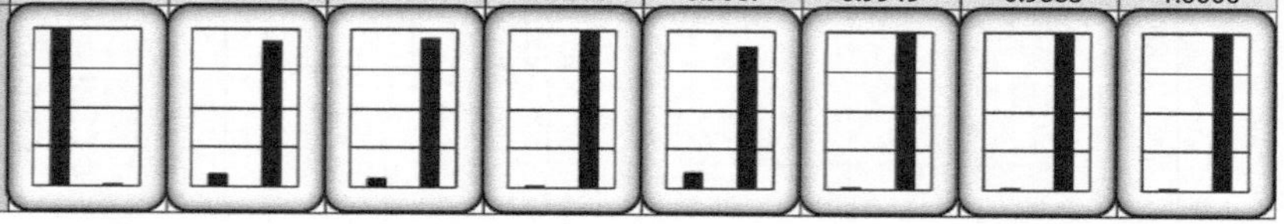							

TABLE 3.127
CPT of *IT*

Arch.		High				Low			
Str.		High		Low		High		Low	
Power		High	Low	High	Low	High	Low	High	Low
I.T.	High	0.9999	0.0137	0.0671	0.0015	0.5550	0.0794	0.0117	0.0000
	Low	0.0001	0.9863	0.9329	0.9985	0.4450	0.9206	0.9883	1.0000
Histograms for conditional probabilities		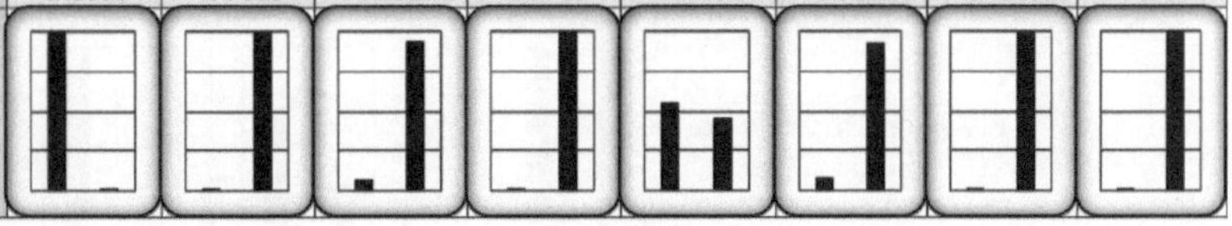							

TABLE 3.128
CPT of *OI_1*

O.I._1	High	0.9
	Low	0.1
Histograms for conditional probabilities		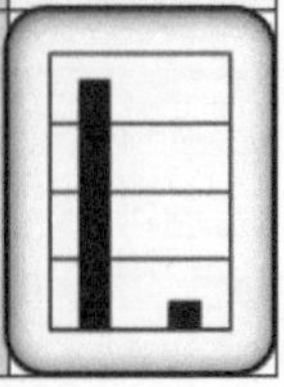

TABLE 3.129
CPT of *OI_2*

O.I._2	High	0.75
	Low	0.25
Histograms for conditional probabilities		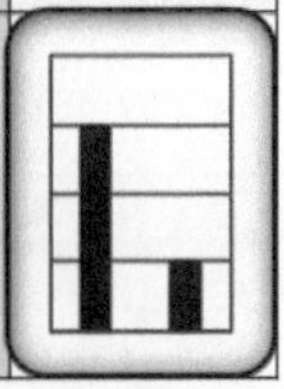

TABLE 3.130
CPT of *OI_3*

O.I._3	High	0.65
	Low	0.35
Histograms for conditional probabilities		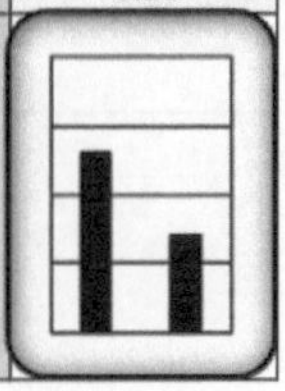

TABLE 3.131
CPT of *Power*

O.I._2		High		Low	
Str.		High	Low	High	Low
Power	High	0.9949	0.3148	0.0117	0.0329
	Low	0.0051	0.6852	0.9883	0.9671
Histograms for conditional probabilities		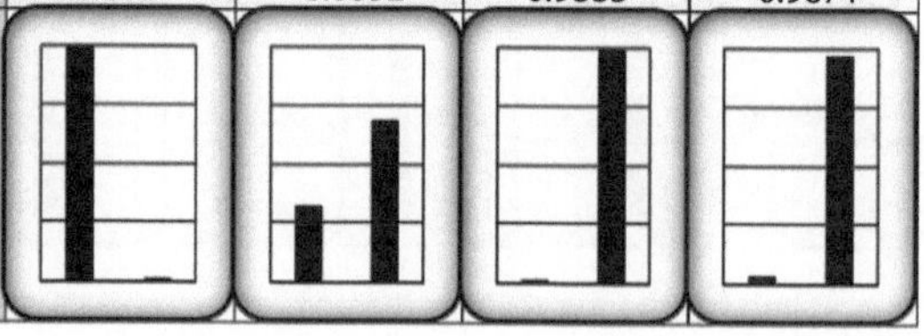			

TABLE 3.132
CPT of *Re*

Time		High			Medium			Low		
CO		High	Medium	Low	High	Medium	Low	High	Medium	Low
Re	High	0.9199	0.2242	0.0585	0.5783	0.1854	0.0207	0.0743	0.1215	0.0056
	Medium	0.0800	0.4626	0.4967	0.4142	0.4525	0.3577	0.7084	0.4148	0.2145
	Low	0.0001	0.3133	0.4448	0.0074	0.3621	0.6216	0.2173	0.4637	0.7799
Histograms for conditional probabilities		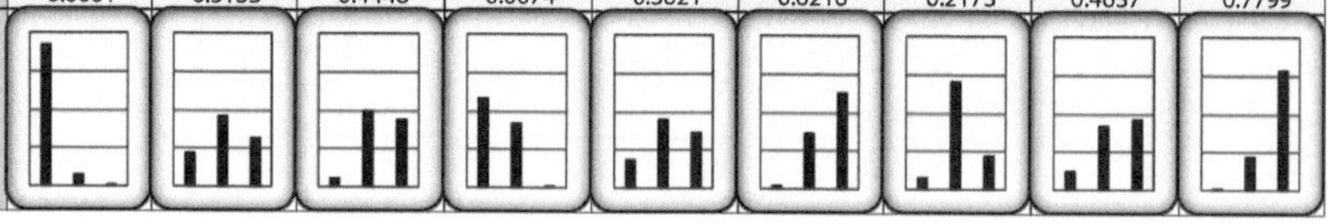								

TABLE 3.133
CPT of *Security*

Str.		High		Low	
Fire		High	Low	High	Low
Security	High	0.9206	0.1698	0.3913	0.1180
	Low	0.0794	0.8302	0.6087	0.8820
Histograms for conditional probabilities		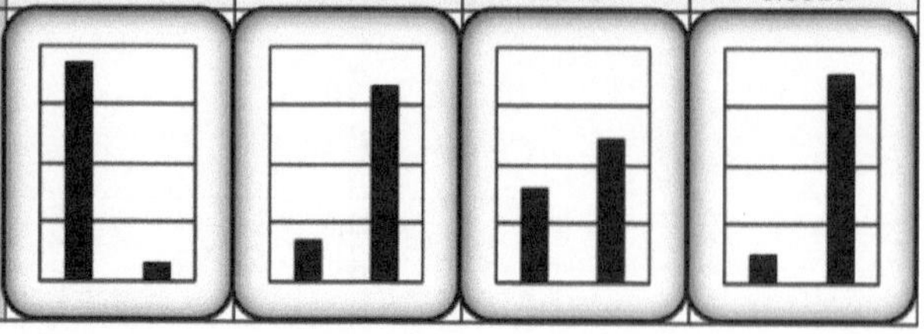			

TABLE 3.134
CPT of *Site*

Site	High	0.85
	Low	0.15
Histograms for conditional probabilities		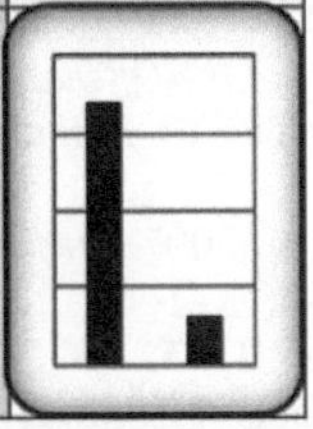

TABLE 3.135
CPT of *Str.*

Site		High	Low
Str.	High	0.9	75
	Low	0.1	0.25
Histograms for conditional probabilities		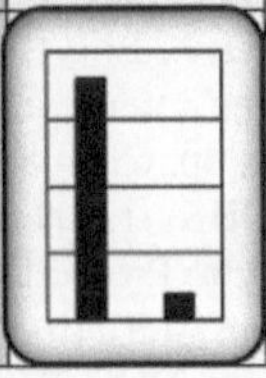	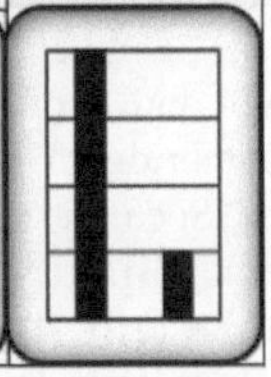

TABLE 3.136
CPT of *Time*

Security		High				High			
Str.		High		Low		High		Low	
Fire		High	Low	High	Low	High	Low	High	Low
Time	High	0.8240	0.4330	0.4963	0.2491	0.4379	0.0963	0.0778	0.0012
	Medium	0.1726	0.4282	0.4282	0.4303	0.4026	0.3894	0.4258	0.1359
	Low	0.0033	0.1387	0.0778	0.3206	0.1595	0.5143	0.4963	0.8629
Histograms for conditional probabilities		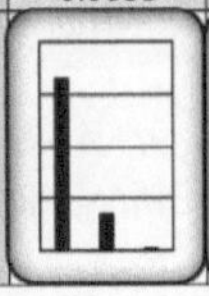	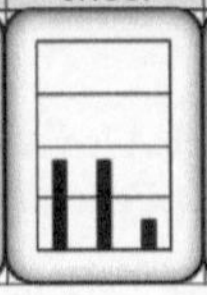	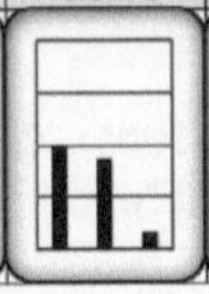	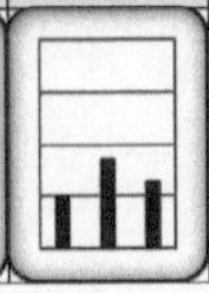	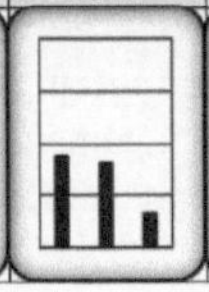	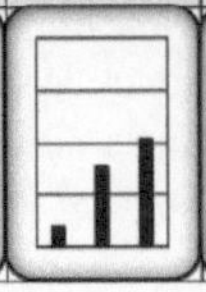	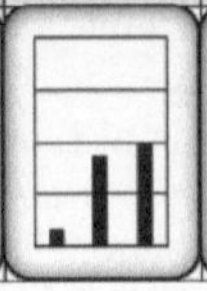	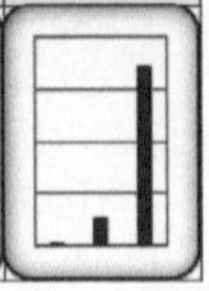

REFERENCES

Aggarwal, R. K. (1991). *Methodology for Assessment by Regulatory Bodies of the Safety of Existing Steel Offshore Platforms*. PhD thesis. University of California, Berkeley, CA.

Agrawal, A., Washer, G., and Xu, G. (2013). *Consistency of the New York State Bridge Inspection Program*. Project # C-07-17. New York State Department of Transportation, Albany, NY.

Anderson, W. D., Silbert, M. N., and Lloyd, J. R. (1982). Reliability Procedures for Fixed Offshore Platforms. *J. Struct. Div.* 108(11), 2517–2537.

ASCE/SEI. (2011). *Seismic Evaluation and Retrofit of Existing Buildings*. American Society of Civil Engineers (ASCE) Standard 41–13, ASCE, Reston, VA.

ASCE. (2013). *2013 Report Card for America's Infrastructure*, American Society of Civil Engineers, Reston, VA.

Ayyub, B. M. (2014). *Risk Analysis in Engineering and Economics*, 2nd ed. CRC, Boca Raton, FL.

Baker, J. W., and Faber, M. H. (2008). Liquefaction Risk Assessment Using Geostatistics to Account for Soil Spatial Variability. *J. Geotech. Geoenviron. Eng.* 134(1), 14–23.

Bea, R. G. (1974). Selection of Environmental Criteria for Offshore Platform Design. *J. Petrol. Technol.* Issue 11, Vol. 26, 1206–1214.

Bea, R. G., and Craig, M. J. K. (1993). Developments in the Assessment and Requalification of Offshore Platforms. *Proceedings of the Offshore Technology Conference*, May 3–6, 1993, Houston, TX, pp. 1–19.

Benjamin, J., and Cornell, A. (1970), *Probability, Statistics, and Decisions for Civil Engineers*, McGraw-Hill, New York.

Bruneau, M., and Reinhorn, A. (2007). Exploring the Concept of Seismic Resilience for Acute Care Facilities. *Earthquake Spectra.* 23(1), 41–62.

Buckle, I. G., and Friedland, I. M. (1995). *Seismic Retrofitting Manual for Highway Bridges,* Rep. No. FHWA-RD-94-052, Federal Highway Administration, Washington, DC.

Canuti, P., Casagli, N., Catani, F., and Fanti, R. (2000). Hydrogeological Hazard and Risk in Archaeological Sites: Some Case Studies in Italy. *J. Cult. Heritage.* 1(2), 117–125.

Celik, O. C., and Ellingwood, B. R. (2009). Seismic Risk Assessment of Gravity Load Designed Reinforced Concrete Frames Subjected to Mid-America Ground Motions. *J. Struct. Eng.* 135(4), 414–424.

Choi, H. H., Cho, H. N., and Seo, J. W. (2004). Risk Assessment Methodology for Underground Construction Projects. *J. Construct. Eng. Manage.* 130(2), 258–272.

Cordona, O. D., van Aalst, M. K., Birkmann, J., Fordham, M., McGregor, G., Perez, R., Pulwarty, R. S., Schipper, E. L. F., and Sinh B. T. (2012). Determinants of Risk: Exposure and Vulnerability. In: *Managing the Risk of Extreme Events and Disasters to Advance Climate Change Adaptation*, Field, C. B., Barros, V., Stocker, T. F., Qin, D., Dokken, D. J., Ebi, K. L., Mastrandrea, M. D., et al. (Eds.). A Special Report of Working Groups I and II of the Intergovernmental Panel on Climate Change (IPCC). Cambridge University Press, Cambridge, UK, pp. 65–108.

Department of Homeland Security (DHS). (2011a). *Integrated Rapid Visual Screening of Mass Transit Stations*, Building and Infrastructure Protection Series, DHS, Washington, DC.

Department of Homeland Security (DHS). (2011b). *Integrated Rapid Visual Screening of Tunnels*, Building and Infrastructure Protection Series, DHS, Washington, DC.

Department of Homeland Security (DHS). (2011c). *Integrated Rapid Visual of Screening of Buildings*, Building and Infrastructure Protection Series. DHS, Washington, DC.

Drdácký, M., Binda, L., Herle, I., Lanza, L. G., Maxwell, I., and Pospísil, S. (2007). *Protecting the Cultural Heritage from Natural Disasters.* Study of the European Parliament IP/B/CULT/IC/2006_163, PE369.029, The European Parliament, Brussels, Belgium.

Ettouney, M.M., and Alampalli, S. (2012a). *Infrastructure Health in Civil Engineering: Theory and Components*, CRC Press, Boca Raton, FL.

Ettouney, M.M., and Alampalli, S. (2012b). *Infrastructure Health in Civil Engineering: Applications and Management*, CRC Press, Boca Raton, FL.

FEMA (Federal Emergency Management Agency). (2003). *NEHRP Recommended Provisions for Seismic Regulations for New Buildings and Other Structures and Commentary*, FEMA 450, Federal Emergency Management Agency, Washington, DC.

FEMA (Federal Emergency Management Agency). (2005). *Risk Assessment: A How-To Guide to Mitigate Terrorist Attacks*, Risk Management Series, FEMA 452, Federal Emergency Management Agency, Washington, DC.

FEMA (Federal Emergency Management Agency). (2009). *Handbook for Rapid Visual Screening of Buildings to Evaluate Terrorism Risks*, Federal Emergency Management Agency (FEMA), Risk Management Series FEMA 455, Federal Emergency Management Agency, Washington, DC.

Fenton, N., and Neil, M. (2013). *Risk Assessment and Decision Analysis with Bayesian Networks*, CRC Press, Boca Raton, FL.

FHWA. (1995). *Recording and Coding for the Structure Inventory and Appraisal of the Nation's Bridges*, FHWA report No. FHWA-PD-96-001, FHWA, Washington, DC.

Fujino, T. (1994). *The Development of a Method for Investigation Construction Site Accidents Using Fuzzy Fault Tree Analysis.* PhD thesis. The Ohio State University, Columbus, OH.

Gonzalo, C. D., Robredo, J. C., and Mintegui, J. A. (2012). Semidistributed Hydrologic Model for Flood Risk Assessment in the Pejibaye River Basin, Costa Rica. *J. Hydrol. Eng.* 17(12), 1333–1344.

Grigg, N. S. (2013). Water Main Breaks: Risk Assessment and Investment Strategies. *J. Pipeline Syst. Eng. Pract.* 4, ISSN 1949-1190/04013001(7)/$25.00.

Hadipriono, F. C., Lim, C. L., and Wong, K. H. (1986). Event Tree Analysis to Prevent Failures in Temporary Structures. *J. Construct. Eng. Manage.* 112(4), 500–513.
Hastak, M., and Shaked, A. (2000). ICRAM-1: Model for International Construction Risk Assessment. *J. Manage. Eng.* 16(1), 59–69.
Holický, M., and Sýkora, M. (2010). Assessment of Flooding Risk to Cultural Heritage in Historic Sites. *J. Perform. Construct. Facil.* 24(5), 432–438.
Huang, D., Chen, T., and Wang, M.-J. J. (2001). A fuzzy Set Approach for Event Tree Analysis. *Fuzzy Sets Syst.* 118, 153–165.
Jorion, P. (2007). *Value at Risk: The New Benchmark for Managing Financial Risk*, McGraw-Hill, New York.
Kanta, L., and Brumbelow, K. (2013). Vulnerability, Risk, and Mitigation Assessment of Water Distribution Systems for Insufficient Fire Flows. *J. Water Resour. Plann. Manage.* 139(6), 593–603.
Koller, D., and Friedman, N. (2009). *Probabilistic Graphical Models: Principles and Techniques*, The MIT Press, Cambridge, MA.
Li, Z., Huang, H., Nadim, F., and Xue, Y. (2010). Quantitative Risk Assessment of Cut-Slope Projects under Construction. *J. Geotech. Geoenviron. Eng.* 136(12), 1644–1654.
Ndiba, P. K., and Axe, L. (2010). Risk Assessment of Metal Leaching into Groundwater from Phosphate and Thermal Treated Sediments. *J. Environ. Eng.* 136(4), 427–434.
Niezgoda, S. L., and Johnson, P. A. (2007). Case Study in Cost-Based Risk Assessment for Selecting a Stream Restoration Design Method for a Channel Relocation Project. *J. Hydraul. Eng.* 133(5), 468–481.
Oudin, L., Andreassian, V., Mathevet, T., Perrin, C., and Michel, C. (2006). Dynamic Averaging of Rainfall-Runoff Model Simulations from Complementary Model Parameterizations. *Water Resour. Res.* 42(7), W07410.
Saaty, T. L. (1980). *The Analytic Hierarchy Process.* McGraw-Hill, New York.
Samadder, S.R., and Subbarao, C. (2007). GIS Approach of Delineation and Risk Assessment of Areas Affected by Arsenic Pollution in Drinking Water. *J. Environ. Eng.* 133(7), 742–749.
Staneff, S. T., Ibbs, C. W., and Bea, R. G. (1995). Risk-Management System for Infrastructure-Condition Assessment. *J. Infrast. Syst.* 1(4), 221–229.
The European Parliament and the Council. (2007). *Directive 2007/60/EC on the Assessment and Management of Flood Risks*, Official Journal of the European Union, Brussels, Belgium.
Tromans, P. S., and Graaf, J. W. (1994). Substantiated Risk Assessment of Jacket Structure. *J. Waterway Port. Coast. Ocean Eng.* 120(6), 535–556.
UNESCO-UNDRO. 1979. *Natural Disasters and Vulnerability Analysis*, Rep. of Expert Group Meeting, Office of the United Nations Disaster Relief Coordinator, Geneva, Switzerland.
Vose, D. (2009). *Risk Analysis a Quantitative Guide.* Wiley, Hoboken, NJ.
Wang, R., Wang, Z., Wang, X., Yang, H., and Sun, J. (2013). Pipe Burst Risk State Assessment and Classification Based on Water Hammer Analysis for Water Supply Networks. *J. Water Resour. Plann. Manage.* doi: 10.1061/(ASCE)WR.1943-5452.0000404.
Wolshon, B., Hamilton, E. U., Levitan, M., and Wilmot, C. (2005b). Review of Policies and Practices for Hurricane Evacuation. II: Traffic Operations, Management, and Control. *Nat. Hazards Rev.* 6(3), 143–161.
Wolshon, B., Urbina, E., Wilmot, C., and Levitan, M. (2005a). Review of Policies and Practices for Hurricane Evacuation. I: Transportation Planning, Preparedness, and Response. *Nat. Hazards Rev.* 6(3), 129–142.
Zadeh, L. A. (1965). *Fuzzy sets. Inf. Contr.* 8, 338–353.
Zatar, W. A., and Harik, I. E. (2008). Bridge Embankments. II: Seismic Risk for I-24 in Kentucky. *J. Perform. Construct. Facil.* 22(3), 181–190.
Zatar, W. A., Harik, I. E., Sutterer, K. G., and Dodds, A. (2008). Bridge Embankments. I: Seismic Risk Assessment and Ranking. *J. Perform. Construct. Facil.* 22(3). 171–180.
Zayed, T. M., and Chang, L. M. (2002). Prototype Model for Build-Operate-Transfer Risk Assessment. *J. Manage. Eng.* 18(1), 7–16.
Zayed, T. M., and Halpin, D. W. (2004). Quantitative Assessment for Piles Productivity Factors. *J. Construct. Eng. Manage.* 130(3), 405–414.
Zayed, T., Minchin, R. E., Jr., Boyd, A. J., Smith, G. R., and McVay, M. C. (2007). Model for the Physical Risk Assessment of Bridges with Unknown Foundation. *J. Perform. Construct. Facil.* 21(1), 44–52.
Zerger, A., and Smith, D. I. (2003). Impediments to Using GIS for Real-Time Disaster Decision Support. *Comput. Environ. Urban Syst.* 27, 123–141.

4 Risk Acceptance

4.1 INTRODUCTION

4.1.1 Overview

In the previous chapter, we explored several methods for risk assessment. We showed that risk can be assessed either in relative terms or in absolute terms. Whenever risk-based processes in civil infrastructures are followed, the assessment usually consumes the greater part of the total effort, and leaves other equally important components of the risk management process with relatively unstudied. Perhaps, the least attention is given to the acceptance threshold, that is, is the just-quantified risk level acceptable when compared to the yet-to-be-assigned risk acceptance threshold? In general, risk acceptance thresholds are not well defined in the early stages and are usually ignored until the later stages of management practice. Thus, if it is assigned an arbitrary value without rigorous studies, the acceptance threshold might render the whole risk management effort less accurate and, in some cases, unnecessarily expensive.

To explain the mismatch between the effort spent in the assessment and acceptance phases in risk management, we quote Jayjock et al. (2000): "Estimating the risk of a particular situation is a scientific pursuit, but determining the acceptability of that risk is a value judgment." This quote illustrates one of the mostly unrecognized facts in the management of infrastructure, whether it is reliability-, exposure-, or risk-based management; the acceptance part is judgment-based, not scientific-based. In risk management, this renders two of the five components judgment-based or qualitative (even though sprinkled with a good portion of qualitative measures themselves), and the other three components are mostly scientific as shown in Figure 4.1.

Acceptance thresholds play a major role in the risk management process. Choosing the optimal risk acceptance threshold not only provides the required performance level but also provides an optimal cost ceiling. Figure 4.2 illustrates this concept. Generally speaking, there are two sources of cost implications for the type of hazards that inflict civil infrastructures: the cost of the hazard itself that obviously increases as the risk level increases and the cost of risk treatment that will be high if we require a lower risk level. The sum of the two costs is the accepted cost at the accepted risk level. An optimal accepted risk level, where these total costs are optimal, is what we strive to use during the risk management process.

4.1.2 Historical Background

Several researchers have explored risk acceptance methods. For example, Reevel (1998) studied the use of probabilistic techniques in risk assessment for sea defenses; that is, the probability of flooding due to wave overtopping and the uncertainties associated with the employed procedure. He demonstrated the use of joint probability function methods (first-order mean value approach and first-order design point approach), including structure function method, direct method, point-estimate method, and Gumbel distribution in calculating the probabilities of exceeding the overtopping threshold for two structures. Based on his investigation, he noted that different methods tend to provide a consensus on weak structures but can differ significantly for structures that are more resilient.

Also, risk management, including risk acceptance, was applied in major construction operations. Zhou and Zhang (2011) note that deep foundation pit engineering projects require risk assessment and management to ensure safety and quality. Based on their literature search, they concluded that most available present methods can only be used as a qualitative or a semiquantitative assessment,

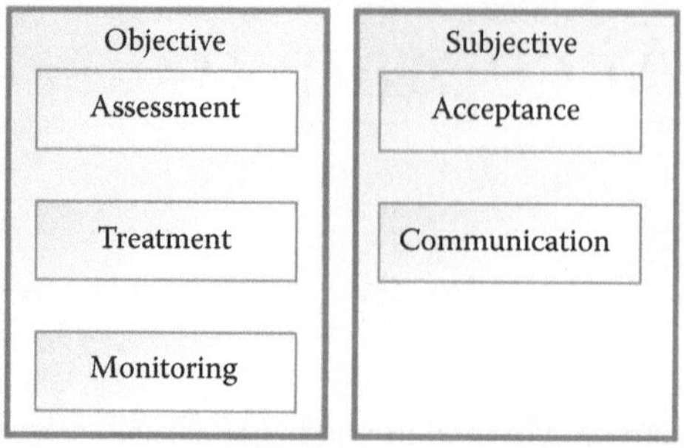

FIGURE 4.1 Subjective vs. objective components of risk management.

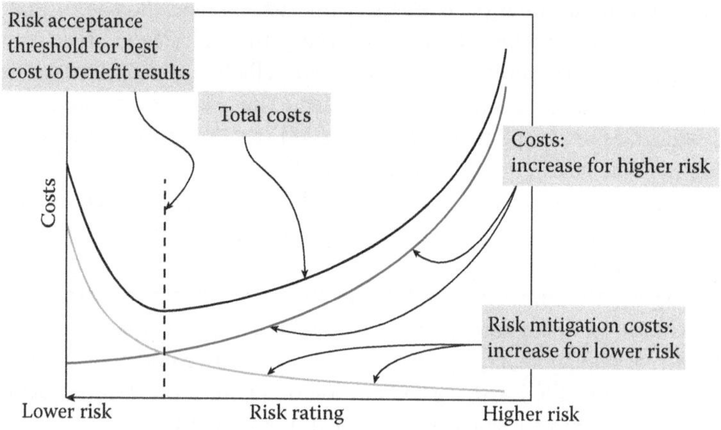

FIGURE 4.2 Optimality of risk acceptance threshold.

fail to take advantage of accumulated historical experiences, and uncertain factors cannot be analyzed precisely. Hence, they proposed a risk assessment method using a fuzzy comprehensive evaluation method based on Bayesian networks (BNs) for deep foundation pit construction to combine the superiority of BN techniques for processing uncertain factors and to integrate historical information with the advantages of the fuzzy comprehensive evaluation method in handling uncertain risks. The BN technique was used to obtain the probabilities of risk events, and the fuzzy comprehensive evaluation method was used to calculate the risk value and determine membership functions for evaluating the risk ratings.

For a case study, Zhou and Zhang used a primary data analysis of a simple accident database from a deep foundation pit construction project in Shanghai to provide the probabilities of basic events and approximate distributions of risk consequences. They obtained the BN model through the transformation of a fault tree analysis. Deductive BN techniques were used to calculate the probability of every risk event. Then, fuzzy analysis (i.e., statistical consequence distributions and weight coefficients of risk events are determined through the database) was used to calculate the consequence of each event. Risk precautions and control measures were suggested on the basis of the risk assessment results and were applied to risk management in deep foundation pit construction.

A mix of subjective and objective acceptance processes in structural failures due to fire following hazards such as earthquakes and blast are needed. Short of adequate acceptance thresholds, progressive collapse of the structure or a reduction of the time available for evacuation operations, thus posing a threat to occupants, may result. There are several studies in the literature on the analysis of structures that have been damaged by an extreme event followed by fire. With an intent to raise awareness of this potential life threatening issue, Quiel and Marjanishvili (2012) studied the effects of fire following an extreme event (i.e., blast or impact) that causes failure of one column on the perimeter of a common steel building frame, designed to satisfy new U.S. Department of Defense (DoD) guidelines, and assumed that the extreme event not only damages one column but also damages active fire protection (i.e., sprinklers) in the vicinity of the structural damage. They used a multihazard approach similar to those of Della Corte et al. (2003) and Chen and Liew (2005), where the performance of the structure is investigated after it had sustained damage due to the primary hazard.

The three phases in this multihazard approach included removal of a key structural element through blast or impact, occurring on a time scale of milliseconds; the redistribution of load and the onset of stable plastic deformations as the progressive collapse resistance is engaged, on a time scale of seconds; and the development of a severe fire near the location of member removal, on a time scale of minutes. Quiel and Marjanishvili assumed that the building sustains no additional structural damage attributable to the initial extreme event beyond the targeted removed members (which is consistent with the unified facilities criteria [UFC] [DoD 2010]) and that the fire initiates immediately after the onset of stable plastic deformations in the structural frame near the removal location. Based on the case study presented, they conclude that the fire following an extreme event represents a significant threat to stability of the structure and that collapse time due to a fire following a major hazard should be considered as a performance tool by building owners and safety officials to determine how much time is realistically available for evacuation before progressive collapse occurs.

Fragility analysis has been also used as a process for an objective acceptance process. For example, studying the vulnerability of an infrastructure to multiple hazards is becoming very important because of the emphasis on mobility along with safety. For example, in most past studies, bridge reliability assessments considered the effects of natural hazards such as earthquakes and flooding independently from traffic loading. But considering the impact of natural hazards along with traffic loading is gaining attention. Realizing this, Ghosh et al. (2013) presented a framework for joint seismic and live-load fragility assessment of highway bridges accounting for variation in bridge parameters, ground motion, and truck load and position to identify the critical truck position that renders the bridge most vulnerable to earthquakes. A case study was presented with a multispan continuous steel girder bridge in the central and southeastern United States. Based on the results, they concluded that it is likely that bridges are more fragile during a seismic event if traffic loading is considered. However, the increased fragility (reduced median peak ground acceleration) can be small once the probabilities of truck presences and gross vehicle weights are taken into account.

4.1.3 Definitions

In this book, we use terminology that closely relates to risk acceptance. First, we differentiate between existing risk and accepted risk. Existing risk, sometimes referred to as an assessed risk, *Ri*, is the current state of risk for the system under consideration. Accepted risk, $Ri|_{Accepted}$, is the risk threshold that is obtained using a value judgment, a subjective method, or a combination of subjective/objective methods. In some infrastructure communities, the term target risk or target risk threshold is also used instead of accepted risk. Another concept in this realm is the concept of achievable risk $Ri|_{Achievable}$, which is the risk that can be achieved after risk treatment. A common example of $Ri|_{Achievable}$ is when a limited budget is available for a particular risk treatment situation. Such a limited budget can only achieve a certain level of risk which is $Ri|_{Achievable}$. In situations when,

$$Ri|_{Achievable} < Ri|_{Accepted} \tag{4.1}$$

we define the remaining risk as

$$Ri|_{Remaining} = Ri|_{Accepted} - Ri|_{Achievable} \tag{4.2}$$

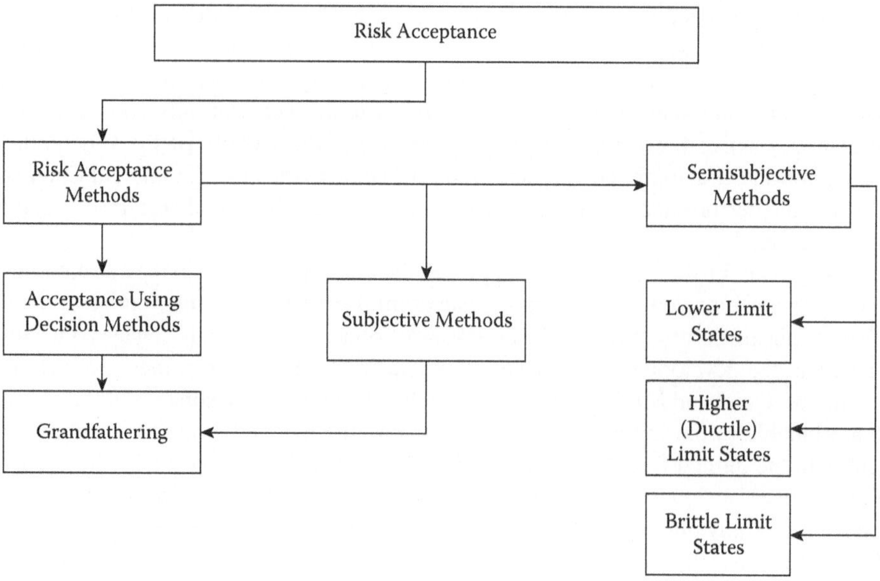

FIGURE 4.3 Contents of this chapter.

Because, as the end, $Ri|_{Accepted}$ is fairly subjective, in situations where $Ri|_{Achievable} < Ri|_{Accepted}$, the acceptance of the presence of a finite and positive $Ri|_{Remaining}$ is also fairly subjective.

4.1.4 This Chapter

We start this chapter with an overview of risk acceptance methods (see Figure 4.3). Even though risk acceptance methods are all subjective in nature, we differentiate between subjective and semisubjective methods. As their titles indicate, subjective methods rely solely on judgments and qualitative processes, whereas semisubjective methods include some measures of objectivity. We then offer several case studies of interest to civil infrastructure stakeholders. These include the acceptance at different limit states: lower limit states (LLS), ductile limit states, or brittle limit states. We also discuss case studies that use decision under uncertainty processes. We end the chapter by exploring the popular grandfathering practice in the field of civil infrastructure from a risk-based viewpoint. We show that the grandfathering process, currently based on subjective judgment, can be looked upon as a risk assessment and acceptance process. Approaching it from such a viewpoint would result in a more objective process that can improve decision-making efficiency (performance and cost).

4.2 OVERVIEW OF RISK ACCEPTANCE METHODS

Several categories of risk acceptance methods are available. We define accepted risk thresholds either in a subjective manner or a semisubjective manner (as we argued earlier, there is no completely objective method for risk acceptance as there will always be a subjective component in the methods of interest). We also categorize methods for risk acceptance by the type of threshold (target): these target thresholds can be target risk, reliability, or safety factor. In addition, the methods for defining acceptance thresholds also depend on the physical infrastructure system under consideration: it can be component, an asset, or a network of assets. Finally, the level of system (or component) response can also affect the acceptance definition method: these levels of responses, or limit states, can be below elastic limit (we call such limit states *lower limit states*. or LLS), above the elastic limit (ductile limit states), ductile failure limit state, or a brittle failure limit state. These different categories are shown in Figure 4.4.

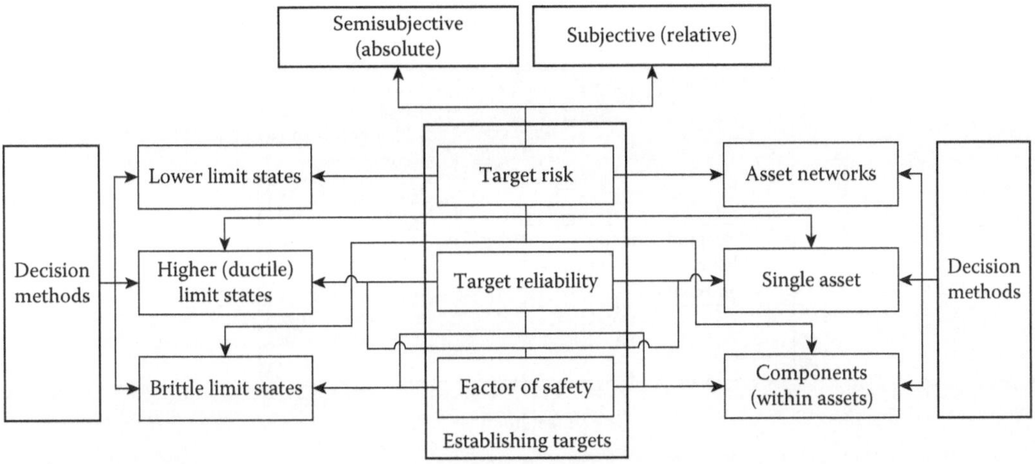

FIGURE 4.4 Categories of acceptance methods in civil infrastructures.

4.3 SUBJECTIVE RISK ACCEPTANCE METHODS

4.3.1 Overview

Several subjective acceptance methods for risk are available. Though some of these methods can be applied to civil infrastructures, others might not be directly applicable. We start by presenting a quick overview of some of the available acceptance methods for subjective risk. We then offer in-depth methods for acceptance of subjective risk that might be more suitable for use in civil infrastructure applications.

4.3.2 Existing Methods

Over the years, many researchers offered methods for subjective risk acceptance. Some of these methods are discussed below.

Risk conversion factors: Let us assume that risks were assessed because of several events that include slow degradation due to the environment the structure is subjected to natural hazards and/or man-made hazards. The risk assessment processes might be different for each of these hazards in their underlying assumptions, and, therefore, the resulting risk values would be difficult to compare to each other. To enable such risk comparisons, conversion factors were proposed for such situations (Ayyub 2003). Use of conversion factors in the field of civil infrastructures might be of limited value because computing risk and its components (threats, vulnerabilities, and consequences) can be handled evenly (e.g., by careful modeling of risk variables) to eliminate the need for them. An example of such careful modeling can be found in the studies by Federal Emergency Management Agency (FEMA) (2009) and Department of Homeland Security (DHS) (2001a, 2011b, 2011c). The authors of those studies noted that risks resulting from failure of bridges should not be directly compared if the type of failures were different. Catastrophic (sudden) failure is considered more risky than gradual failure. By including the types of failure in the risk model (see case studies in Chapter 5), a risk conversion factor is not needed; that is, its equivalent is already built into the risk model.

Farmer's curves: The Farmer curve method (Farmer 1967a, 1967b) describes a function of consequence, C, and likelihood, L, in the space of different hazards in the form

$$f(C, L) = 0 \tag{4.3}$$

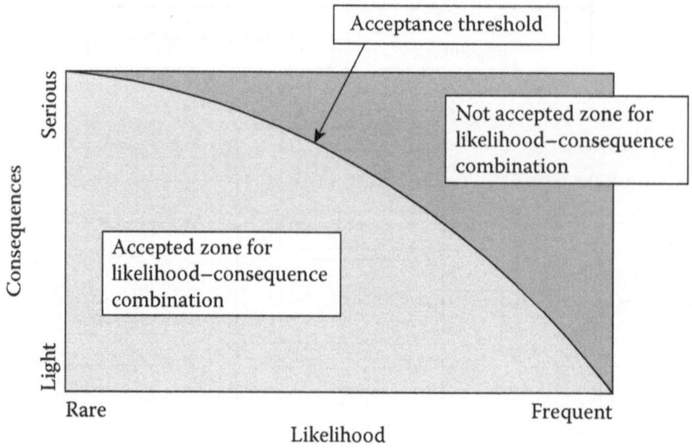

FIGURE 4.5 Concepts of framer curve acceptance method.

This function, within the *C–L* space, shows a zone of acceptable and unacceptable risks as shown in Figure 4.5. Obviously, such functions need to be generated for different types of hazards for each component and for each system. Also equally obvious, the method of generating this function needs to be subjective: it needs some judgment as to the levels of acceptable and unacceptable risks. For the field of civil infrastructure, the value of the use of the method of predetermined acceptance functions for specific components, systems, and network of systems is not clear. A more flexible acceptance system, which accounts for the interactions between different variables in the risk model, should prove more suited and pertinent than fixed/predetermined functions similar to that of Equation 4.3.

Intangibles: In many situations, risk acceptance depends on intangible, that is, nonidentifiable processes. For example, people accept risk in daily routine situations (crossing busy traffic lanes, travel in a plane, visiting a location with dangerous elements, living in a flood zone, etc.) based on their perception, which is an intangible issue. Social or political factors may lead the populace to accept or reject risks, again for not well-defined reasons: an intangible process of acceptance. Ayyub (2003) discussed and analyzed several intangible acceptance methods. Intangible acceptance processes should not be used when addressing risk acceptance for civil infrastructures. Even though we agree to the inevitability of judgment in the risk acceptance process, these judgments need to be based on some reasoning or logic.

4.3.3 Components Based Risk Acceptance

4.3.3.1 Multicomponent Acceptance

Some subjective risk acceptance processes are based on accepting one or more of the risk components. Recall that risk is composed of three components: threat, vulnerability, and consequence. Several civil infrastructure practices include acceptance of consequences, reliability, or exposure only. We discuss briefly these situations and explore how to take advantage of these well-established practices of risk components in a more inclusive acceptance process for risk acceptance. For a two-component risk expression, $Ri = f(x_1, x_2)$, if we know the acceptance thresholds of one of the variables, say x_1 to be $x_1|_{Accepted}$, we need to estimate the acceptance threshold of x_2 as $x_2|_{Accepted}$. As stated earlier, in civil infrastructures, x_1 can be either reliability or exposure, which means that x_2 is the consequences. Thus, in those situations we need to estimate acceptance thresholds for consequences. $x_2|_{Accepted}$ needs to be estimated while accounting for the pertinent interactions between all controlling variables in the risk model. Lack of this all-important step might render the whole process inaccurate. We offer specific observations on the above-suggested approach next.

4.3.3.2 Reliability-Based Acceptance

Accepted reliability thresholds have been studied greatly and, they represent the foundations of Load and Resistance Factor Design (LRFD) design basis (see AISC 1999; ASCE 2010; AASHTO 2014; ACI 2014). Another form of capacity/demand acceptance thresholds in the civil infrastructure community is the factor of safety or the Allowable Stress Design (ASD) design paradigm. Again, acceptance thresholds in ASD paradigms are well established.

When a risk model uses reliability as one of its components in the form, *i*-th = *f* (Reliability, Consequence), it might be possible to establish a relationship between different accepted reliability thresholds for various reliability limit states and estimations of possible consequences at the limits states of those accepted reliability thresholds in the form

$$f\left(Re|_{Accepted}^{i}, C|_{Accepted}^{i}\right) = 0 \tag{4.4}$$

with $Re|_{Accepted}^{i}$ and $C|_{Accepted}^{i}$ representing the accepted reliability and the corresponding accepted consequences at the *i*-th reliability limit state, respectively. This form can be used to produce accepted risk thresholds in any risk model.

4.3.3.3 Exposure-Based Acceptance

We can also develop risk acceptance thresholds through the dependence of risk on exposure and consequences. Recall that, as we argued in Chapter 3, exposure is, similar to reliability, a function of capacity and demand. As such is has several uses in the field of civil infrastructures. One of the main uses of exposure is a subjective definition of condition ratings of structures (see Figures 4.6–4.9). When the risk model uses exposure, *E*, as one of its components in the form *Ri* = *f* (Exposure, Consequence), it might be possible to establish a relationship between accepted exposure thresholds for exposure limit states and estimations of possible consequences at those limit states in the form

$$f\left(E|_{Accepted}^{i}, C|_{Accepted}^{i}\right) = 0 \tag{4.5}$$

with $E|_{Accepted}^{i}$ and $C|_{Accepted}^{i}$ representing the accepted exposure and the corresponding accepted consequences at the *i*-th exposure limit state (e.g., bridge condition rating level), respectively. This form can be used to produce accepted risk thresholds in any risk model.

4.3.3.4 Three-Component Acceptance

We now generalize the above methods for acceptance if risk models include all the three basic risk components, that is, *Ri* = *f*(C, T, V). In such a situation, we have to develop the form

$$f\left(V|_{Accepted}^{i}, T|_{Accepted}^{i}, C|_{Accepted}^{i}\right) = 0 \tag{4.6}$$

with $V|_{Accepted}^{i}$, $T|_{Accepted}^{i}$, and $C|_{Accepted}^{i}$ representing the accepted vulnerability, threat, and the corresponding accepted consequences at an appropriate *i*-th exposure limit state, respectively. This form can be used to produce accepted risk thresholds in any risk model that expresses the three basic components explicitly. Note that this approach is a generalization of the accepted consequence process in Ayyub (2003).

4.3.4 Risk vs. Opportunity or Reward (Cost–Benefit) Acceptance Methods

This method in establishing the risk acceptance threshold is based on the presumption that for every risk (a negative paradigm) there exists a reward (a positive paradigm). If we express risk in monetary

FIGURE 4.6 Example of medium exposure (corrosion on main arch girders) and high consequences bridge.

FIGURE 4.7 Example of low exposure (bridge is in good maintenance condition) and very high consequences bridge.

FIGURE 4.8 Example of low exposure (bridge is in good maintenance condition) and medium consequences bridge.

FIGURE 4.9 Example of low exposure (bridge is in good maintenance condition) and low consequences bridge.

terms, then it essentially becomes a cost–benefit scenario. The acceptance threshold in this situation is expressed by:

$$Ri \leq Reward \tag{4.7}$$

or

$$Cost \leq Benefit \tag{4.8}$$

Computing *Ri* or *Cost* is scientific; most case studies in this book offer one form of such computations. On the other hand, computing *Reward* or *Benefit* is more involved. Ettouney and Alampalli (2012a) offered extensive discussion on the subject of *Benefit* and Ayyub (2003) presented the *Cost/Benefit* paradigm as an acceptance process. Case Study 4.4 presents a study of risk/reward acceptance.

4.4 SEMI-SUBJECTIVE RISK ACCEPTANCE METHODS

4.4.1 Overview

A few methods are a bit less subjective than the methods offered in Section 4.3. This group of methods, though being still subjective in the limit, contains more quantitative components. Each method has its strengths and weaknesses. Ultimately, the stakeholders should use the acceptance method that suits their specific needs.

4.4.2 Utility Theory

4.4.2.1 Overview

We build on our discussion of the cost-benefit (or risk-reward) process in Section 4.3.4 by noting the two main disadvantages of the method:

- It requires too much information that might not be available to the decision maker.
- It lacks the subjective feel that many decision makers have; such subjective feel is extremely difficult to enumerate in a true cost-benefit analysis.

To avoid such limitations, Von Neumann and Morgenstern (1944) developed a new process that is based on developing utility functions. The theory of utility has been used since then as an alternate method for decision-making processes. Table 4.1 shows the main differences between traditional cost-benefit and utility function methods.

In the remainder of this section, we present briefly the theory of utility and types of utility functions. We end the section by discussing how to use utility in the process of risk acceptance.

4.4.2.2 Theory of Utility

According to Trueman (1977), "Utility is the subjective numerical measure of the value of an act to a decision maker when a particular event occurs." In other words, utility is a subjective definition of value.

There are innumerable types of utility metrics as there are decision makers. Every decision maker can establish the type of utility that suits the project at hand. Of course, care must be taken in deciding upon the type of utility used. Some good attributes of utility metrics are:

- *Accuracy*: The utility used must be accurate in describing the type of project. At least, it must be consistent with the *spirit* of the project. For example, monetary or cost utility is suited for projects where costs are an important factor in the decision-making process.

TABLE 4.1
Comparison between Value and Utility Methods in Decision-Making

Cost-Benefit	Utility
Mostly quantitative foundations	Mostly qualitative
Needs mostly objective/mathematical descriptions	Can be adjusted to account for subjective situations
Analytically more rigorous	Less analytical rigor
Due to its rigor, does not need as much care in formulating the problem	Needs more care in formulating the problem
Needs care in interpreting the results	Needs care in interpreting the results

Safety as a metric might be a good utility for a structural security measure. Availability is a good metric for research-oriented structural health monitoring (SHM) project.

- *Simplicity*: The utility metric should be simple to sort out. For example, the monetary utility is fairly simple to understand and apply. In some risk-related decision-making processes, utilities such as vulnerability, asset value, threat, or consequences can be used. They are fairly simply to apply in a qualitative sense.
- *Efficiency*: The utility metric should be efficient in describing as many aspects of the project as possible. Short of this, multiple utilities might be needed to accomplish this objective. Relative risk analysis is a good example of this need. Three utility metrics are used to describe risk levels in many situations: hazard, vulnerability, and consequences.

After deciding upon the type of utility, X, we have to define the utility function, $U(X)$. The choice of the utility value depends on the level of analyst experience. Thus, in using the utility approach, it is prudent for the analyst to be as familiar with the subject as possible.

4.4.2.3 Types of Utility

4.4.2.3.1 Monetary/Cost Utility

The monetary utility is a popular utility commonly used in financial decisions. Well-known qualitative illustrations of different monetary utility function are shown in Figure 4.10

Cost utility describes cost attributes of different aspects of the risk model. It is especially beneficial when exact costs are not well known to the analyst.

4.4.2.3.2 Benefit Utility

We propose another possible utility in the field of risk and risk acceptance: the benefit utility. The benefit utility can be established in a manner that is similar to the cost utility above. The main difference is that the conservative cost utility is similar in shape to the aggressive benefit utility, and the reverse is true.

4.4.2.3.3 Utility of Experimentations

We introduce here another method based on the utility theory to evaluate experiments: utility of experimentation. The utility function can be formed in the same fashion as before. The professional would estimate the maximum and minimum values that are to be obtained from performing an experiment. A maximum and minimum utility of 1.0 and 0.0 is assigned to the function. The midpoint is then estimated following similar procedures as before.

4.4.2.3.4 Other Types of Utility

Other potential utilities can be used. For example, utilities of consequences, vulnerability, and damage can all be established to help the professional in making appropriate decisions.

4.4.2.4 Use of Utility in Risk Acceptance

Upon further reflection, we can approach the utility function as an expression of risk behavior. For example, if we use monetary values (cost) to express utility functions, we can design the function to be the absolute risk, which is the cost of the problem on hand. Similarly, for reward (or benefit) problems, we can assign utility function to represent the benefit/reward and so forth. The user will then assign a particular utility (cost) as an accepted threshold. The assessed risk is then compared to the accepted utility on the utility function curve. Figure 4.11 shows the three utility functions of Figure 4.10 when used in an acceptance problem. Clearly, the risk acceptance threshold can vary as the utility functions vary from conservative to neutral to aggressive. The same approach can be made for any other type of utility as discussed above. We highlight several examples in the remainder of the book that use utility theory in risk acceptance and risk treatment.

4.4.3 Value at Risk

Value at risk (VAR) is a both an assessment and acceptance method that is popular in the financial community (Jorion 2006). It is also known as *financial exposure* and is based on assessing absolute risk of a particular situation, mostly for financial entities such as portfolios, investments, etc. Absolute risk is described using monetary metrics. After assessing the risk, usually called VAR, the manager has to decide whether this VAR is acceptable. Based on such a decision, appropriate actions are taken. Generally speaking, VAR can involve complex analytical processes in assessing the risk. Yet the decision to accept VAR is ultimately subjective. Applying the VAR approach in civil infrastructure is similar: analytical in the assessment phase and subjective in the acceptance phase.

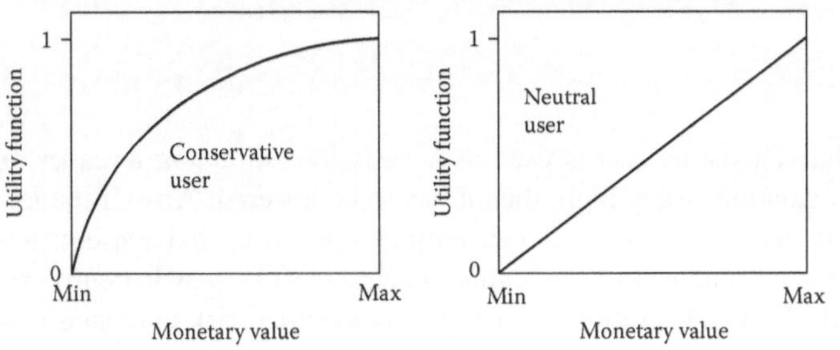

FIGURE 4.10 Monetary utility functions.

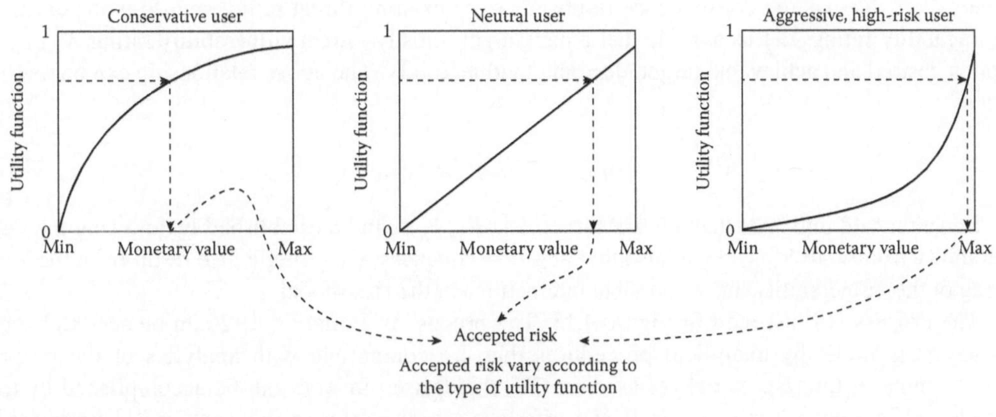

FIGURE 4.11 Acceptance process using utility functions.

The subjectivity in acceptance of VAR can be reduced by following one of two processes, which we make specific for practices in the civil infrastructure community. These are discussed in the next two sections.

4.4.4 Available Budget Method

Available budget method (ABM) is similar to VAR methods in the sense that the acceptable risk is a given budget at hand. The ABM applies to situations where absolute risk is assessed using monetary values. Let us assume that an absolute risk, *Ri*, is assessed for a particular situation. We only ask that the units of *Ri* are in dollars (or any other monetary system).

If there is an allowed fixed budget for this situation, $Ri|_{Allowable}$, then this allowed budget is simply the acceptable risk such that

$$Ri \leq Ri|_{Allowable} \tag{4.9}$$

For civil infrastructure projects, some safety/performance conditions also have to be satisfied such that the function

$$G(T, V) = 0 \tag{4.10}$$

Note that T and V represent hazards and vulnerabilities.

Figure 4.12 illustrates the steps of ABM. The optimal solution of Equations 4.9 and 4.10 is the desired solution. The enforcement of Equation 4.10 can be assured if the risk model that evaluates *Ri* includes adequate representations of T and V. In all of the cases studies in this book, we allow for such representations.

4.4.5 Target Risk Method

In this method, the estimated relative existing risk is used as a basis for estimating a reasonable target risk. The rationale is that if existing risk is high, then it has to be lowered. Also, if the level of risk is moderate, then there is no need to change it. Because threats (hazards) and consequences are generally difficult to modify, it is reasonable to assume that the only realistic way to control risk is to change its vulnerability level. As an illustration, we know that assessed risk is related to its existing components as:

$$Ri = f(C_e, T_e, V_e) \tag{4.11}$$

where C_e is the existing consequence rating, T_e is the existing threat rating, and V_e is the existing vulnerability rating. Let us assume that a new target (mostly lower) vulnerability rating, V_{Target}, is both a desired and achievable target (desirable) vulnerability. The above relationship can be rewritten as:

$$Ri|_{Target} = f(C_{New}, T_e, V_{Target}) \tag{4.12}$$

Equation 4.12 indicates that a lower target risk, $Ri|_{Target}$, can be established by lowering the vulnerability from V_e to V_{Target}. Note also that new consequences, C_{New}, might also result from the lowering of the vulnerability due to possible interactions in the risk model.

The process is illustrated in Figure 4.13. The process in Equation 4.12 can be accomplished in any risk model by analytical procedures that are compatible with analytics of the model. For example, if the risk model is based on GN, the desired inverse can be accomplished by the process of inference (see Chapter 2). We offer case studies later in this section illustrating this process.

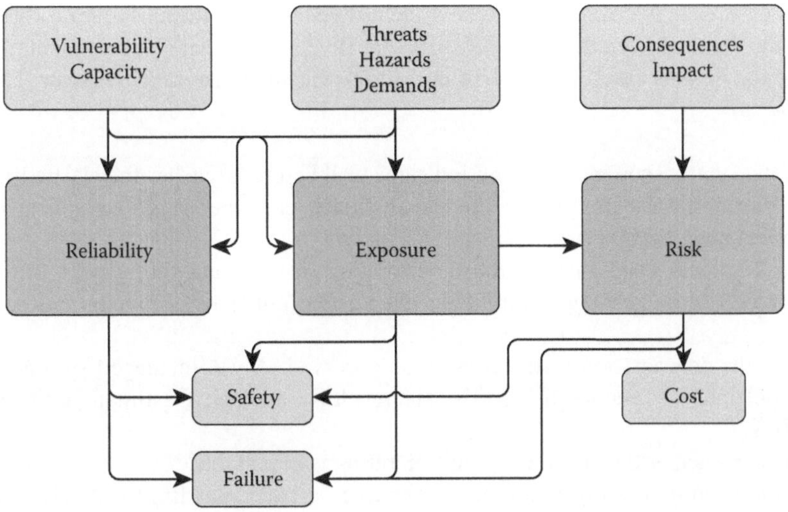

FIGURE 4.12 Available budget method.

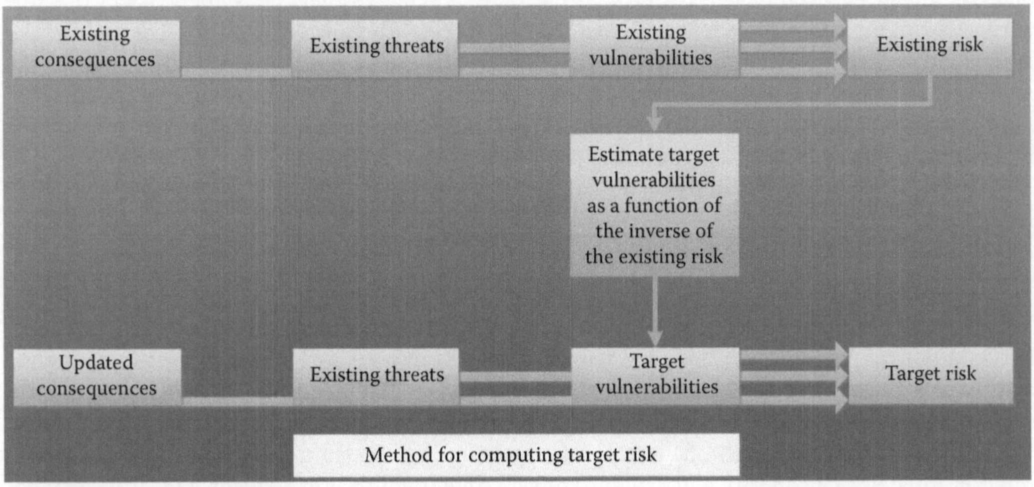

FIGURE 4.13 Target risk method.

4.5 LOWER LIMIT STATES: DETERIORATION OF INFRASTRUCTURES

4.5.1 Overview

Perhaps the most studied limit state in a civil infrastructure community is the elastic limit, that is, when the structure starts behaving in a nonlinear manner. However, most management decisions during the life span of the system are usually made at LLS than the elastic limit. This can be explained by observing the behavior of the civil infrastructure as time marches (see Figure 4.14). As time progresses, the condition rating of the system deteriorates. The deterioration happens because of multitudes of hazards (corrosion, fatigue, thermal freeze/thaw, or simple wear and tear, among many others). As the condition of the system deteriorates, the exposure and risk increase. This increase can be understood by looking at capacity and demand. As time marches, hazards affect the construction material in a gradual manner. This coupled with a healthy factor of safety in the initial design ensures that the capacity of the system remains under the elastic limit for a long time. Meanwhile, the demand from the civil infrastructure (live load, societal demands, etc.) increase in a gradual fashion in most cases. Figure 4.14 also shows this time marching behavior. This results in the following:

1. The system starts showing deterioration signs well before it reaches the elastic limit.
2. Such deterioration, being under the elastic limit, can't be negotiated using common Newtonian mechanics.
3. The stakeholders need to treat the deterioration and prevent the system from further deterioration while keeping it well below the elastic limit. This is done by following three steps:
   a. First the deterioration state needs to be observed. A well-managed system of visual inspection (*V.I.*) accomplishes this. The result is a subjective estimate of the condition rating.
   b. Decide whether the observed condition rating is acceptable.
   c. If the condition rating is not acceptable, i.e., a particular limit state (LLS than the elastic limits) of condition rating is reached, pertinent retrofit measures are taken.

Obviously, each of these LLS thresholds corresponds to a certain accepted exposure threshold and risk threshold. Even though the process currently is subjective and does not consider risk, the acceptance of these LLS thresholds is actually a risk acceptance process at its core. The stakeholder is implicitly accepting a certain degree of risk at each of these LLS of condition ratings. Given the subjective nature of *V.I.* in deciding upon LLS, it is difficult to study the problem objectively without incorporating consequences into the mix. This entails looking at it as a risk-based process. We attempt to do that by offering two case studies next.

4.5.2 Case Study 4.1: Deterioration of Single Asset

4.5.2.1 Overview

We showed above that LLS are defined for civil infrastructures, particularly bridge structures, by condition ratings that are assigned subjectively to the civil assets following a preassigned set of rules (inspection guidelines, for example). We also explained that stakeholders perform certain responses and actions whenever the asset in question reaches each of these limit states. We argued that because each of these limit states implies a certain reliability state (a specific relationship between capacity and demand) or risk state (relationship between capacity, demand, and consequences), we can consider those levels of reliability or risk as an accepted reliability level or risk level at that particular LLS that is subjectively assigned via subjective assignment of condition rating. There are several implications for this LLS acceptance assignment, especially for risk acceptance. We investigate some of these implications in this and the following case studies.

In this example, we show how to evaluate accepted risk at LLS via a simple risk model and then explore some of the implications of the current practices.

4.5.2.2 Model

We use a simple BN risk model as shown in Figure 4.15. We assume that the exposure of the bridge due to deterioration is simulated by condition ratings resulting from visual inspection, *V.I.* We also assume that the total consequences, *C*, is a function of both the exposure of the bridge (as simulated by *V.I.*) and external consequences, *CO*. The states of nodal variables of the model are shown in Table 4.2. We use monetary states for the risk, *Ri*, of this model that make it an absolute risk problem. The definitions of the nodal variables are shown in Table 4.35.

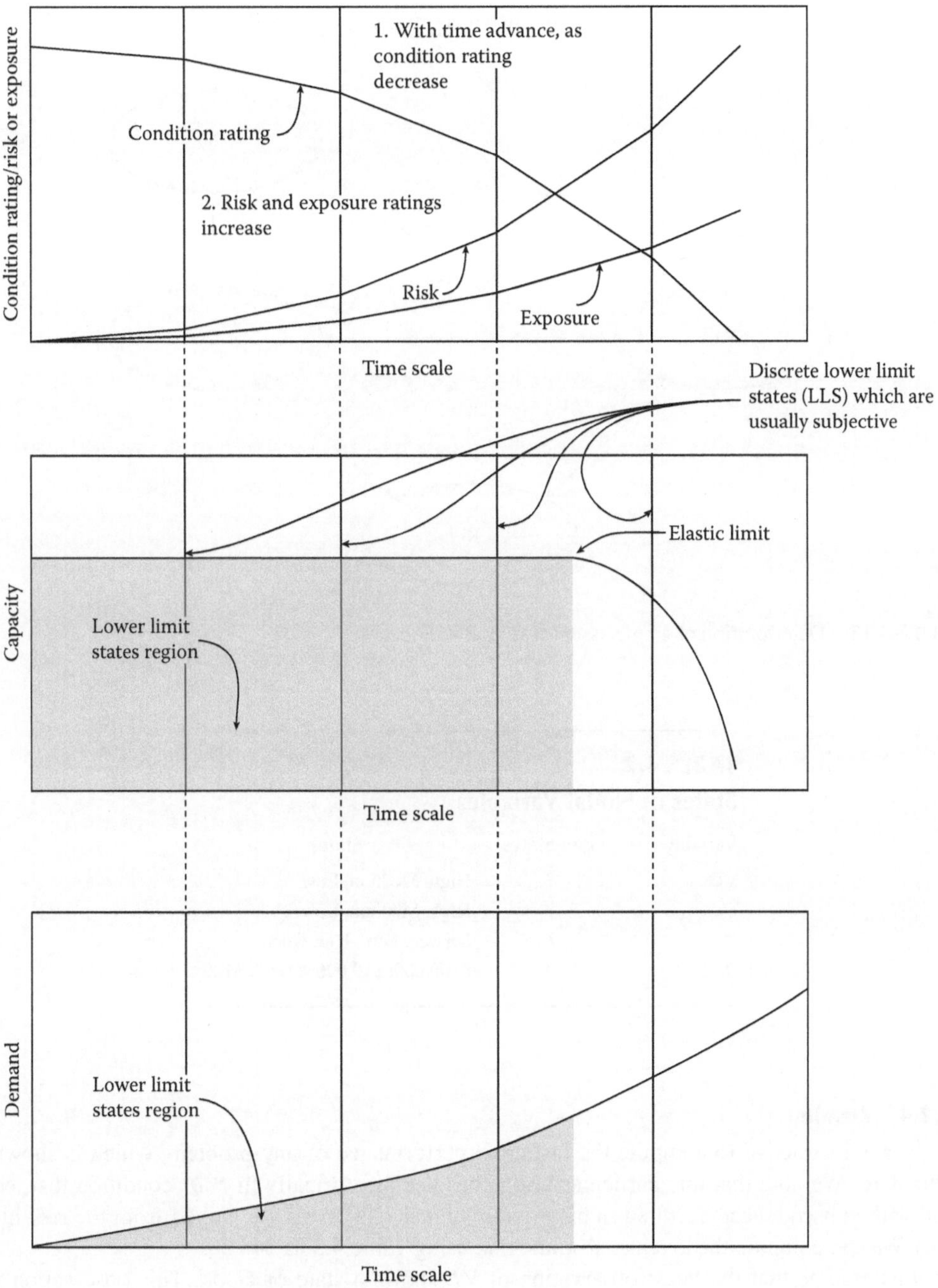

FIGURE 4.14 Time marching and lower limit states.

4.5.2.3 Conditional Probabilities Tables

The different conditional probabilities tables (CPTs) of each variable of the network of Figure 4.15, are shown in Appendix 4.II. As usual, we used our best judgment in filling the different tables in a realistic manner for illustration purposes. For more on different methods of developing CPTs, see Chapter 2.

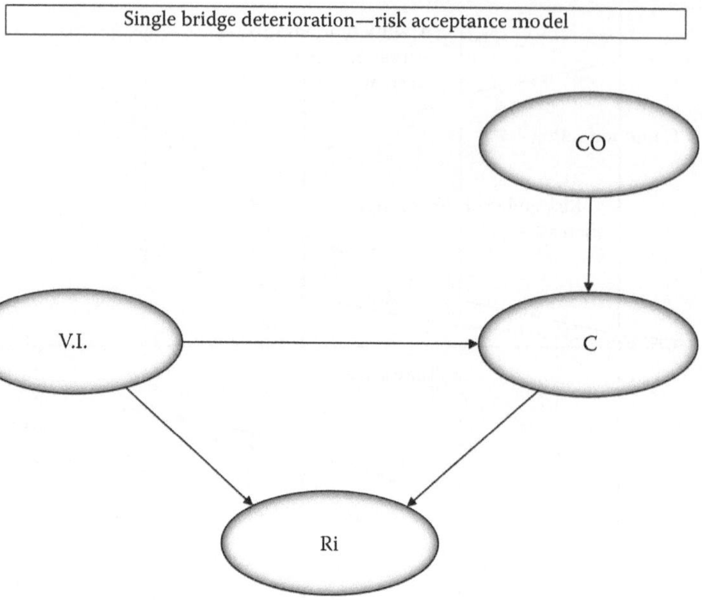

FIGURE 4.15 Deterioration of single asset (bridge) BN model.

TABLE 4.2
States of Nodal Variables

Variable	# of States	States
CO	3	High, Medium, Low
C	3	High, Medium, Low
V.I.	4	Service, Poor, Fair, Good
Ri	4	$100,000, $10,000, $1000, $100

4.5.2.4 Results

It is always instructive to compute the historical performance of any problem, which is shown in Figure 4.16. We note that this particular bridge has been historically in *Poor* condition that, combined with consequences, resulted in historically higher risk levels (as judged from the risk histogram). We can compute the average absolute risk using Table 4.3 as $44,533

Let us assume that the latest observation of *V.I.* shows a state of *Good.* This observation will result in a changed network probabilities as shown in Figure 4.17. The equivalent average risk in this situation can be computed using Table 4.4 as $1976.50. Of course the risk is decreased markedly because the bridge condition is now much better than the historical one. Note that the outside consequences, *CO*, remain the same because it is independent from the bridge condition. Using the LLS argument we presented earlier, we can state that the accepted risk at *Good* limit state is $1976.50. The implication here is that if the condition changed to, say *Poor*, the risk state will increase beyond the accepted $1976.50 threshold, and the stakeholder triggers some kind of response (perhaps in a form of maintenance action). We can find other accepted risk thresholds at different LLS of *Fair*, *Poor*, and *Service* following the same process. The accepted risk thresholds at the four LLS are computed and illustrated in Figure 4.18.

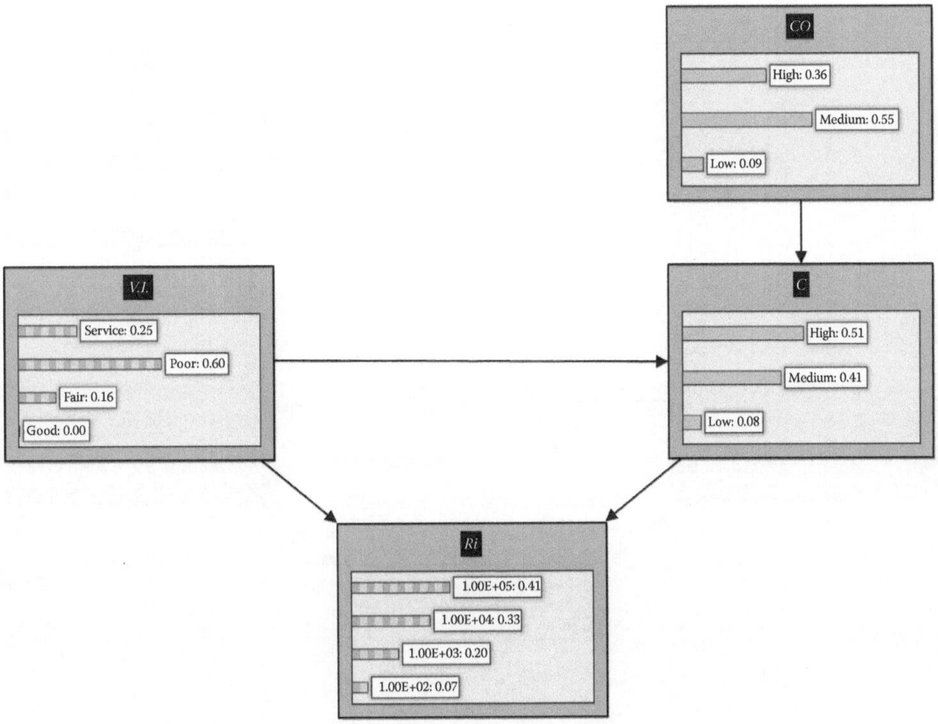

FIGURE 4.16 Deterioration of single asset (bridge): historical probabilities.

TABLE 4.3
Historical Risk for Single Asset (Bridge)

Rank, *i*	$(Ri\,\|_i)$, \$	$p(Ri\,\|_i)$	$(p(Ri\,\|_i))(Ri\,\|_i)$
1	\$100	0.067	\$67
2	\$1000	0.196	\$196
3	\$10,000	0.327	\$3270
4	\$100,000	0.41	\$41,000
Total $(Ri\|_i)$, \$			\$44,533

The above process seems, on the surface, fairly safe and logical. Actually, it is one of the basic foundations of the current bridge management processes: use the changes in condition ratings as observed by the process of visual inspection as the basis for maintenance and retrofit actions in general. On further inspection of the process, we can identify a weakness in this process. When we introduced an observation of *V.I.* = *Good* as a departure from the historical norm, we left the outside consequences constant (as evident from Figures 4.16 and 4.17). Such an assumption is not warranted. Consequences can and will change as easily as bridge conditions may or may not deteriorate. When that happens, the resulting risk will change accordingly. If we assume that in addition to the observation *V.I.* = *Good*, we also observed that the state of the consequences *CO* is now *Low*, the network risk changes as shown in Figure 4.19. The resulting absolute risk is computed in Table 4.5 as \$2577. The absolute risk increased by 30% because of the change in consequences while the state of bridge is still *Good*.

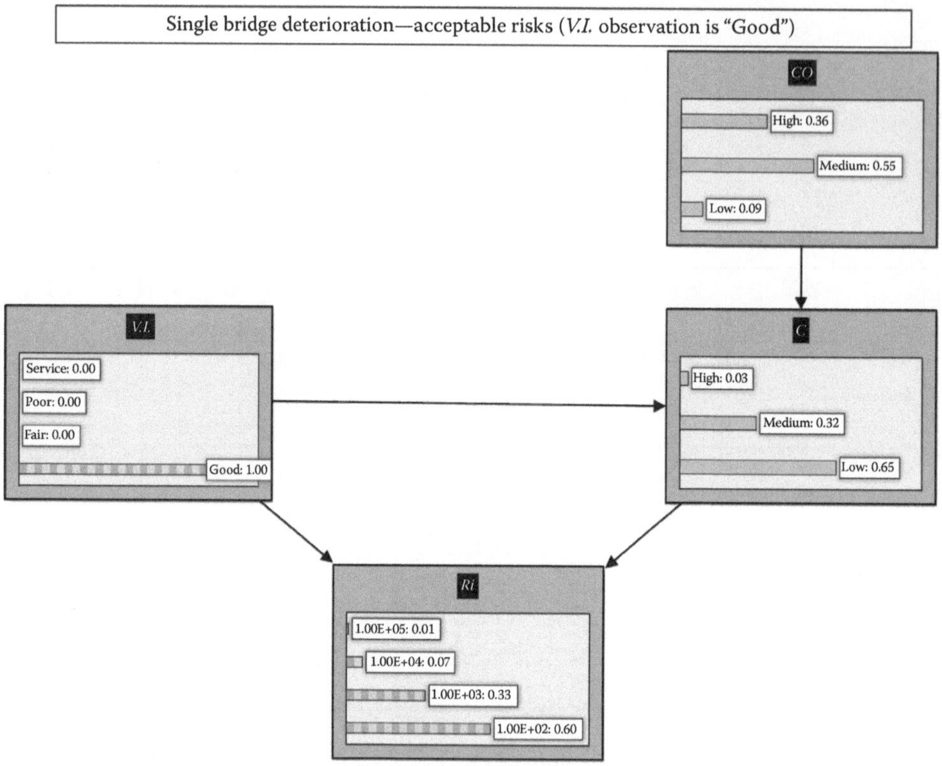

FIGURE 4.17 Deterioration of single asset (bridge): accepted risk as a *Good* observation.

TABLE 4.4
Accepted Risk for Single Asset (Bridge)

Rank, i	$(Ri \mid_i)$, $	$p(Ri \mid_i)$	$(p(Ri \mid_i))(Ri \mid_i)$
1	$100	0.595	$59.50
2	$1000	0.327	$327
3	$10,000	0.069	$690
4	$100,000	0.009	$900
Total $(Ri\mid_i)$, $			$1976.50

4.5.2.5 Summary

We can infer two important lessons from this simple case study:

- The current subjective condition rating of assets, particularly for structures such as bridges and tunnels using visual inspection practices, implies an establishment of an equivalent accepted risk (and reliability) threshold, which can be objectively defined using the method of this case study.
- For any risk-based decision-making management process that relies on decisions based on visual inspection results, care needs to be given to the inevitable changes in consequences. Neglecting consequences when embarking on such a decision would render the foundations of the risk-based processes inaccurate.

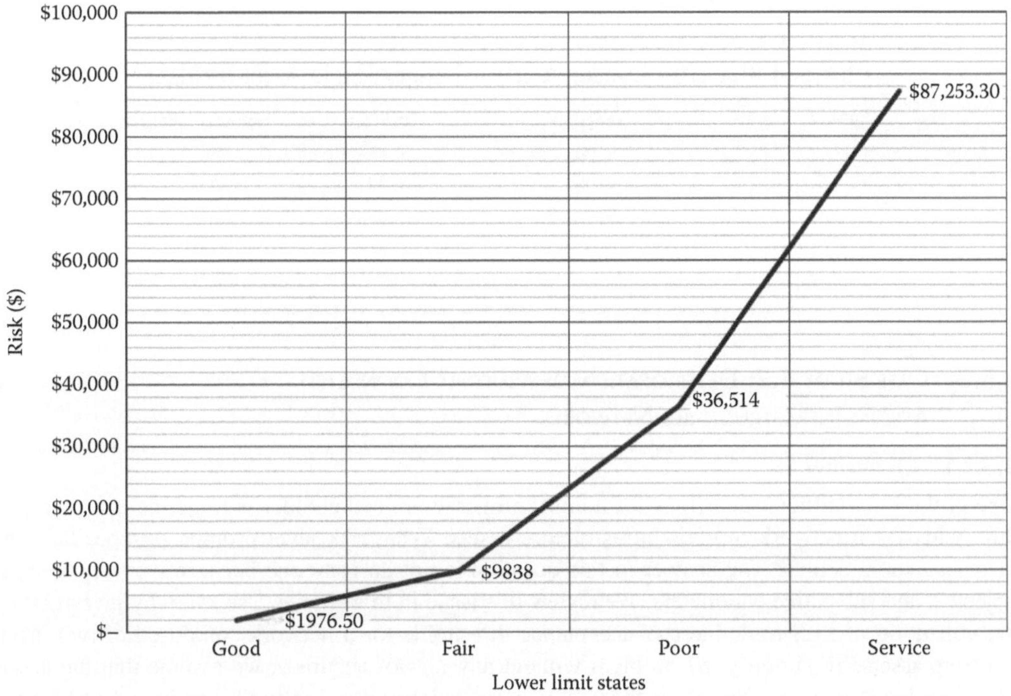

FIGURE 4.18 Accepted risk thresholds at lower limit states.

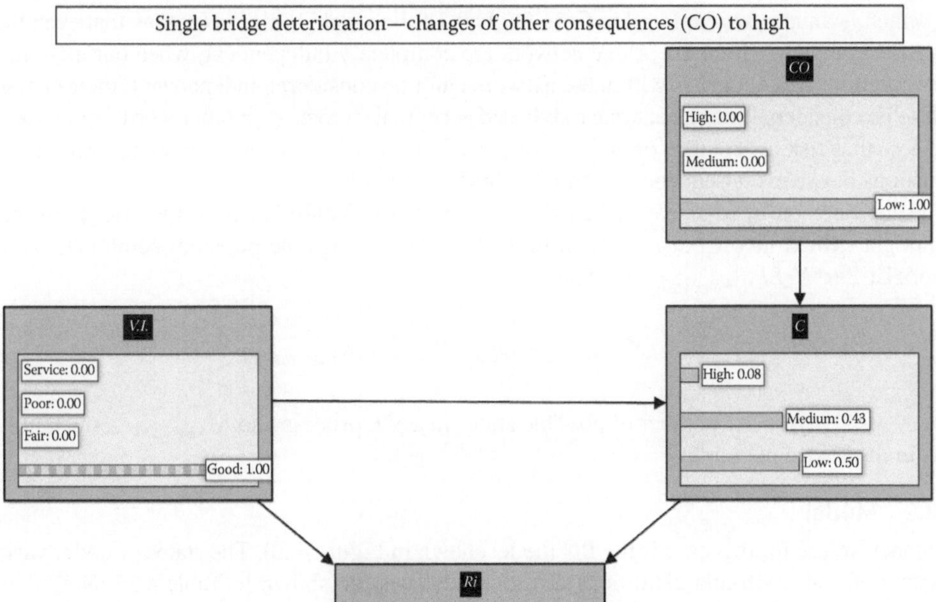

FIGURE 4.19 Deterioration of single asset (bridge): risk changes as consequences (CO) change.

TABLE 4.5
Changed Risk after Consequences Change, Single Asset (Bridge)

Rank, i	$(Ri\,|_i)$, \$	$p(Ri\,|_i)$	$(p(Ri\,|_i))(Ri\,|_i)$
1	\$100	0.558	\$56
2	\$1000	0.341	\$341
3	\$10,000	0.088	\$880
4	\$100,000	0.013	\$1300
Total $(Ri|_i)$, \$			\$2577

4.5.3 Case Study 4.2: Deterioration of Multiple Connected Assets: Infrastructures Networks

4.5.3.1 Overview

One of the most difficult questions for a civil infrastructure stakeholder is to define accepted risk thresholds for a network of civil infrastructure assets. What risk level promotes a specific level of management action? The answer to this question is not an easy one because it involves many objective and subjective parameters. We review this issue in this simple case study by investigating the utilization of LLS to define risk acceptance thresholds for a network composed of two infrastructure assets. To simplify the problem without any loss of accuracy, we assume that the assets are all bridge structures. Our aim is to show how risk acceptance thresholds can be defined for the whole network while accommodating not only exposure (represented in this model via *V.I.*) but also consequences that are modeled separately for each bridge. We use a fairly simple BN model to develop absolute risk. We also illustrate how consequences can interact via modeling the *CO* variable, which is an idea we explored in Case Study 3.1 for a single bridge. We show that even though the *V.I.* observations within the bridge network are completely independent, when our main interest is deterioration, consequences within the network can't be considered independent: their interaction needs to be considered for an accurate risk-based acceptance process. In other words, our case study will show that risk acceptance on a network level is a complex process that should accommodate interactions between consequences within the bridge network.

The approach outlined above still carries one obvious difficulty. Each of the bridges in the network might exhibit its own deterioration limit state. This brings the potential combinations of the network LLS to $N\_LLS|_{NETWORK}$ such that

$$N\_LLS|_{NETWORK} = (N\_LLS|_{ASSET})^2(N|_{NETWORK})^2 \tag{4.13}$$

where $N\_LLS|_{ASSET}$ is the number of possible states in a *V.I.* process and $N|_{NETWORK}$ is the number of assets in the bridge network.

4.5.3.2 Model

The model we use for this case is the BN model shown in Figure 4.20. The states of nodal variables are shown in Table 4.6 and definitions of those variables are shown in Table 4.35. Note that this model is a composite and models each of the bridges in the network in a way that is similar to the bridge in the Case Study 3.1. The main distinction in this bridge network model is that we introduced an external consequences variable, *C*, which is the parent of the two external consequences for each of the two bridges, *CO_1* and *CO_2*. Thus, the two network bridges are now linked not only at the total risk variable, *Ri_NETWORK* but also at the consequences variable *C*. Similar to Case Study 3.1, we assume that the condition rating, as represented by visual inspection state, simulates the bridge exposure. Thus, the individual risks of each bridge, *Ri_1* and *Ri_2*, are the children of the two variables from visual inspection and bridge consequences pairs (*V.I._1*, *C_1*) and (*V.I._2*, *C_2*), respectively. The total network risk, *Ri_NETWORK*, is the child of the individual bridge network risk variables, *Ri_1* and *Ri_2*. All the risk variables in this network are chosen to be absolute risks, that is, their states are monetary.

4.5.3.3 Conditional Probabilities Tables

The different CPTs of each variable of the network of Figure 4.20 are shown in Appendix 4.III. As usual, we used our best judgments in filling the different tables in a realistic manner for illustration purposes. For more on different methods of developing CPTs, see Chapter 2.

4.5.3.4 Results

4.5.3.4.1 Overview

We can explore many aspects of infrastructure network risk assessment and risk acceptance issues with the model of Figure 4.20. First, we look at the historical (prior) risk levels (risk assessment) of the network. We then evaluate the network's risk acceptance levels at LLS of the whole network.

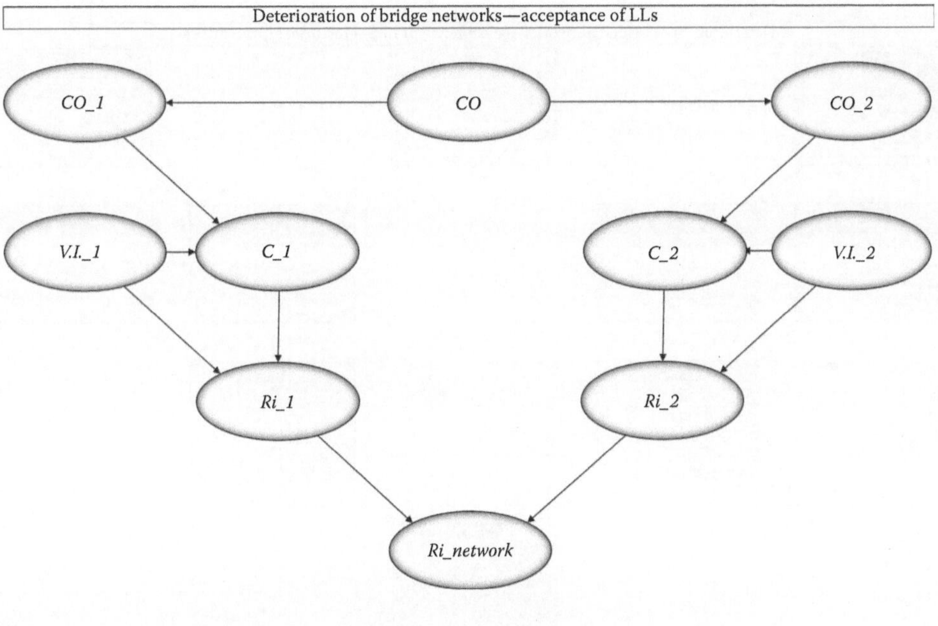

FIGURE 4.20 Risk acceptance BN model for 2-infrastructure assets network.

TABLE 4.6
States of Nodal Variables

Variable	# of States	States
CO_1, CO_2	3	High, Medium, Low
CO	3	High, Medium, Low
V.I._1, V.I._2	4	Service, Poor, Fair, Good
C_1	3	High, Medium, Low
C_2	3	High, Medium, Low
Ri_1, Ri_2	4	\$100,000, \$10,000, \$1000, \$100
Ri_Network	4	\$200,000, \$20,000, \$2000, \$200

We finally explore the issue of consequences and highlight its importance during the process of assigning risk acceptance levels.

4.5.3.4.2 Historical Risk

The network historical total risk can be computed by executing the BN model. The results are shown in Figure 4.21. The average historical risk is then computed to be \$51,082 as shown in Table 4.7.

4.5.3.4.3 Risk Acceptance Thresholds at LLS

There are $N\_LLS|_{NETWORK}$ accepted risk thresholds for the whole network. Each of these thresholds, $(Ri|_{NETWORK}|_A)_{ijkl}$, defines accepted network risk threshold when the i-th bridge in the network is observed at the j-th condition rating state while the k-th bridge in the network is observed at the l-th condition rating state. Obviously, there are a large number of LLS for the network. Even for a 2-bridge network and 4 condition rating states, there are 64 LLS thresholds. We start by exploring the limiting situations where the bridges in the network have identical condition ratings. This will bring down the number of the LLS to only 4. Thus, by enforcing the observations of *V.I._1* = *Good* and *V.I._2* = *Good*, we obtain the probabilities of Figure 4.22. The average accepted risk threshold

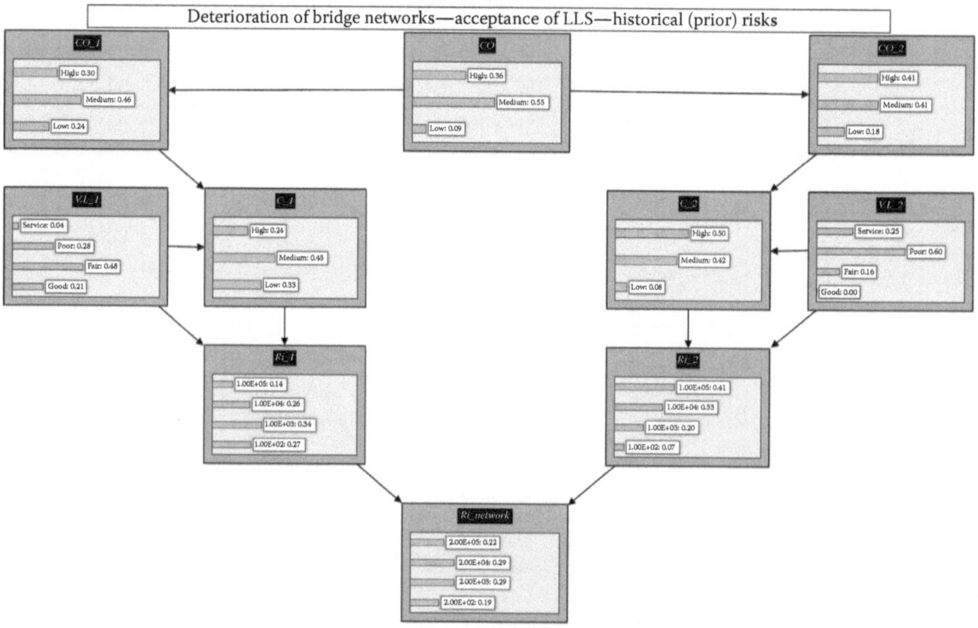

FIGURE 4.21 Risk acceptance BN model for 2-infrastructure assets network: historical probabilities.

TABLE 4.7
Historical Risk for Bridge Network

Rank, i	$((Ri|_{NETWORK})|_i)$, \$	$p((Ri|_{NETWORK})|_i)$	$(p((Ri|_{NETWORK})|_i))((Ri|_{NETWORK})|_i)$
1	\$200	0.192	\$38
2	\$2000	0.292	\$584
3	\$20,000	0.293	\$5860
4	\$200,000	0.223	\$44,600
Total $Ri|_{NETWORK}$, \$			\$51,082

for the network is then computed as in Table 4.8 at $6592. If we computed other accepted risk thresholds assuming the remaining constant condition ratings (*Fair*, *Poor*, and *Service*) across the bridge network, we obtain Figure 4.23. We note that because of the limiting constant condition rating assumption that was used to form Figure 4.23, this figure has only limited, yet important, strategic value to the stakeholders. First, at its bounding points (condition ratings of *Good* and *Service*), we have the total network accepted risks when all bridges in the network are in a *Good* state that presumably indicates that no action is needed across the network or are in a *Service* state that indicates that all bridges in the network are in a worse state. The same implications can be inferred for the intermediate points in the figure.

Figure 4.23 shows interesting and beneficial risk acceptance threshold limits that can guide the stakeholders in deciding different strategic retrofit efforts. However, such benefits are only strategic and cannot be applied to more detailed decisions at an individual bridge level. In such situations, we need to release the assumption of constant condition ratings across the network. The 2D line diagram of Figure 4.23 will change to a 4-D diagram with more demanding computational resources. Such an effort is beyond the scope of this book. We observe, however, that Ettouney and Alampalli (2016) have studied the multihazards interaction between those LLS and

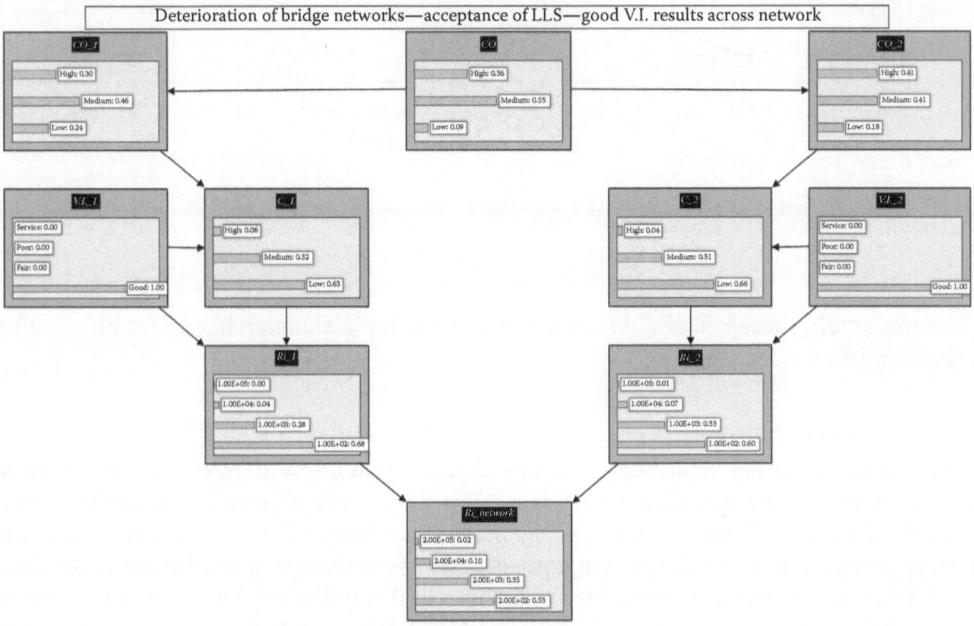

FIGURE 4.22 Risk acceptance BN model for 2-infrastructure assets network: accepted risk when *V.I.* = *Good* observations across network.

TABLE 4.8
Accepted Risk for Bridge Network

Rank, *i*	$(Ri\mid_i)$, $	$p(Ri\mid_i)$	$(p(Ri\mid_i))(Ri\mid_i)$
1	$200	0.529	$106
2	$2000	0.353	$706
3	$20,000	0.099	$1980
4	$200,000	0.019	$3800
Total $(Ri\mid_i)$, $			$6592

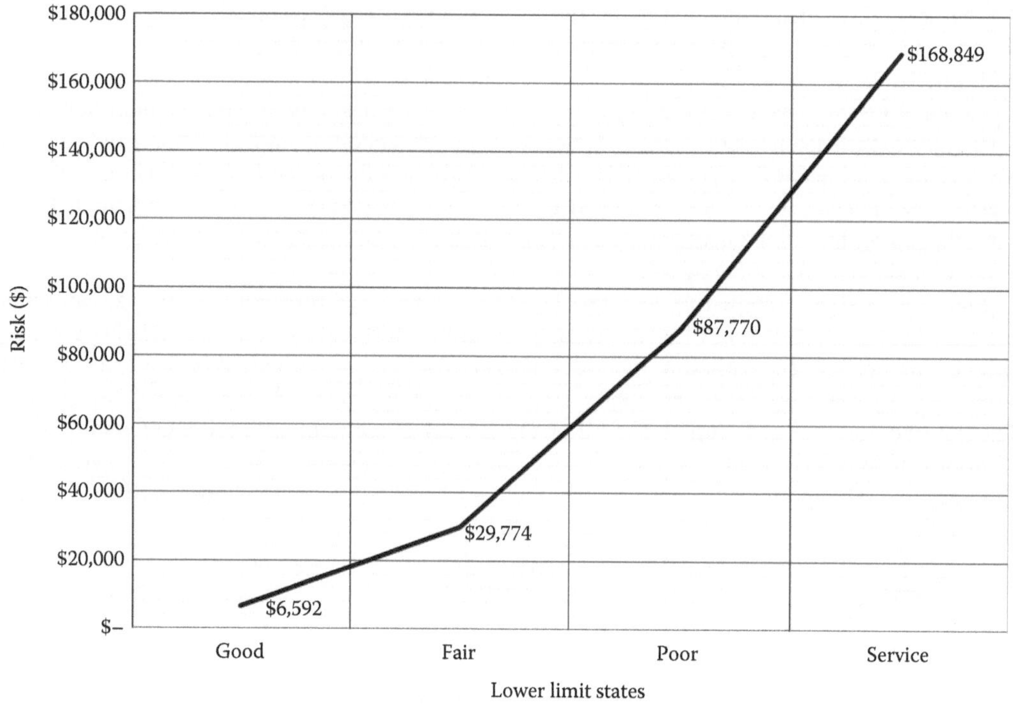

FIGURE 4.23 Accepted total network risk threshold at LLS (assuming constant condition rating across the network).

their corresponding accepted risk thresholds by considering the underlying deterioration of each bridge as an independent hazard.

4.5.3.4.4 Effect of Consequences

Our discussion so far has concentrated on how deterioration of the different bridges in the network (as represented by the condition ratings) affect the implied accepted total network risks as thresholds of the LLS. However, as we pointed out in Case Study 4.1, risk is not only a function of condition ratings; it is also a function of consequences, even though some of those consequences are not a part of the bridge network itself. To illustrate this rather obvious point, let us consider the state of total network risk when all of the condition ratings (*V.I.* states) of the bridges in our current example are *Good.* Let us change the observed state of the external consequences, *CO*, to *High* from its historical state that was kept constant so far. The resulting probabilities are shown in Figure 4.24. The average network risk is computed via Table 4.9 as $7195. This is about 9% higher than the network risk for similar condition ratings with historical consequences (Table 4.8). As concluded before in Case Study 4.1, the implied accepted risks at LLS changes not only according to observed condition ratings but also due to changes in consequences. Neglecting this can lead to inaccurate decisions.

4.6 CASE STUDY 4.3: HIGHER (DUCTILE) LIMIT STATES

4.6.1 Overview

There are many situations in the field of civil infrastructures where the behavior of the asset is allowed to exceed the elastic limit. In such situations, the asset would pass through what is known

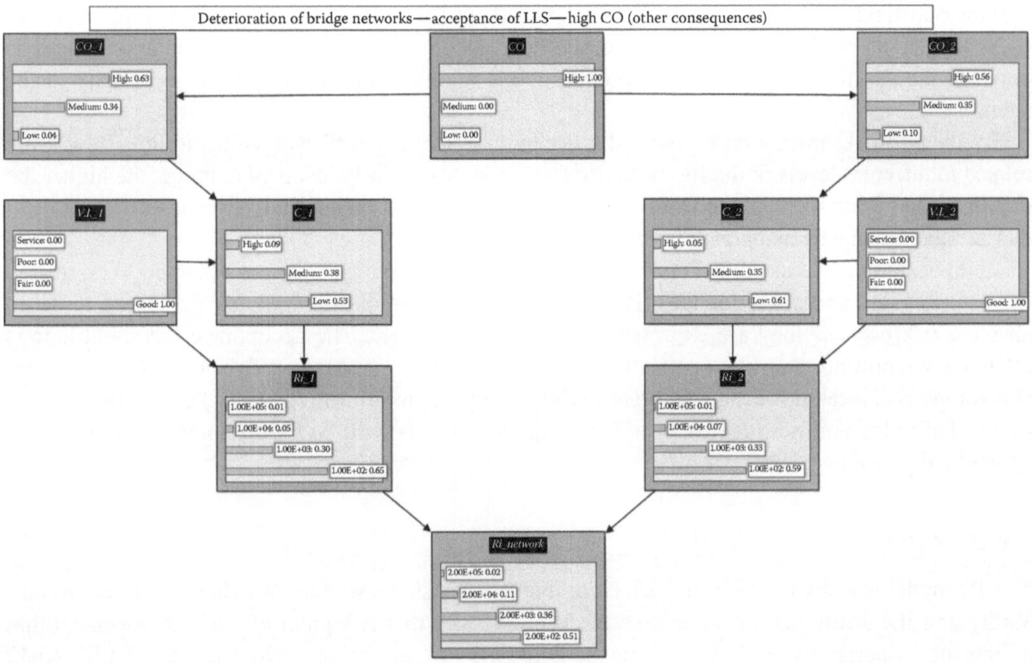

FIGURE 4.24 Risk acceptance BN model for 2-infrastructure assets network: risk changes as consequences (*CO*) change.

TABLE 4.9
Changed Risk after Consequences Change, Bridge Network

Rank, *i*	$(Ri\,|_i)$, $	$p(Ri\,|_i)$	$(p(Ri\,|_i))(Ri\,|_i)$
1	$200	0.514	$103
2	$2000	0.356	$712
3	$20,000	0.109	$2180
4	$200,000	0.021	$4200
Total $(Ri|_i)$, $			$7195

as higher ductile limit states (HDLS). The most obvious examples are assets subjected to earthquakes and/or building security. Specifically, ASCE/SEI (2013) allows for four (continued operations, immediate occupancy, life safety, and collapse prevention) limit states beyond the elastic limit for buildings subjected to earthquake hazard (Ettouney 2015). The common factors among all these situations are: (1) the system is allowed to venture into a regime beyond the elastic limit; (2) the behavior is always ductile behavior, that is, no brittle behavior is allowed; (3) the HDLS stops short of total system failure; and (4) the consequences always increase as the HDLS increases. See Figure 4.25 for a conceptual illustration of the interrelationship among asset vulnerabilities, consequences, and the discrete HDLS. The status of variables vary from 2 to 5, where H = High, L = Low, V.L. = Very low, V.H. = Very high, and M = Medium. Hz is used for Hazard to differentiate *High* and *Hazard*.

The objective of this case study is to quantify the risk acceptance process for civil assets with a finite number of HDLS. We use an influence diagram (ID) process, where the HDLS are used to find the optimal accepted risk level. The use of an ID process allows us to consider graph networks (GN) as a numerical foundation for the solution, accommodating any/all essential interactions between the variables in the decision process while maintaining the simplicity and clarity of the solutions.

By using an ID in this case study, the decisions regarding different vulnerabilities levels are related to different levels of ductile behavior (which are essentially levels of damage: the higher the ductility, the higher the damage levels, thus the higher the cost). These will produce different utilities and the optimal decision is the one that minimizes cost.

In this example, we use the ID utilities to represent risk because they are assumed to be related to both exposure and consequences. Hence, risk in this example is implicitly presumed to be an absolute risk because they are expressed in a monetary metric. The acceptance threshold here is related to the minimum utility (cost) not to a predefined performance (ductility) level. As such, there is no target risk level; however, the target in this example is to minimize (utility) cost. The expectation is that vulnerabilities (ductility) and consequences (costs) will be tightly interrelated as shown symbolically in Figure 4.25.

4.6.2 Model

The ID model is shown in Figure 4.26. Note that the decision variable, *D*, affects both the vulnerability and the utility (as representative of absolute risk); this is logical because the decision that affects the vulnerability level also affects the final utility of the problem. Note also that Utility (risk) is a function of Exposure, *E*, and Consequences, *C*, as usual. Other variables in the network have the usual relationships. This ID model is a shell of the underlying BN model as shown in Figure 4.27. The states of nodal variables are shown in Table 4.10, and the definition of the variables is shown in Table 4.35.

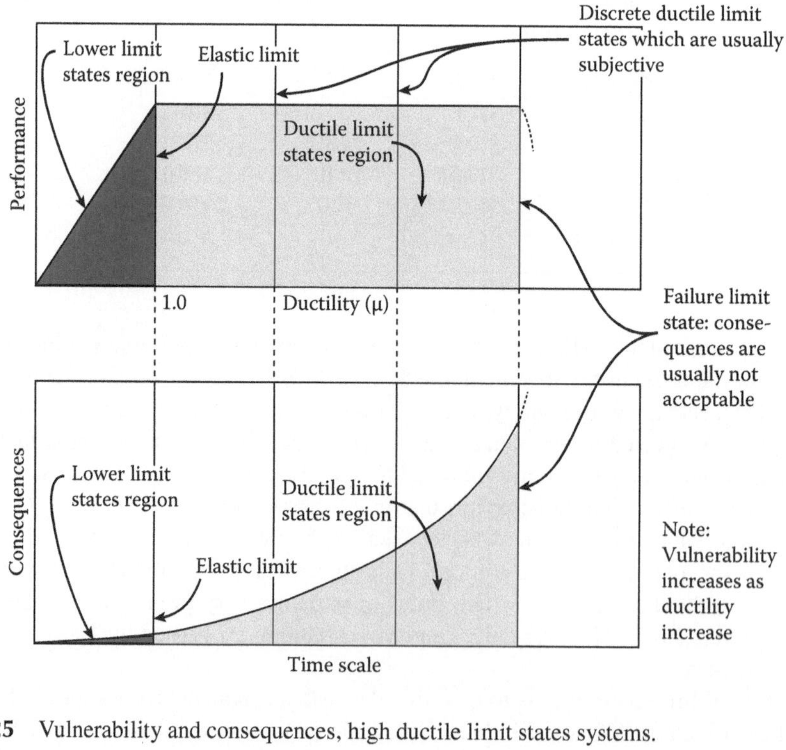

FIGURE 4.25 Vulnerability and consequences, high ductile limit states systems.

This case study, being an ID problem, needs to define the different decision variables as well as the monetary costs for different levels of Exposure, *E*, and Consequences, *C*. We have seen in Chapter 2 that this information is usually given in a utility table. The utility table of this problem is shown in Table 4.11. Note that the decision maker has four available decisions that control the level of vulnerability, that is, different HDLS. The cost of each decision is a function of exposure and consequences levels. Since *V, E,* and *C* are all uncertain, we need to solve the BN using the different decisions as observations to find the pertinent probabilities that correspond to each decision. The problem starts when an assessment showed the current state of the asset vulnerability to be $V = H$. Thus, the decision maker is faced with four possible decisions. We consider these four decisions as the risk acceptance limit states. They are:

- $D = d_1$ – Mitigate from $V = H$ to $V = V.L.$
- $D = d_2$ – Mitigate from $V = H$ to $V = L$
- $D = d_3$ – Mitigate from $V = H$ to $V = M$
- $D = d_4$ – Mitigate from $V = H$ to $V = V.L.$

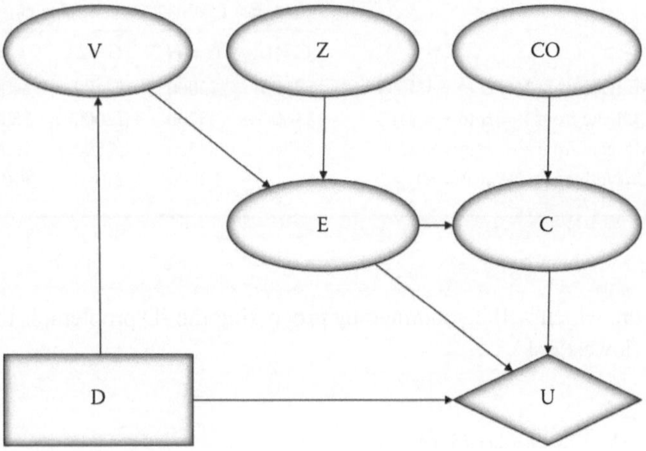

FIGURE 4.26 ID model for high ductile limit states.

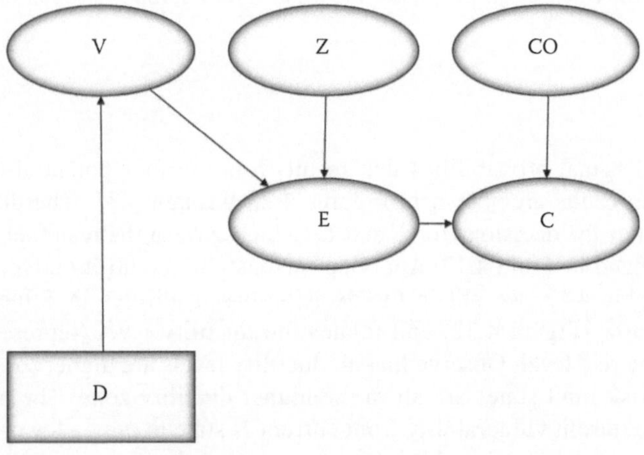

FIGURE 4.27 BN model for high ductile limit states.

TABLE 4.10
States of Nodal Variables

Variable	# of States	States
V.I.	5	*V.H., H., M., L., V.L.*
Hz	2	*H., L.*
E	2	*H., L.*
CO	2	*H., L.*
C	2	*H., L.*
D	4	d_1, d_2, d_3, d_4

TABLE 4.11
Utility Table for HDLS

E	*E* = *L*		*E* = *H*	
C	*C* = *L*	*C* = *H*	*C* = *L*	*C* = *H*
d_1 (Mitigate from *V* = *H* to *V* = *V.L.*)	\$2000	\$3000	\$2500	\$4000
d_2 (Mitigate from *V* = *H* to *V* = *L.*)	\$100	\$3200	\$2000	\$5000
d_3 (Mitigate from *V* = *H* to *V* = *M.*)	\$500	\$3500	\$2500	\$6000
d_4 (Do Nothing: leave at *V* = *H*)	\$–	\$3000	\$1000	\$10,000

The optimal decision, which will be obtained by processing the ID problem, is that which produces the optimal utility (lowest risk).

4.6.3 Conditional Probabilities Tables

The different CPTs for the variables of the network in Figure 4.27 are shown in Appendix 4.IV. These CPTs govern the uncertain behavior of different variables of the BN that is the foundation of the current ID problem. As usual, we used our best judgment in filling the different tables in a realistic manner for illustration purposes. For more on different methods of developing CPTs, see Chapter 2.

4.6.4 Results

We evaluate the marginal probabilities that result from the four potential decisions, $D = d_1$, d_2, d_3, and d_4. The results are shown in Figures 4.28 through 4.31. The different joint probabilities from each of the decisions for *E* and *C* (which govern the resultant *U*, or the absolute risk) are now arranged in Table 4.12. Applying the costs in the utility table, Table 4.11, to the uncertainties in Table 4.12, we obtain the decision-based utilities in Table 4.13. The HDLS can be displayed now (Figure 4.32) and relates the ductility level (represented by degree of vulnerability) to the risk level. Observe that all ductility levels are higher (or equal) than elastic limit; that is, the risk limit states are all in the higher ductility zone. The optimal decision is $D = d_2$: which is to retrofit vulnerability from current *H* state to only a lower state, *V* = *L*. Note that the optimal acceptance level in this case is a midrange ductility, neither the highest nor the lowest ductility.

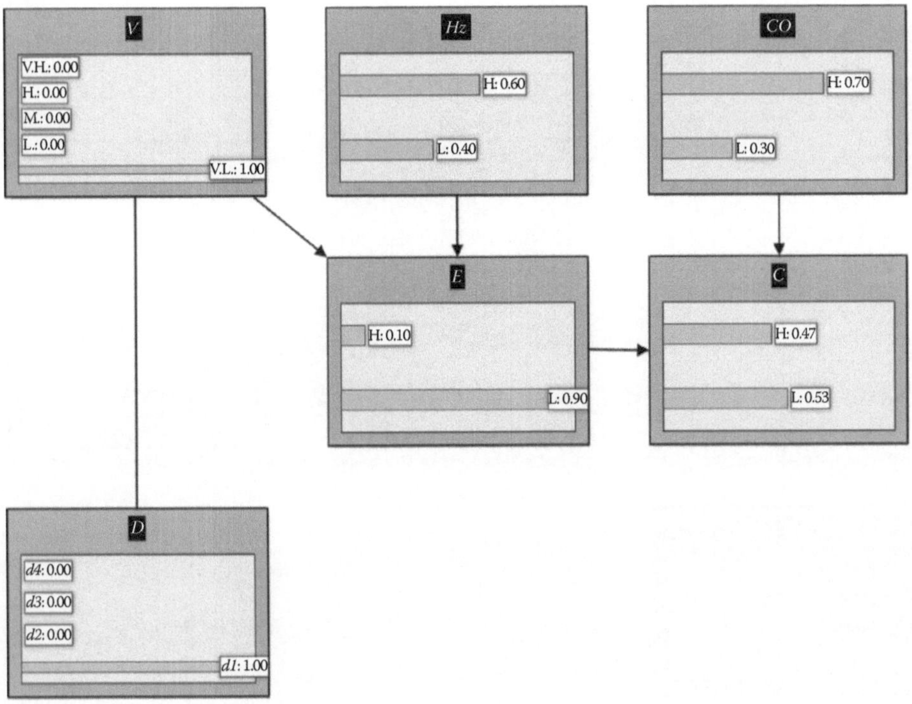

FIGURE 4.28 Ductile limit states: marginal probabilities for $D = d_1$.

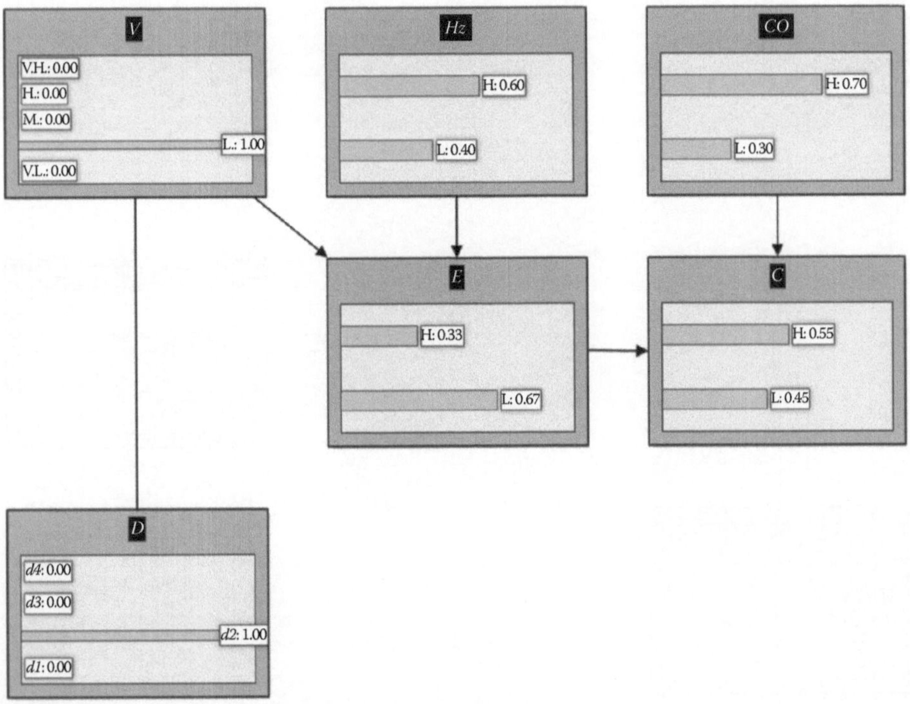

FIGURE 4.29 Ductile limit states: marginal probabilities for $D = d_2$.

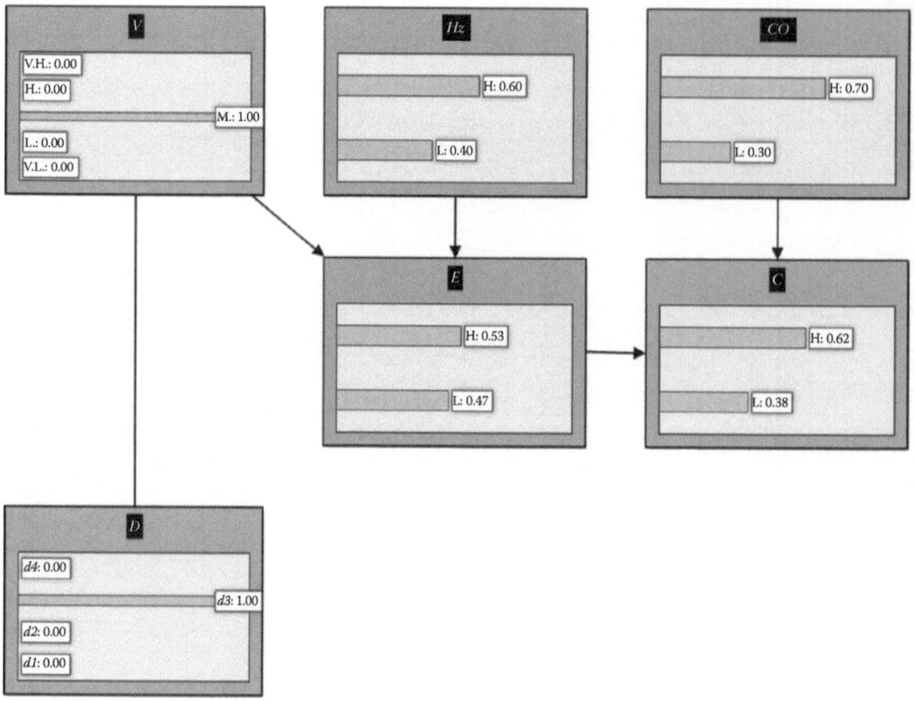

FIGURE 4.30 Ductile limit states: marginal probabilities for $D = d_3$.

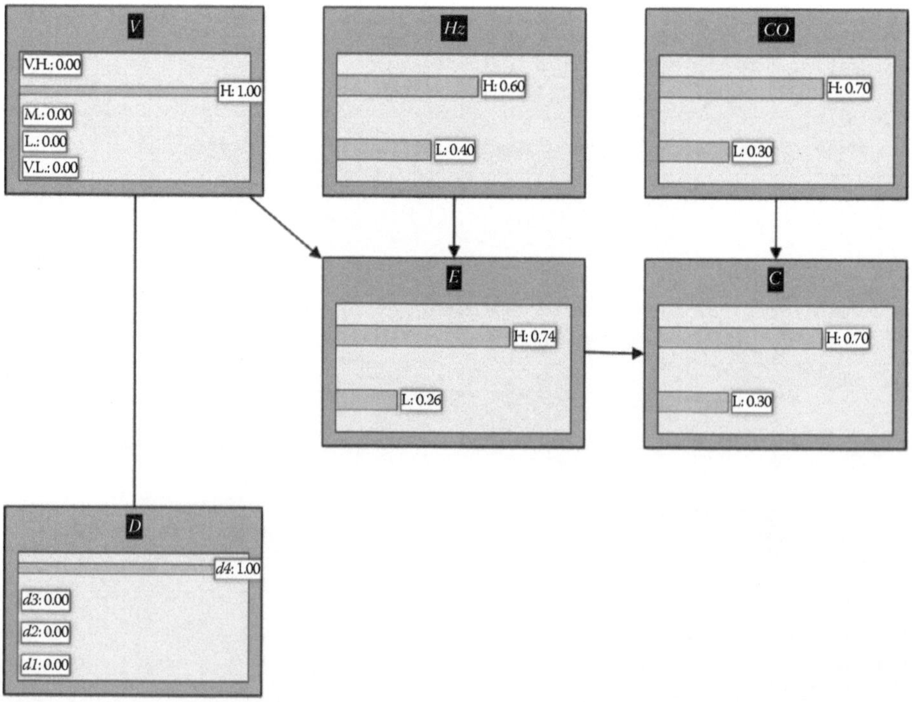

FIGURE 4.31 Ductile limit states: marginal probabilities for $D = d_4$.

TABLE 4.12
Joint Probabilities for Risk Computations (Utility, *U*)

E	*E = L*		*E = H*	
C	*C = L*	*C = H*	*C = L*	*C = H*
d_1 (Mitigate from *V = H to V = V.L.*)	0.4790	0.4180	0.0550	0.0480
d_2 (Mitigate from *V = H to V = L.*)	0.3033	0.3677	0.1487	0.1803
d_3 (Mitigate from *V = H to V = M.*)	0.1796	0.2944	0.1994	0.3266
d_4 (Do Nothing)	0.0791	0.1829	0.2229	0.5151

TABLE 4.13
Decision-Based Utility (Risk)

E	*E = L*		*E = H*		
C	*C = L*	*C = H*	*C = L*	*C = H*	**Total *U***
d_1 (Mitigate from *V = H to V = V.L.*)	$958.00	$1254.01	$137.51	$191.99	$2541.50
d_2 (Mitigate from *V = H to V = L.*)	$30.33	$1176.67	$297.42	$901.46	$2405.87
d_3 (Mitigate from *V = H to V = M.*)	$89.82	$1030.24	$498.39	$1959.88	$3578.32
d_4 (Do Nothing)	$-	$548.63	$222.88	$5151.24	$5922.74

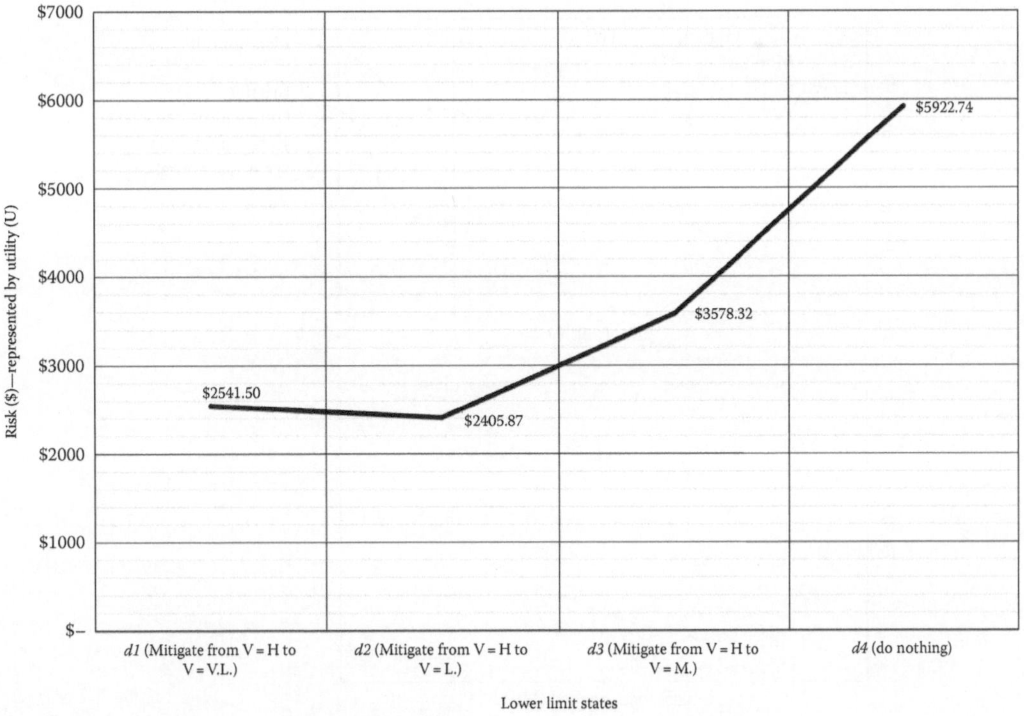

FIGURE 4.32 Accepted risk thresholds at HDLS.

4.7 CASE STUDY 4.4: BRITTLE LIMIT STATES: RISK/REWARD OF PROOF TESTS

4.7.1 Overview

The brittle failure of structural systems occurs when a structure fails suddenly without warning. This can happen when a structure loses its stiffness and becomes unstable. Ettouney and Alampalli (2012a) discussed this subject at length. They pointed out some applications in the field of civil infrastructures, most notably proof testing of bridges. During proof testing of bridges, the stakeholder increases the demand, *Hz*, on the bridge while monitoring its load carrying behavior. In general, as the load increases and approaches its elastic limit point, its shows some indications of distress. This is the point at which the load test should stop because no permanent damage or failure is desired. However, in rare situations, the bridge may fail suddenly without showing any sign of distress if failure mechanisms are not well understood (see Figure 4.33).

This brings an obvious dilemma: on one hand, it is desired to load the bridge to a point as close as possible to the point of failure. On the other hand, without knowing a priory that point of failure (which is the main reason of performing the test in the first place), the stakeholder risks bridge failure.

Generally speaking, in practice, the stakeholder relies on previous experience or structural analysis to guide the loading process decision during proof testing of bridges. In this example, we consider the problem as a risk acceptance problem. We formulate the problem as an ID decision problem. We consider three risk acceptance limit states that are related to three bridge loading, *H*, levels, with, *Hz* = Low (L), *Hz* = Medium (M), and *Hz* = High (H). We compute the utility of the test, as representative of risk, at these three limit states. Note that in this example, utility represents both potential risk and reward or potential cost and potential benefit. The vulnerability is modeled with two states: *Linear* and *Failed*. The consequences can be either *Low*,

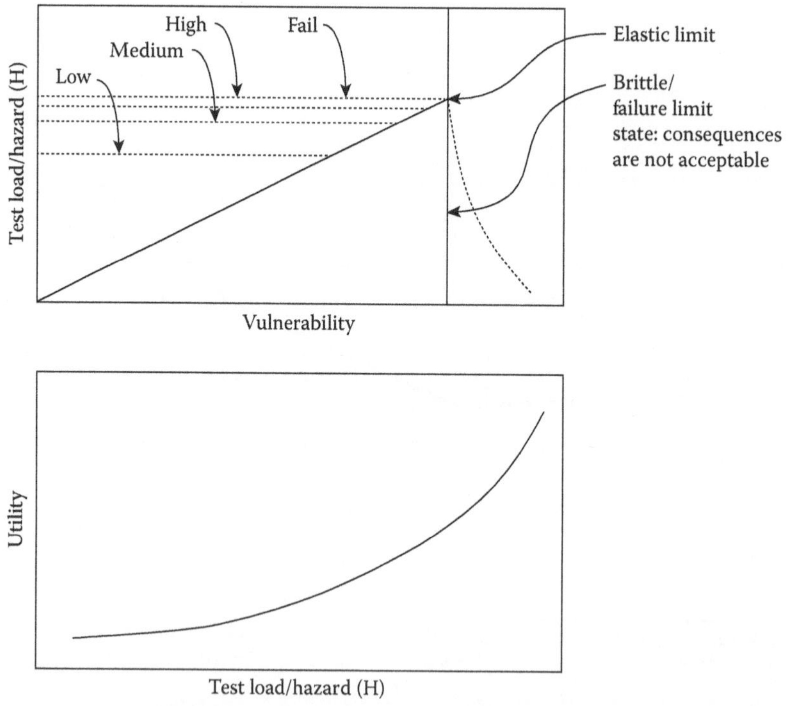

FIGURE 4.33 Brittle limit states and risk acceptance.

High, or *Unacceptable*. By balancing the risk/rewards and uncertainties in this ID way of deciding on the level of loading, the stakeholder will have an objective way of getting the optimal risk/reward from the proof test.

4.7.2 Model

The ID model of this problem is shown in Figure 4.34. It contains a decision variable, D, and a utility variable, U. There are three decisions: $D = d_1$, $D = d_2$, and $D = d_3$. These decisions are as follows:

- $D = d_1$: stop the test at low load level, $Hz = L$
- $D = d_2$: stop the test at medium load level, $Hz = M$
- $D = d_3$: stop the test at high load level, $Hz = H$

Of course, uncertainties correspond to each of these decisions. These are reflected in the structural response as represented by vulnerability: it can either remain linear or fail in a brittle manner. Consequences can be low, high, or unacceptable. The ID model is resolved using the closely related BN model of Figure 4.35. The nodal variable states of the model are presented in Table 4.14. Note that the decision, consequences, and vulnerability variables can affect the utility, as presented in the utility Table 4.15. The table includes relative utility numbers as opposed to the usual monetary values used in ID utility tables so far. This is necessary here because we are aiming at combining the losses incurred due to structural brittle failure or even an undesirable consequence (perhaps due to severe structural behavior that is short of failure) and benefits which might be incurred because of successful execution of the test. The benefits are modeled

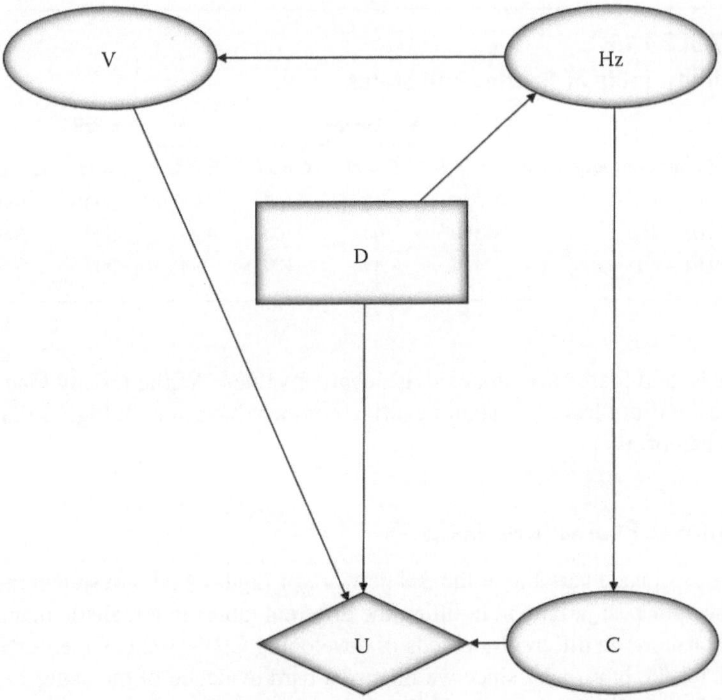

FIGURE 4.34 Proof test ID model (as an example of brittle failure limit state acceptance).

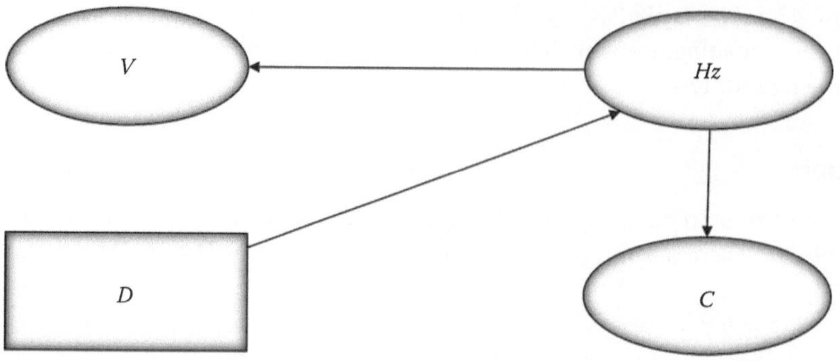

FIGURE 4.35 Proof test BN model (as an example of brittle failure limit state acceptance).

TABLE 4.14
States of Nodal Variables

Variable	# of States	States
Hz	3	*H, M, L*
V	2	Failed, Linear
C	3	Unacceptable, High, Low
D	3	d_1, d_2, d_3
U		

TABLE 4.15
Utility Table of Brittle Limit States

V	V = Linear			V = Fail		
C (other consequences)	C = L	C = H	C = U	C = L	C = H	C = U
d_1 ($Hz = L$)	20	10	−100	−500	−500	−500
d_2 ($Hz = M$)	100	60	−100	−500	−500	−500
d_3 ($Hz = H$)	150	80	−100	−500	−500	−500

as positive values, and losses are modeled as negative values. As the testing load increases, the benefit increases. Failure has a constant negative number that is much higher than any positive benefit the test can offer.

4.7.3 Conditional Probabilities Tables

The different CPTs of each variable of the BN network of Figure 4.35, are shown in Appendix 4.V. As usual, we used our best judgment in filling the different tables in a realistic manner for illustration purposes. For more on different methods of developing CPTs, see Chapter 2. Note that we do not need a CPT for the hazard *Hz* since we impose a hard evidence of the states of the hazard for each of the decision conditions as shown in Figures 4.35, 4.37, and 4.38.

4.7.4 Results

We exercise the BN model for individual decisions $D = d_1$, $D = d_2$, and $D = d_3$ with marginal probabilities shown in Figures 4.36 through 4.38, respectively. From these results, we obtain the joint probabilities needed to compute different decision utilities as shown in Table 4.16. From Tables 4.15 and 4.16, we can compute the utilities (risk/reward) of different decisions as shown in Table 4.17. These utilities represent the acceptance thresholds of risk/reward limits states for the brittle failure problem on hand. Obviously, the optimal decision is $D = d_1$ (highest reward/no risk) indicating that the stakeholder needs to operate the test up to a *Low* test load. This can be explained under the given uncertainties as per the CPTs as well as the high cost of failure (−500) as compared with the relatively limited rewards. If the relative cost of failure is reduced to (−400), the resulting utilities for a different decision will be changed as shown in Table 4.18. The brittle limit states as represented by utilities are changed and the optimal decision is now $D = d_2$: operate the test up to medium test load. Comparison between the acceptance limit states for the two costs of failure is shown in Figure 4.39.

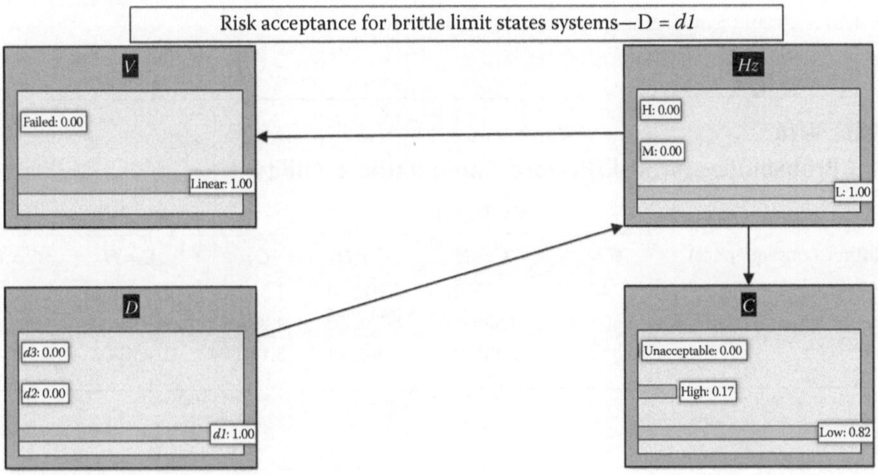

FIGURE 4.36 Proof test marginal probabilities, $D = d_1$.

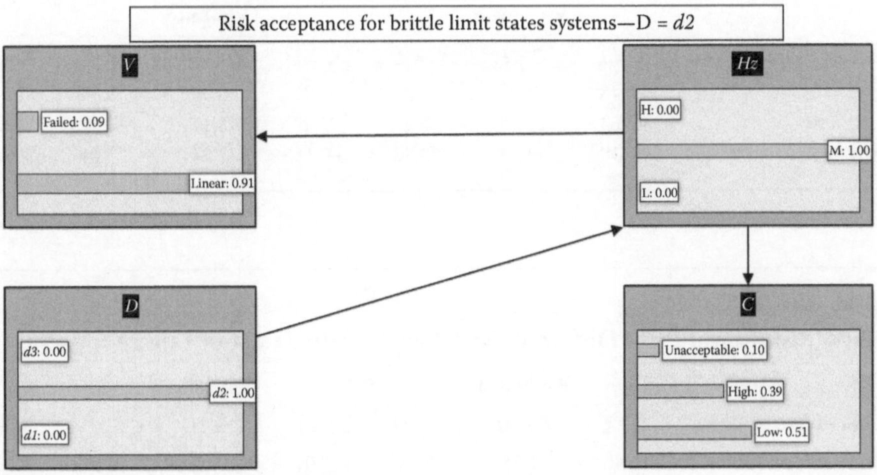

FIGURE 4.37 Proof test marginal probabilities, $D = d_2$.

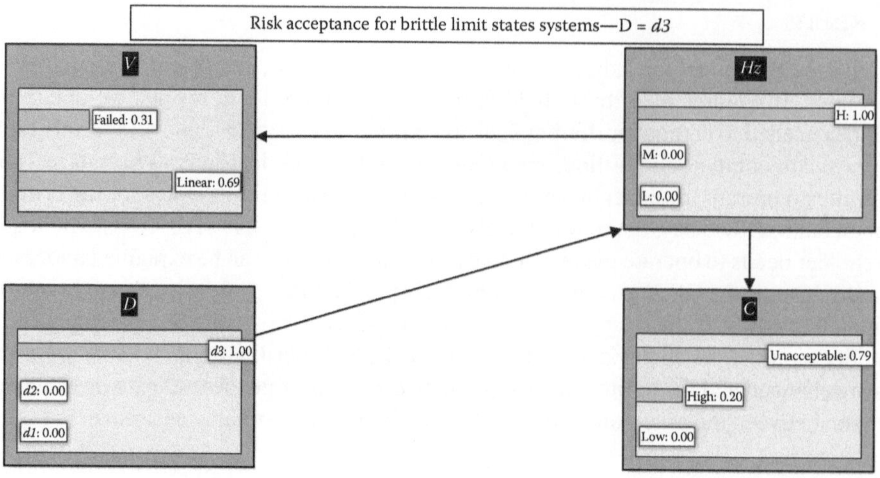

FIGURE 4.38 Proof test marginal probabilities, $D = d_3$.

TABLE 4.16
Joint Probabilities for Risk/Reward Computations (Utility, *U*)

V	V = Linear			V = Fail		
C (other consequences)	*C* = *L*	*C* = *H*	*C* = *U*	*C* = *L*	*C* = *H*	*C* = *U*
d_1 ($Hz = L$)	8.24E-01	1.73E-01	3.30E-03	3.30E-05	6.90E-06	1.32E-07
d_2 ($Hz = M$)	4.66E-01	3.53E-01	8.73E-02	4.80E-02	3.63E-02	8.98E-03
d_3 ($Hz = H$)	6.85E-04	1.40E-01	5.44E-01	3.15E-04	6.45E-02	2.50E-01

TABLE 4.17
Utility (Risk/Reward) for Different Decisions (Relative Cost of Failure = −500)

V	V = Linear			V = Fail			
C (other consequences)	*C* = *L*	*C* = *H*	*C* = *U*	*C* = *L*	*C* = *H*	*C* = *U*	Sum
d_1 ($Hz = L$)	16.48	1.73	−0.33	−0.02	0.00	0.00	**17.86**
d_2 ($Hz = M$)	46.63	21.18	−8.73	−23.99	−18.17	−4.49	**12.43**
d_3 ($Hz = H$)	0.10	11.23	−54.42	−0.16	−32.24	−125.01	**−200.49**

TABLE 4.18
Utility (Risk/Reward) for Different Decisions (Relative Cost of Failure = −400)

V	V = Linear			V = Fail			
C (other consequences)	*C* = *L*	*C* = *H*	*C* = *U*	*C* = *L*	*C* = *H*	*C* = *U*	Sum
d_1 ($Hz = L$)	16.48	1.73	−0.33	−0.01	0.00	0.00	**17.86**
d_2 ($Hz = M$)	46.63	21.18	−8.73	−19.19	−14.53	−3.59	**21.76**
d_3 ($Hz = H$)	0.10	11.23	−54.42	−0.13	−25.79	−100.01	**−169.01**

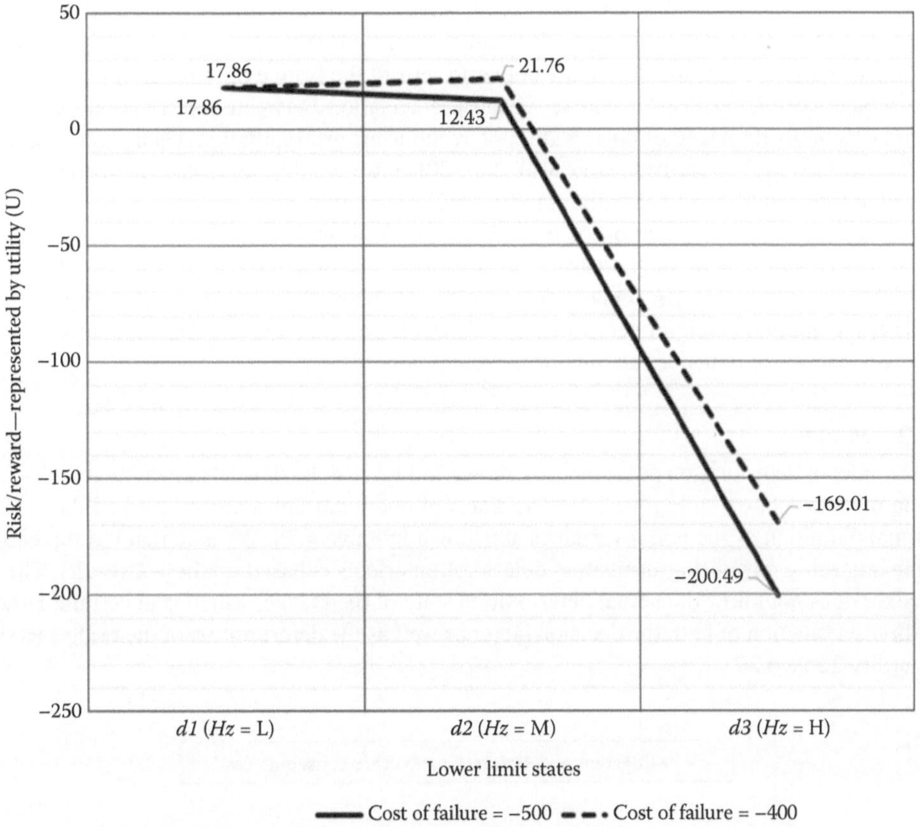

FIGURE 4.39 Risks/rewards limit states for proof load test (brittle failure limit states).

4.8 ACCEPTANCE USING DECISION METHODS

4.8.1 Overview of Decision Methods

Fault tree analysis (FTA) is a tree-based modeling process, which aims at analyzing causes of system failures. Event tree analysis (ETA) has also been used as a decision support process to analyze probabilities of events that result from hazards affecting systems. The ETA model is also a tree-based model similar to that of the FTA. Fenton and Neil (2012) showed that both FTA and ETA can be modeled using BN processes. They also pointed out that because both processes (FTA and ETA) rely on tree structure in their modeling process, they have some limitations that BN does not have. Neapolitan (2004) discussed using tree structure for a decision process; he then generalized the process by using ID while also pointing out the limitation of using tree structure for decision under uncertainty (see Chapter 2). Based on these arguments, we use in this section GN/ID models for illustrating the development of risk acceptance thresholds.

4.8.2 Case Study 4.5: Deterioration (Exposure)-Based Acceptance and Decision

4.8.2.1 Overview

All civil infrastructures deteriorate as time moves on. Numerous authors have studied the deterioration problem from many viewpoints (Agrawal et al. 2009; Ettouney and Alampalli 2012). Current practices in the bridge community is to subjectively estimate deterioration levels of bridges via a system of visual inspection, *V.I.* (Agrawal et al. 2013). We explored this interrelationship in this book in several case studies. In general, *V.I.* can result in different decisions regarding retrofitting deteriorated bridges. Thus, the decision whether to retrofit the bridge or to do nothing is dependent on the subjective *V.I.* results. Of course, because correlation between deterioration (whether the structure is actually damaged or not) and the *V.I.* result is uncertain, though recent studies show that the *V.I.* process is highly accurate (Agrawal et al. 2013), we would like to explore the retrofit decisions due to this underlying uncertainty in this simple case study.

We thus model the deterioration of the bridge (*Det.*) as an uncertain variable with two states (*Pristine* and *Damaged*). We then relate *V.I.* to *Det.* using a reasonable CPT to simulate the underlying subjective nature of the *V.I.* process. We then establish an ID decision model with the decision variable D having two potential decisions (d_1 = *Mitigate* and d_2 = *Do nothing*). The model is fairly simple, yet is realistic in understanding the current practices.

4.8.2.2 Model

The ID model of deterioration acceptance is shown in Figure 4.40. The BN model that underlies the decision model is shown in Figure 4.41. The states of nodal variables are shown in Table 4.19, and the formal definition of the nodal variables are shown in Table 4.35. We note that the stakeholder's decision regarding retrofitting/mitigating deterioration effects is based on the *V.I.* results. The decision maker does not know the actual deterioration state of the bridge, which is uncertain. However, the utility is a function of both the decision factor as well as the deterioration of the bridge, as shown in the utility Table 4.20.

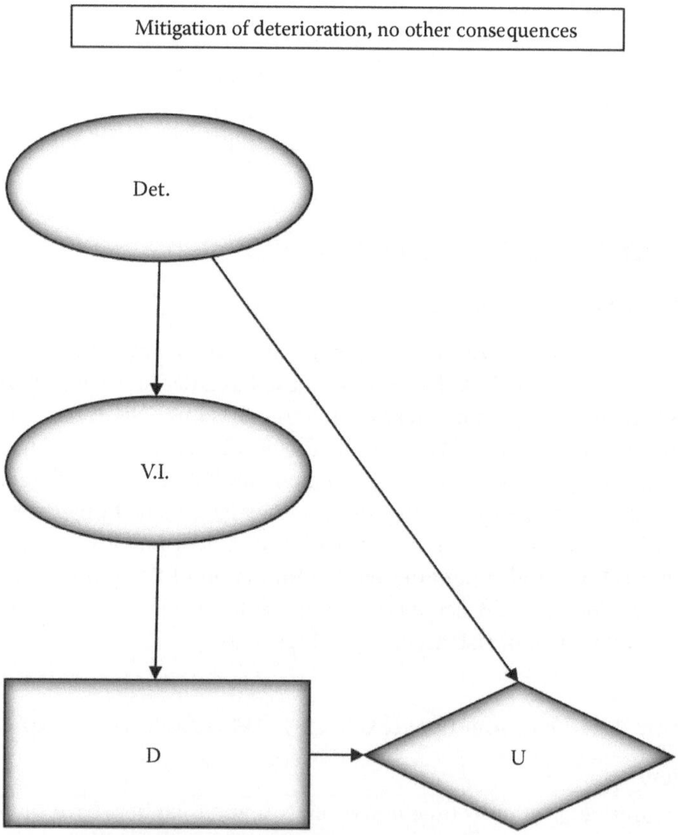

FIGURE 4.40 ID modeling of acceptance: simple model.

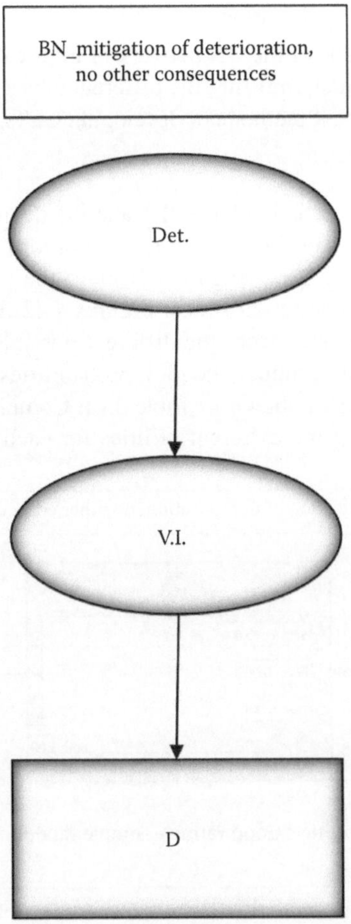

FIGURE 4.41 BN of the ID acceptance model: simple model.

TABLE 4.19
States of Nodal Variables

Variable	# of States	States
Det.	2	Damaged, Pristine
V.I.	4	Service, Poor, Fair, Good
D	2	d_1, d_2

TABLE 4.20
Utility Table for Mitigation Decisions Give Actual Deterioration States

D	d_1 (Mitigate)		d_2 (Do Nothing)	
Det.	Pristine	Damaged	Pristine	Damaged
Utility (U)	$500.00	$500.00	$–	$4000.00

4.8.2.3 Conditional Probabilities Tables

The different CPTs of each variable of the BN network of Figure 4.41 are shown in Appendix 4.VI. As usual, we used our best judgment in filling the different tables in a realistic manner for illustration purposes. For more on different methods of developing CPTs, see Chapter 2.

4.8.2.4 Results

As stated above, assume that the stakeholder will make the decision regarding mitigation versus doing nothing solely based on the results of the *V.I.* observations whereas the actual deterioration state is uncertain. By executing the BN of Figure 4.41 given the *V.I.* observations, we can infer the marginal probabilities of the deterioration states. Figures 4.42 through 4.45 show the deterioration marginal probabilities as inferred from the different possible *V.I.* observations. We can now use these marginal probabilities to produce the joint probabilities that are needed for the decision computations of different utilities as shown in Table 4.21. Combining the joint probabilities with the utilities in Table 4.20, we get the different utilities for each decision and each possible *V.I.*

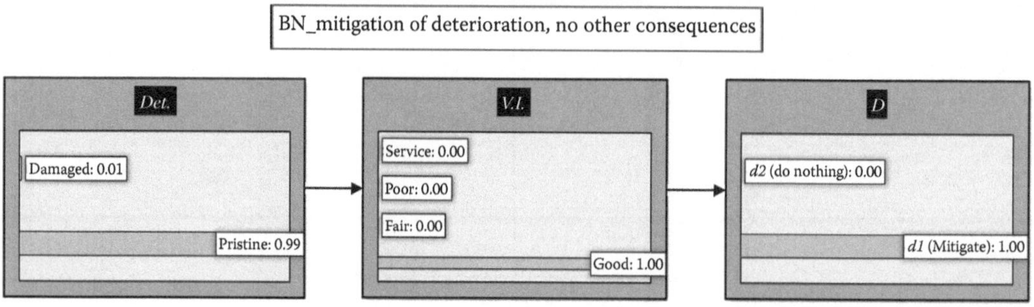

FIGURE 4.42 BN marginal probabilities: good rating—simple model.

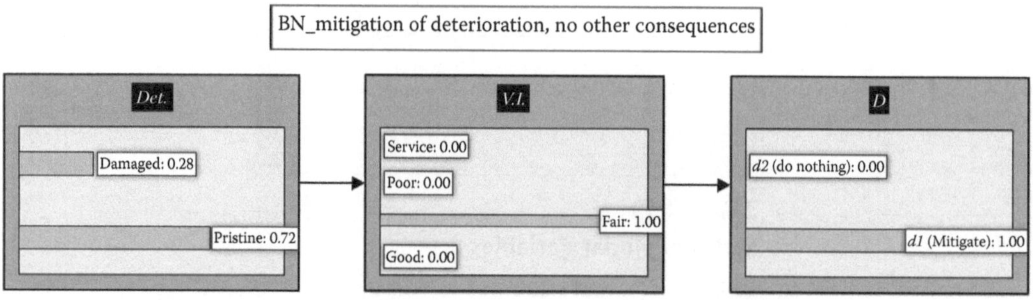

FIGURE 4.43 BN marginal probabilities: fair rating—simple model.

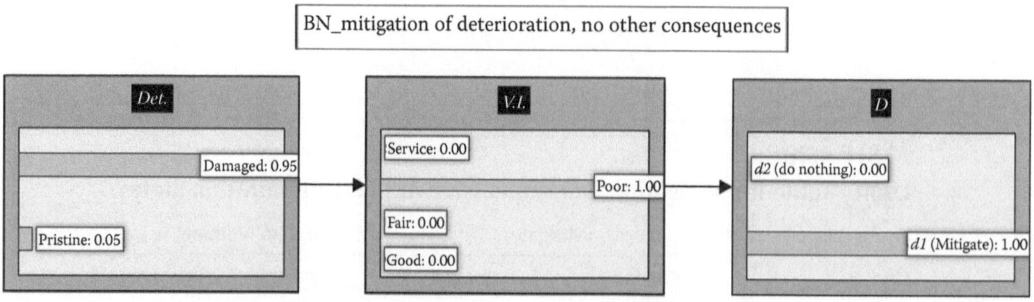

FIGURE 4.44 BN marginal probabilities: poor rating—simple model.

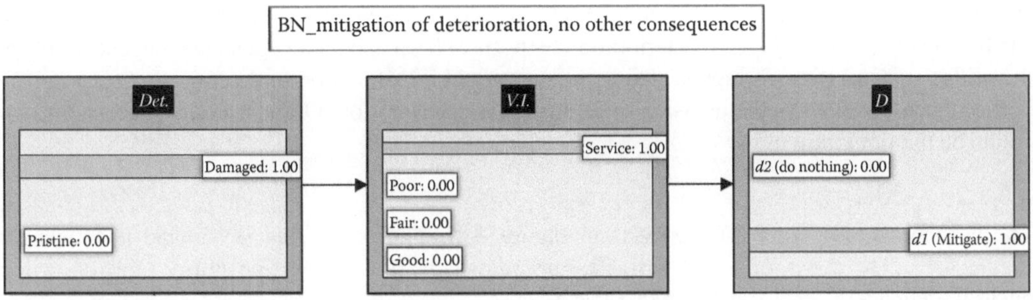

FIGURE 4.45 BN marginal probabilities: service rating—simple model.

TABLE 4.21
Joint Probabilities for Decision-Making

Det.	Pristine	Damaged
V.I. = Good	0.99225	0.00775
V.I. = Fair	0.72147	0.27853
V.I. = Poor	0.05107	0.94893
V.I. = Service	0.00102	0.99898

TABLE 4.22
Optimal Decisions Based on *V.I.* Results

V.I. results	$U(d_1)$	$U(d_2)$	$U(D)$	Optimal Decision, D
V.I. = Good	\$500.00	\$31.00	\$31.00	$D = d_2$ (Do Nothing)
V.I. = Fair	\$500.00	\$1114.12	\$500.00	$D = d_1$ (Mitigate)
V.I. = Poor	\$500.00	\$3795.72	\$500.00	$D = d_1$ (Mitigate)
V.I. = Service	\$500.00	\$3995.92	\$500.00	$D = d_1$ (Mitigate)

observation as displayed in Table 4.22. The different optimal decisions for each possible *V.I.* observation seems to be reasonable: For *V.I. = Good*, the optimal decision is $D = d_1$, which is to do nothing. For all other *V.I.* observations, the optimal decision it to mitigate. Of course, for more realism, a detailed utility table and a case-relevant CPT are also needed in-lieu of the values given in Tables 4.57 and 4.58. Figure 4.46 illustrates different accepted exposure (deterioration) values for each of the *V.I.* observations.

4.8.2.5 Summary

We only considered deterioration in this example (as a form of exposure); we did not formally explore risk-based issues. This means that the decisions discussed above are a form of exposure acceptance. We offer a more complete risk acceptance next.

4.8.3 Case Study 4.6: Deterioration (Exposure) and Consequences Based Acceptance and Decision

4.8.3.1 Overview

We explored acceptance decisions for the deterioration of bridges which is based on exposure (representing the deterioration) utilities. The decision was based only on *V.I.* observations. We expand the process now such that the utility would be representative of risk, not only of exposure.

We can achieve this by adding consequences to the model. As we have seen before, consequences can be a resultant of two factors: deterioration (exposure) level and factors that are independent from the bridge system. Risk, as represented by utility, would be a function of deterioration (exposure), consequences, as well as the decision to mitigate the deterioration. Optimal utility (accepted risk) would be the minimum utility.

4.8.3.2 Model

The decision model is an ID model as in Figure 4.47. The underlying BN model is shown in Figure 4.48. The states of nodal variables are shown in Table 4.23. The utility table showing decision-dependent costs are shown in Table 4.24.

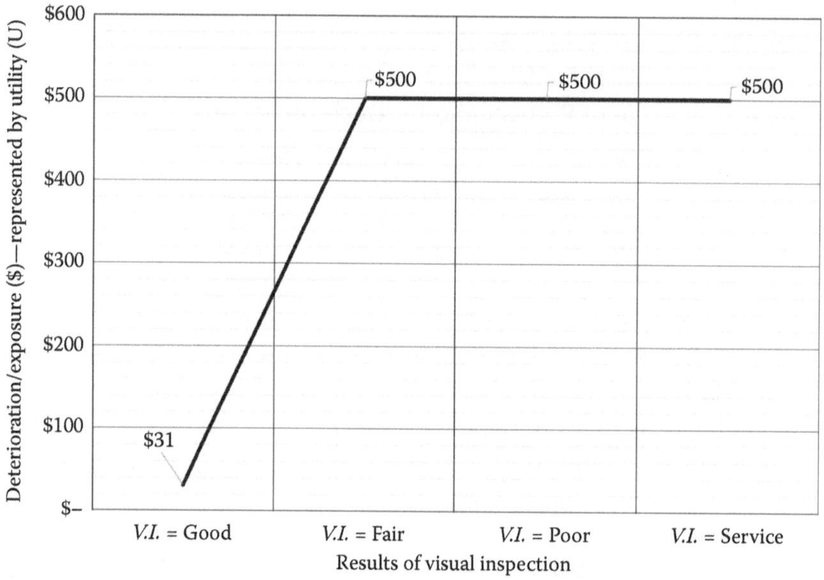

FIGURE 4.46 Accepted deterioration (exposure) as a function of *V.I.* results.

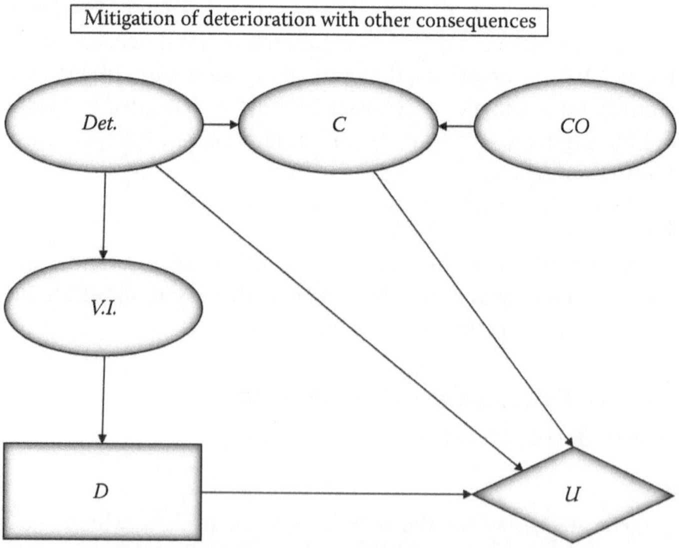

FIGURE 4.47 ID modeling of acceptance: explicitly modeling consequences.

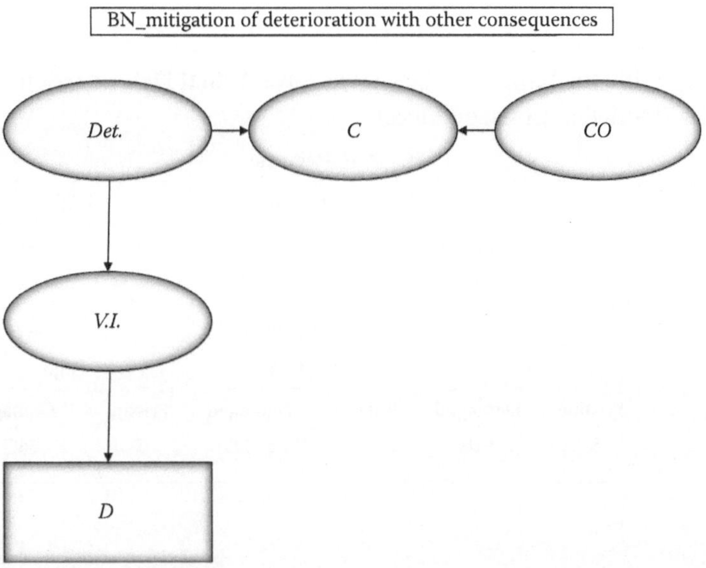

FIGURE 4.48 BN of the ID acceptance model: explicitly modeling consequences.

TABLE 4.23
States of Nodal Variables

Variable	# of States	States
Det.	2	Damaged, Pristine
C	3	High (H), Medium (M), Low (L)
CO	2	High, Low
V.I.	4	Service, Poor, Fair, Good
D	2	d_1, d_2
U		

4.8.3.3 Conditional Probabilities Tables

The different CPTs of each variable of the BN network of Figure 4.48 are shown in Appendix 4.VII. As usual, we used our best judgment in filling the different tables in a realistic manner. For more on different methods of developing CPTs, see Chapter 2.

4.8.3.4 Results

We have to compute joint probabilities for *Det.*, and *C* due to different observations of *V.I.* First, we compute the marginal probabilities for the four individual observations of *V.I.* as shown in Figures 4.49 through 4.52. Using the joint probabilities with the utilities displayed in Table 4.25, we can generate utilities for each decision and each possible *V.I.* observation as displayed in Table 4.26. The optimal decisions for each possible *V.I.* observation seem to be reasonable: For *V.I.* = *Good*, the optimal decision is $D = d_1$, which is to do nothing. We note that by including consequences, the optimal decision for *V.I.* = *Fair* is now changed from $D = d_1$ *(Mitigate)* to $D = d_2$ *(Do Nothing)*. For all other *V.I.* observations, the optimal decision it to mitigate. As in the exposure case 04.05, for more realism, a more detailed utility table as well as a case-relevant CPT are needed in-lieu of the values given in Tables 4.59 through 4.62. Figure 4.53 illustrates accepted exposure (deterioration) values for each of the *V.I.* observations.

TABLE 4.24
Utility Table for Mitigation Decisions Give Actual Deterioration States (Modeling Consequences)

D	d_1 (Mitigate)					
C	Low		Medium		High	
Det.	Pristine	Damaged	Pristine	Damaged	Pristine	Damaged
U	\$500	\$500	\$500	\$500	\$500	\$500
D	d_2 (Do Nothing)					
C	Low		Medium		High	
Det.	Pristine	Damaged	Pristine	Damaged	Pristine	Damaged
U	\$–	\$10	\$–	\$1000	\$–	\$4000

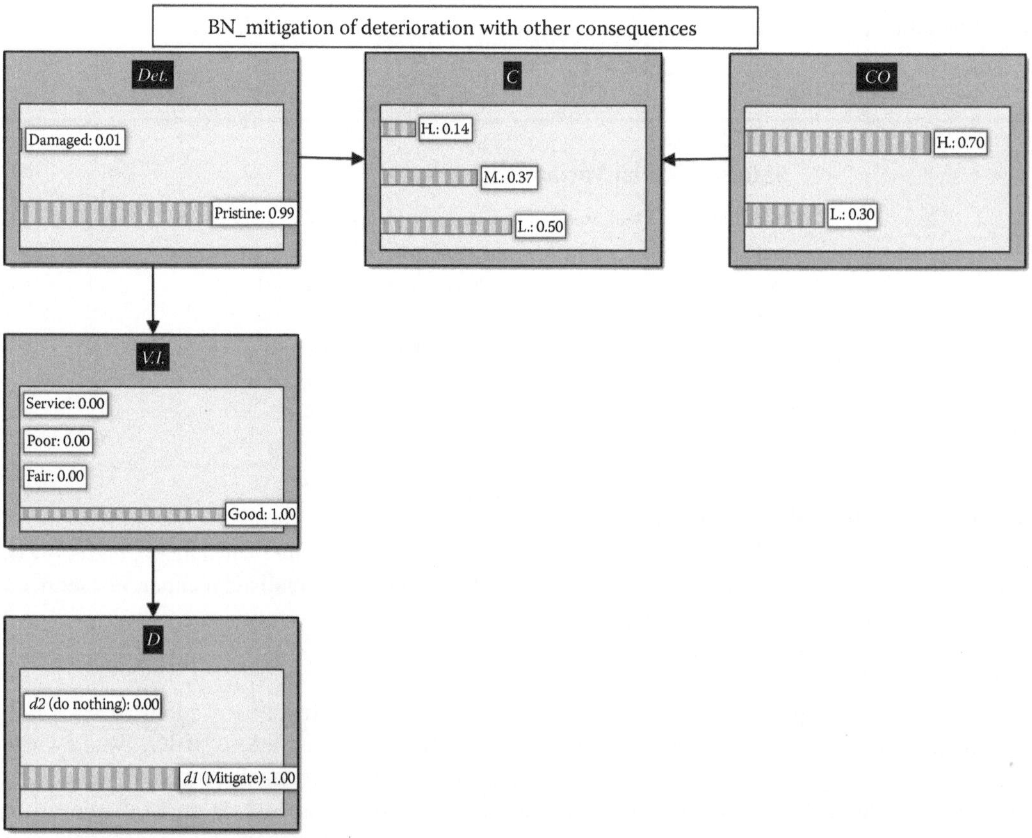

FIGURE 4.49 BN marginal probabilities: good rating—explicitly modeling consequences.

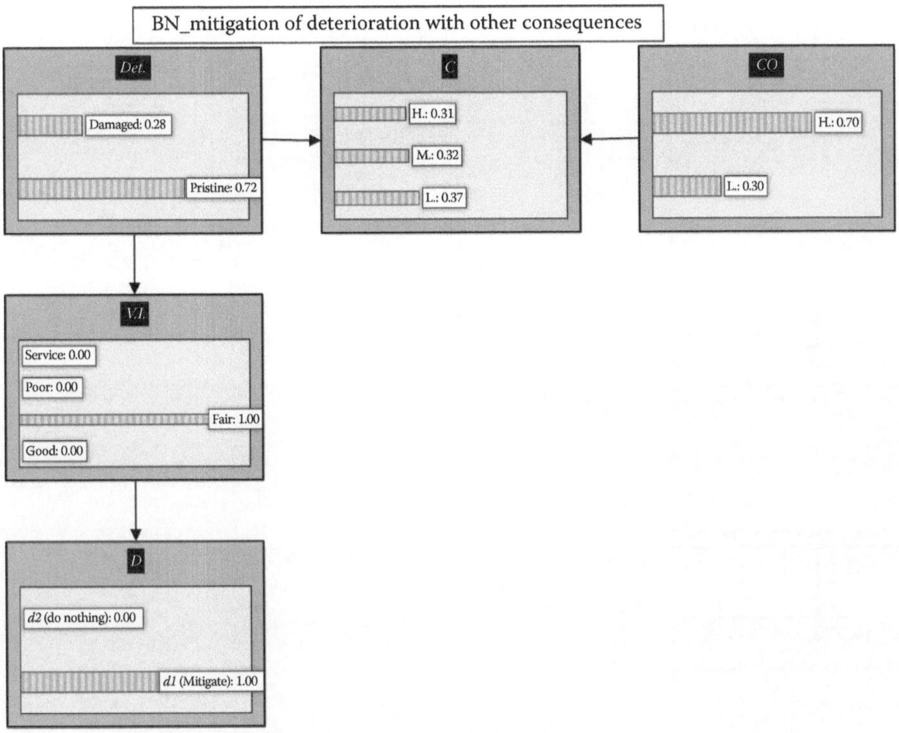

FIGURE 4.50 BN marginal probabilities: fair rating—explicitly modeling consequences.

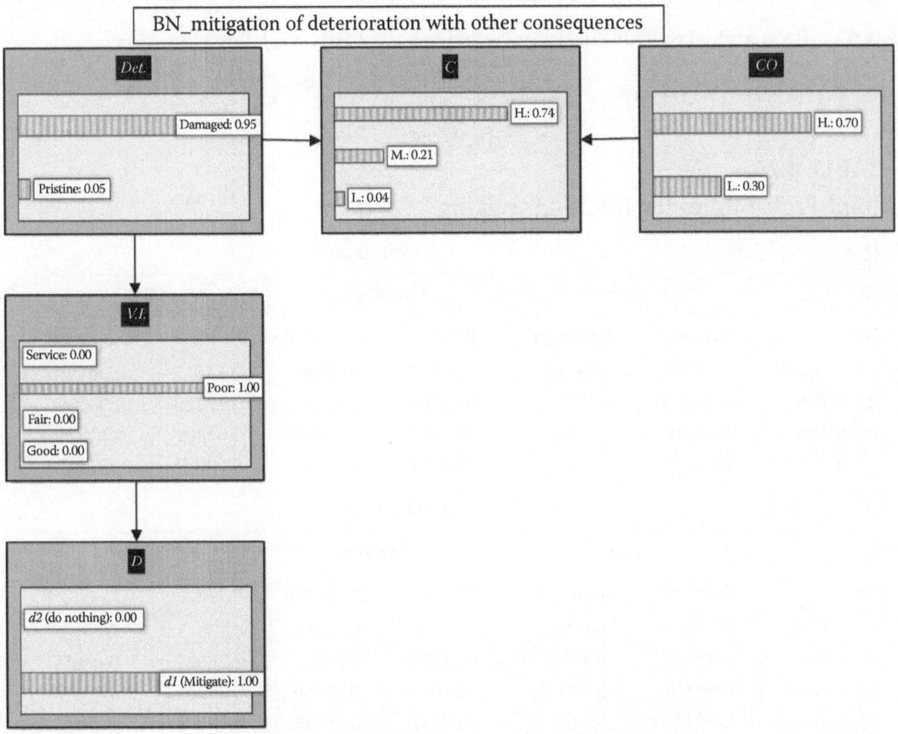

FIGURE 4.51 BN marginal probabilities: poor rating—explicitly modeling consequences.

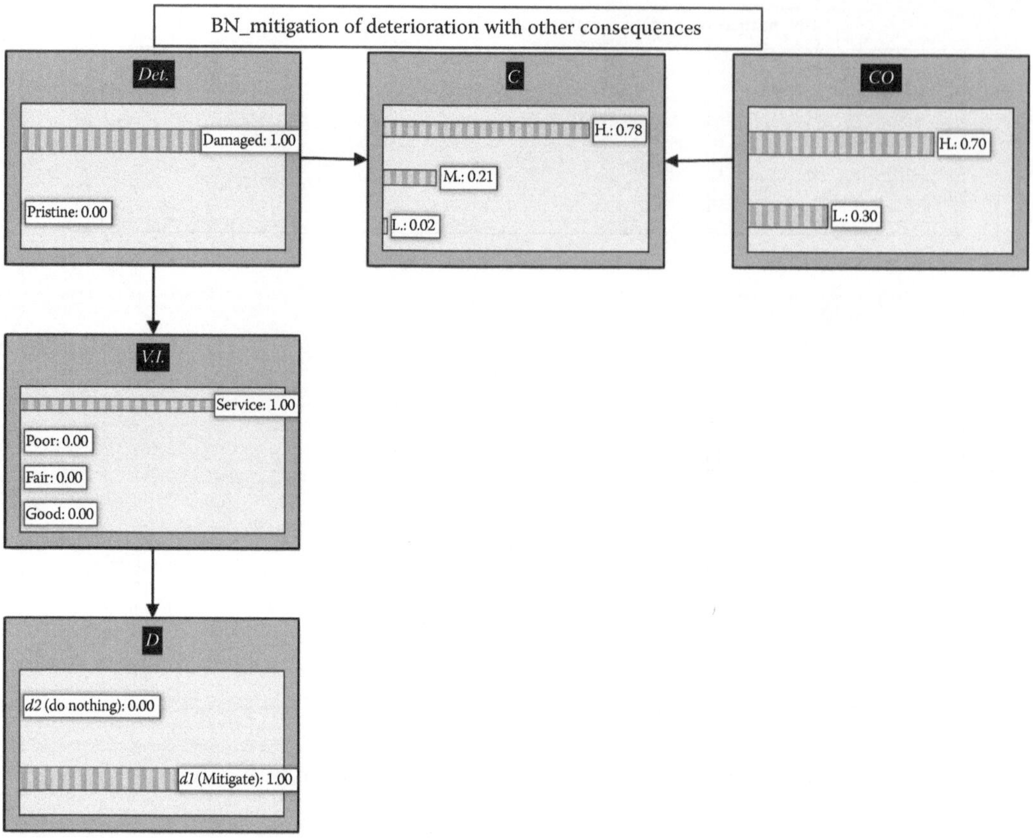

FIGURE 4.52 BN marginal probabilities: service rating—explicitly modeling consequences.

TABLE 4.25
Joint Probabilities for Decision-Making

D	d_1 (Mitigate)					
C	Low		Medium		High	
Det.	Pristine	Damaged	Pristine	Damaged	Pristine	Damaged
V.I. = Good	0.49310	0.00385	0.36442	0.00285	0.13473	0.00105
V.I. = Fair	0.26416	0.10198	0.23303	0.08996	0.22428	0.08659
V.I. = Poor	0.00216	0.04013	0.01090	0.20246	0.03801	0.70634
V.I. = Service	0.00002	0.01809	0.00021	0.20498	0.00079	0.77591
D	d_2 (Do Nothing)					
C	Low		Medium		High	
Det.	Pristine	Damaged	Pristine	Damaged	Pristine	Damaged
V.I. = Good	0.49310	0.00385	0.36442	0.00285	0.13473	0.00105
V.I. = Fair	0.26416	0.10198	0.23303	0.08996	0.22428	0.08659
V.I. = Poor	0.00216	0.04013	0.01090	0.20246	0.03801	0.70634
V.I. = Service	0.00002	0.01809	0.00021	0.20498	0.00079	0.77591

4.8.3.5 Summary

Even at the simplest possible decision modeling, we showed that risk-based decision would produce different results than exposure-only (which is the basis of current practice) decision modeling. Infrastructure stakeholders need to take a note of this risk versus exposure decision policy differences.

4.9 GRANDFATHERING: THE HIDDEN RISK ACCEPTANCE PRACTICE

Nowhere is risk acceptance more important in the field of civil infrastructure than in the asset management arena. Grandfathering situations happen frequently because of changes/upgrading in design codes, changes in government policies and procedures, changes in structural materials/components available in the market, and climate change conditions. Guidelines for building security (Ettouney 2015) and standards for building rehabilitation due to seismic events (ASCE/SEI 2013) contain some popular grandfathering situations. Sometimes personal experiences or subjective code specifications do not allow for grandfathering. We next illustrate that by following a formal risk acceptance process, subjectivity in most grandfathering situations can be eliminated by using an objective approach.

TABLE 4.26
Optimal Decisions Based on *V.I.* Results (with Consequence Modeling)

V.I. Results:	EU(d_1)	EU(d_2)	EU(D)	D
V.I. = Good	$500.00	$7.09	$7.09	$D = d_2$ (Do Nothing)
V.I. = Fair	$500.00	$437.33	$437.33	$D = d_2$ (Do Nothing)
V.I. = Poor	$500.00	$3028.21	$500.00	$D = d_1$ (Mitigate)
V.I. = Service	$500.00	$3308.79	$500.00	$D = d_1$ (Mitigate)

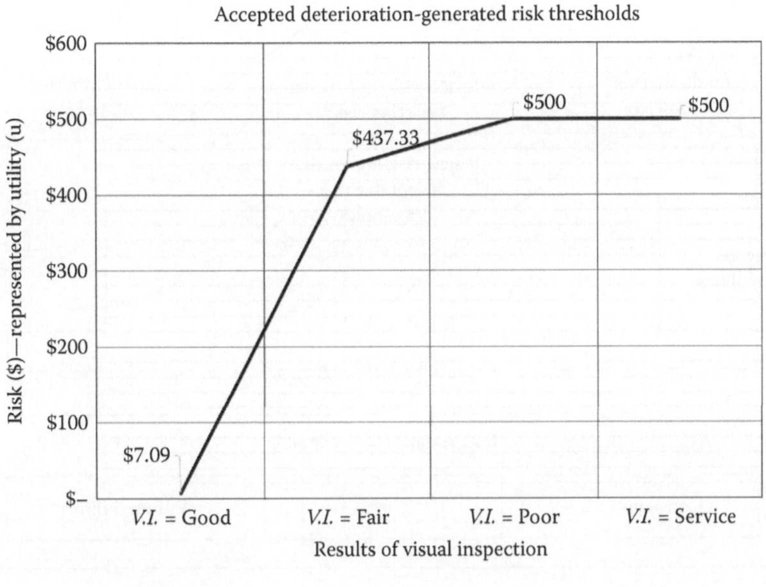

FIGURE 4.53 Accepted deterioration (risk) as a function of V.I. results.

4.9.1 Process

Risk-based approach is useful for making grandfathering an objective rather than a subjective process. This is because grandfathering not only includes reliability issues (vulnerability and threats) but also relies heavily on consequences and also has to accommodate uncertainties. It is really a perfect example for decision in uncertain conditions. See Figure 4.54 for the risk-based grandfathering process. We offer two methods for the Objective Grandfathering process: weighted averages and probabilistic networks.

4.9.2 Case Study 4.7: Objective Grandfathering Process Using Weighted Averages

We reexamine the simple risk approach of Chapter 3. Recall the proposed risk form of

$$Ri = f(C, T, V) \tag{4.14}$$

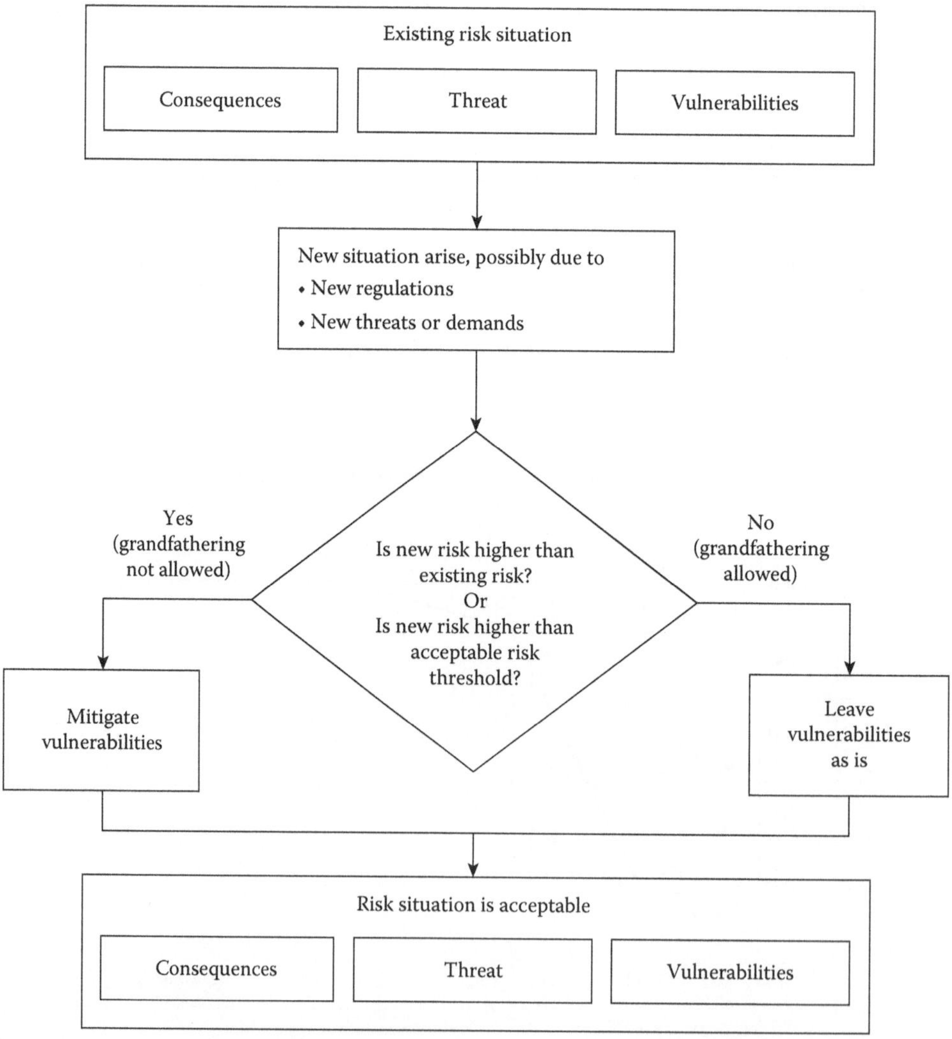

FIGURE 4.54 Risk-based grandfathering process.

As usual, C, T, and V represent the consequence, threat, and vulnerabilities ratings, respectively. We use a simple expression for the function $f(\,)$ as

$$Ri = (C \cdot T \cdot V)^{1/3} \tag{4.15}$$

Furthermore, for the purpose of the grandfathering issue, we assume that C is a function of three components such that

$$C = f(C|_{FUNC}, C|_{CAPEX}, C|_{PHYS}) \tag{4.16}$$

Impact of the loss of function due to threat T is $C|_{FUNC}$. The initial capital expenditure* (in relative terms) is $C|_{CAPEX}$, and the cost of future physical responses to T is $C|_{PHYS}$. For simplicity, we propose

$$C = w_1 C|_{FUNC} + w_2 C|_{CAPEX} + w_3 C|_{PHYS} \tag{4.17}$$

with

$$\sum_{i=1}^{i=3} w_i = 1.0 \tag{4.18}$$

The expressions w_i reflects the relative importance of the three consequence components to the analyst/decision maker.

The analyst/decision maker is now faced with the question of whether to grandfather a particular infrastructure or to apply new requirements. To objectively make the decision, two possibilities affect the decision process (see Figure 4.54):

1. An acceptable risk value, $Ri|_A$ is available, or
2. An acceptable risk value, $Ri|_A$ is not available.

When $Ri|_A$ is available, then the decision to grandfather a particular infrastructure would be when the relationship

$$Ri|_{OLD} \leq Ri|_A \tag{4.19}$$

is true. When $Ri|_A$ is not available, then the decision to grandfather a particular infrastructure would be when the relationship

$$Ri|_{OLD} \leq Ri|_{NEW} \tag{4.20}$$

The analysis can compute $Ri|_{OLD}$ and $Ri|_{NEW}$ from Equations 4.15 through 4.19 for infrastructure with grandfathering condition (old) and the infrastructure where the new requirements are applied (new).

As an example, consider a situation where a new building code is introduced that will affect two buildings: a residential and an office building. We assume first that acceptable risk is known (in the range 0–10) to be

$$Ri|_A = 5.0 \tag{4.21}$$

* This example deals only with initial risks. Risk-based considerations of the life cycle of the grandfathering processes is beyond the scope of this book.

We also assume the analyst decides on relative weights of consequence components as in Table 4.27.

The analyst computes the risk if old conditions of each of the two buildings are grandfathered. The results of the risk assessments are shown in Tables 4.28 (for old conditions of both buildings) and 4.29 (for old and new conditions of both buildings). Note that the vulnerability of both buildings due to an estimated threat rating of 6.00 are fairly low at 2.00 and 3.00 for the new condition. However, given the low consequences at 1.80 and 3.00, the resulting risks are midrange. Still, when the assessed risks for each of the buildings are compared to the acceptable risk, the grandfathering decisions are not the same. For the residential building, it should be grandfathered because $Ri|_{OLD} < Ri|_A$ and the office building need to be upgraded to the new code since $Ri|_{OLD} > Ri|_A$.

In case $Ri|_A$ is not available, then the analyst needs to compare the risks of grandfathering the building (old) versus applying the new building code (new). Assuming that these assessments are made as in Table 4.29, the decisions would be the reverse of the previous decisions. Now the residential building needs to be upgraded to the new code since $Ri|_{OLD} > Ri|_{NEW}$, and the office building needs to be grandfathered because $Ri|_{OLD} < Ri|_{NEW}$.

TABLE 4.27
Relative Consequence Weights

Type of Consequence	$C\|_{FUNC}$	$C\|_{CAPEX}$	$C\|_{PHYS}$	Total
Weight	0.3	0.5	0.2	1.0

TABLE 4.28
Grandfathering Decision Table: $Ri|_A$ Is Available

Type of Infrastructure		Consequences				Current		
		$C\|_{FUNC}$	$C\|_{CAPEX}$	$C\|_{PHYS}$	Total (0–10)	Threat (0–10)	Vulnerability (0–10)	Risk (0–10)
Residential building	Old	3	1	2	1.80	6.00	8.00	4.42
Large office building	Old	7	1	2	3.00	6.00	7.00	5.01

TABLE 4.29
Grandfathering Decision Table: $Ri|_A$ Is Not Available

Type of Infrastructure		Consequences				Current		
		$C\|_{FUNC}$	$C\|_{CAPEX}$	$C\|_{PHYS}$	Total (0–10)	Threat (0–10)	Vulnerability (0–10)	Risk (0–10)
Residential building	Old	3	1	2	1.80	6.00	8.00	4.42
	New	3	7	5	5.40	6.00	2.00	4.02
Large office building	Old	7	1	2	3.00	6.00	7.00	5.01
	New	7	8	8	7.70	6.00	3.00	5.18

4.9.3 Case Study 4.8: Objective Grandfathering Process Using ID

4.9.3.1 Overview

At its core, the grandfathering process is a good example of decision-making when given uncertain input. As such, it is reasonable to solve it using an ID model. We use similar variables as in the previous section because they are the basic variables that affect the decision process. We add a decision variable, *D*, with only two possible decisions: $D = d_{11}$ *(Allow Grandfathering)* and $D = d_{12}$ *(No Grandfathering)*. The utility (representing risk) is a function of the three previously introduced consequences. However, the initial capital expenditure (*CapEx*) is considered deterministic because it is presumed known while the decision is being made. The other two consequences, *C_Phys* and *C_Func.*, are uncertain because they are function of the uncertain hazard, *Threat*, which is yet to happen in some future event. The optimal decision is the decision that produces least risk (utility), which is the accepted risk for this problem.

4.9.3.2 Model

The simple ID model for the grandfathering decision process is shown in Figure 4.55. The underlying risk-based BN model is shown in Figure 4.56. Note that the model variables can easily be classified to the three risk components, as usual. The states of nodal variables are shown in Table 4.30, and the definitions of these variables are shown in Table 4.35. We now have to define the costs of these variables. First, we assume that the *CapEx* costs, which are dependent on the decision *D*, are shown in Table 4.31 with other costs shown in Table 4.32.

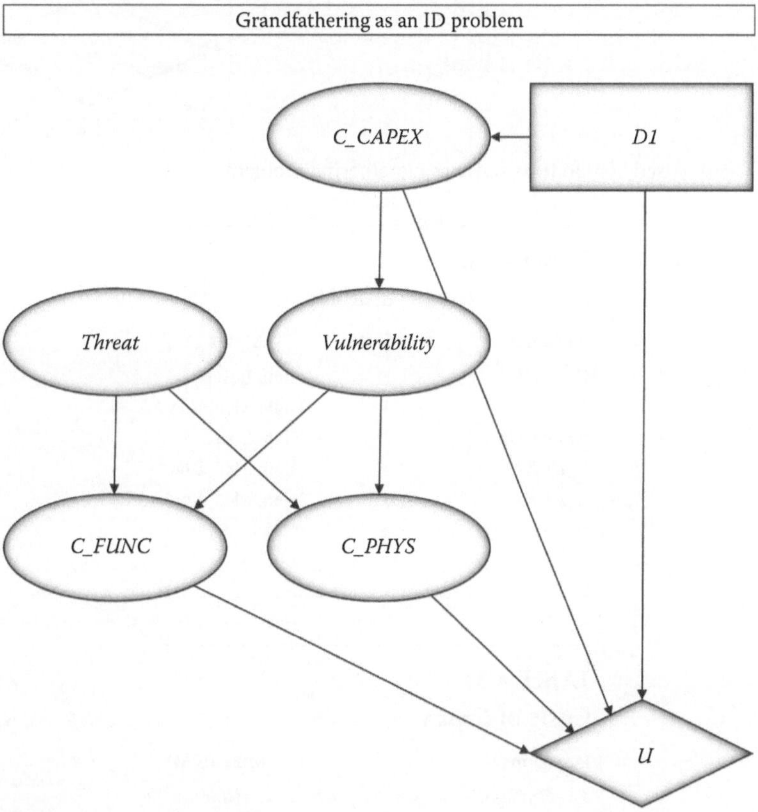

FIGURE 4.55 Risk-based ID model for grandfathering problem.

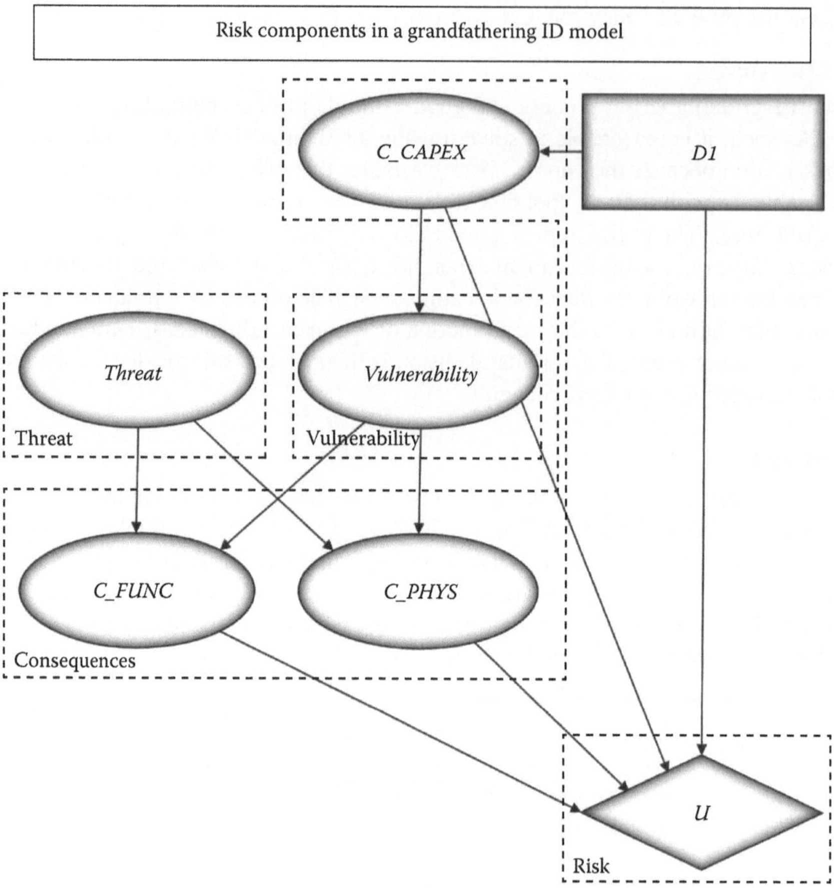

FIGURE 4.56 Risk-based components in ID grandfathering problem.

TABLE 4.30
States of Nodal Variables

Variable	# of States	States
C_CAPEX	2	High, Low
Threat	3	High, Med, Low
Vulnerability	2	High, Low
C_FUNC	3	High, Med, Low
C_PHYS	3	High, Med, Low
D1	2	d_{11},d_{12}
U	—	—

TABLE 4.31
Costs of *CapEx*

Decisions	*CapEx* ($ M)
d_{11} *(Grandfather, C_CapEx = L)*	$0.00
d_{12} *(New, C_CapEx = H)*	$1.20

TABLE 4.32
Costs of *C_Phys* and *C_Func*

Decisions	Costs of *C_Func* and *C_Phys*, $ M					
	C_Phys = *L*	*C_Phys* = *M*	*C_Phys* = *H*	*C_Func* = *L*	*C_Func* = *M*	*C_Func* = *H*
	\$0.20	\$0.50	\$1.00	\$0.20	\$0.50	\$1.00

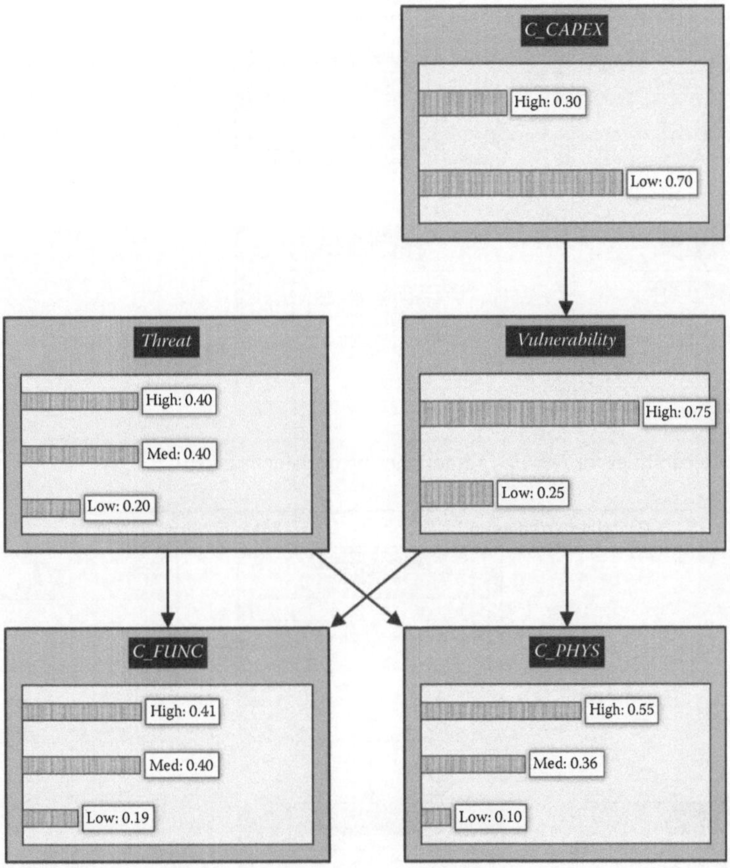

FIGURE 4.57 Grandfathering BN model: historical probabilities.

4.9.3.3 Conditional Probabilities Tables

The different CPTs of each variable of the BN network of Figure 4.56 are shown in Appendix 4.VIII. As usual, we used our best judgment in filling the different tables in a realistic manner for illustration purposes. For more on different methods of developing CPTs, see Chapter 2.

4.9.3.4 Results

The historical probabilities of the different variables of the model can sometimes be of value to the decision maker. In the grandfathering problem, it can act as a valuable background (see Figure 4.57). Of more immediate value are the probabilities for $D = d_{11}$ and $D = d_{12}$. These are shown in Figures 4.58 and 4.59. We now can obtain the needed joint probabilities (Table 4.33) that

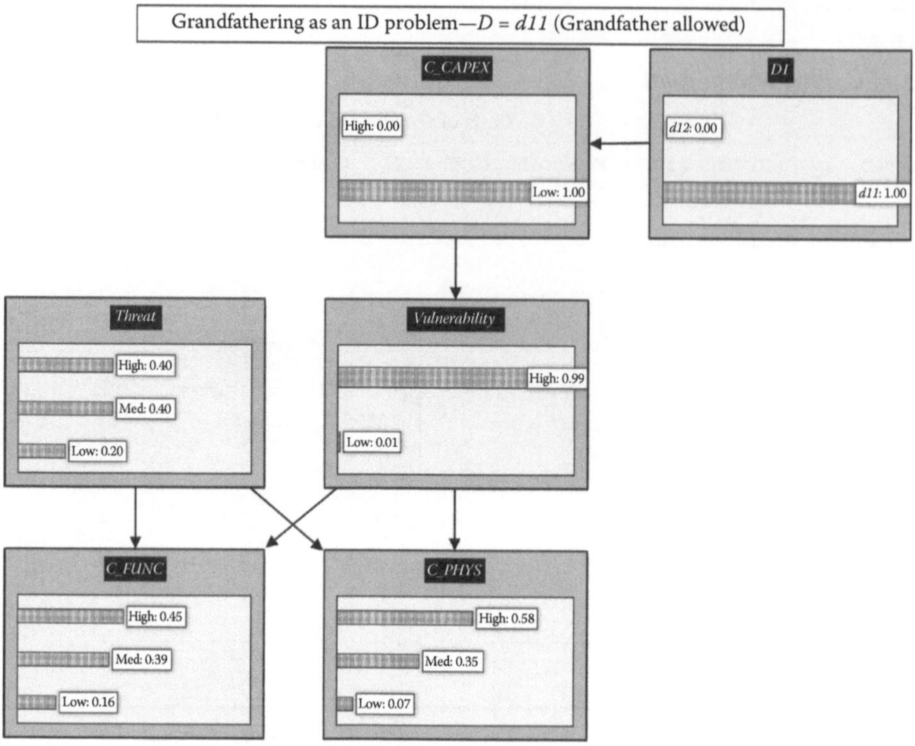

FIGURE 4.58 Probabilities for $D = d_{11}$ (grandfathering decision is taken).

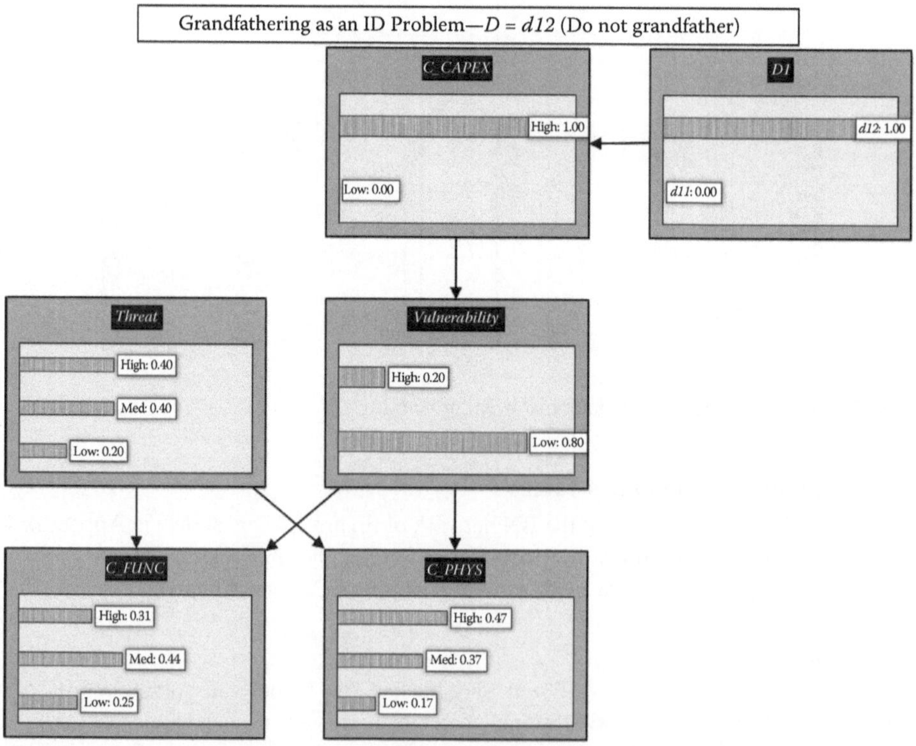

FIGURE 4.59 Probabilities for $D = d_{12}$ (no grandfathering decision is taken).

TABLE 4.33
Joint Probabilities of *C_Phys* and *C_Func* as a Function of Decisions $D = d_{11}$ and $D = d_{12}$

	Joint Probabilities of Consequences					
Decisions	*C_Phys = L*	*C_Phys = M*	*C_Phys = H*	*C_Func = L*	*C_Func = M*	*C_Func = H*
d_{11} *(Grandfather, C_CapEx = L)*	0.0710	0.3510	0.5780	0.1640	0.3870	0.4490
d_{12} *(New, C_CapEx = H)*	0.1660	0.3650	0.4690	0.2500	0.4410	0.3090

TABLE 4.34
Computational Details for Finding Total Utilities (Risk) as a Function of Decisions $D = d_{11}$ and $D = d_{12}$

	Consequences ($M)		
Decisions	*C_Phys*	*C_Func = L*	*C_CapEx*
d_{11} *(Grandfather, C_CapEx = L)*	$0.77	$0.68	$–
d_{12} *(New, C_CapEx = H)*	$0.68	$0.58	$1.20

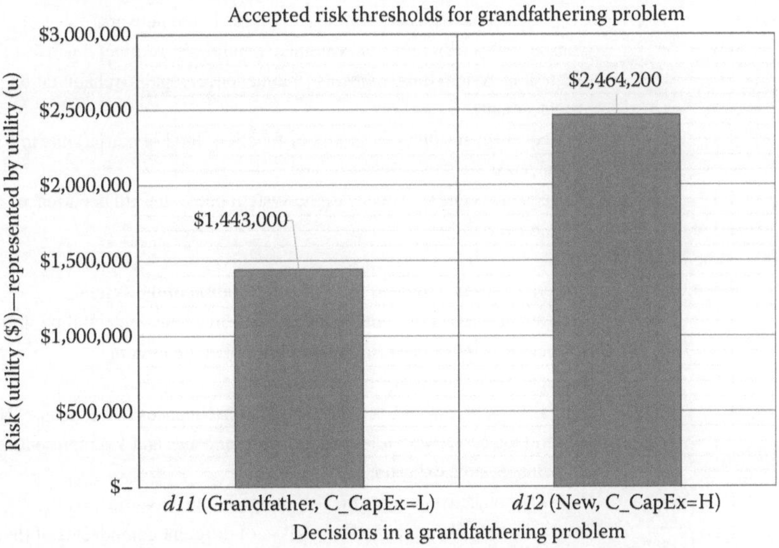

FIGURE 4.60 Total utilities (risk) as a function of decisions $D = d_{11}$ and $D = d_{12}$.

are needed for computing utilities of different decisions as in Table 4.34. Finally, using Tables 4.32 through 4.34, we obtain different cost components (consequences) and total utilities for $D = d_{11}$ and $D = d_{12}$ as shown in Figure 4.60. The optimal risk acceptance threshold is $D = d_{11}$ (*Grandfather*) because it offers the minimum cost (utility).

4.9.4 Summary

Grandfathering is a reasonable approach. We showed how to approach this process from an objective viewpoint by modeling it as a risk process. Employing such a risk-based approach enhances the spirit of the grandfathering process by ensuring an objective balance between safety and costs.

The two methods we offered in this section are simple, yet useful for objective decision-making. Of course, these methods can offer better accuracy if users introduce more pertinent variables into the basic framework that was offered here.

APPENDIX 4.I: DEFINITIONS OF NETWORK NODAL VARIABLES (TABLE 4.35)

TABLE 4.35
Definitions of Nodal Variables

Variable	Used in Case Study #	Definition
C	4.3, 4.6	Objective estimate of total consequences. It is a function of two variables in this example: 1-*Exposure*, or *V.I.* (which approximates system exposure) and their consequences, *CO*. Thus, we establish the presumption that total consequences rely on both the system itself via its exposure and other factors beyond the system itself, *CO*.
C	4.1, 4.4	Objective estimate of consequences. It is assumed here to be a function only of hazard, *H*.
CO.	4.1, 4.2, 4.3, 4.6	Estimate of consequences beyond the system itself, these consequences can be, but are not limited to cost of replacements, cost of traffic delays, symbolic nature, locality, and structure age. In these examples, *CO.* is hazard independent.
CO_1, CO_2	4.2	*CO_i* is consequences beyond the *i*-th bridge in a network.
C_1, C_2	4.2	*CO_i* is the consequences for the *i*-th bridge in the network.
C_CAPEX	4.8	Consequences resulting from capital expenditures. A direct function of decisions regarding system improvements (maintenance, repairs, rehabilitation, or replacement).
C_FUNC	4.8	Consequences resulting from loss of function (total or partial) due to a hazard occurrence.
C_PHYS	4.8	Consequences due to damage to the system due to hazard occurrence.
D, D1	4.3, 4.4, 4.5, 4.6, 4.8	Decision variable
Det.	4.5, 4.6	Objective estimate of the degree of deterioration of the system.
E	4.3	Exposure is a result of both the hazard as well as the vulnerability of the system.
Hz	4.3	Subjective estimate of hazard that might inflict the system.
Hz	4.4	Subjective estimate of the proof load level
Ri	4.1	In this example, risk, *Ri*, is a function of consequences and exposure. Numerous algorithms objectively relate risk to *consequences* and *V.I.* representing an estimate of actual system exposure.
Ri_1, Ri_2	4.2	*Ri_i* is the objective risk for the *i*-th bridge in the network.
Ri_Network	4.2	Total network risk, estimated from risks of different components of the network.
Threat	4.8	Subjective estimate of hazard that might inflict the system.
U	4.3, 4.4, 4.5, 4.6, 4.8	Utility
V	4.4	Vulnerability, see *Vulnerability* below.
V.I.	4.1, 4.3, 4.5, 4.6	Visual inspection results are simulated by this objective variable (*V.I.*). It is a subjective measure of condition rating of the system of interest and is related directly to actual exposure of the system. The number of states has to be consistent with the number of condition ratings the owner uses for visual inspection. Four condition ratings used in this example are only for simplicity.
V.I._1, V.I._2	4.2	*V.I._i* is the visual inspection rating for the *i*-th bridge in the network.
Vulnerability	4.8	Objective estimate of level of vulnerability of the system. In this example, *Vulnerability* is estimated to be a direct function of capital improvements to the system, *C_CAPEX*.

APPENDIX 4.II: CPTs FOR CASE STUDY 4.1 (TABLES 4.36–4.39)

TABLE 4.36
CPT of *C*

CO		Low				Medium				High			
	V.I.	Good	Fair	Poor	Service	Good	Fair	Poor	Service	Good	Fair	Poor	Service
C	Low	0.9199	0.5783	0.0378	0.0355	0.6999	0.3553	0.0359	0.0007	0.4963	0.2407	0.0355	0.0000
	Medium	0.0800	0.4142	0.6364	0.4292	0.2885	0.5527	0.5145	0.1946	0.4258	0.5186	0.4292	0.0055
	High	0.0001	0.0074	0.3258	0.5353	0.0116	0.0921	0.4496	0.8047	0.0778	0.2407	0.5353	0.9945
Histograms for conditional probabilities		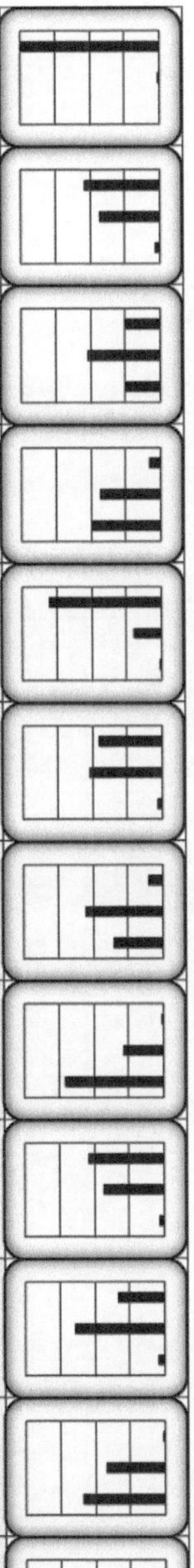											

TABLE 4.37
CPT of *CO*

CO	Low	0.0921
	Medium	0.5527
	High	0.3553
Histograms for conditional probabilities		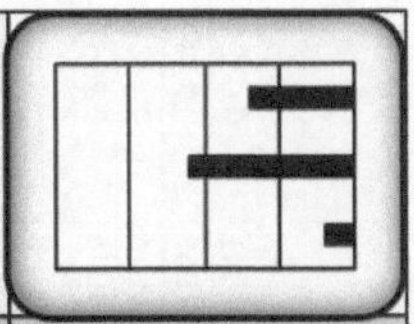

TABLE 4.38
CPT of *Ri*

C		Low				Medium				High			
V.I.		Good	Fair	Poor	Service	Good	Fair	Poor	Service	Good	Fair	Poor	Service
Ri	_1.00E+02	0.6723	0.3585	0.0397	0.0129	0.4666	0.2342	0.0381	0.0001	0.3309	0.1709	0.0378	0.0000
	_1.00E+03	0.2949	0.4717	0.2751	0.1570	0.3882	0.4068	0.2289	0.0249	0.3772	0.3291	0.1994	0.0000
	_1.00E+04	0.0321	0.1570	0.4763	0.4717	0.1285	0.2816	0.4468	0.3815	0.2233	0.3291	0.4164	0.0794
	_1.00E+05	0.0008	0.0129	0.2089	0.3585	0.0168	0.0774	0.2861	0.5935	0.0685	0.1709	0.3464	0.9206
Histograms for conditional probabilities		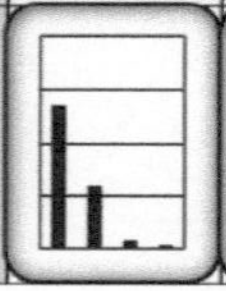		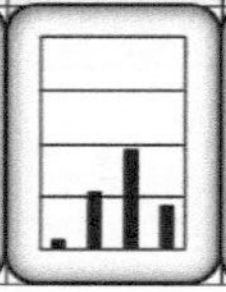	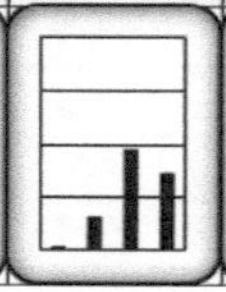	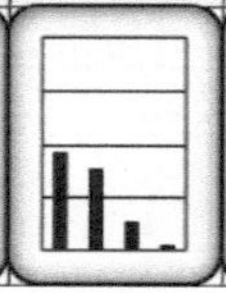	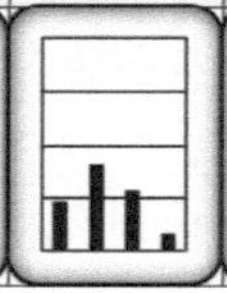	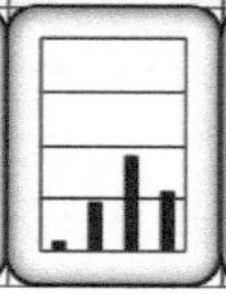	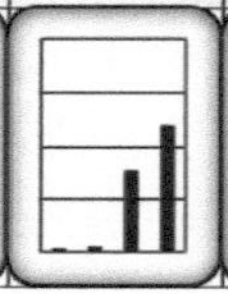	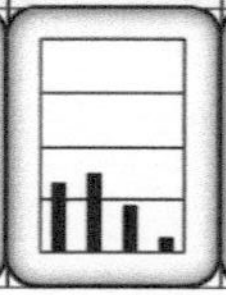	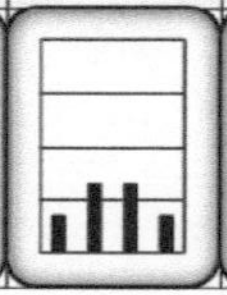	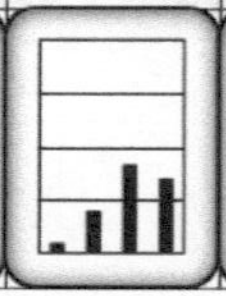	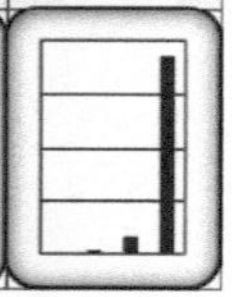

TABLE 4.39
CPT of *V.I.*

V.I.	Good	0.0039
	Fair	0.1564
	Poor	0.5947
	Service	0.2451
Histograms for conditional probabilities		

APPENDIX 4.III: CPTs FOR CASE STUDY 4.2 (TABLES 4.40–4.49)

TABLE 4.40
CPT of *C_1*

CO_1		Low				Medium				High			
VL_1		Good	Fair	Poor	Service	Good	Fair	Poor	Service	Good	Fair	Poor	Service
C_1	Low	0.8240	0.5353	0.0921	0.0355	0.6238	0.3629	0.0828	0.0010	0.4637	0.2670	0.0778	0.0000
	Medium	0.1726	0.4292	0.5527	0.4292	0.3377	0.4931	0.4795	0.2048	0.4148	0.4660	0.4258	0.0117
	High	0.0033	0.0355	0.3553	0.5353	0.0385	0.1439	0.4376	0.7942	0.1215	0.2670	0.4963	0.9883
Histograms for conditional probabilities		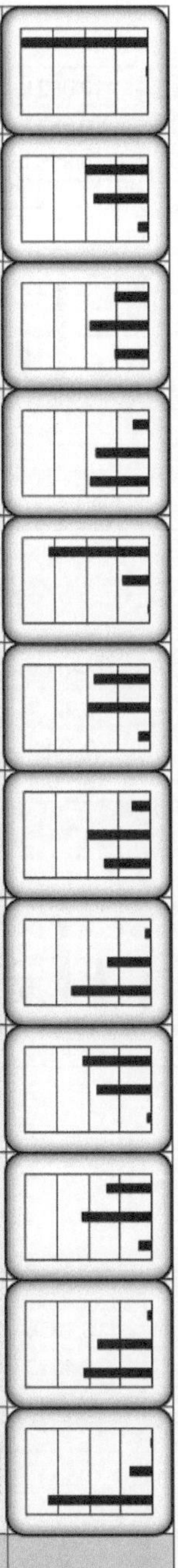											

TABLE 4.41
CPT of *C_2*

CO_2		Low				Medium				High			
VL_2		Good	Fair	Poor	Service	Good	Fair	Poor	Service	Good	Fair	Poor	Service
C_2	Low	0.9199	0.5783	0.0378	0.0355	0.6999	0.3553	0.0359	0.0007	0.4963	0.2407	0.0355	0.0000
	Medium	0.0800	0.4142	0.6364	0.4292	0.2885	0.5527	0.5145	0.1946	0.4258	0.5186	0.4292	0.0055
	High	0.0001	0.0074	0.3258	0.5353	0.0116	0.0921	0.4496	0.8047	0.0778	0.2407	0.5353	0.9945
Histograms for conditional probabilities		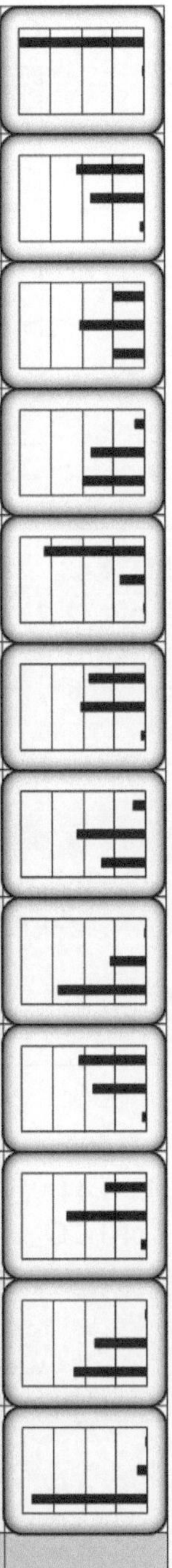											

TABLE 4.42
CPT of *CO*

CO	Low	0.0921
	Medium	0.5527
	High	0.3553
Histograms for conditional probabilities		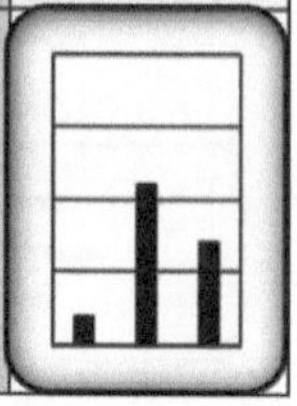

TABLE 4.43
CPT of *CO_1*

CO		Low	Medium	High
CO_1	Low	0.8683	0.2717	0.0385
	Medium	0.1314	0.5898	0.3377
	High	0.0003	0.1385	0.6238
Histograms for conditional probabilities		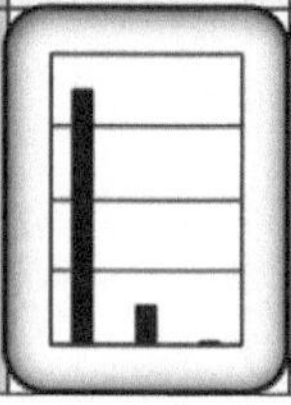	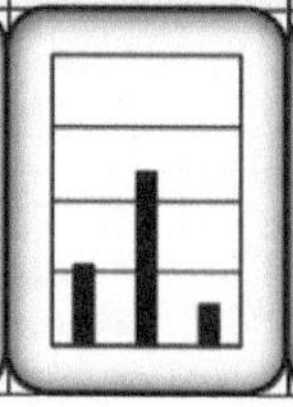	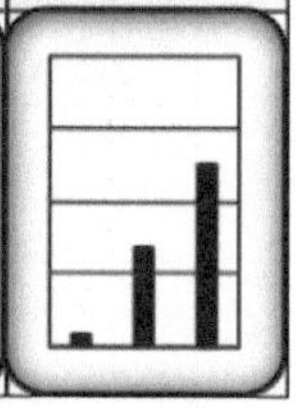

TABLE 4.44
CPT oftt *CO_2*

CO		Low	Medium	High
CO_2	Low	0.5143	0.1854	0.0947
	Medium	0.3894	0.4525	0.3498
	High	0.0963	0.3621	0.5555
Histograms for conditional probabilities		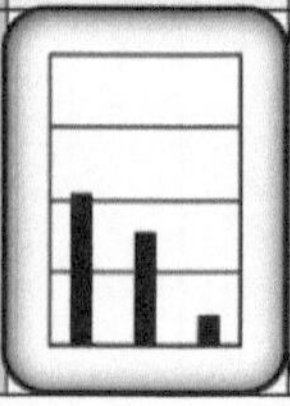		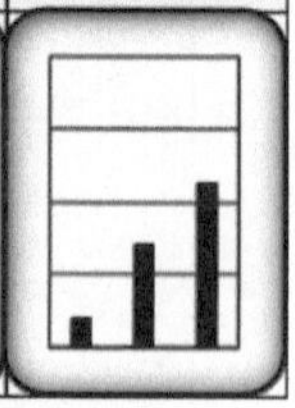

TABLE 4.45
CPT of *Ri_1*

C_1		Low				Medium				High			
	V.I._1	Good	Fair	Poor	Service	Good	Fair	Poor	Service	Good	Fair	Poor	Service
Ri_1	_1.00E+02	0.7878	0.3548	0.0098	0.0129	0.5231	0.2089	0.0113	0.0001	0.3464	0.1424	0.0129	0.0000
	_1.00E+03	0.2071	0.5519	0.2436	0.1570	0.3976	0.4763	0.1889	0.0210	0.4164	0.3576	0.1570	0.0000
	_1.00E+04	0.0051	0.0920	0.5911	0.4717	0.0759	0.2751	0.5326	0.3775	0.1994	0.3576	0.4717	0.0551
	_1.00E+05	0.0000	0.0014	0.1554	0.3585	0.0035	0.0397	0.2672	0.6014	0.0378	0.1424	0.3585	0.9449
Histograms for conditional probabilities		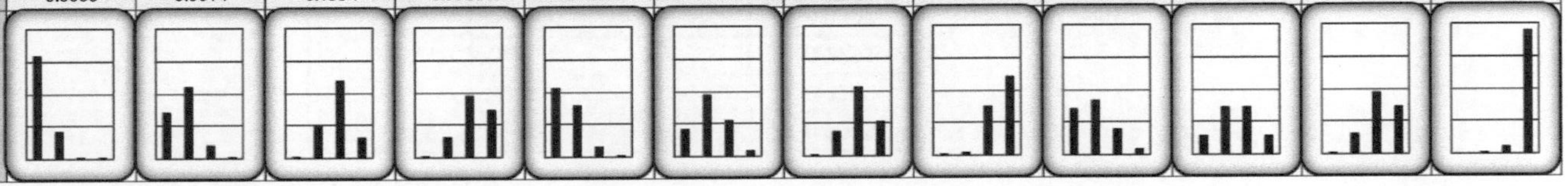											

TABLE 4.46
CPT of *Ri_2*

C_2		Low				Medium				High			
	V.I._2	Good	Fair	Poor	Service	Good	Fair	Poor	Service	Good	Fair	Poor	Service
Ri_2	_1.00E+02	0.6723	0.3585	0.0397	0.0129	0.4666	0.2342	0.0381	0.0001	0.3309	0.1709	0.0378	0.0000
	_1.00E+03	0.2949	0.4717	0.2751	0.1570	0.3882	0.4068	0.2289	0.0249	0.3772	0.3291	0.1994	0.0000
	_1.00E+04	0.0321	0.1570	0.4763	0.4717	0.1285	0.2816	0.4468	0.3815	0.2233	0.3291	0.4164	0.0794
	_1.00E+05	0.0008	0.0129	0.2089	0.3585	0.0168	0.0774	0.2861	0.5935	0.0685	0.1709	0.3464	0.9206
Histograms for conditional probabilities		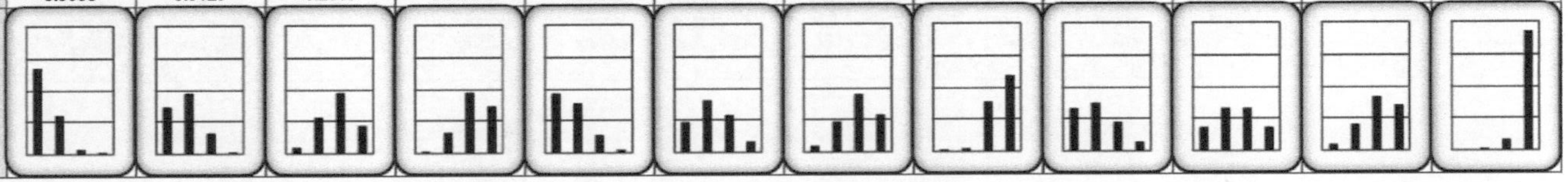											

TABLE 4.47
CPT of *Ri_network*

Ri_2		_1.00E+02				_1.00E+03				_1.00E+04				_1.00E+05			
Ri_1		_1.00E+02	_1.00E+03	_1.00E+04	_1.00E+05	_1.00E+02	_1.00E+03	_1.00E+04	_1.00E+05	_1.00E+02	_1.00E+03	_1.00E+04	_1.00E+05	_1.00E+02	_1.00E+03	_1.00E+04	_1.00E+05
Ri_network	_2.00E+02	0.7268	0.3599	0.0227	0.0129	0.5628	0.2590	0.0229	0.0009	0.4331	0.1966	0.0234	0.0000	0.3387	0.1580	0.0240	4.5703E–12
	_2.00E+03	0.2577	0.5079	0.2641	0.1570	0.3677	0.4702	0.2270	0.0566	0.4057	0.4032	0.2002	0.0057	0.3952	0.3420	0.1810	1.4921E–05
	_2.00E+04	0.0154	0.1270	0.5265	0.4717	0.0663	0.2378	0.4995	0.4415	0.1426	0.3106	0.4696	0.2840	0.2132	0.3420	0.4416	0.0670
	_2.00E+05	0.0001	0.0053	0.1867	0.3585	0.0032	0.0330	0.2506	0.5010	0.0186	0.0896	0.3068	0.7103	0.0530	0.1580	0.3534	0.9330
Histograms for conditional probabilities		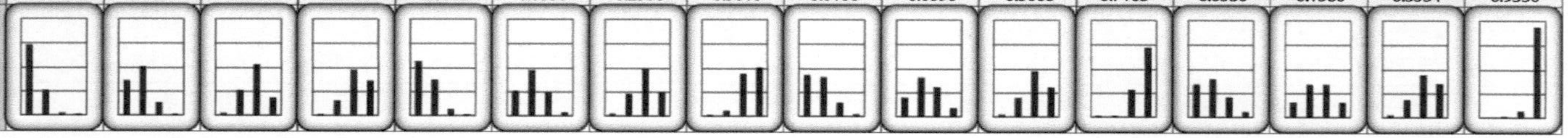															

TABLE 4.48
CPT of *VI_1*

V.I._1	Good	0.2089
	Fair	0.4763
	Poor	0.2751
	Service	0.0397
Histograms for conditional probabilities		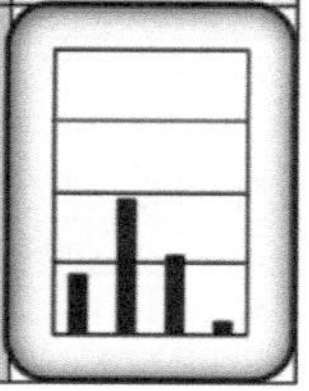

TABLE 4.49
CPT of *VI_2*

V.I._2	Good	0.0039
	Fair	0.1564
	Poor	0.5947
	Service	0.2451
Histograms for conditional probabilities		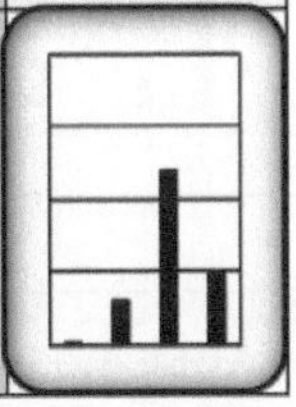

APPENDIX 4.IV: CPTs FOR CASE STUDY 4.3 (TABLES 4.50–4.54)

TABLE 4.50
CPT of *C*

CO		L		H	
E		L	H	L	H
C	L	0.9949	0.6852	0.3913	0.0000
	H	0.0051	0.3148	0.6087	1.0000
Histograms for conditional probabilities		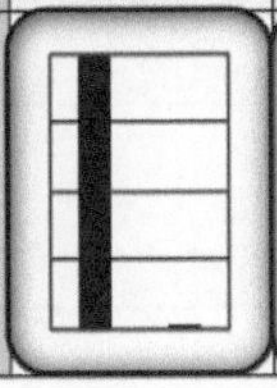	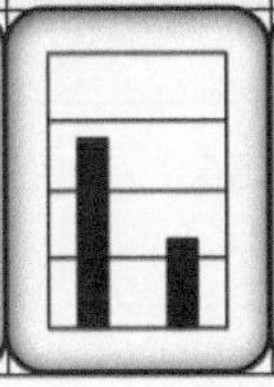	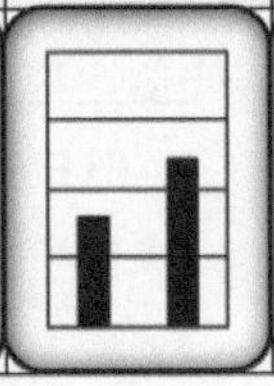	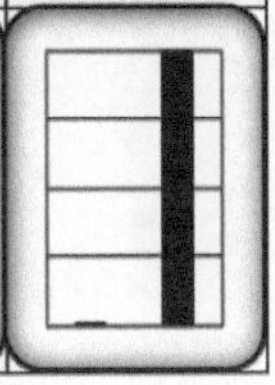

TABLE 4.51
CPT of *CO*

CO	L	0.3
	H	0.7
Histograms for conditional probabilities		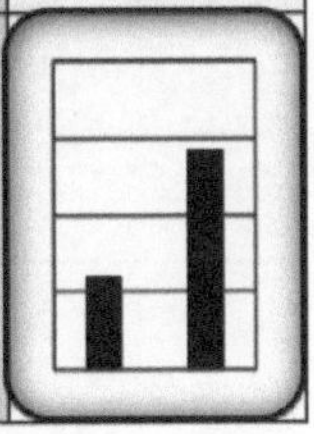

TABLE 4.52
CPT of *E*

Hz		L					H				
V		V.L.	L.	M.	H.	V.H.	V.L.	L.	M.	H.	V.H.
E	L	0.9979	0.8204	0.5981	0.4197	0.0933	0.8302	0.5718	0.3913	0.1583	0.0001
	H	0.0021	0.1796	0.4019	0.5803	0.9067	0.1698	0.4282	0.6087	0.8417	0.9999
Histograms for conditional probabilities											

TABLE 4.53
CPT of *Hz*

Hz	L	0.4
	H	0.6
Histograms for conditional probabilities		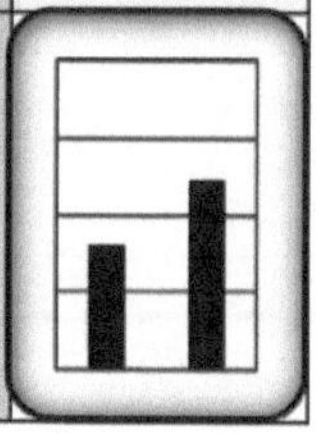

TABLE 4.54
CPT of *V*

V	V.L.	0.6827
	L.	0.2718
	M.	0.0428
	H.	0.0026
	V.H.	0.0001
Histograms for conditional probabilities		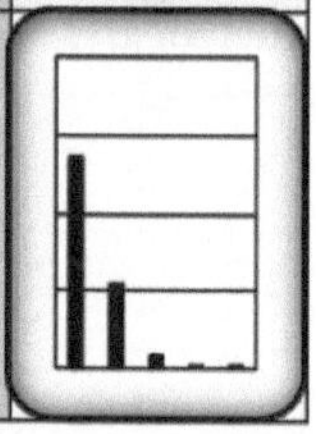

APPENDIX 4.V: CPTs FOR CASE STUDY 4.4 (TABLES 4.55–4.56)

TABLE 4.55
CPT of *C*

Hz		L	M	H
C	Low	0.8240	0.5143	0.0010
	High	0.1726	0.3894	0.2048
	Unacceptable	0.0033	0.0963	0.7942
Histograms for conditional probabilities		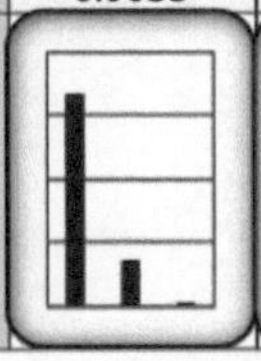	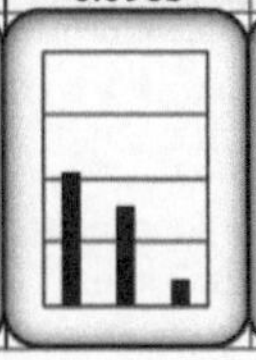	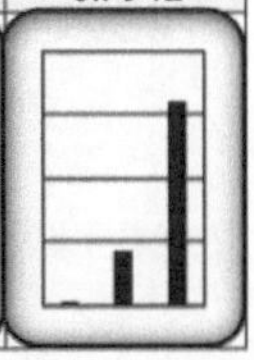

TABLE 4.56
CPT of *V*

Hz		L	M	H
V	Linear	1.00E+00	0.9067	0.6852
	Failed	3.76E–05	0.0933	0.3148
Histograms for conditional probabilities		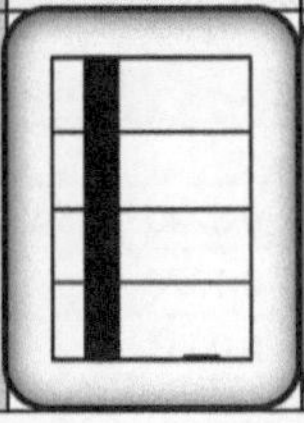	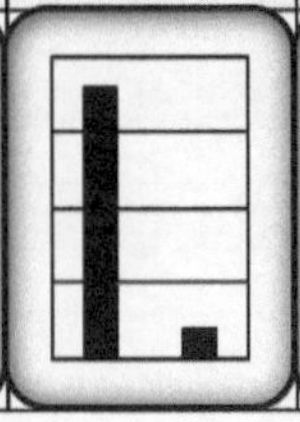	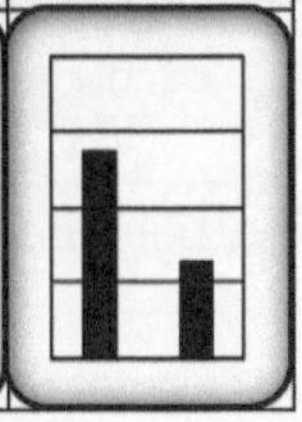

APPENDIX 4.VI: CPTs FOR CASE STUDY 4.5 (TABLE 4.57–4.58)

TABLE 4.57
CPT of *Det.*

Det.	Pristine	0.4
	Damaged	0.6
Histograms for conditional probabilities		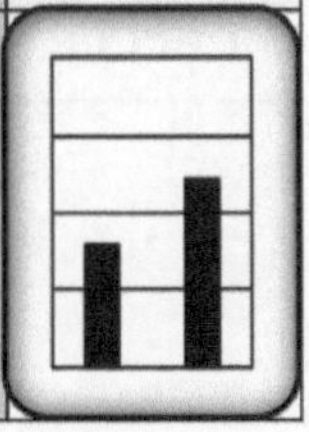

TABLE 4.58
CPT of *V.I.*

Det.		Pristine	Damaged
V.I.	Good	0.6723	0.0035
	Fair	0.2949	0.0759
	Poor	0.0321	0.3976
	Service	0.0008	0.5231
Histograms for conditional probabilities		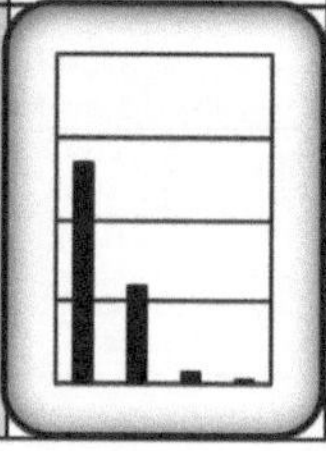	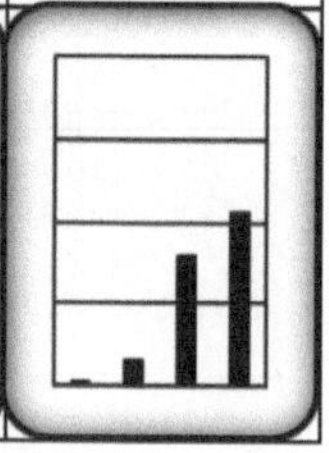

APPENDIX 4.VII: CPTs FOR CASE STUDY 4.6 (TABLES 4.59–4.62)

TABLE 4.59
CPT of *C*

Det.		Pristine		Damaged	
CO		L.	H.	L.	H.
C	L.	0.8240	0.3621	0.0585	0.0001
	M.	0.1726	0.4525	0.4967	0.0800
	H.	0.0033	0.1854	0.4448	0.9199
Histograms for conditional probabilities		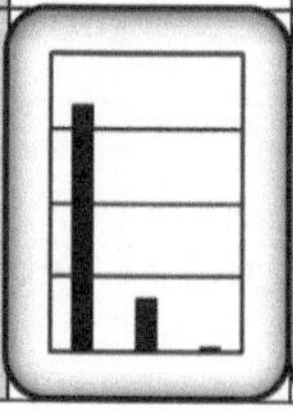	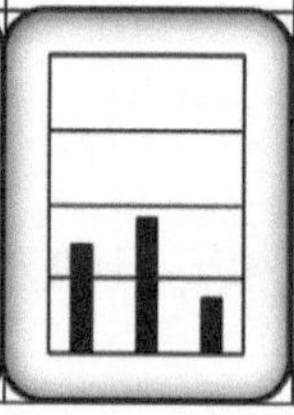	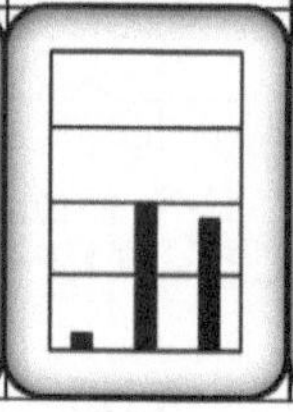	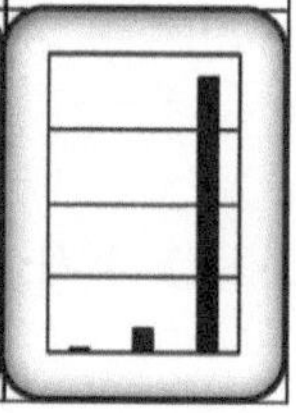

TABLE 4.60
CPT of *CO*

CO	L.	0.3
	H.	0.7
Histograms for conditional probabilities		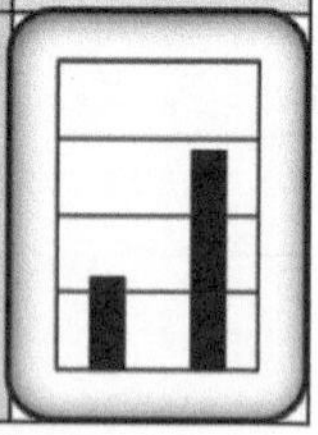

TABLE 4.61
CPT of *Det.*

Det.	Pristine	0.4
	Damaged	0.6
Histograms for conditional probabilities		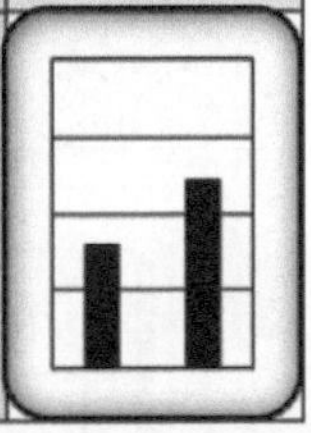

TABLE 4.62
CPT of *V.I.*

Det.		Pristine	Damaged
V.I.	Good	0.6723	0.0035
	Fair	0.2949	0.0759
	Poor	0.0321	0.3976
	Service	0.0008	0.5231
Histograms for conditional probabilities		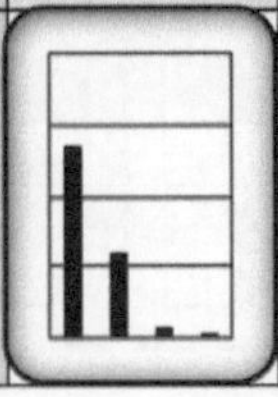	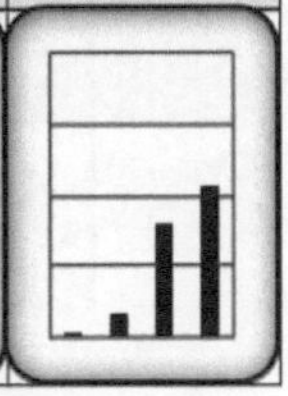

APPENDIX 4.VIII: CPTs FOR CASE STUDY 4.8 (TABLES 4.63–4.67)

TABLE 4.63
CPT of *C_CAPEX*

C_CAPEX	Low	0.7
	High	0.3
Histograms for conditional probabilities		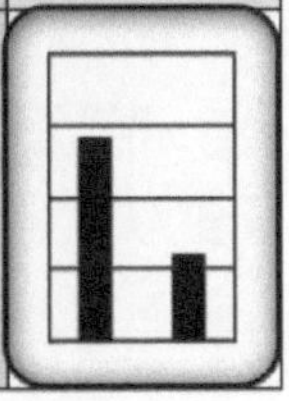

TABLE 4.64
CPT of *C_Func*

Threat		Low		Med		High	
Vulnerability		Low	High	Low	High	Low	High
C_FUNC	Low	0.6999	0.3553	0.3133	0.2242	0.0175	0.0063
	Med	0.2885	0.5527	0.4626	0.4626	0.5321	0.2260
	High	0.0116	0.0921	0.2242	0.3133	0.4504	0.7677
Histograms for conditional probabilities		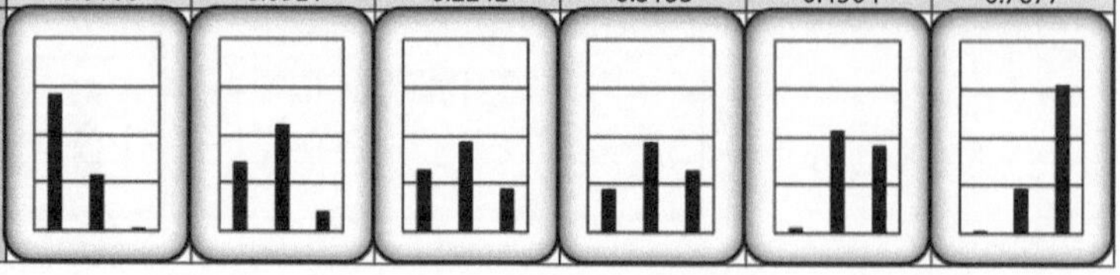					

TABLE 4.65
CPT of *C_Phys*

Threat		Low		Med		High	
Vulnerability		Low	High	Low	High	Low	High
C_PHYS	Low	0.5783	0.1329	0.1886	0.1071	0.0000	0.0000
	Med	0.4142	0.7341	0.5121	0.4635	0.2036	0.0467
	High	0.0074	0.1329	0.2994	0.4293	0.7964	0.9533
Histograms for conditional probabilities		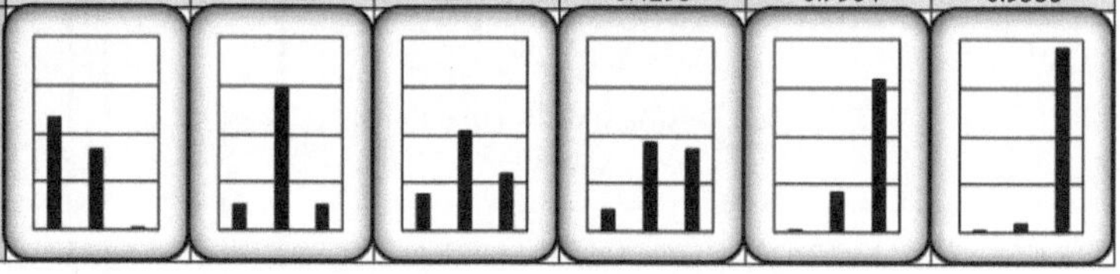					

TABLE 4.66
CPT of *Threat*

Threat	Low	0.2
	Med	0.4
	High	0.4
Histograms for conditional probabilities		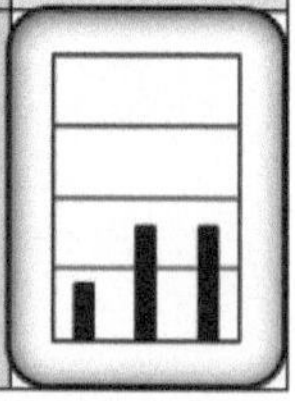

TABLE 4.67
CPT of *Vulnerability*

C_CAPEX		Low	High
Vulnerability	Low	0.01	0.8
	High	0.99	0.2
Histograms for conditional probabilities		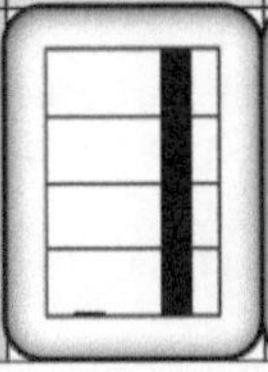	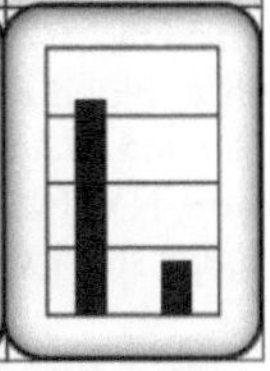

REFERENCES

AASHTO. (2014). *AASHTO LRFD Bridge Design Specifications,* Seventh Edition, American Association of State Highway and Transportation Officials, Washington, DC.

Agrawal, A., Kawaguchi, A., and Qian, G. (2009). *Bridge Element Deterioration Rates: Phase I Report.* Project # C-01-51. New York State Department of Transportation, Albany, NY.

Agrawal, A., Washer, G., and Xu, G. (2013). *Consistency of the New York State Bridge Inspection Program.* Project # C-07-17. New York State Department of Transportation, Albany, NY.

AISC (1999). *Load and Resistance Factor Design Specification for Structural Steel Buildings*, American Institute of Steel Construction, Inc., Chicago, IL.

American Concrete Institute (ACI). (2014). *Building Code Requirements for Structural Concrete* (ACI 318-14) and Commentary (ACI 318R-14), Farmington Hills, MI.

ASCE (2010) *Minimum Design Loads for Buildings and Other Structures* (ASCE/SEI 7-10), American Society for Civil Engineers, Reston, VA.

ASCE/SEI. (2013). *Seismic Evaluation and Rehabilitation of Existing Buildings,* ASCE/SEI Standard ASCE 41–13. American Society of Civil Engineers, Reston, VA.

Ayyub, B. (2003). *Risk Analysis in Engineering and Economics.* Chapman & Hall, Boca Raton, FL.

Chen, H., and Liew, J. Y. R. (2005). Explosion and Fire Analysis of Steel Frames Using Mixed Element Approach. *J. Eng. Mech.* 131(6), 606–616.

Della Corte, G., Landolfo, R., and Mazzolani, F. M. (2003). Postearthquake Fire Resistance of Moment Resisting Steel Frames. *Fire Saf. J.* 38, 593–612.

DoD (Department of Defense). (2010). *Design of Buildings to Resist Progressive Collapse.* Unified Facilities Criteria (UFC) 4-023-03. U.S. DoD, Washington, DC.

DHS. (2011a). *Integrated Rapid Visual Screening of Mass Transit Stations*, Building and Infrastructure Protection Series. DHS, Washington, DC.

DHS. (2011b). *Integrated Rapid Visual Screening of Tunnels*, Building and Infrastructure Protection Series. DHS, Washington, DC.

DHS. (2011c). *Integrated Rapid Visual of Screening of Buildings*, Building and Infrastructure Protection Series. DHS, Washington, DC.

Ettouney, M., Ed. (2015). *Building Security Rating System, Version 2.0 Manual of Practice.* ASCE Press, Reston, VA.

Ettouney, M.M. (2015) *Building Security Rating System: Checklists to Assess Risks, Consequences, and Security Countermeasures*, Ettouney, M.M. (Ed.), American Society for Civil Engineers, Reston, VA.

Ettouney, M.M., and Alampalli, S. (2012a). *Infrastructure Health in Civil Engineering: Theory and Components*, CRC Press, Boca Raton, FL.

Ettouney, M.M., and Alampalli, S. (2016). *Multihazard Considerations in Civil Infrastructure*, CRC Press, Boca Raton, FL.

Farmer, F. R. (1967b). Reactor Safety and Siting: A Proposed Risk Criterion. *Nucl. Saf.* 8(6), 539–548.

Farmer, F. R. (1967a). Siting Criteria—New approach. *Proceedings of Symposium on Containment Siting of Nuclear Power Plants, SM-89/34.* International Atomic Energy Agency, 3–7, 1967, Wieden, Austria.

FEMA. (2009). *Handbook for Rapid Visual Screening of Buildings to Evaluate Terrorism Risks*, Federal Emergency Management Agency (FEMA), Risk Management Series FEMA 455. FEMA, Washington, DC.

Ghosh, J., Caprani, C. C., and Padgett, J. E. (2013). Influence of Traffic Loading on the Seismic Reliability Assessment of Highway Bridge Structures. *J. Bridge Eng.*04013009-1 to 04013009-11

Jayjock, M., Lynch, J., and Nelson, D. (2000). *Risk Assessment Principles for the Industrial Hygienist.* American Industrial Hygiene Association, Falls Church, VA.

Jorion, P. (2006). *Value at Risk.* McGraw Hill, New York.

Neapolitan, R. E. (2004). *Learning Bayesian Networks.* Prentice-Hall, Upper Saddle River, NJ.

Quiel, S. E., and Garlock, M. E. M. (2010). Parameters for ModelingHigh-Rise Steel Building Frame Subject to Fire. *J. Struct. Fire Eng.* 1(2), 115–134.

Reevel, D.E. (1998). Coastal Flood Risk Assessment. *J. Waterway. Port. Coast. Ocean. Eng.* 124(5), 219–228.

Trueman, R.E. (1977). *An Introduction to Quantitative Methods for Decision Making,* 2nd Edition, Holt-Rinehart and Winston, NY, 1977.

Von Neumann, J., and Morgenstern, O. (1944). *Theory of Games and Economic Behavior.* Princeton University Press, Princeton, NJ.

Zhou, H., and Zhang, H. (2011). Risk Assessment Methodology for a Deep Foundation Pit Construction Project in Shanghai, China. *J. Construct. Eng. Manage.* 137(12), 1185–1194.

5 Risk Treatment

5.1 INTRODUCTION

5.1.1 Overview

An integral part of risk management is the risk treatment process. Perhaps the process of risk treatment existed even in a non–risk management environment. Because most perceived safety or deterioration problems in the civil infrastructure community, which have been assessed using paradigms other than risk (such as reliability or exposure), were actually treated using subjective estimates of cost implications; such treatments can arguably be called risk treatments. The main purpose of this chapter is to explore the risk treatment component of risk management and identify a few objectives as well as subjective methods that can be used to increase decision-making efficiency.

Several applications of risk treatment studies are documented in the literature. Balas and Ergin (2002) discussed the first application of reliability-based project management practice for coastal projects in Turkey. Project parameters were estimated by considering the structural damage risk to quantify extreme project risks encountered in coastal construction investments. They then estimated the project completion time by specifying the exceedance probability of the damage level at the design stage and obtaining the delay times of construction activities due to damage at the project-planning phase. For the network evaluation, they derived correlation factors for design parameters and project activities using Monte Carlo and Latin Hypercube sampling. They noted that the application of Monte Carlo simulations (Level III) has the advantage of robustness, provided that the probability distributions of random variables and their correlations are determined from the data compiled over a sufficient number of years.

We note that water mains are an integral part of infrastructure in maintaining the quality of life of any society. So risk treatment of these important infrastructures should be a reasonable approach to ensure accuracy and financial efficiency. Grigg (2013) notes that replacement to sustain the integrity of water mains in the United States is on a cycle of approximately once in 200 years, a slow rate that contributes to the current grade of D– on the Infrastructure Report Card (see ASCE [2013]). The main indicator of physical integrity is frequency of water main breaks, but reducing these is difficult because of the complexity and scale of the problem and the lack of incentives for pipeline replacement. Grigg (2013) reviewed current research to explain the knowledge base about main breaks, the extent to which they are a serious problem deserving more attention, the risk formulation of the decision process for pipeline replacement, the current rates of renewal, and the reasons why utilities renew pipes at low rates.

According to Grigg, based on surveys of larger utilities and conservative estimates, 250,000–300,000 breaks per year occur in the United States because of failures such as structural breaks, joint failures, punctures, or corrosive failure of walls, with an estimated average expected cost of $500,000 due to a large pipe break. The author proposed a general framework for risk assessment considering failure probability and consequences, as well as considering how pipes are exposed to threats that may cause a break, which will cause flooding that may cause damages.

They also identify that the utilities replacement program is not solely based on risk assessment but several other reasons such as gaining efficiencies or coordinating replacement to take advantage of other infrastructure replacements.

Given the resource issues, they note that to reduce the failure rate in the long run, solutions to physical, financial, and institutional problems of water mains will continue to require enlightened and committed professional and political leadership.

Schuster et al. (2005) described another risk treatment strategy relating to the safety of drinking water. According to Schuster et al. (2005), at least 288 such outbreaks occurred in Canada between 1974 and 2001. Following an analytical approach reported in the literature (Sadiq et al. 2004), Lee et al. (2009) described a methodology to assess factors of potential failure in municipal drinking water systems, which may lead to major public health concerns, to assist small utilities. A hierarchical fault tree was used to identify the main contributing factors to water supply failure that may lead to public illness. Because of the unavailability of data to identify frequency of risk factors and the qualitative nature of the problem, the fuzzy logic methodology was used. Thus, fault tree methodology establishes the structure of the risks, and fuzzy logic analysis translates qualitative risk data into probabilities. Sensitivity tests of the fuzzy logic model were performed by changing likelihoods and consequences as well as changing the number of partitions used for the definition of likelihood and consequences for each risk factor. An application of this methodology to the case study of the cryptosporidiosis outbreak in North Battleford showed that waterborne outbreaks can occur and/or be investigated using multiple factors to assess the importance of causative factors.

Historic preservation of buildings can also be applied using risk management strategy. A case in point is the Mansion House, a landmark building in London. It serves as the office and home of the Lord Mayor of the city of London. According to Powderham and Tamaro (1995), a concern of the risk of unacceptable damage arose in 1989 when settlements above the new pedestrian tunnel (constructed in 1988) connecting the Waterloo Line and the City Line appeared to be exceeding the long-term prediction. Tunneling was halted and means of protecting the structure were investigated. These measures included soil improvement to structural underpinning to halting tunnel operations. A system of structural ties through the building was selected as most appropriate, least risky, and most economical. Several tunneling measures were adopted along with development and implementation of the observational method, using reliable instrumentation, to successfully protect the building and complete construction. Trigger levels for building response were established. The building was fully monitored during the subsequent phases of tunneling. Authors noted that the tunneling operations were completed without further delay, in sequence as proposed, and associated settlements remained well within predicted values.

In the field of scour, all bridges over water generally are assessed for scour vulnerability to develop a plan of action in case of flooding and to design appropriate countermeasures. Yanmaz and Apaydin (2012) used a risk-based methodology and software named HYRISK (Pearson et al. 2002), developed by the Federal Highway Administration (FHWA), to evaluate annual risk (R) of scour failure of an existing bridge in monetary values.

$$R = KP_A(Rebuild\_Cost) + (Running\_Cost) + (Time\_Cost) \tag{5.1}$$

where *Rebuild_Cost* is the cost of replacing the bridge as a result of a scour-induced failure, *Running_Cost* is the additional costs associated with vehicles running while detouring during the rebuilding period, *Time_Cost* is the money loss of trucks and people in vehicles while detouring, K is a risk adjustment factor with $K = K_1K_2$, K_1 is the bridge type factor, K_2 is the foundation type factor, P_A is the annual probability of scour failure (1/year). The first estimation of annual probability of scour failure is obtained as a function of overtopping frequency and scour criticality. Annual probability of scour failure regarding bridge scour vulnerability was presented in a tabular format. Scour vulnerability of a bridge can be assessed by the aid of the information about nearby channel characteristics and the substructure condition of the bridge of concern, and methodology such as the one Yanmaz et al. (2007) developed can be used. Authors have described the above methodology in detail and then presented a case study for a bridge crossing Fol Creek in the Black Sea region of Turkey to design a proper scour countermeasure using an economic analysis.

Many researchers have studied using fault tree analysis to prove the efficacy of risk treatment. Specifically, structural systems are made of components, and various components influence the failure differently. Understanding the risk associated with the entire system is an important topic, and Davis-McDaniel et al. (2013) used fault tree analysis as a risk prediction tool to study the whole bridge over its lifetime. A fault tree is a risk assessment tool that depicts a failure event by combining different causal factors and can be used to develop a systematic method of assessing, both qualitatively and quantitatively, failure probability and identifying the initiating events that can lead to system collapse. Qualitative analysis produces a graphical Boolean depiction of the factors (events) leading to bridge failure (top event). Quantitative analysis is performed by determining the occurrence probabilities of all events and by using Boolean algebra and probabilistic mathematics to solve for the occurrence probability of the pertinent events. The occurrence probability of a unique combination of events leading to failure provides insight into the most likely scenarios.

Researchers are prudent in noting that general knowledge of the bridge and its structural components is necessary to develop a fault tree, and a list of plausible failures and their causes must be compiled using sources such as expert opinion, failure case studies, and inspection reports. Probabilistic data for the basic events is required to transition from the qualitative fault tree for the quantitative analysis.

Davis-McDaniel et al. (2013) applied the fault tree–based bridge risk analysis process to the James B. Edwards Bridge in Charleston, South Carolina, constructed in 1989 and consisting of two post-tensioned side-by-side concrete segmental box girders. They suggested that the top five critical factors leading to segmental concrete box girder bridge failure, from most to least critical, are as follows: flood, scour, overloading, corrosion of posttensioning tendons, and earthquake. Their analysis showed the bridge approaching an unsafe structural state at 50 years, with a 1.21×10^{-3} probability of failure. They do note that the causal analysis of failure through the fault tree can help identify countermeasures to minimize failure risks but suggest using it in combination with current methodologies, such as visual inspections and structural health monitoring sensors. They also note that lack of research on occurrence probabilities of in-depth basic events, such as those related to corrosion, as a limitation to this methodology.

5.1.2 Objectives of Risk Treatment

The risk treatment process originates when, for a given hazard and a given limit state, assessed risk, $Ri|_{Assessed}$, exceeds acceptable risk, $Ri|_{Accepted}$, such that

$$Ri|_{Assessed} > Ri|_{Accepted} \tag{5.2}$$

The purpose of the risk treatment process is simply to reduce the risk to, $Ri|_{New}$, such that

$$Ri|_{New} \leq Ri|_{Accepted} \tag{5.3}$$

If and when it is not possible to achieve condition (5.3), then the objective of the treatment process would produce an adequate residual risk, $Ri|_{Residual}$, which is defined by

$$Ri|_{Residual} = Ri|_{Accepted} - Ri|_{New} \tag{5.4}$$

As we observed that setting, the level of $Ri|_{Accepted}$ is ultimately subjective and hence the level of residual risk would also be ultimately subjective. In setting $Ri|_{Residual}$, the stakeholder may use one of the several techniques offered in Chapter 4. For the remainder of this chapter though, we comment on the processes of risk treatment from a decision-making viewpoint. Our main goal is to lessen the subjective nature of risk treatment decisions to make it a more objective process. We show that by following an objective decision process in risk treatment, stakeholders can achieve cost savings as well as an efficient risk treatment process.

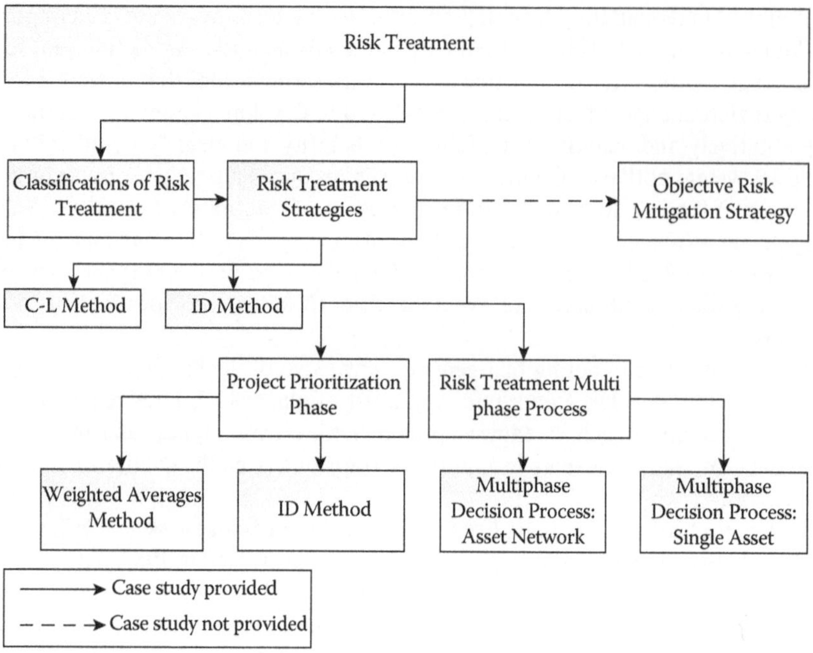

FIGURE 5.1 Risk treatment chapter layout.

5.1.3 This Chapter

We present an overview of the nature of risk as it relates to civil infrastructure followed by categories of risk treatment. The remainder of the chapter looks at the different processes of risk treatment categories through specific, simple, yet realistic examples. Figure 5.1 shows an overview of the sections of this chapter.

5.2 PHASES OF RISK TREATMENT

5.2.1 Overview

When assessed risk is greater than acceptable risk, the stakeholders of the infrastructure asset or community of assets may decide to embark on a risk treatment process. Historically, this process has been fairly subjective, but in recent years, more objective processes have been evolving, mainly in the infrastructure construction phase. It is not usual to use objective processes during *all* risk treatment phases. However, as we will see later, greater efficiencies and cost savings may be realized by implementing objectivity in as much of the process as possible.

Three fairly distinct phases compose a risk treatment process. All of the phases are important, and some of the phases can be interlinked. For an accurate and efficient process, such linking between the phases needs to be considered: a fact that will become apparent in some of the case studies later in this chapter. The three phases of risk treatment are choosing treatment strategy, planning and prioritizing for projects, and executing the projects. The process is shown in Figure 5.2 and will be explored next in detail.

5.2.2 Phase I: Choose Strategy

This phase is well discussed in risk management communities (finance, insurance, engineering, etc.). In this phase, decision makers need to choose a general strategy for risk treatment. Overall, there are four possible strategies, as follows.

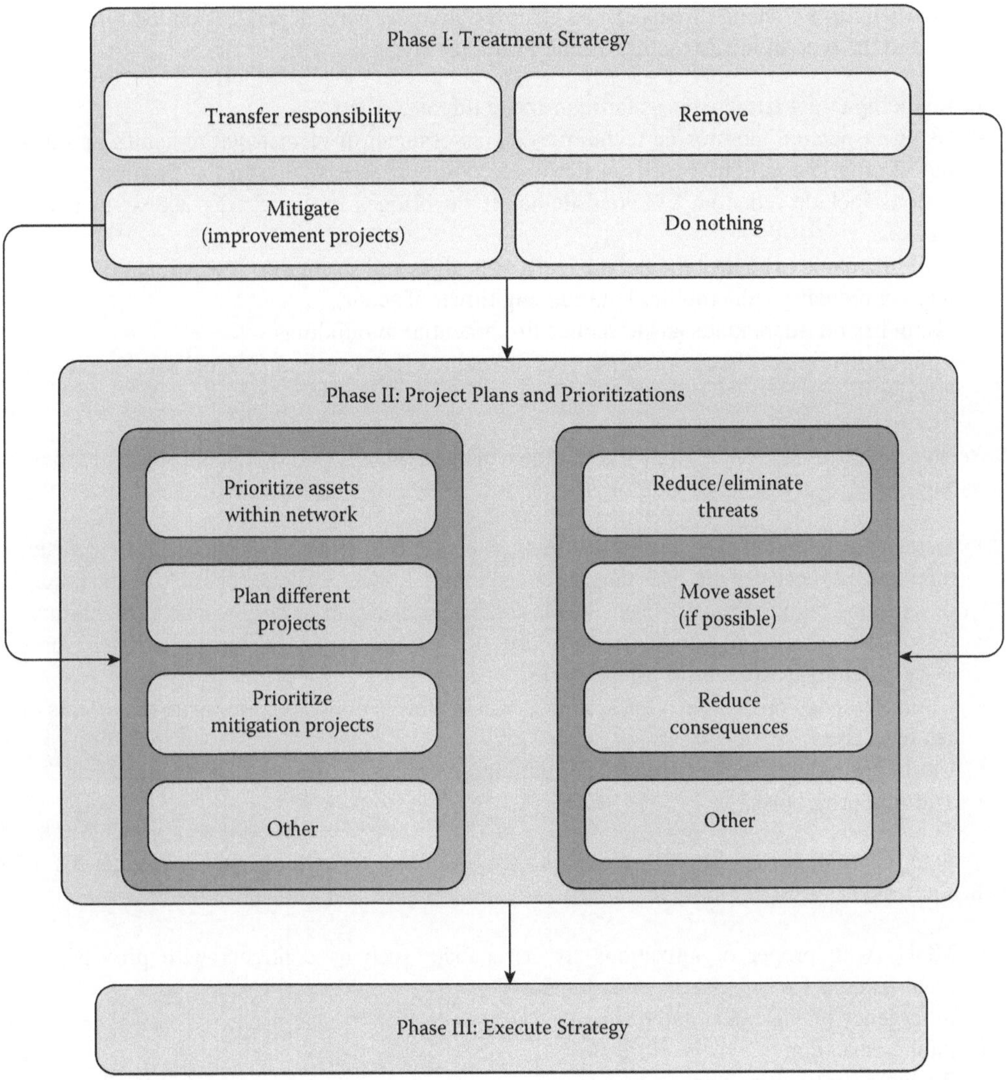

FIGURE 5.2 Risk treatment processes.

5.2.2.1 Risk Transfer (or Sharing)

In this strategy, the transfer can be made by some provisions (contracting, joint ventures, etc.) with other organizations, such as insurance companies. Owners of civil infrastructures might resort to flood insurance or earthquake insurance. In certain situations, insurance against man-made activities might be available. Seeking professional help also might transfer certain types of the risks.

5.2.2.2 Risk Mitigation (Modification)

This strategy is perhaps the most common strategy within the civil infrastructure community. It is based on reducing one or more of the risk components by retrofitting, maintenance, or rehabilitation efforts. Risk treatment is distinct from risk mitigation. Risk treatment is the general process of improving the risk picture, whereas risk mitigation is one of four possible strategies within the risk treatment process. Mitigation can have several facets as described below.

Reduce Threat: Reducing threats can be achieved in many ways depending on the type of threat (hazard) and the type of infrastructure. Some examples are:

- Reducing target attractiveness for man-made threats.
- Using inspection/monitoring techniques for assessment of infrastructure health and taking appropriate actions based on periodic condition assessment and evaluation. These actions include reducing live load demands on bridges by restricting access to heavy trucks.
- Building dams to control the flood hazard of bridges and structures downstream (note that it is not reducing vulnerability but reducing threat of scour).
- Using fire retardant materials to reduce fire potential in buildings.
- Long-term climate change prevention or reduction efforts such as CO_2 emission reduction efforts.

Reduce Vulnerability: Reducing vulnerabilities includes many operational and physical processes including:

- Infrastructure maintenance and rehabilitation decisions to reduce the vulnerability of structures, such as not using pin-and-hanger connections in bridges to reduce possible brittle failure or replacing high rocker bearings with elastomeric bearings to reduce seismic vulnerability
- Corrosion prevention efforts through use of new materials or appropriate measures (e.g., cathodic protection for bridge decks)
- Climate change preventive efforts, such as raising levels of assets to mitigate the effects of sea level rise
- Use of sprinkler systems to reduce the damaging effects of fire
- Training programs

Reduce Consequences: Several methods can be used to reduce the consequences of a particular risk:

- MOUs with proper organizations for assistance, such as contracts with professional organizations for help during emergencies
- Emergency management and training
- Public education
- Business continuity plans
- Recovery plans

5.2.2.3 Accept/Tolerate Risk

This strategy is generally the least desired strategy. It is found in one of two situations:

1. It is decided that the residual risk, $Ri|_{Residual}$, can be accepted. It is a judgmental call and should be taken only with great caution.
2. There is no available budget to embark on a risk treatment effort.

When this strategy is adopted, some contingent emergency plans need to be in place just in case the risky situation materializes. Such contingencies may include: reserve emergency funding, personnel training, and so forth.

5.2.2.4 Remove/Circumvent/Avoid Risk

Sometimes it is possible to remove the risk entirely. For example, in case of flood risk, the stakeholder may choose not to place the asset in a flood plain. In case of a man-made hazard, the stakeholder might decide to move the facility to a more secure location.

5.2.3 Phase II: Project Plans and Prioritization

The effort required by decision makers depends on the chosen risk treatment strategy as noted below:

- *Risk transfer*: After the needed operational risk transfer efforts are taken (such as getting appropriate insurance), no additional steps are generally required except for periodic follow-up efforts.
- *Mitigation*: This strategy involves several additional steps. First, if the domain of the decision maker is a community of assets, then prioritizing these assets within the community is needed. This should be followed by planning for different risk improvement projects for each asset. Prioritization of projects for each asset is next. Finally, any needed additional social, political, or economic steps should be taken in preparation for Phase III.
- *Accept risk*: No further effort is generally needed when this strategy is chosen except for appropriate periodic follow-up.
- *Remove risk*: Studies regarding optimal risk efforts need to be performed. Among the possible efforts are reducing or eliminating threats, moving the asset from its current location, and/or reducing or eliminating consequences.

5.2.4 Phase III: Execute Strategy

Finally, this step involves executing the chosen strategy of Phase I and any efforts that are needed as a result in Phase II. Periodic evaluation is required as conditions generally change with time.

5.3 NEED FOR OBJECTIVE APPROACH WHEN CHOOSING RISK TREATMENT STRATEGY

5.3.1 Overview

We now turn our attention to the practical process of choosing between the four major risk treatment methods of the previous sections. A decision maker, given a risk assessment and acceptance results, has to decide how to treat the risk that was just assessed in the most optimal manner. Traditionally, a subjective approach, for example (Vose 2009 and Ayyub 2003), is to use the exposure (likelihood)–consequence, or *E–C* diagram (also known as *L–C* diagram), as shown in Figure 5.3. The *L–C* diagram has four quadrants:

- Low exposure (likelihood)—Low consequence
- Low exposure (likelihood)—High consequence
- High exposure (likelihood)—Low consequence
- High exposure (likelihood)—High consequence

The generally recommended risk treatment method, according to the *L–C* diagram, varies as shown in Figure 5.3. The subjectivity of the *L–C* diagram limits its general usefulness greatly. Because of this, a more objective process can be used to decide upon the optimal risk treatment method. We will revisit this issue in Section 5.3.4. For now, we explore a more objective decision making process that utilizes the influence diagram (ID) method for finding optimal risk treatment methods.

5.3.2 Modeling Processes

We begin the risk treatment decision examples with a relatively simple case. There is an uncertain hazard, *H* (for illustration, we assume that the hazard is a seismic hazard). The decision maker

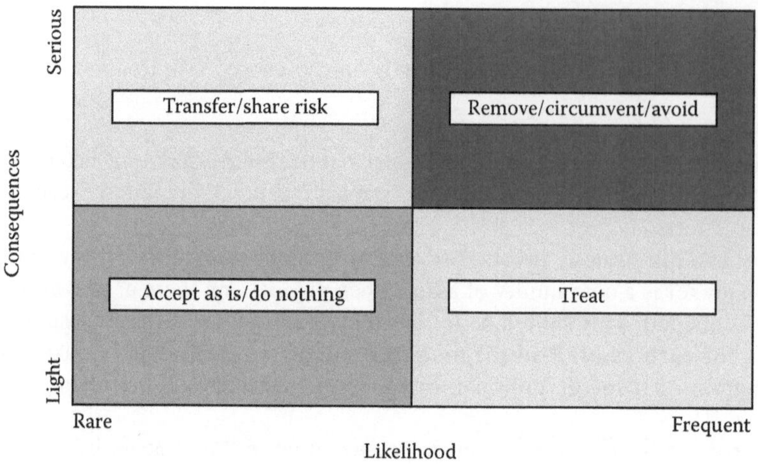

FIGURE 5.3 Likelihood–consequence (L–C) diagram.

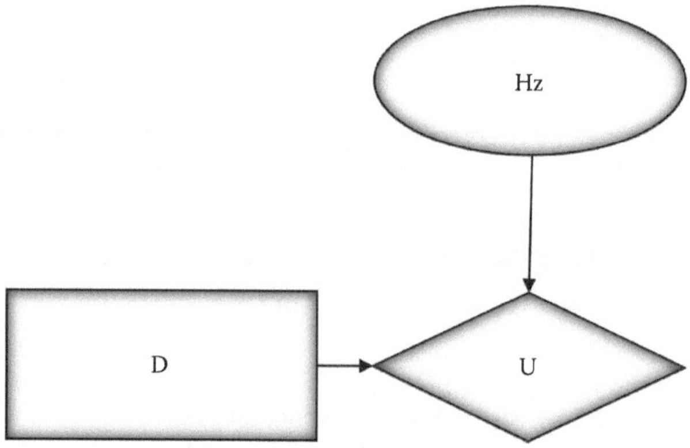

FIGURE 5.4 ID model for simple risk treatment problem.

has to make a decision, D, to treat this uncertain hazard. The optimal utility, U, of the decision will determine which decision offers the best value for risk treatment. Note that, as described in Chapter 2, the utility used in the ID examples is an expression of absolute risk. Figure 5.4 shows the ID model for this simple DM case. The basic assumption in this model is that the hazard, Hz, and the decision, D, are mutually independent. Because the decision maker does not have control over the occurrence of the hazard, there is no arrow from D to Hz. Also, the decision is made without any knowledge of the outcome of Hz, so there is no an arrow from Hz to D. We examine these assumptions in later examples. For this simple case, the probabilities of the hazard states are shown in Table 5.1. Decision variable, D, has four states (possibilities) as shown in Table 5.2. The utilities (costs) of these decisions for each of the possible realizations of the hazard are shown in Table 5.3. We note that the cost of removing the risk, which in case of an earthquake hazard would involve building the facility in an area with low seismic threat. The cost of risk acceptance would cost nothing in the case of $Hz = H_2$; however, the cost can be fairly high if $Hz = H_1$. Cost of mitigation, $D = d_2$, is more reasonable. Though the cost of insurance is moderate on an annual basis, it can accumulate on a life cycle basis. For simplicity, we assume that Hz is computed for a return period of 50 years and the life cycle of the facility of interest is also 50 years. We ignore discount rates for this example.

TABLE 5.1
Probabilities and States of Hazard, *H*

State of *H*	Probability
$Hz = H_1$, Yes—Hazard happens	0.1
$Hz = H_2$, No—Hazard doesn't happen	0.9

TABLE 5.2
Decision Set

D	Risk Treatment Process
$D = d_1$	Transfer Risk (insurance)
$D = d_2$	Mitigate/Modify Risk
$D = d_3$	Accept Risk/Do Nothing
$D = d_4$	Remove/Avoid/Circumvent Risk

TABLE 5.3
Decisions and Their Costs (Utility Table)

D	*U*, Utility (Cost, $)	
	$Hz = H_1$	$Hz = H_2$
$D = d_1$	1.00E + 04	1.00E + 04
$D = d_2$	1.00E + 05	1.00E + 05
$D = d_3$	1.00E + 09	0
$D = d_4$	1.00E + 08	1.00E + 08

TABLE 5.4
Total Utilities of Different Decisions

D	Risk Treatment Process	Cost ($)
$D = d_1$	Transfer Risk (insurance)	1.00E + 04
$D = d_2$	Mitigate/Modify Risk	1.00E + 05
$D = d_3$	Accept Risk/Do Nothing	1.00E + 09
$D = d_4$	Remove/Avoid/Circumvent Risk	1.00E + 08

Applying the ID process of Chapter 2 to this problem, we get the costs of different decisions as shown in Table 5.4. The optimal decision is to mitigate the risk by possibly retrofitting the asset. Clearly doing nothing, $D = d_3$, is not acceptable.

The problem we just discussed is fairly simple but is based on several unreasonable assumptions. We try to address some of these assumptions in a more realistic example next.

5.3.3 Case Study 5.1: Simple Risk Treatment Model: Interaction between Exposure and Consequences

5.3.3.1 Overview

In Section 5.3.2, we made the assumption that none of the four risk treatment decisions would affect the exposure to the hazard. This assumption, however, might not be accurate on most occasions. Consider the ID model of Figure 5.5. It shows that the risk treatment decision can affect some aspects of *both* the exposure and the utility. As an example of this situation, consider the security hazard of a particular facility.

As usual, four decision variables represent the four risk treatment options in decision node, *D*, with utilities (representing absolute risk costs) shown in Table 5.5.

5.3.3.2 Bayesian Network Model

The foundation of the ID model of Figure 5.5 is the Bayesian network (BN) model of Figure 5.6. Note that we removed the utility variable while retaining the uncertain variables and the decision variables. The states of the nodal variables are shown in Table 5.6, and the definitions of those variables are given in Appendix 5.I (Table 5.31). There are four potential decisions/strategies for risk treatment of this problem. For simplicity, we choose the basic four risk treatment categories discussed earlier to model the decision choices shown in Table 5.7. Of course, the decision set can be more detailed. Other ID case studies in this book have more detailed decision sets.

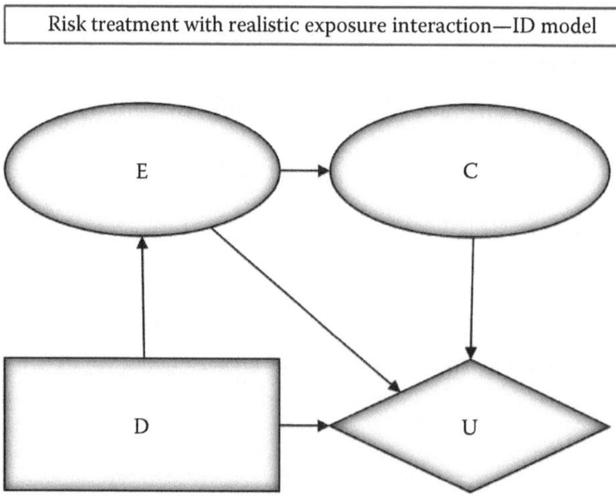

FIGURE 5.5 Risk treatment ID model for man-made risk hazard.

TABLE 5.5
Utilities as Function of Decisions, Exposure, and Consequences for Man-made Hazards

E	*E* = *L*		*E* = *H*	
C	*C* = *L*	*C* = *H*	*C* = *L*	*C* = *H*
d_1 (Transfer)	\$2000.00	\$2000.00	\$2000.00	\$2000.00
d_2 (Mitigate)	\$1500.00	\$1500.00	\$1500.00	\$1500.00
d_{13} (Accept/Do Nothing)	\$–	\$3000.00	\$1000.00	\$5000.00
d_4 (Remove)	\$7000.00	\$7000.00	\$7000.00	\$7000.00

Note: *H* = High, and *L* = Low.

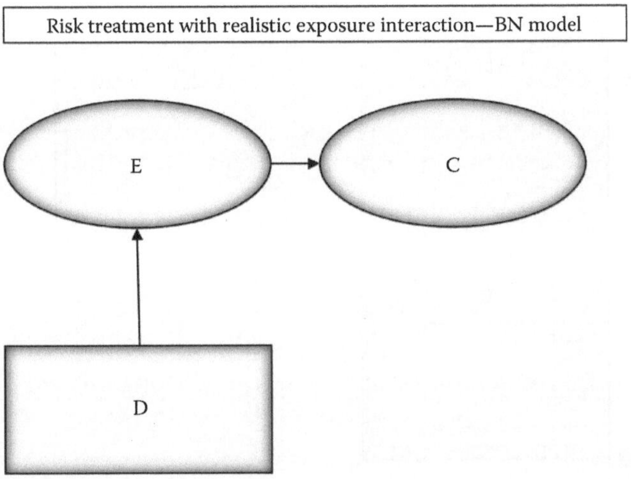

FIGURE 5.6 BN base model for the ID problem.

TABLE 5.6
States of Nodal Variables

Variable	# of States	States
E	2	$H = High$, $L = Low$
C	2	$H = High$, $L = Low$
D	4	d_1, d_2, d_3, d_4
U	—	—

TABLE 5.7
Decision Set

Decision	Activity
$D = d_1$	Transfer
$D = d_2$	Mitigate
$D = d_3$	Accept/Do Nothing
$D = d_4$	Remove

5.3.3.3 Conditional Probabilities Tables

The different conditional probabilities Tables (CPTs) of each variable of the network of Figure 5.6 are shown in Appendix 5.II. As usual we used our best judgment in filling the different tables in a realistic manner for illustration purposes. For more on different methods of developing CPTs, see Chapter 2.

5.3.3.4 Results

We process the BN model for the four decisions within the decision set. The marginal probabilities of the four decisions are shown in Figures 5.7 through 5.10. From the realizations of the BN, we can form the joint probabilities table for different decisions as shown in Table 5.8.

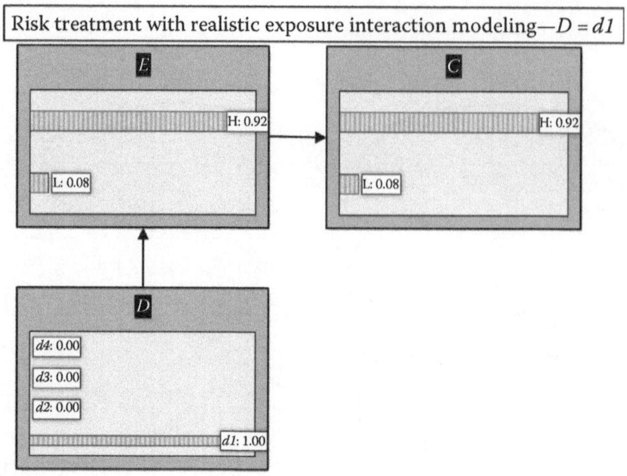

FIGURE 5.7 Marginal probabilities for $D = d_1$.

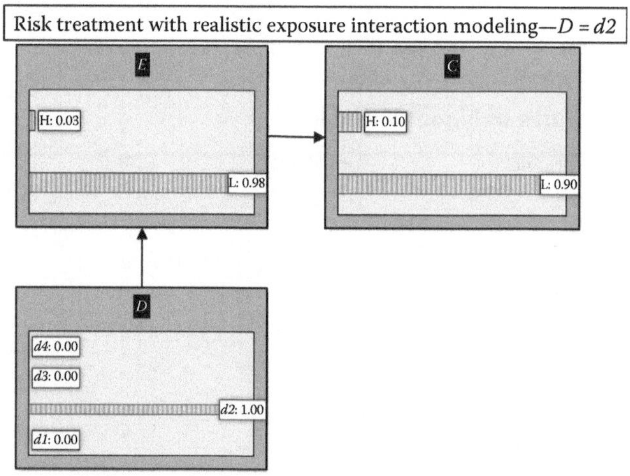

FIGURE 5.8 Marginal probabilities for $D = d_2$.

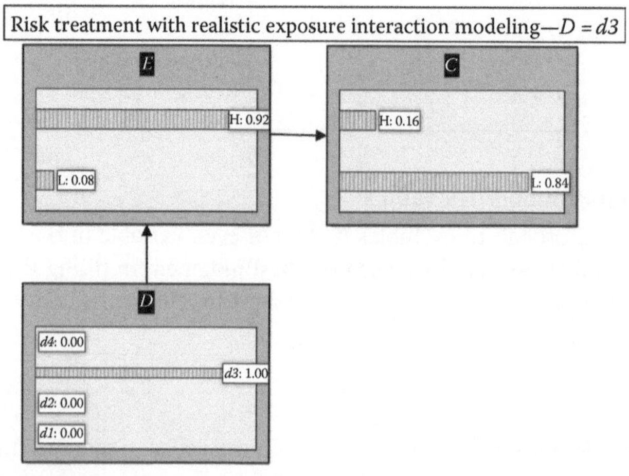

FIGURE 5.9 Marginal probabilities for $D = d_3$.

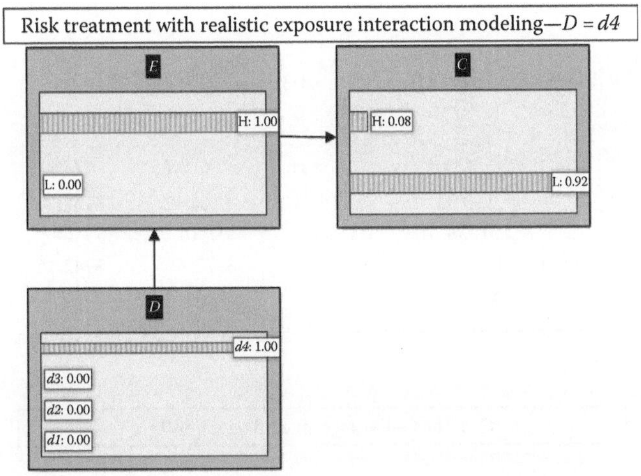

FIGURE 5.10 Marginal probabilities for $D = d_4$.

TABLE 5.8
Joint Probabilities

E	*E* = *L*		*E* = *H*	
C	*C* = *L*	*C* = *H*	*C* = *L*	*C* = *H*
d_1 (Transfer)	0.0067	0.0727	0.0772	0.8434
d_2 (Mitigate)	0.8754	0.0996	0.0224	0.0026
d_{13} (Accept/Do Nothing)	0.0666	0.0128	0.7721	0.1485
d_4 (Remove)	0.0000	0.0000	0.9206	0.0794

TABLE 5.9
Utilities of Different Decisions

E	*E* = *L*		*E* = *H*		
C	*C* = *L*	*C* = *H*	*C* = *L*	*C* = *H*	**Utility**
d_1 (Transfer)	$13.32	$145.48	$154.42	$1686.78	$2000.00
d_2 (Mitigate)	$1313.15	$149.35	$33.67	$3.83	$1500.00
d_{13} (Accept/Do Nothing)	$–	$38.42	$772.11	$742.46	$1552.99
d_4 (Remove)	$–	$–	$6444.20	$555.80	$7000.00

5.3.3.5 Optimal Decision/Strategy for Risk Treatment

Combining Tables 5.3 and 5.8, we can easily generate the utilities (absolute risk) of each decision as in Table 5.9. Comparing the resulting utilities (absolute risks), it appears that the optimal decision is $D = d_2$, which is to mitigate the risk because it is the decision which produces the minimum utility (minimum absolute risk). If, on the other hand, the cost of mitigation were $1700.00 (instead of $1500.00), then the optimal utilities table would be $D = d_3$ (Accept the Risk, Do Nothing). The details of the computations are shown in Table 5.10.

TABLE 5.10
Alternate Utilities of Different Decisions

E	*E* = *L*		*E* = *H*		
C	*C* = *L*	*C* = *H*	*C* = *L*	*C* = *H*	Utility
d_1 (Transfer)	$13.32	$145.48	$154.42	$1686.78	$2000.00
d_2 (Mitigate)	$1488.24	$169.26	$38.16	$4.34	$1700.00
d_{13} (Accept/Do Nothing)	$–	$38.42	$772.11	$742.46	$1552.99
d_4 (Remove)	$–	$–	$6444.20	$555.80	$7000.00

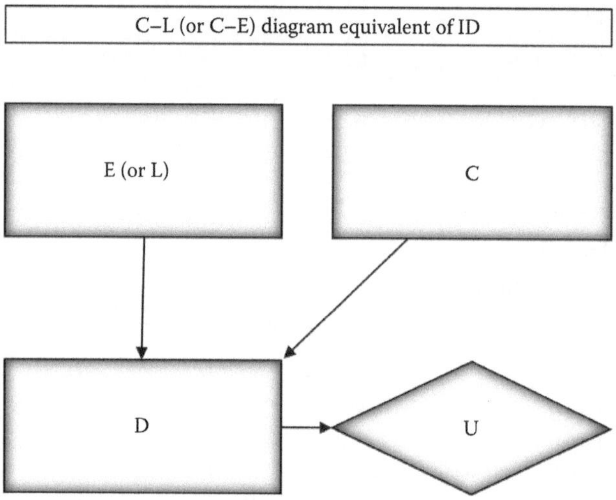

FIGURE 5.11 ID representations of the *C–L* risk treatment process.

5.3.4 Revisiting the *C–L* Diagram Method

We explored different examples of how to decide upon the optimal risk treatment method for some civil infrastructure hazards in a risk-based setting, that is, when *C–L* information is known to the decision maker. We implied, through the CPT of different nodes in the ID models in all of these examples, that the optimal risk treatment decision depends on the values of *C* and *L*. A less obvious implication is that the optimal risk treatment decision is not as static as the *C–L* diagram method of Figure 5.3 indicates.

Let us consider once more the CPTs of Appendix 5.II for the example in Section 5.3.3. These CPTs indicate clearly that the optimal decision is not only a function of both *L* and *C*, but that any risk treatment decision can also change the nature and value of *L* and/or *C*. As such, the risk treatment decision will effectively change the very fabric of the *C–L* diagram. The not too obvious conclusion is: using the *C–L* diagram to obtain a risk treatment decision method can lead to a less than optimal risk treatment decision. It is safer to use a more objective approach that accommodates the complex interactions and dependencies between *L*, *C*, and *D* as shown in this section (See Figure 5.11).

We also observe that even though our simple examples concentrated on civil infrastructure hazards risks, there is no reason to limit the above conclusion regarding the potential inaccuracy of the *C–L* diagram in any other community (such as insurance or finance) that used risk-based management.

5.4 RISK VS. RELIABILITY TREATMENT

We described the differences between risk and reliability in several locations in this book. The main difference, conceptually, between risk and reliability, is that risk is a combination of both reliability and consequences. This simple addition causes many differences in the treatment of both management paradigms as can be seen in Table 5.11.

Studying the table carefully, we reach the conclusion that we can perhaps achieve some parts of the reliability treatment in Phase I; however, in the equally important Phase II of treatment, we can't really prioritize projects based on the information reliability components (capacity and demand) provide. We always need a measure of consequences to prioritize reliability mitigation projects. In those cases, the stakeholder would revert to subjective judgments. Hence, we conclude that there is, in fact, nothing that can be called *reliability-based mitigation*, and all mitigations, no matter on which management paradigm they are based, are risk-based mitigations.

Therefore, we devote the rest of this chapter in exploring the risk process in more detail.

5.5 PROJECT PRIORITIZATION STRATEGIES

5.5.1 Overview

We noted that prioritizations of available mitigation projects is one of the essential components of Phase II in risk treatment (see Figure 5.20). Over the years, several authors have addressed this issue. For example, in port management projects, Hashemi et al. (2013) noted that a few studies on risk management are found in the related literature (Newland 1998; Balas et al. 2000; Kakimoto and Seneviratne 2000; Trbojevic and Carr 2000; Peris-Mora et al. 2005; Tsai and SU 2005; Yip Suedel et al. 2008; 2008) on port projects even though about 90% of the cargo transportation in Asian countries is carried out by marine boundaries. According to researchers, this comparative neglect is because of the complexity of the risk assessment process with nonparametric resampling methods and interval computations. In risk assessment, required data and experts' judgments often are limited. Moreover, statistical distributions of parameters, which play a significant role, are usually unknown. Hence, traditional approaches do not have enough accuracy to assist such problems. Therefore, Hashemi et al. (2013) presented a compromise ranking approach based on bootstrap confidence intervals for risk scores to data obtained from experts' judgments in the port industry.

Main features of the proposed approach, in contrast with other methods for assessing risk in port projects, include a new compromise solution approach based on the decision makers' preferences by employing linguistic terms, a bootstrap method applied to estimate confidence intervals for the risk assessment criteria (i.e., probability and impact) and influential parameters for the ranking process, and a new ranking index with interval numbers for precise and accurate analysis. Based on an

TABLE 5.11
Treatment Strategies of Risk and Reliability Paradigms

Treatment Strategy	Risk	Reliability
Transfer	Yes	Can't be done directly
Remove: Reduce threats	Yes	Yes
Remove: Move asset	Yes	Yes
Remove: Reduce consequences	Yes	Not applicable
Mitigate: Design and plan mitigation projects	Yes	Can't be done directly
Mitigate: Prioritize projects	Yes	Can't be done directly
Accept	Yes	Yes

empirical study using a real case in the port industry in developing countries, it was concluded that the bootstrap method exhibits high performance when the sample size was small and the distributions of the original data were unknown.

We next explore several views of some popular project prioritization processes and also how to incorporate risk methods to improve the processes.

5.5.2 The Dilemma of Project Prioritization

Using initial capital expenditure, *CapEx*, as the prioritization metric is the simplest and by far the most popular process for project prioritization. Another popular method is to use a cost/benefit function as the metric. Finally, each of these two methods can be extended to accommodate life cycle cost (LCC) or life cycle analysis (LCA) prioritization metrics. We discuss each of these methods and their limitations next.

5.5.2.1 CapEx as a Prioritization Metric

Using *CapEx* as a prioritization metric is simple. If two capital expenditures for two projects are $CapEx|^1$ and $CapEx|^2$, assuming

$$CapEx|^1 < CapEx|^2 \tag{5.5}$$

the stakeholder would generally choose Project 1 because it is simply less expensive than Project 2. The reverse is also true.

Obviously this approach has several limitations:

1. It doesn't account for the potential benefits (current and future) that will be gained from each of the two projects.
2. It doesn't account for costs that are not related directly to each of the projects.
3. It only accounts for a snapshot of the cost and ignores future costs during the lifespan of the asset at hand.

5.5.2.2 Cost/Benefit as a Prioritization Metric

Another method for project prioritization is to estimate the benefits of each project and compare them to project costs such that

$$CB|^i = \frac{CapEx|^i}{Benefit|^i} \tag{5.6}$$

where $CB|^i$ is the cost to benefit ratio and $Benefit|^i$ is the benefit estimate for the i-th project. The prioritization process proceeds so that the projects are prioritized according to their $CB|^i$ values, with the highest ranking project the one with the lowest $CB|^i$ value.

Several problems with this method are found:

1. Estimating benefits is fairly difficult and not as straightforward as estimating costs.
2. Smaller projects will usually yield smaller benefits, and projects with higher costs will yield higher benefits. This results in a confusing, or perhaps misleading $CB|^i$ values that may have a built-in bias, depending on how the benefits are computed.
3. Similar to the *CapEx* method, this method does not have a mechanism to accommodate costs and benefits that might occur beyond the domain of the project itself.
4. It only accounts for a snapshot of the cost and ignores future costs during the lifespan of the asset at hand.

5.5.2.3 LCC and LCA as a Prioritization Metric

The *CapEx* and Cost/Benefit models can be improved by using a life cycle prioritization process (see Ettouney and Alampalli 2012a). This resolves the snapshot versus time marching limitations of both methods. However, the other limitations, as detailed above, still restrict their accuracy.

5.5.3 The Case for Risk as Prioritization Metric

Reflecting on the limitations of the prioritization processes of Section 5.5.2, we argue that these limitations can be eliminated by adopting a risk-based approach based on the following:

1. The benefits can be estimated for different projects, both direct and indirect, by accommodating as many variables in the risk model as needed. (We assume that a graph network modeling technique is followed. Using the decision-tree model might prove to be difficult to build and maintain as well as being computationally inefficient.) The links and interactions between these variables can also be easily modeled using either directional or nondirectional link types. In many situations, the benefits can be simulated as cost reductions for pertinent scenarios. These risk models can be as simple or as complex as the stakeholders desire. In addition, if an ID model is followed, it would be possible to model all projects within a single ID model on an equal footing without any bias in modeling thus, resolving naturally one of the most difficult issues in modeling benefits.
2. Parameters (whether costs and/or benefits) beyond the direct project domain that might affect the solutions can easily be included in the model.
3. Risk models can be solved for snapshot or life cycle situations.
4. The cost/benefit dilemma of smaller, *CapEx*–smaller, benefits versus larger, *CapEx*–larger, benefits is also resolved in situations where an absolute risk model is followed.

Using risk models for prioritization is fairly simple. For example, if it is decided to follow a risk reduction plan with two feasible mitigation projects, a straightforward solution would be to compare the two risks after mitigation and choose the project that offers the lowest new risk. Formally, if the two new risks are $Ri|_{New}^{1}$ and $Ri|_{New}^{2}$ for each of the two projects, and if

$$Ri|_{New}^{1} < Ri|_{New}^{2}, \tag{5.7}$$

then Project 1 is the optimal project. The reverse is also true. We can extend the logic to a multitude of projects because we can prioritize them by the order of their new risk status. This approach seems simple. We only need to pay attention to Item 4 above if we use a relative risk metric because the prioritization scheme may offer either biased and/or unreasonable results. Hence, we recommend using an absolute risk model, perhaps with monetary metrics, for the prioritization situation.

5.6 RISK TREATMENT: MULTIPHASE DECISION PROCESS

5.6.1 Overview

Figure 5.2 shows that risk treatment is a multiphase process. Because decisions made at later phases depend on decisions made in earlier phases, an optimal risk treatment decision process needs to account for all phases in an integrated manner. To illustrate this important concept, we consider in this section two case studies to show how ID and BN can be used for optimal decision-making in a multiphase risk treatment process.

5.6.2 Case Study 5.2: Multiphase Risk Treatment Decisions for a Single Asset

Upon further reflection on the model of Case Study 5.1, we can immediately make the following observations:

1. Some of the consequences are a function of decisions made.
2. The risk is none other than the utility value, in this case it is the absolute risk.

Based on these observations and in order to recast the ID model using the exposure, consequences, and vulnerability representations in this book, we offer the ID model shown in Figure 5.12. We model the phases of risk treatment decisions in a simplified manner by using three decision variables, *D1*, *D2*, and *D3*, as follows:

1. *D*1 represents Phase 1 (treatment strategy) with the exception of risk removal.
2. *D*2 represents Phase 2 (mitigation strategies).
3. *D*3 represents risk removal (which is the remainder of the risk treatment options in Phase 1).

For *D*3, risk removal is achieved via either reducing the level of hazard, *Hz*, or reducing consequences, *C.O.* Table 5.12 shows the details of the decision set. Note that the relationships of the ID model of Figure 5.12 reflect how different decision variables affect other variables in the model.

5.6.2.1 Model

The base BN model that controls the ID process is shown in Figure 5.13. The states of the nodal variables of the model are shown in Table 5.13, and the definitions of these variables are shown in Appendix 5.I (Table 5.31). We embark now on developing the utility (cost) tables that are consistent with the ID model of Figure 5.12 as well as the decision set of Table 5.12. We assume that the current state of the vulnerability is *V* = *H*. Based on this, we come up with estimated utilities for different states of variables and different decisions as shown in Table 5.14.

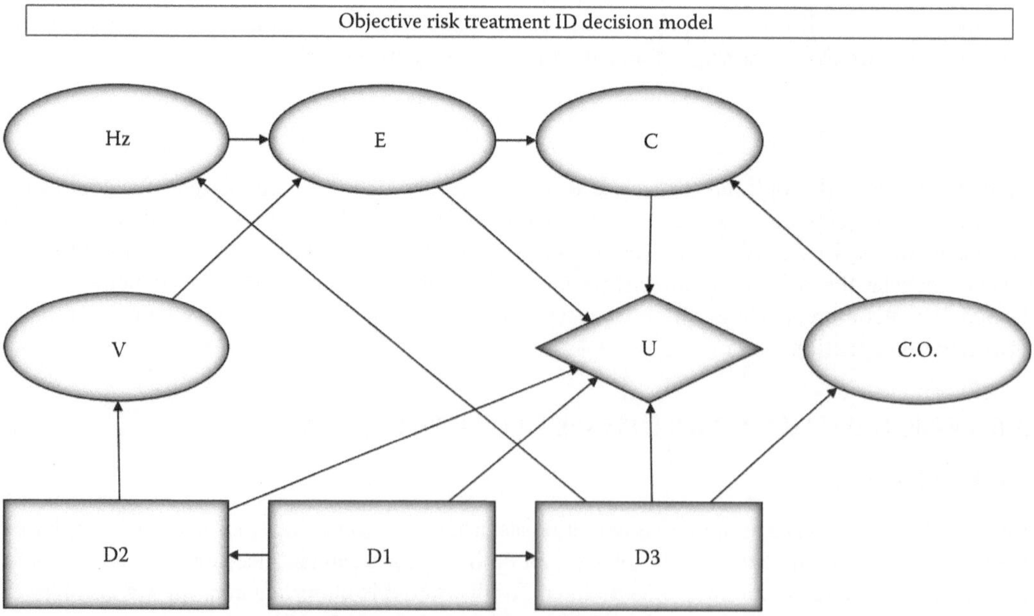

FIGURE 5.12 ID model for risk treatment, mitigation, and removal.

TABLE 5.12
Decision Set

Decision Variables	Decision Subsets
$D1$	d_{11} (Transfer)
	d_{12} (Mitigate) = $D2$
	d_{13} (Accept/Do Nothing)
	d_{14} (Remove) = $D3$
$D2$	d_{21} (Mitigate from $V = H$ to $V = V.L.$)
	d_{22} (Mitigate from $V = H$ to $V = L.$)
	d_{23} (Mitigate from $V = H$ to $V = M.$)
	d_{24} (Accept/Do Nothing)
$D3$	d_{31} (Reduce H to L)
	d_{32} (Reduce $C.O.$ to L)

Note: L. = Low, H. = High, V.L. = Very Low, V.H. = Very High, and M. = Medium.

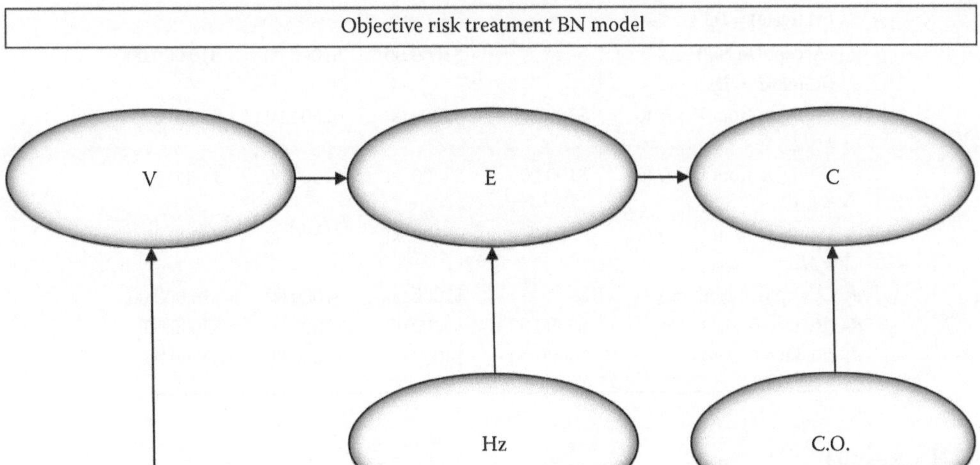

FIGURE 5.13 BN model which supports the ID model of Figure 5.12.

5.6.2.2 Conditional Probabilities Tables

The different CPTs of each variable of the network in Figure 5.13 are shown in Appendix 5.III. As usual, we used our best judgment in filling the different tables in a realistic manner for illustration purposes. For more on different methods of developing CPTs, see Chapter 2.

TABLE 5.13
States of Nodal Variables

Variable	# of States	States
Hz	2	H, L
E	2	H, L
C	2	H, L
V	5	$V.H., H., M., L., V.L.$
$C.O.$	2	H, L
D_1, D_2, D_3	10 (Decisions)	$d_{11}, d_{12}, d_{13}, d_{14}, d_{21}, d_{22}, d_{23}, d_{24}, d_{31}, d_{32}$
U		

TABLE 5.14
Utilities of Treatment, Mitigation, and Transfer

E	*E* = *L*		*E* = *H*	
C	*C* = *L*	*C* = *H*	*C* = *L*	*C* = *H*
d_{11} (Transfer)	\$5000.00	\$5000.00	\$5000.00	\$5000.00
d_{12} (Mitigate) = $D2$				
d_{13} (Accept/Do Nothing)	\$–	\$3000.00	\$1000.00	\$10,000.00
d_{14} (Remove) = $D3$				
d_{21} (Mitigate from $V = H$ to $V = V.L.$)	\$2000.00	\$3000.00	\$2500.00	\$4000.00
d_{22} (Mitigate from $V = H$ to $V = L.$)	\$100.00	\$3200.00	\$2000.00	\$5000.00
d_{23} (Mitigate from $V = H$ to $V = M.$)	\$500.00	\$3500.00	\$2500.00	\$6000.00
d_{24} (Accept/Do Nothing)	\$–	\$3000.00	\$1000.00	\$10,000.00
d_{31} (Reduce H to L)	\$3000.00	\$3000.00	\$3000.00	\$3000.00
d_{32} (Reduce $C.O.$ to L)	\$4000.00	\$4000.00	\$4000.00	\$4000.00

5.6.2.3 Results

The marginal probabilities for different BN observations (which are none other than the deterministic decisions from the decision set of Table 5.12) are shown in Figures 5.14 through 5.19. Note that the proper phasing between decisions is accommodated in each realization of BN. The joint probabilities needed for computing expected utilities of different decisions can now be computed using these marginal probabilities and are shown in Table 5.15.

5.6.2.4 Optimal Decisions

Using the utilities table of Table 5.14 and the joint probabilities of Table 5.15, we can compute the utilities (costs, or absolute risk) of different decisions as shown in Table 5.16. The optimal decision is $D_2 = d_{21}$, which is to mitigate via improving the vulnerability state from $V = H$ to $V = V.L.$ Obviously, the optimal decision can change if the parameters and utilities of the problem change.

5.6.3 Case Study 5.3: Multiphase Risk Treatment Decisions for Asset Networks

5.6.3.1 Overview

We expand Case Study 5.2 into the more complex situation of an asset network. For simplicity, we consider only two assets. However, the potential decision set will multiply very quickly.

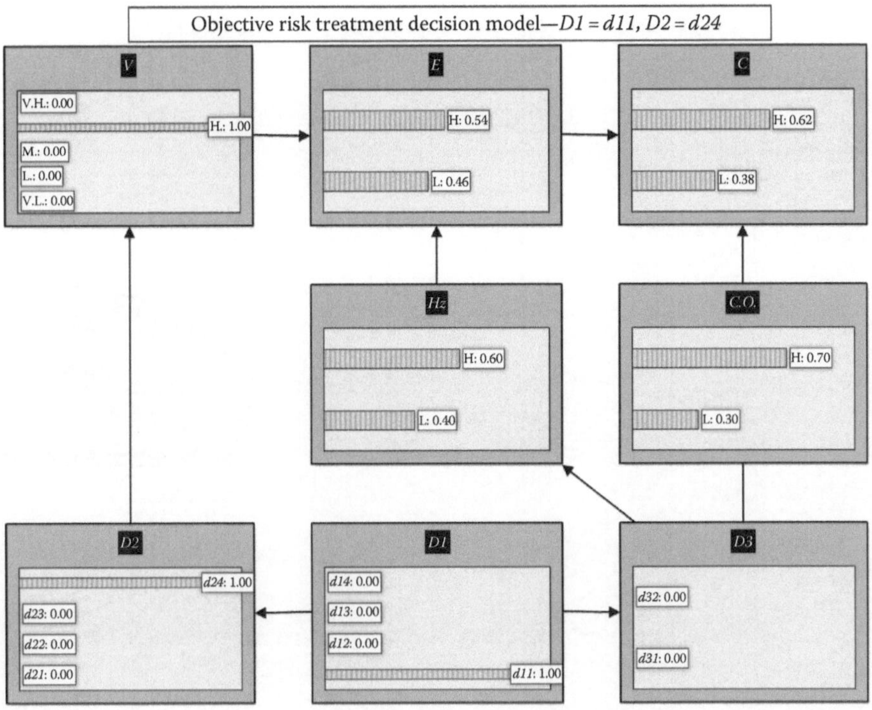

FIGURE 5.14 Marginal probabilities of the BN model: $D_1 = d_{11}$, $D_2 = d_{24}$ (Do Nothing decision).

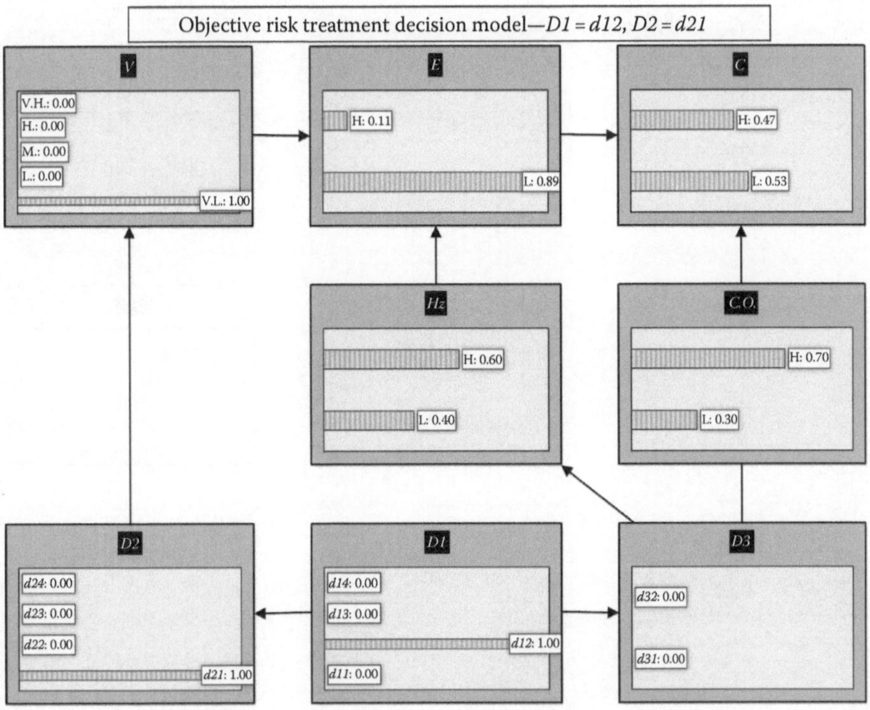

FIGURE 5.15 Marginal probabilities of the BN model: $D_1 = d_{12}$, $D_2 = d_{21}$.

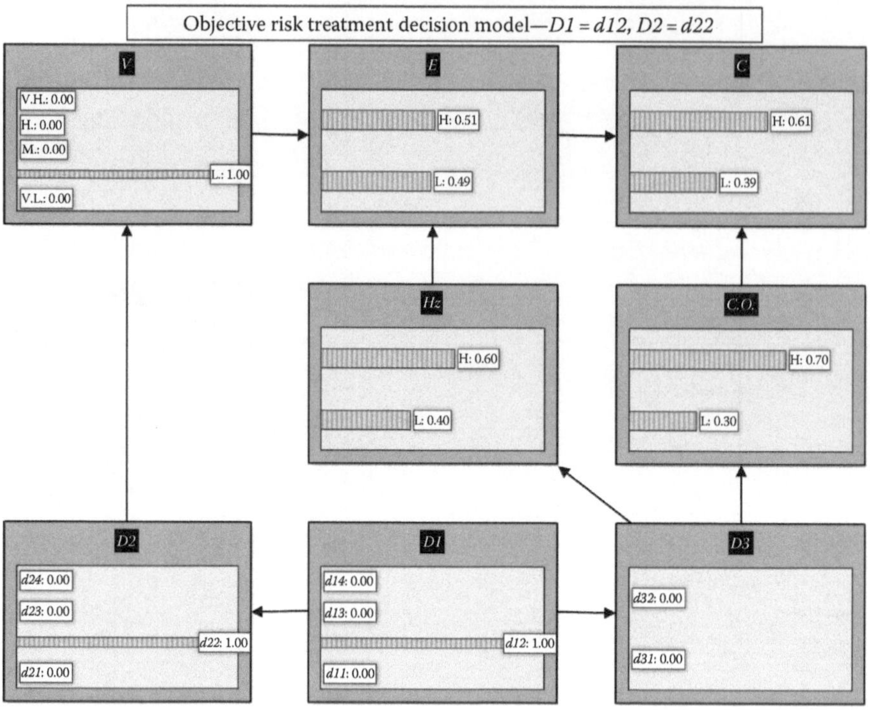

FIGURE 5.16 Marginal probabilities of the BN model: $D_1 = d_{12}$, $D_2 = d_{22}$.

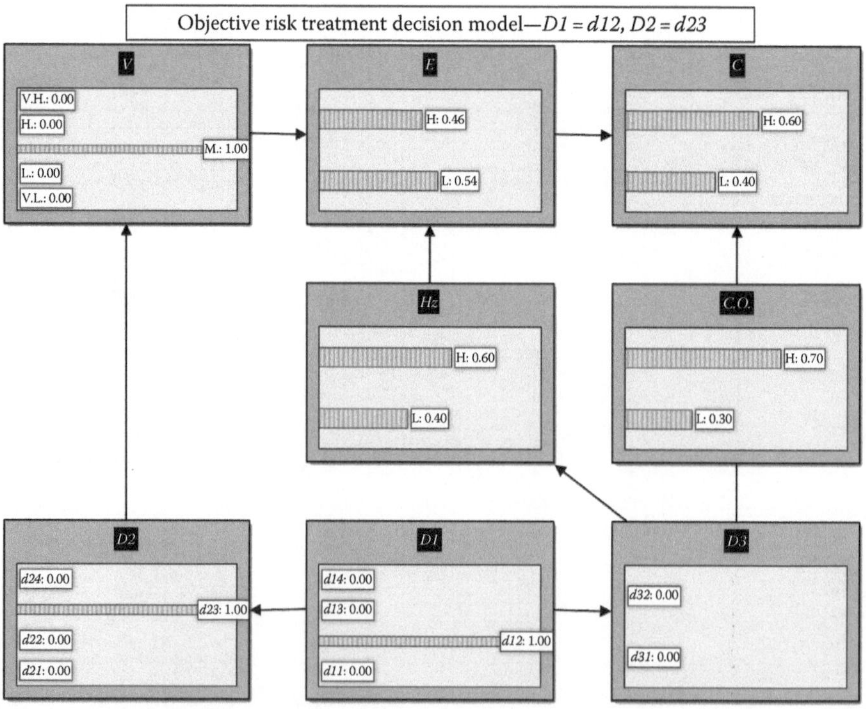

FIGURE 5.17 Marginal probabilities of the BN model: $D_1 = d_{12}$, $D_2 = d_{23}$.

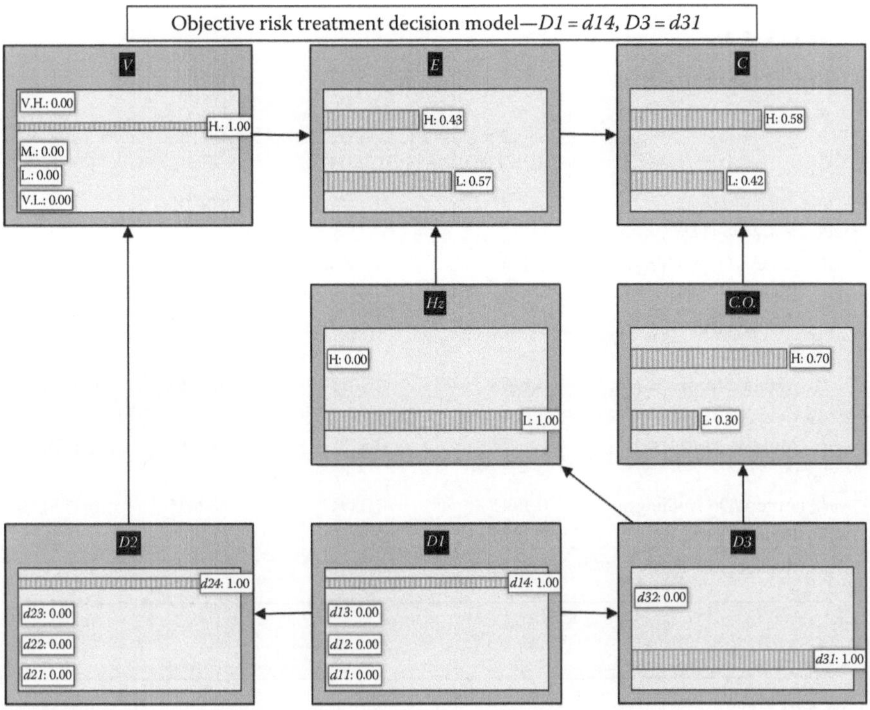

FIGURE 5.18 Marginal probabilities of the BN model: $D_1 = d_{14}$, $D_3 = d_{31}$.

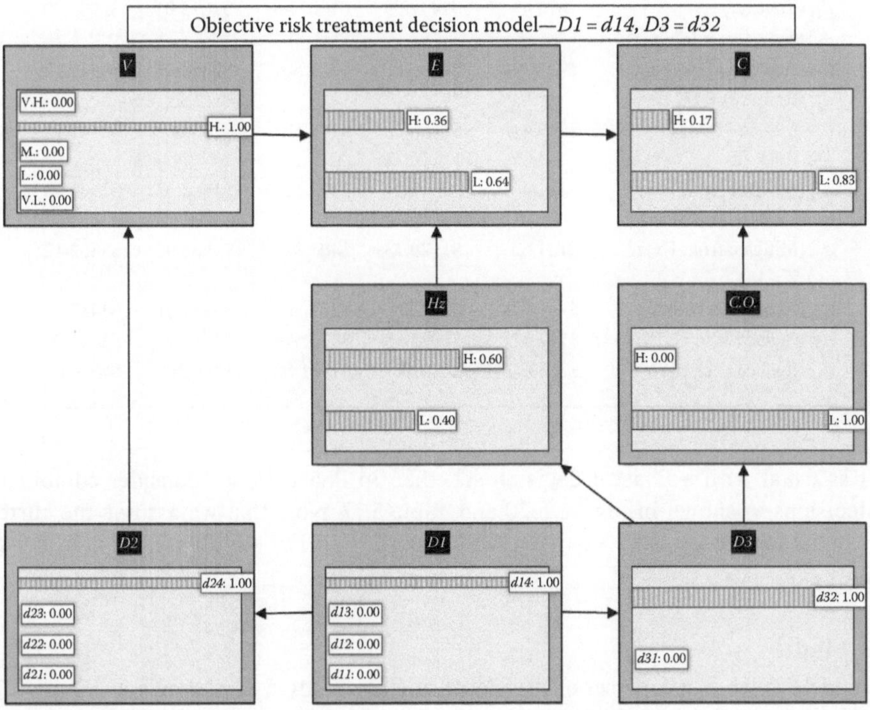

FIGURE 5.19 Marginal probabilities of the BN model—$D_1 = d_{14}$, $D_3 = d_{32}$.

TABLE 5.15
Joint Probabilities

E	*E = L*		*E = H*	
C	*C = L*	*C = H*	*C = L*	*C = H*
d_{11} (Transfer)	0.1741	0.2894	0.2015	0.3350
d_{12} (Mitigate) = *D2*				
d_{13} (Accept/Do Nothing)	0.1741	0.2894	0.2015	0.3350
d_{14} (Remove) = *D3*				
d_{21} (Mitigate from $V = H$ to $V = V.L.$)	0.4747	0.4167	0.0578	0.0508
d_{22} (Mitigate from $V = H$ to $V = L.$)	0.1894	0.3017	0.1963	0.3126
d_{23} (Mitigate from $V = H$ to $V = M.$)	0.2152	0.3202	0.1867	0.2778
d_{24} (Accept/Do Nothing)	0.1741	0.2894	0.2015	0.3350
d_{31} (Reduce *H* to *L*)	0.2375	0.3343	0.1778	0.2504
d_{32} (Reduce *C.O.* to *L*)	0.5333	0.1102	0.2954	0.0611

TABLE 5.16
Utilities (Absolute Risk) of Different Decisions

E	*E = L*		*E = H*		**Total**
C	*C = L*	*C = H*	*C = L*	*C = H*	**Utility**
d_{11} (Transfer)	$870.41	$1447.09	$1007.49	$1675.01	$5000.00
d_{12} (Mitigate) = *D2*					
d_{13} (Accept/Do Nothing)	$–	$868.26	$201.50	$3350.01	$4419.77
d_{14} (Remove) = *D3*					
d_{21} (Mitigate from $V = H$ to $V = V.L.$)	$949.41	$1250.08	$144.58	$203.06	$2547.14
d_{22} (Mitigate from $V = H$ to $V = L.$)	$18.94	$965.32	$392.59	$1563.23	$2940.08
d_{23} (Mitigate from $V = H$ to $V = M.$)	$107.62	$1120.73	$466.85	$1666.92	$3362.12
d_{24} (Accept/Do Nothing)	$–	$868.26	$201.50	$3350.01	$4419.77
d_{31} (Reduce *H* to *L*)	$712.42	$1002.98	$533.51	$751.09	$3000.00
d_{32} (Reduce *C.O.* to *L*)	$2133.20	$440.80	$1181.80	$244.20	$4000.00

We start as usual with a Phase I decision set, then in Phase II, we consider combinations of retrofit decisions as shown in Figure 5.20 and Table 5.17. Note that we assume the current vulnerability state for the two assets as $V1 = V.H.$ and $V2 = H$. For simplicity, we consider a single hazard only.

5.6.3.2 Model

The ID model (which is a superset of the model of Case Study 5.2) is shown in Figure 5.20. The base BN model that controls the ID process is shown in Figure 5.21. The states of nodal variables of the model are shown in Table 5.18, and the definitions of these variables are shown in

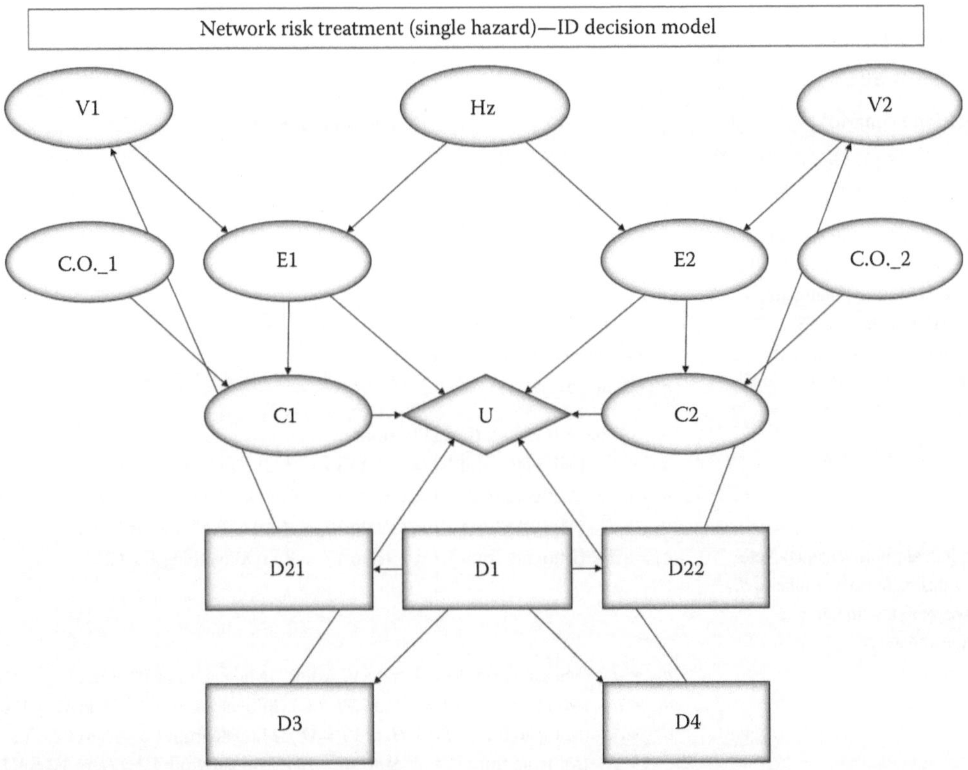

FIGURE 5.20 Multiphase ID decision model for a two-asset network.

Appendix 5.I (Table 5.31). Developing the utility (cost) Tables should be consistent with the ID model of Figure 5.20 as well as the decision set of table 5.17. Based on this, we come up with estimated utilities for different states of variables and different decisions as shown in Table 5.19.

5.6.3.3 Conditional Probabilities Tables

The different CPTs of each variable of the network in Figure 5.20 are shown in Appendix 5.IV. As usual, we used our best judgment in filling the different tables in a realistic manner for illustration purposes. For more on different methods of developing CPTs, see Chapter 2.

5.6.3.4 Results

The marginal probabilities for different decision scenarios are shown in Figures 5.44 through 5.59 in Appendix 5.V. Note that the proper phasing between decisions is accommodated in each realization of BN. The joint probabilities needed for computing expected utilities of different decisions can now be computed using these marginal probabilities and are shown in Table 5.20.

5.6.3.5 Optimal Decisions

Using the utilities table of Table 5.19 and the joint probabilities of Table 5.20, we can compute the utilities (costs or absolute risk) of different decisions shown in Table 5.21. The optimal decision is $D_1 = d_{12}$ (risk mitigation decision), which leads to decisions regarding the two-asset retrofits: $D_{21} = d_{212}$ (retrofit asset #1 via improving the vulnerability state from $V1 = V.H.$ to $V1 = L.$) and $D_{22} = {}_{d221}$ (retrofit asset #2 via improving the vulnerability state from $V21 = H.$ to $V2 = M.$). Obviously, the optimal decision can change if the parameters and utilities of the problem changes.

TABLE 5.17
Decision Set

Decision Scenario	Decision Subsets
	Phase I Decisions
0	d_{11} (Transfer)
This decision to mitigate, which is part of Phase I treatment, will lead to further Phase II mitigation scenarios (1 through 15)	d_{12} (Mitigate) = D_{21} and D_{22}
0 (minimal improvements)	d_{13} (Accept/Do Minimum Retrofit)
0	d_{14} (Remove)
	Phase II Mitigation Decisions
1	$d_{211} + d_{224}$ (Mitigate from $V1 = V.H.$ to $V1 = V.L.$) (Do Nothing for $V2$)
2	$d_{212} + d_{224}$ (Mitigate from $V1 = V.H.$ to $V1 = L.$) (Do Nothing for $V2$)
3	$d_{213} + d_{224}$ (Mitigate from $V1 = V.H.$ to $V1 = M.$) (Do Nothing for $V2$)
0 (minimal improvements) Note that this scenario is similar to d_{13} (Accept risk with minimal improvements)	$d_{214} + d_{224}$ (Mitigate from $V1 = V.H.$ to $V1 = H.$) (Do Nothing for $V2$)
4	$d_{211} + d_{223}$ (Mitigate from $V1 = V.H.$ to $V1 = V.L.$) (Mitigate from $V2 = H.$ to $V2 = V.L.$)
5	$d_{212} + d_{223}$ (Mitigate from $V1 = V.H.$ to $V1 = L.$) (Mitigate from $V2 = H.$ to $V2 = V.L.$)
6	$d_{213} + d_{223}$ (Mitigate from $V1 = V.H.$ to $V1 = M.$) (Mitigate from $V2 = H.$ to $V2 = V.L.$)
7	$d_{214} + d_{223}$ (Mitigate from $V1 = V.H.$ to $V1 = H.$) (Mitigate from $V2 = H.$ to $V2 = V.L.$)
8	$d_{211} + d_{222}$ (Mitigate from $V1 = V.H.$ to $V1 = V.L.$) (Mitigate from $V2 = H.$ to $V2 = L.$)
9	$d_{212} + d_{222}$ (Mitigate from $V1 = V.H.$ to $V1 = L.$) (Mitigate from $V2 = H.$ to $V2 = L.$)
10	$d_{213} + d_{222}$ (Mitigate from $V1 = V.H.$ to $V1 = M.$) (Mitigate from $V2 = H.$ to $V2 = L.$)
11	$d_{214} + d_{222}$ (Mitigate from $V1 = V.H.$ to $V1 = H.$) (Mitigate from $V2 = H.$ to $V2 = L.$)
12	$d_{211} + d_{221}$ (Mitigate from $V1 = V.H.$ to $V1 = V.L.$) (Mitigate from $V2 = H.$ to $V2 = M.$)
13	$d_{212} + d_{221}$ (Mitigate from $V1 = V.H.$ to $V1 = L.$) (Mitigate from $V2 = H.$ to $V2 = M.$)
14	$d_{213} + d_{221}$ (Mitigate from $V1 = V.H.$ to $V1 = M.$) (Mitigate from $V2 = H.$ to $V2 = M.$)
15	$d_{214} + d_{221}$ (Mitigate from $V1 = V.H.$ to $V1 = H.$) (Mitigate from $V2 = H.$ to $V2 = M.$)

5.7 CASE STUDY 5.4: PRIORITIZATION METHODS IN RISK TREATMENT

5.7.1 Overview

Phase III of risk treatment processes (see Figure 5.2) involves the prioritizations of projects within any asset network. This is a frequently encountered situation in infrastructure management. Generally speaking, various indexes that can be considered risk score are used for this prioritization process based on network characteristics, asset conditions, asset characteristics, and consequences. In most cases, consequences are indirectly accounted for based on stakeholders' perception of asset importance or the services the asset provides. Specifically, for a network that contains N_{ASSET} that requires mitigation to reduce its respective assessed risk, $Ri|_i$ with $i = 1,2, \ldots, N_{ASSET}$, we may use those assessed risks $Ri|_i$ as a prioritization metric to decide upon the order at which each asset, i, receives a risk mitigation effort.

We present in this section a practical procedure that some asset owners use for evaluating risk for the prioritization of projects in an asset network with limited resources. We start by offering the variables that affects $Ri|_i$; we then analyze the popular risk procedure of using weighted averages. We then explore a more involved process using the chain graph (CG) method. We use bridges for the example. Of course, the methodologies apply to any other type of infrastructure. One only needs to change the variables according to the type of infrastructure of interest.

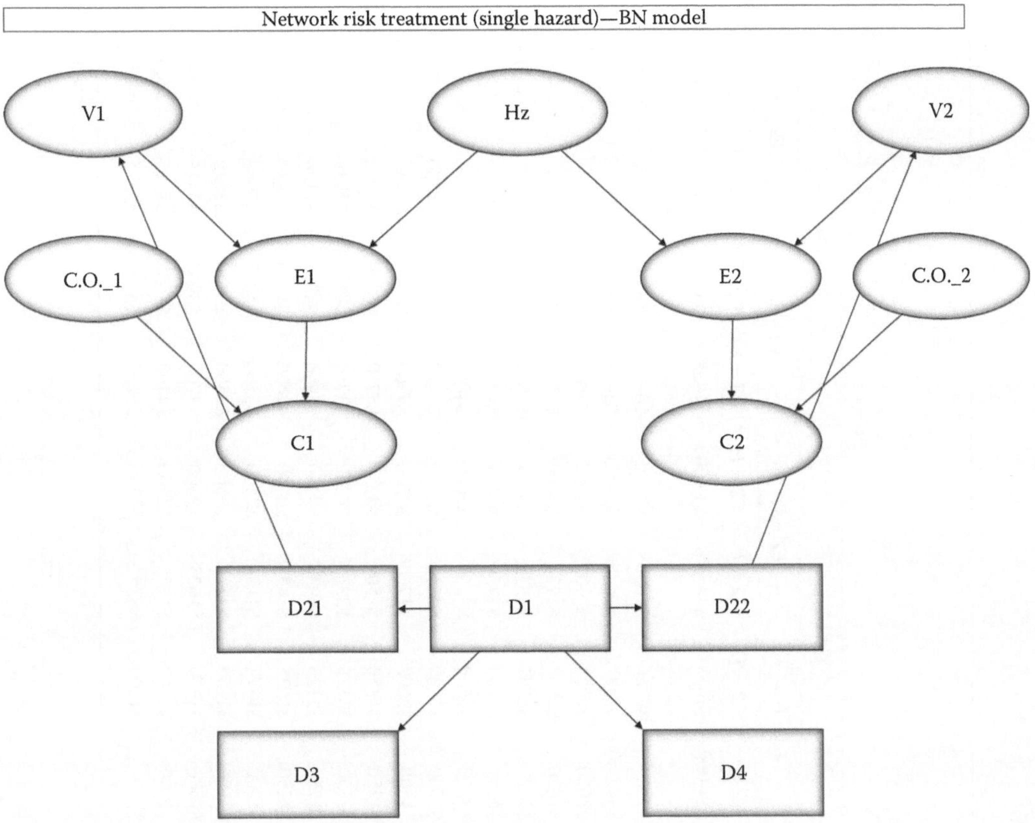

FIGURE 5.21 BN model of the ID of Figure 5.20.

TABLE 5.18
States of Nodal Variables

Variable	# of States	States
V_1, *V2*	5	*V.H., H., M., L., V.L.*
Hz	2	*H, L*
C.O._1, C.O._2	2	*H, L*
E1, E2	2	*H, L*
C1, C2	2	*H, L*
D_1, D_{21}, D_{22}, D_3, D_4	12 (Decisions)	d_{11}, d_{12}, d_{13}, d_{14}, d_{211}, d_{212}, d_{213}, d_{214}, d_{221}, d_{222}, d_{223}, d_{224}
U		

5.7.2 Risk Variables

The main risk contributors for bridge mitigation needs can be subdivided into four general categories: capital needs, importance, restrictions, and vulnerabilities. The capital needs category involves descriptions of physical conditions and retrofit needs for those physical conditions based on agency asset management objectives. The importance category is a description of the relative importance of the bridge under consideration based on its location and the stakeholders it serves. Restrictions describe any restrictions that have already been imposed on the use or operations of the bridge.

TABLE 5.19
Utilities of Treatment, Mitigation, and Transfer

E1	*E1* = *L*							
C1	*C1* = *L*				*C1* = *H*			
E2	*E2* = *L*		*E2* = *H*		*E2* = *L*		*E2* = *H*	
C2	*C2* = *L*	*C2* = *H*	*C2* = *L*	*C2* = *H*	*C2* = *L*	*C2* = *H*	*C2* = *L*	*C2* = *H*
d11 (Transfer)	$14,000	$14,000	$14,000	$14,000	$14,000	$14,000	$14,000	$14,000
d12 (Mitigate) = *D21* and *D22*								
d13 (Accept/Do Minimum Retrofit)	$250	$2750	$1000	$7250	$6000	$8500	$6750	$13,000
d14 (Remove)	$34,000	$34,000	$34,000	$34,000	$34,000	$34,000	$34,000	$34,000
d211 + *d224* (Mitigate from *V1* = *V.H.* to *V1* = *V.L.*) (Do Nothing for *V2*)	$2000	$2667	$2333	$4667	$3000	$3667	$3333	$5667
d212 + *d224* (Mitigate from *V1* = *V.H.* to *V1* = *L.*) (Do Nothing for *V2*)	$1000	$2467	$1733	$4600	$3200	$4667	$3933	$6800
d213 + *d224* (Mitigate from *V1* = *V.H.* to *V1* = *M.*) (Do Nothing for *V2*)	$500	$2767	$1633	$5367	$3900	$6167	$5033	$8767
d214 + *d224* (Mitigate from *V1* = *V.H.* to *V1* = *H.*) (Do Nothing for *V2*)	$250	$3250	$1750	$6417	$4750	$7750	$6250	$10917
*d*211 + *d223* (Mitigate from *V1* = *V.H.* to *V1* = *V.L.*) (Mitigate from *V2* = *H.* to *V2* = *V.L.*)	$3000	$3500	$3250	$5000	$4000	$4500	$4250	$6000
d212 + *d223* (Mitigate from *V1* = *V.H.* to *V1* = *L.*) (Mitigate from *V2* = *H.* to *V2* = *V.L.*)	$2000	$2500	$2250	$4000	$4200	$4700	$4450	$6200
d213 + *d223* (Mitigate from *V1* = *V.H.* to *V1* = *M.*) (Mitigate from *V2* = *H.* to *V2* = *V.L.*)	$1500	$2000	$1750	$3500	$4900	$5400	$5150	$6900
d214 + *d223* (Mitigate from *V1* = *V.H.* to *V1* = *H.*) (Mitigate from *V2* = *H.* to *V2* = *V.L.*)	$1250	$1750	$1500	$3250	$5750	$6250	$6000	$7750
d211 + *d222* (Mitigate from *V1* = *V.H.* to *V1* = *V.L.*) (Mitigate from *V2* = *H.* to *V2* = *L.*)	$2750	$3125	$2938	$4250	$3750	$4125	$3938	$5250
d212 + *d222* (Mitigate from *V1* = *V.H.* to *V1* = *L.*) (Mitigate from *V2* = *H.* to *V2* = *L.*)	$1750	$2125	$1938	$3250	$3950	$4325	$4138	$5450
d213 + *d222* (Mitigate from *V1* = *V.H.* to *V1* = *M.*) (Mitigate from *V2* = *H.* to *V2* = *L.*)	$1250	$1625	$1438	$2750	$4650	$5025	$4838	$6150
d214 + *d222* (Mitigate from *V1* = *V.H.* to *V1* = *H.*) (Mitigate from *V2* = *H.* to *V2* = *L.*)	$1000	$1375	$1188	$2500	$5500	$5875	$5688	$7000
d211 + *d221* (Mitigate from *V1* = *V.H.* to *V1* = *V.L.*) (Mitigate from *V2* = *H.* to *V2* = *M.*)	$2500	$2750	$2625	$3500	$3500	$3750	$3625	$4500
d212 + *d221* (Mitigate from *V1* = *V.H.* to *V1* = *L.*) (Mitigate from *V2* = *H.* to *V2* = *M.*)	$1500	$1750	$1625	$2500	$3700	$3950	$3825	$4700
d213 + *d221* (Mitigate from *V1* = *V.H.* to *V1* = *M.*) (Mitigate from *V2* = *H.* to *V2* = *M.*)	$1000	$1250	$1125	$2000	$4400	$4650	$4525	$5400
d214 + *d221* (Mitigate from *V1* = *V.H.* to *V1* = *H.*) (Mitigate from *V2* = *H.* to *V2* = *M.*)	$750	$1000	$875	$1750	$5250	$5500	$5375	$6250

(Continued)

TABLE 5.19 *(Continued)*
Utilities of Treatment, Mitigation, and Transfer

E1	*E1 = H*							
C1	*C1 = L*				*C1 = H*			
E2	*E2 = L*		*E2 = H*		*E2 = L*		*E2 = H*	
C2	*C2 = L*	*C2 = H*	*C2 = L*	*C2 = H*	*C2 = L*	*C2 = H*	*C2 = L*	*C2 = H*
d11 (Transfer)	$14,000	$14,000	$14,000	$14,000	$14,000	$14,000	$14,000	$14,000
d12 (Mitigate) = *D21* and *D22*								
d13 (Accept/Do Minimum Retrofit)	$4000	$6500	$4750	$11,000	$18,000	$20,500	$18,750	$25,000
d14 (Remove)	$34,000	$34,000	$34,000	$34,000	$34,000	$34,000	$34,000	$34,000
d211 + *d224* (Mitigate from *V1* = *V.H.* to *V1* = *V.L.*) (Do Nothing for *V2*)	$2500	$3167	$2833	$5167	$6000	$6667	$6333	$8667
d212 + *d224* (Mitigate from *V1* = *V.H.* to *V1* = *L.*) (Do Nothing for *V2*)	$2100	$3567	$2833	$5700	$6400	$7867	$7133	$10,000
d213 + *d224* (Mitigate from *V1* = *V.H.* to *V1* = *M.*) (Do Nothing for *V2*)	$2200	$4467	$3333	$7067	$7800	$10,067	$8933	$12,667
d214 + *d224* (Mitigate from *V1* = *V.H.* to *V1* = *H.*) (Do Nothing for *V2*)	$2500	$5500	$4000	$8667	$9500	$12,500	$11,000	$15,667
d211 + *d223* (Mitigate from *V1* = *V.H.* to *V1* = *V.L.*) (Mitigate from *V2* = *H.* to *V2* = *V.L.*)	$3500	$4000	$3750	$5500	$7000	$7500	$7250	$9000
d212 + *d223* (Mitigate from *V1* = *V.H.* to *V1* = *L.*) (Mitigate from *V2* = *H.* to *V2* = *V.L.*)	$3100	$3600	$3350	$5100	$7400	$7900	$7650	$9400
d213 + *d223* (Mitigate from *V1* = *V.H.* to *V1* = *M.*) (Mitigate from *V2* = *H.* to *V2* = *V.L.*)	$3200	$3700	$3450	$5200	$8800	$9300	$9050	$10,800
d214 + *d223* (Mitigate from *V1* = *V.H.* to *V1* = *H.*) (Mitigate from *V2* = *H.* to *V2* = *V.L.*)	$3500	$4000	$3750	$5500	$10,500	$11,000	$10,750	$12,500
d211 + *d222* (Mitigate from *V1* = *V.H.* to *V1* = *V.L.*) (Mitigate from *V2* = *H.* to *V2* = *L.*)	$3250	$3625	$3438	$4750	$6750	$7125	$6938	$8250
d212 + *d222* (Mitigate from *V1* = *V.H.* to *V1* = *L.*) (Mitigate from *V2* = *H.* to *V2* = *L.*)	$2850	$3225	$3038	$4350	$7150	$7525	$7338	$8650
d213 + *d222* (Mitigate from *V1* = *V.H.* to *V1* = *M.*) (Mitigate from *V2* = *H.* to *V2* = *L.*)	$2950	$3325	$3138	$4450	$8550	$8925	$8738	$10,050
d214 + *d222* (Mitigate from *V1* = *V.H.* to *V1* = *H.*) (Mitigate from *V2* = *H.* to *V2* = *L.*)	$3250	$3625	$3438	$4750	$10,250	$10,625	$10,438	$11,750
d211 + *d221* (Mitigate from *V1* = *V.H.* to *V1* = *V.L.*) (Mitigate from *V2* = *H.* to *V2* = *M.*)	$3000	$3250	$3125	$4000	$6500	$6750	$6625	$7500
d212 + *d221* (Mitigate from *V1* = *V.H.* to *V1* = *L.*) (Mitigate from *V2* = *H.* to *V2* = *M.*)	$2600	$2850	$2725	$3600	$6900	$7150	$7025	$7900
d213 + *d221* (Mitigate from *V1* = *V.H.* to *V1* = *M.*) (Mitigate from *V2* = *H.* to *V2* = *M.*)	$2700	$2950	$2825	$3700	$8300	$8550	$8425	$9300
d214 + *d221* (Mitigate from *V1* = *V.H.* to *V1* = *H.*) (Mitigate from *V2* = *H.* to *V2* = *M.*)	$3000	$3250	$3125	$4000	$10,000	$10,250	$10,125	$11,000

TABLE 5.20
Joint Probabilities

E1	*E1* = *L*							
C1	*C1* = *L*				*C1* = *H*			
E2	*E2* = *L*		*E2* = *H*		*E2* = *L*		*E2* = *H*	
C2	*C2* = *L*	*C2* = *H*	*C2* = *L*	*C2* =*H*	*C2* = *L*	*C2* = *H*	*C2* = *L*	*C2* = *H*
d11 (Transfer)	0.00	0.01	0.02	0.08	0.00	0.02	0.02	0.11
d12 (Mitigate) = *D21* and *D22*								
d13 (Accept/Do Minimum Retrofit)	0.00	0.01	0.02	0.08	0.00	0.02	0.02	0.11
d14 (Remove)	0.00	0.01	0.02	0.08	0.00	0.02	0.02	0.11
d211 + *d224* (Mitigate from *V1* = *V.H.* to *V1* = *V.L.*) (do nothing for *V2*)	0.02	0.07	0.08	0.39	0.01	0.04	0.05	0.23
d212 + *d224* (Mitigate from *V1* = *V.H.* to *V1* = *L.*) (Do Nothing for *V2*)	0.01	0.05	0.05	0.25	0.01	0.04	0.05	0.21
d213 + *d224* (Mitigate from *V1* = *V.H.* to *V1* = *M.*) (Do Nothing for *V2*)	0.01	0.03	0.03	0.16	0.01	0.03	0.04	0.17
d214 + *d224* (Mitigate from *V1* = *V.H.* to *V1* = *H.*) (Do Nothing for *V2*)	0.00	0.01	0.02	0.08	0.00	0.02	0.02	0.11
d211 + *d223* (Mitigate from *V1* = *V.H.* to *V1* = *V.L.*) (Mitigate from *V2* = *H.* to *V2* = *V.L.*)	0.02	0.06	0.01	0.02	0.03	0.09	0.01	0.02
d212 + *d223* (Mitigate from *V1* = *V.H.* to *V1* = *L.*) (Mitigate from *V2* = *H.* to *V2* = *V.L.*)	0.12	0.34	0.03	0.08	0.07	0.20	0.02	0.05
d213 + *d223* (Mitigate from *V1* = *V.H.* to *V1* = *M.*) (Mitigate from *V2* = *H.* to *V2* = *V.L.*)	0.08	0.22	0.02	0.05	0.06	0.18	0.02	0.04
d214 + *d223* (Mitigate from *V1* = *V.H.* to *V1* = *H.*) (Mitigate from *V2* = *H.* to *V2* = *V.L.*)	0.05	0.14	0.01	0.03	0.05	0.15	0.01	0.04
d211 + *d222* (Mitigate from *V1* = *V.H.* to *V1* = *V.L.*) (Mitigate from *V2* = *H.* to *V2* = *L.*)	0.01	0.05	0.01	0.03	0.02	0.07	0.02	0.05
d212 + *d222* (Mitigate from *V1* = *V.H.* to *V1* = *L.*) (Mitigate from *V2* = *H.* to *V2* = *L.*)	0.07	0.25	0.05	0.18	0.04	0.15	0.03	0.11
d213 + *d222* (Mitigate from *V1* = *V.H.* to *V1* = *M.*) (Mitigate from *V2* = *H.* to *V2* = *L.*)	0.05	0.16	0.04	0.12	0.04	0.13	0.03	0.10
d214 + *d222* (Mitigate from *V1* = *V.H.* to *V1* = *H.*) (Mitigate from *V2* = *H.* to *V2* = *L.*)	0.03	0.10	0.02	0.07	0.03	0.11	0.02	0.08
d211 + *d221* (Mitigate from *V1* = *V.H.* to *V1* = *V.L.*) (Mitigate from *V2* = *H.* to *V2* = *M.*)	0.01	0.03	0.01	0.05	0.01	0.05	0.02	0.08
d212 + *d221* (Mitigate from *V1* = *V.H.* to *V1* = *L.*) (Mitigate from *V2* = *H.* to *V2* = *M.*)	0.05	0.18	0.07	0.27	0.03	0.11	0.04	0.16
d213 + *d221* (Mitigate from *V1* = *V.H.* to *V1* = *M.*) (Mitigate from *V2* = *H.* to *V2* = *M.*)	0.03	0.12	0.05	0.18	0.02	0.10	0.04	0.15
d214 + *d221* (Mitigate from *V1* = *V.H.* to *V1* = *H.*) (Mitigate from *V2* = *H.* to *V2* = *M.*)	0.02	0.07	0.03	0.11	0.02	0.08	0.03	0.12

(Continued)

TABLE 5.20 *(Continued)*
Joint Probabilities

E1	*E1 = H*							
C1	*C1 = L*				*C1 = H*			
E2	*E2 = L*		*E2 = H*		*E2 = L*		*E2 = H*	
C2	*C2 = L*	*C2 = H*	*C2 = L*	*C2 = H*	*C2 = L*	*C2 = H*	*C2 = L*	*C2 = H*
d11 (Transfer)	0.01	0.04	0.05	0.21	0.01	0.06	0.06	0.30
d12 (Mitigate) = D21 and D22								
d13 (Accept/Do Minimum Retrofit)	0.01	0.04	0.05	0.21	0.01	0.06	0.06	0.30
d14 (Remove)	0.01	0.04	0.05	0.21	0.01	0.06	0.06	0.30
d211 + *d224* (Mitigate from *V1* = *V.H.* to *V1* = *V.L.*) (Do Nothing for *V2*)	0.00	0.01	0.01	0.04	0.00	0.00	0.01	0.03
d212 + *d224* (Mitigate from *V1* = *V.H.* to *V1* = *L.*) (Do Nothing for *V2*)	0.00	0.02	0.03	0.12	0.00	0.02	0.02	0.10
d213 + *d224* (Mitigate from *V1* = *V.H.* to *V1* = *M.*) (Do Nothing for *V2*)	0.01	0.03	0.04	0.17	0.01	0.04	0.04	0.19
d214 + *d224* (Mitigate from *V1* = *V.H.* to *V1* = *H.*) (Do Nothing for *V2*)	0.01	0.04	0.05	0.21	0.01	0.06	0.06	0.30
d211 + *d223* (Mitigate from *V1* = *V.H.* to *V1* = *V.L.*) (Mitigate from *V2* = *H.* to *V2* = *V.L.*)	0.06	0.18	0.01	0.04	0.09	0.26	0.02	0.06
d212 + *d223* (Mitigate from *V1* = *V.H.* to *V1* = *L.*) (Mitigate from *V2* = *H.* to *V2* = *V.L.*)	0.01	0.04	0.00	0.01	0.01	0.02	0.00	0.01
d213 + *d223* (Mitigate from *V1* = *V.H.* to *V1* = *M.*) (Mitigate from *V2* = *H.* to *V2* = *V.L.*)	0.04	0.11	0.01	0.03	0.03	0.09	0.01	0.02
d214 + *d223* (Mitigate from *V1* = *V.H.* to *V1* = *H.*) (Mitigate from *V2* = *H.* to *V2* = *V.L.*)	0.05	0.15	0.01	0.04	0.06	0.16	0.01	0.04
d211 + *d222* (Mitigate from *V1* = *V.H.* to *V1* = *V.L.*) (Mitigate from *V2* = *H.* to *V2* = *L.*)	0.04	0.13	0.03	0.10	0.06	0.20	0.04	0.15
d212 + *d222* (Mitigate from *V1* = *V.H.* to *V1* = *L.*) (Mitigate from *V2* = *H.* to *V2* = *L.*)	0.01	0.03	0.01	0.02	0.01	0.02	0.00	0.01
d213 + *d222* (Mitigate from *V1* = *V.H.* to *V1* = *M.*) (Mitigate from *V2* = *H.* to *V2* = *L.*)	0.02	0.08	0.02	0.06	0.02	0.07	0.01	0.05
d214 + *d222* (Mitigate from *V1* = *V.H.* to *V1* = *H.*) (Mitigate from *V2* = *H.* to *V2* = *L.*)	0.03	0.11	0.02	0.08	0.04	0.12	0.03	0.09
d211 + *d221* (Mitigate from *V1* = *V.H.* to *V1* = *V.L.*) (Mitigate from *V2* = *H.* to *V2* = *M.*)	0.02	0.09	0.04	0.14	0.04	0.14	0.05	0.21
d212 + *d221* (Mitigate from *V1* = *V.H.* to *V1* = *L.*) (Mitigate from *V2* = *H.* to *V2* = *M.*)	0.01	0.02	0.01	0.03	0.00	0.01	0.00	0.02
d213 + *d221* (Mitigate from *V1* = *V.H.* to *V1* = *M.*) (Mitigate from *V2* = *H.* to *V2* = *M.*)	0.01	0.06	0.02	0.09	0.01	0.05	0.02	0.07
d214 + *d221* (Mitigate from *V1* = *V.H.* to *V1* = *H.*) (Mitigate from *V2* = *H.* to *V2* = *M.*)	0.02	0.08	0.03	0.12	0.02	0.09	0.03	0.13

TABLE 5.21
Utilities (Absolute Risk) of Different Decisions

E1	*E1 = L*							
C1	*C1 = L*				*C1 = H*			
E2	*E2 = L*		*E2 = H*		*E2 = L*		*E2 = H*	
C2	*C2 = L*	*C2 = H*	*C2 = L*	*C2 = H*	*C2 = L*	*C2 = H*	*C2 = L*	*C2 = H*
d11 (Transfer)	$42.41	$197.28	$228.09	$1060.90	$59.69	$277.64	$320.99	$1493.03
d12 (Mitigate) = *D21* and *D22*								
d13 (Accept/Do Minimum Retrofit)	$0.76	$38.75	$16.29	$549.40	$25.58	$168.57	$154.76	$1386.38
d14 (Remove)	$103.01	$479.12	$553.93	$2576.48	$144.96	$674.27	$779.55	$3625.93
d211 + *d224* (Mitigate from *V1* = *V.H.* to *V1* = *V.L.*) (Do Nothing for *V2*)	$31.19	$193.46	$195.71	$1820.59	$27.90	$158.59	$166.68	$1317.96
d212 + *d224* (Mitigate from *V1* = *V.H.* to *V1* = *L.*) (Do Nothing for *V2*)	$10.18	$116.74	$94.84	$1170.75	$27.04	$183.39	$178.71	$1437.03
d213 + *d224* (Mitigate from *V1* = *V.H.* to *V1* = *M.*) (Do Nothing for *V2*)	$3.13	$80.63	$55.04	$841.10	$26.86	$197.52	$186.39	$1509.98
d214 + *d224* (Mitigate from *V1* = *V.H.* to *V1* = *H.*) (Do Nothing for *V2*)	$0.76	$45.80	$28.51	$486.25	$20.25	$153.69	$143.30	$1164.21
d211 + *d223* (Mitigate from *V1* = *V.H.*to *V1* = *V.L.*) (Mitigate from *V2* = *H.* to *V2*= *V.L.*)	$65.76	$221.27	$17.08	$75.77	$130.67	$424.01	$33.28	$135.52
d212 + *d223* (Mitigate from *V1* = *V.H.* to *V1* = *L.*) (Mitigate from *V2* = *H.* to *V2* = *V.L.*)	$233.48	$841.77	$62.96	$322.85	$292.31	$943.46	$74.24	$298.33
d213 + *d223* (Mitigate from *V1* = *V.H.* to *V1* = *M.*) (Mitigate from *V2* = *H.* to *V2* = *V.L.*)	$114.24	$439.32	$31.95	$184.29	$309.86	$984.91	$78.07	$301.68
d214 + *d223* (Mitigate from *V1* = *V.H.* to *V1* = *H.*) (Mitigate from *V2* = *H.* to *V2* = *V.L.*)	$58.62	$236.72	$16.86	$105.38	$296.36	$929.11	$74.13	$276.17
d211 + *d222* (Mitigate from *V1* = *V.H.*to *V1* = *V.L.*) (Mitigate from *V2* = *H.* to *V2*= *L.*)	$38.05	$145.93	$30.23	$147.59	$77.33	$287.09	$60.38	$271.73
d212 + *d222* (Mitigate from *V1* = *V.H.* to *V1* = *L.*) (Mitigate from *V2* = *H.* to *V2* = *L.*)	$128.96	$528.49	$106.18	$601.10	$173.53	$641.26	$135.17	$600.94

(Continued)

TABLE 5.21 ***(Continued)***
Utilities (Absolute Risk) of Different Decisions

E1	*E1 = L*								
C1	*C1 = L*				*C1 = H*				
E2	*E2 = L*		*E2 = H*		*E2 = L*		*E2 = H*		
C2	*C2 = L*	*C2 = H*	*C2 = L*	*C2 = H*	*C2 = L*	*C2 = H*	*C2 = L*	*C2 = H*	
d213 + *d222* (Mitigate from *V1* = *V.H.* to *V1* = *M.*) (Mitigate from *V2* = *H.* to *V2* = *L.*)	$60.09	$263.65	$51.39	$331.82	$185.61	$676.96	$143.60	$616.15	
d214 + *d222* (Mitigate from *V1* = *V.H.* to *V1* = *H.*) (Mitigate from *V2* = *H.* to *V2* = *L.*)	$29.60	$137.38	$26.14	$185.76	$178.93	$645.09	$137.61	$571.60	
d211 + *d221* (Mitigate from *V1* = *V.H.* to *V1* = *V.L.*) (Mitigate from *V2* = *H.* to *V2*= *M.*)	$21.66	$91.43	$34.53	$176.69	$45.19	$185.82	$71.07	$338.58	
d212 + *d221* (Mitigate from *V1* = *V.H.* to *V1* = *L.*) (Mitigate from *V2* = *H.* to *V2* = *M.*)	$69.21	$309.88	$113.85	$672.17	$101.78	$416.99	$159.76	$753.37	
d213 + *d221* (Mitigate from *V1* = *V.H.* to *V1* = *M.*) (Mitigate from *V2* = *H.* to *V2* = *M.*)	$30.10	$144.40	$51.42	$350.81	$109.97	$446.03	$171.73	$786.47	
d214 + *d221* (Mitigate from *V1* = *V.H.* to *V1* = *H.*) (Mitigate from *V2* = *H.* to *V2* = *M.*)	$13.90	$71.14	$24.63	$189.02	$106.95	$429.98	$166.26	$741.91	
E1	*E1 = H*								
C1	*C1 = L*				*C1 = H*				
E2	*E2 = L*		*E2 = H*		*E2 = L*		*E2 = H*		**Total**
C2	*C2 = L*	*C2 = H*	*C2 = L*	*C2 = H*	*C2 = L*	*C2 = H*	*C2 = L*	*C2 = H*	**Utility**
d11 (Transfer)	$118.94	$553.24	$639.63	$2,975.10	$167.39	$778.59	$900.16	$4,186.91	$14,000.00
d12 (Mitigate) = *D21* and *D22*									
d13 (Accept/Do Minimum Retrofit)	$33.98	$256.86	$217.02	$2337.58	$215.22	$1140.08	$1205.57	$7476.63	$15,223.42
d14 (Remove)	$288.86	$1343.59	$1553.38	$7225.24	$406.52	$1890.86	$2186.09	$10,168.21	$34,000.00
d211 + *d224* (Mitigate from *V1* = *V.H.* to *V1* = *V.L.*) (Do Nothing for *V2*)	$4.46	$26.30	$27.21	$230.75	$6.39	$33.01	$36.25	$230.76	$4507.20
d212 + *d224* (Mitigate from *V1* = *V.H.* to *V1* = *L.*) (Do Nothing for *V2*)	$10.47	$82.68	$75.93	$710.53	$26.48	$151.41	$158.74	$1035.04	$5469.94

(Continued)

TABLE 5.21 *(Continued)*
Utilities (Absolute Risk) of Different Decisions

E1	*E1 = H*								
C1	*C1 = L*				*C1 = H*				
E2	*E2 = L*		*E2 = H*		*E2 = L*		*E2 = H*		Total
C2	*C2 = L*	*C2 = H*	*C2 = L*	*C2 = H*	*C2 = L*	*C2 = H*	*C2 = L*	*C2 = H*	Utility
d213 + *d224* (Mitigate from *V1* = *V.H.* to *V1* = *M.*) (Do Nothing for *V2*)	$15.30	$144.45	$124.63	$1228.94	$59.60	$357.78	$367.07	$2420.87	$7619.27
d214 + *d224* (Mitigate from *V1* = *V.H.* to *V1* = *H.*) (Do Nothing for *V2*)	$21.24	$217.35	$182.75	$1841.73	$113.59	$695.17	$707.27	$4685.35	$10,507.20
d211 + *d223* (Mitigate from *V1* = *V.H.*to *V1* = *V.L.*) (Mitigate from *V2* = *H.* to *V2*= *V.L.*)	$215.14	$709.16	$55.25	$233.74	$641.29	$1981.76	$159.21	$570.06	$5668.99
d212 + *d223* (Mitigate from *V1* = *V.H.* to *V1* = *L.*) (Mitigate from *V2* = *H.* to *V2* = *V.L.*)	$41.43	$138.77	$10.73	$47.12	$58.96	$181.54	$14.61	$51.78	$3614.35
d213 + *d223* (Mitigate from *V1* = *V.H.* to *V1* = *M.*) (Mitigate from *V2* = *H.* to *V2* = *V.L.*)	$119.36	$398.07	$30.85	$134.11	$272.56	$830.78	$67.19	$231.27	$4528.50
d214 + *d223* (Mitigate from *V1* = *V.H.* to *V1* = *H.*) (Mitigate from *V2* = *H.* to *V2* = *V.L.*)	$182.14	$600.38	$46.78	$197.89	$600.50	$1814.47	$147.37	$494.26	$6077.14
d211 + *d222* (Mitigate from *V1* = *V.H.*to *V1* = *V.L.*) (Mitigate from *V2* = *H.* to *V2* = *L.*)	$126.10	$474.70	$99.19	$462.58	$390.34	$1390.59	$298.35	$1197.44	$5497.61
d212 + *d222* (Mitigate from *V1* = *V.H.* to *V1* = *L.*) (Mitigate from *V2* = *H.* to *V2* = *L.*)	$24.04	$91.82	$19.06	$92.10	$35.96	$127.73	$27.44	$109.19	$3442.97
d213 + *d222* (Mitigate from *V1* = *V.H.* to *V1* = *M.*) (Mitigate from *V2* = *H.* to *V2* = *L.*)	$69.46	$264.22	$54.94	$262.98	$167.15	$588.90	$127.04	$493.15	$4357.12
d214 + *d222* (Mitigate from *V1* = *V.H.* to *V1* = *H.*) (Mitigate from *V2* = *H.* to *V2* = *L.*)	$106.76	$401.88	$83.97	$391.62	$370.02	$1294.53	$280.21	$1064.65	$5905.75
d211 + *d221* (Mitigate from *V1* = *V.H.*to *V1* = *V.L.*) (Mitigate from *V2* = *H.* to *V2*= *M.*)	$72.89	$303.02	$115.28	$566.28	$235.36	$937.99	$364.24	$1582.48	$5142.52
d212 + *d221* (Mitigate from *V1* = *V.H.* to *V1* = *L.*) (Mitigate from *V2* = *H.* to *V2* = *M.*)	$13.73	$57.77	$21.86	$110.81	$21.73	$86.41	$33.59	$144.97	$3087.88
d213 + *d221* (Mitigate from *V1* = *V.H.* to *V1* = *M.*) (Mitigate from *V2* = *H.* to *V2* = *M.*)	$39.81	$166.91	$63.24	$317.87	$101.61	$401.68	$156.60	$663.40	$4002.03
d214 + *d221* (Mitigate from *V1* = *V.H.* to *V1* = *H.*) (Mitigate from *V2* = *H.* to *V2* = *M.*)	$61.70	$256.54	$97.60	$479.41	$226.04	$889.17	$347.51	$1448.89	$5550.67

Finally, the vulnerability category describes potential vulnerabilities of the bridge that can be improved based on existing state-of-the-art and state-of-the-practice. Each category can be subdivided into subcategories. The discrete variables are defined in Appendix 5.I (Table 5.31). Note that this is purely for illustration purposes and the factors and categories vary significantly from organization to organization.

5.7.3 Weighted Averages

5.7.3.1 Overview

The currently popular method for project prioritization is to assess risk via the weighted averages method (see Chapter 3). As noted earlier, the method is fairly simple but sometimes might lead to inconsistent results. It is used extensively in risk assessment procedures. For further information, see, for example, Fenton and Neil (2013), FEMA (2009), and DHS (2011a, 2011b, and 2011c). It estimates the risk rating as

$$Ri = \sum_{i=1}^{i=4} w_i \sum_{j=1}^{j=N_i} w_{ij} \tag{5.8}$$

with N_j representing the number of subcategories of the i-th category. The proposed weights, w_i and w_{ij}, are shown in Table 5.22.

TABLE 5.22
Suggested Weights for Weighted Averages Method

Category	Category Weights (w_i)	Subcategory	Subcategory Weights (w_{ij})	# of States
Condition-based needs (*Cap. Needs*)	1.0	Condition rating (*CR*)	24	4
		Structural deficiencies (*SDB*)	6	2
		Asset needs (*Needs*)	12	3
Importance (*Importance*)	1.0	Traffic service class (*Func. Class*)	8	4
		National commerce value (*NCV*)	6	2
		Traffic count (*ADT*)	6	3
		Truck volume (*Trucks*)	6	3
		Detour length (*Detour*)	6	3
Functional restrictions (*Restrictions*)	1.0	Functional restrictions (*Restrictions*). See Figures 5.22 through 5.24	4	2
Asset vulnerability characteristics (*Vulnerability*)	1.0	Fracture/fatigue susceptibility (*F.C.*)	4	2
		Main material (*Material*)	6	3
		Design type (*Design*)	6	3
		Other vulnerabilities (*Other*)	6	3

FIGURE 5.22 Example of load posting (detour).

FIGURE 5.23 Example of load posting.

5.7.3.2 Results

To explore the behavior of the weighted averages method, let us consider the two situations shown in Tables 5.23 and 5.24 for two bridges, A and B. The bridges are fairly similar except for two attributes: condition rating (*CR*) and importance (see Table 5.27 for condition states). Bridge A obtained a *CR* of 4 because it was judged to be in service condition, and Bridge B received a *CR* of 3 because it was judged to be in a poor condition. The importance of the bridges also varies. Bridge A received lower scores than Bridge B in the importance category. The estimated risks of Bridge A are about the same as that of Bridge B (72 vs. 77) even though the two bridges are about the same in *CR* (with only a single point difference) and Bridge B is more important than Bridge A. The heavily weighted *CR* subcategory obviously played a role and balanced the importance of Bridge B. In this case, mobility and bridge condition were weighed similarly and, therefore, this result. Thus, weights for each category are very important and any change is weights without a good rationale can give different results.

Another example might shed some light on another limitation of the weighted average method; that is, it does not account for interactions between parameters. Let us consider the risk scores for bridges C and D as shown in Tables 5.25 and 5.26. The bridges again differ in *CR* rating,

FIGURE 5.24 Example of load posting (major bridge).

TABLE 5.23
Risk Computations with Weighted Averages Method: Bridge A

Category	Category Weights (w_i)	Subcategory	Subcategory Weight (w_{ij})	# of States	Assessed Value	Points
Condition-based needs (*Cap. Needs*)	1.0	Condition rating (*CR*)	24	4	4	24
		Structural deficiencies (*SDB*)	6	2	1	3
		Asset needs (*Needs*)	12	3	2	8
Importance (*Importance*)	1.0	Traffic service class (*Func. Class*)	8	4	4	8
		National commerce value (*NCV*)	6	2	1	3
		Traffic count (*ADT*)	6	3	1	2
		Truck volume (*Trucks*)	6	3	1	2
		Detour length (*Detour*)	6	3	1	2
Functional restrictions (*Restrictions*)	1.0	Functional restrictions (*Restrictions*)	4	2	1	2
Asset vulnerability characteristics (*Vulnerability*)	1.0	Fracture/fatigue susceptibility (*F.C.*)	4	2	2	4
		Main material (*Material*)	6	3	2	4
		Design type (*Design*)	6	3	2	4
		Other vulnerabilities (*Other*)	6	3	3	6
						72

Note: Lower assessed value indicates better condition. Points = (Subcategory Weight/# of States) * Assessed Value.

TABLE 5.24
Risk Computations with Weighted Averages Method: Bridge B

Category	Category Weights (w_i)	Subcategory	Subcategory Weights (w_{ij})	# of States	Assessed Value	Point
Condition-based needs (*Cap. Needs*)	1.0	Condition rating (*CR*)	24	4	3	18
		Structural deficiencies (*SDB*)	6	2	1	3
		Asset needs (*Needs*)	12	3	2	8
Importance (*Importance*)	1.0	Traffic service class (*Func. Class*)	8	4	4	8
		National commerce value (*NCV*)	6	2	2	6
		Traffic count (*ADT*)	6	3	3	6
		Truck volume (*Trucks*)	6	3	3	6
		Detour length (*Detour*)	6	3	1	2
Functional restrictions (*Restrictions*)	1.0	Functional restrictions (*Restrictions*)	4	2	1	2
Asset vulnerability characteristics (*Vulnerability*)	1.0	Fracture/fatigue susceptibility (*F.C.*)	4	2	2	4
		Main material (*Material*)	6	3	2	4
		Design type (*Design*)	6	3	2	4
		Other vulnerabilities (*Other*)	6	3	3	6
						77

TABLE 5.25
Risk Computations with Weighted Averages Method: Bridge C

Category	Category Weights (w_i)	Subcategory	Subcategory Weights (w_{ij})	# of States	Assessed Value	Point
Condition-based needs (*Cap. Needs*)	1.0	Condition rating (*CR*)	24	4	4	24
		Structural deficiencies (*SDB*)	6	2	2	6
		Asset needs (*Needs*)	12	3	1	4
Importance (*Importance*)	1.0	Traffic service class (*Func. Class*)	8	4	4	8
		National commerce value (*NCV*)	6	2	1	3
		Traffic count (*ADT*)	6	3	1	2
		Truck volume (*Trucks*)	6	3	1	2
		Detour length (*Detour*)	6	3	2	4
Functional restrictions (*Restrictions*)	1.0	Functional restrictions (*Restrictions*)	4	2	1	2
Asset vulnerability characteristics (*Vulnerability*)	1.0	Fracture/fatigue susceptibility (*F.C.*)	4	2	1	2
		Main material (*Material*)	6	3	1	2
		Design type (*Design*)	6	3	1	2
		Other vulnerabilities (*Other*)	6	3	1	2
						63

TABLE 5.26
Risk Computations with Weighted Averages Method: Bridge D

Category	Category Weights (w_i)	Subcategory	Subcategory Weights (w_{ij})	# of States	Assessed Value	Point
Condition-based needs (*Cap. Needs*)	1.0	Condition rating (*CR*)	24	4	3	18
		Structural deficiencies (*SDB*)	6	2	1	3
		Asset needs (*Needs*)	12	3	1	4
Importance (*Importance*)	1.0	Traffic service class (*Func. Class*)	8	4	4	8
		National commerce value (*NCV*)	6	2	1	3
		Traffic count (*ADT*)	6	3	3	6
		Truck volume (*Trucks*)	6	3	3	6
		Detour length (*Detour*)	6	3	3	6
Functional restrictions (*Restrictions*)	1.0	Functional restrictions (*Restrictions*)	4	2	1	2
Asset vulnerability characteristics (*Vulnerability*)	1.0	Fracture/Fatigue susceptibility (*F.C.*)	4	2	2	4
		Main material (*Material*)	6	3	1	2
		Design type (*Design*)	6	3	1	2
		Other vulnerabilities (*Other*)	6	3	1	2
						66

where Bridge C is rated to be slightly in worse condition than Bridge D. This resulted in *CR* ratings for the two bridges of 4 and 3, respectively. However, it is known that Bridge C is less fracture prone and has low traffic and truck volume. Bridge D, on the other hand, is fracture critical and supports heavy traffic and truck loads. Yet, the method results in similar scores for Bridge C (63) and Bridge D (66). Essentially, the bridge condition was offset by the vulnerabilities because the intersections between *CR* and other vulnerability subcategories are not well accounted for in this method.

5.7.4 Probabilistic Graph Network

5.7.4.1 Overview

We turn our attention now to a network-based method for assessing the same risk and its components. We keep the same main variables of the previous example while adding a few more internal variables to increase the efficiency of computations and ease of input to the network model. We re-examine the examples in the previous section in order to have a practical view on the advantages and limitations of the weighted averages versus the network approaches.

5.7.4.2 Model

The proposed network model is shown in Figure 5.25. We note that the model is composed of four parts, each representing a subnetwork. The three upstream subnetworks each form a Markov network, and the downstream subnetwork forms a Bayes network. The four subnetworks are connected to form a chain graph network. See more descriptions on Markov, Bayes, and chain graph networks in Chapter 2. A brief description of each subnetwork is given below.

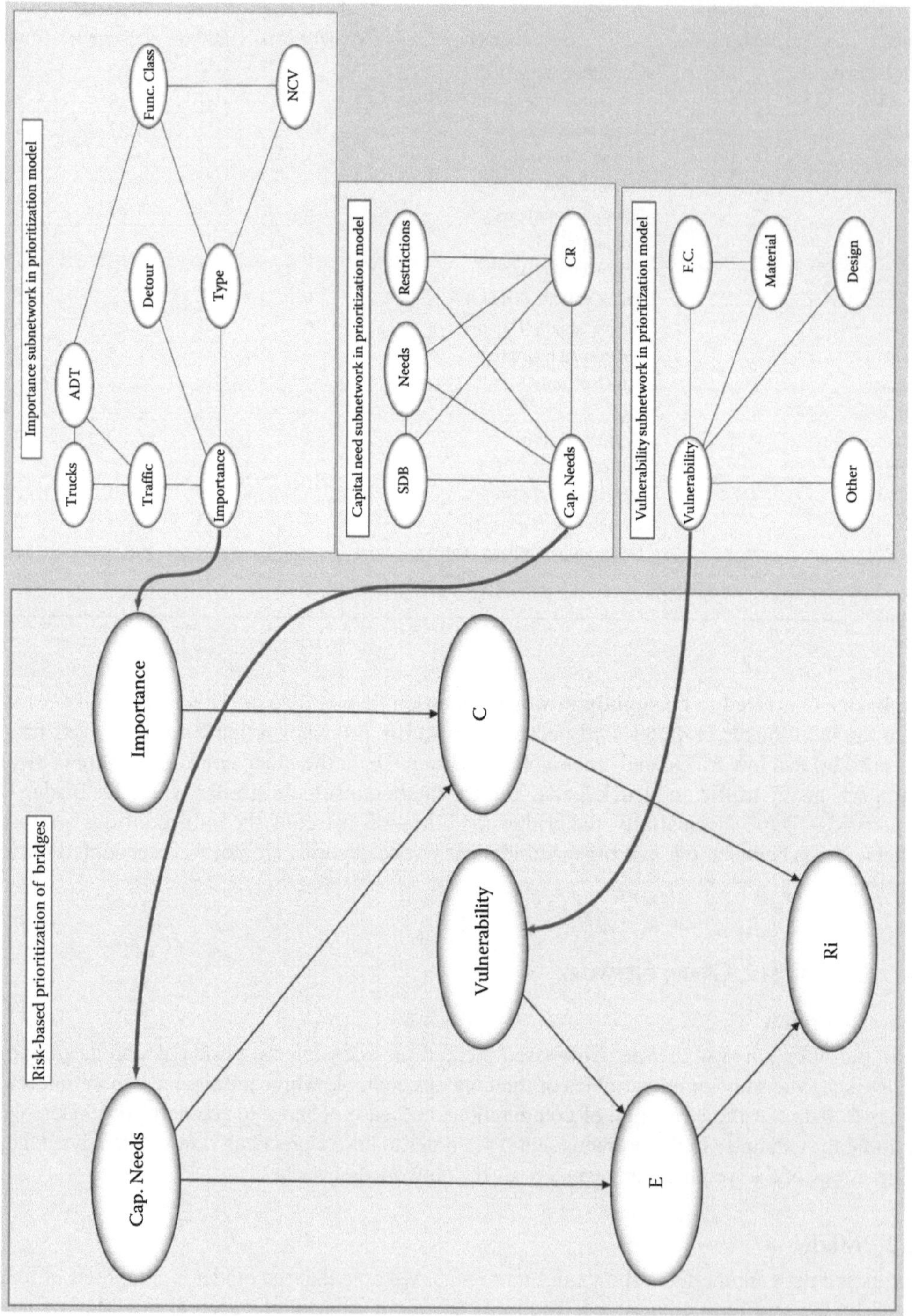

FIGURE 5.25 Overall chain graph model for project prioritization using risk as a metric.

5.7.4.3 Subnetworks

Importance Network: This network is composed of the variables in Table 5.22 that control the importance parameters of the bridge. Note that we added two intermediate (hidden) nodes: type and traffic nodes. The type node is related to the National Commerce Value (NCV) and the function class variables, and the traffic node is related to the Trucks and the Average Daily Traffic/Traffic Count (ADT) variables. The purpose of the hidden nodes is to simplify and reduce the size of the potential tables, thus increasing the computational efficiency of the network. Note that all the edges between the nodes in this network have no directions, which indicates that the relationships between the variables are dependent and the dependencies can flow either way. This results in a Markov-type network. Objectively, the potentials can describe the correlations (affinities) between the variables. Figure 5.26, Relationships of potentials and observation layer in the importance subnetwork, shows the network topology—the assumed potentials between the nodes. The figure also shows the observations nodes/layer. These are the observation nodes that will lead in the downstream to the risk score of the bridge. Note that the observation nodes are similar to those of the weighted averages independent variables in Table 5.22. Upon the execution of this subnetwork, the marginal probabilities of the importance variable are computed.

Vulnerability Network: Similarly, Figure 5.27 shows a Markov network that computes the vulnerability marginal probabilities. The observation nodes/layer are consistent with those of Table 5.22. The potentials of this network also controls the correlations/affinities between the nodal variables and are also shown in Figure 5.27.

Capital Needs Network: Again, a Markov network with its potentials is used to describe the topology of this group of variables. Figure 5.28 illustrates this topology.

Risk Network: Upon computing the marginal probabilities of importance, vulnerabilities, and capital needs, we can proceed to compute risk. The inputs to this downstream subnetwork, Figure 5.29, are the marginal probabilities of the importance, vulnerability, and capital needs variables. The output is the marginal probability of the risk variable, *Ri*. As usual, we need to add two hidden/intermediate nodes that represent exposure, *E*, and consequences, *C*. Note that all edges in this subnetwork are directional and that the network does not produce any circularity which, therefore, makes it a Bayes network. We only have to define the conditional probabilities of the different nodes.

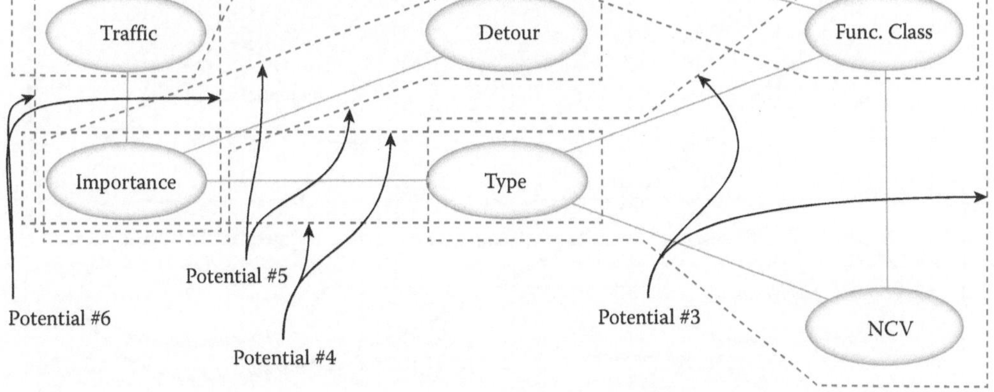

FIGURE 5.26 Relationships of potentials and observation layer in the importance subnetwork.

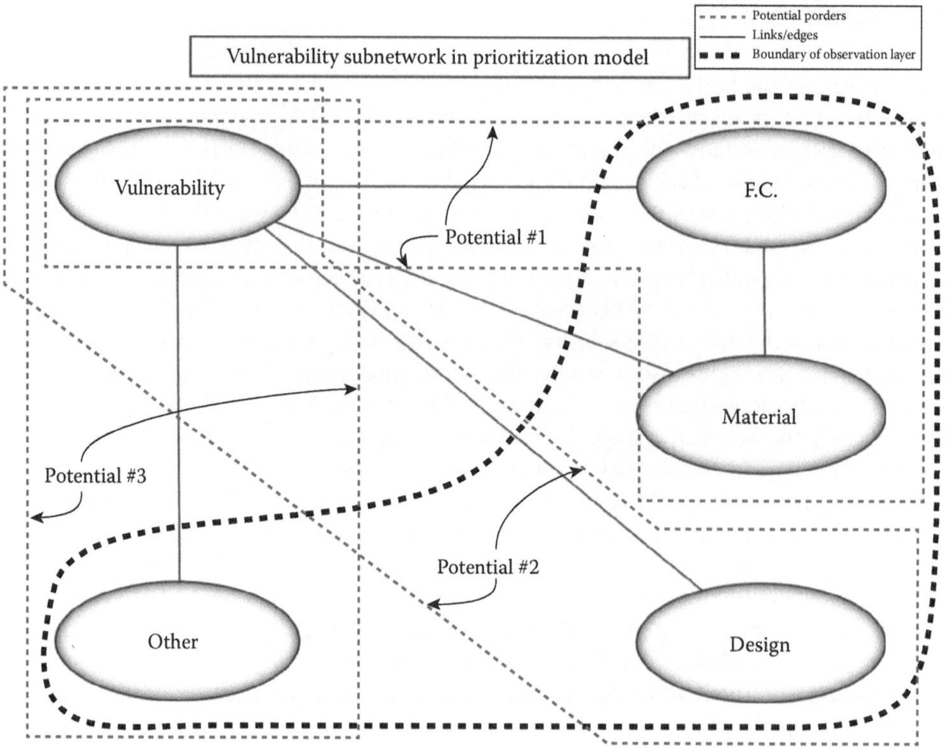

FIGURE 5.27 Relationships of potentials and observation layer in the vulnerability subnetwork.

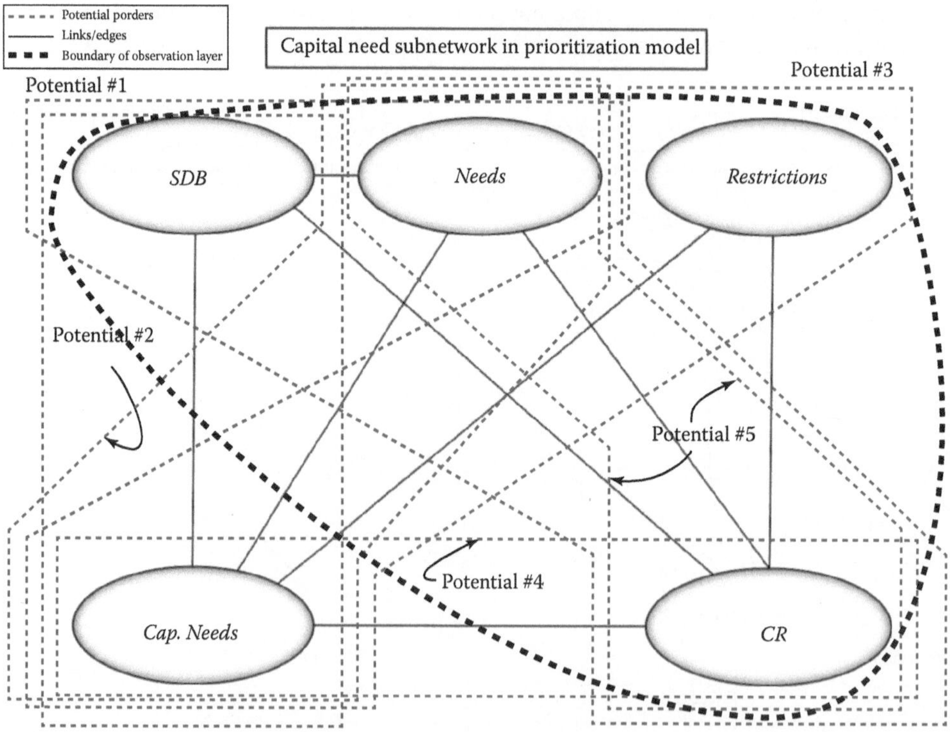

FIGURE 5.28 Relationships of potentials and observation layer in the capital needs subnetwork.

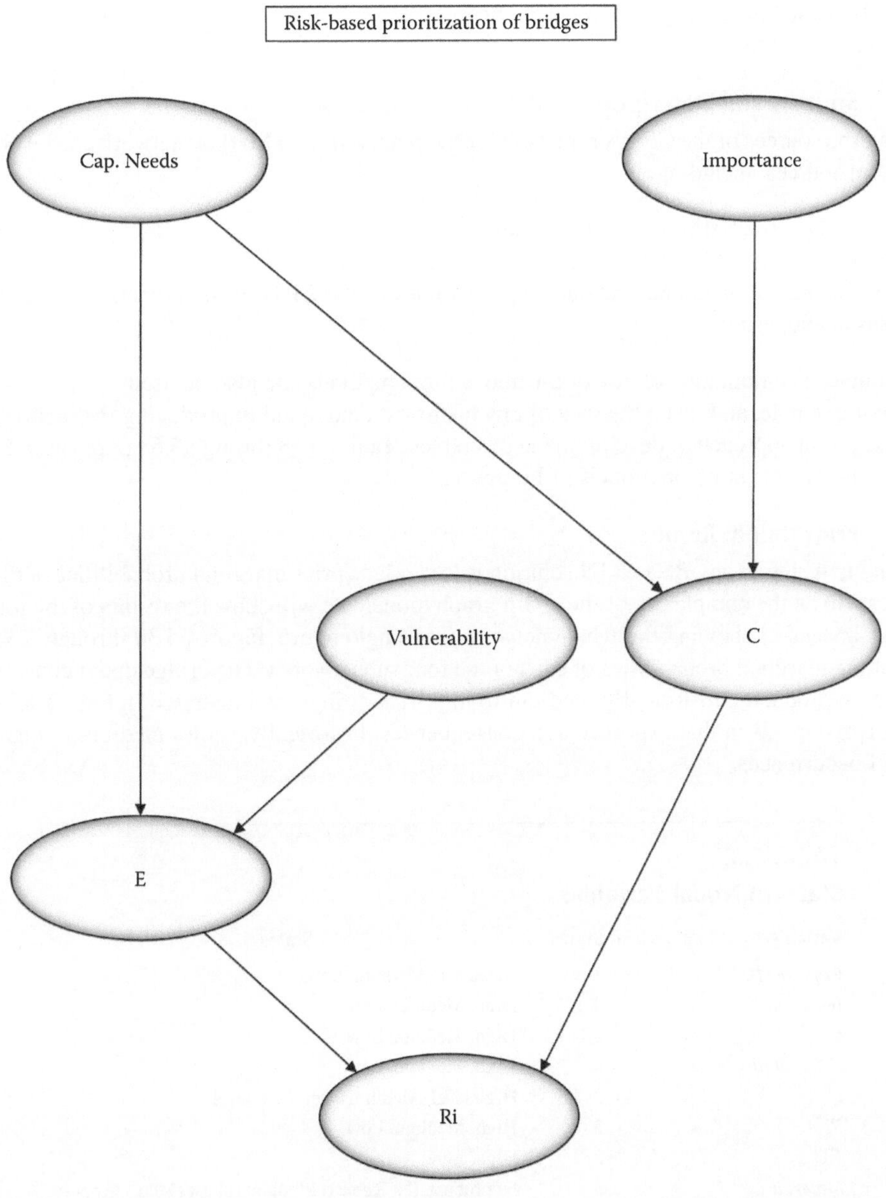

FIGURE 5.29 Risk network (downstream of the three subnetworks of Figure 5.26 through Figure 5.28).

Solution Process: As described above, the network of this model is composed of three Markov subnetworks at the upstream and a downstream Bayes network, which results in a chain graph network. We first solve the three upstream Markov networks for prior or any observation scenarios, then use the resulting marginal probabilities for importance, vulnerability, and capital needs variables to solve the downstream Bayes network for evaluating the marginal of E, C, and Ri.

5.7.4.4 Nodes and Their States

We choose a simple set of nodal states to describe the different network variables. For more realistic situations, the reader can use a higher number of states for variables. The variables for C and Ri can even be described in an absolute or monetary manner if desired. Table 5.27 shows the nodal states and their corresponding states. Note that these are the same variables used while solving with the weighted averages process in Section 5.7.3. The variables are defined in Appendix 5.I (Table 5.31).

5.7.4.5 Sources and Description of Potentials and CPTs

As usual, the sources of the CPTs or the BNs, or the potentials of a Markov network can be obtained via several sources, including:

1. Historical observations, records, or data
2. Subjective best judgment or experience of the users or officials
3. Exercising the model and readjusting the values of the CPTs or the potentials to produce reasonable results

Of course, a combination of any of the above three methods can also be used.

For our example, and given the lack of any historical data to aid in producing the tables, we use a best judgment approach to develop the needs tables. Tables 5.48 through 5.63 (Appendix 5.V) are used to solve the chain graph models of Figure 5.25.

5.7.4.6 Prior (Initial Results)

The usual first step in any BN or MN solution is to produce prior marginal probabilities of different variables. Given the complexity of the chain graph model, we will show the results of the four subnetworks instead of showing the whole network in a single figure. Figures 5.30 through 5.33 show the resulting marginal probabilities of each of the four subnetworks. The bridge under consideration seems to be producing historically medium to high-risk ratings, as illustrated in Figure 5.33. This is not surprising, given that exposure and consequences, historically, shows medium to high probability of occurrences.

TABLE 5.27
States of Nodal Variables

Variable	# of States	States
Cap. Needs	3	Extensive, Medium, Low
Importance	3	High, Medium, Low
C	3	High, Medium, Low
Vulnerability	3	High, Medium, Low
E	3	High level, Medium level, Low level
Ri	3	High, Medium, Low
F.C.	2	High, Low
Material	3	Not robust, (N. Robust) Medium robust (Med.), Robust
Other	3	High, Medium, Low
Design	3	Not robust, Medium robust, Robust
Trucks	3	High, Medium, Low
ADT	3	High, Med., Low
Traffic	3	High, Medium, Low
Detour	3	Long, Medium, Short
Func. Class	4	Interstate, Arterial, Collector, Local
Type	3	High impact, Medium impact, Low impact
NCV	2	High, Low
SDB	2	High, Low
Needs	3	Extensive, Medium, Low
Restrictions	2	High, Low
CR	4	Service, Poor, Fair, Good

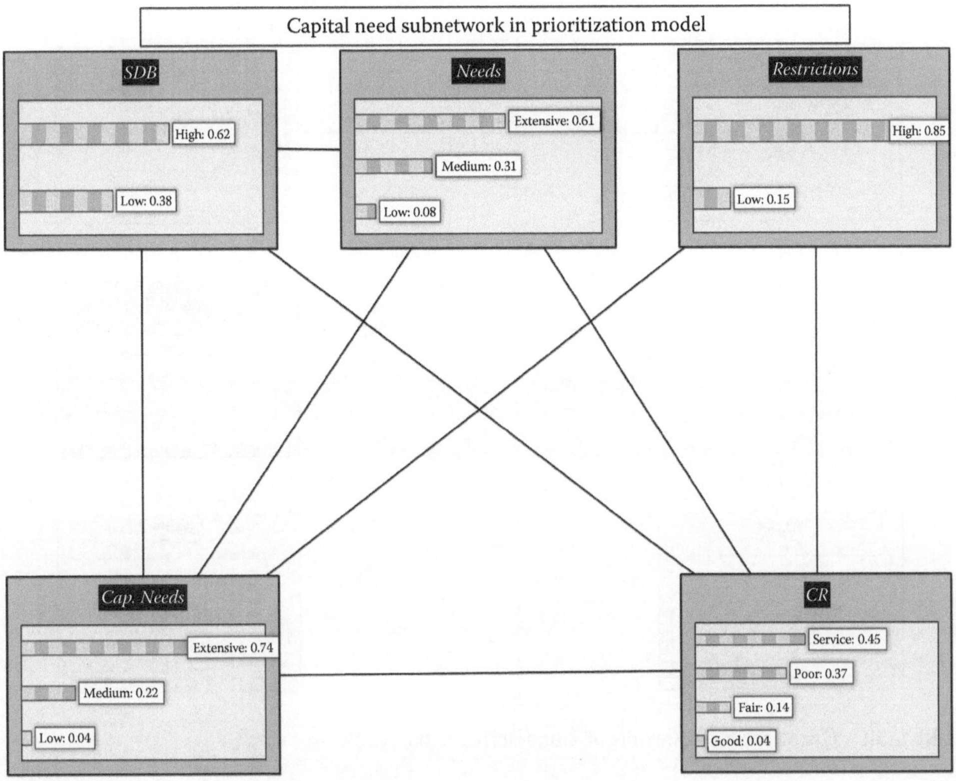

FIGURE 5.30 Historical (prior) results of capital needs subnetwork.

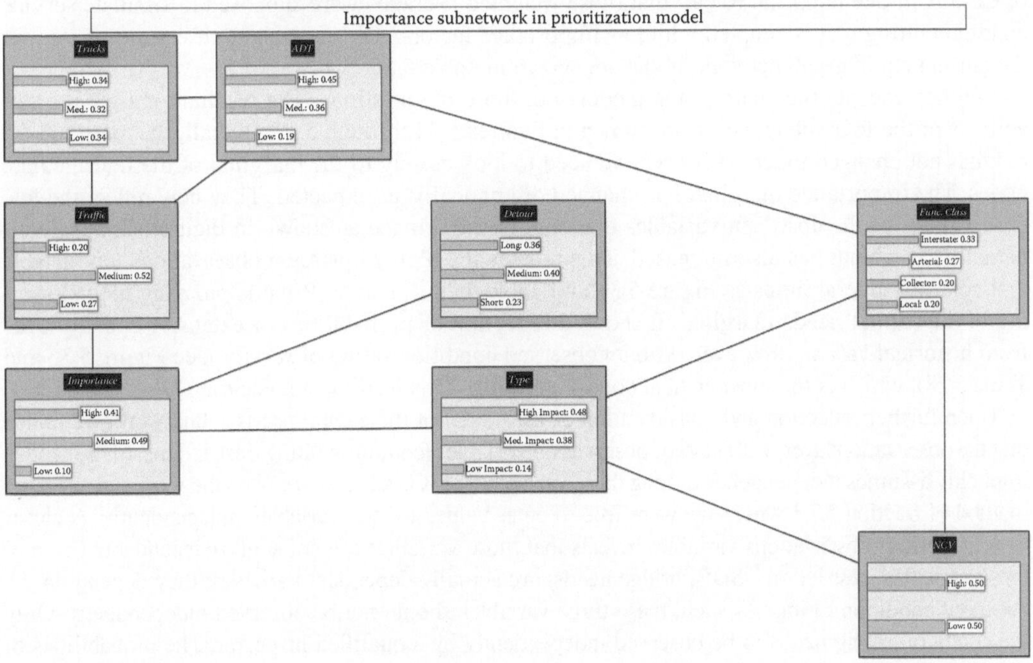

FIGURE 5.31 Historical (prior) results of importance subnetwork.

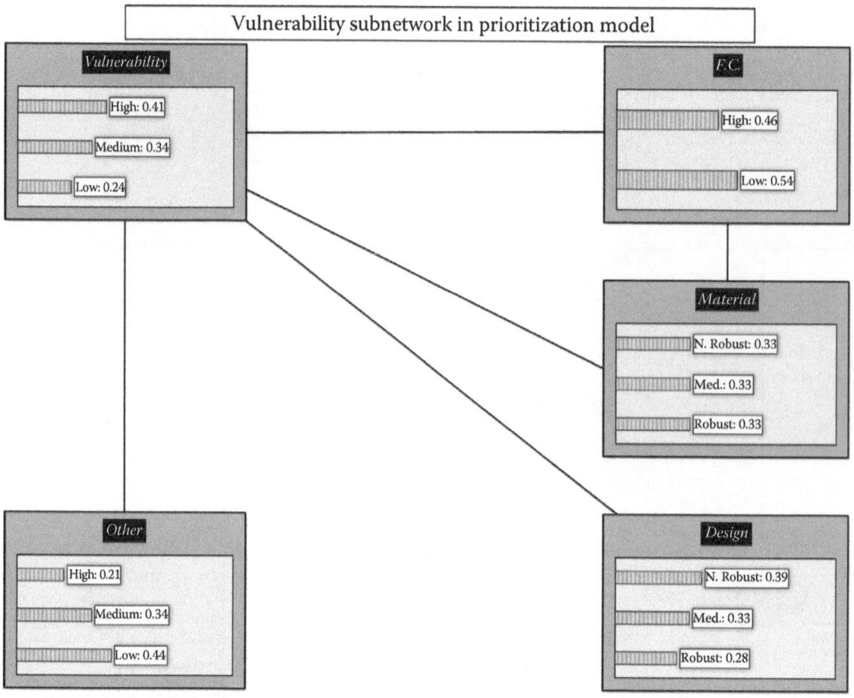

FIGURE 5.32 Historical (prior) results of vulnerability subnetwork.

5.7.4.7 Evidence (Observation) Introductions

We now proceed to evaluate risk value that is based on actual observations to update the prior historically based values of the previous section. We use observations similar to the weighted averages of Case A in Section 5.7.3. Recall that those weighted averages were supposed to simulate service condition rating that is coupled with low importance factors. The equivalent observation states for the current chain graph network model are shown in Table 5.28.

We then execute the chain graph model with those observations. The resultant marginal probabilities of the four subnetworks are shown in Figures 5.34 through 5.37. Overall, the marginal for risk has not changed much. However, we need to look closely to the marginal of the intermediate nodes. The importance marginal has changed dramatically, as expected. They now reflect the low observations of the upstream variables that control importance as shown in Figure 5.35. The vulnerability marginal has also increased. This reflects the higher upstream observations for variable that control vulnerabilities as Figure 5.36 and Table 5.28 illustrate. What is puzzling is the lowering of the capital needs marginal. It shows a reduction of probabilities of extensive capital needs from historical 74% to 30% even with an observed condition rating of *service* (see Figure 5.36 and Table 5.28), which is the most critical condition rating. This result is not accurate.

Upon further reflection and consideration of the model of the capital needs subnetwork, we notice that the observation layer, with its four observation variables (condition rating, restrictions, SDB, needs), implicitly assumes independence among these variables. This is a carryover from the weighted averages method of Section 5.7.3 where we were free to enter values for the variables independently. A closer look at the four observations variables reveals that those variables are not at all independent. Three of these variables (restrictions, SDB, bridge needs) are actually dependent variables; they depend on the observed condition rating. As such, these three variables should not be observed independently. Only the condition rating needs to be observed independently by a qualified inspector. The probabilities of the three variables are then computed using the potentials during the solution phase of the network.

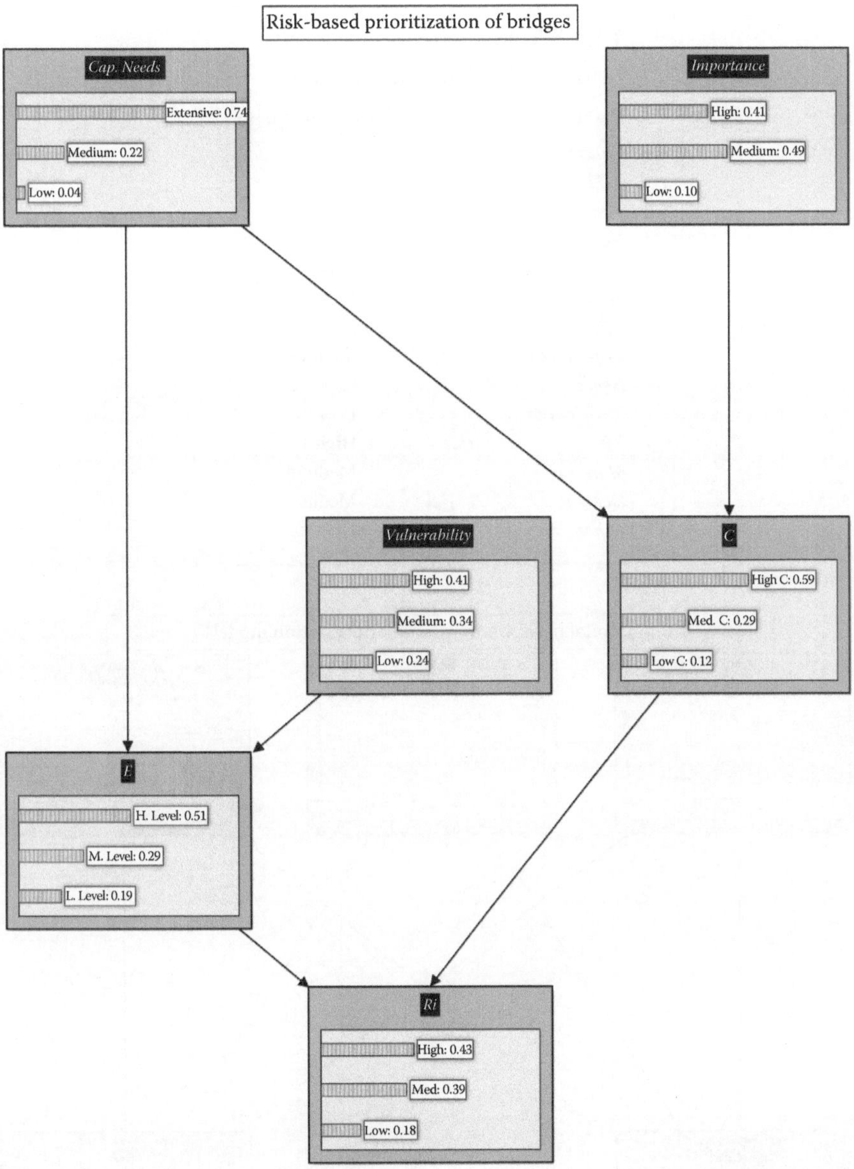

FIGURE 5.33 Historical (prior) results of risk subnetwork.

Based on the above, we need to remodel the capital needs subnetwork. Specifically, its observations layer is shown in Figure 5.38. Now the observation layer has only a single observation variable: condition rating (*CR*). Note also that there is no need to make any changes in the potentials topology or potentials tables of the subnetwork.

Upon resolving the chain graph network using the modified observations, the resulting marginal probabilities are now shown in Figures 5.39 and 5.40. Note that the marginal probabilities of both vulnerability and importance subnetworks are the same as in Figures 5.35 and 5.36 because the changes in the capital needs subnetwork have no effects on either of the two subnetworks. But the probability of extensive capital needs have now increased to 83%, a reasonable value given the severity of the *CR* observation. The high-risk probability has now increased to 52% from a

TABLE 5.28
Bridge "A" Observation States

Variable	Observation State
Trucks	Low
ADT	Low
Func. Class	Interstate
NCV	Low
Detour	Short
SDB	Low
CR	Service
Cap. Needs	Medium
Needs	Low
Restrictions	Low
FC	High
Material	Medium
Design	Medium
Other	High

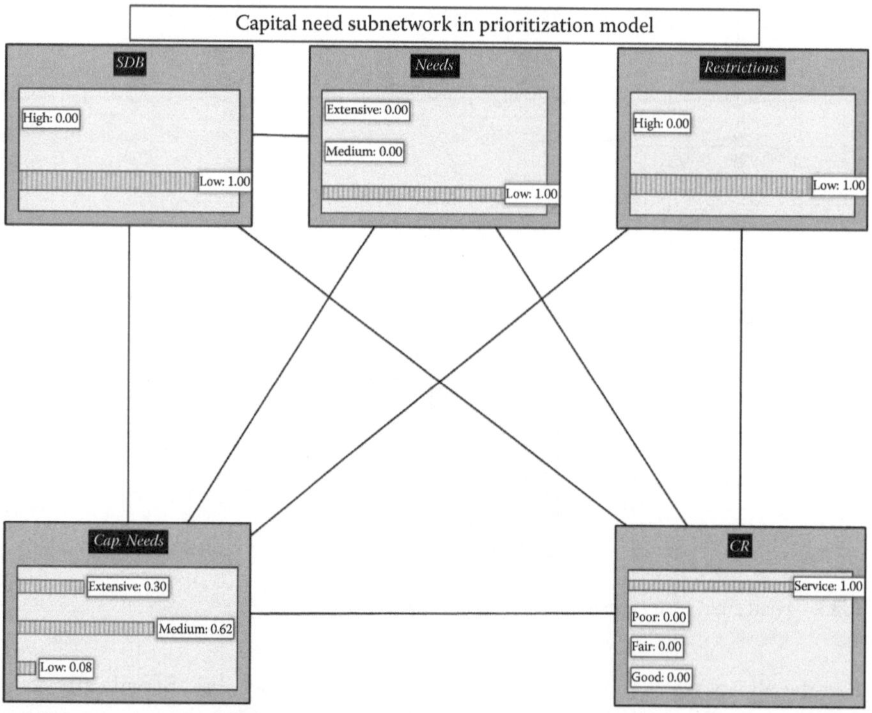

FIGURE 5.34 Results using bridge "A" observations: capital needs subnetwork.

historical rating of 43%. Given the extremely high exposure (91% probability) and the relatively moderate to high consequences (33% and 48%, respectively), this risk rating is not too surprising.

We now re-execute the chain graph model with a different set of observations consistent with the B model of Section 5.7.3. Recall that the model aims at simulating a lesser critical *CR* that is coupled with higher importance observations. The equivalent set of observations is shown in Table 5.30.

Executing this high importance chain graph model produces a set of marginal probabilities that are shown in Figures 5.41 through 5.43. Note that the vulnerability subnetwork is the same as in

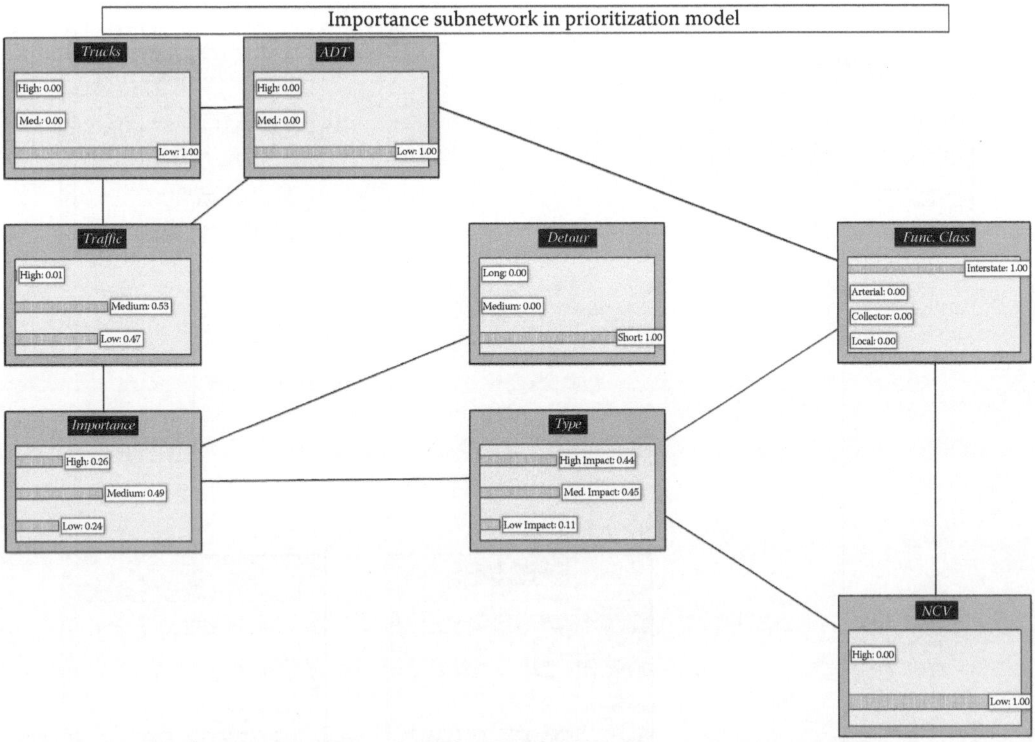

FIGURE 5.35 Results using bridge "A" observations: importance subnetwork.

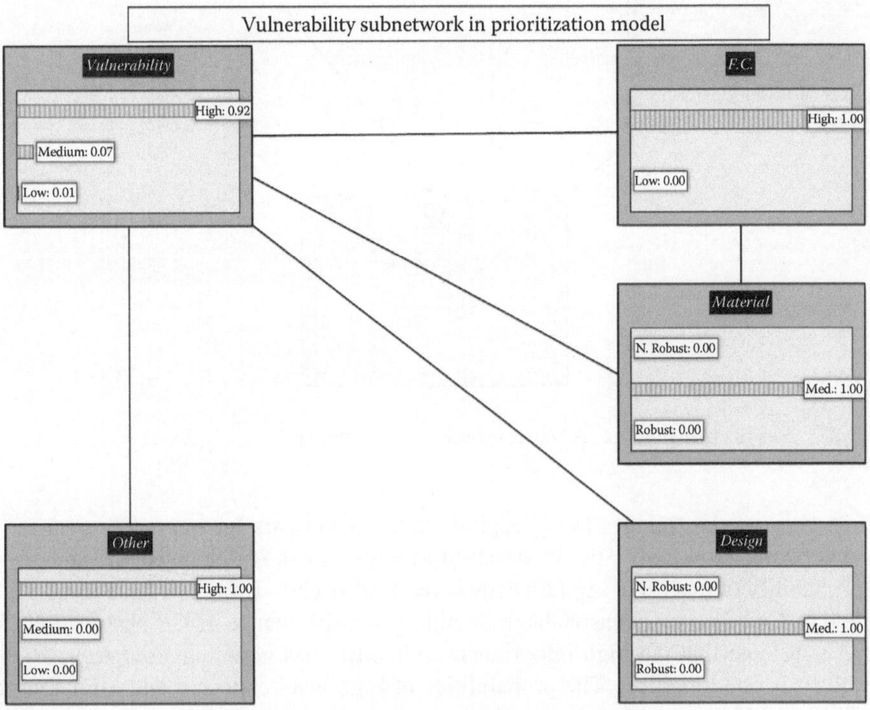

FIGURE 5.36 Results using bridge "A" observations: vulnerability subnetwork.

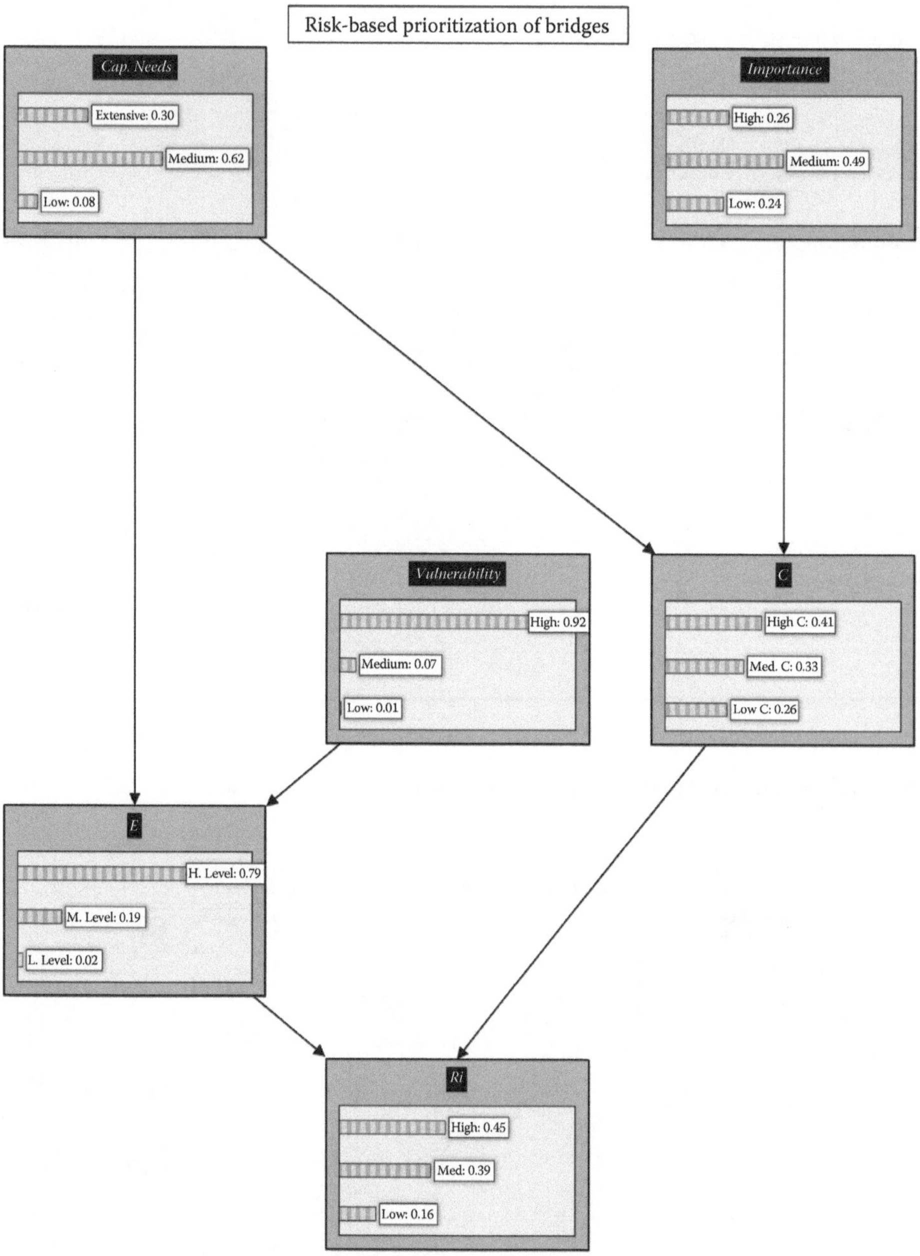

FIGURE 5.37 Results using bridge "A" observations: risk subnetwork.

Figure 5.36 because the changes in the capital needs subnetwork have no effects on the vulnerability subnetwork. Expectedly, the vulnerability marginal has not changed. Of course, we note that the probability of extensive capital needs is reduced slightly from 83% (due to *CR* of service) to 75% (due to *CR* of poor): a reasonable reduction given the change of *CR* observations. Equally interesting is to note that the high importance probability has now increased from 26% to 58% (historically it is rated at 40%). The probabilities of high-level exposure and high consequences are now 89% and 72%, which produces a high-risk probability of 63%, again reasonable and not too surprising results.

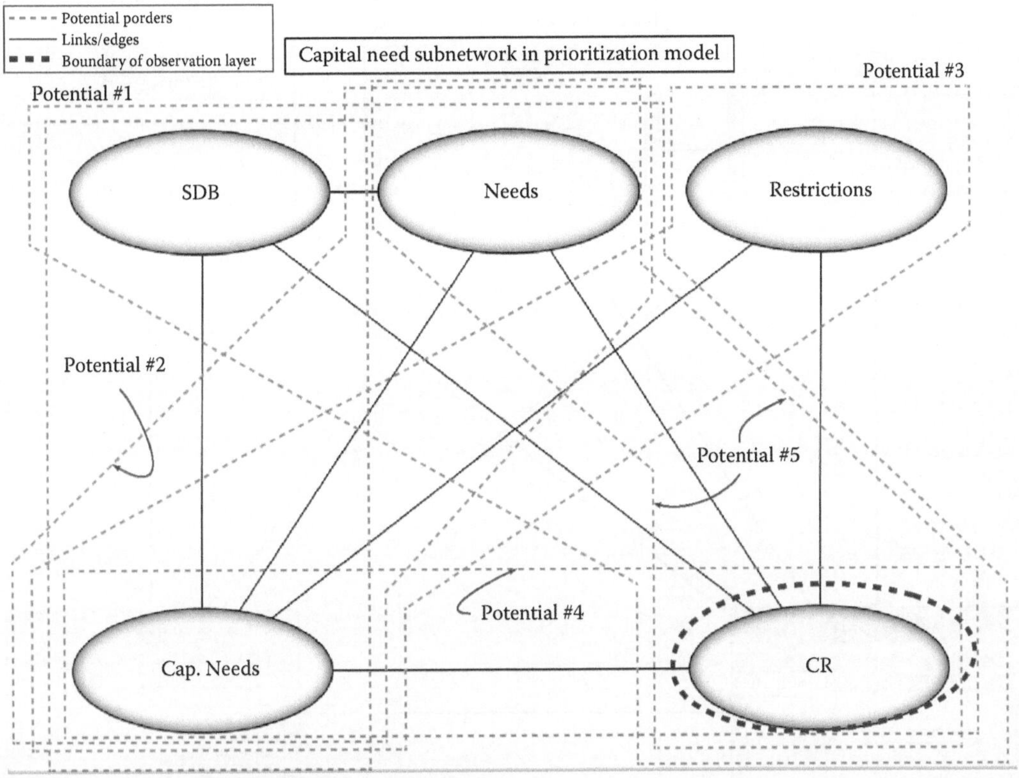

FIGURE 5.38 Modified capital needs subnetwork.

TABLE 5.29
Bridge "A" Observation States (Modified)

Variable	Observation State
Trucks Low	
ADT Low	
Func. Class Interstate	
NCV Low	
Detour Short	
CR Service	
FC High	
Material Medium	
Design Medium	
Other High	

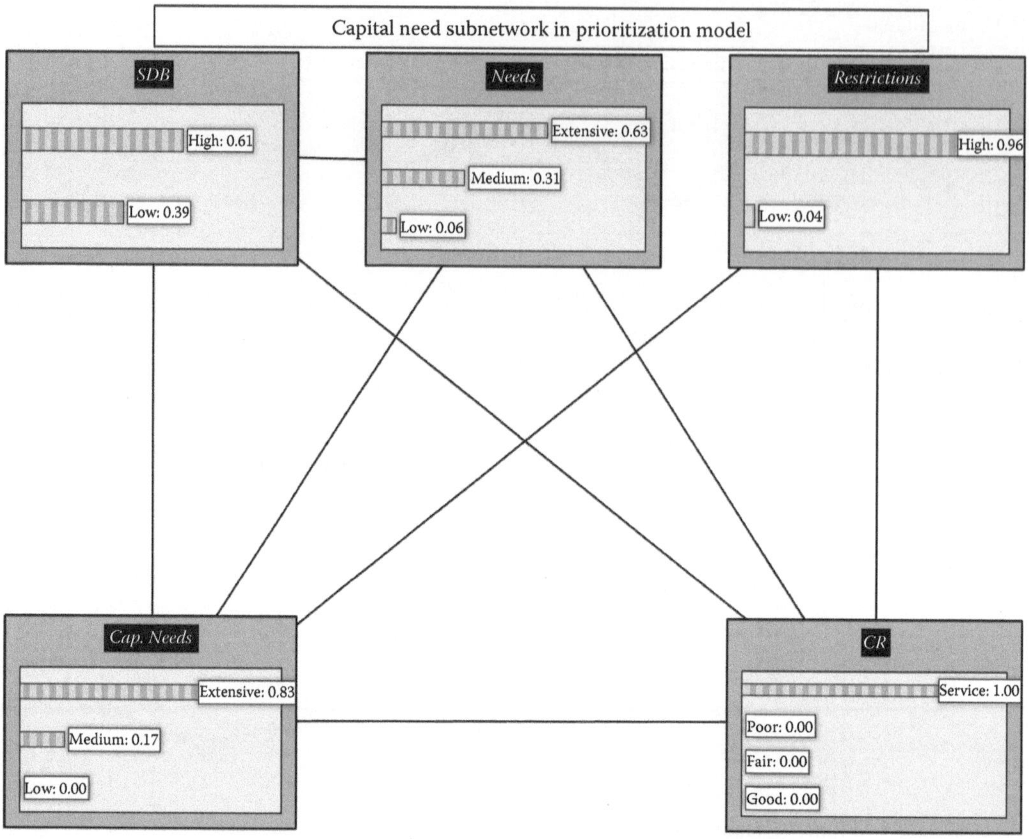

FIGURE 5.39 Results using bridge "A" observations: modified capital needs subnetwork.

The changes in *CR* and importance observations have produced different results that are reasonable, whereas the reader might recall that the weighted averages method, Section 5.7.3, produced relative risk ratings that were not too convincing.

5.7.5 Concluding Remarks

We offered prioritization models for risk ratings using two methods: weighted averages and chain graph network. Both methods have certain advantages and disadvantages. The weighted averages method features simplicity, and the chain graph method is obviously complex and can be computationally demanding. On the other hand, we showed that the weighted averages method cannot accommodate variable dependencies and intersections. The network method can accommodate such intersections and dependencies by default. Other advantages of the network method that were exhibited by this example are:

- *Scalability*: when there was a need to change observations or dependence between variables, such changes were possible with minimal efforts.
- Accommodating uncertainties/considerations are built in.
- *Ease of understanding*: it is easy to explore the roles of different variables in the network by studying the marginal probabilities and the network topology.
- Based on historical observations by default.

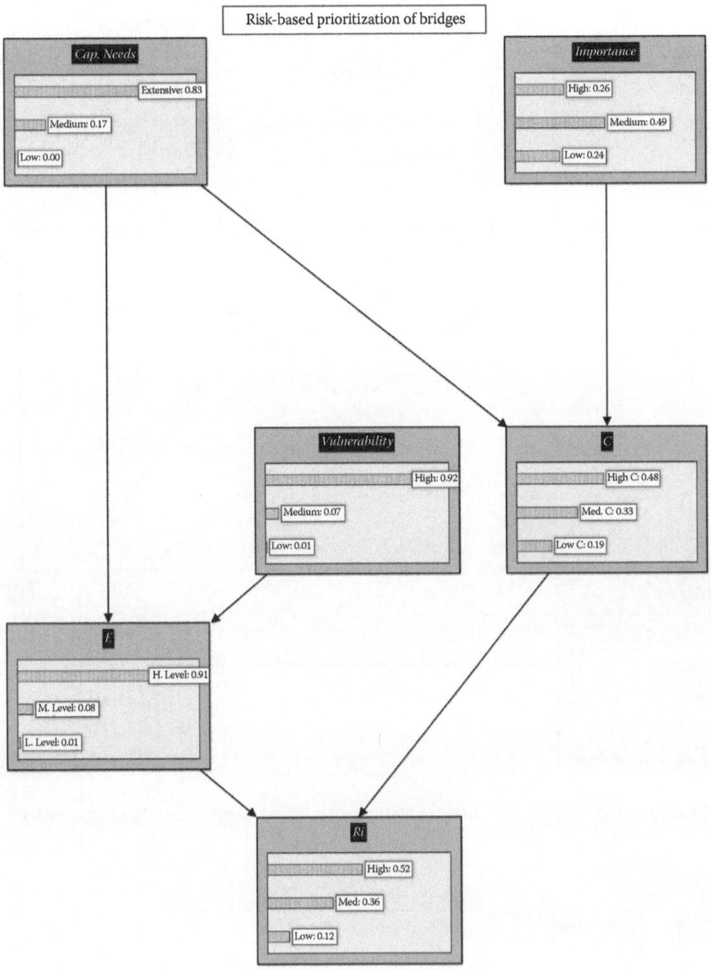

FIGURE 5.40 Results using bridge "A" observations: risk subnetwork.

TABLE 5.30
Bridge "B" Observation States

Variable	Observation State
Trucks	High
ADT	High
Func. Class	Interstate
NCV	High
Detour	Short
CR	Poor
FC	High
Material	Medium
Design	Medium
Other	High

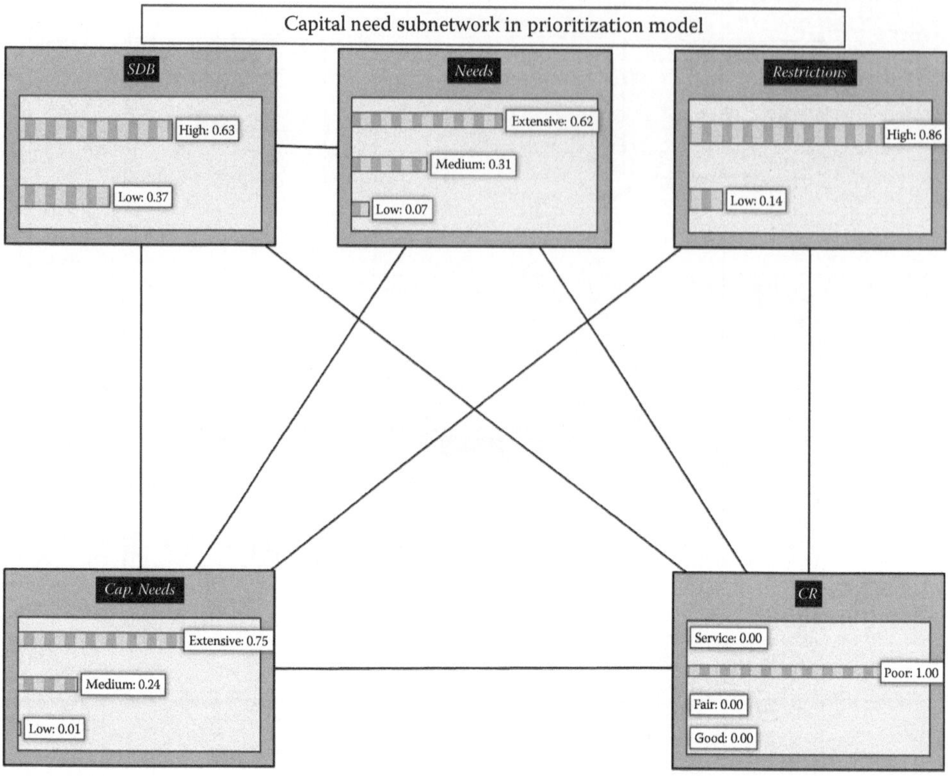

FIGURE 5.41 Results using bridge "B" observations: modified capital needs subnetwork.

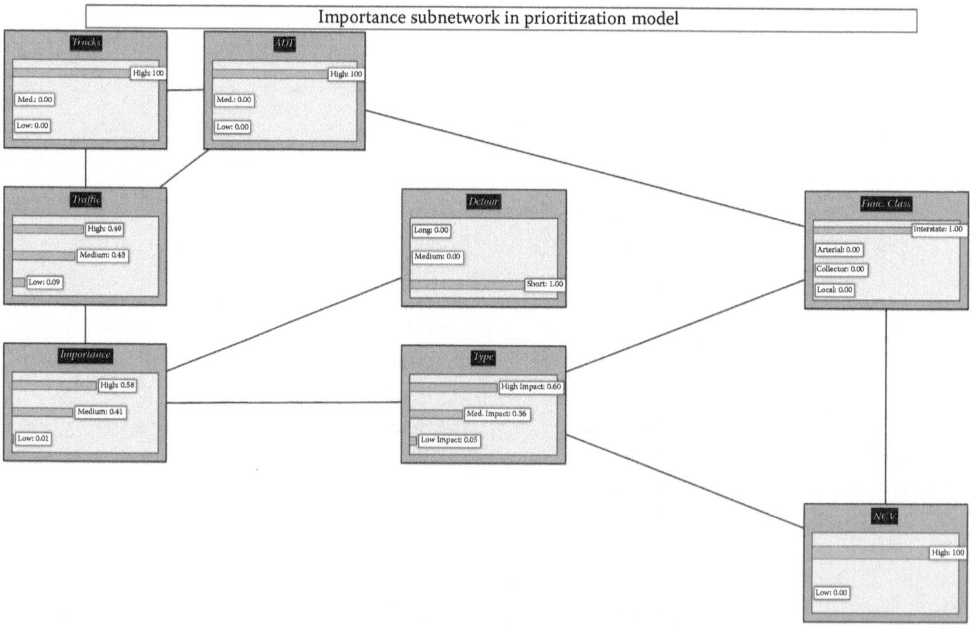

FIGURE 5.42 Results using bridge "B" observations: modified importance subnetwork.

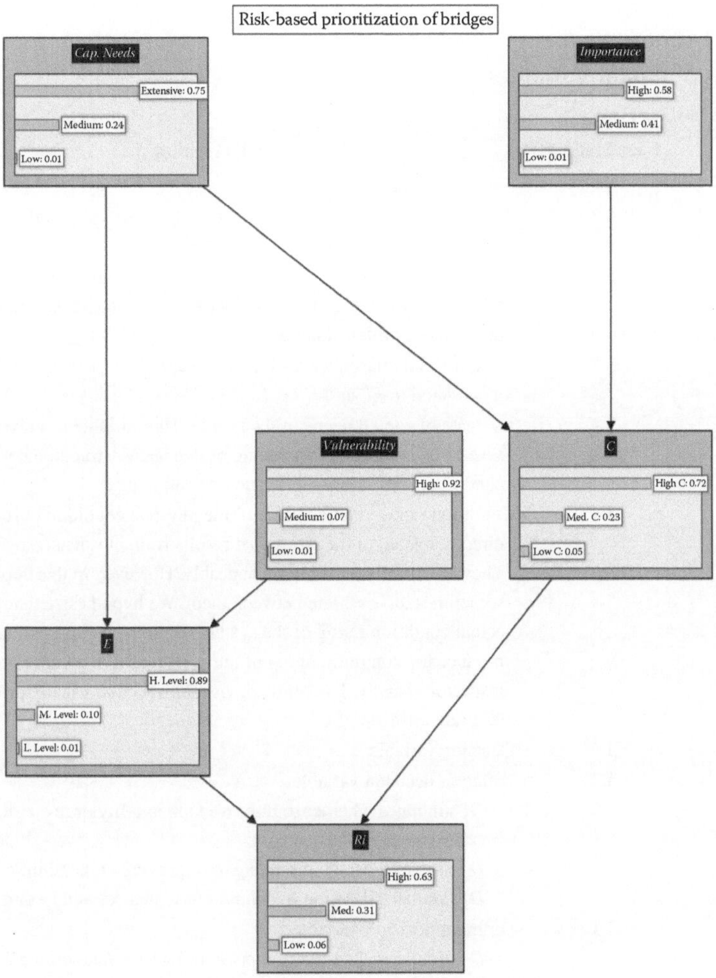

FIGURE 5.43 Results using bridge "B" observations: risk subnetwork.

APPENDIX 5.I: DEFINITIONS OF NETWORK NODAL VARIABLES (TABLE 5.31)

TABLE 5.31
Definitions of Nodal Variables

Variable	Used in Case Study #	Definition
ADT	5.4	Average daily traffic on the bridge of interest
(Bridge) Needs	5.4	Describes bridge physical maintenance, rehabilitation, or replacement needs. It relates to condition rating and structural deficiency status.
C	5.1	Consequences of this example is only a function of bridge deterioration. It does not include any factors beyond the bridge physical system such as detours or any other social factors.
C	5.2	Consequences of this example is a function of 1-bridge deterioration as represented by *Exposure*, and 2-*C.O.*, which simulates factors beyond the bridge physical system such as detours and/or any other social factors.

(Continued)

TABLE 5.31 *(Continued)*
Definitions of Nodal Variables

Variable	Used in Case Study #	Definition
C	5.4	Total consequences
C.O.	5.2	This variable simulates factors beyond the bridge physical system such as detours and/or any other social factors.
C.O._1, C.O._2	5.3	*C.O._i* simulates factors beyond the *i*-th bridge physical system such as detours and/or any other social factors. Their impact varies, depending on each *i*-th bridge situation.
C1, C2	5.3	*Ci* is the total consequences for *i*-th bridge in the network. In this example, it is related to *Ei,* and *C.O._i.*
Cap. Needs	5.4	Estimate of capital needs of the bridge. This variable is a function of several other variables including bridge needs, structural deficiency condition, restrictions, and condition rating.
CR	5.4	Condition rating is an estimate of the physical condition of the bridge. It is directly related to the rating that results from physical inspections. They are usually used interchangeably. However, in this book, we sometimes differentiate between them. We hypothesize that *CR* is the actual condition rating of the system, while visual inspection would produce the condition rating of the system as it is observed by the inspector visually. Even though visual inspection usually produces an accurate estimate of *CR*, they are not identical.
D	5.1	Decision variable
D1, D2, D3	5.2	Different decision variables: D_1 simulates whether to improve vulnerability or to control consequences and hazards. D_2 simulates decision relating to improving vulnerability. D_3 simulates decision to control consequences and hazards.
D_1, D_{21}, D_{22}, D_3, D_4	5.3	Different decision variables: D_1 simulates whether to improve bridge 1 or/and bridge 2 in the network. D_{12} simulates decision relating to improving bridge 1 in the network. D_{22} simulates decision relating to improving bridge 2 in the network. D_3 simulates decision to transfer risk. D_4 simulates decision to do nothing.
Design Type	5.4	Some bridge design types offer more robustness than others—some are harder to replace than others and durability may vary.
Detour	5.4	Estimate of detour length, if applicable.
E	5.1	Exposure in this example indicates the extent of deterioration of the bridge.
E	5.2	Exposure of the bridge in this example is a function of both hazard and vulnerability of the bridge.
E	5.4	Total exposure
E1, E2	5.3	*Ei* is the exposure the indicates the extent of deterioration of the *i*-th bridge.
F.C.	5.4	Fracture critical status of the bridge.
Func. Class	5.4	This variable describes the character of traffic service of the route carried by the bridge.
Hz	5.2	Subjective variable which indicated magnitude of postulated hazard.
Hz	5.3	This hazard, *Hz,* affects all bridges in the network. However, its effects on each bridge are different.

(Continued)

TABLE 5.31 ***(Continued)***
Definitions of Nodal Variables

Variable	Used in Case Study #	Definition
Hydraulic Vuln.	5.4	This variable indicates vulnerability of bridge to hydraulic-related hazards. This includes scour, overtopping, storm surge, etc.
Importance	5.4	Depends on factors such as truck traffic, average daily traffic, detour length, etc.
Material	5.4	Material of construction of bridge. Maintenance costs and durability vary depending on the materials.
NCV	5.4	Value of the asset contributing to commerce.
Restrictions	5.4	Describes whether there are any restrictions on bridge use. For example, restricted to heavy loads or low under-bridge clearance.
Ri	5.4	Risk
SDB	5.4	Describes whether the structure is deficient.
Traffic	5.4	This variable is a function of *Trucks* and *ADT*.
Trucks	5.4	Truck traffic level on the bridge of interest.
Type	5.4	This variable simulates the impact of physical loss of the bridge of interest.
U	5.1	Utility
U	5.2	Utility
U	5.3	Utility
V	5.2	Vulnerability in this example is a function of the decision to the level of mitigation, D_2.
V1, V2	5.3	*Vi* simulates the *V*, vulnerability, of the *i*-th bridge in the network.
Vulnerability	5.4	*Vulnerability* in this example is an indication of bridge's physical vulnerability, which is a function of the fracture critical condition of the bridge, material of construction, vulnerability status, and the design type.

APPENDIX 5.II: CPTs FOR CASE STUDY 5.1 (TABLES 5.32–5.33)

TABLE 5.32
CPT of *C*

E		L	H
C	L	0.9206	0.0117
	H	0.0794	0.9883
Histograms for conditional probabilities		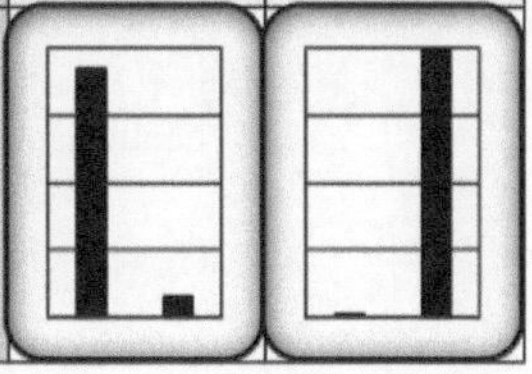	

TABLE 5.33
CPT of *E*

D		d1	d2	d3	d4
E	L	0.0794	0.9750	0.0794	1.0000
	H	0.9206	0.0250	0.9206	0.0000
Histograms for conditional probabilities		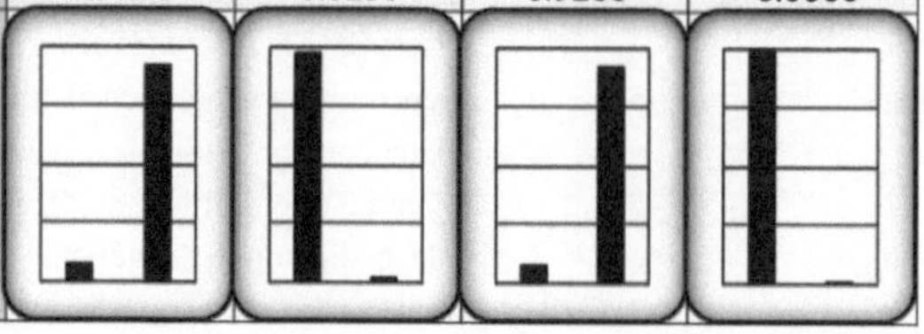			

APPENDIX 5.III: CPTs FOR CASE STUDY 5.2 (TABLES 5.34–5.38)

TABLE 5.34
CPT of *C*

C.O.		L		H	
E		L	H	L	H
C	L	0.9949	0.6852	0.3913	0.0000
	H	0.0051	0.3148	0.6087	1.0000
Histograms for conditional probabilities		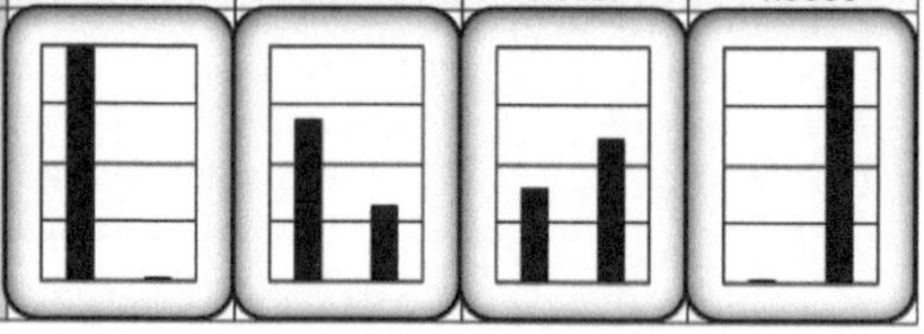			

TABLE 5.35
CPT of *CO*

C.O.	L	0.3
	H	0.7
Histograms for conditional probabilities		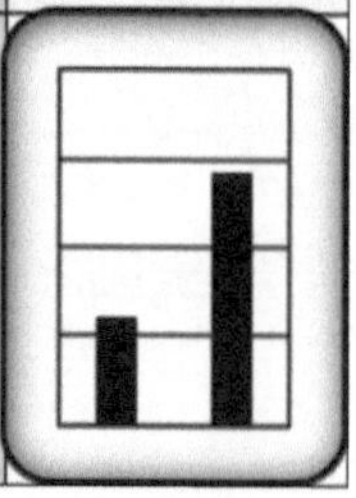

TABLE 5.36
CPT of *E*

Hz		L					H				
V		V.L.	L.	M.	H.	V.H.	V.L.	L.	M.	H.	V.H.
E	L	0.9979	0.8204	0.5981	0.4197	0.0933	0.8302	0.5718	0.3913	0.1583	0.0001
	H	0.0021	0.1796	0.4019	0.5803	0.9067	0.1698	0.4282	0.6087	0.8417	0.9999
Histograms for conditional probabilities			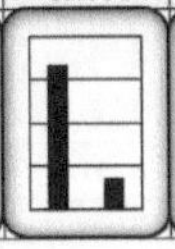	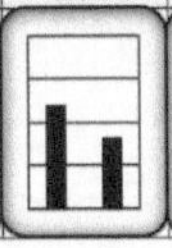	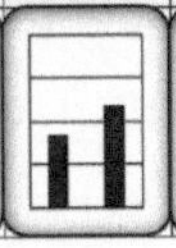	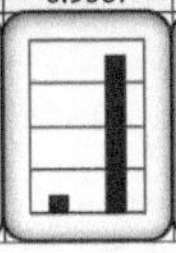	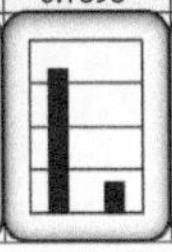	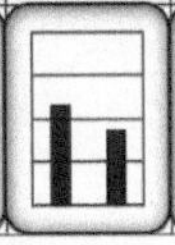	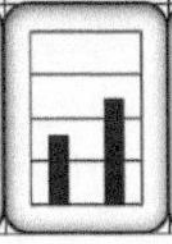	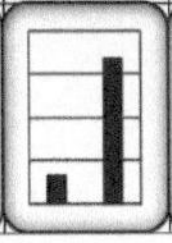	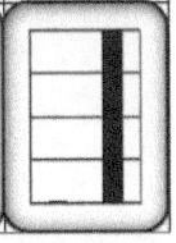

TABLE 5.37
CPT of *Hz*

Z	L	0.4
	H	0.6
Histograms for conditional probabilities		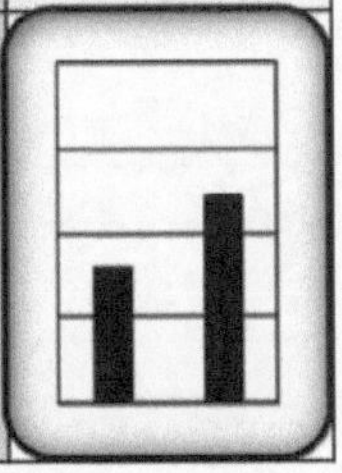

TABLE 5.38
CPT of *V*

D2		d21	d22	d23	d24
V	V.L.	0.6827	0.6827	0.6827	0.6827
	L.	0.2718	0.2718	0.2718	0.2718
	M.	0.0428	0.0428	0.0428	0.0428
	H.	0.0026	0.0026	0.0026	0.0026
	V.H.	0.0001	0.0001	0.0001	0.0001
Histograms for conditional probabilities		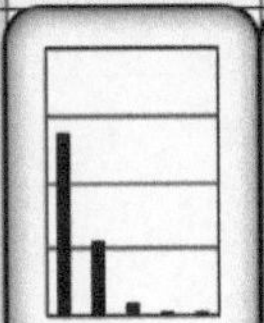	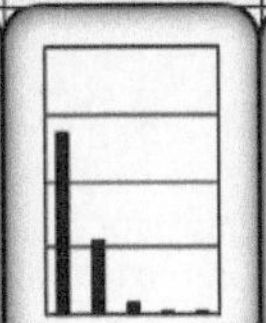	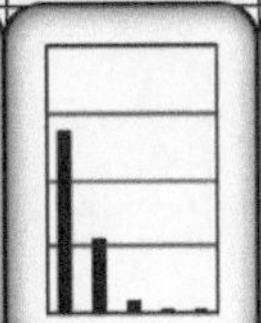	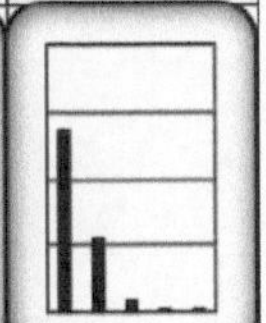

APPENDIX 5.IV: CPTs FOR CASE STUDY 5.3 (TABLES 5.39–5.47)

TABLE 5.39
CPT of *C*1

C.O._1		L		H	
E1		L	H	L	H
C1	L	0.9949	0.6852	0.3913	0.0000
	H	0.0051	0.3148	0.6087	1.0000
Histograms for conditional probabilities		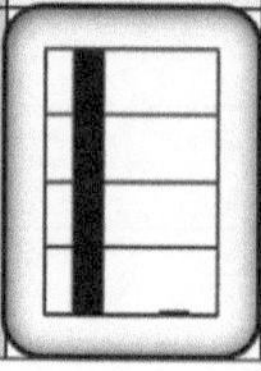	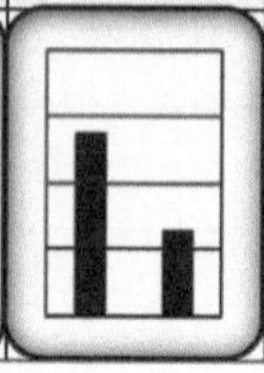	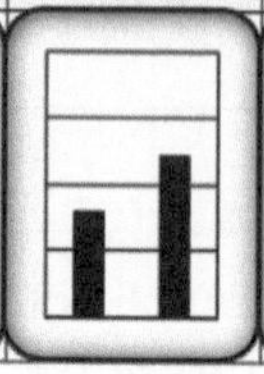	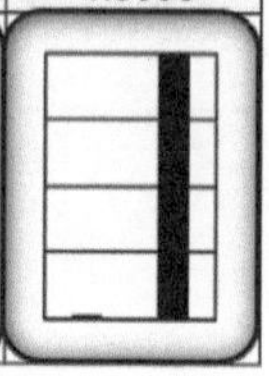

TABLE 5.40
CPT of *C*2

C.O._2		L		H	
E2		L	H	L	H
C2	L	0.9750	0.6301	0.0502	0.0000
	Hz	0.0250	0.3699	0.9498	1.0000
Histograms for conditional probabilities		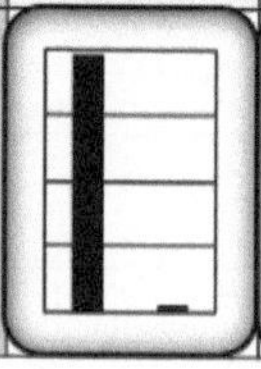	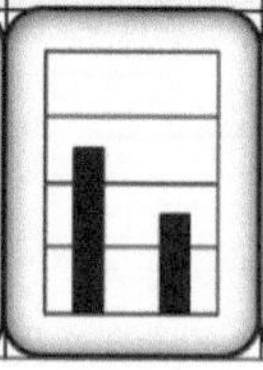	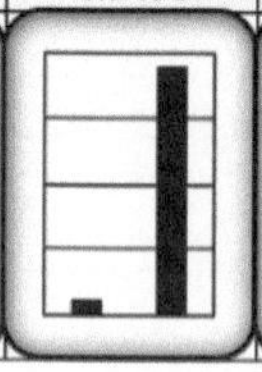	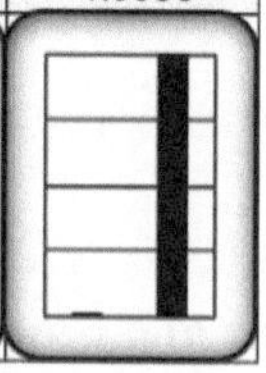

TABLE 5.41
CPT of *CO*_1

C.O._1	L	0.45
	H	0.55
Histograms for conditional probabilities		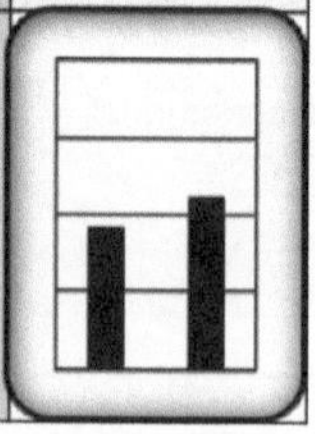

TABLE 5.42
CPT of *CO_2*

C.O._2	L	0.25
	H	0.75
Histograms for conditional probabilities		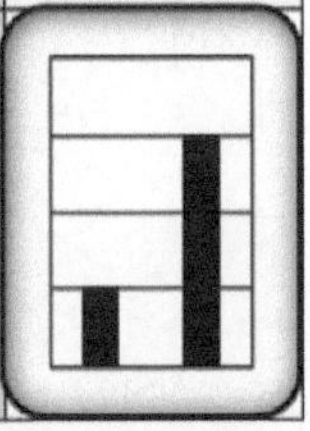

TABLE 5.43
CPT of *E*1

H		L					H				
V1		V.L.	L.	M.	H.	V.H.	V.L.	L.	M.	H.	V.H.
E1	L	0.9979	0.8204	0.5981	0.4197	0.0933	0.8302	0.5718	0.3913	0.1583	0.0001
	H	0.0021	0.1796	0.4019	0.5803	0.9067	0.1698	0.4282	0.6087	0.8417	0.9999
Histograms for conditional probabilities		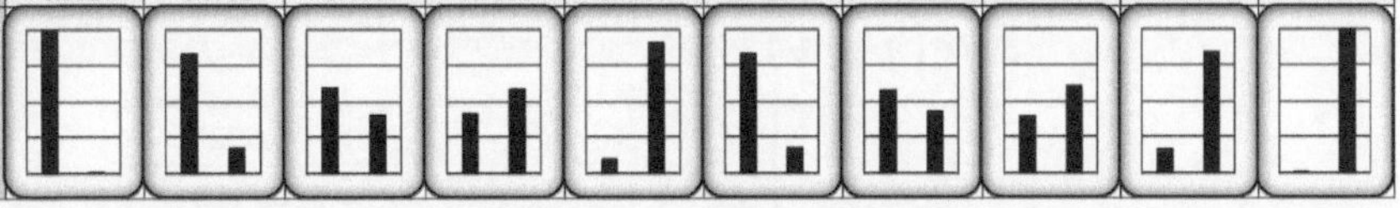									

TABLE 5.44
CPT of *E*2

H		L					H				
V2		V.L.	L.	M.	H.	V.H.	V.L.	L.	M.	H.	V.H.
E2	L	0.9888	0.7271	0.5550	0.2729	0.0502	0.6852	0.4711	0.2918	0.0794	0.0005
	H	0.0112	0.2729	0.4450	0.7271	0.9498	0.3148	0.5289	0.7082	0.9206	0.9995
Histograms for conditional probabilities											

TABLE 5.45
CPT of *H*

Hz	L	0.4
	H	0.6
Histograms for conditional probabilities		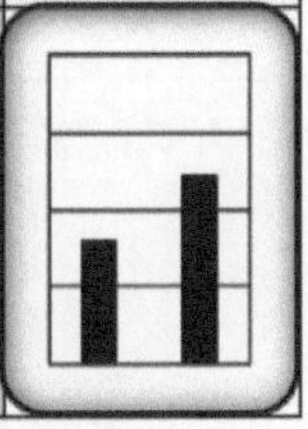

TABLE 5.46
CPT of *V*1

V1	V.L.	0.6827
	L.	0.2718
	M.	0.0428
	H.	0.0026
	V.H.	0.0001
Histograms for conditional probabilities		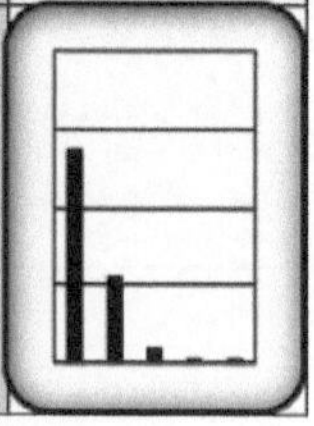

TABLE 5.47
CPT of *V*2

V2	V.L.	0.6827
	L.	0.2718
	M.	0.0428
	H.	0.0026
	V.H.	0.0001
Histograms for conditional probabilities		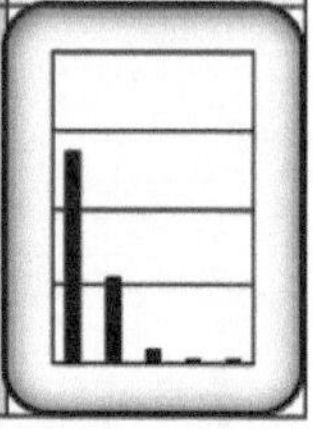

APPENDIX 5.V: MARGINAL PROBABILITY FOR DECISION SCENARIOS OF CASE STUDY 5.3 (FIGURES 5.44–5.59)

This section contains the marginal probability results for decision scenarios 00 through 15 for Case Study 5.3 of Section 5.6.3.

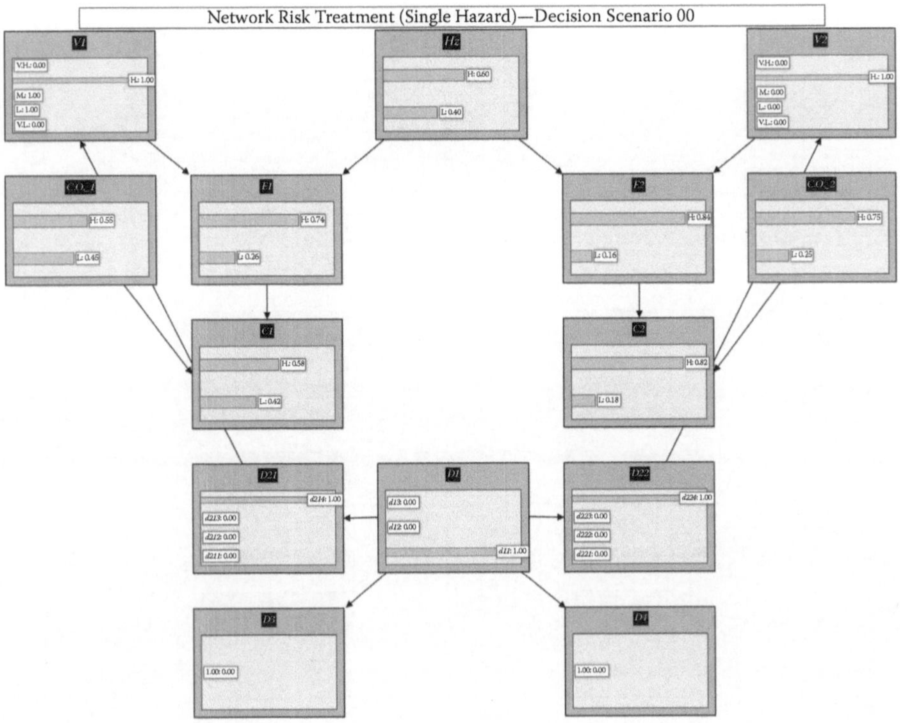

FIGURE 5.44 Marginal probabilities for design scenario 00.

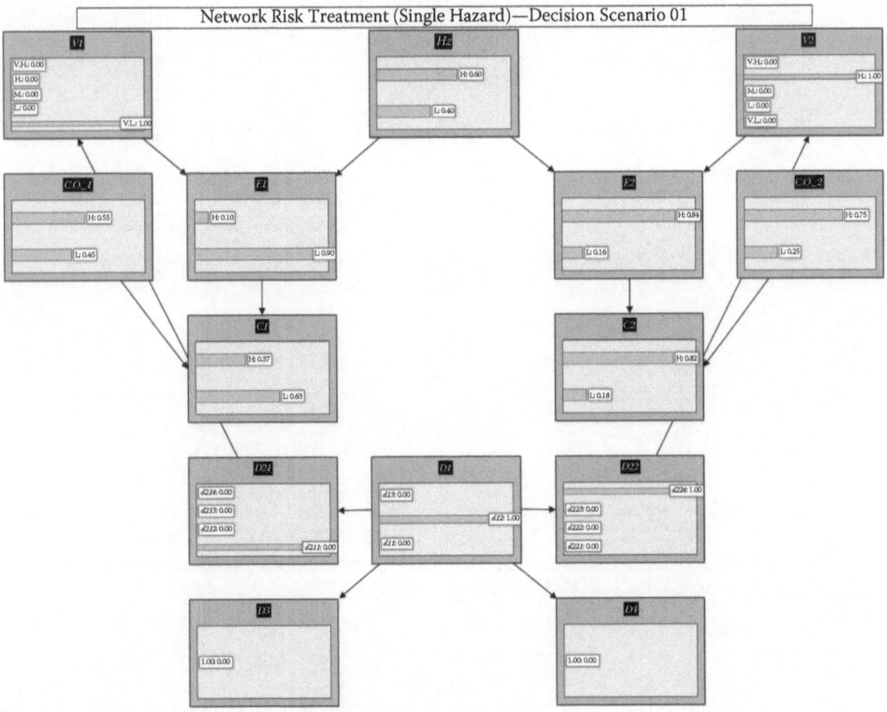

FIGURE 5.45 Marginal probabilities for design scenario 01.

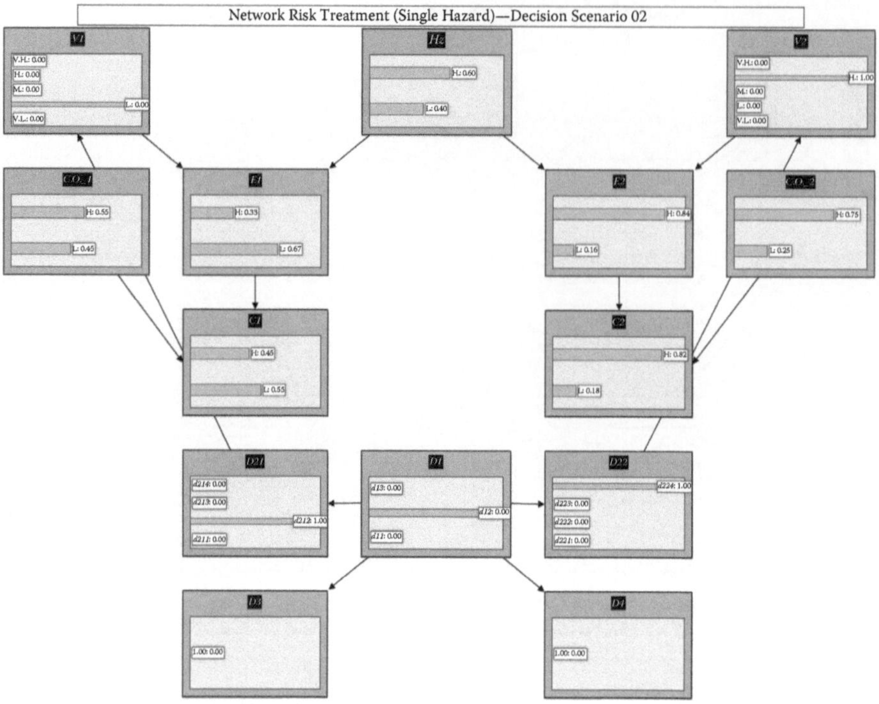

FIGURE 5.46 Marginal probabilities for design scenario 02.

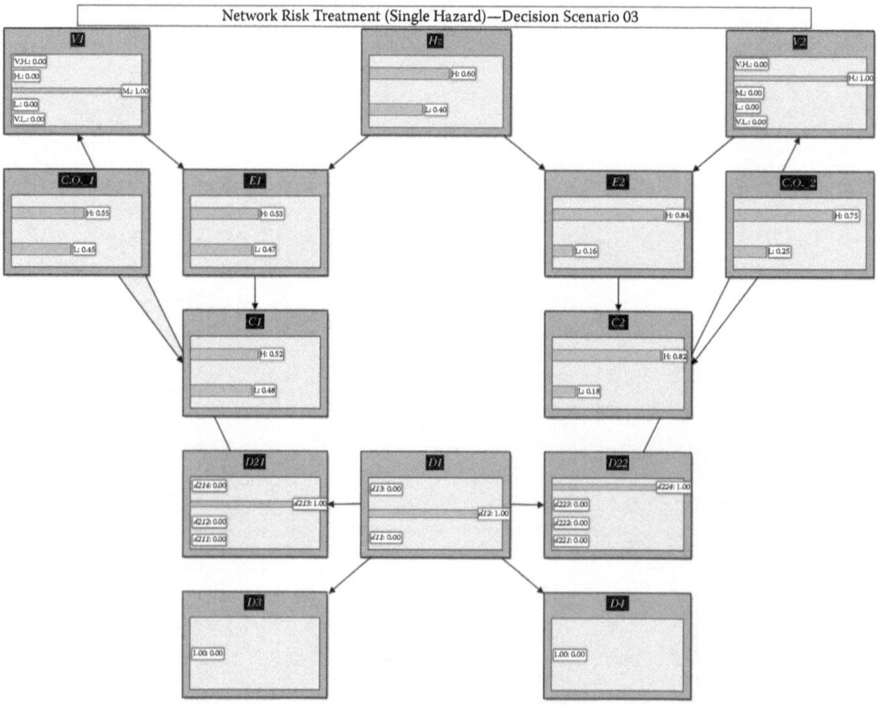

FIGURE 5.47 Marginal probabilities for design scenario 03.

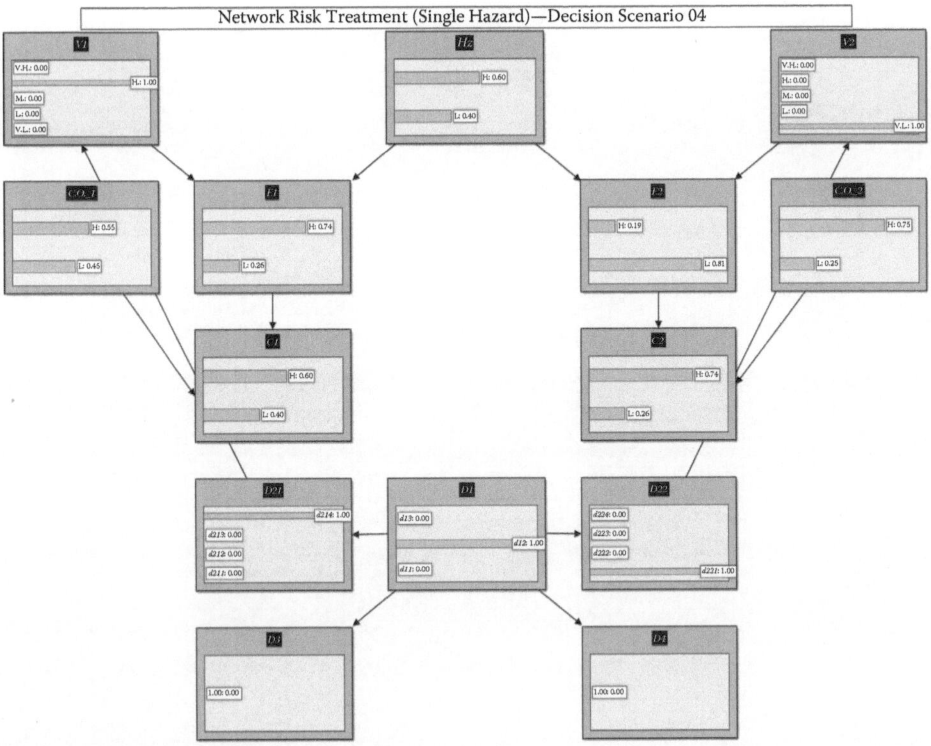

FIGURE 5.48 Marginal probabilities for design scenario 04.

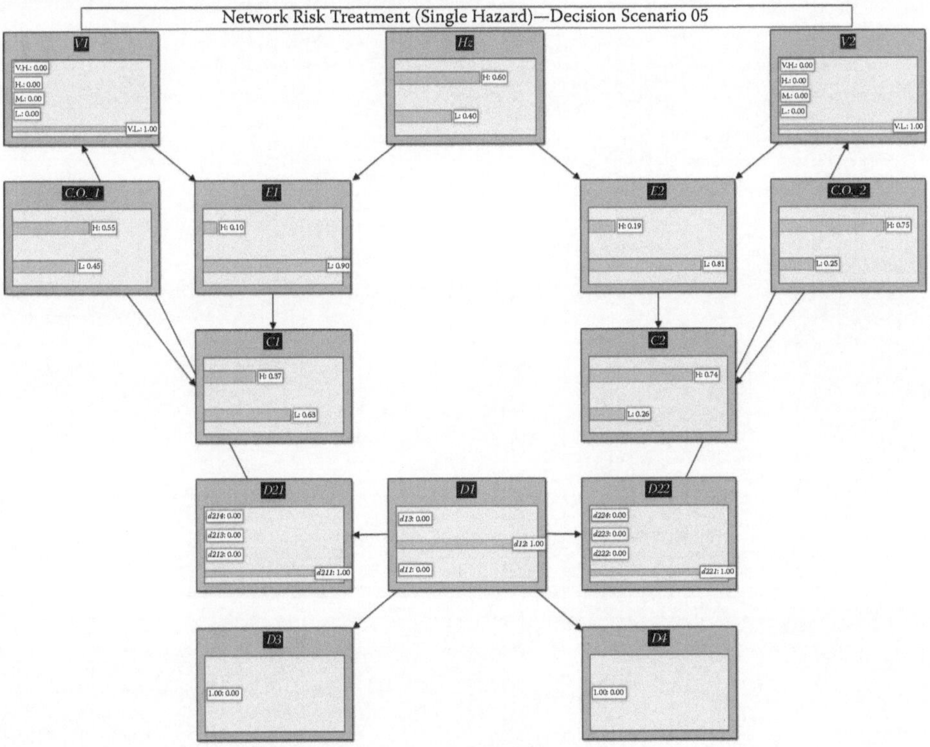

FIGURE 5.49 Marginal probabilities for design scenario 05.

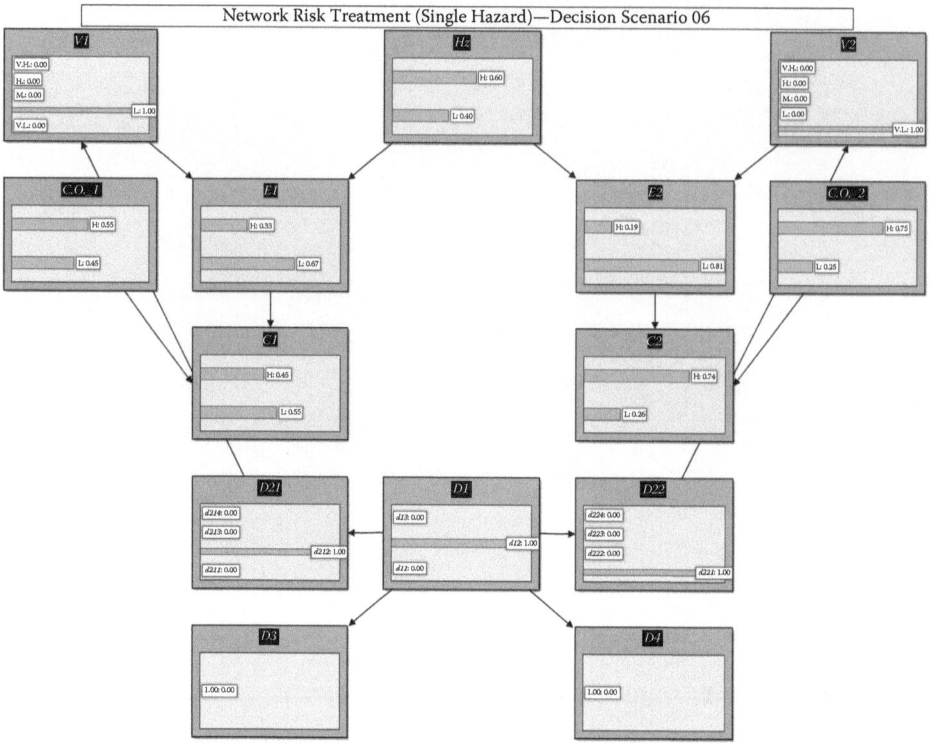

FIGURE 5.50 Marginal probabilities for design scenario 06.

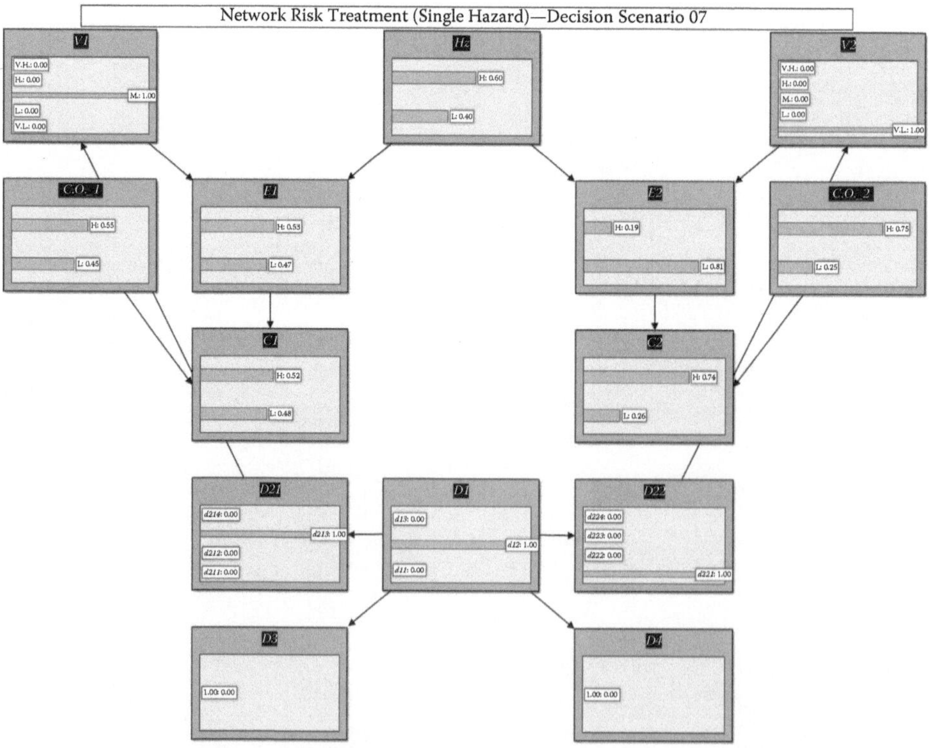

FIGURE 5.51 Marginal probabilities for design scenario 07.

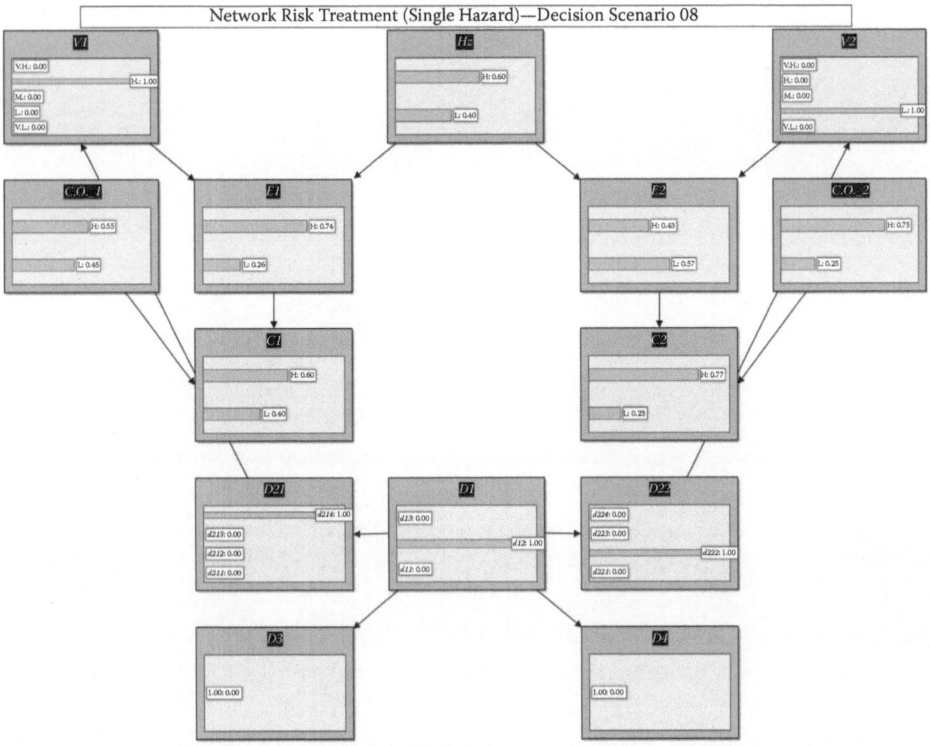

FIGURE 5.52 Marginal probabilities for design scenario 08.

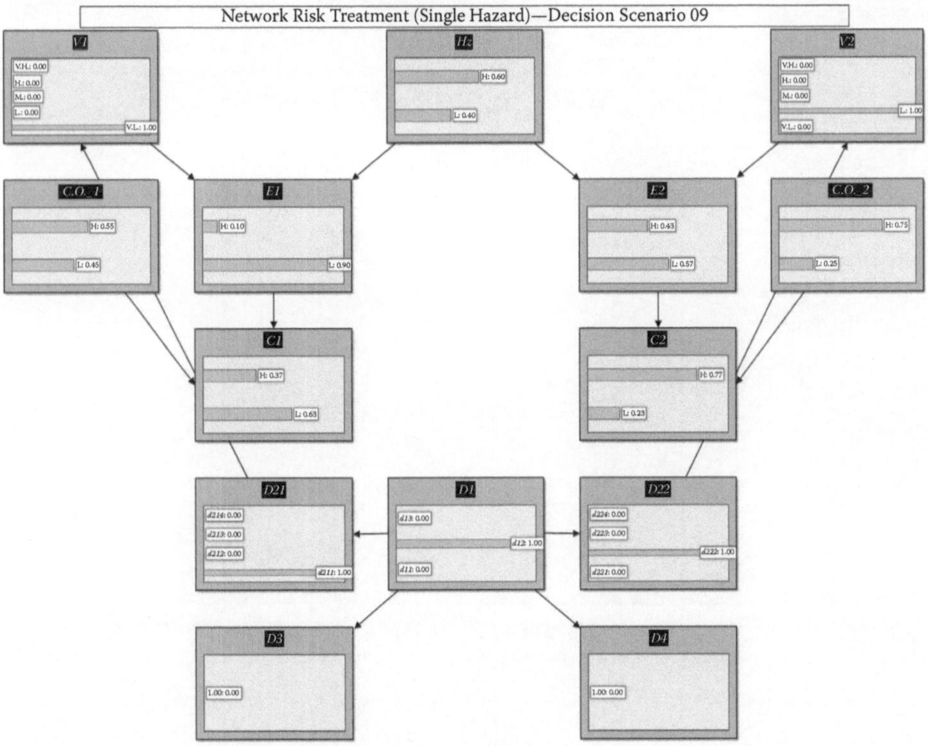

FIGURE 5.53 Marginal probabilities for design scenario 09.

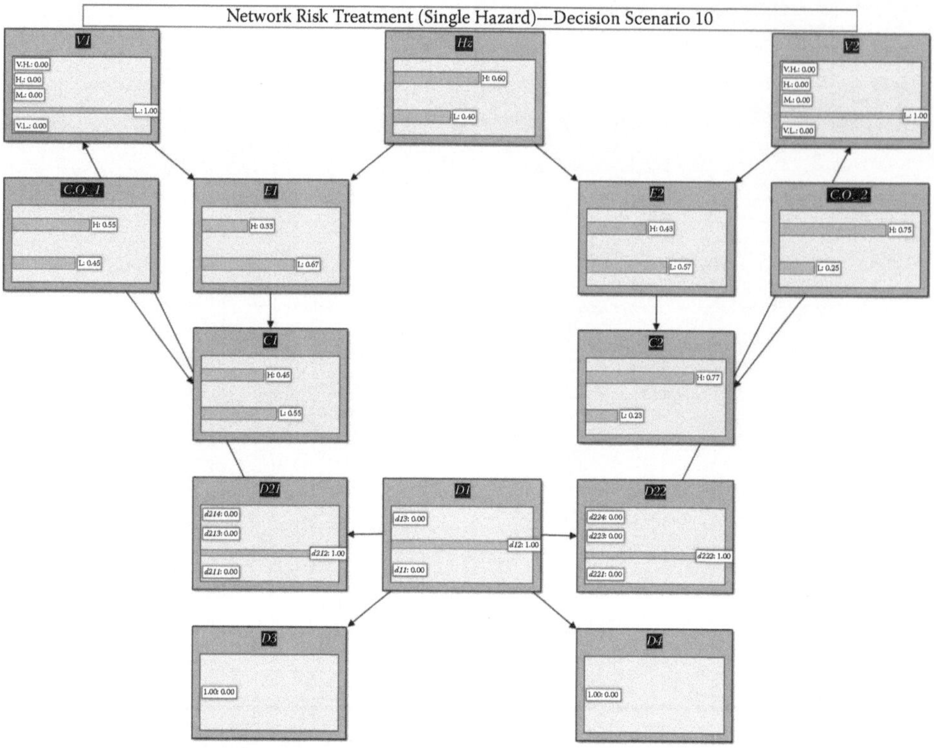

FIGURE 5.54 Marginal probabilities for design scenario 10.

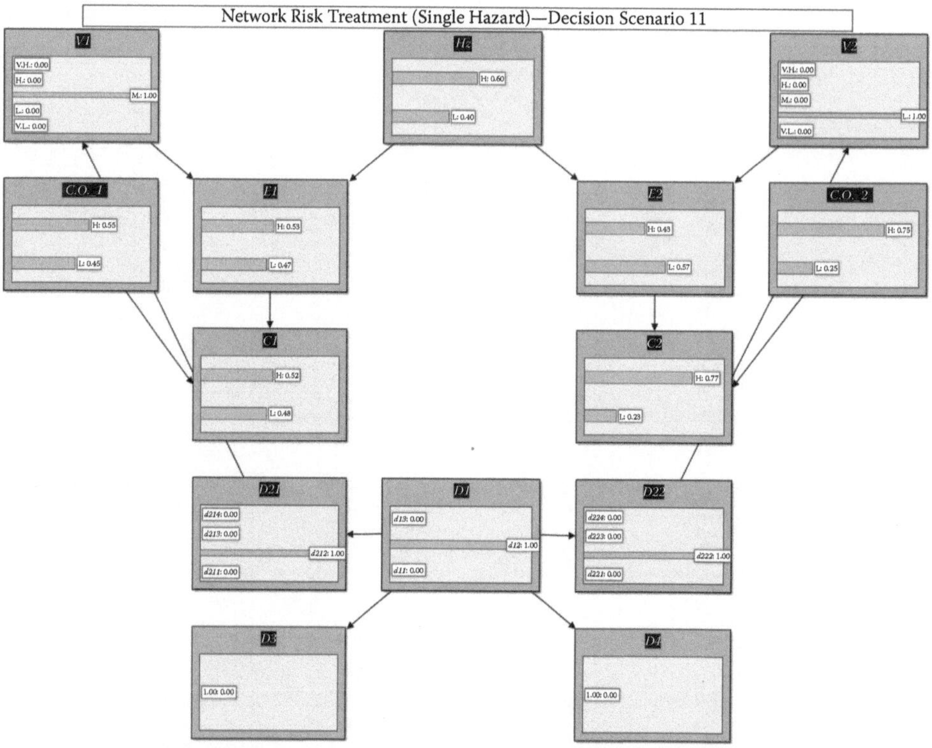

FIGURE 5.55 Marginal probabilities for design scenario 11.

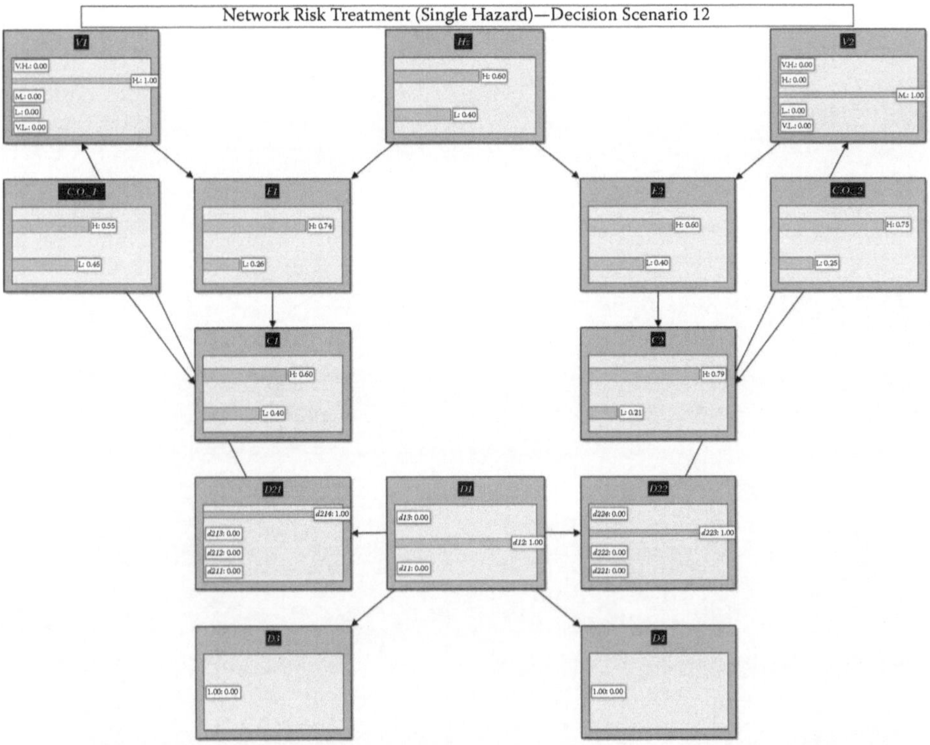

FIGURE 5.56 Marginal probabilities for design scenario 12.

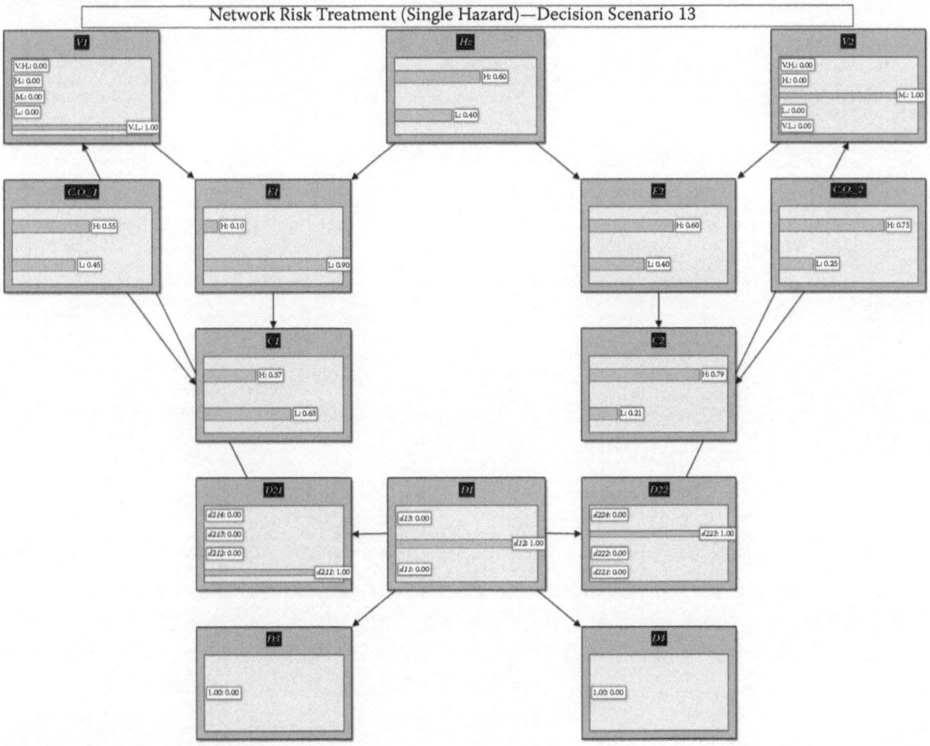

FIGURE 5.57 Marginal probabilities for design scenario 13.

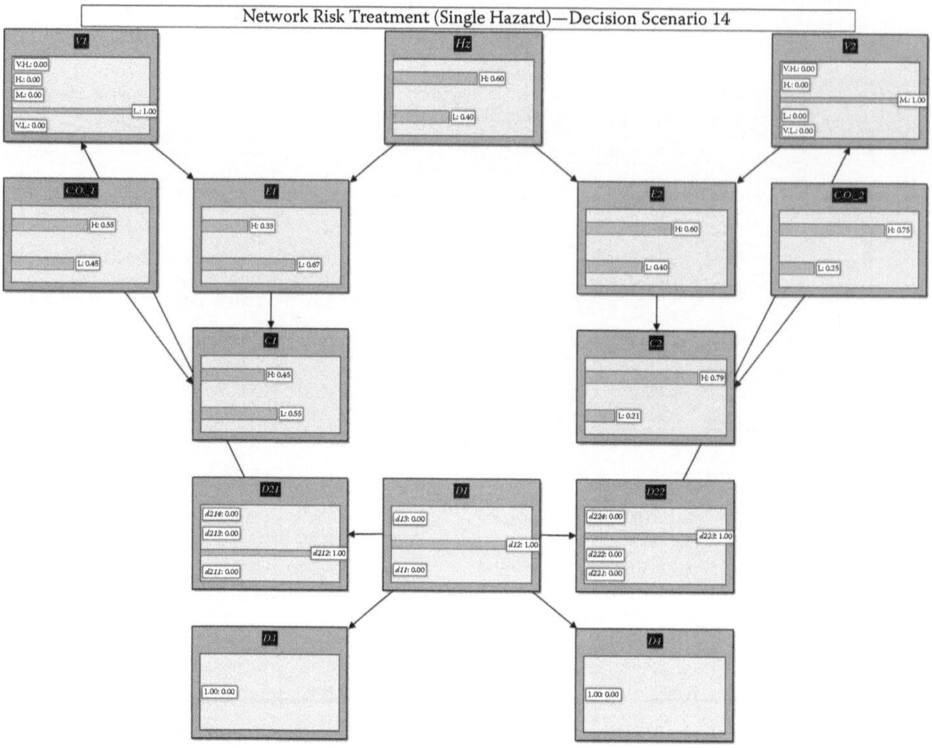

FIGURE 5.58 Marginal probabilities for design scenario 14.

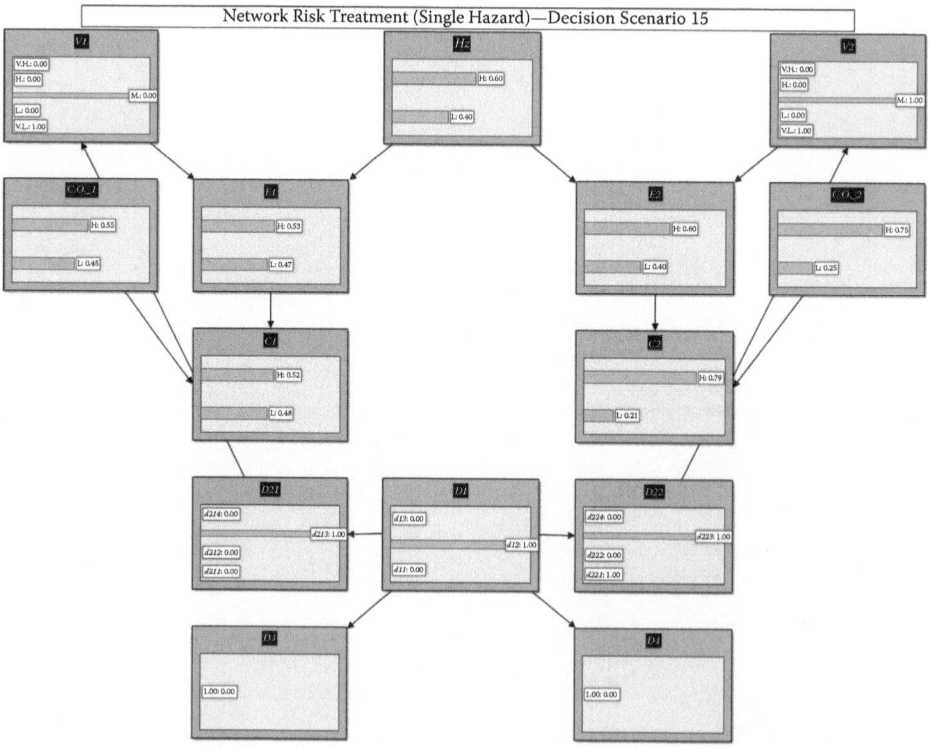

FIGURE 5.59 Marginal probabilities for decision scenario 15.

APPENDIX 5.VI: CPTs AND POTENTIALS FOR CASE STUDY 5.4 (TABLES 5.48–5.63)

TABLE 5.48
Potential ϕ (*Needs, CR, SDB*)

Bridge Needs		Low				Medium				Extensive			
CR		Good	Fair	Poor	Service	Good	Fair	Poor	Service	Good	Fair	Poor	Service
SDB	No	100	110	130	150	125	143	170	200	150	175	210	250
	Yes	150	160	180	200	163	185	220	250	175	210	260	300

TABLE 5.49
Potential ϕ (*Needs, Cap. Needs*)

Bridge Needs		Low	Medium	Extensive
Cap. Needs	None	100	60	50
	Some	50	75	150
	Extensive	10	100	200

TABLE 5.50
Potential ϕ (*Restrictions, Cap. Needs*)

Restrictions		High	Low
Cap. Needs	None	50	100
	Some	100	50
	Extensive	150	40

TABLE 5.51
Potential ϕ (*Cap. Needs, CR*)

CR		Good	Fair	Poor	Service
Cap. Needs	None	300	150	50	10
	Some	100	125	250	200
	Extensive	10	100	250	300

TABLE 5.52
Potential ϕ (*Vulnerability, Material, F.C.*)

Material		Robust		Med.		N. Robust	
F.C.		High	Low	High	Low	High	Low
Vulnerability	Low	30	60	20	30	10	20
	Medium	35	45	30	35	25	30
	High	40	30	50	40	60	40

TABLE 5.53
Potential ϕ (*Vulnerability, Design Type*)

Design Type		**Robust**	**Med.**	**N. Robust**
Vulnerability	Low	70	80	90
	Medium	55	65	75
	High	40	50	60

TABLE 5.54
Potential ϕ (*Vulnerability,* Other.)

Other		**Low**	**Medium**	**High**
Vulnerability	Low	70	10	1
	Medium	60	80	10
	High	50	70	100

TABLE 5.55
Potential ϕ (*Traffic, Importance*)

Traffic		**Low**	**Medium**	**High**
Importance	Low	100	10	1
	Medium	50	120	90
	High	20	100	150

TABLE 5.56
Potential ϕ (*Detour, Importance*)

Detour		**Short**	**Medium**	**Long**
Importance	Low	50	40	10
	Medium	45	80	70
	High	40	90	100

TABLE 5.57
Potential ϕ (*Type, Importance*)

Type		**Low Impact**	**Medium Impact**	**High Impact**
Importance	Low	100	70	10
	Medium	50	100	70
	High	10	50	100

TABLE 5.58
Potential ϕ (NCV, *Func. Class, Type*)

NHS+		No				Yes			
Func. Class		**Local**	**Collector**	**Arterial**	**Interstate**	**Local**	**Collector**	**Arterial**	**Interstate**
Type	Low impact	60	48	36	24	40	32	24	16
	Medium impact	43	42	40	42	67	70	72	82
	High impact	48	56	64	80	72	84	96	120

TABLE 5.59
Potential ϕ (*Func. Class, ADT*)

Func. Class		**Local**	**Collector**	**Arterial**	**Interstate**
ADT	Low	50	40	30	20
	Medium	55	60	65	70
	High	60	70	80	100

TABLE 5.60
Potential ϕ (*ADT, Trucks, Traffic*)

ADT		Low			Medium			High		
Trucks		**Low**	**Med.**	**High**	**Low**	**Med.**	**High**	**Low**	**Med.**	**High**
Traffic	Low	100	60	20	70	45	10	30	20	1
	Medium	70	60	50	60	100	55	50	60	70
	High	1	4	10	15	50	70	50	70	100

TABLE 5.61
CPT of C

Importance		Low			Medium			High		
Cap. Needs		None	Some	Extensive	None	Some	Extensive	None	Some	Extensive
C	Low C	0.8240	0.7128	0.5828	0.4637	0.2242	0.0585	0.0355	0.0003	0.0000
	Med. C	0.1726	0.2665	0.3491	0.4148	0.4626	0.4967	0.4292	0.1314	0.0117
	High C	0.0033	0.0207	0.0681	0.1215	0.3133	0.4448	0.5353	0.8683	0.9883
Histograms for conditional probabilities		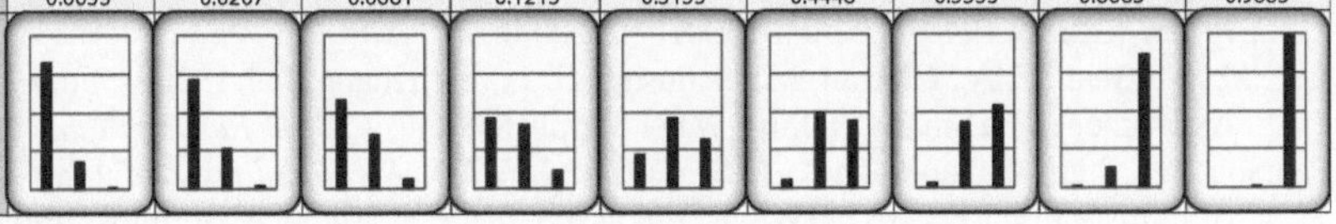								

TABLE 5.62
CPT of *E*

Vulnerability		Low			Medium			High		
Cap. Needs		None	Some	Extensive	None	Some	Extensive	None	Some	Extensive
E	L. Level	0.9199	0.6811	0.5636	0.4379	0.2407	0.0921	0.0207	0.0010	0.0000
	M. Level	0.0800	0.2927	0.3610	0.4026	0.5186	0.5527	0.3577	0.2048	0.0117
	H. Level	0.0001	0.0263	0.0754	0.1595	0.2407	0.3553	0.6216	0.7942	0.9883
Histograms for conditional probabilities										

TABLE 5.63
CPT of *Ri*

C		Low C			Med. C			High C		
E		L. Level	M. Level	H. Level	L. Level	M. Level	H. Level	L. Level	M. Level	H. Level
Ri	Low	0.7677	0.4562	0.3621	0.4562	0.1986	0.1071	0.4024	0.1071	0.0063
	Med	0.2260	0.4493	0.4525	0.4493	0.6028	0.4635	0.4399	0.4635	0.2260
	High	0.0063	0.0945	0.1854	0.0945	0.1986	0.4293	0.1576	0.4293	0.7677
Histograms for conditional probabilities										

REFERENCES

Ayyub, B. M. (2003). *Risk Analysis in Engineering and Economics, Risk Analysis in Engineering and Economics.*Chapman & Hall/CRC, Boca Raton, FL.

ASCE. (2013). 2013 Report Card for America's Infrastructure, *American Society of Civil Engineers*, Reston, VA.

Balas, C. E., and Ergin, A. (2002). Reliability-Based Risk Assessment in Coastal Projects: Case Study in Turkey. *J. Waterway Port. Coast. Ocean Eng.* 128(2), 52–61.

Davis-McDaniel, C., Chowdhury, M., Pang, W., and Dey, K. (2013). Fault-Tree Model for Risk Assessment of Bridge Failure: Case Study for Segmental Box Girder Bridges. *J. Infrastruct. Syst.* 19(3), 326–334.

DHS. (2011a). *Integrated Rapid Visual Screening of Mass Transit Stations*, Building and Infrastructure Protection Series. DHS, Washington, DC.

DHS. (2011b). *Integrated Rapid Visual Screening of Tunnels*, Building and Infrastructure Protection Series. DHS, Washington, DC.

DHS. (2011c). *Integrated Rapid Visual of Screening of Buildings*, Building and Infrastructure Protection Series. DHS, Washington, DC.

Ettouney, M., and Alampalli, S. (2012a). *Infrastructure Health in Civil Engineering: Theory and Components*, CRC Press, Boca Raton, FL.

FEMA. (2009). *Handbook for Rapid Visual Screening of Buildings to Evaluate Terrorism Risks*, Federal Emergency Management Agency (FEMA), Risk Management Series FEMA 455. FEMA, Washington, DC.

Fenton, N., and Neil, M. (2013). *Risk Assessment and Decision Analysis with Bayesian Networks.* CRC Press, Boca Raton, FL.

Grigg, N. S. (2013). Water Main Breaks: Risk Assessment and Investment Strategies. *J. Pipeline Syst. Eng. Pract.* 4, ISSN 1949-1190/04013001(7)/$25.00.

Lee, M., McBean, E. A., Ghazali, M., Schuster, C. J., and Huang, J. J. (2009). Fuzzy-Logic Modeling of Risk Assessment for a Small Drinking-Water Supply System. *J. Water Resour. Plann. Manage.* 135(6), 547–552.

Pearson, D., Stein, S., and Jones, J. S. (2002). *HYRISK Methodology and User Guide.* Federal Highway Administration Rep. No.FHWA-RD-02-XXX. Federal Highway Administration, Washington, DC.

Powderham, A. J., and Tamaro, G. J. (1995). Mansion House London: Risk Assessment and Protection. *J. Construct. Eng. Manage.* 121(3), 266–272.

Sadiq, R., Kleiner, Y., and Rajani, B. (2004). Aggregative Risk Analysis for Water Quality Failure in Distribution Networks. *J. Water Supply Res. Technol.* 53(4), 241–261.

Vose, D. (2009). *Risk Analysis a Quantitative Guide.* Wiley, Hoboken, NJ.

Yanmaz, A. M., and Apaydin, M. (2012). Bridge Scour Risk Assessment and Countermeasure Design. *J. Perform. Construct. Facil.* 26(4), 499–506.

Yanmaz, A. M., Caner, A., and Berk, A. (2007). Renovation of a Safety-Inspection Methodology for River Bridges. *J. Perform. Constr. Facil.* 21(5), 382–389.

6 Risk Monitoring

6.1 INTRODUCTION

Risk monitoring is perhaps one of the least discussed components of risk management. Structural inspection, nondestructive testing (NDT), and structural health monitoring (SHM) have been advocated and used for decades as a means of risk monitoring (Ettouney and Alampalli 2012a, 2012b). These methods have been used as the main means for monitoring capacity, demands, and/or exposure, and thus can be used to partially monitor risk. However, these three popular processes are not entirely sufficient or have not been used in a systematic fashion for proper and complete risk monitoring. This is partly because of their use and ability in monitoring only one of the needed parameters, that is, either capacity (or vulnerability), demands (or threats/hazards), or exposure. These processes are generally not well suited for monitoring consequences, which is needed for proper and accurate risk monitoring

6.1.1 Overview

Several researchers from various fields have explored risk monitoring. This section briefly describes some of these studies. Panu and Sharma (2002) explored consequence monitoring, which is an integral part of risk monitoring. According to them, the definition of agricultural drought hovers around soil moisture deficiency in relation to climatic factors (water supply) and crop water consumption and its impact on agricultural production. Mannocchi et al. (2009) contend that an objective definition of agricultural drought, for example, the critical level of imbalance that determines the critical impact on agricultural production has not yet been univocally defined. Therefore, agricultural drought is not clearly differentiated from the other phenomenon of water shortage, and consequently, it is not easy to identify or to quantify the corresponding risk. Hence, they proposed a new methodology for the risk analysis and economic impact assessment of agricultural droughts (ADERA) based on an original objective definition of *agricultural drought* that accounts for its peculiar characteristics. Authors applied this methodology to a rain-fed sunflower crop in the Papiano (central Italy) experimental plots for which the 26-year time series for yield and the corresponding phenological, climatic, and economic data were available. They conclude, based on a case study, that the methodology makes it possible to evaluate the intensity of the agricultural drought phenomenon and to evaluate whether the hazard will occur within or after the growing period and, therefore, to assess whether the vulnerability is low or high in terms of the amount of water eventually needed to mitigate the phenomena. The main limitation of this method is that it requires the experimental crop yield time series, which is typically unavailable.

Also, according to Batchelor and Valdes (1998), although quantitative risk assessments play an important role in developing remedial plans at sites contaminated with hazardous materials. Stochastic risk assessment techniques offer the potential for being more objective by explicitly describing the effects of site variability and uncertainty on determination of risk. Subjective judgment of the extent of safety required to prudently protect public health can be made by selecting a degree of assurance for setting risk reduction goals, that is, by selecting the point on the distribution of calculated risk to be used as a risk reduction goal. Authors developed such a model for a site contaminated with polychlorinated biphenyls by representing the parameters used in the risk assessment as probability distribution functions rather than single values.

As noted earlier, consequence monitoring is an integral part of any safety risk study. The U.S. Mine Safety and Health Administration (MSHA) reports provide evidence that construction operations associated with earthmoving and surface mining activities are generally associated with higher risk. According to MSHA (2009), heavy machinery caused more than 30% of the total fatalities in coal and metal/nonmetal mining from 2004 to 2008. Manual inspections and internal controls such as checklists are predominantly used for minimizing risk. These procedures are not only expensive and time consuming, but the success of these procedures depends on the competency of the personnel deployed (Ahmad and Gibb 2004). With recent advances, spatial-based technologies have been gaining attention to reduce these risk factors. Hence, Chi and Caldas (2012) investigated the use of an automated image-based safety assessment method for earthmoving and surface mining activities.

Based on a literature review, Chi and Caldas (2012) focused on three risk factors associated with operator errors: excessive operation speeds, dangerous access to the unsafe areas, and close proximity between objects such as heavy machinery and workers. Then they identified best practices and spatial data needs for a safety assessment in relation to these risk factors to investigate image-based data collection devices and algorithms for spatial safety assessment. The analyzed algorithms were designed to systematically collect and interpret safety-related data. Based on a case study, they concluded that the applied safety rules enabled automated violation detection.

Risk monitoring was also considered in construction management because risk management is essential and is used commonly in major construction operations. Zhou and Zhang (2011) noted that deep foundation pit engineering projects require risk assessment and management to ensure safety and quality. Based on their literature search, the authors concluded that most available present methods can only be used as either a qualitative or semiquantitative assessment and failed to take advantage of accumulated historical experiences and the fact that uncertain factors cannot be analyzed precisely. Therefore, Zhou and Zhang (2011) proposed a risk assessment method using a fuzzy comprehensive evaluation method based on Bayesian networks (BNs) for deep foundation pit construction to combine the superiority of BN techniques for processing uncertain factors and integrating historical information with the advantages of the fuzzy comprehensive evaluation method in handling uncertain risks. The BN technique was used to obtain the probabilities of risk events, and the fuzzy comprehensive evaluation method was used to calculate the risk value and determine membership functions for evaluating the risk ratings. For a case study, primary data analysis of a simple accident database from a deep foundation pit construction project in Shanghai was used to provide the probabilities of basic events and approximate distributions of risk consequences. Authors obtained the BN model through the transformation of a fault tree analysis. The probability of every risk event is calculated by using deductive BN techniques. Then the consequence of each event is calculated by using fuzzy analysis (i.e., statistical consequence distributions and weight coefficients of risk events are determined through the database). Risk precautions and control measures were suggested on the basis of the risk assessment results and were applied to risk management in deep foundation pit construction.

Sewer or wastewater pipes are a significant portion of wastewater systems forming the pathway between points of wastewater generation and treatment plants or outfalls. More than 1 million miles of these pipes are supposed to exist in the United States and require considerable resources to maintain them in operating condition. Given constrained resources, prioritization is required, and risk-based methodologies are gaining attention. Salman and Salem (2012) note that even though identifying probability of failure for sewer pipes through deterioration models has attracted the attention of several researchers, determining the consequences of failure and incorporating this information to prioritize sewers has not been fully examined. Hence, researchers attempted to generate a risk assessment tool to prioritize sewer pipes considering consequences-of-failure values using a Weighted Scoring System. Authors used the data sets from the Metropolitan Sewer District of Greater Cincinnati (MSDGC) that serves approximately 230,000 residential and commercial users by managing a sewer network that has an approximate total length of 3,000 miles. Each pipe was evaluated with respect to a set of factors and a weighted sum of the performance of the pipe was calculated. ArcGIS software was used to obtain the required information for each factor and to geographically represent the consequence of failure values for the overall sewer network. Thus, one limitation of the proposed methodology is the lack of high quality data sets that can be imported into a GIS software package. Three different risk assessment prioritization methods were used: multiplication of probability and consequences of failure, risk matrices, and fuzzy inference systems. Authors report that using fuzzy inference systems resulted in a more flexible method to combine probability and consequences of failure because it allowed modeling of the fuzziness around boundary values for different levels of probability. One interesting aspect researchers identified is the lack of data sets that include other utility lines such as water pipes, power lines, and gas pipes in the proximity of sewer pipes to consider interaction in estimating consequences of failure.

6.1.2 Objectives of Risk Monitoring

6.1.2.1 Monitoring Risk Components

We discussed the need for risk monitoring in the framework of a risk management process earlier in Section 1.4. Given most of the risk components are time dependent, we argued that risk assessment is a snapshot in time and, hence, requires continuous or periodic evaluation as conditions affecting the assessment change. In addition, as noted in Chapter 4 that discussed risk acceptance, risk acceptance thresholds can be time dependent and need periodic evaluation as conditions change. Similarly state-of-the-practice in technology and communications are also constantly evolving, and, thus, solutions not available today may be feasible in the future. Cost-effectiveness of current methodologies also significantly change with advancing technologies. In short, because of the time-dependent nature of risk and its components, all risk components, risk thresholds, risk treatment techniques, and risk communication require periodic or constant evaluation based on changes affecting the analyzed situation.

6.1.2.2 Monitoring Temporal Behavior of Risk: Trending and Forecasting

To accomplish periodic risk monitoring, we need adequate information about past and future developments. This can be accomplished through understanding how risk and risk components behaved in the past as a function of time and how they are most likely to behave in the future as a function of time. In this book, we call this time-dependent behavior of risk and its components collectively as risk monitoring. Specifically, effective risk monitoring should enable us to follow trends of risk and risk components over time past and forecast them to aid in decision-making processes. We follow the definitions for trending (past performance) and forecasting (future performance) given in Chapter 2.

We note that the above two objectives are linked together with risk acceptance, risk treatment, and risk communications efforts as shown in Figure 6.1.

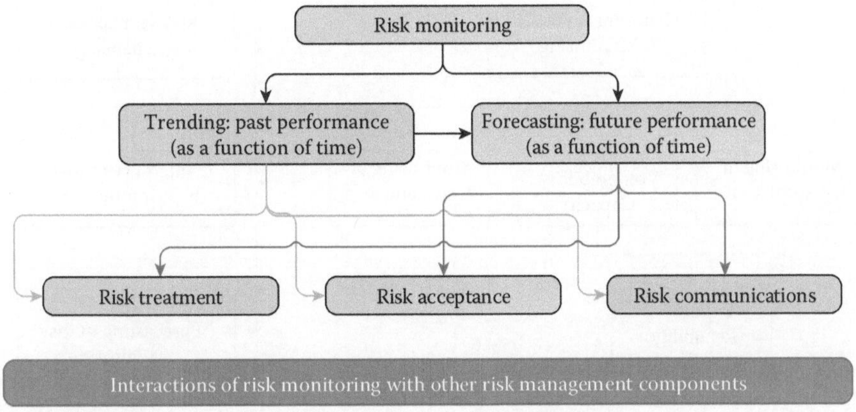

FIGURE 6.1 Risk monitoring and other components of risk management.

6.1.2.3 Other Risk Monitoring Objectives

Other important risk monitoring objectives improve overall assessment from a practice point-of-view and include the following:

- Validation of modeling assumptions
- Subjective evaluations of conditional probabilities tables (CPTs) and/or potentials (see Chapter 2)
- Improving overall accuracy of subjective or objective risk-based decisions
- Helping communications with stakeholders (see Chapter 7)

6.1.3 This Chapter

This chapter covers some of the salient aspects of risk monitoring. We discuss briefly the monitoring attributes of risk components, with special attention to consequence monitoring as it is the least studied and most overlooked component. We then offer three practical case studies to illustrate various aspects of risk monitoring. We review monitoring risk versus monitoring exposure (or reliability) first and then show the importance of using inference in risk monitoring practice. Finally, we illustrate how to forecast risk with time using an application of dynamic graph network (DGN). Figure 6.2 shows the layout of this chapter.

6.2 COMPONENTS OF RISK MONITORING

6.2.1 Overview

Risk monitoring is a fairly complex issue because of the diversified aspects it encompasses. It can be looked at from three different viewpoints:

- *Events of interest*: Events of interest can be man-made or natural. We refer to them in this book, without any loss of accuracy, as demands, hazards, or threats. The temporal aspect and magnitude of these events vary greatly; that is, these events can be almost instantaneous (such as a bomb blast), slow (such as cumulative fatigue or corrosion, see Figure 6.3), or extremely slow (such as climate change effects).

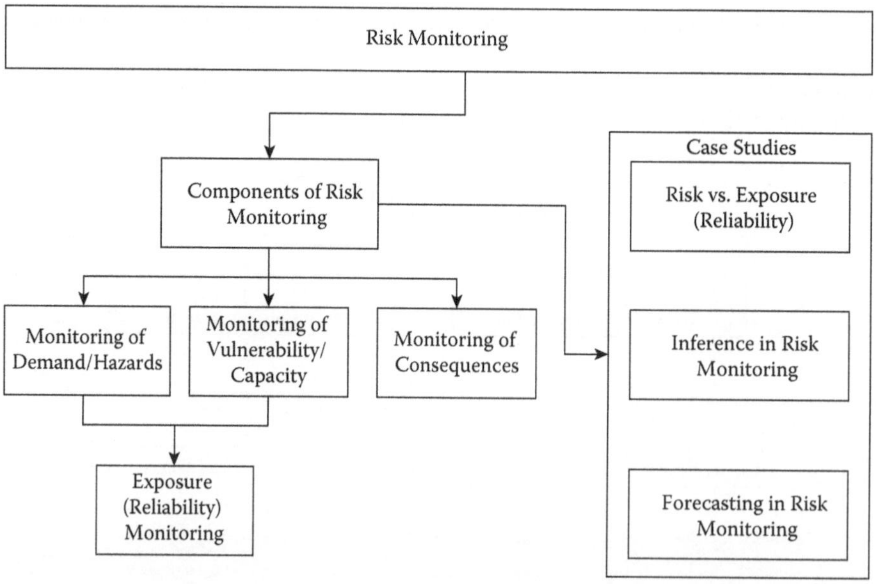

FIGURE 6.2 Layout of this chapter.

- *Risk components*: Risk has three distinct components (see Chapter 1): hazards, vulnerability, and consequences. Effects of hazards and vulnerability are combined in some occasions, when convenient. Such combined effect is referred to as exposure, reliability, or likelihood.
- *Monitoring*: The methods of monitoring (measurements) are extensive and depend on the nature of events as well as the risk component under consideration. As stated earlier, visual inspection (*V.I.*), structural health monitoring (*SHM*), and NDT are often used as a framework for risk monitoring (see Figure 6.4). *SHM* or NDT techniques are popular monitoring methods for demands and/or capacity, whereas *V.I.* is popular for exposure measurement. *V.I.* is generally subjective, and *SHM* and NDT can be both objective and subjective based how they are executed (see Alampalli and Ettouney 2012a, 2012b).

Figure 6.5 illustrates the interrelationships between some of the above viewpoints of risk monitoring. Many authors, including Ettouney and Alampalli (2012a, 2012b), explored monitoring methods for capacity and demands. Similarly, other authors explored monitoring and observing events (hazards/threats). For the remainder of this section, we discuss briefly risk monitoring from a risk components viewpoint, with some detailed discussion of consequence monitoring.

FIGURE 6.3 Onset of corrosion in bridge floor system.

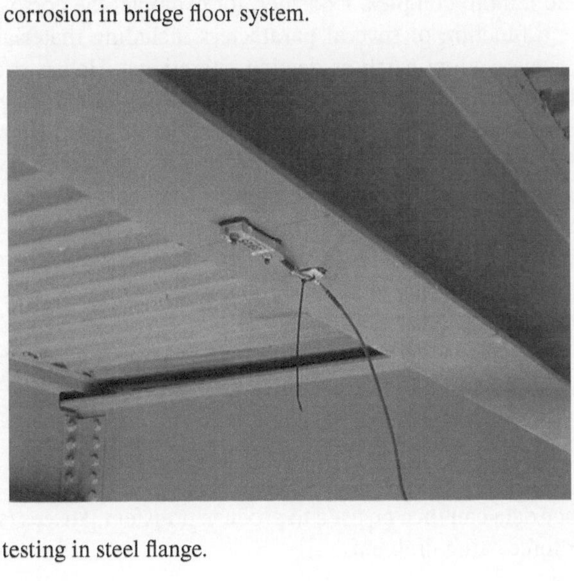

FIGURE 6.4 Fatigue testing in steel flange.

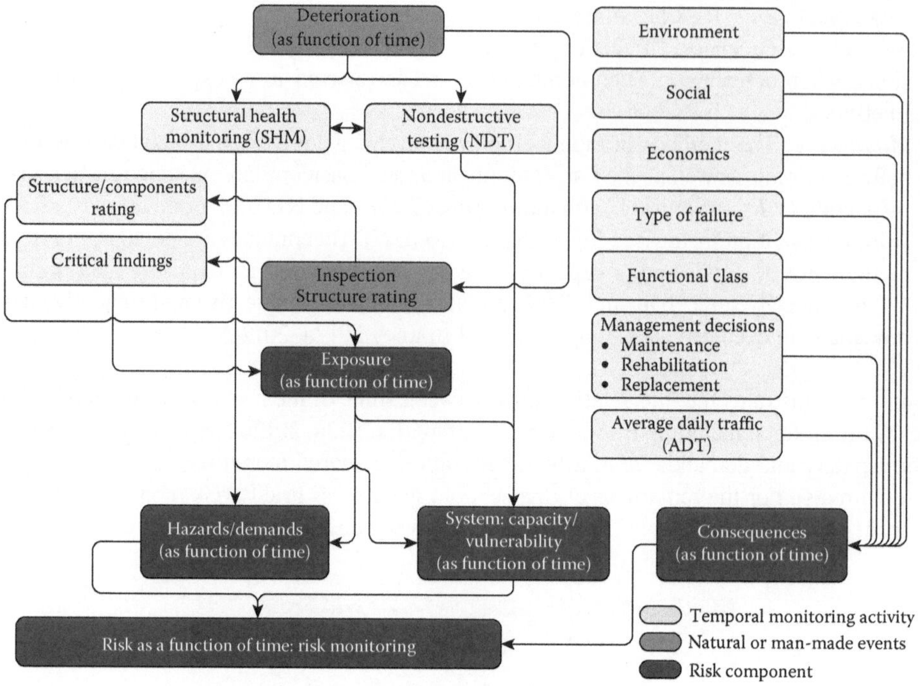

FIGURE 6.5 Components of risk monitoring.

6.2.2 Hazards/Demand Monitoring

Demand monitoring, besides exposure monitoring, is perhaps the most practiced of all of the risk components, and this is attributed to its simplicity. Almost all types of demand can be monitored, either directly or indirectly. The monitoring can be subjective or objective. We summarize some of the hazards and their monitoring attributes in Table 6.1.

Monitoring demands with time can be complex and requires a thorough understanding of its behavior, influencing factors, and its effects on the structure that is being assessed. A very slow rate of change can make it more complex. Consider, for example, the corrosion demand. It is well know that corrosion is a function of several parameters including material type, environmental factors, and practices owners adopt (such as deicing operations). Hence, a detailed model with a good understanding of all influencing factors is needed for proper monitoring of corrosion demand. An example is shown in Figure 6.6. We note that this model might be computationally demanding. Depending on the procedure adopted, influencing factors can be grouped (see Figure 6.7) and produce a simpler model. At the least, a simple corrosion model, Figure 6.8, may even produce adequate results. Either of the three models can be used by observing the levels of one (or more) of the variables and inferring the probabilities of the remaining variables. The CPTs that governs the relationships between different variables can be estimated either subjectively or objectively following the methods illustrated in the Chapter 2.

We note that monitoring demands produce information regarding the behavior of the hazard variable of interest, H, as a function of time such that

$$H = H(t) \tag{6.1}$$

This time-dependence data can then be used to produce transfer matrices (see Chapter 2) that can be used for trending or forecasting problems.

TABLE 6.1
Monitoring Demands (Hazards)

Hazard (Demand)	Monitoring Method
Live load on bridges	• Direct monitoring of loads using weigh-in-motion systems • Indirect monitoring via structural response, e.g., strain or acceleration monitoring
Deterioration (corrosion)	• Direct corrosion rate monitoring • Indirect monitoring via measuring (or estimating) section losses
Collision (bridges)	• Indirect via monitoring structural response in real time (mostly accelerations) • Indirect via inference based on postevent observations
Earthquakes	• Direct monitoring of ground motions and/or structural response in real time • Indirect monitoring after event observation
Bomb blast	• Indirect monitoring via air pressure sensors away from the event • Indirect inference after event
Temperature	• Direct monitoring in real time is a possibility • Indirect monitoring using weather station data nearby, regional weather centers, etc.

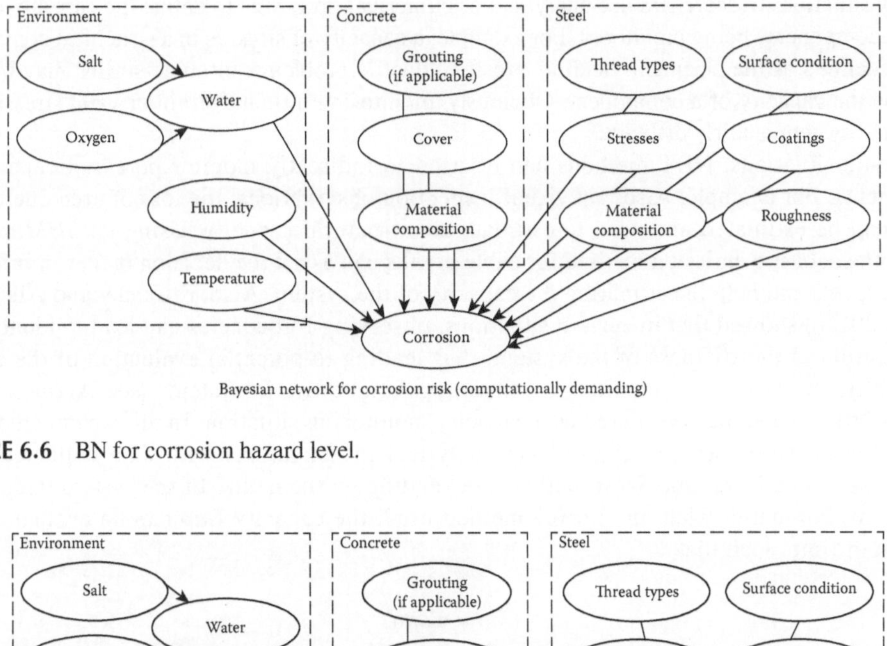

FIGURE 6.6 BN for corrosion hazard level.

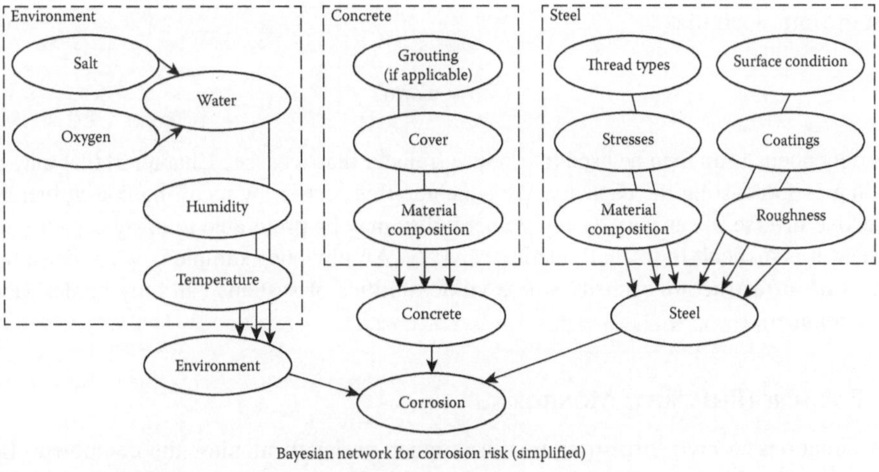

FIGURE 6.7 Simplified BN for corrosion hazard level.

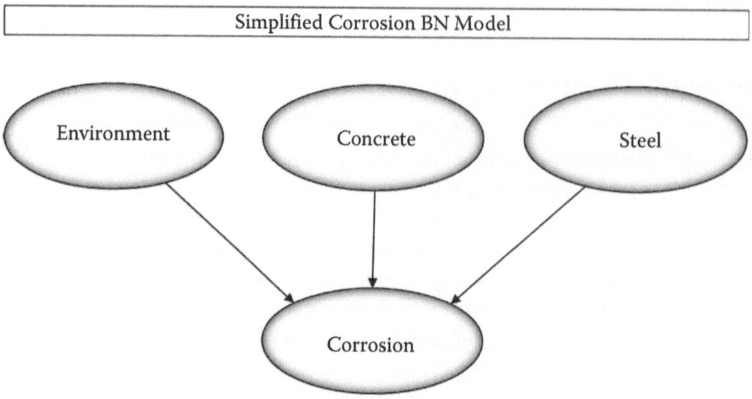

FIGURE 6.8 BN of corrosion hazard level.

6.2.3 Capacity Monitoring

In many situations, researchers use analytical or numerical processes to define the capacity of systems or components being monitored. For example, a capacity of stresses in a structural component may be defined as the pertinent yielding stress. In ductile problems, ultimate strains may be used to define the capacity of a component. Obviously, monitoring ultimate strain or yield stresses in a functional system is fairly difficult.

In some situations, NDT methods can be used to indirectly monitor potential capacity of components. For example, based on visual inspection observations, the loss of area due to corrosion can be estimated and used to estimate the structural capacity. Using an *SHM* method such as the acoustic emission method for cable breaks can assist the decision maker in inferring damage levels and help in estimating the capacity of the system. Also, Ettouney and Alampalli (2012a, 2012b) showed that in certain situations, observing frequencies can lead to identifying the reduction in the stiffness of the system, thus leading to potential evaluation of the capacity. Finally, the process of monitoring remaining fatigue life in systems (see Alampalli and Lund 2006) can also be considered as a capacity monitoring situation. In all, we conclude that monitoring capacity can be achieved via analytical investigations with data collected from visual inspection, NDT, and *SHM* methods, depending on the nature of the system under considerations. No matter what monitoring method used, the capacity *Cap* can be evaluated as a function of time, such that

$$Cap = Cap(t) \tag{6.2}$$

Time dependence can then be used to produce transfer matrices (see Chapter 2) that may be used for trending or forecasting the capacity. We conclude this section by recalling that vulnerability is the subjective inverse of capacity. Using vulnerability may be preferable to using capacity in many cases where it is difficult to formally define capacity. An obvious example is when defining many parameters of infrastructure security where vulnerabilities of systems can only be described in a subjective manner.

6.2.4 Exposure (Reliability) Monitoring

In many situations in civil infrastructures, it is much easier to monitor the combined effects of hazards and vulnerabilities. Recall that we defined this combined effect as *Exposure*. Perhaps the most popular exposure monitoring activity is observing condition ratings using visual inspection.

Monitoring in this case is a subjective monitoring that sometime may be improved by using measurement tools, *SHM*, and NDT.

Reliability is another popular metric that relates capacity and demands. Monitoring reliability is generally objective (assuming that both capacity and demands are estimated objectively). In general, exposure is the subjective inverse of reliability.

Similar to capacity and demands, it might be of interest to evaluate exposure (or reliability) as a function of time such that

$$Exposure = Exposure(t) \tag{6.3}$$

6.2.4.1 Exposure Monitoring at Lower Limit States

An important, yet largely ignored issue in the community as of the writing of this book is the subject of lower limit states (LLS), which we discussed earlier in Chapters 3 and 4. We note that traditionally the most widely used limit state in civil infrastructures is the elastic limit state. In general, elastic limit state is synonymous with failure limit state. However, for safety assurance and cost-effective management, most decisions are made at limit states below elastic limit or failure (recall the corrosion example). Unfortunately, all traditional structural analysis methods are incapable of addressing or defining those LLS and, hence, monitoring processes are needed to identify them.

Though the rest of the discussion will be focused on bridge structures, it is applicable to most civil infrastructures. In general, there are three generic monitoring methods in bridge engineering. The three methods are *V.I.*, *SHM*, and NDT. For simplicity and without loss of generality, we combine *SHM* and NDT and refer it as *SHM*. The primary question is can either or both of these generic monitoring techniques identify LLS?

SHM allows for a wealth of measurements/responses but there is no objectively or subjectively known scale that would clearly identify using *SHM*. *SHM* generally can only identify failure limit state, a state that is near failure. Such knowledge is of course extremely beneficial. Unfortunately, most bridge management decisions occur at LLS (corrosion, fatigue, painting, deck rehabilitations, component replacements, etc.). Simply stated, bridge managers do not wait for a bridge or its component to fail or approach failure to make a decision.

This leaves us with *V.I.*, which we have observed is a form of exposure monitoring. Note that different LLS will be used in the case studies later in this chapter, and they all are below the elastic limit of the system under considerations. *V.I.* output generally is the rating of a bridge or one of its components on a discrete scale. The scale varies depending on the methods used by owners. For example, NYSDOT (2014) uses a 1–7 scale, whereas the Federal Highway Administration (1995) uses 0–9 scale. We also note that most of these steps are below failure states, that is, LLS. The *V.I.* practice has been around for many years and the resulting ratings have proved to be fairly accurate and beneficial, and bridge managers use these ratings to reach many of their decisions. This infers an important finding—*V.I.* condition ratings essentially are the sought-after LLS. Moreover, the acceptance of these condition ratings, from a well-established practice, represents the corresponding acceptance thresholds. We now have to establish a process to relate *V.I.* condition ratings with risk ratings to properly place *V.I.* in a true risk-based management system. Let us revisit now the value of using *SHM*, or remote sensing, in LLS. If a well-designed *SHM*/remote sensing effort is performed in LLS in coordination with *V.I.* efforts, it is expected that the net value of the process will be increased accuracy, as can be witnessed from the illustration shown in Figure 6.9.

6.2.5 Consequences Monitoring

6.2.5.1 Overview

Consequence monitoring is another complex issue within the risk monitoring subject. It is not well studied even though it is an essential component. The complexity of consequence monitoring is

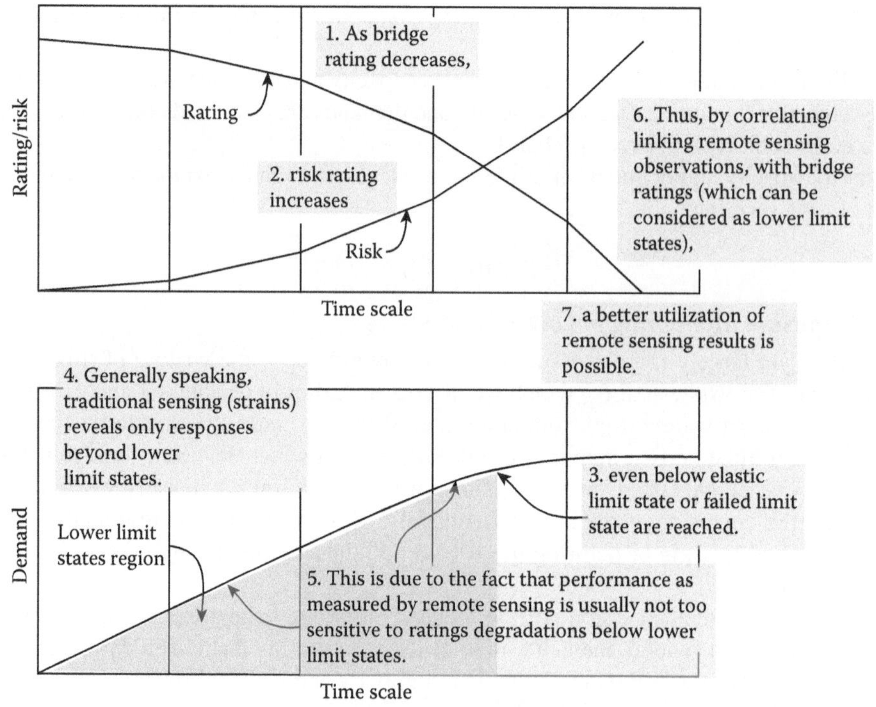

FIGURE 6.9 Lower limit states in relationship with remote sensing.

TABLE 6.2
Types and Potential Sources of Consequences

Type of Consequences	Examples
System Related	Parameters such as direct costs of failure, cost of retrofit, management costs, type of failures, functional class, traffic (for bridges or tunnels), number of occupants (for buildings), and/or operational costs
Outside System Realm	Social/economics factors such as location, indirect cost of failure, symbolic value, environmental costs, costs of detours, partial closings, etc.

due to the many facets of consequences as shown in Table 6.2. In addition, consequences can also be either subjective (relative) or objective (absolute). Sometimes consequences can be expressed monetarily.

An example of a BN for computing consequences is shown in Figure 6.10. Note how parameters affecting consequences can also interact with each other. The definitions of these variables are given in Appendix 6.I (Table 6.10).

An even more involved consequence-monitoring model is shown in Figure 6.11 for illustration purposes. Note that *Functional Class* does not directly affect *Consequences* as in the previous model, but its effects are now indirect through other layers. An intermediate variable, *Social-Economics*, is also used to gather all social and economic factors into a single variable. This modeling technique helps simplify the CPTs and compartmentalize the modeling efforts, thus reducing the computational requirements of the model.

Given that consequences can vary with time, we use the model of Figure 6.10 as template for a dynamic Bayesian network (DBN) model as shown in Figure 6.12. We used the traffic information,

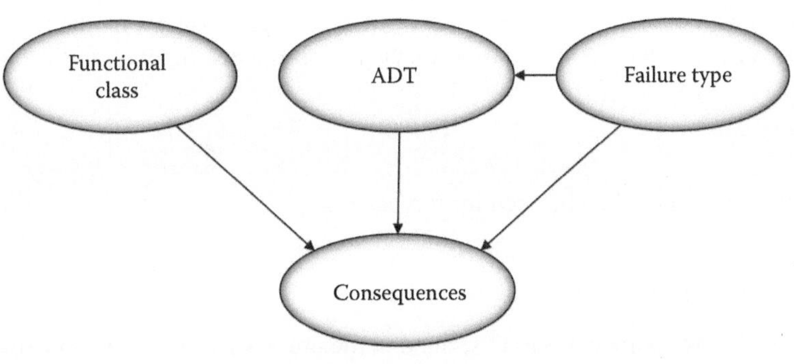

FIGURE 6.10 Simple BN model for estimating consequences.

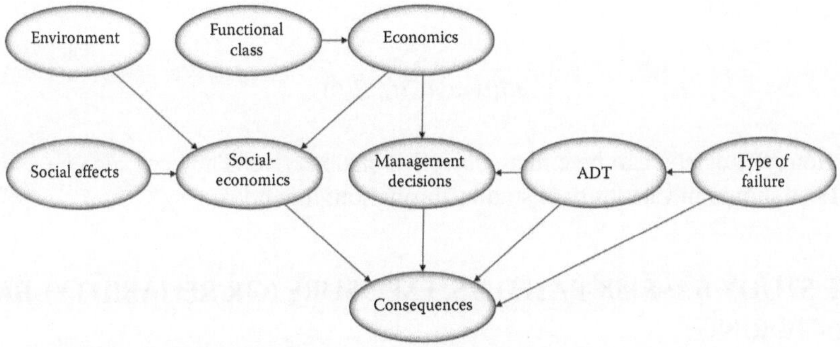

FIGURE 6.11 Advanced BN model for modeling consequences.

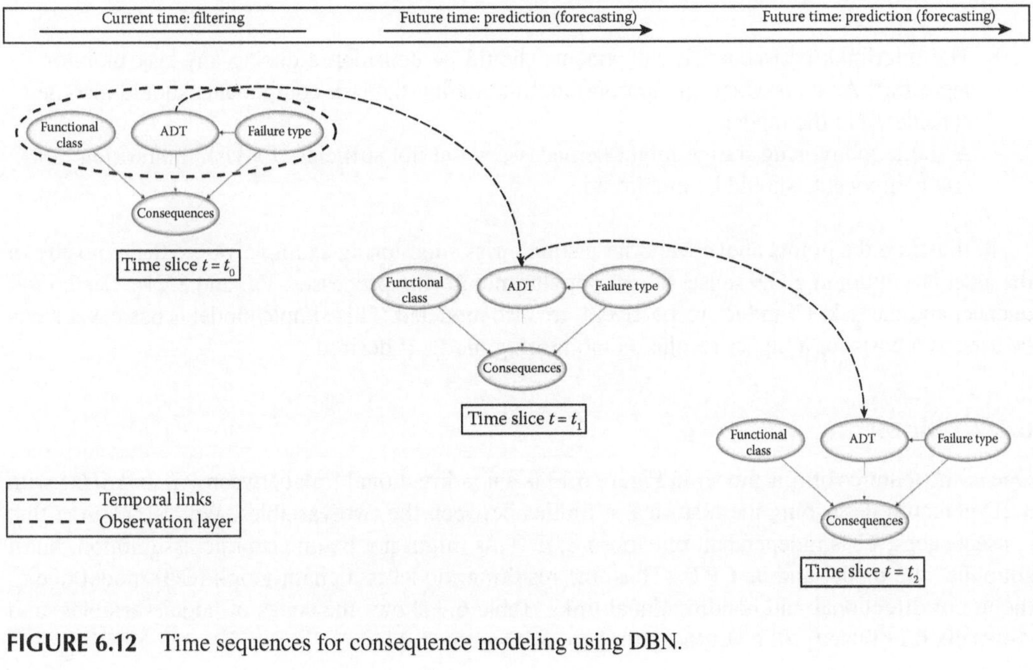

FIGURE 6.12 Time sequences for consequence modeling using DBN.

average daily traffic (ADT), as the interface between time slices. Using a model similar to the one shown in Figure 6.12, we can establish a time varying consequence function in the form of:

$$C = C(t) \tag{6.4}$$

The time dependence can then be used for forecasting.

6.2.6 Putting It All Together: Risk Monitoring

After monitoring all risk components as described in the previous sections, we can now monitor the risk, *Ri*, given the functional relationships introduced in Chapter 2, which can be displayed in its time-dependence form as:

$$Ri(t) = f(C(t), H(t), V(t)) \tag{6.5}$$

or

$$Ri(t) = g(C(t), E(t)) \tag{6.6}$$

The functions *f*( ) or *g*( ) can be either subjective or objective. They can also be estimated as a form of CPTs as shown in various case studies throughout this book.

6.3 CASE STUDY 6.1: RISK-BASED VS. EXPOSURE (OR RELIABILITY)-BASED MONITORING

6.3.1 Overview

We argued that risk associated with a civil infrastructure asset is a time varying entity as most of its underlying components change with time. Given the essential interlinking between these components, we can make two important observations:

- The interlinks between risk components should be considered during any risk monitoring effort. As a corollary, an appropriate link among the risk components should also be considered in the model.
- A single monitoring source might be necessary but not sufficient for risk monitoring: all risk components should be monitored.

To illustrate the points above, we offer a simple risk monitoring example where the exposure of the asset is a child, in a GN sense, of two popular monitoring processes: *V.I.* and *SHM*. The consequences and the risk, *C*, and *Ri*, respectively, are also modeled. This simple model is basic, yet it can be used as a basis for a higher resolution monitoring model if desired.

6.3.2 Model

One of the features of this model in Figure 6.13 is a nondirectional link between *V.I.* and *SHM* with a 2D potential describing the historical affinities between the two variables. We also assume that consequences, *C*, is independent of exposure, *E*. This might not be an accurate assumption, but it simplifies the model and its CPTs. Thus, the resulting model is a chain graph (CG) model due to the use of directional and nondirectional links. Table 6.3 shows the states of nodal variables, and Appendix 6.I (Table 6.10) gives the definitions of those variables.

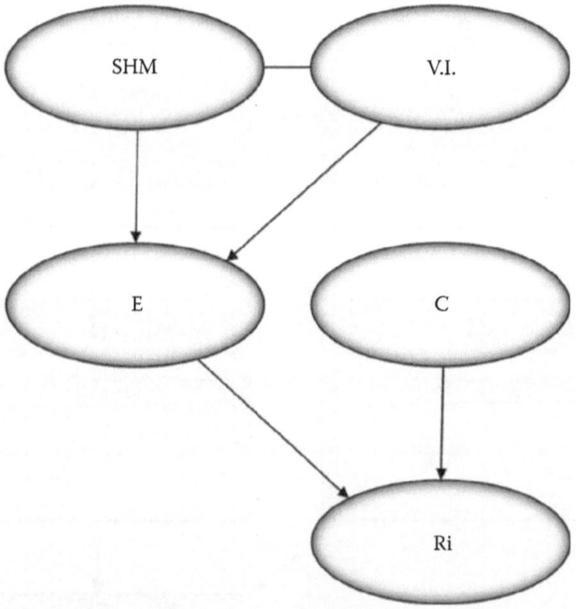

FIGURE 6.13 Risk monitoring using both *V.I.* and *SHM:* chain graph model.

TABLE 6.3
States of Nodal Variables

Variable	# of States	States
SHM	2	*Poor, Good*
V.I.	2	*Poor, Good*
E	2	*High, Low*
C	2	*High, Low*
Ri	2	*High, Low*

6.3.3 CPTs and Potentials

The CPTs and potentials of each variable of the network of Figure 6.13 are shown in Appendix 6.II. Because this is a CG model, it includes a mix of CPTs (Tables 6.12 through 6.14) and potentials (Table 6.11). As usual, we used our best judgment in filling the different tables in a realistic manner.

6.3.4 Results

We start this study by computing the historical (prior) probabilities of the variables (see Figure 6.14) that indicate the asset of interest having an above average exposure, yet given its low historical consequences, it shows the resulting risk to be about average (neither high nor low). Let us assume that an *SHM* experiment, with no equivalent *V.I.* observation, resulted in a *Poor* observation. The resulting network variables are computed and are shown in Figure 6.15. The exposure is high now with risk leaning toward the *High* state. With an equivalent and supporting *V.I.* observation of *Poor*, as shown in Figure 6.16, the exposure, *E*, becomes as expected very high. It is of interest to note that the risk has increased but not considerably. This can be attributed to the consequences remaining

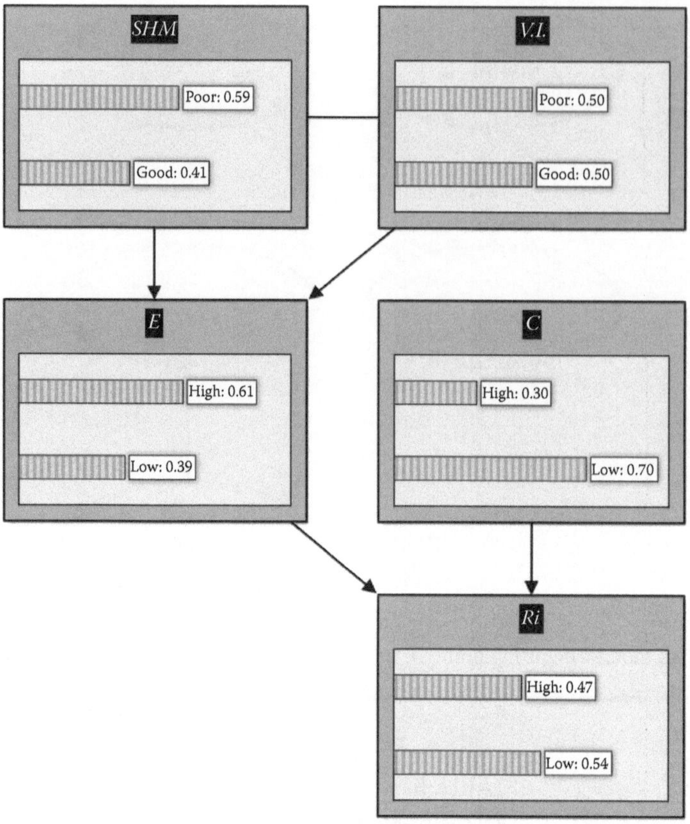

FIGURE 6.14 Risk monitoring historical (prior) observations: chain graph model.

low and, thus, moderated the effects of fairly high exposure on the overall risk performance. This simple monitoring example shows the power of GN modeling and its value. It is also another demonstration of the essential difference between exposure-based (or reliability-based) management and risk-based management. In exposure-based management, this particular asset would have been fairly high on the retrofit list because of the high exposure state. In risk-based management, this asset would not be as high on the retrofit list, because of its average risk rating attributable to its low consequence rating.

6.4 CASE STUDY 6.2: INFERENCE IN RISK MONITORING

6.4.1 Overview

In Case Study 6.1, we made the implicit assumption that the observations during monitoring efforts (whether *V.I.* or *SHM*) are synonymous with exposure, *E*. Such an assumption is not exactly correct because the actual exposure (or reliability) as a representation of the relationship between capacity and demands is not the same as what is being monitored. There are numerous reasons for this inequality, including inherent subjectivity and limitations associated with all the monitoring methods such as *V.I.*, *SHM*, and *NDT* (Ettouney and Alampalli 2012a, 2012b). The major limitation of methods such as *SHM* or *NDT* is that, in most situations, they concentrate on either capacity or demand monitoring, not exposure (or reliability) monitoring. In most cases, the missing components are either assumed or computed.

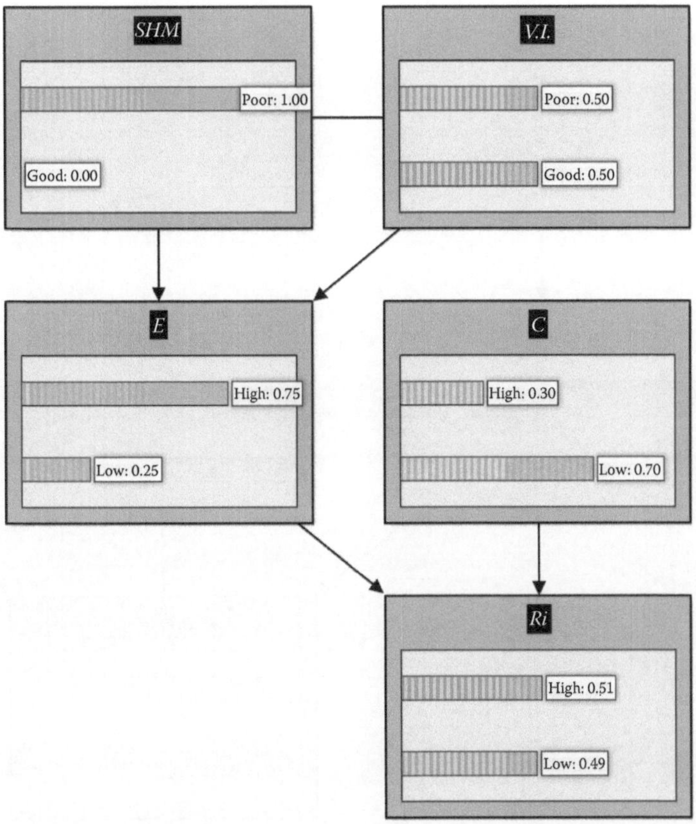

FIGURE 6.15 Risk monitoring with *SHM* observe state of *Poor*: chain graph model.

To accommodate the distinction between monitoring results and the actual state of exposure, these two variables need to be modeled separately in the GN. We can then introduce the monitored results as observations (evidence) in the network and use the inference processes of graph networks (GN) to estimate the actual exposure. Next, we provide an example to illustrate this process.

6.4.2 Model

We model the actual level of damage by the variable *Corr-Fat* as in Figure 6.17 graph network. Because this variable is hidden, that is, it cannot be estimated directly, its effect can be observed through visual inspection activity that is modeled as another variable, *V.I.* Both of these variables can be used to estimate the actual exposure, *E*. Also, we assume that the mitigation level is a function of the *V.I.* observation, see Figure 6.18. This is because mitigation is only performed based on the observation *V.I.*, not the actual state of *Corr-Fat*. The relationships between other model variables are similar to the case studies we offered earlier in the book. The states of the nodal variables are given in Table 6.4, and the definitions of the variables are given in 6.I Table 6.10.

6.4.3 Conditional Probabilities Tables

The different CPTs of each variable of the network of Figure 6.17 are shown in Appendix 6.III. Expert judgment or historic data are needed to complete these CPTs in a realistic manner.

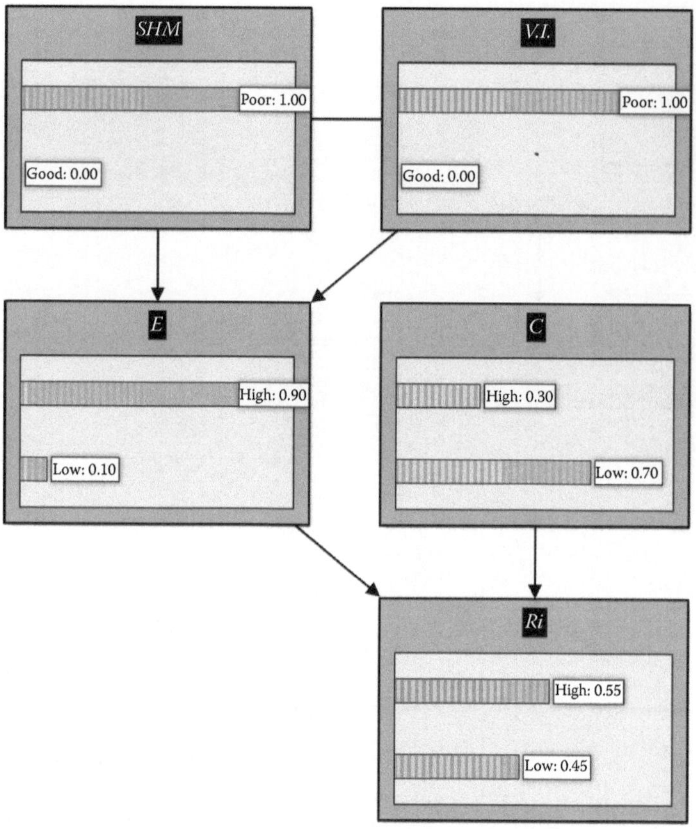

FIGURE 6.16 Risk monitoring with both *SHM* and *V.I.* observed state of *Poor*: chain graph model.

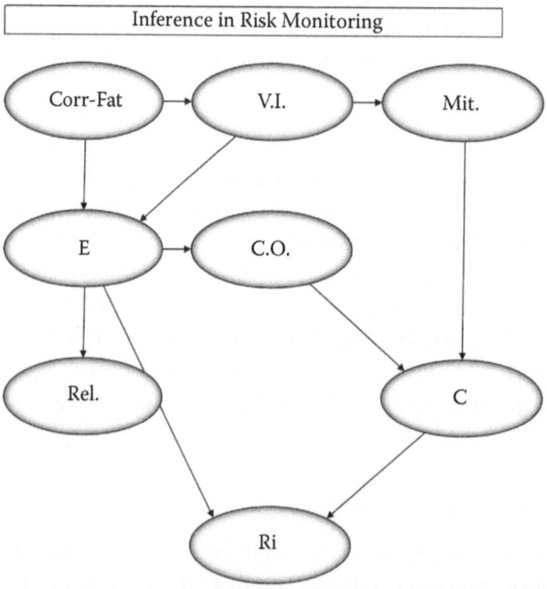

FIGURE 6.17 BN for inference and estimating actual corrosion-fatigue damage, reliability, and risk.

FIGURE 6.18 An engineer under a bridge for inspection.

TABLE 6.4
States of Nodal Variables

Variable	# of States	States
Corr_Fat	2	*Damaged, Pristine*
V.I.	2	*Damaged, Pristine*
Mit.	2	*Not Adequate, Adequate*
E	3	*High (H.) Level, Medium (M.) Level, Low (L.) Level*
C.O.	2	*High Con., Low Con.*
Rel.	2	*Low (unsafe), High (safe)*
C	3	*High C, Medium C, Low C*
Ri	3	*High, Medium, Low*

6.4.4 Results

The historical (prior) probabilities of the variables are obtained by solving the BN of Figure 6.19. The historical risk, *Ri*, is fairly uniformly distributed. However, the historical reliability, *Rel*, is more on the *high* side. When an actual inspection is performed, we can enter the result as evidence in the model. Let us assume that the observed inspection state is *V.I.* = *Pristine*. The resulting variable probabilities are as shown in Figure 6.20, and, as expected, both *Ri* and *Rel* improved. If the observed inspection state is *V.I.* = *Damaged*, then the resulting variable probabilities as shown in Figure 6.21 shows the worsened picture of both *Ri* and *Rel*, as expected.

The model reveals more than the probabilities of *Ri* and *Rel*. It shows the actual levels of the *Corr_Fat* damage which cannot be observed directly. This is possible because the BN can infer these values from the observations of *V.I.*, one of the fundamentals of Bayes Theorem. Hence, when we observe *V.I.* = *Pristine*, the BN infers that the probability of *Corr-Fat* = *Pristine* is 83% and the probability of *Corr-Fat*= *Damaged* is 17%. Similarly, when we observe a *V.I.* = *Damaged*, the BN infers that the probability of *Corr-Fat* = *Pristine* is 19% and the probability of *Corr-Fat* = *Damaged* is 81%. This is one of the most powerful capabilities of GNs, and its utility in this type of monitoring (estimating the actual, nonobservable, value of a variable) is fairly helpful to decision makers. To gain more understanding of the utility of this technique, let us assume that *Corr-Fat* = *Damaged*. The resulting values of the variables are shown in Figure 6.22. Note that the *V.I.* histogram has now finite values for probabilities of the two states (*Pristine*

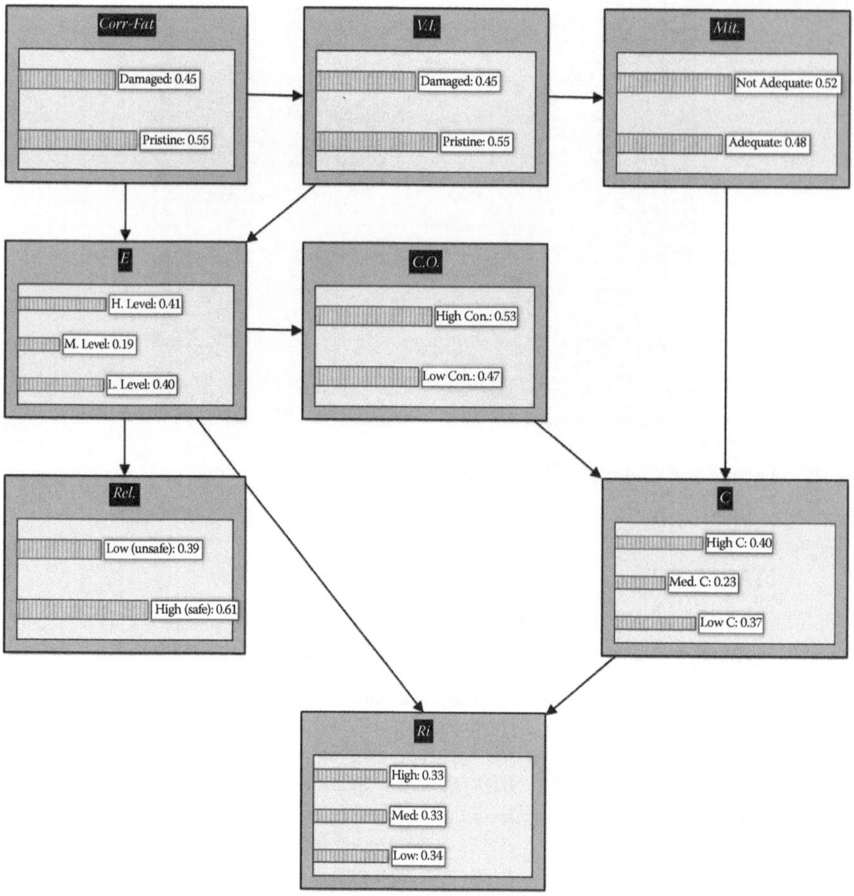

FIGURE 6.19 BN for inference and estimating actual corrosion-fatigue damage, reliability, and risk: historical performance.

and *Damaged*). This particular situation is not possible, since as argued earlier, observing the exact state of *Corr-Fat* is generally not possible. An interesting condition to investigate would be the role of the *Mit* in controlling *Ri*. If, for example, we observe a *V.I.* = *Damaged* and *Mit* = *Not Adequate*, the resulting network is shown in Figure 6.23. As expected, the risk states are worse than earlier situations.

6.5 CASE STUDY 6.3: FORECASTING FUTURE RISK OF DETERIORATION AND CONDITION RATINGS

6.5.1 Overview

Bridge deterioration forecasting is one of the major components of bridge management. Because of this, issues that relate to bridge deterioration have received considerable interest from researchers, practicing engineers, and bridge managers. To start with, we need to define deterioration of bridges. Ettouney and Alampalli (2012b) defined deterioration as the reduction of the *state* of the bridge (or any type of infrastructure) from a given *baseline*. Usually, such a *baseline* is defined as the initial desired state of the bridge, for example, at conclusion of construction. Definition of the *state* is more difficult. Many define the *state* as the *capacity* to resist specific *demands*. Such a definition

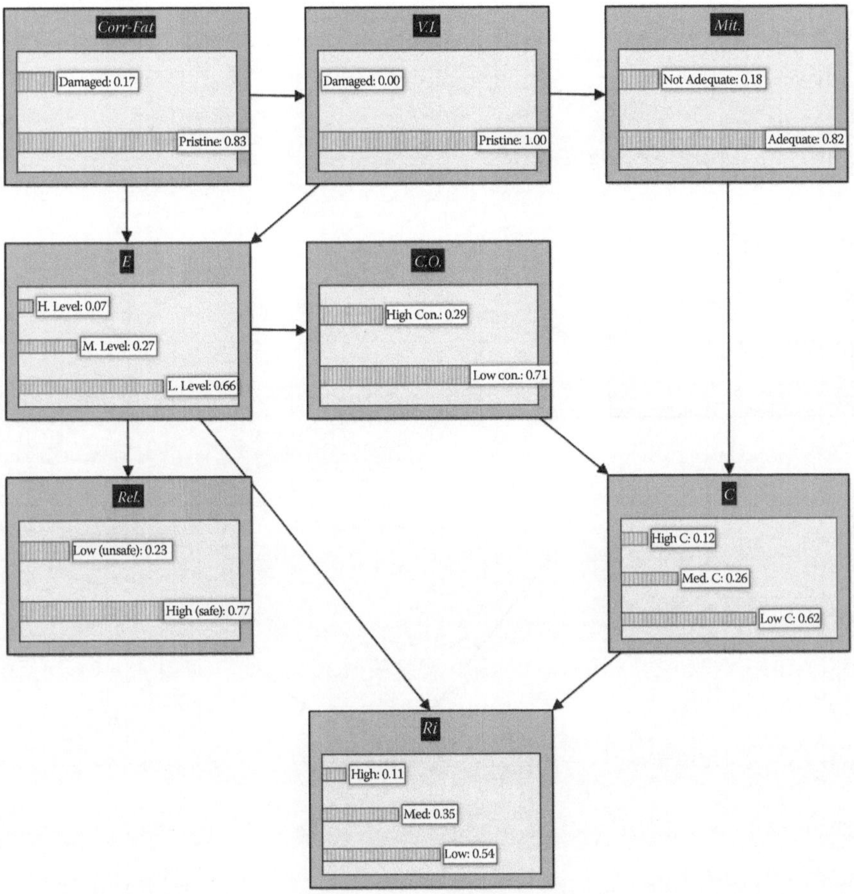

FIGURE 6.20 BN for inference and estimating actual corrosion-fatigue damage, reliability, and risk: pristine *V.I.* observation.

can indicate that the desired *state* varies as *capacity* and/or *demand* varies. Following such logic indicates that deterioration relates to current bridge capacity and demand. Thus, determining deterioration can be an involved process.

Given the complexity of the considerations of bridge deterioration, we summarize quickly in this section some modeling techniques of deterioration as well as sources of deterioration before offering a risk-based methodology for forecasting deterioration. Further detailed discussion of deterioration can be found in Ettouney and Alampalli (2012b).

6.5.1.1 Sources of Deterioration

Sources of bridge deterioration are numerous. Some deterioration sources are given in Table 6.5. Obviously, effects of these sources are cumulative, and many of them have cascading effects. As such, quantitative estimates of such effects are difficult. However, many models are used to estimate objectively the resulting deterioration: these models are described next.

6.5.1.2 Modeling Deterioration

Bridge managers need to estimate future states of bridges for appropriate maintenance decisions. To do an objective analysis, bridge deterioration as a function of time is needed. We assume that the bridge (system) is composed of one or more subcomponents. Obviously, the deteriorated state

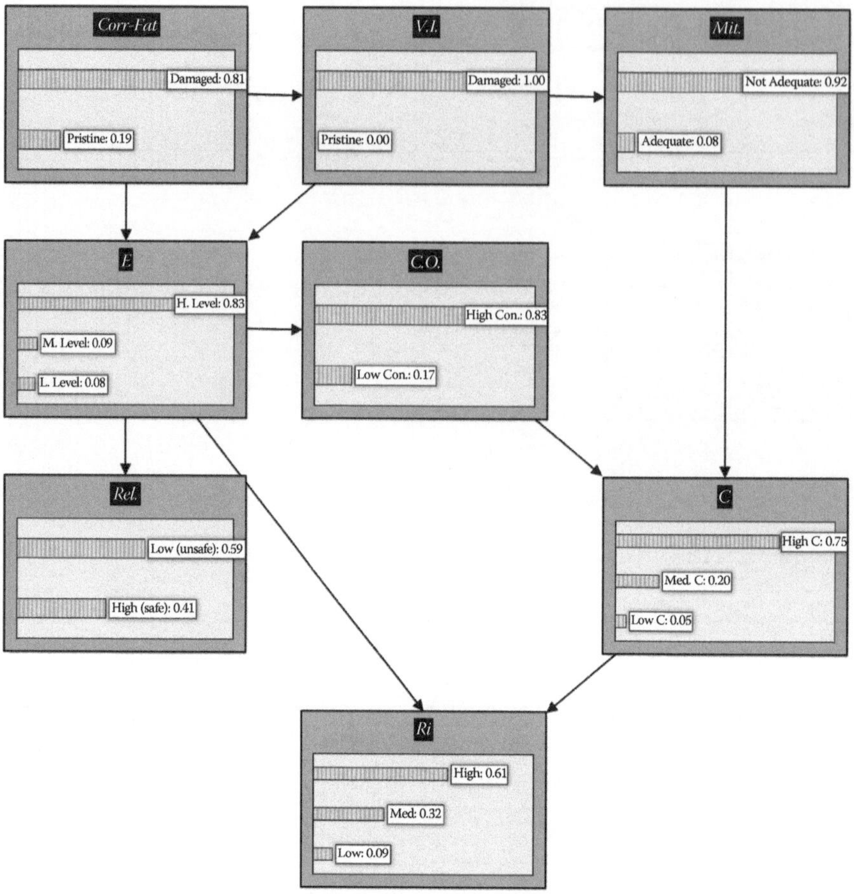

FIGURE 6.21 BN for inference and estimating actual corrosion-fatigue damage, reliability, and risk: damaged *V.I.* observation.

of the bridge (or any other type of infrastructure) is a function of the deteriorated states of its subcomponents such that

$$D(t) = f(s_1, s_2, \ldots, s_n) \tag{6.7}$$

where

$D(t)$ = A measure of the deteriorated state of the desired component, or system, at time t
s_i = deteriorated state of the i-th subcomponent of the system, with $i = 1,2\ldots, n$
n = Number of components of the system or component under consideration

Note that $n = 1$ if only one component in the system is under consideration. The function $f(\ )$ is a suitable function.

Further, the deteriorated state of the i-th component can be described as

$$s_i = g_i\left(x_1, x_2, \ldots, x_{m_i}\right) \tag{6.8}$$

where

x_j = The j-th source of the deterioration of i-th component, with $j = 1,2,\ldots, m_i$

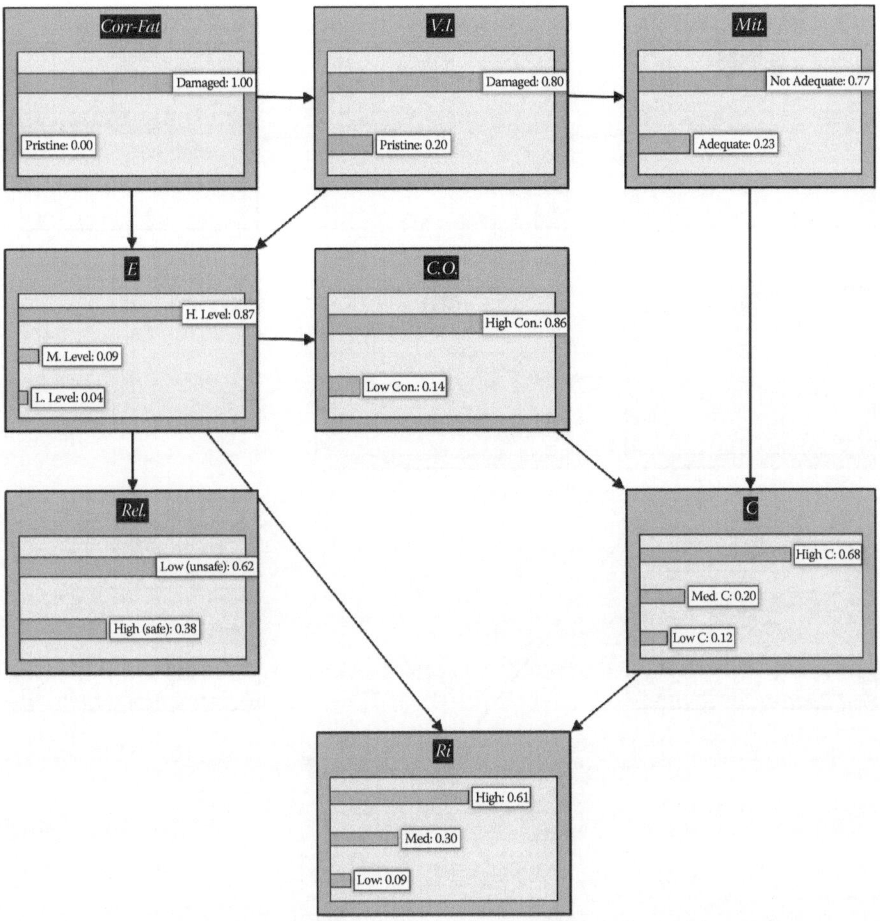

FIGURE 6.22 BN for inference and estimating actual corrosion-fatigue damage, reliability, and risk: actual *Corr-Fat* damage.

The sources for deterioration of different components can be found from studying Table 6.5. Obviously, the function $g_i(\,)$ is a difficult function to quantify. Because of this, evaluating $D(t)$ using formal relations between the deteriorating state and the sources of deterioration is beyond the current state-of-the-art.

A simple method to estimate the deteriorating state of bridges is possible by using the concept of bridge rating, R, as a deterioration measure such that

$$D(t) = R(t) \tag{6.9}$$

Thus,

$$R(t) = f_1(R_1, R_2, \ldots R_n) \tag{6.10}$$

where R_i is the rating of the i-th component of the system. Rating bridge components is carried out by using visual inspection in a qualitative manner. The function $f_1(\,)$ is a simple function that accounts for all major components in the system under consideration.

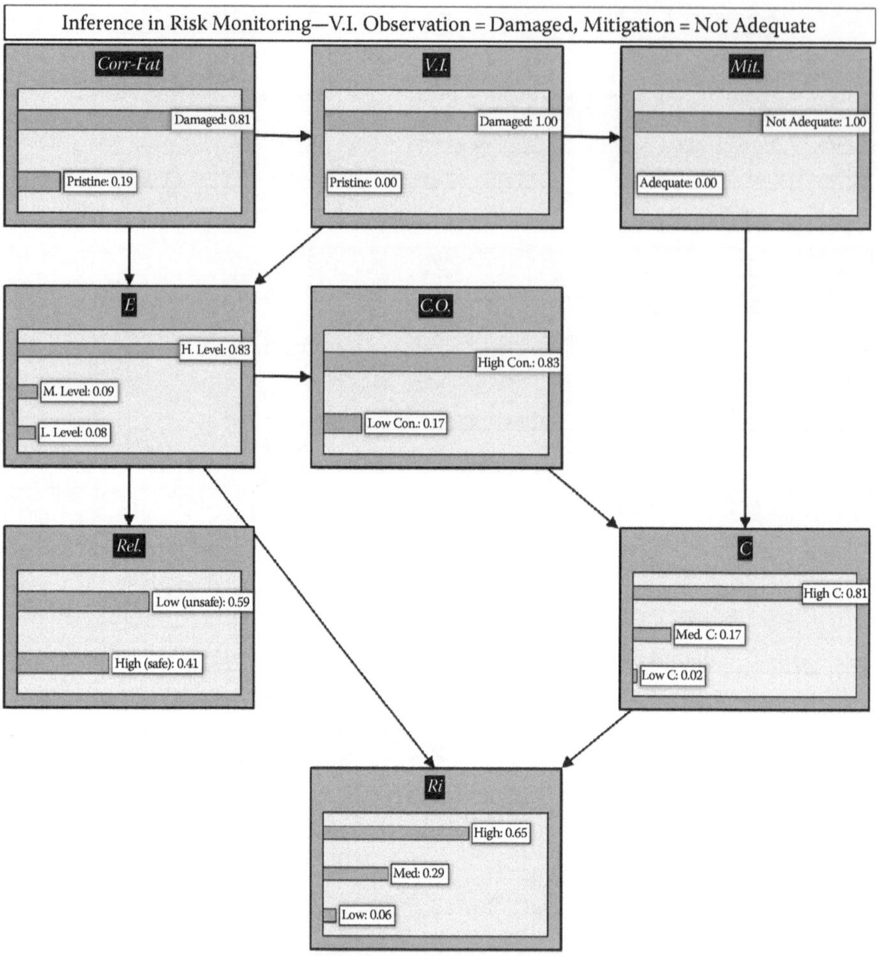

FIGURE 6.23 BN for inference and estimating actual corrosion-fatigue damage, reliability, and risk: risk and reliability for *V.I.* observed damage with damage not mitigated.

For the bridge manager, it is important to know the current rating, $R(t = \text{Now})$, as well as the future rating, $R(t > \text{Now})$. To achieve this extrapolation to the future, the deterioration rate (assuming that $D(t) \simeq R(t)$) needs to be estimated. Bridge managers use several models for estimating such a rating. For example, Yanev (1998) explored regression modeling for evaluating deterioration rates. These methods are deterministic in nature. We describe in detail two of the stochastic-based deterioration rate models next.

6.5.1.3 Markov Model

The Markovian model to predict deterioration is simple. First, it uses *CR* states as the measures for bridge deterioration. Depending on the application, the number of the condition ratings for a component N can differ. Benjamin and Cornell (1970) used five states for a pavement condition. In the Arizona Pavement Management System, 45 condition states were used in the Markovian modeling of pavement deterioration. A transition probability matrix $[P]_k$ is then defined for the component under consideration. The matrix $[P]_k$ is a square matrix of size N. The components of the matrix p_{ijk} (i-th row and j-th column) are defined as the probability that the condition of the component will change from the i-th condition state to the j-th condition state at the end of the k-th time period. The final piece in deterioration Markov modeling is defining the state probability vector $\{S\}_k$; the size of

TABLE 6.5
Sources of Deterioration of Bridges

Type	Time Scale	Source
Man-made source	Long	• Construction defect (e.g., inadequate grouting of post tensioned tendons or inadequate waterproofing, inadequate expansion joints) • Inadequate designs • Overloads
	Sudden/short	• Construction defects • Overload • Vehicle/barge collision • Fire
Natural source	Long	• Corrosion • Fatigue • Thermal effects • Freeze-thaw • Chemical processes such as alkali–silica interaction or carbonation • Inadequate drainage • Paint deterioration • Creep/shrinkage • Settlement of supports/foundations
	Sudden/short	• Earthquake • Wind • Flood/scour/debris

$\{S\}_k$ is also N. The components of the vector s_{ik} (i-th row) are defined as the probability that the condition rating is $R_k = i$ at the end of the k-th time period. More details on the generic Markov process method as well as its generalization into a DBN were offered in Chapter 2.

Another practical use of Markovian models in estimating deterioration is a simple, yet effective, computer code that was developed for the City and County of Denver. The computer code has a built-in database of information regarding different bridges that are managed by the City and County of Denver Department of Public Works. The database includes such information as the year the structure was built, ADT, available historical condition ratings of the deck, superstructure, and substructure. The computer code uses the built-in Markovian model method to extrapolate the deterioration of different components in the future. The deterioration of the components is represented by condition ratings of the component. Two types of deterioration models can be used by this method: the age of the component, and the ADT.

We also note that Markov processes are also used in bridge management systems such as Pontis (2005a, 2005b).

6.5.1.4 Probability Distributions

Probability Density Function (PDF) models are another technique used to quantify and study deterioration rates of bridges or bridge components. In this technique, available historical records of bridge or bridge component ratings are arranged in logical histograms. The histograms are then approximated using appropriate PDF. The resulting PDFs can then be used in decision-making processes that involve the changes (deterioration) of the bridge or component ratings. Agrawal et al. (2008) used Weibull distributions to model deterioration for a statewide bridge deterioration trending and forecasting.

6.5.1.5 Reliability Index Model

The use of condition rating as a measure of deterioration has been criticized for its qualitative nature (of course, by utilizing *SHM* techniques, the condition rating can be more quantitative, as discussed elsewhere in this book). Instead of using deterioration to describe the adequacy of the bridge state, the bridge capability to carry load, that is, safety, is used. The time-changing reliability index of the bridge β is then used as a measure for the adequacy of the bridge to perform its function as it ages. Frangopol and Yang (2004) suggested states for the bridge based on its reliability index β.

6.5.1.6 Concluding Remarks

The above mentioned deterioration models aim at modeling behavioral trends then move on to forecasting potential future trends. We follow this approach by offering a GN approach, which builds on the common threads of all of those methods while adding some needed features such as:

1. Infusing risk-based capability by accommodating consequences-type variables
2. Combine/aggregate appropriate metrics so that the resulting statistics are more accurate
3. Be specific as to which component is being handled during processes
4. Acknowledge interactions between different variables

Thus, by following a GN approach, we can take advantage of the strengths of different deterioration models and assure even more accuracy in the trending and forecasting processes.

6.5.2 CGN Process in Forecasting

Temporal changes in risk and risk components are of interest to civil infrastructure stakeholders. For example, the changes in bridge condition rating as time progresses forms one of the basic tenets in bridge management. Previously observed condition ratings of structures and its components are used to forecast expected future behavior, mostly deterioration, of condition ratings. The bridge managers then make appropriate decisions, based in part on these forecasts, regarding maintenance and rehabilitation.

One of the basic tools for these forecasts are Markov chains [see Chapter 2 and Ettouney and Alampalli (2012)], which relate the current condition rating with forecasts of the probabilities of different states of condition ratings in the form:

$$\{X\}_{t=i+NPERIOD} = [T]^{NPERIOD}\{X\}_{t=i} \tag{6.11}$$

where $\{X\}_{t=i}$ is the vector of marginal probabilities of occurrence of different states, x_j, at time $t = i$, $[T]$ is a transition matrix for marginal probabilities in a single period, and $NPERIOD$ is the number of periods. $\{X\}_{t=i+NPERIOD}$ is the vector of marginal probabilities of occurrence of different states, x_j, at time $t = i + NPERIOD$. The different discrete states of the condition ratings are x_j and $j = 1,2,\ldots, N_{STATE}$, where N_{STATE} is the rank. To keep the developments of this section general, we do not specify the actual length of a period but assume that it is constant without any loss of generality. In practice, a period is chosen, generally, to be a multiple of years. In Chapter 2, we showed that the Markov chain system is a simplified form of DBN. For example, if we consider a component condition rating to be perfectly aligned with its exposure and observe that the condition rating of the component to be *poor*, then the marginal probability vector at present time, $t = 0$ is:

$$\{X\}_{t=0} = \begin{Bmatrix} 0 \\ 0 \\ 1 \\ 0 \end{Bmatrix} \tag{6.12}$$

Here we assume the condition ratings rank $N_{STATE} = 4$, with different states defined as *good*, *fair*, *poor*, and *service*. Let us assume a transition matrix for this particular component as

$$[T]=\begin{bmatrix} .6 & 0 & 0 & 0 \\ .3 & .55 & 0 & 0 \\ .08 & .35 & .6 & 0 \\ .02 & .1 & .4 & 1 \end{bmatrix} \quad (6.13)$$

Applying Equation 6.11, we get the marginal probabilities of different states of condition rating at $t = t_2$ (after two periods) as

$$\{X\}_{t=2}=\begin{Bmatrix} 0 \\ 0 \\ .36 \\ .64 \end{Bmatrix} \quad (6.14)$$

Assuming perfect alignment between exposure and condition rating, CR such that

$$\{E\} = \{CR\} \quad (6.15)$$

we estimate that, according to Equation 6.14, the probability that the exposure of the component degrading to state of *poor*, after two periods, to be 36% and that the probability that the exposure of the component is requiring *service* is 64%.

Applying the above Markov chains method is obviously simple, yet it might not be accurate enough or adequate enough for proper decision-making. It has two major deficiencies:

- It presumes complete alignment between observation, *CR*, and exposure, *E*.
- It ignores consequences, *C*, in the decision process. As discussed throughout this book, the use of *E* in any decision process is necessary but not sufficient to reach an optimal decision. Ultimately, it is desirable to use risk, *Ri*, in the decision process, which accommodates both *E* and *C*.

DBN offers capability of using observations (evidence) of more than one variable at any time slice, then inferring marginal probabilities for the rest of the model at any present (process of filtering), past (process of smoothing), or future (process of prediction or forecasting) time slice. Hence, DBN may be used to rectify the above deficiencies.

6.5.3 Model

We reconsider the above problem by first realizing one of its basic assumptions—presumption of perfect alignment between observation, *CR*, and exposure, *E*. We also need to introduce consequences and risk into the problem. Thus, we end up with a static BN. We then apply the DBN method to explore temporal risk monitoring, that is, the fact that there are changes in many of the variables of the problem with time. A simple DBN template incorporating these elements is shown in Figure 6.24. The states of the nodal variables are given in Table 6.6, and the definitions of the variables are given in Table 6.10. The variables of this simple BN template are *CR*, *E*, *C*, and *Ri*. We also use two parent variables for consequence, *C*. These are *C_1* and *C_2*, and they respectively represent consequences of increased exposure of the bridge such as maintenance and rehabilitation efforts and consequences due to other factors such as abnormal hazards, social/economic factors, or user costs. Figure 6.25 shows three time slices of the template, $t = t_0, t_1, t_2$. We assume that as time progresses, the exposure at time $t = i + 1$ depends on exposure at time $t = i$. Similarly, we also assume that the consequence C_2 at time $t = i + 1$ depends on consequences C_2 at time $t = i$. Let us assume the transfer matrix for these consequences to be:

$$[T]=\begin{bmatrix} .55 & .15 \\ .45 & .85 \end{bmatrix} \quad (6.16)$$

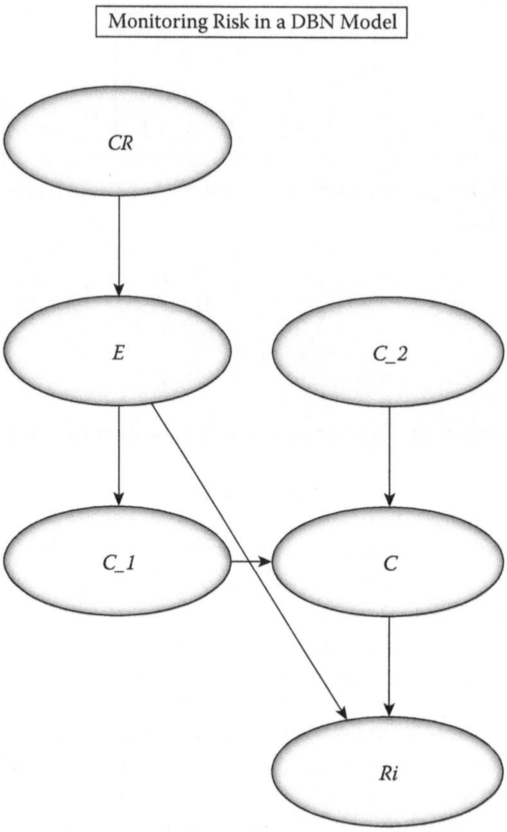

FIGURE 6.24 Template for DBN risk monitoring.

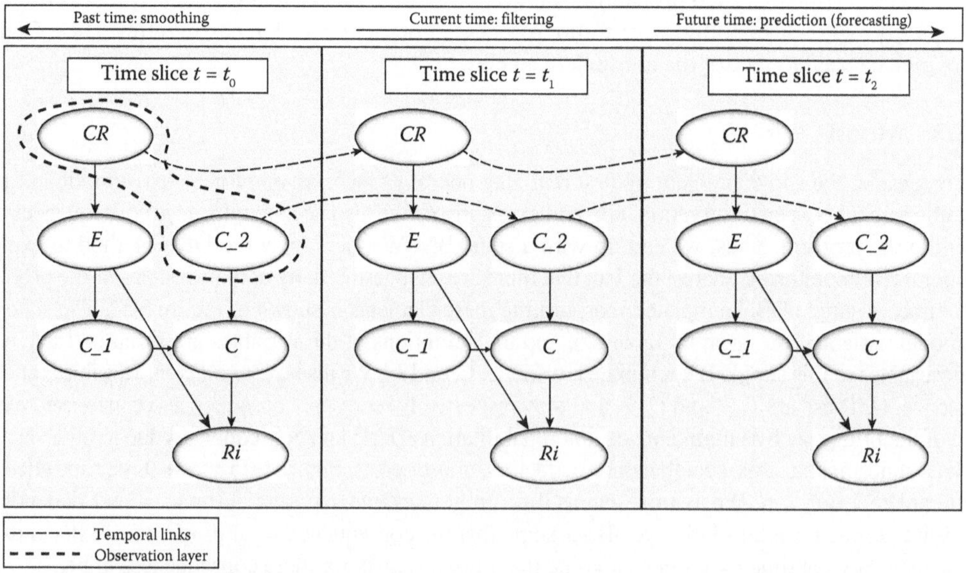

FIGURE 6.25 DBN for risk monitoring.

TABLE 6.6
States of Nodal Variables

Variable	# of States	States
CR	4	*Service, Poor, Fair, Good* (See Figures 6.26 through 6.31 for examples of different damage states)
E	3	*High (H.) Level, Middle (M.) Level, Low (L.) Level*
C_1, C_2	2	*High Con., Low Con.*
C	3	*High C, Med. C, Low C*
Ri	3	*High, Med, Low*

FIGURE 6.26 Steel girder in *Good* condition rating.

FIGURE 6.27 Concrete girder in *Good* condition rating.

FIGURE 6.28 Steel girder in *Fair* condition rating.

FIGURE 6.29 Steel girder in *Poor* condition rating.

We use the same transfer matrix shown earlier in Equation 6.13 to describe the temporal dependence of the exposure. These dependencies can be described in the model as temporal links, as shown in Figure 6.25. As illustrated in Chapter 2, temporal links used in DBN are similar in nature to Markov chains transfer matrices.

6.5.4 Conditional Probabilities Tables

The different CPTs of each variable of the BN network of the template of Figure 6.24 are shown in Appendix 6.IV. We used our best judgment in filling the different tables in a realistic manner for illustration purposes. For more on different methods of developing CPTs, see Chapter 2.

FIGURE 6.30 Concrete girder in *Poor* condition rating.

FIGURE 6.31 Steel girder in *Service* condition rating.

6.5.5 Prior (Initial) Marginal Probabilities (Initial Time Slice)

We use the CPTs in Appendix 6.IV and the BN of the template in Figure 6.24 to compute the historical probabilities of the model that are shown in Figure 6.32. Of course, these historical probabilities are only indicative of past performances; we need now to impose initial observations, i.e., CR = poor, to the template that will result in the probabilities of Figure 6.33. These initial probabilities are then used with the DBN time-marching process to obtain the desired network forecast, which we perform next.

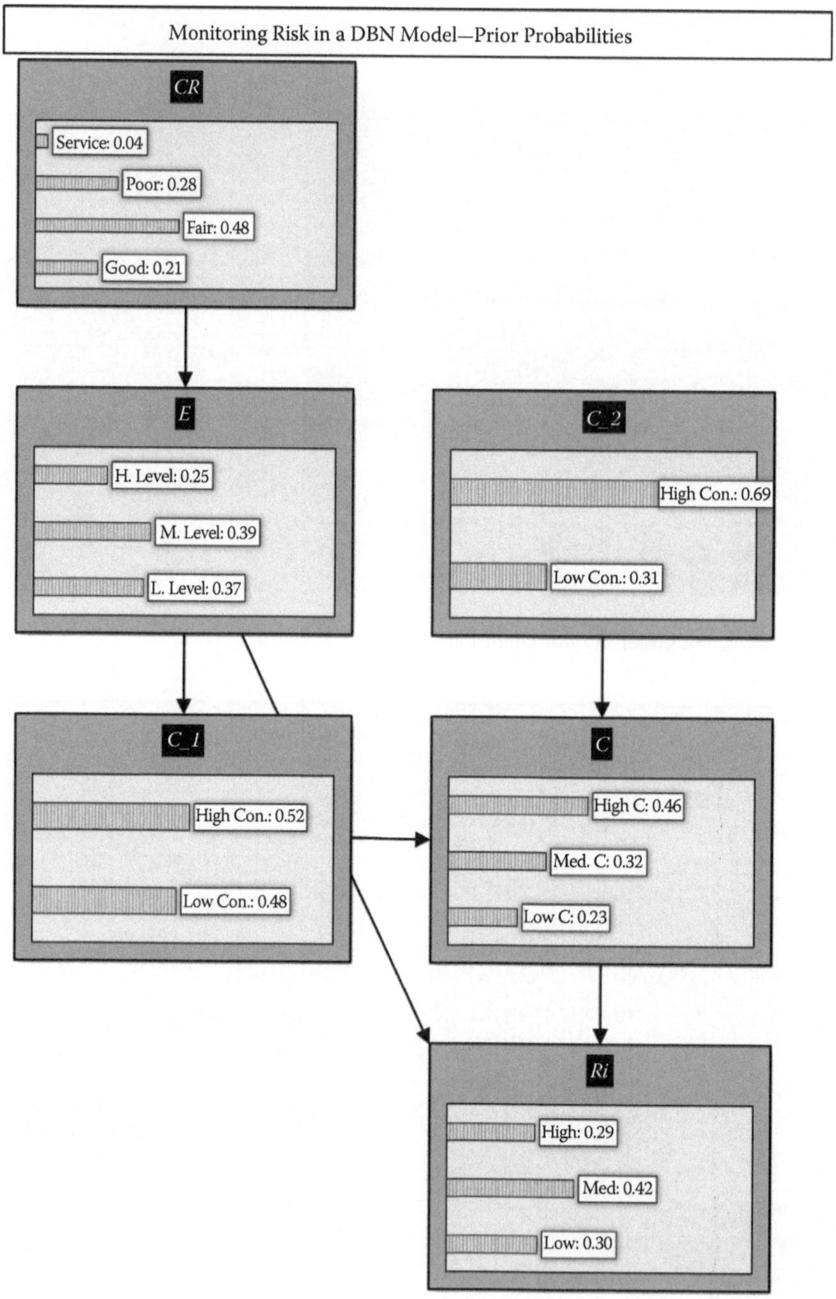

FIGURE 6.32 Historical (prior) performance.

6.5.6 Filtering (First Time Slice)

To obtain marginal probabilities of the network variables after one time period, we apply transfer matrices of Equations 6.13 and 6.14 on all temporal links of Figure 6.25. There are two temporal links in the BN template that apply to *CR* and *C_2*. The resulting marginals of the two variables present soft evidence of the BN template at period 1: *P*(*CR*) and *P*(*C_2*). We then use the method in Chapter 2 to apply these soft evidences on the network to resolve for the marginals of

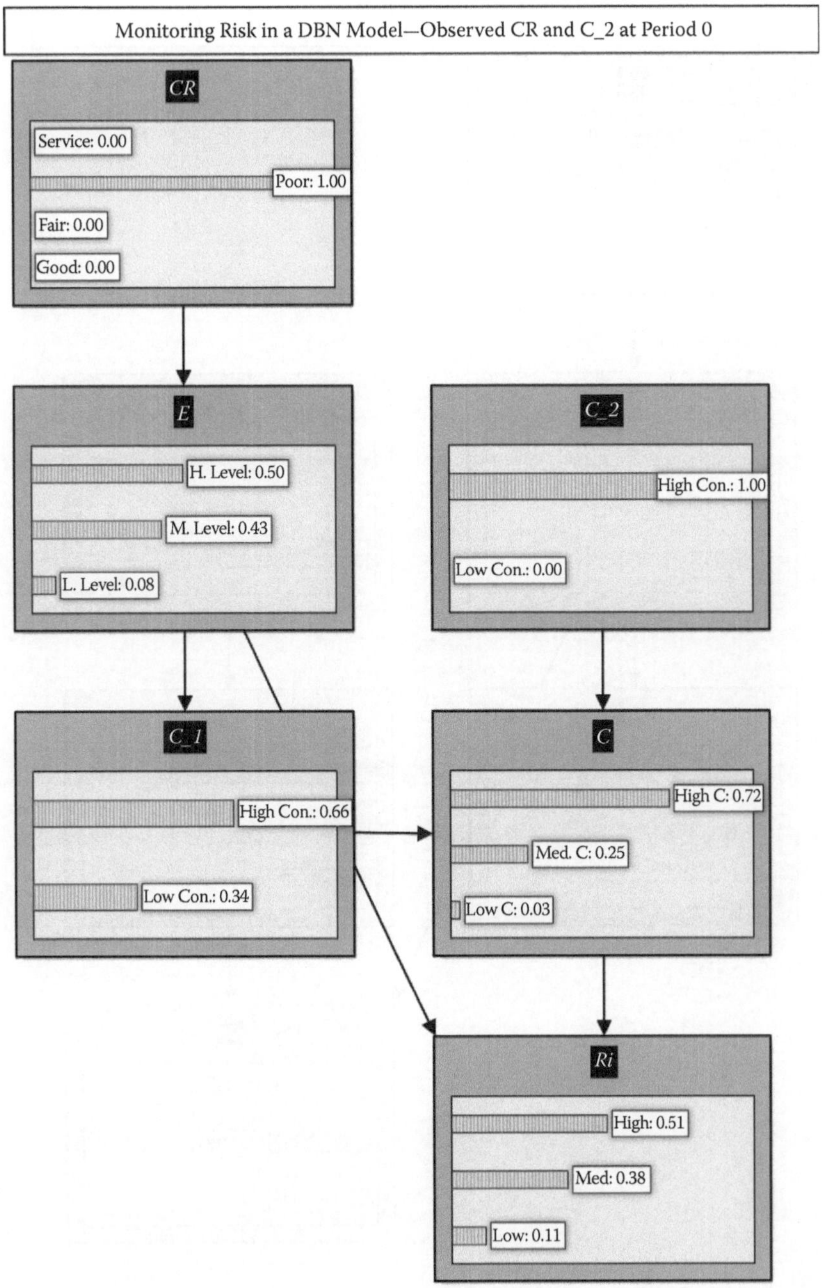

FIGURE 6.33 Probabilities at initial time period.

the other variables in the network. The resulting marginals are shown in Figure 6.34. Note that an observation of CR = 100% *poor* has changed to 60% *poor* and 40% *service.* While an observation of *C_2* = 100%, consequences has now changed to 85% high consequences and 15% low consequences. Both observations are considered soft evidence in the network because they were obtained using transfer matrices from the initial period to this period. After solving the network for these soft evidences, probabilities of the rest of the variables have changed. For example, the probability of high level exposure raised to 69% (from 50% in period 0), and the probability of high risk raised to 54%

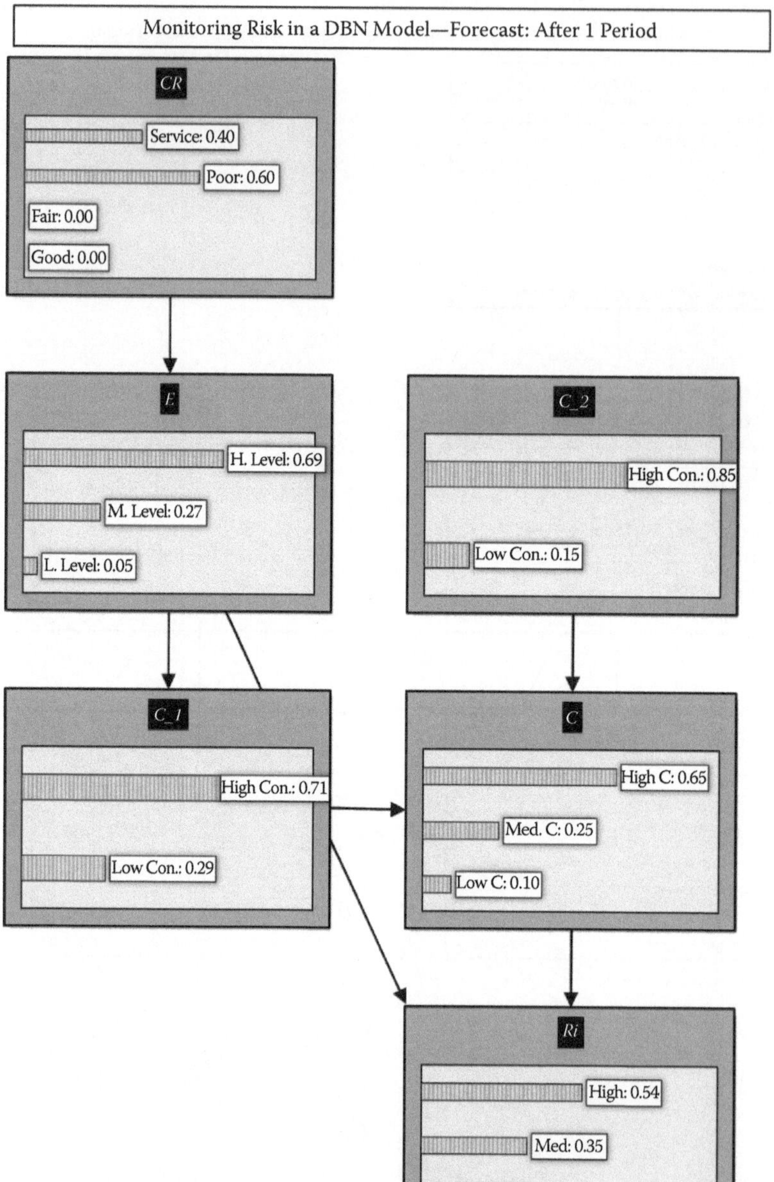

FIGURE 6.34 Projections after one time period.

(from 51% in period 0). The changes in exposure are expected to be higher than the changes in risk because all the changes in the CR are worst in period 1 compared to period 0, while the high consequences of *C_2* have been reduced from 100% to 85%. Obviously, the transfer matrices play a key role in how the variables of the BN change from one period to the next.

6.5.7 Forecasting (Prediction) at Future Time Slice

Continuing the forecasting effort of the network template, we obtain the marginal probabilities shown in Figure 6.35. We again extrapolated the two temporal links *CR* and *C_2* from period 1 to period 2,

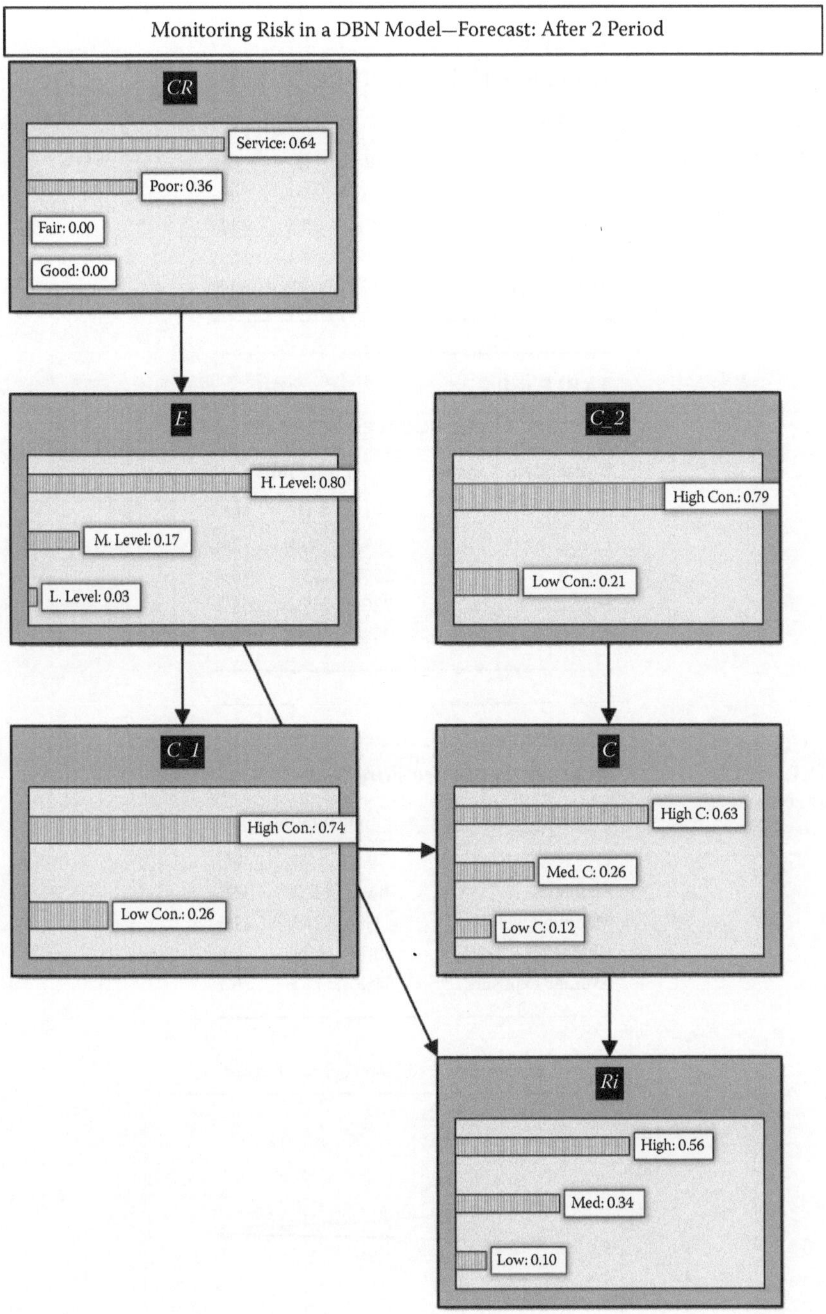

FIGURE 6.35 Projections after two time periods.

using the transfer matrices of Equations 6.13 and 6.14, respectively. We also apply the resulting probabilities as soft evidence to the BN using methods shown in Chapter 2. The exposure and risk have continued their respective increases as in period 1. Also, the rate of exposure increase is more than the rate of risk increase because of the differences in the transfer matrices of *CR* and *C_2*, as discussed earlier. These results can be illustrated by computing forecasts of average risk, average exposure, and average consequences as in Tables 6.7 through 6.9 and Figure 6.36.

TABLE 6.7
Average Risk Forecast

Risk State	Probability		
	t0	t1	t2
Low	11%	11%	10%
Med	38%	35%	34%
High	51%	54%	56%
Average risk	63%	64%	65%

TABLE 6.8
Average Consequences Forecast

Consequence State	Probability		
	t0	t1	t2
Low C	3%	10%	12%
Med. C	25%	25%	26%
High C	72%	65%	63%
Average consequences	73%	68%	67%

TABLE 6.9
Average Exposure Forecast

Exposure State	Probability		
	t0	t1	t2
Low level	8%	5%	3%
Medium level	43%	27%	17%
High level	50%	69%	80%
Average exposure	64%	71%	76%

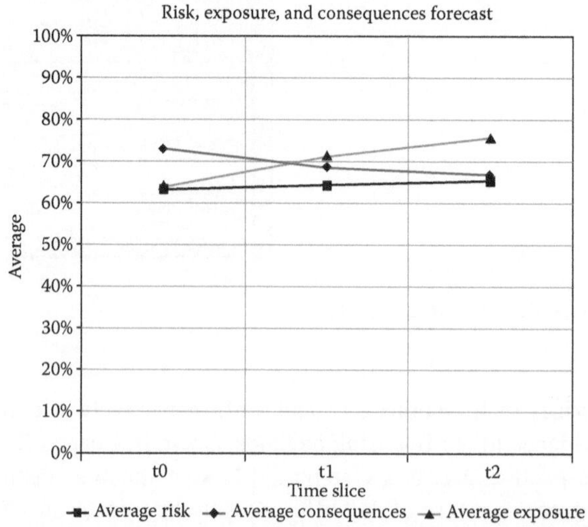

FIGURE 6.36 Forecasting risk, exposure, and consequences.

APPENDIX 6.I: DEFINITIONS OF NETWORK NODAL VARIABLES (TABLE 6.10)

TABLE 6.10
Definitions of Nodal Variables

Variable	Used in Case Study #	Definition
C	6.3	Total consequences is a combination of C_1, C_2.
C	6.2	Measure of consequences in this example is a function of the degree of repair (*Mit*) as well as different social and economic factors as represented by *C.O.* In this example, consequences are estimated subjectively. Obviously, consequences are higher when the system has adequate *Mit*. Similarly, different social and economic factors, *C.O.*, can affect levels of consequences.
C	6.1	Consequences in this example is assumed to be an independent variable (no parent in the network). This assumption is taken only for simplicity though it is not an accurate assumption.
C.O.	6.2	Estimate of consequences beyond the system itself, these consequences can be, but are not limited to, cost of replacements, cost of traffic delays, symbolic nature of the structure, structure locality, and structure age. In these examples, C.O. is hazard independent.
C_1, C_2	6.3	*C_1* is an estimation of direct consequences resulting from damages resulting from *CR*. These include costs of maintenance, repairs, rehabilitation, or replacements. C_2 is a subjective estimation of consequences beyond the system itself, these consequences can be, but are not limited to, cost of replacements, cost of traffic delays, symbolic nature of the structure, structure locality, and structure age.
Corr_Fat	6.2	Even though corrosion or fatigue are different hazards, they have in common the fact that their effects might be observed via visual inspection. As such, this variable estimates the deterioration of the system(or component) resulting from either of the two hazards.
CR	6.3	Condition rating is an estimate of the physical condition of the bridge. It is directly related to the rating that results from physical inspections. They are usually used interchangeably. However, in this book we sometimes differentiate between them. We hypothesize that *CR* is the actual condition rating of the system, and visual inspection produces the condition rating of the system as the inspector observes it visually.
E	6.3	Estimate of the exposure of the system. In this example, *Exposure* relates to *CR*. In many situations, *CR* is directly related to observations resulting from visual inspections.
E	6.2	Estimate of the exposure of the system. In this example, *Exposure* relates to both *V.I.* and *Corr_Fat*.
E	6.1	Actual *Exposure* of the system is estimated here from the results of *V.I.* and *SHM*. This produces a better exposure estimate than reliance on a single monitoring/observation method.
Mit.	6.2	Probability of mitigation and its potential adequacy.
Rel.	6.2	This is a metric of the capacity-demands relationship. It is an objective inverse of the exposure of the system. In this example, it is assigned the state of *High* (Safe) if capacity exceeds demands or *Low* (Unsafe) if reverse is true.
Ri	6.3	Risk, Ri, is a function of exposure and consequences.
Ri	6.2	Risk, Ri, is a function of exposure and consequences.
Ri	6.1	Risk, Ri, is a function of exposure and consequences.
SHM	6.1	This objective variable is a measure of conditions rating of the system that result from S.H.M. In this chain graph (CG) model, it is assumed that there is a nondirectional link (affinity) between *V.I.* and S.H.M.

(Continued)

TABLE 6.10 ***(Continued)***
Definitions of Nodal Variables

Variable	Used in Case Study #	Definition
V.I.	6.2	Visual inspection results are simulated by this objective variable (*V.I.*). It is a subjective measure of condition rating of the system of interest. *V.I.* estimates are function of both hazards and vulnerability. Thus, it is related directly to actual response of the system to *Corr_Fat*. The number of states has to be consistent with the number of condition ratings the owner uses for visual inspection. Three condition ratings used in this example are only for simplicity.
V.I.	6.1	Visual inspection results are simulated by this objective variable (*V.I.*). It is a subjective measure of condition rating of the system of interest. *V.I.* estimates are function of both hazards and vulnerability. Thus, it is related directly to actual exposure of the system. The number of states has to be consistent with the number of condition ratings the owner uses for visual inspection. Three condition ratings used in this example are only for simplicity. In this chain graph (CG) model, it is assumed that there is a nondirectional link (affinity) between *V.I.* and *SHM*.

APPENDIX 6.II: CPTs AND POTENTIALS FOR CASE STUDY 6.1 (TABLES 6.11–6.14)

TABLE 6.11
Potential ϕ(*V.I.*, *SHM*)

SHM		*Good*	*Poor*
V.I.	*Good*	0.7	0.4
	Poor	0.2	0.9

TABLE 6.12
CPT of *C*

C	Low	0.7
	High	0.3
Histograms for conditional probabilities		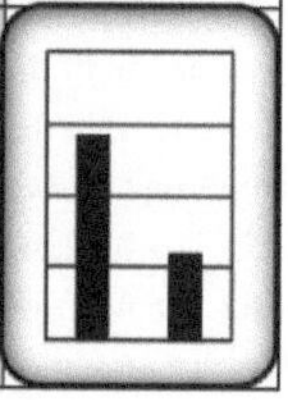

TABLE 6.13
CPT of *E*

SHM		Good		Poor	
V.I.		Good	Poor	Good	Poor
E	Low	0.9	0.3	0.4	0.1
	High	0.1	0.7	0.6	0.9
Histograms for conditional probabilities					

TABLE 6.14
CPT of *Ri*

C		Low		High	
E		Low	High	Low	High
Ri	Low	0.9	0.55	0.3	0.1
	High	0.1	0.45	0.7	0.9
Histograms for conditional probabilities					

APPENDIX 6.III: CPTs FOR CASE STUDY 6.2 (TABLES 6.15–6.22)

TABLE 6.15
CPT of *C*

C.O.		Low Con.		High Con.	
Mit.		Adequate	Not Adequate	Adequate	Not Adequate
C	Low C	0.9067	0.0921	0.3553	0.0000
	Med. C	0.0933	0.5527	0.5527	0.0933
	High C	0.0000	0.3553	0.0921	0.9067
Histograms for conditional probabilities		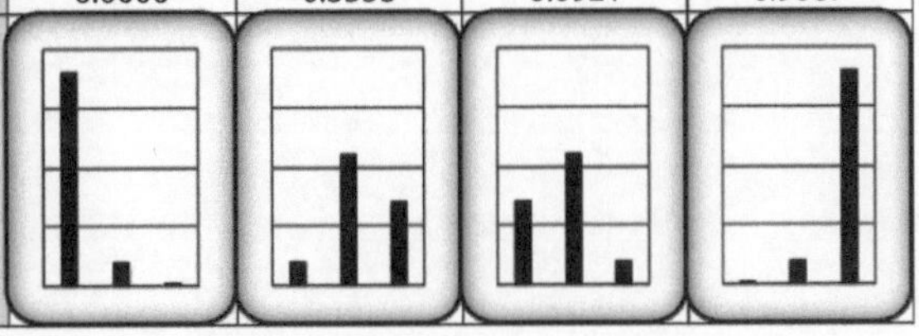			

TABLE 6.16
CPT of *C.O.*

E		L. Level	M. Level	H. Level
C.O.	Low Con.	0.7082	0.4319	0.2082
	High Con.	0.2918	0.5681	0.7918
Histograms for conditional probabilities		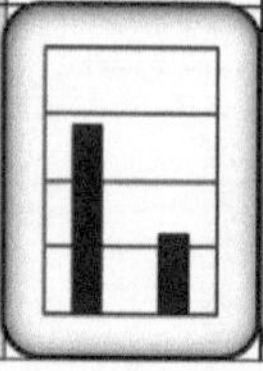	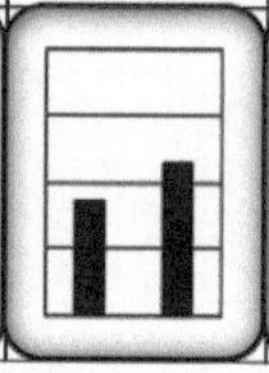	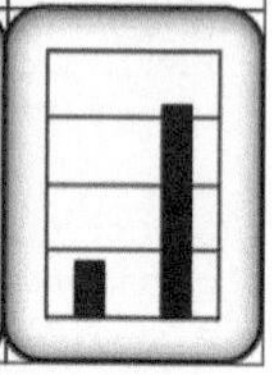

TABLE 6.17
CPT of *Corr-Fat*

Corr-Fat	Pristine	0.5501
	Damaged	0.4499
Histograms for conditional probabilities		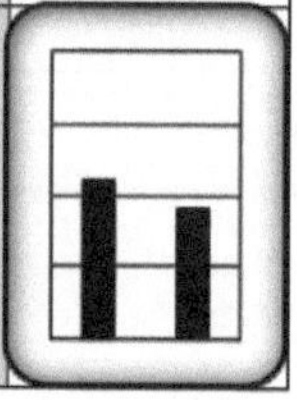

TABLE 6.18
CPT of *E*

V.I.		Pristine		Damaged	
Corr-Fat		Pristine	Damaged	Pristine	Damaged
E	L. Level	0.7491	0.1904	0.4181	0.0000
	M. Level	0.2418	0.3916	0.3916	0.0117
	H. Level	0.0091	0.4181	0.1904	0.9883
Histograms for conditional probabilities		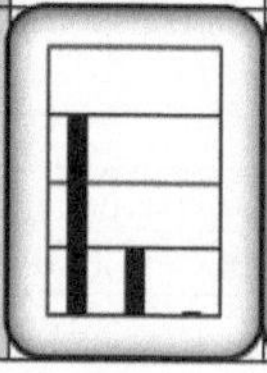		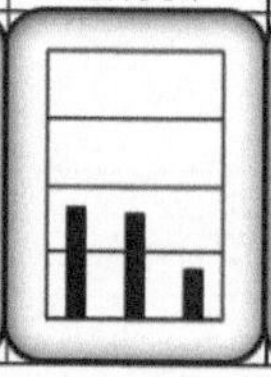	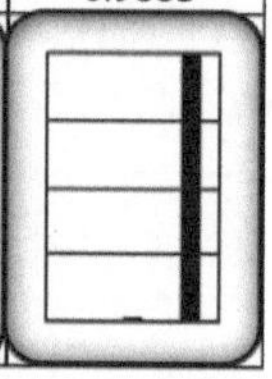

TABLE 6.19
CPT of *Mit.*

V.I.		Pristine	Damaged
Mit.	Adequate	0.8127	0.0794
	Not Adequate	0.1873	0.9206
Histograms for conditional probabilities		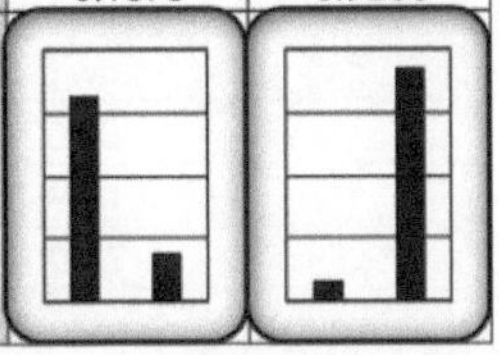	

TABLE 6.20
CPT of *Rel*

E		L. Level	M. Level	H. Level
Rel.	High (safe)	0.9206	0.5550	0.3399
	Low (unsafe)	0.0794	0.4450	0.6601
Histograms for conditional probabilities		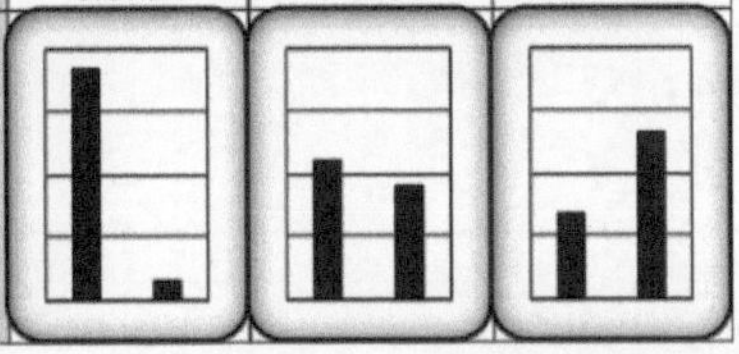		

TABLE 6.21
CPT of *Ri*

E		L. Level			M. Level			H. Level		
C		Low C	Med. C	High C	Low C	Med. C	High C	Low C	Med. C	High C
Ri	Low	0.7677	0.4562	0.4024	0.4562	0.1986	0.1071	0.3621	0.1071	0.0063
	Med	0.2260	0.4493	0.4399	0.4493	0.6028	0.4635	0.4525	0.4635	0.2260
	High	0.0063	0.0945	0.1576	0.0945	0.1986	0.4293	0.1854	0.4293	0.7677
Histograms for conditional probabilities										

TABLE 6.22
CPT of *V.I.*

Corr-Fat		Pristine	Damaged
V.I.	Pristine	0.8413	0.2024
	Damaged	0.1587	0.7976
Histograms for conditional probabilities		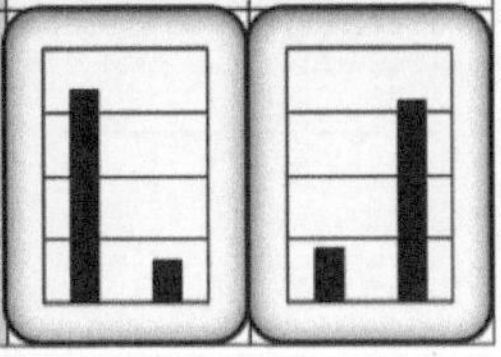	

APPENDIX 6.IV: CPTs FOR CASE STUDY 6.3 (TABLES 6.23–6.28)

TABLE 6.23
CPT of *C*

C_2		Low Con.		High Con.	
C_1		Low Con.	High Con.	Low Con.	High Con.
C	Low C	0.9067	0.3553	0.0921	0.0000
	Med. C	0.0933	0.5527	0.5527	0.0933
	High C	0.0000	0.0921	0.3553	0.9067
Histograms for conditional probabilities		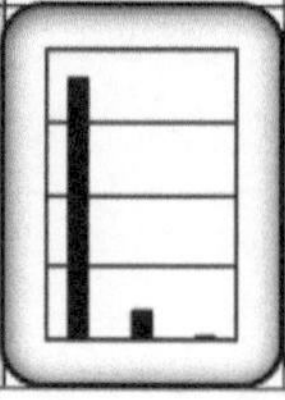	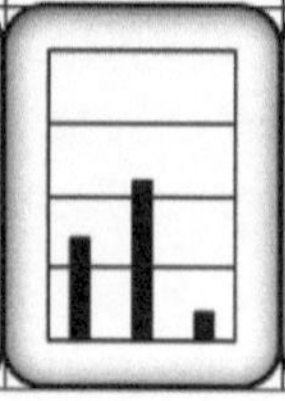	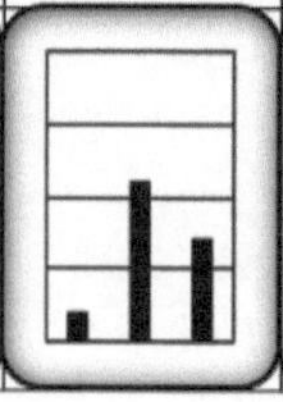	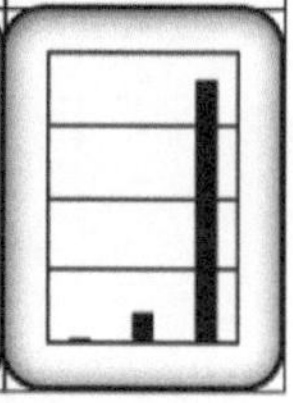

TABLE 6.24
CPT of *C_1*

E		L. Level	M. Level	H. Level
C_1	Low Con.	0.7082	0.4319	0.2082
	High Con.	0.2918	0.5681	0.7918
Histograms for conditional probabilities		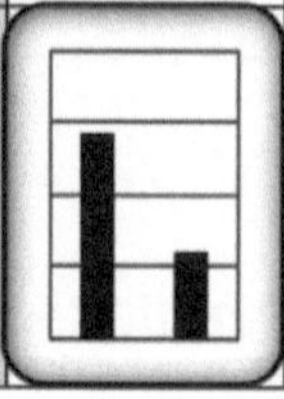	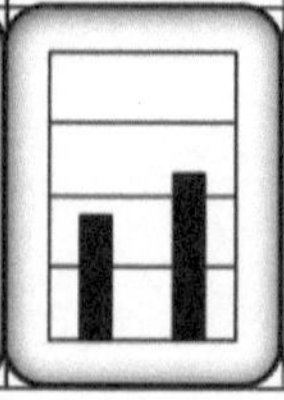	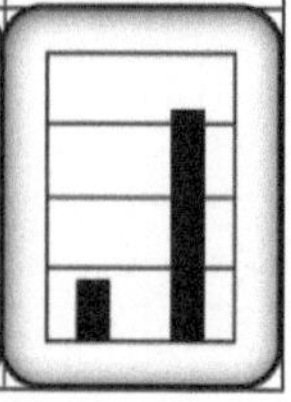

TABLE 6.25
CPT of *C_2*

C_2	Low Con.	0.3148
	High Con.	0.6852
Histograms for conditional probabilities		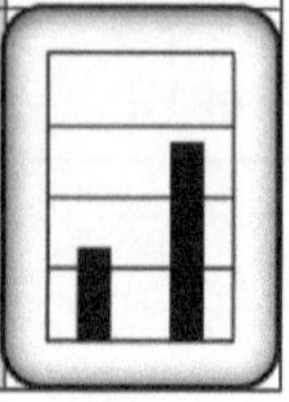

TABLE 6.26
CPT of *CR*

CR	Good	0.2089
	Fair	0.4763
	Poor	0.2751
	Service	0.0397
Histograms for conditional probabilities		

TABLE 6.27
CPT of *E*

CR		Good	Fair	Poor	Service
E	L. Level	0.8240	0.3629	0.0778	0.0000
	M. Level	0.1726	0.4931	0.4258	0.0117
	H. Level	0.0033	0.1439	0.4963	0.9883
Histograms for conditional probabilities		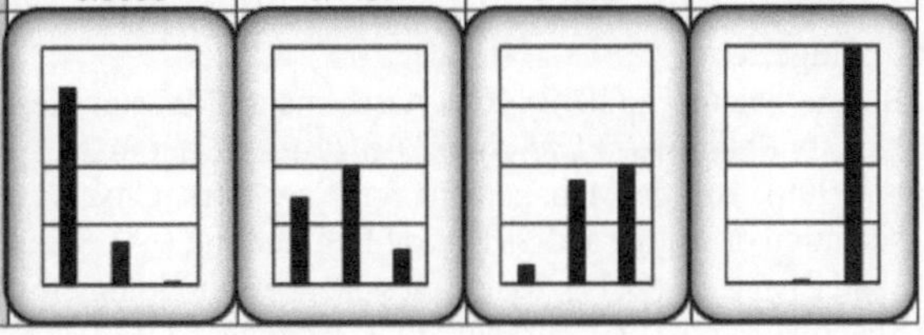			

TABLE 6.28
CPT of *Ri*

C		Low C			Med. C			High C		
E		L. Level	M. Level	H. Level	L. Level	M. Level	H. Level	L. Level	M. Level	H. Level
Ri	Low	0.7677	0.4562	0.3621	0.4562	0.1986	0.1071	0.4024	0.1071	0.0063
	Med	0.2260	0.4493	0.4525	0.4493	0.6028	0.4635	0.4399	0.4635	0.2260
	High	0.0063	0.0945	0.1854	0.0945	0.1986	0.4293	0.1576	0.4293	0.7677
Histograms for conditional probabilities										

REFERENCES

Agrawal, A., Kawaguchi, A., Chen, Z., Lagace, S., and Delisle, R. (2008). *Case Study on Factors Affecting Deterioration of Bridge Elements in New York*, NDE/NDT for Highways and Bridges: Structural Materials Technology (SMT). ASNT, Oakland, CA.

Ahmad, K., and Gibb, A. (2004). *Towards effective safety performance measurement—evaluation of existing techniques and proposals for the future.* Construction Management Systems, Spon, London.

Alampalli, S., and Lund, R. (2006). Estimating Fatigue Life of Bridge Components Using Measured Strains. *J. Bridge. Eng.* 11(6), 725–736.

Batchelor, B., Valdes, J., and Araganth, V. (1998). Stochastic Risk Assessment of Sites Contaminated by Hazardous Wastes. *J. Environ. Eng.* 124(4), 380–388.

Benjamin, J and Cornell, A. (1970), *Probability, Statistics, and Decisions for Civil Engineers*, McGraw-Hill, New York.

Chi, S., and Caldas, C. H. (2012). Image-Based Safety Assessment: Automated Spatial Safety Risk Identification of Earthmoving and Surface Mining Activities. *J. Construct. Eng. Manage.* 138(3), 341–351.

Ettouney, M., and Alampalli, S. (2012a). *Infrastructure Health in Civil Engineering: Theory and Components.* CRC Press, Boca Raton, FL.

Ettouney, M. and Alampalli, S. (2012b). *Infrastructure Health in Civil Engineering: Applications and Management*, CRC Press, Boca Raton, FL.

FHWA. (1995). *Recording and Coding for the Structure Inventory and Appraisal of the Nation's Bridges*, FHWA report No. FHWA-PD-96-001, FHWA, Washington DC.

Frangopol, D., and Yang, C. (2004). Health and Safety of Civil Infrastructures: A Unified Approach. *Proceedings of 2nd International Workshop on Structural Health Monitoring of Innovative Civil Engineering Structures.* September 22–23, 2004, ISIS Canada Corporation, Manitoba, Canada.

Mannocchi, F., Todisco, F., and Vergni, L. (2009). New Methodology for the Risk Analysis and Economic Impact Assessment of Agricultural Droughts. *J. Irrigat. Drain. Eng.* 135(5), 643–653.

MSHA (Mine Safety and Health Administration). (2009). *MSHA Fatality Statistics.* Available from: http://www.msha.gov/stats/charts/chartshome.htm [accessed May 14, 2009].

NYSDOT. (2014). Bridge Inspection Manual, New York State Department of Transportation, Albany, NY.

Panu, U. S., and Sharma, T. C. (2002). Challenge in Drought Research: Some Perspectives and Future Directions. *Hydrol. Sci. J.* 47S, S19–S30.

Pontis 4.4. (2005a). *Pontis Release 4.4: User's Manual.* Prepared for AASHTO by Cambridge Systematics, Cambridge, MA.

Pontis 4.4. (2005b). *Pontis Release 4.4: Technical Manual.* Prepared for AASHTO by Cambridge Systematics, Cambridge, MA.

Salman, B., and Salem, O. (2012). Risk Assessment of Wastewater Collection Lines Using Failure Models and Criticality Ratings. *J. Pipel. Syst. Eng. Pract.* 3(3), 68–76.

Yanev, B. (1998). Bridge Management for New York City. *Struct. Eng. Int.* 8(4), 211–215. International Association of Bridge and Structural Engineering (IABSE), Zurich, Switzerland.

Zhou, H., and Zhang, H. (2011). Risk Assessment Methodology for a Deep Foundation Pit Construction Project in Shanghai, China. *J. Construct. Eng. Manage.* 137(12), 1185–1194.

7 Risk Communication

7.1 OVERVIEW

7.1.1 What Is Risk Communications?

Our journey through risk management components in civil infrastructures has showed us one important theme, albeit in an implicit manner: there is a need to interact with different types of stakeholders throughout the risk management process. These stakeholders can be the risk managers themselves, infrastructure owners, public officials, insurance companies, funding agencies or groups, and members of the public at large. Without successful interaction, the very objectives of risk management can be at peril. Because of this, we offer two important points:

1. We group all of these interactions as a separate risk management component: risk communication.
2. This risk communication component may be the most important component in the risk management process. Therefore, we devote this chapter to discuss its different aspects.

As a start, we look for a suitable definition of risk communication. Infanti et al. (2013) defined health risk communications as an *exchange of information*. Covello and Sandman (2001) specified such an exchange to be about health risks caused by environmental, industrial, or agricultural processes, policies, or products among individuals, groups, and institutions. For our immediate purposes (in civil infrastructure risk management), we allow risk communication to be an exchange of information during all phases of risk management among all potential stakeholders, individuals, groups, and institutions.

It is helpful to make a distinction between risk communications as we defined above and crisis communication. Risk communication covers exchange of information, and crisis communication, as Infanti et al. (2013) define, is a spontaneous and reactive process that often occurs in unexpected emergency situations. Of course we can argue that this definition of crisis communication can be considered a part of our more encompassing definition of risk communication as a component of risk management.

7.1.2 Lasswell Model

The Lasswell model for risk communication is based on segmenting the communication process into five basic steps (Lasswell 1948; Gutteling and Wiegman 1996). Gutteling and Wiegman (1996) summed up the model as: "Who says what to whom, how, and with what effects." The model as illustrated in Figure 7.1 and is based on identifying:

1. The *source* of communication, the originator (who?)
2. The *message* of the communication process (what is being communicated?)
3. The *medium* (sometimes called the *channel*) of communication (how is the message being channeled?)
4. The *receivers* of the message (to whom is the communications aimed at?)
5. The *effects* of the communications (what are the *desired/achieved* effects of this communication process?). By comparing the achieved effects with the desired effect, the success of the communication process can be measured.

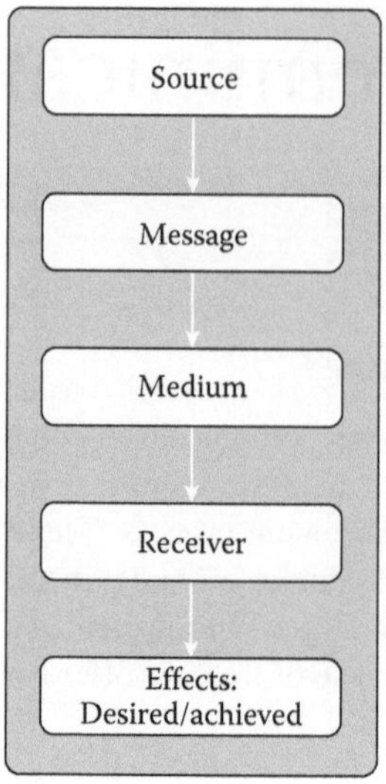

FIGURE 7.1 Lasswell model for risk communications.

TABLE 7.1
General Risk Communications Objectives

Risk Communications Objective	Typical Receivers
Increase public trust	P
Cultivate partnerships with different stakeholders	Pro, P, and O
Understand and accept different management decisions	P and O
Improve education programs	Pro, P, and O
Increase participation of all stakeholders	Pro, P, and O

Note: O: officials/decision makers; P: public: Pro: professionals.

The Lasswell model is fairly simple and logical, yet powerful and encompassing. Because of this, we use it as the basis of our exploration of risk communications in civil infrastructures.

7.1.3 Objectives of Risk Communications

Many authors have addressed the objectives of risk communications; see, for example, Gutteling, and Wiegman (1996) and Sheppard, Janoske, and Liu (2012). In our discussion of risk communications objectives, we recognize that communications objectives can be general (see Table 7.1). Some communication objectives can also be specific to the risk management component and to the receivers of the message (some of these specific objectives are shown in Table 7.2).

7.1.4 This Chapter

We start this chapter by exploring three important tenets of risk communications: (a) its centrality in the risk management process, (b) the different theories that aim at understanding the nature of risk communications, and (c) the rich and varied objectives of risk communications. We follow that by delving at depth into the Lasswell model of risk communications, especially as it applies to civil infrastructures. We show that each of the components of the model are closely interrelated with the other components of the model, and only by understanding such interrelations, can the risk manager

TABLE 7.2
Specific Risk Communications Objectives

Risk Management Component	Objective	Typical Receiver
Risk assessment	Relay information about different important issues during assessment process	Pro, P, and O
Risk acceptance	Help in setting different acceptance thresholds	Pro, P, and O
Risk treatment	Secure funding for different treatment projects	O
	Prioritize treatment projects	Pro
	Improve public acceptance for any inconveniences while executing projects	P
Risk monitoring	Allow for participation of different organizations while forecasting risk variables	Pro
	Secure funding for different treatment projects (especially when monitoring consequences)	O
	Improve public acceptance for any inconveniences while executing projects	P

Note: O: officials/decision makers; P: public: Pro: professionals.

produce an effective communication process. Then we offer an objective communications model to produce an objective assessment of the effectiveness of the communications; the process is based on the Lasswell model. We end the chapter with a discussion of the specifics of communications of risk components and risk management components. We argue that communicating each risk component (vulnerability, threat/hazard, consequences, reliability, and/or exposure) might require different considerations. Similarly, we show that communicating different risk management components (assessment, acceptance, treatment, and/or monitoring) may require different communications strategies.

7.2 ISSUES OF IMPORTANCE IN COMMUNICATING RISK

7.2.1 Centrality of Risk Communications within Risk Management

We argued earlier that communicating risk is central to the whole process of risk management (see Chapter 1). This idea is not new and was illustrated by Chorus and Bartram (1999), Tabbaa (2010), and Infanti et al. (2013), for communicating risk within a health risk management process as illustrated in Figure 7.3. Our point here is to highlight the centrality of communicating risk in both communities (health and civil infrastructure). Of course, communication issues vary greatly as evident by a quick comparison of Figures 1.16, 1.17, and 7.3.

7.2.2 Theories of Risk Communications

Numerous theories exist regarding risk communication aspects. Gutteling and Wiegman (1996) and Sheppard et al. (2012) summarized and discussed many of these theories. We note that the vast majority of existing theories of risk communications involve one of the following attributes:

1. They seldom involve lower limit states. They concentrate mainly on failures as the subject of the communications message. Given the fact that lower limit states represent most everyday activities, there is an immense knowledge gap in the current state of risk communications theory.

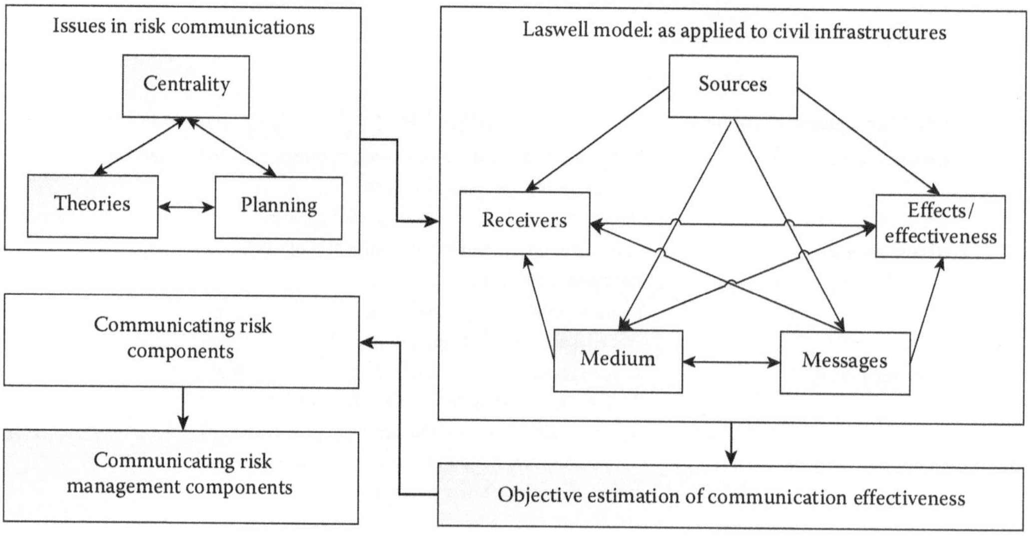

FIGURE 7.2 Contents of risk communications chapter.

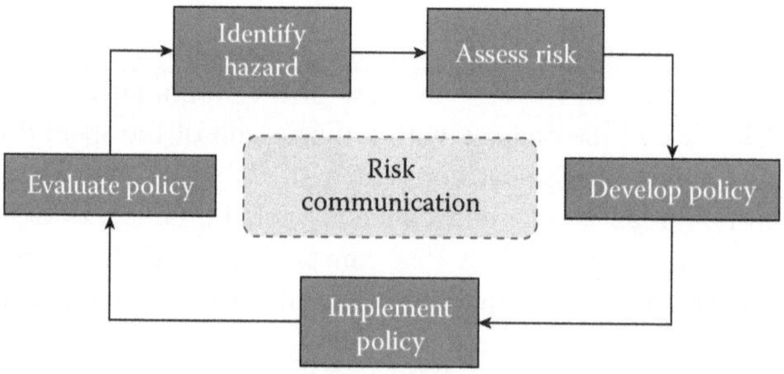

FIGURE 7.3 Centrality of risk communication in health risk management. (Based on Infanti, J., et al., *A Literature Review on Effective Risk Communication for the Prevention and Control of Communicable Diseases in Europe,* ECDC, Stockholm, Sweden, 2013; Tabbaa, D., *Int. J. Antimicrob. Ag.*, 36(1), 80–83, 2010; Chorus, I., and Bartram, J., *Toxic Cyanobacteria in Water: A Guide to Their Public Health Consequences, Monitoring and Management,* E & FN Spon, London, 1999.)

2. They relate to the public as a receiver in the communication efforts.
3. They involve, primarily, professionals as a source.
4. They center on major disasters, specifically, health, natural, or man-made disasters.

We discuss different risk communications theories in different parts of this chapter. Figure 7.4 illustrates two of the more generic communications theories that Gutteling and Wiegman (1996) describe in detail.

7.2.3 Planning Risk Communications

Risk communication efforts invariably start with the source. From that point, planning efforts should follow similar steps shown in Figure 7.5. Closer inspection of the planning steps in Figure 7.5 show that it follows closely the Lasswell's model described in Section 7.1.2. The planning effort will first identify the receiver or receivers. Immediately follows the identification of the message that has to be communicated. The next step is the plan design step. Some of the criteria for communication

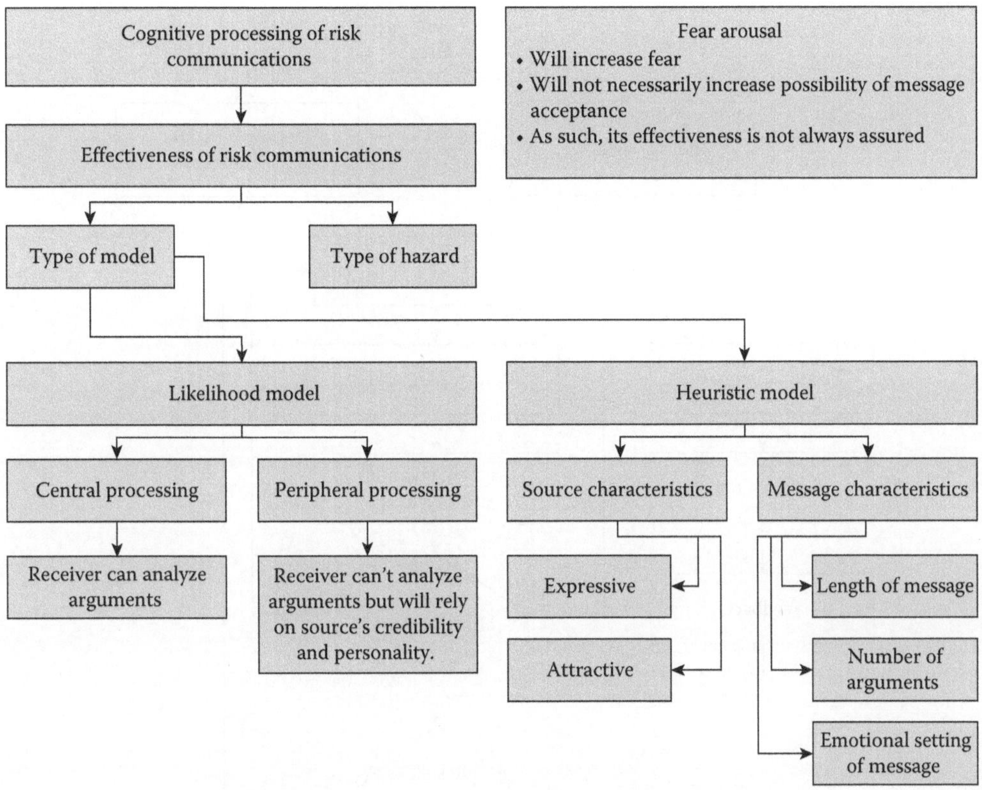

FIGURE 7.4 Attributes of two generic risk communications theories.

adequacy as well as the adequacy of the communication methods are shown in Tables 7.3 and 7.4, respectively. The designing of the plan will then commence by identifying the medium followed by analyzing for potential effectiveness (see Section 7.7 for an objective methodology to estimate the effectiveness of risk communication). If the estimated effectiveness is acceptable, then the communication plan will be produced and implemented. In most situations, a feedback from the receivers and perhaps all others involved in the plan should be sought. There are two decision points in the communication plan in Figure 7.5. The first decision point occurs when evaluating the effectiveness of the plan as designed. If the effectiveness is deemed unacceptable, then a revision of the plan is needed. The second decision point is after receiving a feedback from the implementation of the plan. If the feedback shows an unacceptable result from the communication efforts, then a redesign of the plan is required.

The plan shown in Figure 7.5 is generic. Of course, the degree of sophistication and the steps as shown need changes, as the situation requires. For example, an interoffice communication might not be as detailed or as extensive as a risk communication effort to warn the public prior to an impending hazard or a communication effort that aims at securing funding for a major civil infrastructure risk mitigation project.

7.3 SOURCES OF RISK COMMUNICATIONS

As noted earlier, according to the Lasswell model of risk communications, the whole process starts with the source of the communications. We can categorize the sources into two generic categories: individuals and organizations. Each of these is categorized further into a set of subcategories as shown in Figure 7.6.

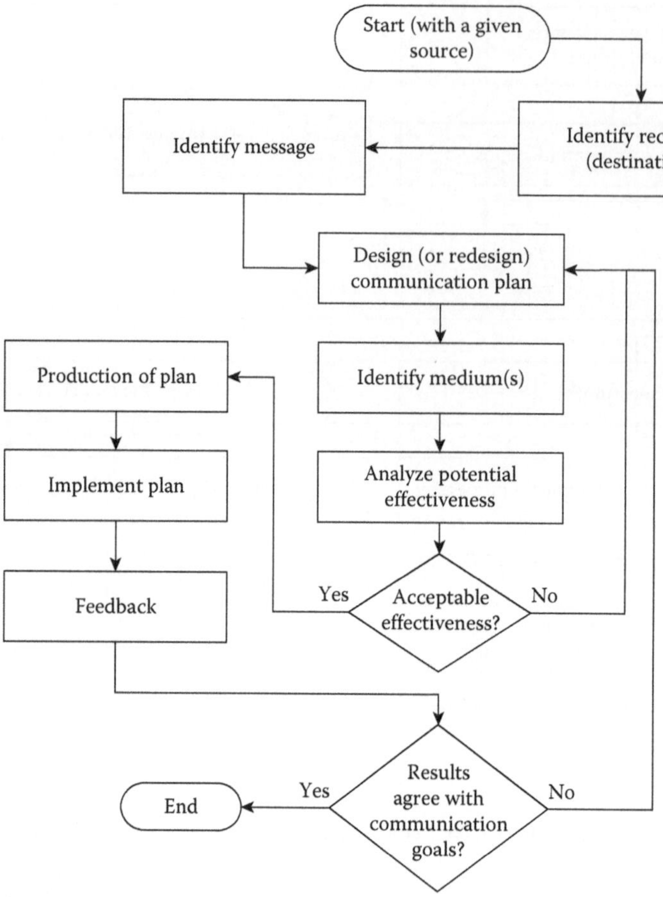

FIGURE 7.5 Systematic risk communication planning.

TABLE 7.3
Criteria for Adequate Risk Communications

Criteria	Comments
Clear motivation/goals	This is a very important first step and all involved must understand it clearly.
Is it misleading?	Misleading messages can hurt credibility and can hamper future communication efforts.
Can it raise any doubts?	Choice of words is important, as professionals and the public may perceive them very differently. For example, a structurally deficient bridge does not mean the bridge is unsafe but public perception and understanding is very different from that of professionals.
Completeness	Increases effectiveness and reduces efforts/resources required in the long-term.
Which component of risk management can it be used for?	Conveying vulnerabilities, consequences, or mitigation efforts, approach and usage of words has to be different based on the goals of the message.

TABLE 7.4
Adequacy of Communication Methods

Attributes of Communication Method		Comment
General	Comprehensive	Should give full picture to receivers.
	Suitable to communication objective?	The method needs to be consistent with the stated objectives.
	Evaluation	The method needs to be helpful, accurate, and clear.
Specific	Single hazard or multihazards?	Unintended consequences of different hazards need to be addressed, whenever pertinent.
	Uniformity	If there are other communications regarding the same issue, all methods/subjects have to be uniformly designed.
	Contingency for communications failure	Redundant communication methods need to be planned, whenever pertinent.

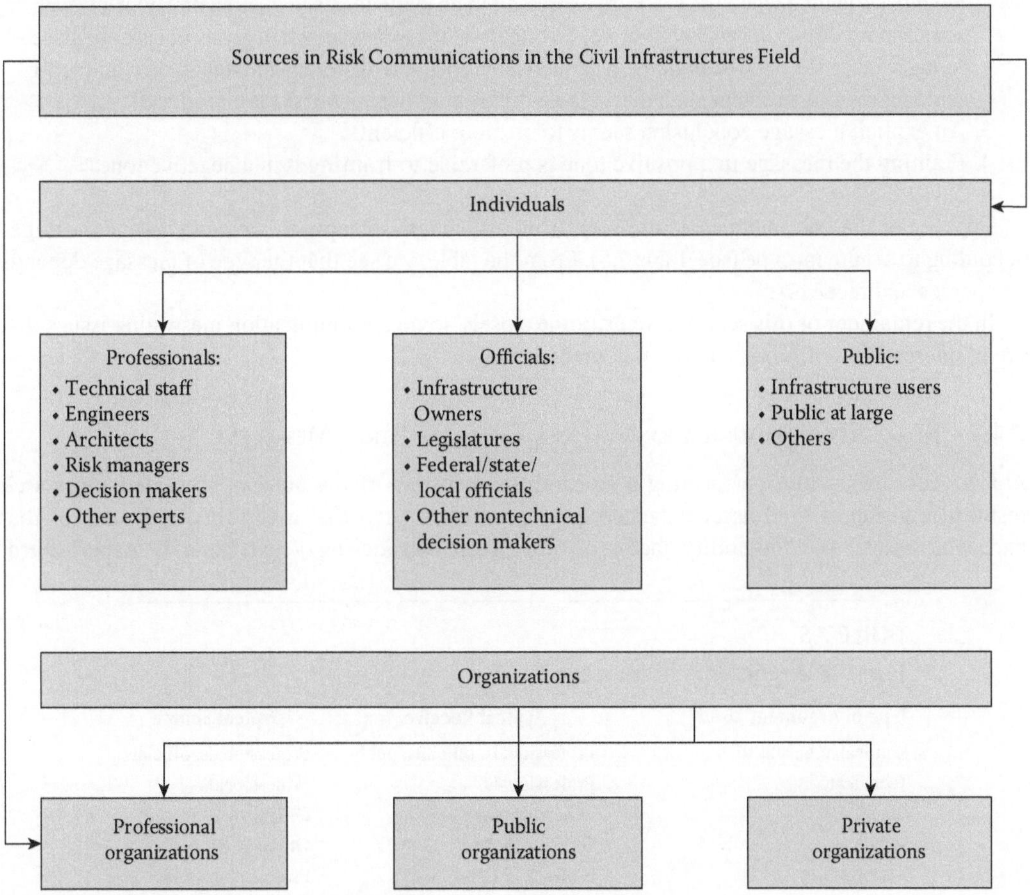

FIGURE 7.6 Sources in risk communications.

7.4 COMMUNICATION MESSAGE: HOW TO COMMUNICATE RISK

7.4.1 Overview

There are several attributes to effective risk communication messages. For example, Chorus and Bartram (1999) pointed to the need of the message to be:

1. Dynamic, that is, changing according to situation.
2. Interactive, that is, the exchange between the *receivers* and *sources* must be part of the message.
3. Inclusive, that is, all potential stakeholders should be considered in the risk communication plan.

Looking at what provides for an effective message, Gutteling and Wiegman (1996) provided several criteria, such as:

1. The effectiveness of defining risk as a probabilistic value, for example, 75% probability, as compared with defining risk in subjective terms, such as high risk, depends on several factors including the problem at hand (catastrophic hazard or corrosion hazard) and the receiver (public or professionals).
2. Comparing different levels of risk can be an effective message in itself. Such comparisons are most effective when the risks are expressed in absolute terms (e.g., monetarily). However, sometimes comparing risks might not be effective, for example, when levels of risk are close to each other. In such situations, it is advisable to use a different messaging technique or remodel the risk problem such that sizable differences between risks are produced.
3. An explicit message conclusion seems to be more efficient.
4. Framing the message in a positive tone is preferable to framing it in a negative tone.

Looking at the communication message from a different viewpoint, we can classify messages according to argument type (see Table 7.5). From the table, we see that the type of message depends on sources and receivers.

In the remainder of this section, we examine closely some communication messaging issues that are of interest to civil infrastructure risk management.

7.4.2 Reliability Communications vs. Risk Communication Messages

A few years ago, while performing a research project, one of the authors presented a research result to a group of civil asset managers. The presenter stated that a certain component of that particular asset has a reliability factor of 5.32 when considering a particular demand/hazard.

TABLE 7.5
Types of Arguments in Message

Type of Argument Used	Typical Receiver	Typical Source
Rational, e.g., statistics	Professionals, officials, public	Professionals, officials
Technical	Professionals	Professionals
Emotional	Officials, public	Officials
Vivid	Officials, public	Officials
Narrative	Professionals, officials	Professionals, officials

Source: From Gutteling and Wiegman (1996).

TABLE 7.6
Communicating Risk and Reliability

Aspect of Communication	Reliability	Risk
Generic form	*f(Cap, Dem)*	*f(C,T,V)*
Failure limit state?	Yes	Yes
Lower limit states (LLS)	Not well represented	Yes
Metric of communication	Analytical, e.g., reliability factor or factor of safety	Analytical, monetary
Amenability to LCA	Can be complicated	Yes

The group commented that such an expression was not too useful. The presenter then stated that this exact condition is equivalent to a cost of $3.50 million over the next two years. The group remarked that such an expression is more useful for their asset management practice.

We can make few useful observations from the above experience. First, we note that the reliability factor is high and indicated that safety is not an issue for the particular component under consideration. This means that the presenter was dealing with one of the Lower Limit States (LLS), which we covered at length in earlier chapters. Because safety was not an issue, using the reliability factor as a communication metric was not too successful. Only when the presenter moved into a risk metric (cost) did the audience relate to the situation at hand.

We conclude then that there are differences between communicating reliability and risk. Table 7.6 shows some of these differences. Because risk has consequences built in as one of its basic components, it is far easier to communicate to different audiences than reliability, which can be difficult to communicate at times, especially at the much more prevailing and encountered LLS.

7.4.3 Communication Messages with the Public

When communicating with the public, there are certain qualitative rules that the risk manager might want to follow. Chess et al. (1994) addressed these rules. Among them is to address the reasons for communicating risk, the audience and the type of information they need/desire to know, and the type of information that the risk manager wants to communicate to the intended audience. Also, the method of communication, the feedback from the audience, and information regarding risk treatment plans including: how they are to be implemented, schedules for completion, and the goals of these plans. Chess et al. (1994) summarized these issues concisely in ten questions that can be used as a checklist in any risk communication effort with the public.

7.4.4 Obstacles to Efficient Communication Messages

Covello and Sandman (2001) observed that several obstacles to communicating risk exist. But in general, they specified four obstacles as follows:

1. The underlying nature of risk itself offers an obstacle to communication. Remember that risk, by nature and as we have covered its many attributes in this book, has the following qualities:
   a. Uncertain/probabilistic nature
   b. Complex: Its underlying variables are many and the interactions between these variables can be fairly complex
   c. Incomplete: Due to the many variables and immense complexities involved, risk models can be incomplete

   The message needs to recognize and acknowledge the potential of any/all of the above risk attributes and address them accordingly.

2. Distrust is another obstacle to efficient risk communication message. The origins of such distrust can be:
   a. Disputes between different communication sources
   b. Conflicting messages from different sources
   c. Unskilled communicators/sources
   d. Lack of sensitivity within the message itself
   e. Incomplete participation of all stakeholders, including the public
   f. Historical mismanagement and/or neglect
   g. Historical inefficient or misleading messages

   Effective messages need to overcome the trust issue. This might be possible by avoiding or negating some of the above mentioned sources. Covello and Sandman (2001) offered a more detailed strategy to overcome the trust obstacle.
3. Selectivity of message: For example, when news media become selective in their reporting of a particular event. Another example would be when a risk manager is selective when reporting a particular situation to decision makers/funders (say overstating benefits and understating costs). The remedy for this obstacle is self-evident: try to be as comprehensive in the message as possible.
4. Different psychological/social factors such as:
   a. Erroneous calculations of probabilities of damaging hazards and their consequences
   b. Prevalence of apathy in some situations
   c. Overconfidence
   d. Difficulty in absorbing or understanding the message
   e. The unreasonable demand of some message receivers (e.g., the public) of objective certainty, which, given the very nature of risk, is not possible to provide
   f. Difficulties in overcoming strongly held beliefs, even in the face of presence of strong evidence

Detailed discussion of obstacles to risk communications and how to overcome them can be found in Covello and Sandman (2001).

7.5 MEDIUMS (CHANNELS) OF COMMUNICATIONS

7.5.1 Overview

Effectiveness of risk communication messages also depends largely on the medium used by the source in communicating the message. The advent of the information and electronic age provided risk managers with additional mediums that were not available only a few years ago. As such, a rich mix of traditional (audio, verbal, statistical, etc.) and modern (video, presentations, television, Internet, social media, etc.) mediums are at the beckon of the risk managers. A judicious choice of the medium can help greatly in increasing the effectiveness of the communications. Next, we explore in detail some of the mediums available to risk managers for communicating their risk messages.

7.5.2 Visual Mediums

Perhaps the most used form of messaging in risk communications are visual messaging techniques. Among these methods are:

- *Network visuals*: These can be risk models, which we used in many of our case studies.
- *Activity networks and diagrams*: These are used in risk treatment projects that can relate different project activities in a simple manner to different stakeholders as well as the project team.

- *Different charting types*: These include bar charts, line and scatter diagrams, and trend lines. There are numerous other types of charting methods, which we used in different case studies in this book.
- *Tree-style visuals*: These include decision trees, fault trees, etc. These models are specialized cases of network visuals. They offer simplicity; however, they are not suitable in dealing with complex interactions between nodes.
- *Flow charts and flow processes*: Very useful in relating different aspects of risk-related problems and situations. These can be used in any risk management phase or component with great effectiveness.
- *Gantt charts*: Well-known visual technique that is used in project management processing. Can be adapted for use in risk management problems that use DBN and/or LCA.

7.5.3 Statistical/Analytical Mediums

Communicating risk through analytical messaging is most effective when both the *source* and the *receiver* are both reasonably familiar with the analytical method used in the *message*. When there is discrepancy in the degree of familiarity of the analytical method used in the message, the communication may lose its effectiveness. For example, this book relies heavily on the use of histograms to communicate different messages and lessons of the case studies. Readers who are familiar with histograms may get more benefit from the different case studies than those readers who are not very familiar with histograms. In addition to histograms, other statistical metrics such as averages, variances, coefficients of variation (COVs), and so forth, may also prove to be reasonable risk communications methods. This is not too surprising because uncertainties, after all, are among the foundations of risk.

More advance analytical or statistical methods for communicating risks also may be used for complex risk communications messaging. Some of those methods are the Kano model, integrated definition methodology with function modeling (IDEFO), kappa statistics, matrix data and diagrams, prioritization matrix, process capability, process precision program simulation, tables, or value analysis. Detailed discussion of these methods is beyond the scope of this book. The reader might want to consult other references such as Ayyub (2003), Jorion (2007), and Vose (2009) for more details on analytical risk communication methods.

7.5.4 Qualitative Mediums

Qualitative methods of communicating the message are as important as any other method. In some cases, they can be the only available means of risk communications because of their timeliness, cost effectiveness, and efficiency. These methods include:

- *Direct verbal communications*: Perhaps the most efficient of all communication methods. It can be delivered in real time at almost no cost.
- *Groups*: Using groups of different stakeholders can help in focusing the different risk management messages while executing any of the components (assessment, acceptance, monitoring, and treatment).
- *Surveys*: These are particularly effective messaging methods because they can be used in:
  - Preparing conditional probabilities tables (CPTs)
  - Relating assessment results in the form of survey questions
  - Helping in prioritizing risk treatment projects
  - Explaining and justifying risk monitoring efforts
  - Help in setting risk acceptance thresholds
- *Voting*: In certain situations, voting results can communicate the fairness of the different decisions to stakeholders. This can be especially effective in risk treatment decision processes.

TABLE 7.7
Communications Objectives and Receivers of Messages

	Suitability to Receiver		
Potential Objectives of Communications	**Technical/ Professional**	**Officials (Local, State, Federal)**	**Public**
Reduction of fear	L	M	H
Change of attitude/ behavior	M	M	H
Information transfer	H	H	M
Budget prioritizations/ allotments	M	H	L
Needs for treatment	H	H	L
Treatment strategy	H	H	L
Life cycle	H	M	M

Note: L = Low, M = Medium, and H = High

7.5.5 Others Communication Mediums

Several other communication mediums are available: audio, verbal, and electronic mediums such as found in the social media field and its wealth of applications. The suitability to the task is mostly situational. Demographics also play a major role, depending on to whom the message is being delivered. As usual, the effectiveness of the communications effort depends on the proper application of the medium used in communicating the risk message.

7.6 RECEIVERS OF RISK COMMUNICATIONS

Traditionally, risk communication receivers are mostly presumed to be the public at large, especially when communicating health risks or when there is an impending natural or man-made hazard. For risk in the civil infrastructure, we generalize the potential of risk communication receivers to all potential stakeholders. Thus, all potential sources shown in Figure 7.6 can also be potential receivers, depending on the situation. Table 7.7 shows that for different types of risk communication objectives, the potential receivers can be fairly varied. Following this generalization of receivers, any source type can be aiming their communication message to any receiver type. The communication message, medium, and effectiveness need adjustment accordingly.

7.7 CASE STUDY 7.1: EFFECTS OF COMMUNICATIONS: QUANTIFICATION OF LASSWELL MODEL

7.7.1 Overview

The very reason of the existence of risk communications is to produce an *effect*. Traditionally, that elusive effect or effects have been measured/evaluated subjectively. We introduce in this section, an objective process to measure the effectiveness of communications. We note that this process is specifically tailored for the risk management process as we have presented it in this book: fairly applicable to civil infrastructures. This process might need some modifications before it is applied to other risk management communities.

7.7.2 Model

We use a BN model to simulate the risk communications process with the Lasswell model as its foundation. The model is shown in Figure 7.7. Note that the nodal variables in the model represent the five components of the Lasswell model as we discussed them in Sections 7.1.2, and 7.3 through 7.6, given the close interrelation between *Medium* and *Message*, we introduce a variable, *Mess-Med*, to accommodate the net effects of these two variables. The states of nodal variables of the model are shown in Table 7.8. The definitions of those variables are given in Appendix 7.I (Table 7.14).

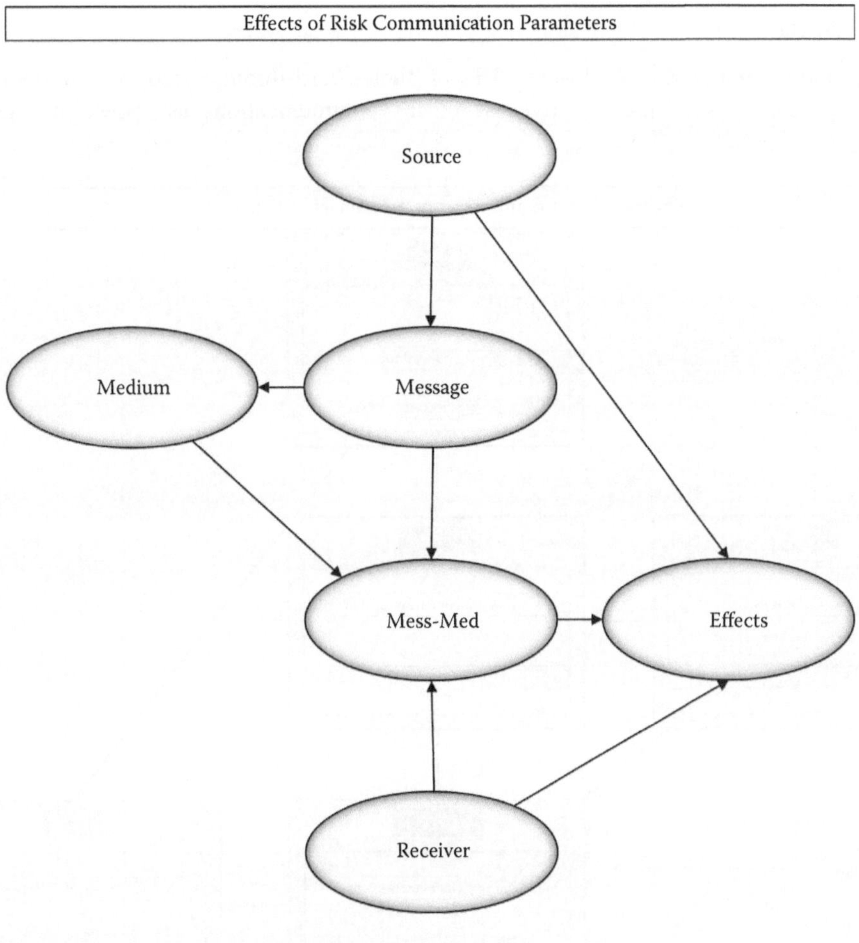

FIGURE 7.7 BN for Lasswell model of risk communication.

TABLE 7.8
States of Nodal Variables

Variable	# of States	States
Source	3	P, O, Pro
Message	2	Nontechnical, Technical
Medium	2	Nonvisual, visual
Mess-Med	2	Not effective, effective
Receiver	3	P, O, Pro
Effects	2	Not effective, effective

Of course, several states can be added to each nodal variable; however, we opted for only two states per variable to simplify the model.

7.7.3 Conditional Probabilities Tables

The CPTs of this model should be based on surveys the risk managers used to measure the relationships between different variables in the model. We provided some estimates of potential CPTs, to illustrate the functionality of the model, in Appendix 7.II.

7.7.4 Results

Using the model of Figure 7.7 and the CPTs of Tables 7.15 through 7.20, we can compute the historical probabilities of the effectiveness of the communications as shown in Figure 7.8.

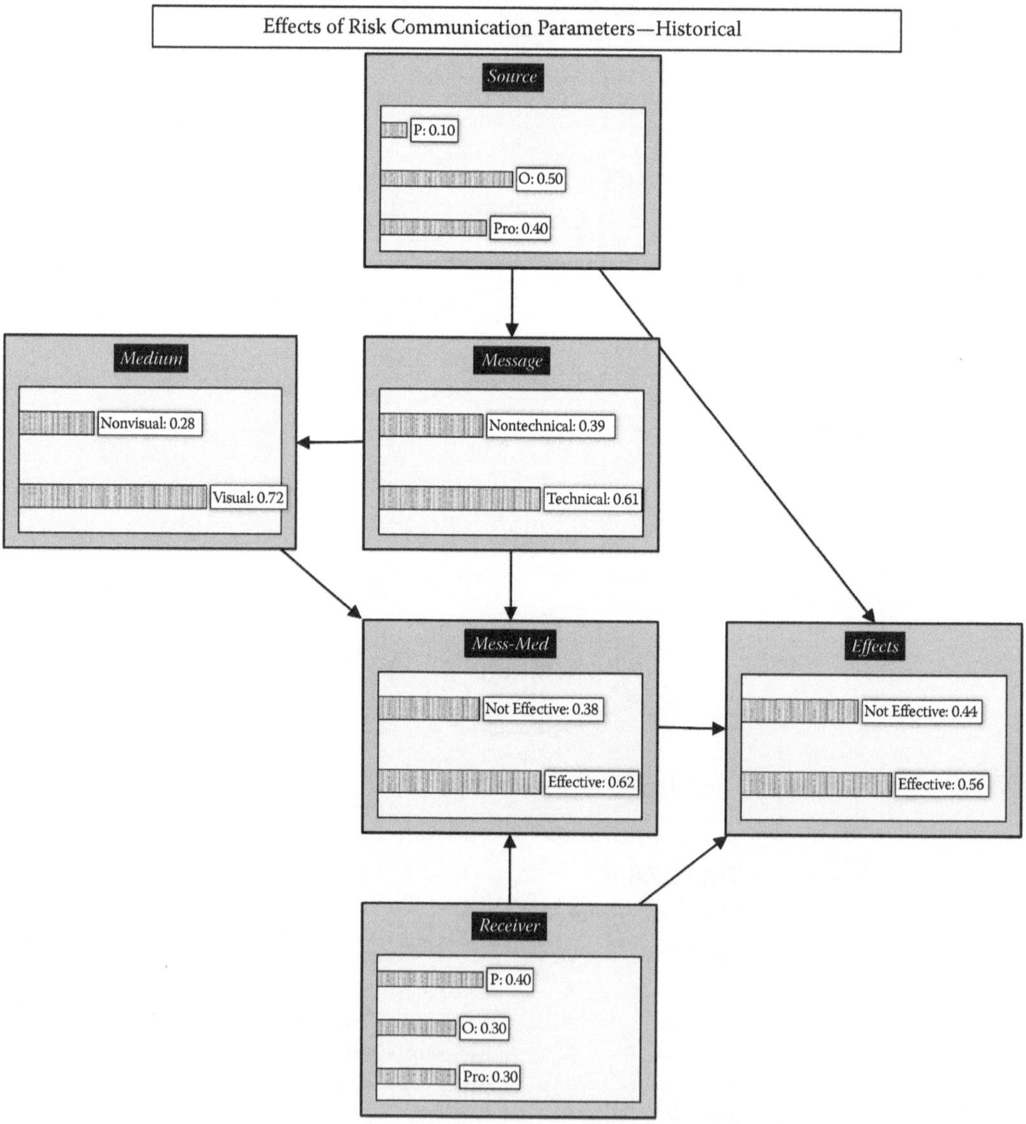

FIGURE 7.8 Historical probabilities of risk communications effectiveness.

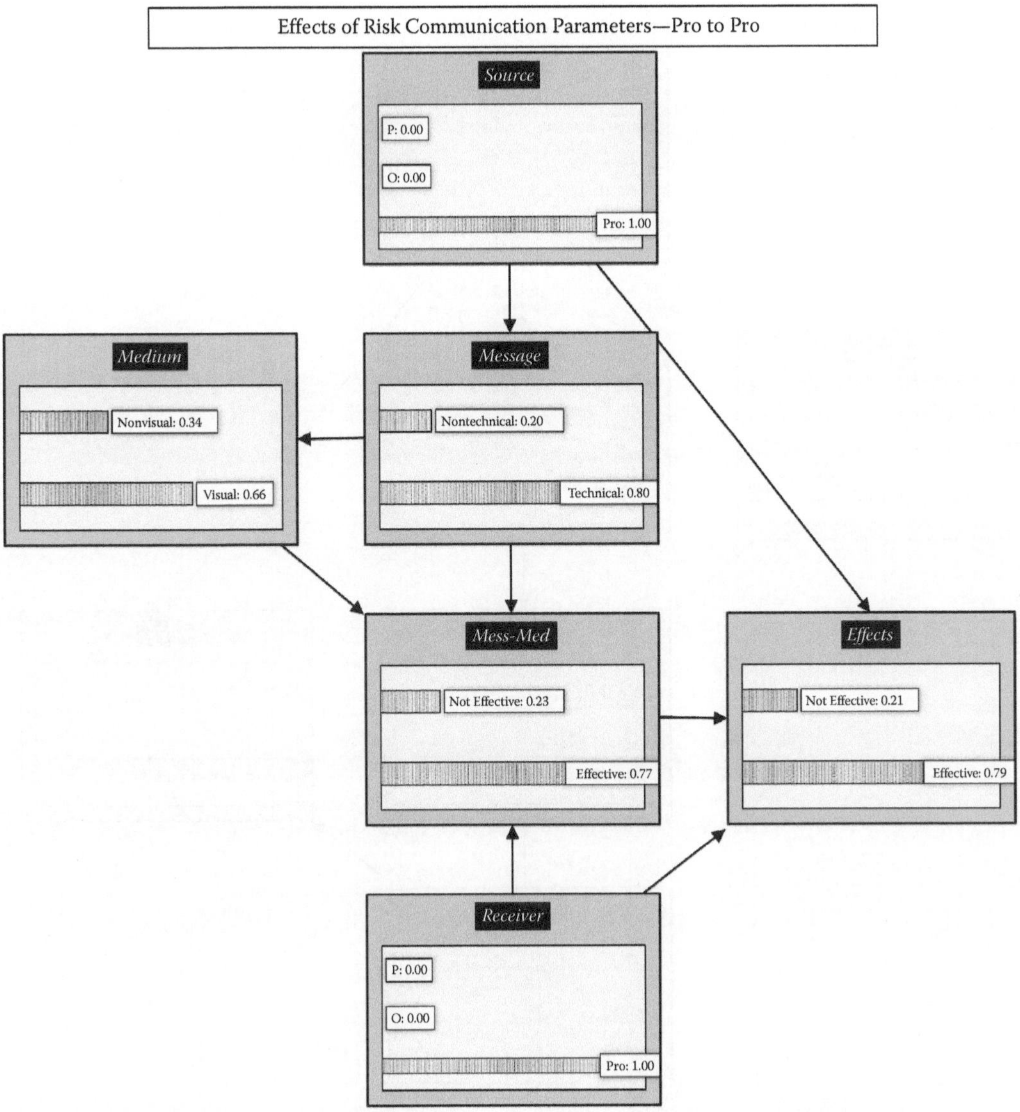

FIGURE 7.9 Risk communications effectiveness when both sources and receivers are *Pro*.

For this particular model, the communications are effective 56% of the time. In situations where the risk communications are specifically conducted from professionals (as sources) to professionals (as receivers), the effectiveness of the communications increases to 79%, as shown in Figure 7.9. In situations where the sources are professionals, the receivers are the public, and the professionals used nontechnical messages and a visual medium (such as colorful presentations or video graphics), the effectiveness of the communications are reduced to 58%, as shown in Figure 7.10. However, when the message and medium changes to technical message and nonvisual medium (such as verbal only medium), the effectiveness of the communications dropped to 42%, as shown in Figure 7.11.

7.7.5 Remarks

Risk managers can use a similar process, shown in this case study, to evaluate effectiveness of their risk communications processes and to optimize the results, while perhaps reducing communication costs.

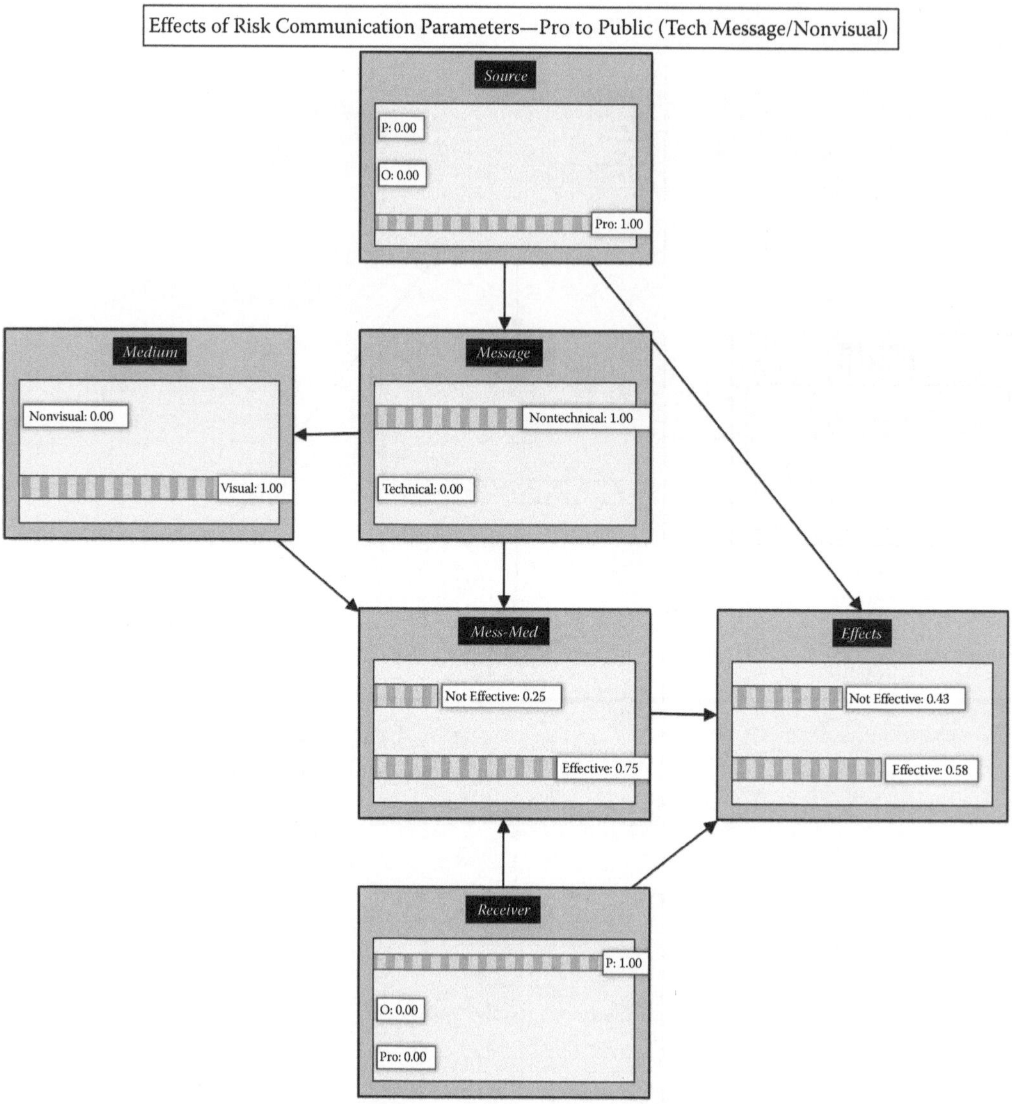

FIGURE 7.10 Risk communications effectiveness when sources are *Pro* and receivers are *Public* (nontechnical message/visual medium).

7.8 COMMUNICATING BASIC RISK COMPONENTS

7.8.1 Overview

Our opinion throughout the book has been that risk (as applied to the civil infrastructures field) is composed of several basic components. As a result, we expect risk communications to rely heavily on communicating some or all of these components in addition to communicating the risk itself. By communicating the risk components, instead of or in addition to communicating risk itself, the desired communication messages might be easier to compose and the communications effects might be easier for the receivers to grasp. The net result would be an effective risk communication effort with obvious benefits to all stakeholders.

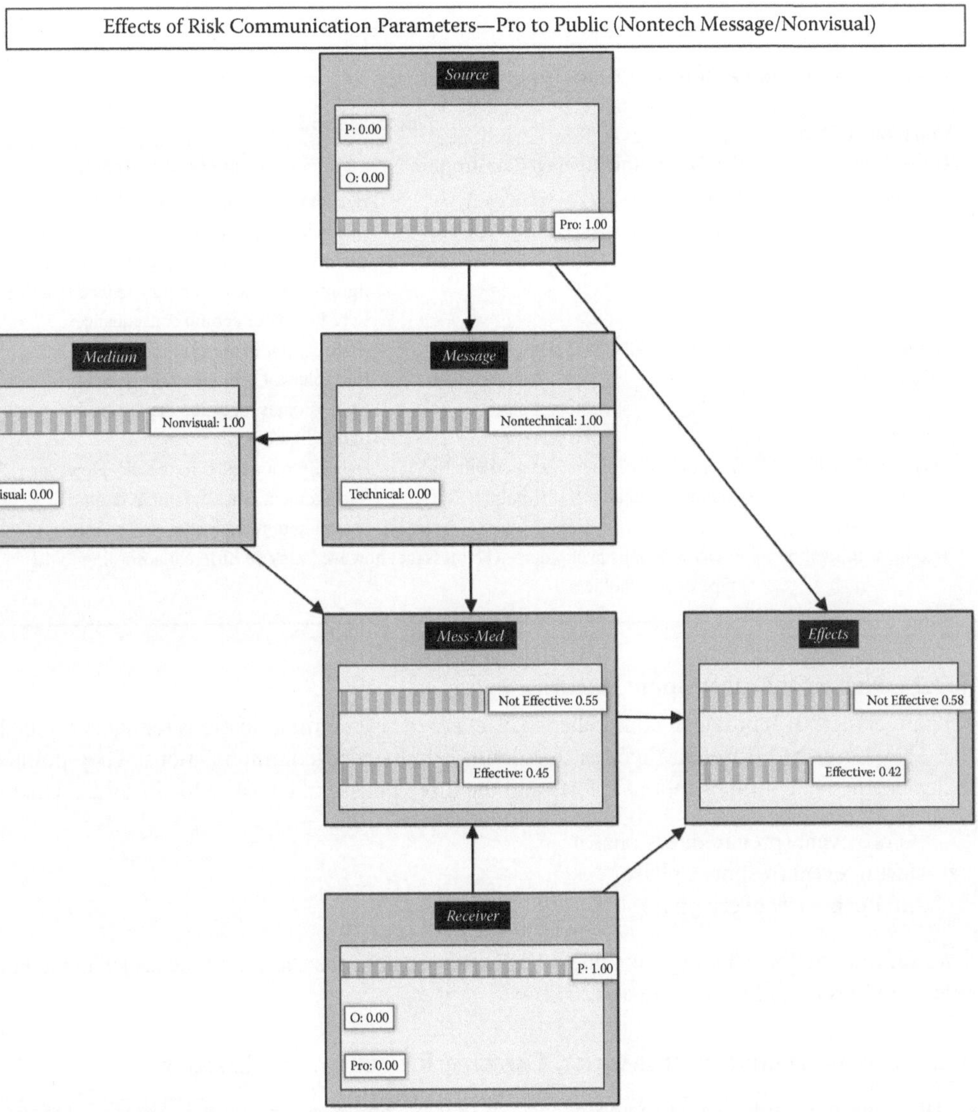

FIGURE 7.11 Risk communications effectiveness when sources are *Pro* and receivers are *Public* (nontechnical message/nonvisual medium).

7.8.2 Communicating Hazards

7.8.2.1 Type of Hazards

Types of communications are functions of the type of hazards. As such, we categorize civil infrastructure hazards into two types:

- Catastrophic or life threatening hazards such as earthquakes, floods (including potential of sudden scour effects), high wind (hurricanes/tornadoes), collision, failure, etc.
- LLS/not life threatening: wear and tear, corrosion, scour (gradual), fatigue, security (man-made hazards), etc.

Table 7.9 shows how communications objectives can vary based on the type of hazards.

TABLE 7.9
Communications Dependence on Type of Hazards

Communications Objectives	Type of Hazard: Safety/Life Threatening/Catastrophic	Type of Hazard: LLS/Not Life Threatening
Education	Essential, but varies in its intensity depending on the receiving audience.	Educating the public might not be as essential as educating professionals, decision makers, and officials (such as funders and public figures). But education may reduce risk due to reduction of certain consequences.
Warnings	Essential to all stakeholders; should be timely.	Generally not needed for the public/nontechnical. Important for owners, professionals, decision makers, funders, and public figures.
Behavior change	Essential to communicate methods for behavioral change to the public.	Generally not needed for public. However, management practices might require changing to reduce or avoid hazards risks.
Treatment/mitigation	Essential to all stakeholders. The message, however, may be different according to the type of stakeholders.	

7.8.2.2 Phases of Catastrophic Hazards

Sheppard et al. (2012) provided comprehensive details of risk communications for catastrophic hazards. They described different theories that control the needed communications. They pointed to the dependence of communications efforts and the three phases of catastrophic hazards, which are:

- Before event (preparedness phase)
- During event (response phase)
- After event (recovery phase)

We summarize their findings in Table 7.10. For a complete description of the issues involved, the reader should refer to Sheppard et al. (2012).

7.8.3 Communicating Vulnerability, Capacity, Reliability, and Exposure

We have seen in Chapter 1 that the propensity of any system or components in the civil infrastructure portfolio to degrade when subjected to a hazard is described by its vulnerability. Another expression for describing the inverse of vulnerability, albeit in a more objective manner, is capacity. Another subclass of risk components is reliability, which in a general sense expresses a relationship between capacity and demand/hazard (see Chapters 1 and 2) and exposure that expresses the relationship between vulnerability and hazard. There are several methods of communicating these risk components, which we used throughout this book.

- *Vulnerability*: Vulnerability has been used objectively and subjectively in risk-based efforts, see FEMA (2005). This is a powerful communication tool because it can be simple and can be comprehended by almost everyone, yet powerful enough to be used in complex risk management models as can be witnessed from FEMA (2005), as well as the different examples in this book.
- *Capacity*: Capacity is expressed in an objective manner. Because of this, its use is universal within professionals and technical stakeholders. Unfortunately, its usefulness as a communications metric to receiving stakeholders without a technical background might be limited.

TABLE 7.10
Communications before, during, and after Catastrophic Events

Phase of Catastrophic Hazard	Applicable Communication Theory/ Model	Specific Communications Steps
Before event/ preparedness	Actionable risk communication model, Wood et al. (2012) Theory of reasoned action (risk information seeking and processing model), Ajzen and Fishbein (1980)	• Circulate different preparedness messages • Use multiple and repeated messaging methods • Ensure that the type of circulated messages to the public is consistent with the public perception and understanding of the risks involved

During event/response	Image Restoration and Repair model, Benoit (1997) and Situational Crisis Communication Theory (SCCT), Ulmer (2001), Coombs (2004), and Coombs (2012)	• Use different pertinent response strategies
		• Use the most effective strategies depending on conditions on the ground. Some of the issues needing consideration are:
		  • Type and extent of catastrophic hazard
		  • Type and experience of responding organization
		  • Public perception of the responding organization

After event/recovery	• CAUSE model, Rowan et al. (2010) • Precaution adoption process model, National Cancer Institute (2003) • Social Amplification of Risk Framework (SARF) model, Rogers et al. (2007)	• Rapid recovery • Recognize issues affecting how to recover from the specific catastrophe • Accommodate potential unintended consequences • Try to completely eliminate vulnerabilities • Account for potential multihazard effects • Accommodate social/economic issues as well as any potential unintended consequences

Source: Excerpted from Sheppard et al. 2012, *Understanding Risk Communication Theory: A Guide for Emergency Managers and Communicators*, Report to Human Factors/Behavioral Sciences Division, Science and Technology Directorate, U.S. Department of Homeland Security, College Park, MD.

- *Reliability*: Reliability, as a relationship between capacity and demand, is mostly an objective metric. Similar to capacity, this objectivity is highly desired by professional and technical stakeholders, while fairly limiting for other stakeholders. However, for experienced professionals, it can also be used as a communication tool in a subjective manner, e.g., safe/unsafe, or highly reliable, limited reliability, etc. This makes reliability a general communication tool for professionals/technocrats.
- *Exposure*: Usually communicated in a subjective manner as a condition rating of the system of interest. Condition ratings are popular in the bridge community. It is the basis of most decisions for bridge management; see Ettouney and Alampalli (2012a, 2012b). Sometimes it is also used to communicate states of infrastructures to other stakeholders, including the public at large (ASCE 2013).

7.8.4 Communicating Consequences

Communicating consequences is much more complex than communicating threats or vulnerabilities. This complexity is because communicating consequences depends on type of sources, message and/or medium, and type of receivers. We can easily contend that all the potential sources and receivers of Sections 7.3 and 7.6, respectively, can communicate consequences. In addition, any of the mediums in Section 7.5 can be used to communicate consequences. The consequence messages fall into two generic categories: monetary (absolute) and relative (subjective). Several case studies of both monetary and subjective consequence messages were covered in different case studies throughout this book.

The consequence communication message can also be either detailed or inclusive. Inclusive consequences would be the integral consequences of the problem.* Detailed consequences can be the values of the components of the inclusive consequences. For example, see Table 3.3 of Case Study 3.1.

Due to the above mentioned complexities as well as the importance of communicating consequences, careful planning is essential before execution. A good practice would be to perform an effectiveness study similar to the one we offered in Section 7.7 to ensure optimal results.

7.9 COMMUNICATING RISK MANAGEMENT COMPONENTS

7.9.1 Overview

We now turn our attentions to risk communications from the risk component view point. As per Figure 7.3, risk communication activities are closely linked to all components of risk management. In what follows, we explore the different flavors of risk communications for each of the four risk management components. We hasten to say that while making our presentations, we only cover a small portion of the rich and varied communication aspects here.

7.9.2 Assessment

The risk assessment process includes several steps. Each of the steps would require different communications activities. Figure 7.12 illustrates a partial listing of those steps and different potential sources, receivers, and messages.

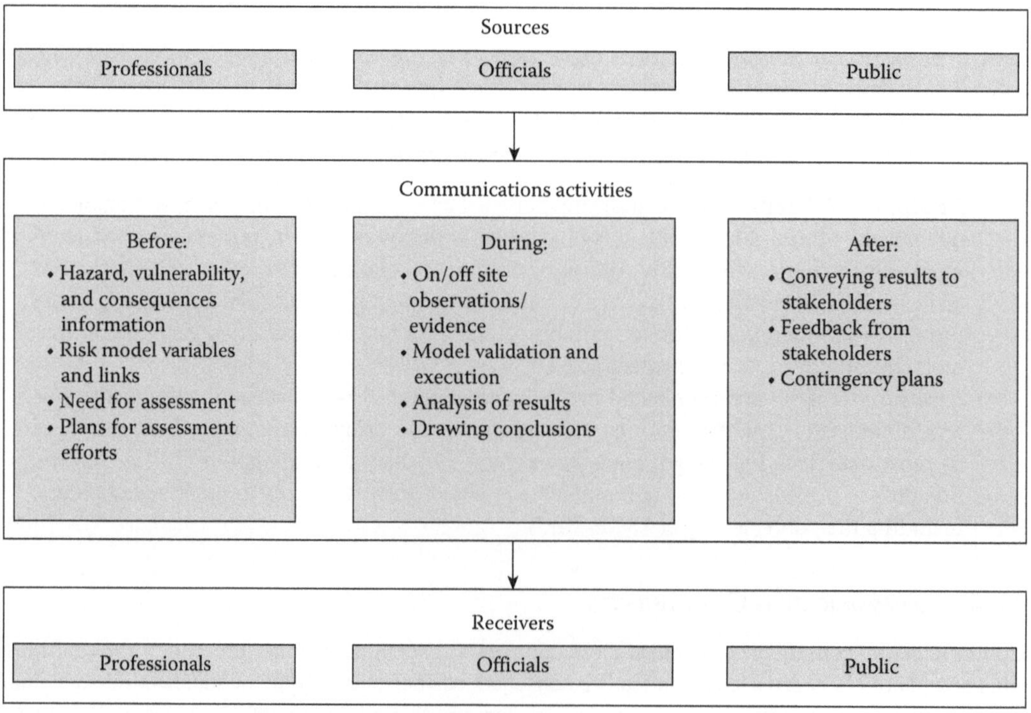

FIGURE 7.12 Risk communication during risk assessment phase.

* E.g., the values of variable C in most of the case studies in this book.

7.9.3 Acceptance

Fischoff et al. (1981) addressed how risk perception and acceptance changes according to the type of undesirable events in a nonobjective manner, as shown in Table 7.11. This means that when communicating risk acceptance thresholds the risk manager should be careful in communicating different thresholds. For example, the table indicates that corrosion risk acceptance thresholds (gradually occurring) may be higher than earthquake risk acceptance thresholds (sudden and catastrophic). For an objective and cost effective risk-based policy, this may not be the most efficient policy. Clearly, such a discrepancy needs to be communicated clearly to the stakeholders to achieve a reasonable and efficient risk management policy with balanced treatment of corrosion versus earthquake risks.

7.9.4 Treatment

While discussing communications for risk treatment components, it is helpful to use the three phases of risk treatment described in Chapter 5. As expected, several types of communications are needed during different risk treatment phases. Table 7.12 shows some of these communications. We note that the communication types shown in the tables are only a small sample of the communications needed. The reader is invited to add to this sample other types of communications as an exercise.

7.9.5 Monitoring

If monitoring processes are reviewed from only a physical viewpoint, their communications issues should be similar to the risk mitigation (Phase II of risk treatment) issues. Some of these communication issues are described in Table 7.12. However, several risk monitoring properties differentiate them from their risk mitigation and, therefore, require different communication handling. Some of these monitoring communication issues are summarized in Table 7.13. Note that numerous other communications issues are not mentioned in the table because of space limitations.

TABLE 7.11
Undesirable Events Influence on Risk Acceptance Thresholds

Source of Undesirable Event	Risk Acceptance Threshold Is High (Accepting More Risk) If:	Risk Acceptance Threshold Is Low (Accepting Less Risk) If:
Controlled events	Risk taken is voluntary	Risk taken is imposed
Risk vs. rewards	Taking the risk has obvious rewards/benefits	Taking the risk has no obvious benefits
Risk applies in a fair manner to stakeholders	Damages apply fairly	Damages apply in an unfair manner
Nature of undesirable events—1	Natural hazards	Man-made hazards
Nature of undesirable events—2	Gradually occurring	Sudden catastrophes
Source of undesirable events (if man-made hazards)—1	Trusted sources	Untrusted sources
Source of undesirable events (if man-made hazards)—2	Familiar sources	Unfamiliar sources
Age of affected persons	Adults	Children

TABLE 7.12
Communications during Different Risk Treatment Phases

Risk Treatment Phase	Step	Communication Source	Receiver	Objective
Phase I: Treatment strategy	Choose overall risk strategy	Pro (risk manager, analyst, etc.)	Officials (decision makers)	Decide on overall treatment strategy
		Officials (Decision makers)	Officials (owners, legislature)	Afford budgets
Phase II: Risk mitigation	Prioritization	Pro (risk manager, analysts, designers, etc.)	• Pro (risk manager, analysts, designers, etc.) • Officials (Decision makers, etc.)	Choose optimal asset and project prioritization order
	Planning	Officials (PR, decision makers, etc.)	Public	Convey specifics about the project and how it might impact the public (benefits, inconveniences such as detours, etc.)
Phase III	Executing the projects	Pro (construction managers)	Owners	Projects progress (technical/financial)
		Officials	Public	Projects progress (from public's viewpoint)

TABLE 7.13
Some Communications Issues for Risk Monitoring

Issue	Communications Source	Receiver	Objective
Monitoring demands (use of the infrastructure, such as current and future traffic demands)	Public	Pro (designers, analysts, or risk managers)	Provide accurate picture of demands
Monitoring consequences	Pro (economists, social scientists)	Pro (risk analysts) Officials (owners, legislature)	Provide accurate data to compute consequences
Forecasting	Pro (economists, social scientists)	Pro (risk analysts)	Accurate LCA
Life cycle costs	Officials (decision makers, risk managers)	Officials (owners, legislature)	Accurate current and future budgeting and funding
Life cycle benefits	Officials (decision makers, risk managers)	Public and officials (owners, legislature)	Justify current and future expenditures

APPENDIX 7.I: DEFINITIONS OF NODAL VARIABLES (TABLE 7.14)

TABLE 7.14
Definitions of Nodal Variables

Variable	Used in Case Study #	Definition
Effects	7.1	Objective measure of the effectiveness of the risk communication effort. Can also be described in a subjective manner
Medium	7.1	Types of mediums used
Message	7.1	Type of message used
Mess-Med	7.1	Effectiveness of the combined *Messages* and *Medium* used in the process
Receiver	7.1	Receiver categories. In the example provided they are: Public (*P*), Officials (*O*), and Professionals (*Pro*)
Source	7.1	Sources categories. In the example provided they are: Public (*P*), Officials (*O*), and Professionals (*Pro*)

APPENDIX 7.II: CPTs FOR CASE STUDY 7.1 (TABLES 7.15–7.20) (OBJECTIVE LASSWELL MODEL TABLES 7.15–7.20)

TABLE 7.15
CPT of *Effects*

Mess-Med		Effective								
Receiver		Pro			O			P		
Source		Pro	O	P	Pro	O	P	Pro	O	P
Effects	Effective	0.9	0.85	0.8	0.8	0.75	0.7	0.7	0.65	0.6
	Not Effective	0.1	0.15	0.2	0.2	0.25	0.3	0.3	0.35	0.4

Histograms for conditional probabilities

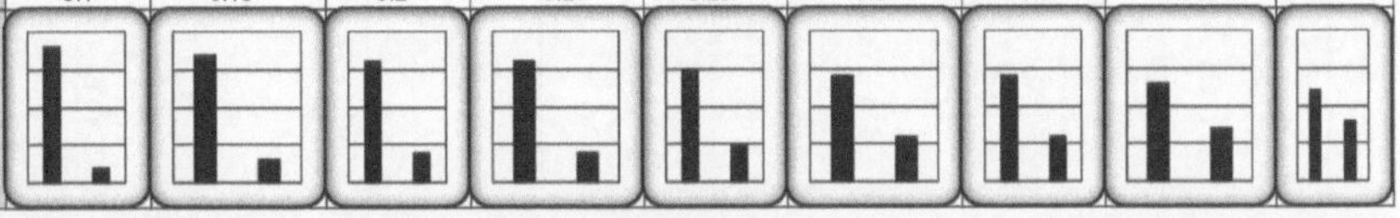

Mess-Med		Non Effective								
Receiver		Pro			O			P		
Source		Pro	O	P	Pro	O	P	Pro	O	P
Effects	Effective	0.4	0.35	0.3	0.3	0.25	0.2	0.2	0.15	0.1
	Not Effective	0.6	0.65	0.7	0.7	0.75	0.8	0.8	0.85	0.9

Histograms for conditional probabilities

TABLE 7.16
CPT of *Medium*

Message		Technical	Nontechnical
Medium	Visual	0.6	0.9
	Nonvisual	0.4	0.1
Histograms for conditional probabilities		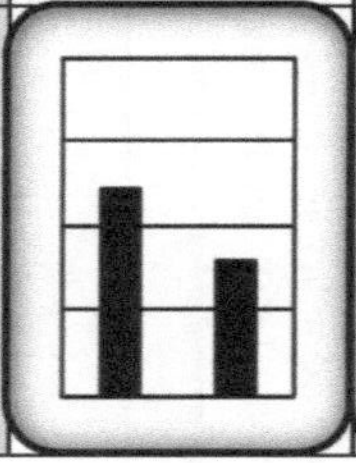	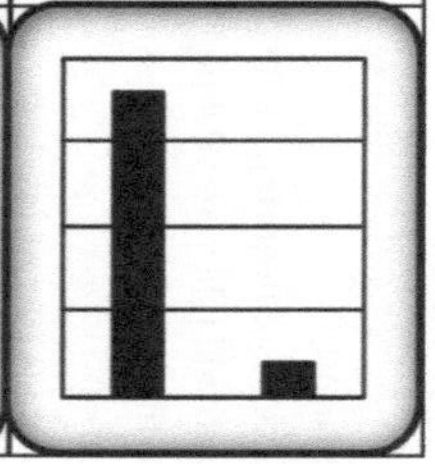

TABLE 7.17
CPT of *Mess-Med*

Receiver		Pro				O				P			
Medium		Visual		Nonvisual		Visual		Nonvisual		Visual		Nonvisual	
Message		Technical	Nontechnical	Technical	Nontechnical	Technical	Nontechnical	Technical	Nontechnical	Technical	Nontechnical	Technical	Nontechnical
Mess-Med	Effective	0.8	0.85	0.7	0.55	0.6	0.7	0.4	0.55	0.45	0.75	0.3	0.45
	Not Effective	0.2	0.15	0.3	0.45	0.4	0.3	0.6	0.45	0.55	0.25	0.7	0.55
Histograms for conditional probabilities		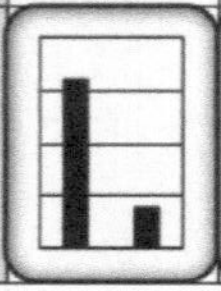		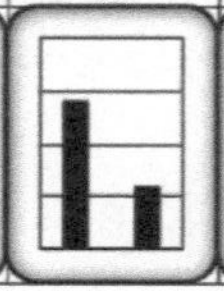		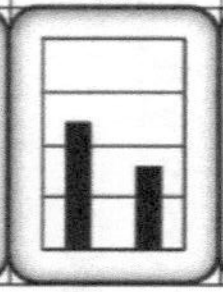		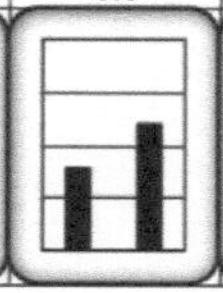		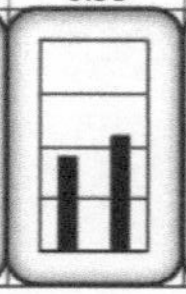		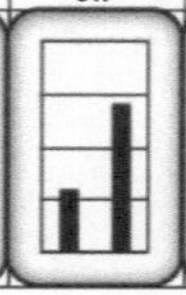	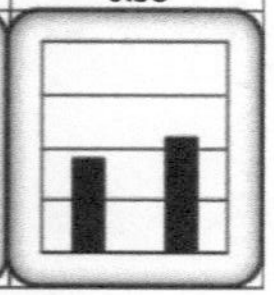

TABLE 7.18
CPT of *Message*

Source		Pro	O	P
Message	Technical	0.8	0.4	0.9
	Nontechnical	0.2	0.6	0.1
Histograms for conditional probabilities		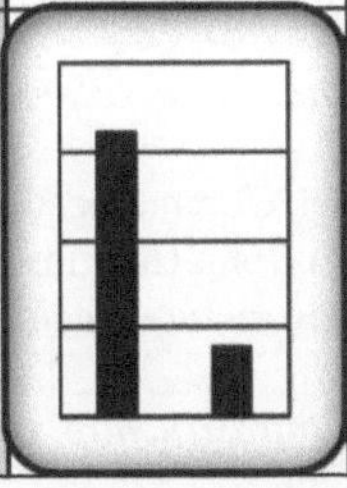	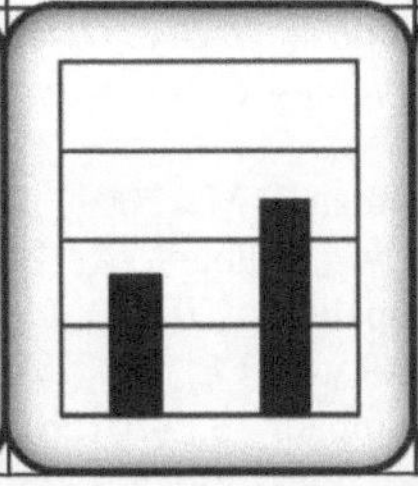	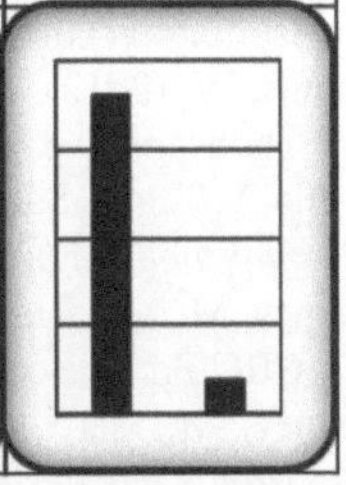

TABLE 7.19
CPT of *Receiver*

Receiver	Pro	0.3
	O	0.3
	P	0.4
Histograms for conditional probabilities		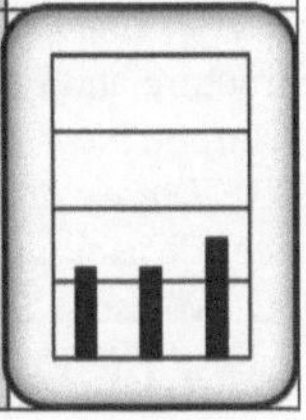

TABLE 7.20
CPT of *Source*

Source	Pro	0.4
	O	0.5
	P	0.1
Histograms for conditional probabilities		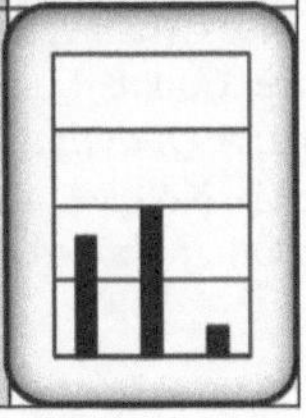

REFERENCES

Ajzen, I., and Fishbein, M. (1980). *Understanding Attitudes and Predicting Social Behavior*. Prentics Hall, Englewood Cliffs, NJ.

ASCE. (2013). *2013 Report Card for America's Infrastructure*. American Society of Civil Engineers, Reston, VA.

Ayyub, B. M. (2003). *Risk Analysis in Engineering and Economics*. Chapman & Hall/CRC, Boca Raton, FL.

Benoit, W. (1997). Image Repair Discourse and Crises Communication. *Public Relat. Rev*. 23(2), 177–186.

Chess, C., and Hance, B.-J. (1994). *Communicating with the Public: Ten Questions Environmental Managers Should Ask*. The Center for Environmental Communication, Cook College, Rutgers University, New Brunswick, NJ.

Chorus, I., and Bartram, J. (1999). *Toxic Cyanobacteria in Water: A Guide to Their Public Health Consequences, Monitoring and Management*. E & FN Spon, London.

Coombs, W. (2004). Impact of Past Crises on Current Crises Communication: Insights from Situational Crises Communication Theory. *J. Bus. Comm*. 41(3), 265–289.

Coombs, W. (2012). *Ongoing Crises Communication: Planning, Managing, and Responding*, 3rd ed. Sage, Thousand Oaks, CA.

Covello, V. T., and Sandman, P. M. (2001). Risk Communication: Evolution and Revolution. In: *Solutions to an Environment in Peril*, Wolbarst. A. (Ed.). John Hopkins University Press, Baltimore, MD, pp. 164–178.

Ettouney, M., and Alampalli, S. (2012a). *Infrastructure Health in Civil Engineering: Theory and Components*. CRC Press, Boca Raton, FL.

Ettouney, M., and Alampalli, S. (2012b). *Infrastructure Health in Civil Engineering: Applications and Management*. CRC Press, Boca Raton, FL.

FEMA. (2005). *Risk Assessment: A How-To Guide to Mitigate Terrorist Attacks*, Risk Management Series, FEMA 452. Federal Emergency Management Agency, Washington, DC.

Fischoff, B., Lichtenstein, S., Slovic, P., Derby, S., and Keeney, R. (1981). *Acceptable Risk*. Cambridge University Press, New York, NY.

Gutteling, J., & Wiegman, O. (1996). *Exploring Risk Communications*. Kluwer Academic Publishers Group, Norwell, MA.

Infanti, J., Sixsmith, J., Barry, M. M., Núñez-Córdoba, J., Oroviogoicoechea-Ortega, C., and Guillén-Grima, F. (2013). *A Literature Review on Effective Risk Communication for the Prevention and Control of Communicable Diseases in Europe*. ECDC, Stockholm, Sweden.

Jorion, P. (2007). *Value at Risk: The New Benchmark for Managing Financial Risk*. McGraw-Hill, New York.

Lasswell, H. D. (1948). The Structure and Function of Communication in Society. In: *The Communications of Ideas*, Bryson, L. (Ed.). Harper, New York, pp. 37–51.

National Cancer Institute. (2003). *Theory at a Glance: A Guide for Health Promotion Practice*. Department of Health and Human Services, Washington, DC.

Rogers, M., Amlot, R., Rubin, G., Wessely, S., and Krieger, K. (2007). Mediating the Social and Psychological Impacts of Terrorist Attacks: The Role of Risk Perception and Communication. *Int Rev. Psychiatry*. 19(3), 279–288.

Rowan, K., Botan, C., Kreps, G., Samoilenko, S., and Fransworth, K. (2010). Risk Communication Education for Local Emergency Managers: Using the CAUSE Model for Research, Education, and Outreach. In: *Handbook of Risk and Crisis Communication*, Heath, R. and O'Hair, H. (Eds.). Routledge, New York, pp. 168–191.

Sheppard, B., Janoske, M., and Liu, B. (2012). *Understanding Risk Communication Theory: A Guide for Emergency Managers and Communicators*. Report to Human Factors/Behavioral Sciences Division, Science and Technology Directorate. U.S. Department of Homeland Security, College Park, MD.

Tabbaa, D. (2010). Emerging Zoonoses Responsible Communication with the Media-Lessons Learned and Future Perspectives. *Int. J. Antimicrob. Ag*. 36(1), 80–83.

Ulmer, R. (2001). Effective Crises Management Through Established Stakeholder Relationships: Malden Mills as a Case Study. *Manage. Comm. Q*. 14(4), 590–615.

Vose, D. (2009). *Risk Analysis: A Quantitative Guide*. Wiley, Hoboken, NJ.

Wood, M., Mileti, D., Kano, M., Kellyey, M., Regan, R., and Bourque, L. (2012). Communicating Actionable Risk for Terrorism and Other Hazards. *Risk Anal*. 32(4), 601–615.

8 Risk Management Applications

8.1 INTRODUCTION

8.1.1 Overview

Previous chapters in this book examined relevant aspects of risk management for civil infrastructures. Risk management offers an objective approach to addressing numerous types of problems the community faces in a comprehensive and efficient manner. It has potential to offer improved performance at reasonable costs. In many situations, we have seen that risk management offers objective solutions to problems, which have only been addressed, thus far, mostly in a subjective manner. This final chapter provides more important applications that showcase these capabilities and the promise of risk management as a comprehensive management method.

Many researchers have already offered studies in risk management, or some pertinent components of it, in addressing various applications in civil infrastructures. For example, Elhag and Wang (2007) argue that approaches to risk assessment for bridge maintenance use in literature (Gralund and Puckett 1996; Abudayyeh and Al-Battaineh 2003; Lounis 2004) generally have been based on conventional statistical and computing techniques and most are incapable of handling qualitative factors that have a significant impact on the risk assessment of bridges. Hence, they propose an alternative method, using an artificial neural network (ANN), suitable for bridge risk assessment, which involves qualitative and quantitative expert judgments. According to the authors, ANN is an effective approach for bridge risk assessment because it has the capability of learning from past experiences or examples, generalizing the captured knowledge for future solutions, and self-updating.

Elhag and Wang used back-propagation neural networks developed to model bridge risk score and risk categories, using 506 bridge maintenance projects from the British Highways Agency. Besides ANN, they compared the results with multiple regression analysis and also proposed two hybrid models that combine neural networks and regression analysis. Risk was defined as a product of likelihood and consequences. All maintenance needs were classified as essential, preventive, and upgrading. Risk events that would occur if no action was taken in 3 to 4 years were identified—protective coating failure, deterioration, failure to meet obligation for required operational capacity, concrete spalling, equipment failure, nonstructural component failure, structural component failure, element failure, or structural failure.

They then broke the risk events down further into a risk event chain. For these risks, four likelihood ratings (certain, high, medium, and low) and four consequence ratings (High*, high, medium, and low) based on factors related to safety, functionality, sustainability, and environment were obtained using engineering judgment. Overall risk associated with a risk event was determined by a combination of likelihood and consequence. High* was assigned only for safety-related risks, and these were assigned a value of 100 to give them high priority, whereas all other projects were scored under 100 in terms of the benefits and detriments that each project was going to deliver.

Based on their study, it was concluded that risk score and risk category for bridge maintenance projects can be modeled well using neural networks, neural networks performed better than multiple regression models, and ANN models combined with regression models performed much better.

Evans et al. (2004) applied risk assessment processes to erosion hazard at the Magela Creek and floodplain, which are located within Kakadu National Park in Australia. These are listed as Wetlands of

International Importance and recognized under the World Heritage Convention. Kakadu National Park is one of the largest and most environmentally diverse national parks worldwide and is managed for conservation, tourism, and natural and cultural features and values (Kakadu Board of Management and Parks Australia 1998). The Jabiluka Uranium Mine site is on the Ngarradj, the downstream waterway that flows into Kakadu National Park. Evans et al. (2003) assessed the impact of constructing future waste rock dumps (WRDs) at the Jabiluka Uranium Mine site by estimating the possible erosion, that is, soil loss. Based on the extensive monitoring data available, they used Monte Carlo Simulations, with 500,000 sets of randomized variable values, using relatively simple equations for soil loss, sediment delivery, and discharge. Besides estimating the percentage of increase in mean suspended-sediment (mud) concentrations for 32 years due to WRDs construction, they also modeled sensitivity analysis that has identified pertinent factors that contribute to the uncertainty of erosion prediction.

Many authors have studied security-related applications. Based on recent domestic and international terrorist activities, transportation infrastructures have been considered attractive targets because of their accessibility and potential impacts on human lives and economic activity. Hence, bridge owners are concerned about the safety of bridges against man-made hazards, such as blast. Blast considerations are incorporated into major or important bridges and are slowly being incorporated in the design of new bridges. But the performance of existing bridges against blast loading is a serious concern given constrained resources and issues with providing adequate standoff distances. NCHRP 645 report titled "Blast-Resistant Highway Bridges: Design and Detailing Guidelines" (Williamson et al. 2010) does present some simplified design guidelines against blast loads, but not much information is available on failure modes of different bridge components during blast loads, and this has been an important topic of research in recent years.

To address this gap in the literature, Yi et al. (2014a) researched an effective approach for the application of blast loads on structural components, verification of the reliability of finite-element models in LSDYNA in simulating blast load effects, and a detailed investigation of the behavior of various bridge components during blast loads. Authors developed a high-fidelity three-dimensional finite-element model of a typical three-span simply supported bridge in such detail that it can simulate both global and local failure modes. Low, medium, and high-level blast load effects on the reinforced concrete bridge were investigated and concluded that bridge components such as elastomeric bearings are subjected to excessively high levels of stresses during a high level of blast loads.

Realizing the value of multihazard design and because many existing bridges are already designed for seismic loads, the authors (Yi et al. 2014b) also developed a unified approach for multihazard design and presented the correlation of failure modes of the bridge under both seismic and blast loads. They identified the failure modes of bridge components based on three-dimensional (3D) simulation of a full bridge subjected to three levels of underdeck blast loads and quantified seismic resistance–blast load correlations to identify a possible trade-off between mitigating these two hazards. This information can be used for a cost-effective retrofit of structures to resist both seismic and blast loads.

Another application of risk-based processes is prioritization of different issues in civil infrastructure. Prioritization is an essential component of infrastructure management for optimal allocation of constrained resources. Maintenance plays a prominent role in infrastructure management in order to maintain the required level of service and durability in a cost-effective manner. Risk assessment based on risk score has been recommended as a basis for prioritization of bridge maintenance activities (see Adey et al. 2003; Chassiakos et al. 2005; Elhag and Wang 2007; Wang et al. 2007). Risk score (RS) is an aggregation of safety-risk rating (SRR), functionality-risk rating (FRR), sustainability risk rating (SUR), and environment-risk rating (ERR) of a bridge.

Based on existing models and RSs based on prioritization philosophy, Cheng and Hoang (2012) proposed an artificial intelligence (AI) approach known as evolutionary fuzzy least-squares support-vector machine (LSSVM) inference model (EFLSIM) that they contend can operate automatically without human intervention and domain knowledge as it can predict RSs from a new input pattern. Fuzzy logic (FL) was used to enhance the capability of approximate reasoning and to deal with subjective information, which was obtained from human judgment. The inference model employed

LSSVM as a supervised learning technique to infer the fuzzy input-output mapping relationship. Differential evolution (DE) was integrated into the model to optimize its tuning parameters. The model was validated based on a single database of 55 evaluation cases and, hence, requires more validation before it can be used. At the same time, availability of such databases that can be readily used with models such as these is very limited.

In structural dynamics, considerations of possible structural responses are a necessity for probabilistic performance assessment of structures and risk analysis. Incremental dynamic analysis (IDA) (Vamvatsikos and Cornell 2002) can assist in this aspect for more detailed seismic performance predictions of structures subjected to different seismic excitation levels. This can be used for fragility curves development and to predict failure probabilities. Tehrani and Mitchell (2013) note that this method has been successfully used for buildings and also apply this method to predict the seismic risk assessment of four-span bridges, designed using Canadian bridge design code with different column heights and diameters in Montreal. They studied the seismic risk associated with exceeding different damage states in the columns, including yielding, cover spalling, bar buckling, and structural collapse (i.e., dynamic instability). Based on the estimated mean annual probability of exceeding different damage states in the columns, they concluded that the probabilities of exceeding different damage states were reasonably low.

In another structural failure-related study that recognizes the need for the integration of structural failure probabilities with the probabilities of the events/hazards that cause the failure, Saydam et al. (2013) proposed a methodology for quantifying lifetime risk associated with the component failure and risk-based robustness of bridge superstructures. They quantified risk in terms of expected direct and indirect losses. The AASHTO/PONTIS element condition rating system (AASHTO 1993, PONTIS 1995) is used for identifying expected losses. A reliability-based approach is applied to compute the component and system failure probabilities, given the condition states of components.

The expected direct risk is defined as that associated only with the component failure and is computed by combining the condition state probabilities, the component failure probabilities in different condition states, and the direct consequences of the component failure. The total expected direct loss was calculated as the sum of the expected direct losses for all components. Indirect risk was defined as the risk associated with the system failure initiated by the component failure. A scenario-based approach regarding the condition states of the components was applied and the total expected indirect loss was calculated as the summation of the expected indirect losses for all scenarios. The time-variant risk-based robustness index was computed as the ratio of the expected direct loss to the sum of the expected direct and indirect losses. The major issues with proposed methodology, as identified by Saydam et al. (2013), is the availability of reliable data regarding the transition probabilities between condition states of each component within a system.

8.1.2 This Chapter

The remainder of this chapter aims at exploring the rich and varying nature of risk management applications in the field of civil infrastructures (see Figure 8.1). Analytically, the applications we provide in this chapter will cover different graph network types. We also cover different types of infrastructure: bridge security, critical findings practice, bridge network management, cascading effects (CEs), redundancy, and community resilience. We also present the risk management viewpoint for two issues that have gained attention lately: severely deteriorated (deficient) and functionally deficient condition of structures. In many of these applications, similar to many of the other case studies that we offered throughout this book, we show that risk management offers an objective solution to problems that have been generally approached subjectively. For all of the following case studies, depending on the nature of the application, we discuss pertinent components of risk management. Note that even though most case studies are bridge management examples, except for one case study (CEs), all the case studies can be easily applicable to any type of infrastructures with appropriate changes to pertinent parameters.

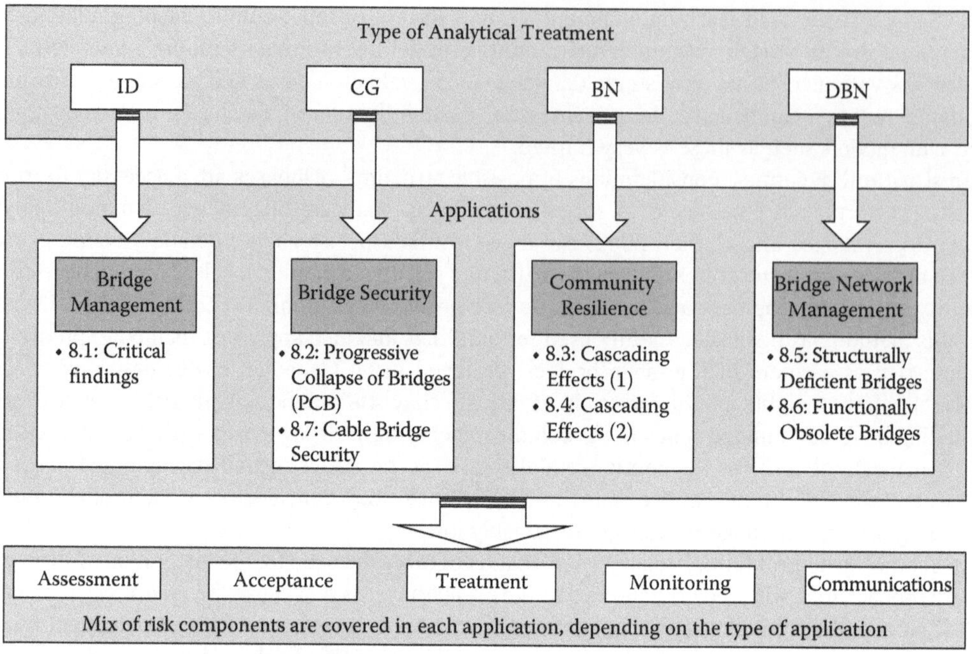

FIGURE 8.1 Composition of this chapter.

8.2 CASE STUDY 8.1: CRITICAL FINDINGS PRACTICE: IS IT A COMPLETE RISK-MANAGEMENT PROCESS?

8.2.1 Overview

8.2.1.1 Critical Findings Practice

The primary purpose of structural inspection is to assure safety of users through identification of critical findings, both structural and functional. The inspection process is also instrumental in maintaining an uninterrupted level of service through timely identification of corrective maintenance items. These findings can be used for appropriate planning and cost-effective infrastructure management. In the Highway bridge field, National Bridge Inspection Standards (NBIS) requires not only most highway bridges in the United States to be inspected at least once every 24 months to ensure public safety but also the establishment of a statewide procedure to ensure that critical findings are addressed in a timely manner (FHWA 2004). It further requires periodically notifying the Federal Highway Administration (FHWA) of the actions taken to resolve or monitor critical findings. Inspections or monitoring at higher frequencies may be required based on the structural condition observed through these inspections and subsequent evaluations.

Critical findings procedures vary depending on organizational structure, management philosophy, and organizational history. For example, in New York State, the bridge inspectors role is generally limited to reporting the critical findings to the bridge owner, and they do not make any work recommendations. At the California Department of Transportation (Caltrans), the bridge inspectors do make work recommendations to assist the maintenance crews located in their districts (Alampalli 2014). But in general, a good procedure will not only identify the critical findings but also set forth written procedures for communicating and addressing these deficiencies in a timely fashion.

Critical findings can also represent various levels of issues, structural or nonstructural, based on the identified issues and how they affect the structure or the users. For example, NYSDOT divides the structural critical findings (flags) into Red Structural Flags and Yellow Structural Flags. Red Structural Flags (see Figure 8.2) are used to report the failure of a critical primary structural component or a failure that is likely to occur before the next scheduled inspection. Yellow Structural Flags, see Figure 8.3, are used to report a potentially hazardous condition that if left unattended beyond the next anticipated inspection, would likely become a clear and present danger. Nonstructural Safety Flags, see Figures 8.4 and 8.5, are used to report a condition presenting a clear and present danger to vehicle or pedestrian traffic but is in no danger of structural

FIGURE 8.2 Red flag condition example.

FIGURE 8.3 Yellow flag condition example.

FIGURE 8.4 Nonstructural life line in good maintenance condition example.

FIGURE 8.5 Safety flag condition example.

failure or collapse. The notification and response procedures vary depending on the flag type (NYSDOT 2010).

8.2.1.2 Critical Findings Process (Flagging) as a Risk Management Process

We build on the critical findings process described above. We argue that it is essentially a form of risk assessment and acceptance in one step and is a direct result of the inspection process that is a form of exposure assessment. We use *flagging* to describe the *critical finding process* as it is easy to use. When a flag is raised, the implication is that the inspector has decided that the consequences (bridge damage or even failure) of the observed exposure has reached a risk threshold which is not acceptable in his/her judgment based on established policies and procedures of the agency/owner. Also, by raising a red flag, certain responsive steps are taken, which is the start of the risk treatment process. The flag itself can be considered as a visual communication of the risk state. Also, the flagging protocols institute certain monitoring actions. Thus, we can conclude that the flagging process is a complete subjective risk management process.

One immediate effect of placing an appropriate flag on a bridge is that it might reduce demands, if deemed appropriate by professionals, that in turn reduce the potential of failure. On the other hand, the detours that might arise will increase user costs. The net effects are usually reduced overall risks as summarized in Figure 8.6. However, we are interested in addressing the risk-related critical findings process in a more objective manner by using an ID approach.

The ID approach provides a method for estimating costs of placing a flag during a bridge inspection or after any other inspection. For example, when inspectors place a red tag for deficient buildings after an earthquake event, it might be necessary to estimate the cost of this tagging. First, we recast the concept of flagging that was discussed so far in this section in a form of a Bayesian network as shown in Figure 8.7. However, we immediately note that by assessing different variable probabilities only, we are not addressing many of the other essential elements of the critical findings process including the many different decision steps that have to be accommodated during the process. These decision steps are summarized in Table 8.1. Thus, the use of an ID process seems to be a reasonable way to address not only the assessment of risks but also risk thresholds and risk treatment as an objective way to apply the critical findings process.

8.2.2 Model

We first explore the problem as a simple BN model, which is shown in Figure 8.8. Note that the model includes, implicitly, the five components of risk management as shown in Figure 8.9, again showing how the critical findings (flagging) process is a complete risk management process indeed. The proposed states of nodal variables are shown in Table 8.2 and the definitions of these variables are shown in Appendix 8.I (Table 8.36).

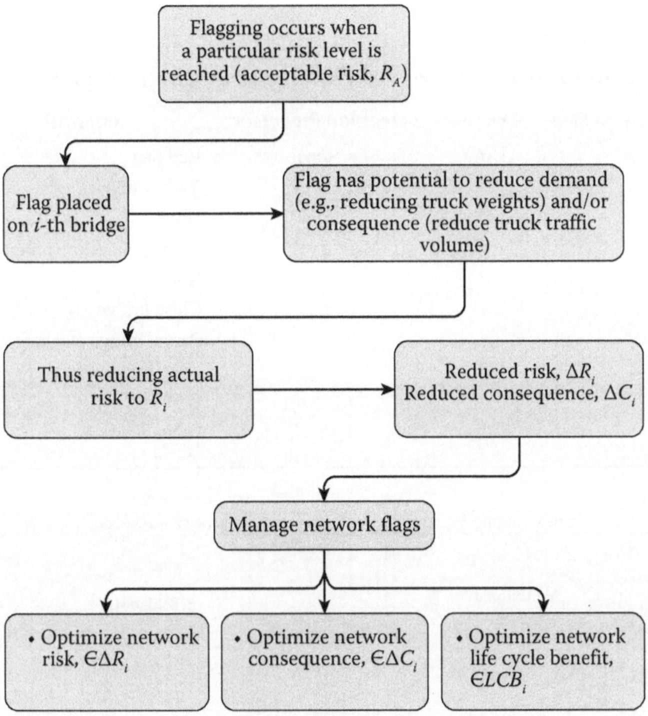

FIGURE 8.6 Process of controlling risk through critical findings procedures.

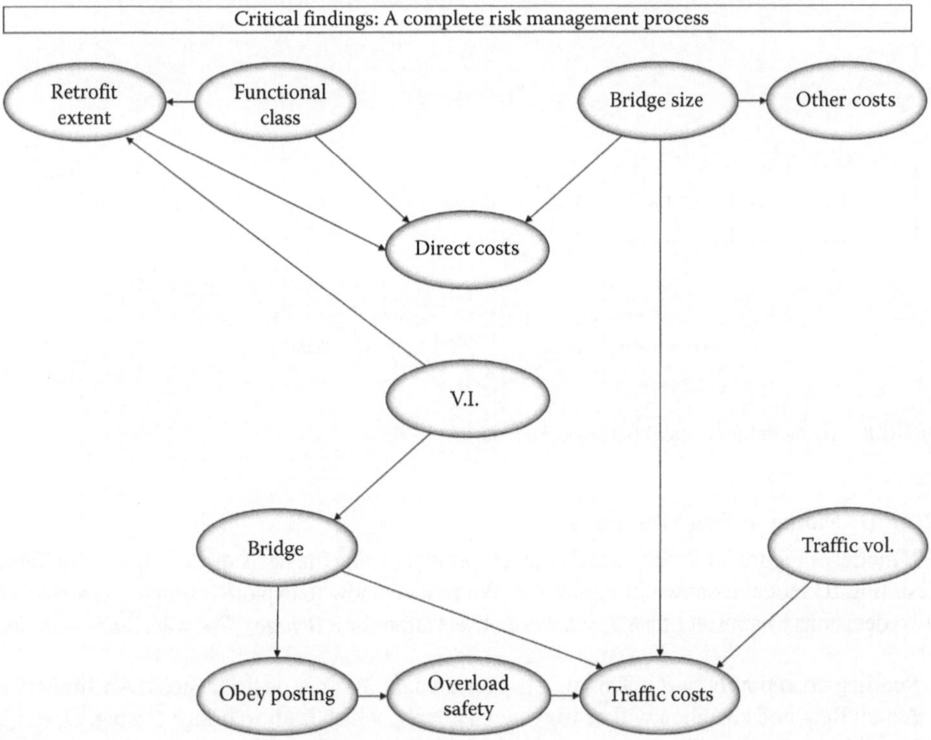

FIGURE 8.7 BN model for critical findings risk assessment only.

TABLE 8.1
Decision Process in Critical Findings (Flagging) Process

Decision Step	Decision	Decision Precursor	Comment
D_1	d_{11}	None	Red flag
	d_{12}	None	Yellow flag
	d_{13}	None	None
D_2	d_{21}	d_{11}	Immediate action required
	d_{22}	d_{11}	None
D_3	d_{31}	d_{21}	Close bridge
	d_{32}	d_{21}	Partial bridge closing
	d_{33}	d_{21}	Keep bridge open
D_4	d_{41}	d_{32} or d_{33}	Post
	d_{42}	d_{32} or d_{33}	Other action

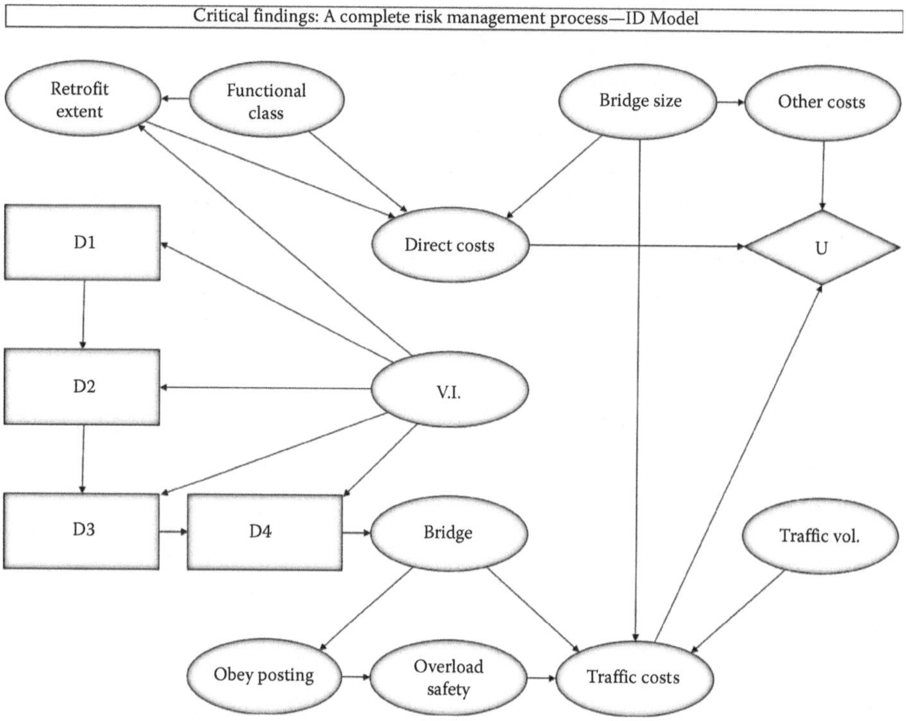

FIGURE 8.8 ID model for critical findings procedures.

8.2.2.1 ID Model of Flag Situations

The BN model of Figure 8.7 is now recast as an ID problem using the decision descriptions of Table 8.1. The resulting ID model is shown in Figure 8.8. We proceed now to explore a simple example to compare two decisions as a result of a *V.I. = Severe* observation for a *Bridge_Size = Medium* situation as:

1. Red flag condition $D_1 = d_{11}$. This will trigger a chain in other decision steps. An Immediate Action Required condition will be triggered, $D_2 = d_{21}$, which leads to bridge closing, $D_3 = d_{31}$,
2. No action flags $D_1 = d_{13}$, $D_2 = d_{22}$, $D_3 = d_{33}$, and $D_4 = d_{42}$. This case illustrates not having a critical findings process even though the conditions for it exists.

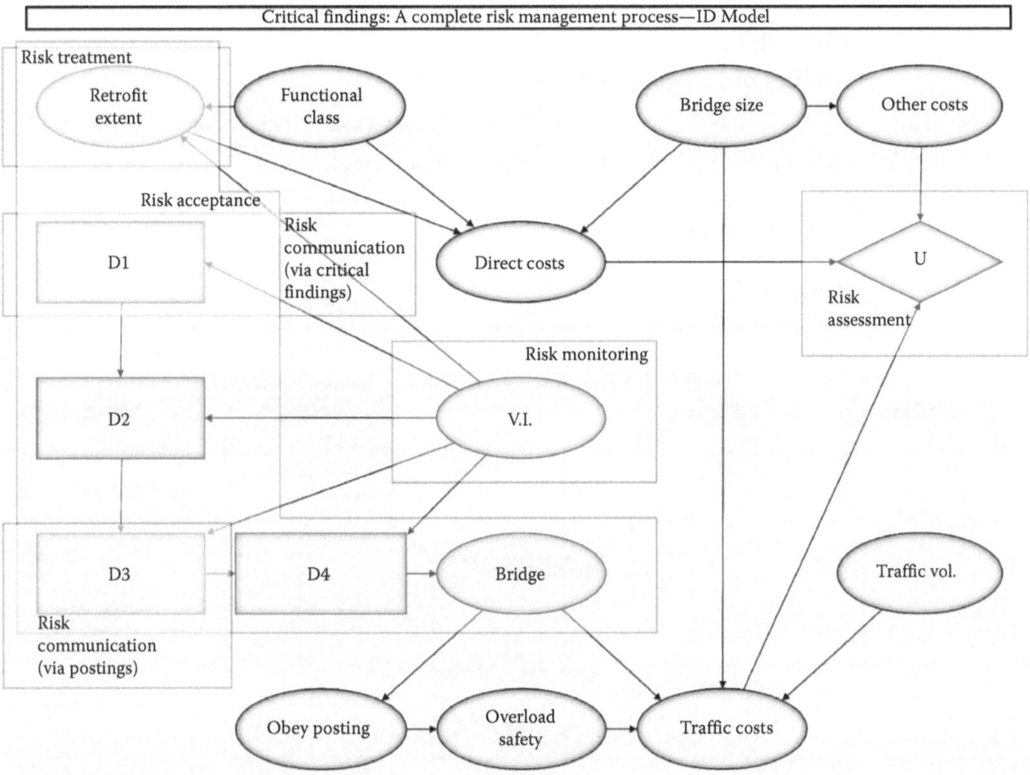

FIGURE 8.9 Critical findings as a risk management process.

TABLE 8.2
States of Nodal Variables

Variable	# of States	States
Retrofit Extent	3	High, Medium, Little
Functional Class	4	Interstate, Arterial, Collector, Local
Bridge Size	3	Large, Medium, Small
Other Costs	2	High, Low
Direct Costs	2	High, Low
V.I.	4	Service, Poor, Fair, Good
Traffic Vol.	3	High, Medium, Low
Obey Posting	2	No, Yes
Overload Safety	2	Low, High
Traffic Costs	2	High, Low
Bridge	3	Close, Partial, Open
D1, D2, D3, D4	Varies	*d11, d12, d13, d21, d22, d31, d32, d33*

There are several other possible decision combinations, and those will be left to the reader as an exercise.

8.2.2.2 Utility of Different Decisions

The utilities in this problem include costs from the sum of three sources: *Traffic_Costs*, *Direct_Costs*, and *Other_Costs*. A hypothetical utilities is shown in Table 8.3.

TABLE 8.3
Utilities of Flagging Problem

	Traffic_Costs		Direct_Costs		Other_Costs	
Scenarios	**Low**	**High**	**Low**	**High**	**Low**	**High**
$d_{11} + d_{21} + d_{31} + d_{41}$ Red flag and close bridge	\$100	\$1000	\$100	\$500	\$100	\$500
$d_{13} + d_{22} + d_{32}$ Do nothing	\$100	\$1000	\$100	\$500	\$100	\$500

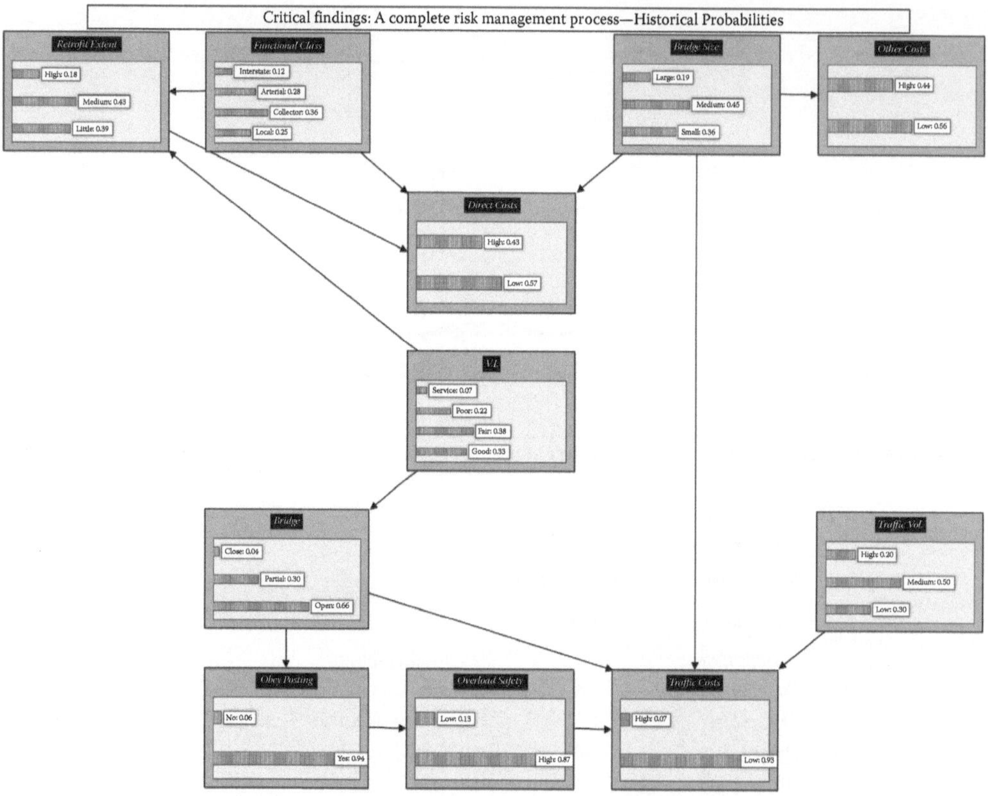

FIGURE 8.10 Historical (prior) probabilities.

8.2.3 CPTs

The different CPTs of each variable of the BN network of Figure 8.7 are shown in Appendix 8.II. As usual, we used our best judgment in filling the different tables in a realistic manner for illustration purposes. For more on different methods of developing CPTs, see Chapter 2.

8.2.4 Results

As a frame of reference, we execute the BN model for historical probabilities that are shown in Figure 8.10. We then execute the ID cases that are needed to generate different joint probabilities for the two decisions that are considered for this problem ($d_{11} + d_{21} + d_{31} + d_{41}$ and $d_{13} + d_{22} + d_{32}$). The two cases are shown in Figures 8.11 and 8.12.

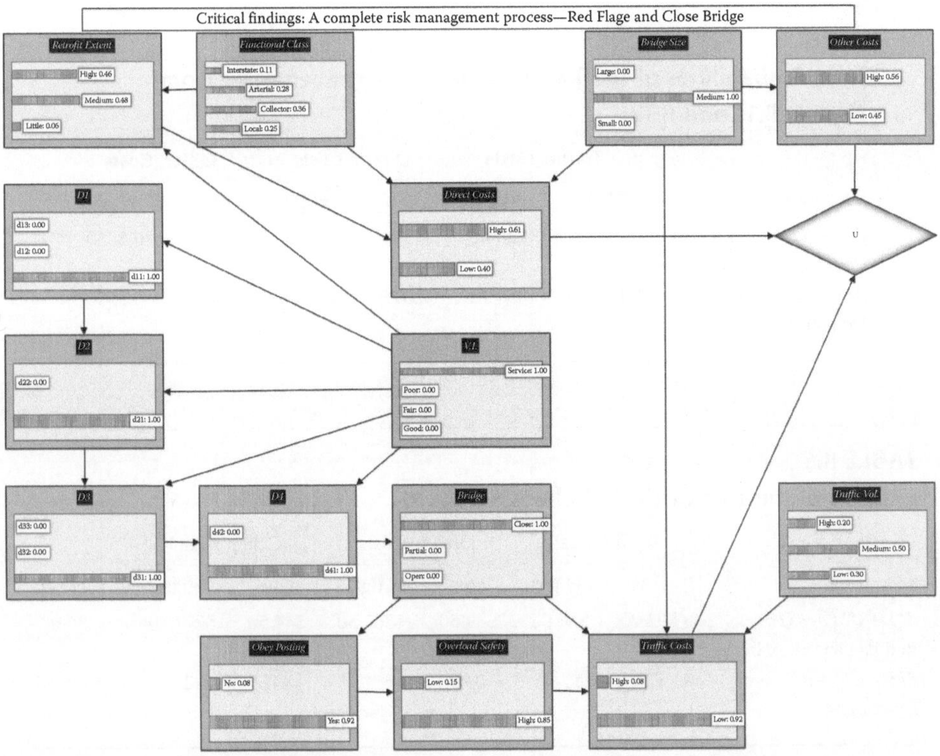

FIGURE 8.11 ID probabilities for bridge closing after a critical finding.

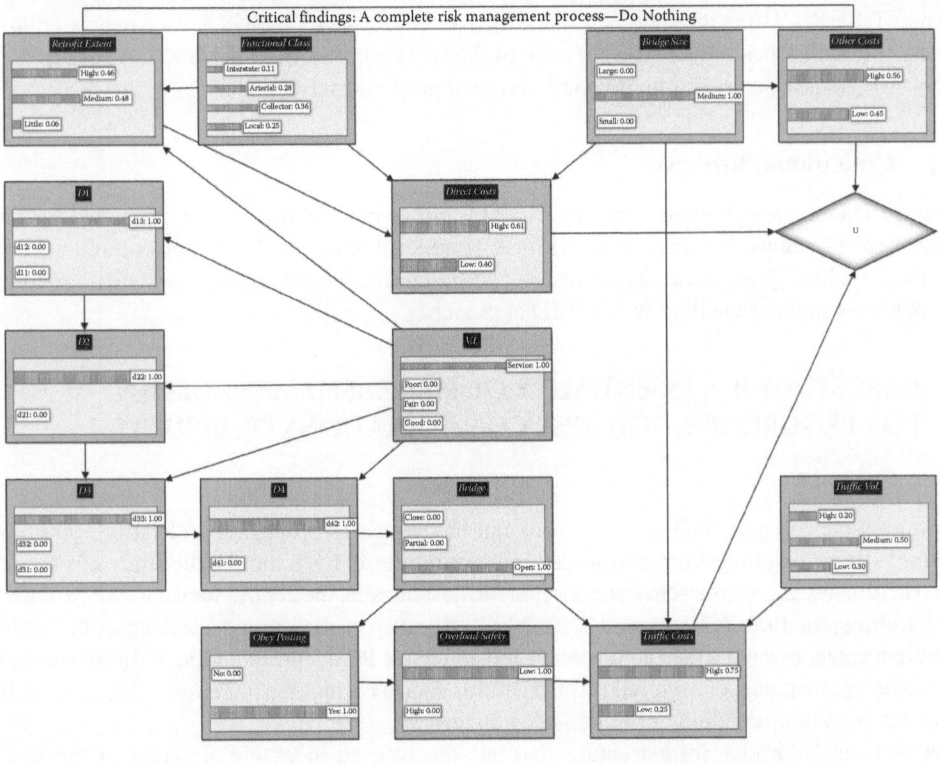

FIGURE 8.12 ID probabilities for doing nothing after a critical finding condition is noticed.

TABLE 8.4
Joint Probabilities of the Two Critical Findings Scenarios (from Figures 8.11 and 8.12)

	Traffic_Costs		Direct_Costs		Other_Costs	
Scenarios	**Low**	**High**	**Low**	**High**	**Low**	**High**
d11 + d21 + d31 Red flag and close bridge	0.9200	0.0800	0.3950	0.6050	0.4450	0.5550
d13 + d22 + d32 Do nothing	0.2530	0.7470	0.3950	0.6050	0.4450	0.5550

TABLE 8.5
Utilities of the Two Critical Findings Scenarios

	Traffic_Costs		Direct_Costs		Other_Costs		
Scenarios	**Low**	**High**	**Low**	**High**	**Low**	**High**	**Total *U***
d11 + d21 + d31 Red flag and close bridge	$92.00	$80.00	$39.50	$302.50	$44.50	$277.50	$836.00
d13 + d22 + d32 Do nothing	$25	$747	$40	$303	$45	$278	$1436.30

We are now able to summarize the needed joint probabilities of the two decision scenarios as shown in Table 8.4. Using these probabilities and from the costs of Table 8.3, we can now obtain the utilities, *U* (total costs/absolute risks) of each of the two scenarios as in Table 8.5. As expected, it is far less risky to issue a critical finding and taking appropriate action.

8.2.5 Concluding Remarks

We just provided a simple model for the critical findings process that can be expanded for bridge management situations. However, even with the simplicity of the model, the power and benefits of the critical findings process can be shown, in addition to the benefits of casting such a process in a more objective manner via the use of the ID approach.

8.3 CASE STUDY 8.2: ESSENTIALITY OF RISK-BASED MANAGEMENT FOR PROGRESSIVE COLLAPSE CONSIDERATIONS OF BRIDGES

8.3.1 Overview

The progressive collapse (PC) of a structure can be defined as propagation of failure through the structural system resulting from a localized structural failure. Even though the study of PC of civil infrastructure has gained immense attention in the past decade, the central focus of these studies has been building structures (Ellingwood et al. 2007). Because of the limited similarities in responses of different categories of infrastructure when subjected to a PC event, extending buildings-based PC practices to another category of civil infrastructures such as bridges is not easy. If not careful, these studies can lead to misleading or even unsafe conclusions.

Let us consider bridge infrastructure that has received considerable attention in the last few years due to the notable failure of bridges such I-35W Bridge in the State of Minnesota in 2007

(NTSB 2008). Unlike buildings, standardized design procedures or guidelines to meet the PC hazard is nonexistent for bridge structures. In addition, there are numerous bridge PC behavioral issues that are fundamentally different from their building counterparts. These include the nature of failure front propagation through the structure after the initiation of a PC event, varied potential PC failure modes depending on the type of bridge and connection details, and relative nature of the importance of the system vs. component behavior during a PC event in bridges. In the authors' opinion, the question is: "Can we study and then resolve PC issues by solely, or even mostly, observing/researching components and ignoring systems?" It is hardly addressed by the civil infrastructure community.

Another crucial difference between buildings and bridges when it comes to the PC response is the decision-making (DM) process. When PC is considered from the DM lens, the relative importance of analysis, design, and mitigation processes in both communities would change considerably. Costs (both initial capital and life cycle) of PC analysis, design, mitigation, etc. need to be considered while deciding on PC performance. Adequacy (or inadequacies) of empiricism, limits of bridge hardening, and the nature of bridge redundancies play a role; thus, scale differences between buildings and bridges in time, geometry, and amplitude need accommodation. Because of this, reasonable bridge PC considerations need to accommodate DM practices from the ground up. This might include, but not be limited to, multihazards, life cycle considerations and/or risk, and resilience management processes (Ettouney and Alampalli 2012a, 2012b).

This case study addresses the subject of progressive collapse in bridges (PCB) with the aim of showing that the subject matter needs to be addressed via a risk-based viewpoint. We first review the fracture critical bridges (FCBs) and fracture critical members (FCMs) because of the attention they have received in recent years. We then follow with subjective comparisons between FCBs and PC of buildings. Next, we make the case for a risk-based PC process for managing PCBs. We end the section with a simple example of risk-modeling of the PCB as related to FCB and a potential abnormal hazard that might trigger a PC event.

8.3.2 Historical Overview of PCB

Because of the earlier definition of PC, the failure of a bridge member that would trigger a complete or partial failure of the bridge has always been of intense interest to the bridge community and reminds us of the FCBs that have been at the forefront of discussions around the world due to failures, such as the I-35W Bridge in the State of Minnesota in 2007 (see Figure 8.13). There are well-documented cases of FCB failures (Ettouney and Alampalli 2012a), including the Silver Bridge (LeRose 2001) in 1967, the Mianus River Bridge (NTSB 1984) in 1983, the I-35W Bridge (NTSB 2008) in 2007, and the I-5 Skagit River Bridge (Leavitt and Straight 2013) in 2013. FCBs are seriously addressed by the bridge community in all aspects of bridge management (NYSDOT 1997; NYSDOT 2000; FHWA 2004; NYSDOT 2008; Federal Highway Administration 2012; Alampalli 2014). Figure 8.14 shows these different aspects of FCBs.

FIGURE 8.13 Collapsed I-35W Bridge. (From NTSB Report.)

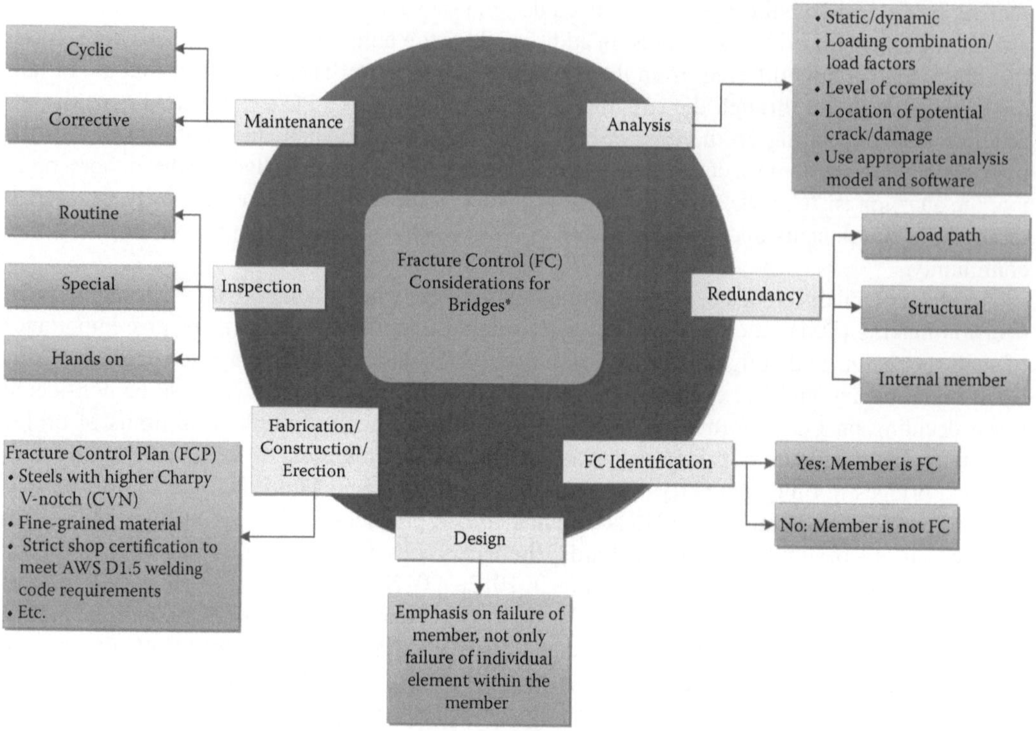

FIGURE 8.14 Fracture critical bridge management processes in bridge structures.

FIGURE 8.15 Schoharie Creek Bridge collapse. (From United States Geological Survey website http://sc.water.usgs.gov/projects/bridgescour/.)

8.3.3 Fracture Critical Condition and the Susceptibility to PC

Perhaps the most important difference between FCB collapse and PCB is in applicability. FCB is limited only to bridges with FCM while PCB applies to all types of bridges. For example, the failure of the Schoharie Creek Bridge in 1987 occurred because of scour beneath the bridge foundation (NTSB 1988; Ettouney and Alampalli 2012a) (see Figure 8.15). This failed bridge is not considered an FCB. In addition to the above difference, FCB is generally applied only to steel bridges,

FIGURE 8.16 Failure of a bridge during construction.

while PCB can be attributed to all types of bridges. Another difference is that FCB failures, when designed and maintained properly, consider only deterioration and design-based code demands. PCB would venture beyond those normal design demands to consider abnormal man-made (such as construction accidents and bomb blasts) or natural hazards (see Figure 8.16 for failure of a nonredundant bridge during construction). There are several other differences between FCBs and PCBs, such as inspection and maintenance practices. In general, FCBs enjoy a higher level of care than other PC susceptible bridges.

Based on the above, we observe that FCB is a subset of PCB. We can make the following conclusion: All FCBs are susceptible to PC, but not all PC susceptible bridges are FCBs. We devote the rest of this work to address PCB. We take advantage of the fact that FCBs represent only part of the PCB community.

8.3.4 Bridge-Specific Generalized Definition of PC

Without departing from the earlier PC definition, we can obtain a general and adequate definition of PC by generalizing the FCB definition described in FHWA (2012) memorandum. A PC event in bridges occurs when a local failure initiates a propagation of large-scale failures throughout the bridge to such an extent that it will interfere with the acceptable functionality of the bridge and/or it will result in unacceptable consequences beyond the immediate environs of the bridge system. Note that we did not make any exception to the cause of the PC, the type of the initial local failure, or the material of construction. Note also that we included in our definition the potential consequences of the PC—an important issue that needs consideration when risk-based PCB issues are discussed in the later section.

8.3.5 PC: Buildings vs. Bridges

PC considerations in the building community have enjoyed wide interest in the past few decades (Ellingwood et al. 2007). There are, however, numerous differences between PC of buildings, FCBs, and PCBs. For completion, we present some of those differences in Table 8.6.

8.3.6 Assessment of PC Potential in Bridges

The first step in understanding and avoiding PC requires assessing a bridge to PC susceptibility. Conventionally, this requires determining the effect of the loss of a bridge component due to any postulated event. Two key issues that should be addressed for this analysis are the bridge (component)

TABLE 8.6
Comparison between PC for Buildings, FCBs, and PCBs

Issue	Buildings	Bridges: F.C.	Bridges: P.C.
Hazard	Usually hazard independent	Hazard independent	Not well defined
Analysis	Can be static or dynamic, linear or nonlinear	Can be static or dynamic, linear or nonlinear	Not well defined
Service-life performance	None	Well controlled and specified	None
Construction material	All	Mostly steel	Should be all materials, but no controlling mechanism
Triggering event	Loss of any structural member (usually columns)	Loss of fracture critical member (which leads to partial or total bridge failure)	Not codified. Should be similar to the building definition in its generality
Design loading after triggering event	DL + reduced LL (some accommodation for dynamic effects)	DL + postulated truck load. Dynamic effects not well prescribed	Not well defined. Should be similar to F.C. design loading
Acceptance limit states	Components: Damage (ductile behavior) allowed. Brittle behavior not well prescribed. System: No postdamage behavior is prescribed	Components: Damage (ductile behavior) allowed. Brittle behavior not well prescribed. System and postdamage behavior is prescribed	Not well defined
Mitigation strategies	Specific load resistance (hardening) Alternate path (redundancy) Indirect (tying/bridging)	Try to avoid FC members by providing redundancy If F.C.M. is present, provide special attention to them (equivalent to hardening in building)	Not well defined
Reliability design	Yes	Yes	Not well defined
Risk-based design	No	No	No
Multihazards decisions?	None available	Partial	No
Life cycle analysis	No	Partial	No

that may be compromised and the event that can cause it, if any. Answering these questions requires good understanding of the following:

- Both the local and global behavior of the structure because of the need for an appropriate mix of detailed local models and balanced global models
- Careful modeling of current and future demands
- Risk-based estimations of abnormal demands
- Static and dynamic modeling of compression members that is very difficult
- Overall stability
- Postdamage modeling
- Modeling existing field condition instead of as-designed condition
- 3D analysis

8.3.7 Risk Management of PCB Potential

8.3.7.1 General

After identifying the susceptibility of the bridge to PC, the issue of mitigation/treatment needs to be addressed. In general, PC treatment processes of PC in buildings include providing an alternate load path (improving redundancy), adding local resistance (hardening), and indirect (empirical) methods. Mitigation of FCB failures in bridges is focused in providing adequate redundancy through one or more methods: load path redundancy, structural redundancy, and internal member redundancy (Federal Highway Administration 2012). Another aspect of mitigation in the bridge community is the higher level of attention given to the FCMs from fabrication to maintenance. Thus, the mitigation is exemplified in a higher level of QA/QC during manufacturing and construction, higher degree of inspection, and priorities given to addressing FCMs during maintenance and rehabilitation. This is analogous to specific load resistance procedures in the building community even though the actual steps are fairly different. Finally, we observe that there is no analogous process for the building PC process of the indirect (empirical) mitigation method in the bridge community's FC mitigation methods. One possible reason might be that bridges are more high consequence facilities than buildings.

The FCB mitigation processes are then fairly compatible with the building PC mitigation processes. What remains for us is to propose generalizations to these processes for the wider subject of PCB care.

8.3.7.2 Essential Components of Dealing with Potential of PCBs

Studying Figure 8.17 and Table 8.6 reveal the main differences between PCBs and FCBs. Based on this, additional needed mitigation considerations for PCBs can be introduced. These include, but are not limited to:

- Generalized hazard-demand studies to accommodate all potential abnormal hazards.
- Generalized analysis for assessments to accommodate all construction materials for all types of bridges.
- Whenever a PCB potential is identified from the above two steps, develop a mitigation strategy that includes hardening the relevant bridge members, providing for additional redundancy, and allowing for an enhanced inspection/maintenance process for the identified bridge members.

The above proposed PCB process is comprehensive and is compatible with the current FCB practices. Hence, its implementation within the bridge community is straightforward. However, such an implementation is too costly to apply in practice. A risk-based policy may provide a reasonable balance between the PC consequences costs and the costs of implementing corresponding mitigation measures.

8.3.7.3 Risk-Based PCB Treatment

Risk-based policies provide a link between exposure to hazards and the consequences. It is recommended to adopt a risk-based policy in managing hazards that are rare in occurrence yet with high consequences. We recall that risk can be evaluated as:

$$R = f(T, V, C) \tag{8.1}$$

where T, V, and C are the threat (hazard), vulnerabilities, and consequence ratings, respectively (see Ettouney and Alampalli 2012a). Of these three risk components, computing the PCB vulnerability rating is perhaps the most well-defined procedure (NYSDOT 2000; FHWA 2012). A PCB threat rating T is not an easy parameter to compute; however, some guidelines can be

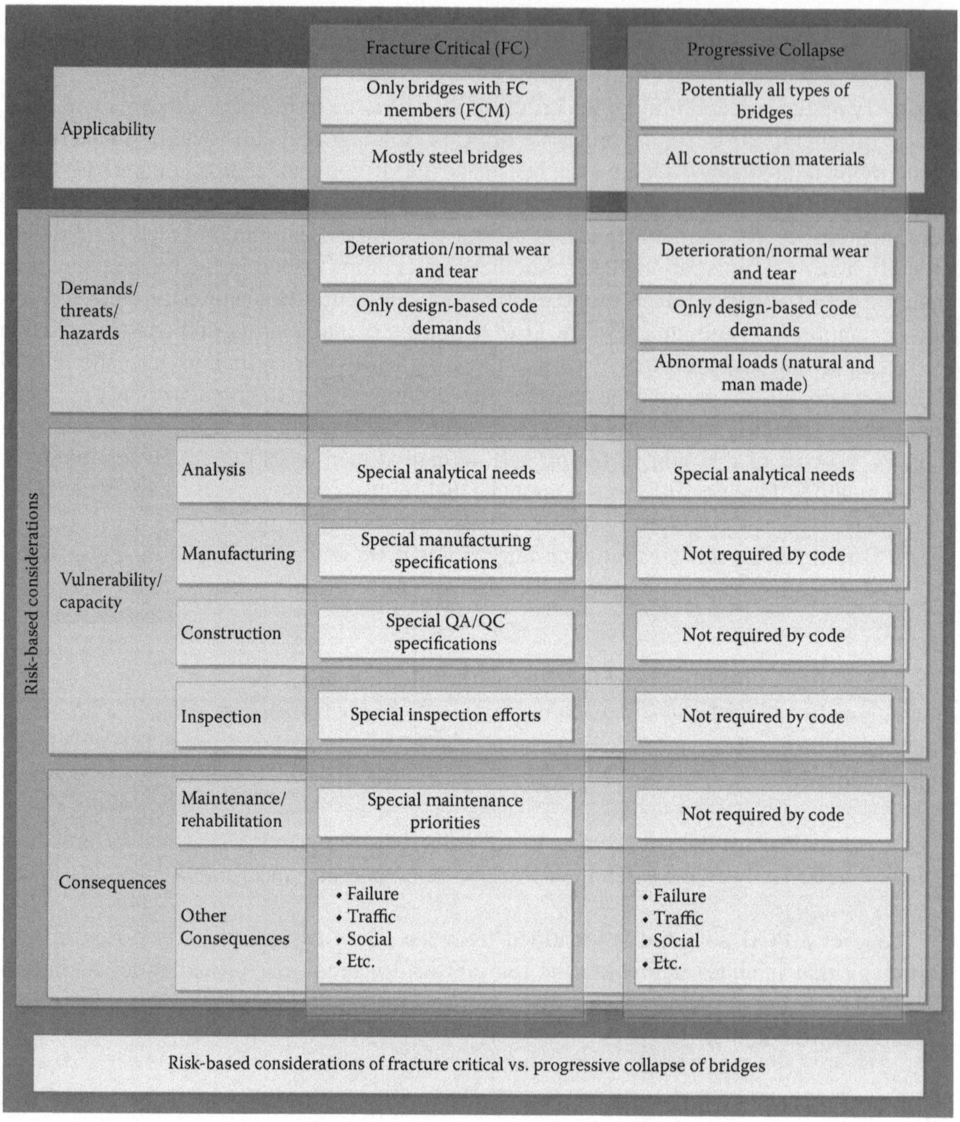

FIGURE 8.17 Comparison between FCBs and PCBs.

found in the literature (NYSDOT 2000; DHS 2011d). The consequences *C* can be either relative or absolute (DHS 2011d).

When using risk in a PCB mitigation process, we have to compute the in-service risk *Ra* (sometime referred to as assessed risk), a reasonable threshold (target) risk *Rt*, and then compare the two. If *Ra* is less than *Rt*, in-service PCB situation is acceptable. If not, the in-field PCB situation is not acceptable, and, thus, PC mitigation is needed. Such mitigation can be provided by reducing one of the three risk components *T*, *V*, or *C*. Strategies for reducing risk components in general are provided in the literature (DHS 2008). Current methods for dealing with FCB can be used to reduce *V* whenever applicable. However, we note that reducing *V* might not be the most efficient or cost-effective method for reducing PC risk. Depending on the bridge situation, reducing *T* or *C*, or a combination of reducing all the three risk components might provide a more efficient risk reduction method. Unfortunately, specific strategies for bridge structures, especially for reducing *T* or *C* are not well developed. Table 8.7 shows qualitative estimates of contributions of different bridge management

TABLE 8.7
Risk-Based PCB Management Reduction Parameters

Management Process	Decrease Threat?	Decrease Vulnerability?	Decrease Consequences?
Design	Low	High	Medium
Construction/ manufacturing	Low	Medium	Medium
Inspection	Low	High	Low
Maintenance	Medium	Medium	Medium
Other (site, traffic, etc.)	High	Low	High

processes to the reduction of risk components. In-depth studies of the contributions of all of these components to risk reduction are needed.

8.3.8 PCB Risk Assessment Using BN

8.3.8.1 Overview

We have argued so far that all FCBs are susceptible to PC, but not all PC susceptible bridges are FCBs. Similarities and differences between PCBs and FCBs from several viewpoints including design, manufacturing, construction, inspection, and maintenance as well as the triggering hazard were presented. We also observed that though FCB is well understood and regulated, PCB is not. Among the several possible ways of dealing with the potential of PCB, we suggested that a risk-based management should be followed in order to achieve the desired safety at reasonable costs. In the remainder of this section, we offer a simple BN model that aims at assessing PCB risk. We include in the model several important variables such as the fracture critical nature of the bridge, the state of vulnerability mitigations, and the role of visual inspection as well as other risk components such as exposure and consequences.

8.3.8.2 Model

The proposed BN model for assessing PCB risk is shown in Figure 8.18. The states of nodal variables are shown in Table 8.8 and the formal definition of these variables is shown in Appendix 8.I (Table 8.36). Note that in this model we defined exposure, *E*, as a function of *F.C.B.*, *Abn.Hz.*, and *S.V.I.* We added a reliability only variable, *Rel*, which is only a function of exposure to contrast the role of reliability vs. risk in a PCB problem. For simplicity, we are using only consequences that are a direct function of the abnormal hazard and the bridge system itself; that is, any external/independent consequences are ignored in this model. This is not exactly an accurate assumption and needs to be rectified. We leave this to the reader as an exercise. The risk, *Ri*, is a subjective risk in this model. It is a function of exposure and consequences. Expressing absolute risk in monetary value may add further insight to the process.

8.3.8.3 Conditional Probabilities Tables

The different conditional probabilities tables (CPTs) of each variable of the BN network in Figure 8.19 are shown in Appendix 8.III. In this example, when assigning the CPTs, we made the following assumptions:

1. The abnormal hazard is scour.
2. Exposure is increased due to the abnormal hazard (scour). This is a fair assumption since FCB's lack of redundancy would increase the exposure, but scour will also have a great effect on non-FCBs.
3. Decisions to mitigate or not is also slightly affected by the FCB condition of the bridge.

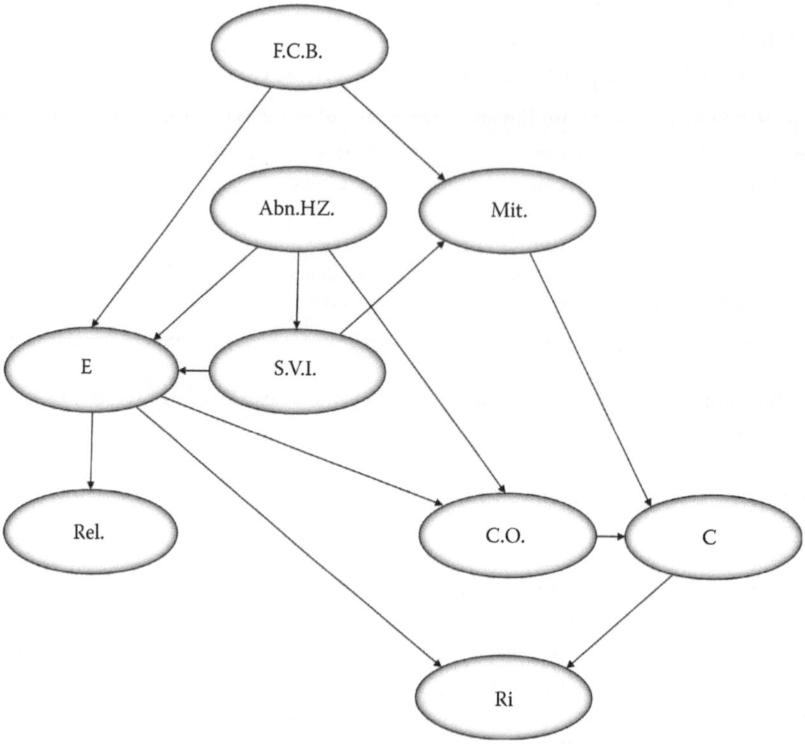

FIGURE 8.18 Progressive collapse (due to abnormal hazard) risk BN model.

TABLE 8.8
States of Nodal Variables

Variable	# of States	Ranks (States)
F.C.B.	2	No, Yes
Abn.Hz.	2	H. Level, L. Level
Mit.	3	None (Hazard Occurs before Mit.), Not Adequate, Adequate
E	3	H. Level, M. Level, L. Level
S.V.I.	2	Damaged, Pristine
Rel.	2	Low (unsafe), High (safe)
C	3	High C, Med. C, Low C
C.O.	2	High Cons., Low Con.
Ri	3	High, Med. Low

As usual, we used our best judgment in filling the different tables in a realistic manner for illustration purposes. For more on different methods of developing CPTs, see Chapter 2.

8.3.8.4 Results

We can assess PCB risk due to abnormal hazards that might produce PCB using the BN model shown in Figure 8.18. Since the FCB status is also included as a variable, we can assess the

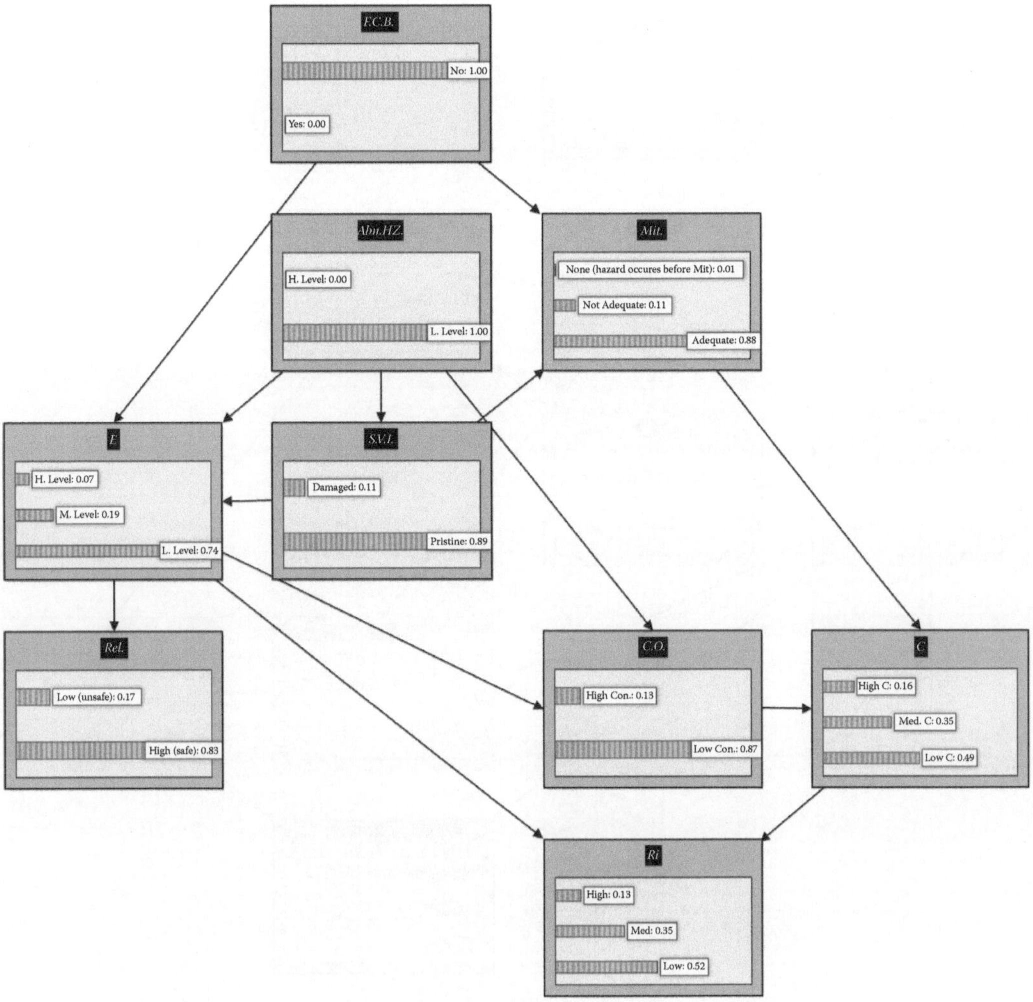

FIGURE 8.19 Progressive collapse (due to abnormal hazard) risk for *F.C.B.* = *No* and *Abn.Hz.* = *L.*

interrelationship between such a status and the FCB risk. Figures 8.19 through 8.22 show the limits of the states of both *FCB* and *Abn.Hz* as they control the risk states. The summary is shown in Table 8.9. We see that the average risk is controlled mainly by the state of the abnormal hazard while the effects of the *FCB* state have minimal effects on the risk state (see Table 8.10).

8.3.8.5 Summary

Throughout this section, we explored the PCB phenomenon. We showed that for the most effective treatment, risk management policy needs to be considered. We then presented a BN model for assessment of risk potential due to PCB and FCB. We showed that abnormal hazard might control the PCB more, with smaller effects resulting from the FCB state in this example. Similar behavior was also observed for bridge reliability. Current practices in bridge management put a major emphasis on the FCB state and much less emphasis on PCB. Obviously, the current bridge management status regarding this issue needs to be recalibrated as needed.

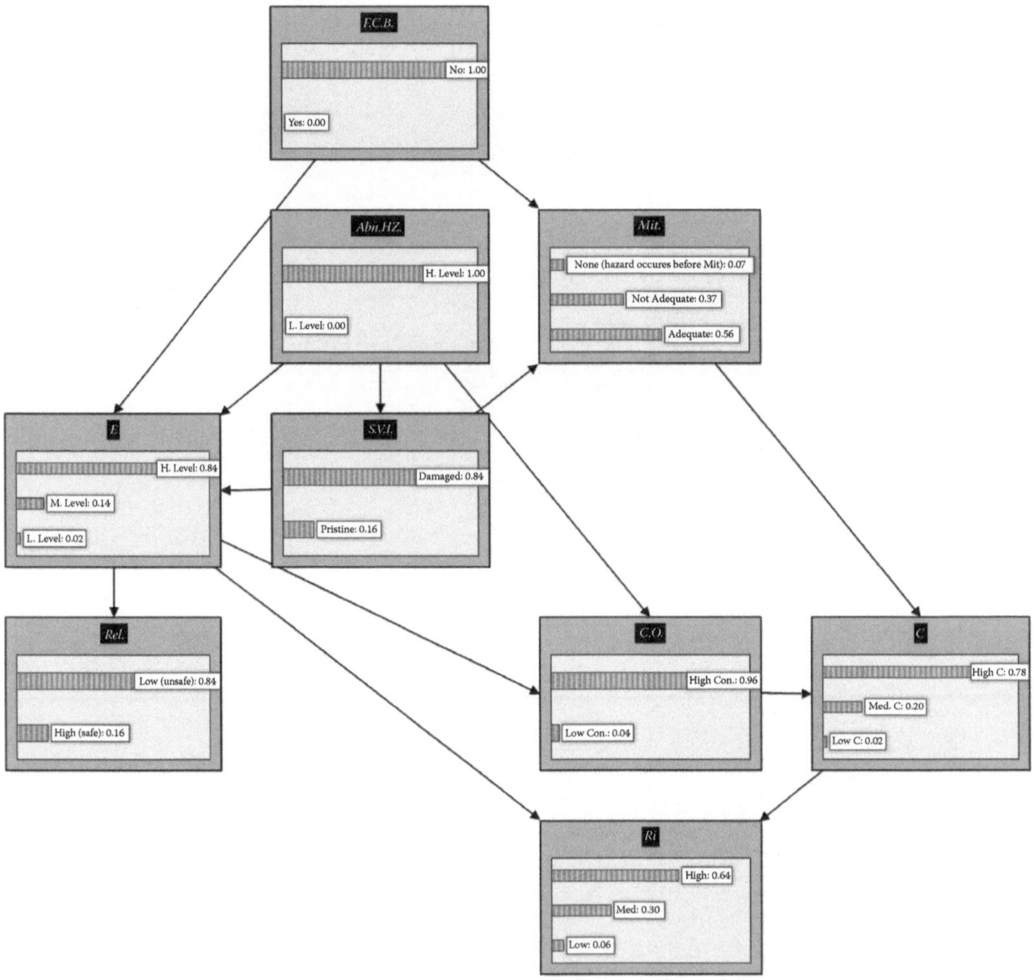

FIGURE 8.20 Progressive collapse (due to abnormal hazard) risk for *F.C.B.* = *No* and *Abn.Hz.* = *H.*

8.4 CASE STUDIES 8.3 AND 8.4: SUPER STORMS: CASCADING EFFECTS (CE) AND RESILIENCE (RE)

8.4.1 Overview

The concept of resilient infrastructures has emerged in the past few years as an important goal for their performance and operations. DHS identifies two types of resilience: asset resilience and community resilience. Asset represents a single facility such as a bridge, building, tunnel, a power station, or a dam. A community represents a group of assets that are interconnected. The assets within a community can be of varying types. For example, a community can be composed of several assets (subway stations, bridges, power stations, police precincts or other facilities, fire stations, and schools), and the links that connect those assets including roadways, railways, tunnels, power lines, and communication lines. A community can be as small as two assets linked together or it can be as large as a whole geographical or operational region. Furthermore, DHS identifies four basic components that comprise resilience: robustness, resourcefulness, recovery, and redundancy. These four components are commonly referred to as the 4Rs. A common mode of failure (or severe degradation of performance) that can affect one or more of the 4Rs is known as *cascading effect*. A CE can

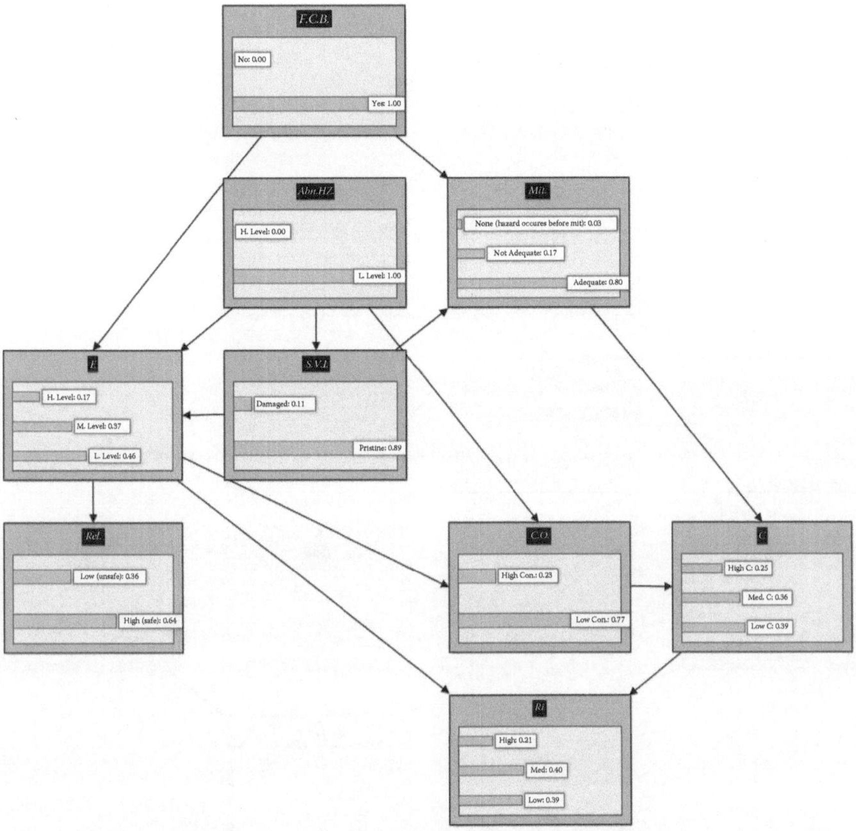

FIGURE 8.21 Progressive collapse (due to abnormal hazard) risk for *F.C.B.* = *Yes* and *Abn.Hz.* = *L.*

be subjectively defined as a failure (or severe degradation in performance) of single consideration within an asset or community that will propagate throughout the asset or community causing more failures as it propagates.

Let us consider in more detail how CE affects an asset or a community. Whether it is an asset or a community, each is composed of different physical or operational considerations (nodes). These considerations (nodes) are interconnected/linked together in a manner that is unique to the asset or community. Note that each link can also be considered as a consideration or node. Figure 8.23 shows a simple example of asset interconnectivity. Figure 8.24 shows an example of community interconnectivity (network). A degradation of performance in any of these considerations (nodes) might lead to a degradation of the performance of the neighboring considerations (nodes). This degradation might continue to propagate throughout the whole network (asset or community). This phenomenon is known as cascading effect, or CE. The degradation of performance in nodes can vary from limited degradation to complete failure of the consideration (nodes).

There are numerous examples of CE. PC of buildings is a classic example of CE. We can find examples of CEs in the aftermath of any natural or man-made disaster. A recent unfortunate example of CE is the 2012 super storm Sandy (see Figures 8.25 and 8.26) in the New York-New Jersey region.

Obviously, resilience of assets or communities is directly related to CEs. A large degree of CE failures within an asset or community is a sign of low resilience and the reverse is true. We note that CE can have an effect on the ratings of all 4Rs. For example a structural failure of a roadway (*robustness*) can have an effect on *recovery* efforts. Lack of *redundancy* of power sources can

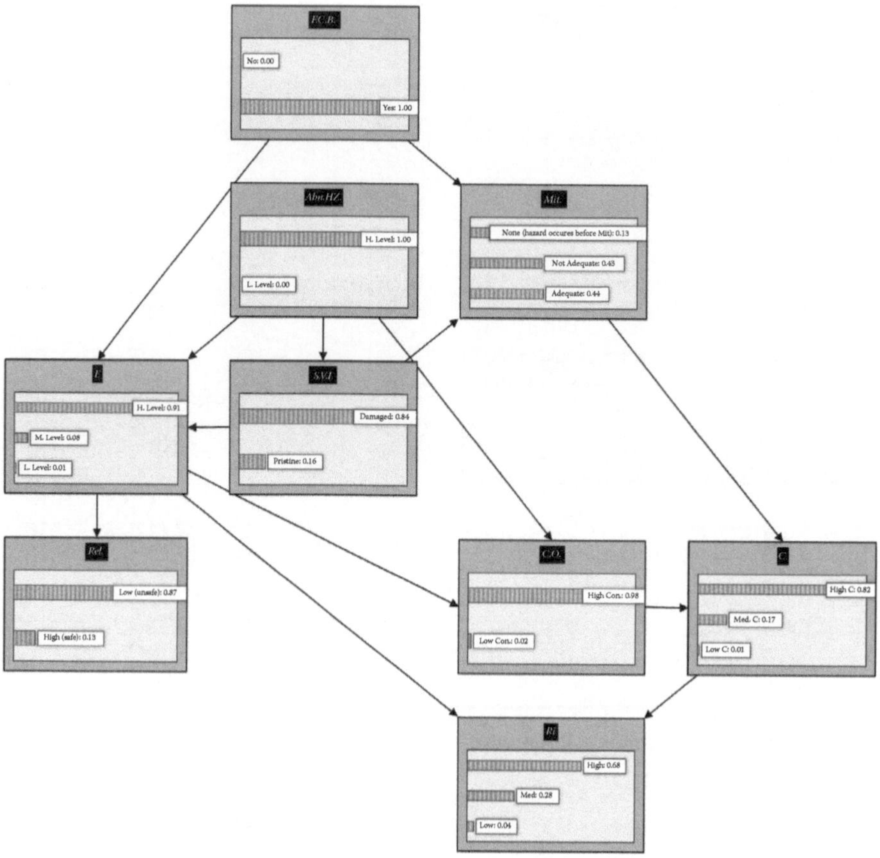

FIGURE 8.22 Progressive collapse (due to abnormal hazard) risk for *F.C.B.* = *Yes* and *Abn.H.* = *Hz.*

TABLE 8.9
Effects of Hazards, Fracture Critical Status on Risk

	F.C.B. = No		***F.C.B. = Yes***	
***Ri* State**	***Abn.Hz. = H***	***Abn.Hz. = L***	***Abn.Hz. = H***	***Abn.H. = L***
Low	6%	52%	4%	39%
Med	30%	35%	28%	40%
High	64%	13%	68%	21%
Average *Ri*	69%	37%	71%	44%

TABLE 8.10
Effects of Hazards, Fracture Critical Status on Reliability

	F.C.B. = No		***F.C.B. = Yes***	
***Rel* State**	***Abn.Hz. = H***	***Abn.Hz. = L***	***Abn.Hz. = H***	***Abn.Hz. = L***
High (safe)	16%	83%	13%	64%
Low (Unsafe)	84%	17%	87%	36%
Average *Rel*	33%	67%	32%	57%

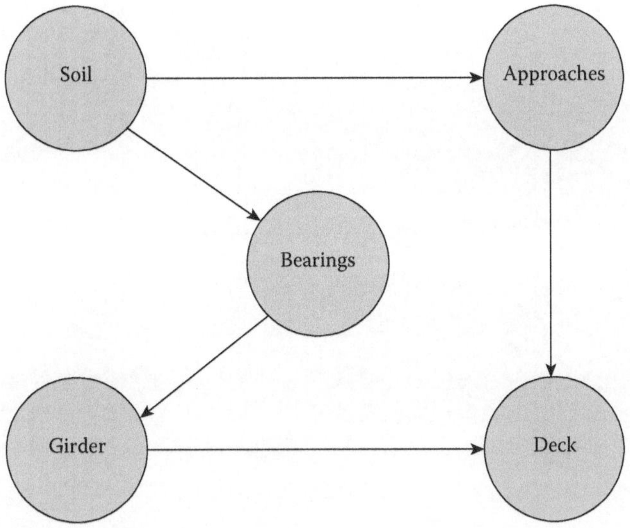

FIGURE 8.23 Simple asset (bridge) interconnectivity.

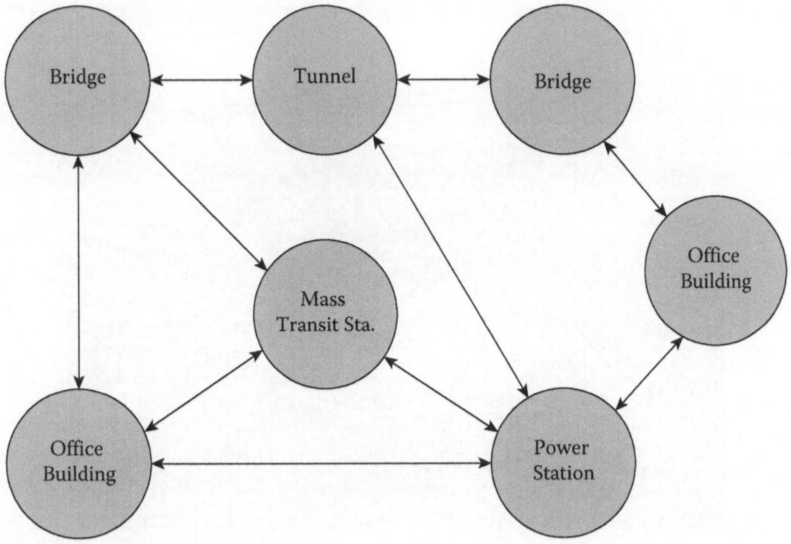

FIGURE 8.24 Simple community interconnectivity.

have an effect on control of flood gates, which might lead to flooding related failures (*robustness*). Thus, there is a need to understand CE in both assets and communities of interest. Such understanding should include ways to assess and mitigate CE. In the remainder of this section, we explore further different aspects of CE.

8.4.2 Metrics and Methodologies of CE

To objectively compute CE properties of an asset or a community, we have to identify the metrics of CE performance. First, starting from a particular consideration within an asset or a community, we define cascading effects extent (*CEE*) as a measure of the extent of the effects a particular

FIGURE 8.25 Cascading effect: Storm surge leading to interruption of subway services. (Courtesy of Albert DiBernardo.)

FIGURE 8.26 Cascading effect: Wind and flood damage leading to fire. (Courtesy of Marco Coco.)

consideration will propagate within the asset or community. We might want to opt for a metric that describe the relative importance of considerations that are affected by the consideration of interest, their condition ratings, as well as the geometric extent of the propagation of effects of such a consideration throughout the facility or community. We call this effect *CES*, or cascading effects strength. Note that while *CEE* is a purely geometric metric, *CES* is a metric that combines geometry, importance, and condition ratings all in once. For simplicity, *CEE* and *CES* are taken to be in the range 0–100. Because *CEE* is merely a geometric measure, its value is not an indication of higher or lower performance. But if *CES* is low, it would indicate lower performance, and higher performance is inferred for higher values of *CES*. Figure 8.27 shows schematically different performances of *CEE*.

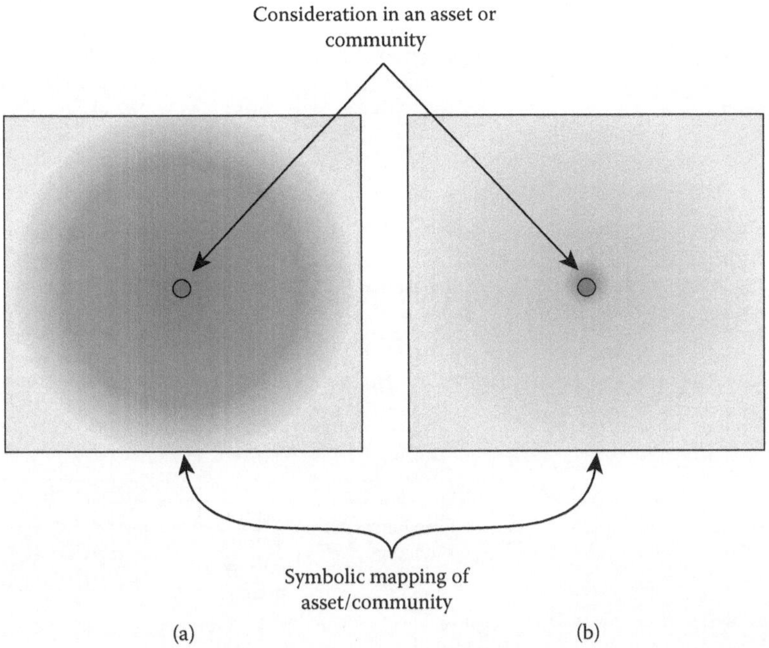

FIGURE 8.27 Examples of cascading effects extent (CEE). (a) Large CEE; and (b) small CEE.

There are numerous ways to compute *CEE* and *CES*. Graph theory, Bayesian networks methods, and/or Markov random fields (Markov networks) offer a wealth of processes for those computational tasks. Of course, simpler procedures can be followed from any of those methods with the appropriate assumptions.

8.4.3 Assessment and Mitigation

As was mentioned earlier, infrastructures stakeholders need to consider CE from two viewpoints: assessment and mitigation. To assess CE on single assets, or community of assets, the foremost important point is to enumerate a complete and accurate list of considerations/issues that affect the physical and operational performance of the asset (node) or community (network). Following the completion of such a list, an accurate estimate of how these considerations/issues are linked, and whether these links are directional is needed. The directionality between two nodes indicates which node affects the other and whether the link between the two nodes is directionless. For example, in Figure 8.23 it is obvious that the soil supports the approaches and the bearings, thus the behavior of the soil affects the behavior of both the approaches and bearings. Whereas the bearings and the girder both affect each other, there should not be any directionality between both nodes (indicating bidirectional nature). The final important step in assessing CE is to accurately estimate the properties of both nodes and links within the network. Following these three important steps, evaluations of the CE metrics can be possible either probabilistically or deterministically.

If a CE assessment revealed that mitigation is warranted, the effects of different mitigation options on CE should be included. It is important to note that CE can have an effect on both costs and benefits of different mitigations projects. Because of this, changes of CE patterns and metrics need to be evaluated if a mitigation project is performed. Without evaluating CE changes and effects (positive or negative), unintended consequences might occur after performing mitigation projects as shown in Figure 8.28.

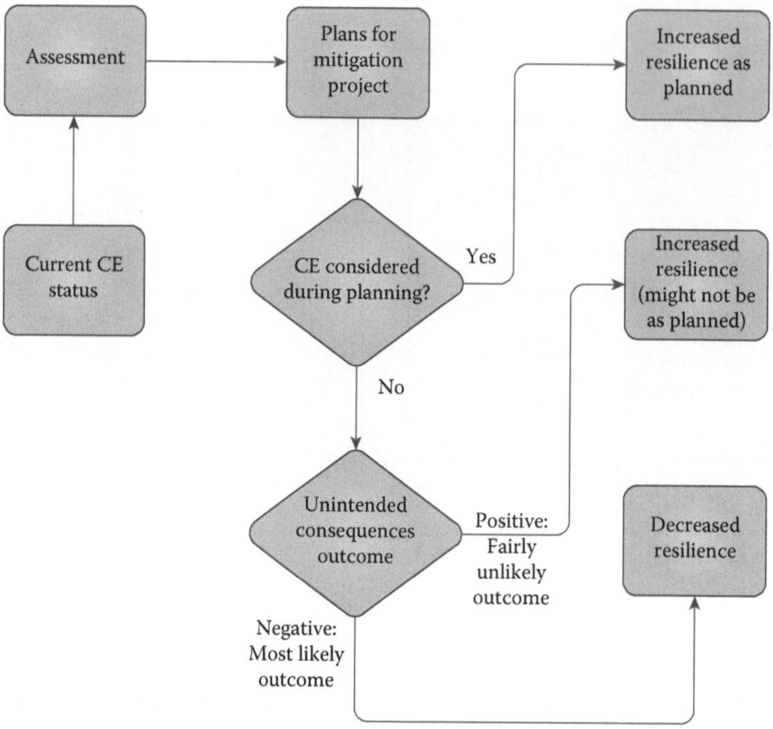

FIGURE 8.28 Cascading effects and unintended consequences in planning mitigation projects.

8.4.4 State of the Art

Currently there are limited available methodologies and tools that objectively include asset (node)-based or community (network)-based CE in either assessments or mitigation efforts. Any successful CE tool or methodology needs to include a healthy mix of subjective and objective procedures. Subjective procedures include understating how different disciplines work in a facility, multihazards interactions, and respect and inclusion of operators and managers of the facility in building up any CE model. Objective procedures include appropriate modeling of networks (Bayes networks and/or Markov network offer a great base to start), balance resolution of the network geometry and the ultimate objectives (assessment, mitigation, and prioritizations), improved functionality and balanced choices of the 4Rs, and balanced choices of TTR (Time to Recovery) and CO (Continued Operations).

An example of publicly available simple tools that apply a simplified application of CE principles is the DHS sponsored Integrated Rapid Visual Screening (IRVS tools). Similar tools and methodologies are definitely needed to ensure higher performance and resilient infrastructures.

8.4.5 Case Study 8.3: Community Resilience and Cascading Effects

8.4.5.1 Overview

We have seen earlier resilience models which utilize both time and continued operations as the governing factors for resilience assessment or the 4Rs as the governing factors for resilience assessment. In this case study, we are modeling CE on resilience of the different assets within a community network. We can, of course, model each asset in as much detail and then connect them to other assets in the network with appropriate links. However, for simplicity, we model each asset with a single variable that denotes the state of its operational efficiency. This is an indirect way to express its resilience. Within the community network, we also model the links between the assets via either

CPTs (for directional links) or potentials (for nondirectional or two-way links). The resulting model is a chain graph (CG) model. These links simulate the desired CEs after an abnormal event. Again for clarity, we assume the model to be hazard independent. We accommodate the effects of hazards by imposing a reduction of operational efficiency on one or more of the assets within the network. The desired CE can then be studied through the propagation of reduced operational efficiencies, or lack of them, throughout the assets within the network. Figure 8.29 shows the simple community of assets we use in this study. It is composed of two office buildings (OB1 and OB2), a single tunnel (T), two bridges (B1 and B2), a single mass transit station (M.T.), and a single power station (P.S.). The links between the different assets are as shown in the figure. There is a river crossing the community. The river is prone to flooding, which when it happens, can reduce the operational efficiency of some of the assets on the right bank of the river.

8.4.5.2 Model Description

The model that simulates CE is shown in Figure 8.30. The states of nodal variables are shown in Table 8.11. Note that we used a simple 2-state descriptor for the operational efficiency of each asset as either *Low O.E* or *High O.E*. The definitions of the nodal variables are in Appendix 8.I (Table 8.36). Note that the links in this model are as follows:

- *Directional*: For example, the power station operation is obviously a parent to the tunnel and one of the office buildings.
- *Nondirectional (two-way)*: For example, the roadway links between the tunnel, the two bridges, and the rest of the network are nondirectional. Similarly, the telecommunication link between the two office buildings is nondirectional.

8.4.5.3 CPTs and Potentials

The different CPTs and potentials of each variable of the network of Figure 8.30 are shown in Appendix 8.IV. Because this is a CG model, it includes a mix of potentials (Tables 8.57 through 8.63)

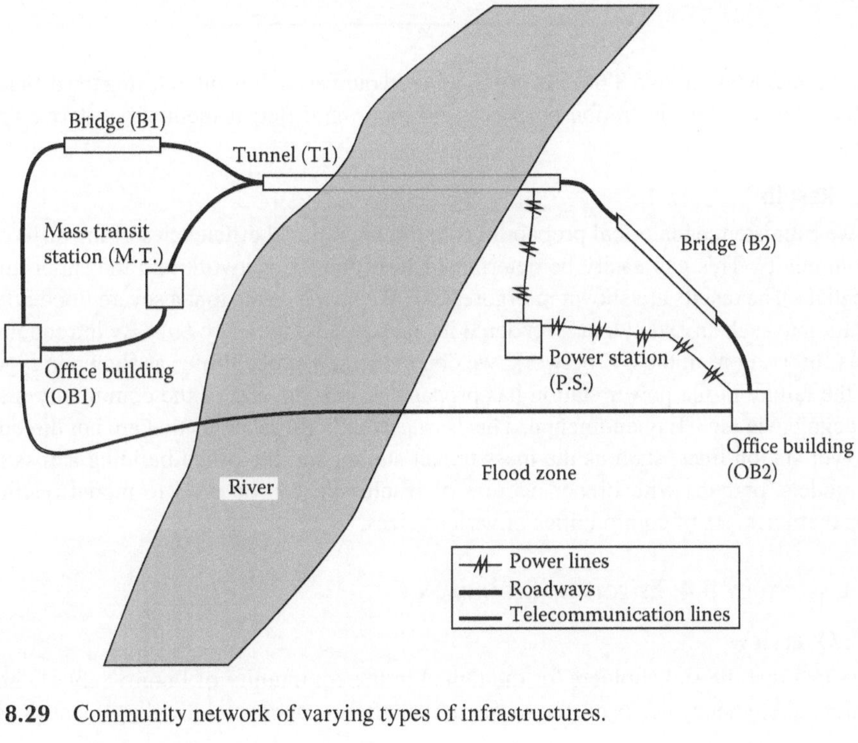

FIGURE 8.29 Community network of varying types of infrastructures.

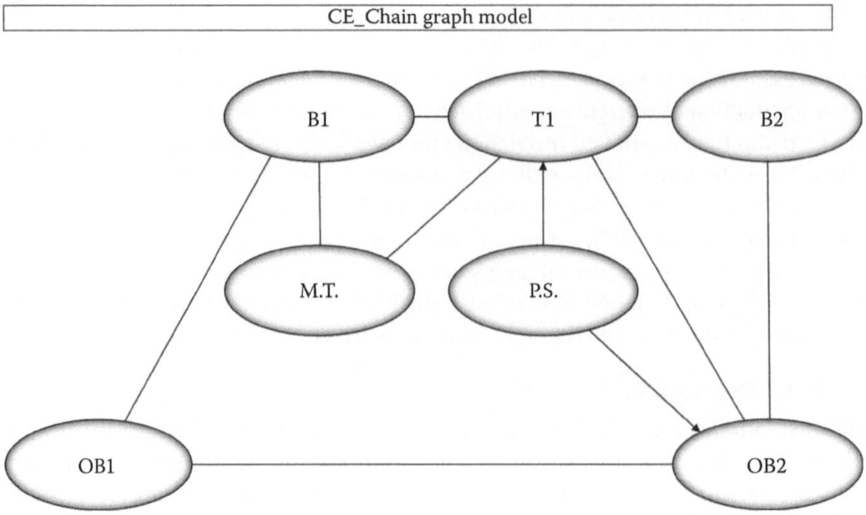

FIGURE 8.30 Chain graph model of community network nonredundant power stations.

TABLE 8.11
States of Nodal Variables

Variable	# of States	Ranks (States)
B1, B2	2	*Low O.E., High O.E.*
T1	2	*Low O.E., High O.E.*
M.T.	2	*Low O.E., High O.E.*
P.S.	2	*Low O.E., High O.E.*
OB1, OB2	2	*Low O.E., High O.E.*

and CPTs (Tables 8.64 through 8.66). As usual, we used our best judgment in filling the different tables in a realistic manner for illustration purposes. For more on different methods of developing CPTs, see Chapter 2.

8.4.5.4 Results

At first, we estimate the historical probabilities of the operational efficiencies of the different assets in the community. This can easily be determined from the CG network and the supporting CPTs and potentials. The results are shown in Figure 8.31. We now assume that a severe flood affected the power station to such an extent that it rendered its operational efficiency *Low.* By introducing this as evidence (observation) in the CG network, we obtain the new probabilities as shown in Figure 8.32. Clearly, the failure in the power station has propagated into the rest of the community assets, to a varying degree, via the CE phenomenon. This is true even to those assets that are not directly linked to the power station lines, such as the mass transit station and the office building across the river. Similar models, perhaps with higher degrees of resolution, can be used to model resilience and operational efficiencies of communities of various sizes.

8.4.6 Case Study 8.4: Effects of Redundancy

8.4.6.1 Overview

Let us assume that the stakeholders for the tunnel in the community of Figure 8.29 did not accept the operational efficiency performance of the tunnel after an abnormal flooding event as computed

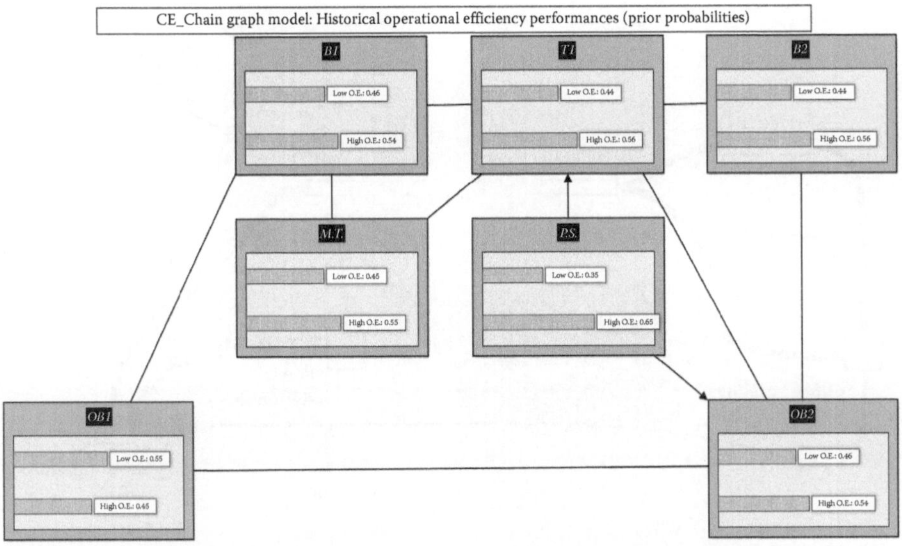

FIGURE 8.31 Historical (prior) probabilities for the redundant network.

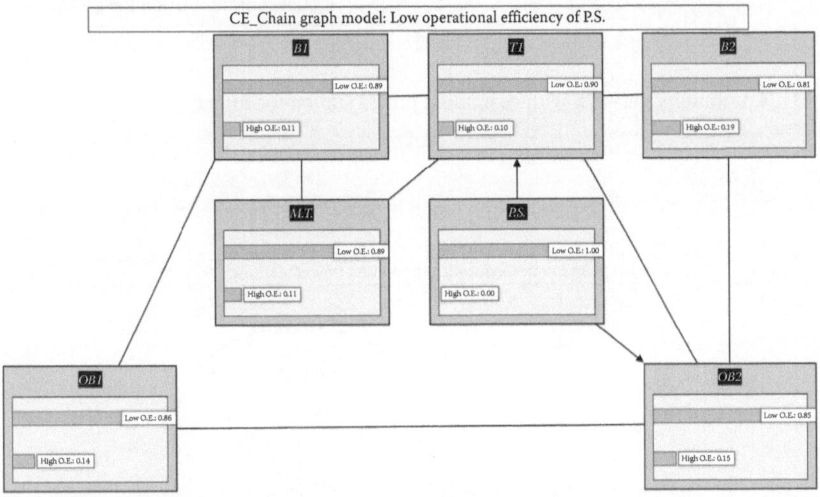

FIGURE 8.32 Probabilities after the failure of power station *P.S.*

in Case Study 8.3. One possible way to treat such poor performance is by introducing a redundant power station (Red_P.S.) that is not affected by flooding events as shown in Figure 8.33. Now we are interested in examining the effects of this added redundancy on the operational efficiencies of different assets in the community.

8.4.6.2 Model

The new CG model is shown in Figure 8.34. The states of nodal variables are shown in Table 8.12. As usual, the definition of the nodal variables are shown in Appendix 8.I (Table 8.36). The model is fairly similar to the model in Case Study 8.3 with the addition of a single new variable, *Red_P.S.*, to simulate the added redundant power station that will power the tunnel if a flooding event causes failure of the main power station. The link to the tunnel is obviously directional; the redundant *Red_P.S.* is the parent to the tunnel, *T1*. Given that the new station is meant only to serve a single

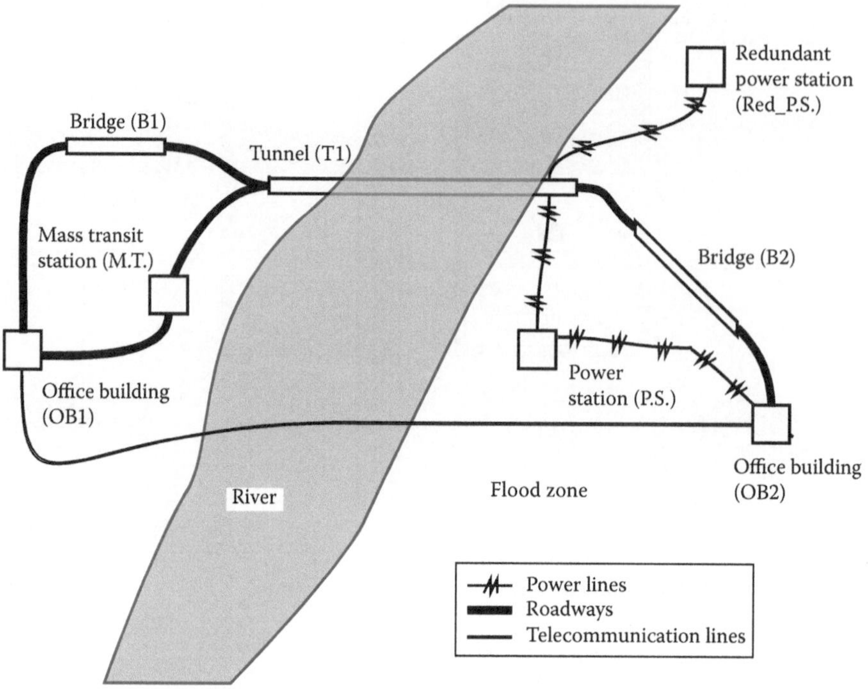

FIGURE 8.33 Community network with redundant power station for tunnel.

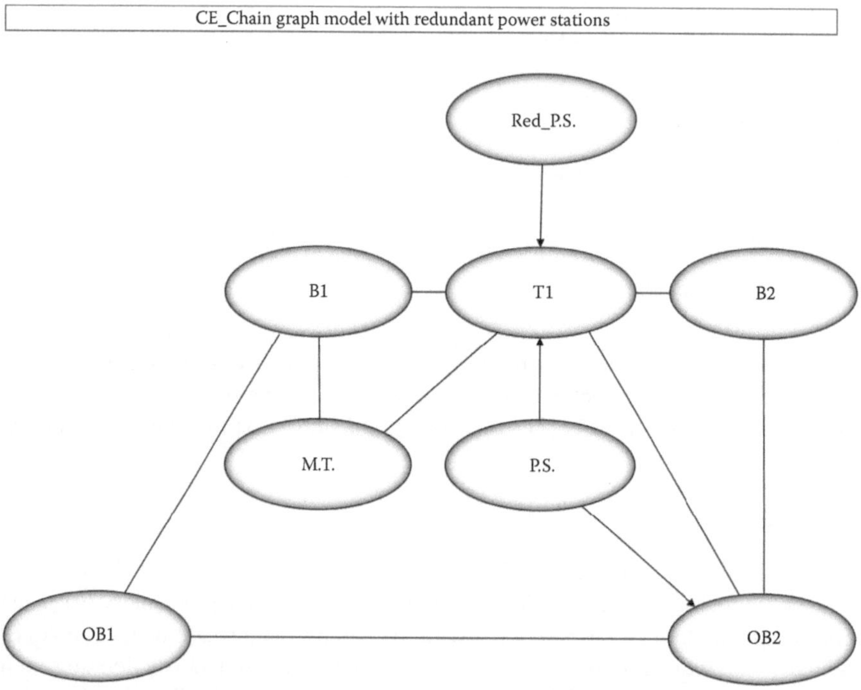

FIGURE 8.34 Chain graph model of community network with redundant power stations.

TABLE 8.12
States of Nodal Variables

Variable	# of States	Ranks (States)
B1, B2		*Low O.E., High O.E.*
T1		*Low O.E., High O.E.*
M.T.		*Low O.E., High O.E.*
P.S.		*Low O.E., High O.E.*
OB1, OB2		*Low O.E., High O.E.*
Red_P.S.		*Low O.E., High O.E.*

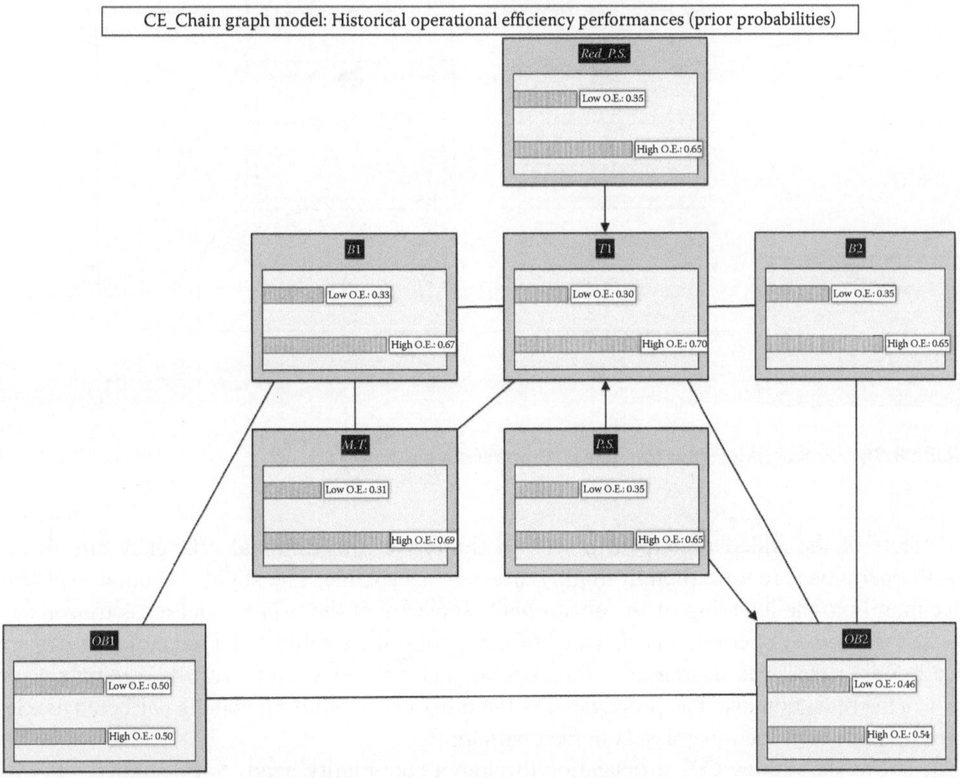

FIGURE 8.35 Historical (prior) probabilities for the redundant network.

asset, the tunnel in case of an emergency, the costs of adding such a station are obviously low compared to potential costs if the new station is meant to power more assets.

8.4.6.3 CPTs and Potentials

All the CPTs and potentials of this model are identical to those of Case Study 8.3 as displayed in Appendix 8.IV. Two additional CPTs that have to be defined to accommodate the new redundant power station: the CPT for the new variable, *Red_P.S.*, and the CPT for the child variable, *T.* Both are displayed in Appendix 8.V.

8.4.6.4 Results

With the addition of the new redundant power station, new prior probabilities are computed as shown in Figure 8.35. Of more interest is to study the operational efficiency when the flood causes a massive failure in the operational efficiency of the main power station as shown in Figure 8.36.

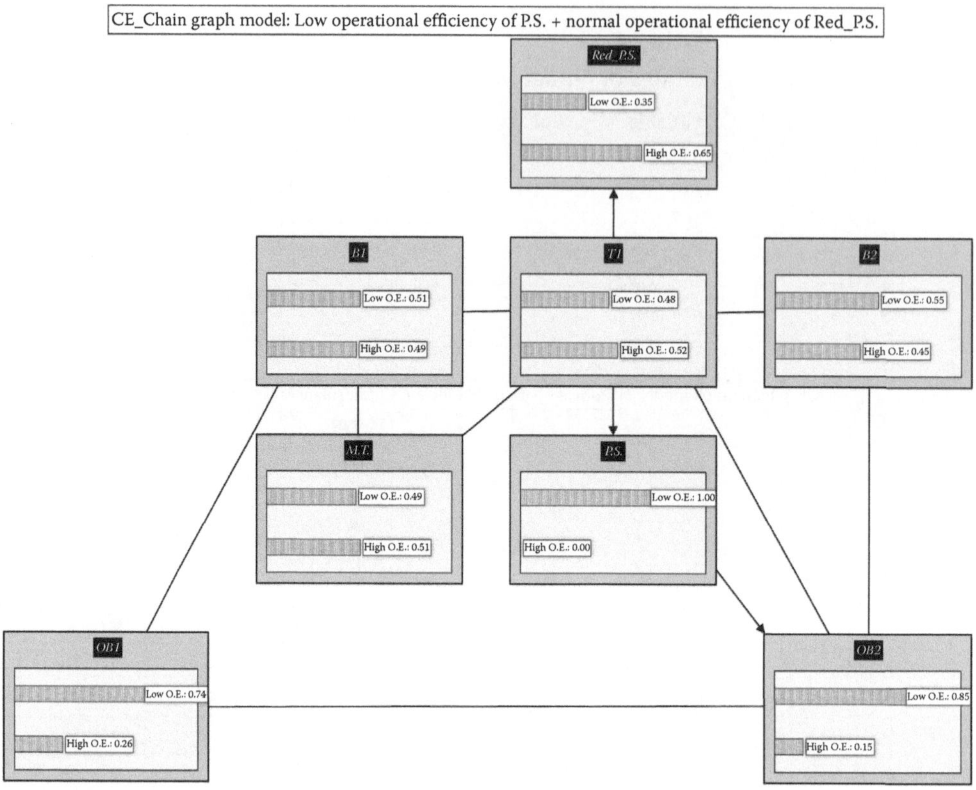

FIGURE 8.36 Probabilities after the failure of power station *P.S.*

The effects of the added redundant power on the tunnel's operational efficiency are clear: the tunnel's operations are now, though slightly affected, acceptable. The slight reduction in efficiency is due mainly to the lowering of the operational efficiency of the bridges and mass transit station. However, the loss of operational efficiency of the main station with no redundancy that can power office building *OB2* has decreased *OB2*'s operational efficiency even with the redundant power station. This reduction also has propagated to the other office building and also affected to a lesser degree operations of the other assets in the community.

The above shows how CEs propagation through a community needs to be studied and treated carefully. Because the community acts as a network, making changes in a form of mitigation might result in an unexpected, and perhaps undesired effect, in different parts of the network. The savings in the cost of adding a redundant power station might improve the performance of parts of the community, but such improved performance might not be globally realized.

8.5 CASE STUDY 8.5: RISK-BASED STRUCTURAL DEFICIENCY OF BRIDGES

8.5.1 Overview

In general, structural deficiency of bridges is caused due to deterioration of its components. When structural deterioration reaches a severe stage where the structure becomes deficient to carry legal loads or provides a level of service less than expected, the owner may have to spend significant resources to bring the structure back to its fully functional, nondeficient condition.

In this case study, we look at the severely deteriorated bridge condition (SDBC) leading to structural deficiency as a risk management problem where stakeholders make decisions on a risk-based underpinning. To do this, we need to put the SDBC in a risk framework, which is consistent with the nomenclature of this book. Upon some reflection on the very state of *structural deficiency*, we come to the conclusion that it is none other than a bridge exposure, following the definition we have used throughout the book. Structural deficiency is a condition that results from two factors: the bridge vulnerability state and various hazards the bridge is exposed to. There is no structural deficiency without being vulnerable and no deficiency without being subjected to hazard (or multitude of hazards). Thus, exposure is directly related to the severity of bridge deficiency.

Because of this we can relate risk, *Ri*, to *SDBC* as

$$Ri = f(SDBC, C) \tag{8.2}$$

We chose to model the risk as a temporal entity using a DBN process. We need to model the time dependency of Consequences *C* and *SDBC* in order to forecast the future risk of *SDBC*.

8.5.2 Model

8.5.2.1 Model Topology

One of the features of this model is the incorporation of independent component ratings as the effect of the *SDBC*. Bridge structures can be generally divided into three major components: deck, superstructure, and substructure. Thus, it can be generally characterized that *SDBC* is a function of the independent ratings of the deck, superstructure, and substructure. Bridge deficiency can be caused by severe deterioration of any of these conditions to reduce the level of service of the bridge below the intended level. Accordingly, the variables of this model are shown in Figure 8.37 and their states are shown in Appendix 8.I (Table 8.13). A summary of the definitions of nodal variables is shown in Table 8.36. As stated earlier, we will consider the temporal behavior of the risk by modeling the problem in a DBN process. To achieve that, we use the model in Figure 8.37 as the template (slice) as explained in Chapter 2. The temporal links will be assumed to be the condition ratings of the three components, *CR_1*, *CR_2*, and *CR_3*. The DBN scheme which can be used for time-marching forecasting is shown in Figure 8.38.

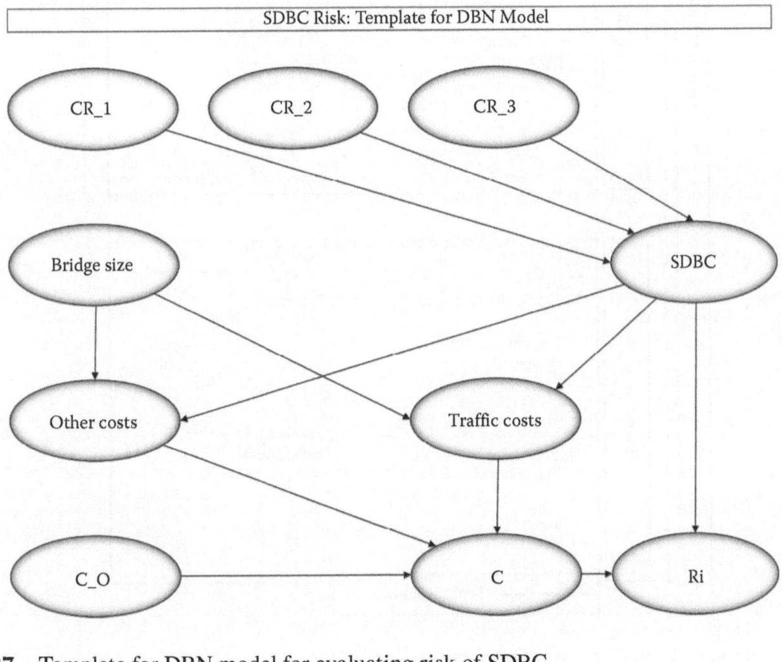

FIGURE 8.37 Template for DBN model for evaluating risk of SDBC.

TABLE 8.13
States of Nodal Variables

Variable	# of States	Ranks (States)
CR_1, CR_2, CR_3		Damaged, Pristine
Bridge Size		Large, Medium, Small
SDBC		Yes, No
Other Costs		High, Low
Traffic Costs		High, Low
C_O		High, Low
C		High, Med., Low
Ri		High, Med., Low

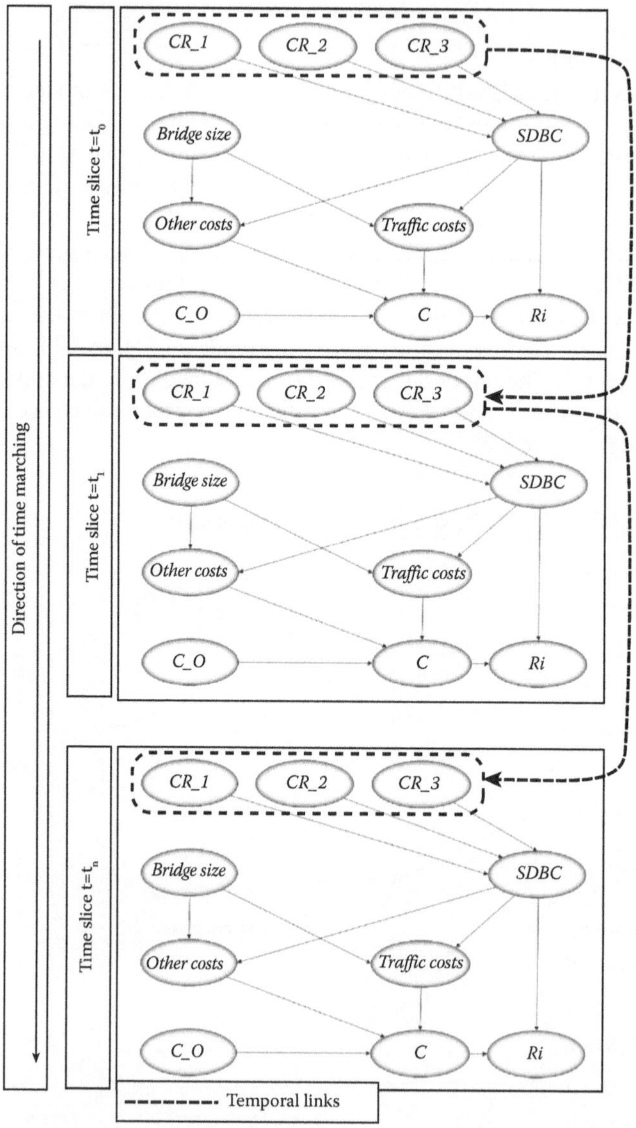

FIGURE 8.38 Time-marching DBN for SDBC model.

8.5.2.2 Transfer Matrices

As stated earlier, *SDBC* depends on one of potential three condition ratings. The purpose of this example is to explore the time dependency of these changes. In order to simulate these time changes in a DBN setting, we utilize transfer matrices in extrapolating time-dependent nodes in the time template (slice) of Figure 8.37. For our current purpose, we assume that only three condition ratings are going to change with time. The resulting time-marching DBN model is shown in Figure 8.38.

The deck is assumed to be the fastest to deteriorate. We further assume that the superstructure deterioration rate is the slowest, followed by the substructure. Thus, for a given time period, the transfer matrix of the superstructure, *CR_1* is assumed to be

$$[T]_{CR\_1} = \begin{bmatrix} 0.9 & 0.0 \\ 0.1 & 1.0 \end{bmatrix} \tag{8.3}$$

For the substructure, the transfer matrix for *CR_2* is assumed to be

$$[T]_{CR\_2} = \begin{bmatrix} 0.85 & 0.0 \\ 0.15 & 1.0 \end{bmatrix} \tag{8.4}$$

Finally, the deck transfer matrix for *CR_3* is assumed to be

$$[T]_{CR\_3} = \begin{bmatrix} 0.8 & 0.0 \\ 0.2 & 1.0 \end{bmatrix} \tag{8.5}$$

We will assume that the condition rating at the start of the excursive, $t = t_0$, is 100% pristine (0% damaged) for all three bridge components. Applying the transfer matrices (8.3) through (8.5), we get the probabilities of different condition ratings after five time periods as shown in Table 8.14.

8.5.3 Conditional Probabilities Tables

The different CPTs of each variable of the BN network of Figure 8.37 are shown in Appendix 8.VI. As usual we used our best judgment in filling the different tables in a realistic manner for illustration purposes. For more on different methods of developing CPTs, see Chapter 2.

8.5.4 Results

Historical (prior) probabilities are estimated using the model and are shown in the Figure 8.39 and the average historic risk is calculated using the method described in Appendix B. The average historical risk is then $Ri|_{\text{average}} = 46\%$. For a specific new bridge, if we apply the observations of *CR_1*, *CR_2*, and *CR_3* from Table 8.14 at $t = t_0$, we obtain the probability values for the network as in Figure 8.40. The average risk value of that specific bridge is $Ri|_{\text{average}} = 39\%$, a bit smaller than the historical value due to the assumed pristine condition of the bridge under consideration.

We are now confronted with the task of identifying objectively the time at which the bridge is declared as a severely deteriorated bridge (SDB). We first recall that an SDB condition is reached when one of the three condition ratings is found below a threshold rating. Based on this, and using Table 8.14, we can identify the expected time of interest based on a threshold probability of damaged CR. We choose a probability of damage of 60% as a suitable threshold for illustration purposes. By examining Table 8.14, it seems that the deck, *CR_3*, will reach this threshold probability at approximately $t = t_4$. We thus compute the BN probabilities at that time instant. Those marginal probabilities are shown in Figure 8.41. We note that the average risk that corresponds to this instant of time is $Ri|_{\text{average}} = 68\%$: this is the implied threshold acceptable risk of being deficient for this particular bridge

We had assumed all along that other consequences, *C_O*, are time independent for this example. This is of course not a very accurate assumption because consequences change with time. We now release that assumption and assume that *C_O* increases via a transfer matrix as

$$[T]_{C\_O} = \begin{bmatrix} 0.95 & 0.0 \\ 0.05 & 1.0 \end{bmatrix} \tag{8.6}$$

The changes in *C_O* probabilities are now computed as in Table 8.15. The resulting time-marching DBN model is shown in Figure 8.42.

TABLE 8.14
Changes of Probabilities of CR of Bridge Components

Bridge Component		Probabilities at Start of Time Periods					
		$t = t_0$	$t = t_1$	$t = t_2$	$t = t_3$	$t = t_4$	$t = t_5$
CR_1	*pristine*	100%	90%	81%	73%	66%	59%
	damaged	0%	10%	19%	27%	34%	41%
CR_2	*pristine*	100%	85%	72%	61%	52%	44%
	damaged	0%	15%	28%	39%	48%	56%
CR_3	*pristine*	100%	80%	64%	51%	41%	33%
	damaged	0%	20%	36%	49%	59%	67%

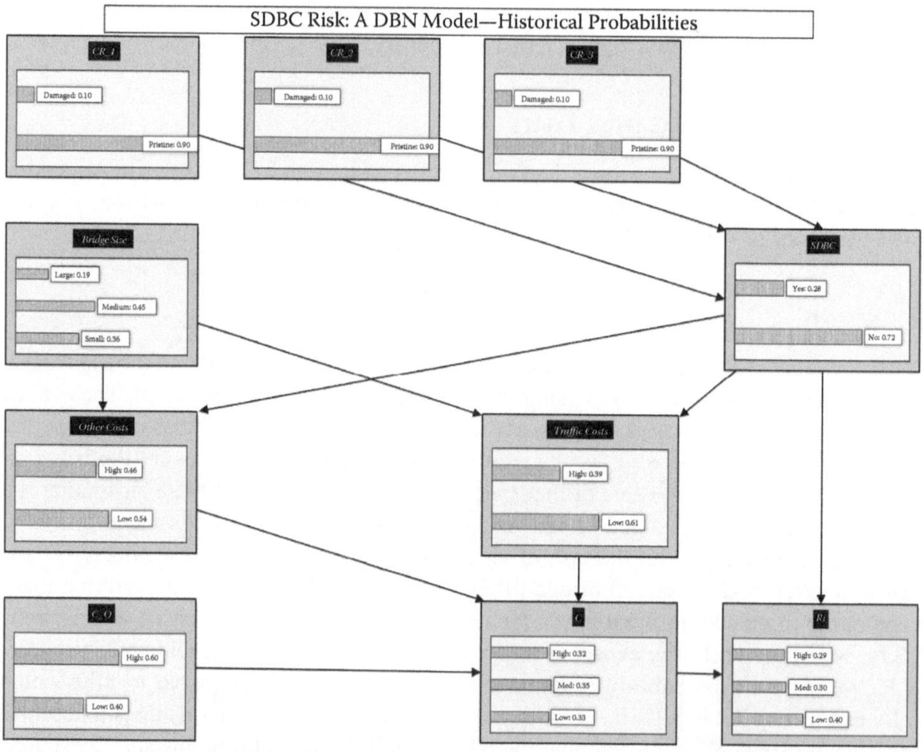

FIGURE 8.39 Historical probabilities of SDBC.

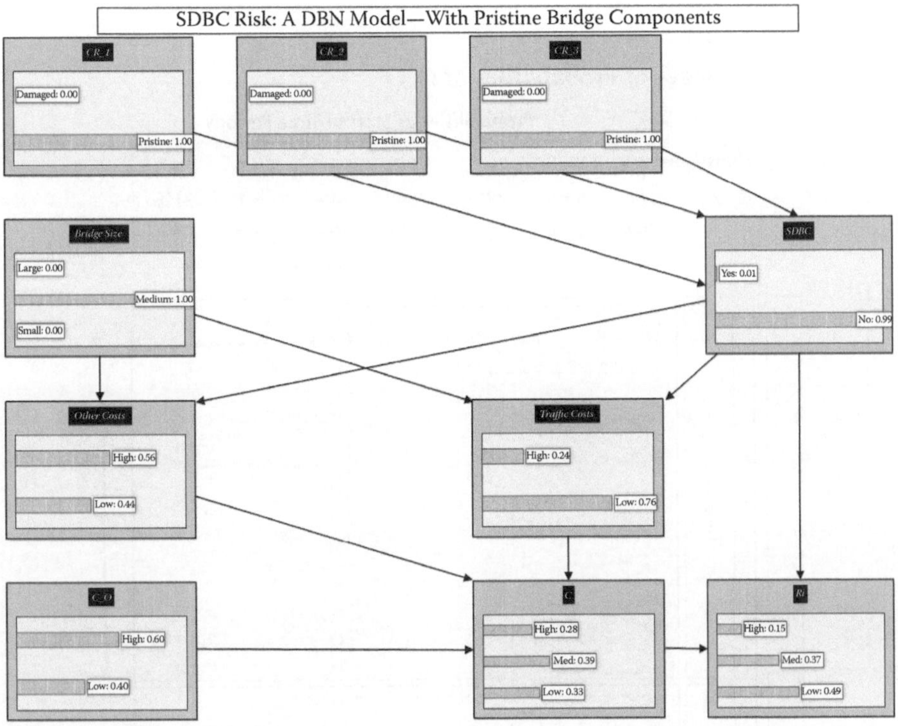

FIGURE 8.40 Marginal probabilities at $t = t_0$ for medium size pristine bridges.

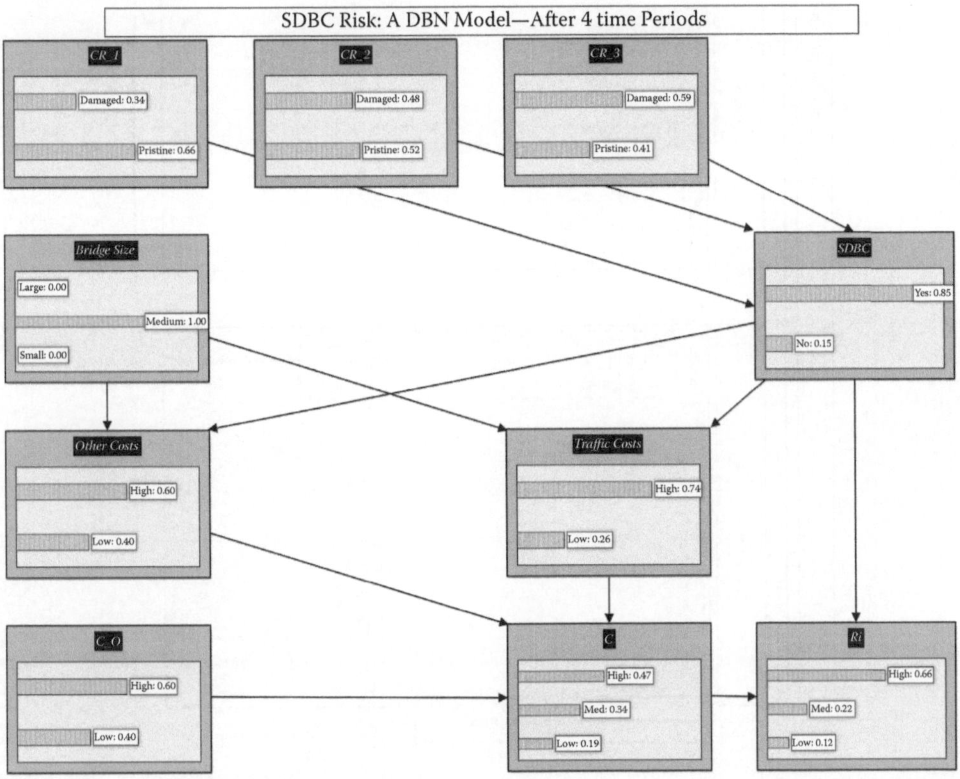

FIGURE 8.41 Marginal probabilities at $t = t_4$.

TABLE 8.15
Changes of Probabilities of *C_O*

Consequences		Probabilities at Start of Time Periods					
		$t = t_0$	$t = t_1$	$t = t_2$	$t = t_3$	$t = t_4$	$t = t_5$
C_O	*Low*	40%	38%	36%	34%	33%	31%
	High	60%	62%	64%	66%	67%	69%

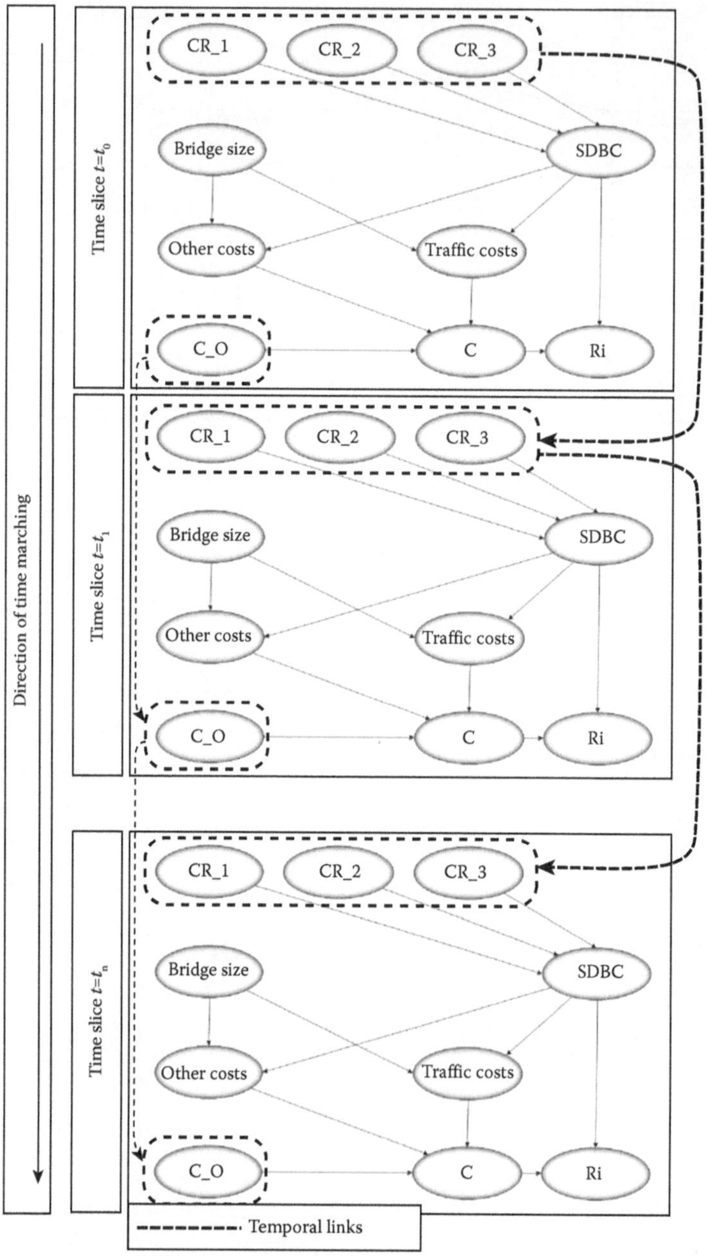

FIGURE 8.42 Time-marching DBN for SDB model with both condition ratings and consequences are time dependent.

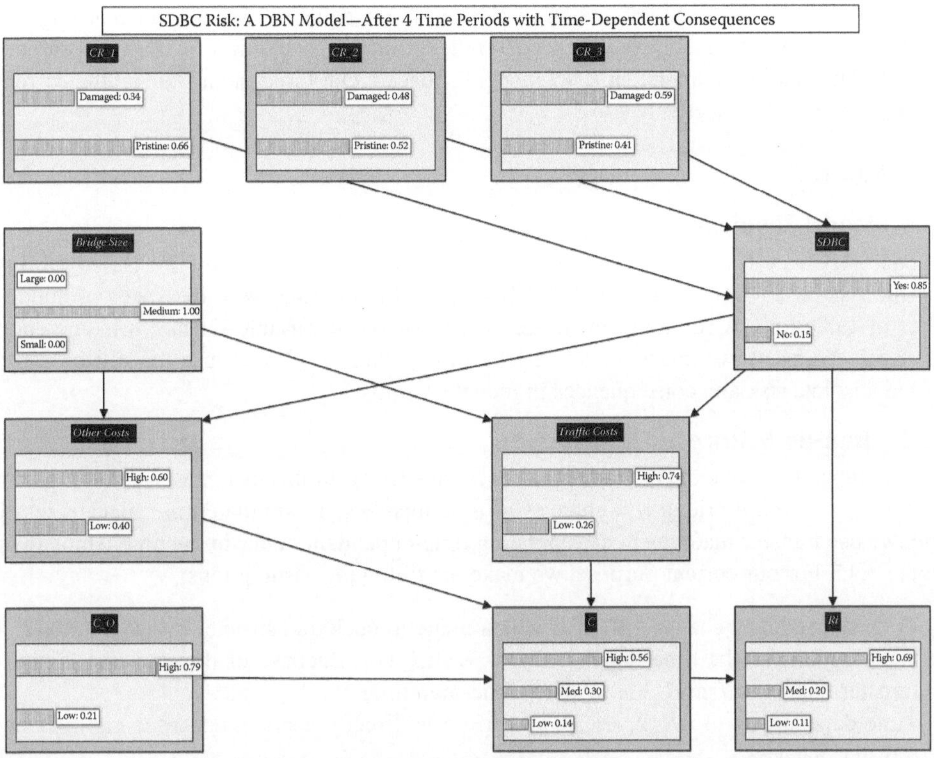

FIGURE 8.43 Marginal probabilities at $t = t_4$ and time-varying C_O.

Applying these changes in the DBN up to $t = t_4$, we obtain the marginal probabilities as shown in Figure 8.43. The average risk in this instant is $Ri|_{\text{average}} = 69\%$: slightly higher than when we didn't consider consequences changes (Figure 8.41). This indicates that the bridge reached the risk threshold a bit earlier. To find the exact time when the acceptable risk threshold of 60% is reached, a series of DBN time-marching analyses have to be undertaken.

8.5.5 Concluding Remarks

One of the main conclusions of the example should be that risk-based *SDBC* management can help provide a cost-effective and safer approach because it can forecast early on the incipient SDBC, as well as accommodate cost (consequences) in the forecasting process.

8.6 CASE STUDY 8.6: RISK-BASED FUNCTIONAL DEFICIENCY

8.6.1 Overview

Structures can be in a good condition but can be functionally inadequate (functional deficiency condition, FDC) to meet the current demands compared to when they were built. Several bridges built in major cities 30 to 40 years ago fall into this category. Despite the maintenance and rehabilitation efforts to keep them in good condition, they may not be able to provide the required service due to factors such as geometry, increased traffic volumes, inadequate under clearance, approach road way alignment, waterway inadequacy, and very poor structural condition. The FHWA uses appraisal ratings to document functional issues of the bridges. These ratings are based on an evaluation of bridge characteristics relative to the current standards used for highway and bridge design (FHWA 1995) and are used to designate functionally deficient/obsolete condition.

FDC merely represents a combination of a type of consequence and a factor that might affect exposure. So in our risk model, we will need to reflect this reality. Similar to *SDBC* considerations, we model FDC in a time-dependent DBN model so that we can forecast and budget the current and future effects of FDC objectively.

8.6.2 Model

8.6.2.1 Model Topology

We model FDC as a function of five appraisal ratings as shown in Table 8.16. Based on this, we model the risk-based DBN time template (slice) as in Figure 8.45, with the states of nodal variables as in Table 8.17. A summary of the definitions of nodal variables is shown in Appendix 8.I (Table 8.36). We used two intermediate variables *FO_1* and *FO_2* to reduce the size of CPTs. We also used absolute risk and consequences in monetary units.

8.6.2.2 Transfer Matrices

FDC rank depends on one of the five appraisal ratings (AR). In this example, we investigate some issues with how overall bridge risk changes as time marches. To simulate time changes in a DBN setting, we use transfer matrices in extrapolating time-dependent nodes in the time template (slice) of Figure 8.45. For our current purpose, we make the following assumptions:

- Time-dependency changes of ARs, which relate to deck and structures, *AR_1* and *AR_4*, are functions of the time dependency of *Travel_Vol*. Because of this, we only assign a transfer matrix to *Travel_Vol* to affect time-marching of *AR_1* and *AR_4*.
- Time dependency of *AR_2*, *AR_3*, and *AR_5* is directly accommodated using their own transfer matrices.

TABLE 8.16
Factors Affecting Functional Deficiency

Variable	Component	Function of Traffic Volume?
AR_1	Deck geometry	Yes
AR_2	Under clearance	No
AR_3	Approach roadway alignment (see Figure 8.44)	No
AR_4	Structural condition	Yes
AR_5	Waterway adequacy	No

FIGURE 8.44 Example of functionally deficient road approach alignment.

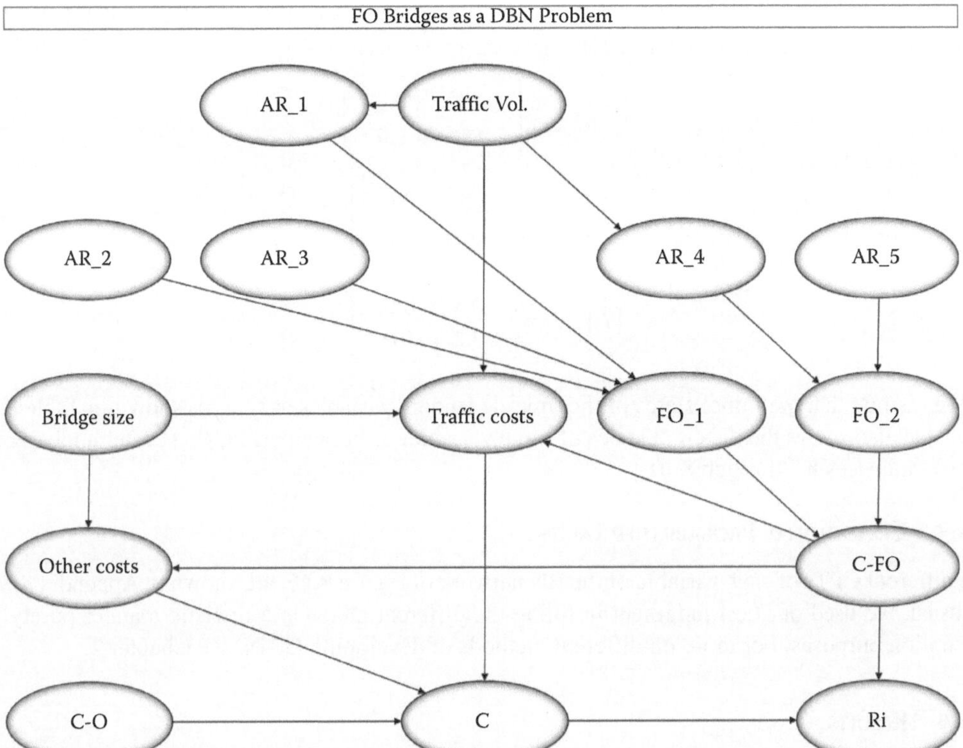

FIGURE 8.45 Template for functional deficiency DBN model.

TABLE 8.17
States of Nodal Variables

Variable	# of States	Ranks (States)
AR_1, AR_2, AR_3, AR_4, AR_5	2	Obsolete, Acceptable
Traffic Vol.	3	High, Medium, Low
Bridge Size	3	Large, Medium, Small
Traffic Costs	2	High, Low
FO_1	2	Obsolete, Acceptable
FO_2	2	Obsolete, Acceptable
Other Costs	2	High, Low
C-FO	2	Obsolete, Acceptable
C-O	2	High, Low
C	3	\$100 M, \$10 M, \$0.0 M
Ri		\$100 M, \$10 M, \$0.0 M

The resulting time-marching DBN model is shown in Figure 8.48.

Assuming that *Traffic_Vol* changes from one period to another according to the transfer matrix

$$[T]_{Traffic\_Vol} = \begin{bmatrix} 0.7 & 0.0 & 0.0 \\ 0.2 & 0.9 & 0.0 \\ 0.1 & 0.1 & 1.0 \end{bmatrix} \quad (8.7)$$

The changes in the probabilities of *AR_1* and *AR_4* are then conditioned by CPTs relating them to *Traffic_Vol*. Furthermore, we assume additional transfer matrices for *AR_2*, *AR_3*, and *AR_5* as follows:

$$[T]_{AR\_2} = \begin{bmatrix} 0.85 & 0.0 \\ 0.15 & 1.0 \end{bmatrix} \tag{8.8}$$

$$[T]_{AR\_3} = \begin{bmatrix} 0.9 & 0.0 \\ 0.1 & 1.0 \end{bmatrix} \tag{8.9}$$

$$[T]_{AR\_5} = \begin{bmatrix} 0.8 & 0.0 \\ 0.2 & 1.0 \end{bmatrix} \tag{8.10}$$

We assume that the initial states of the time-dependent variables at $t = t_0$ as shown in Table 8.18. The table also shows the different states of same variables after one period, $t = t_1$, after applying the transfer matrices 8.7 through 8.10.

8.6.3 Conditional Probabilities Tables

The different CPTs of each variable of the BN network of Figure 8.46, are shown in Appendix 8.VII. As usual, we used our best judgment in filling the different tables in a realistic manner purely for illustration purposes. For more on different methods of developing CPTs, see Chapter 2.

8.6.4 Results

We first execute the template of Figure 8.45 using different CTPs to obtain the historical probabilities for different variables. Those historical probabilities are shown in Figure 8.46. Using the methodology of Chapter 2, the *Ri* histogram results in a historical average of \$40.27 M. Note that the risk value is expressed in absolute monetary manner. We then introduce an observation of *Bridge_Size* = *Medium* to indicate the specific size of the bridge of interest. We assume that the bridge is designed for a low traffic condition, *Traffic_Vol* = *Low*. The remainder of the appraisal conditions at $t = t_0$ are shown in Table 8.18. Executing BN with those observations results in the histograms of Figure 8. 47. The average risk value of that specific bridge at its initial phase of operation is \$13.32 M, smaller than the historical value because the assumed excellent condition of the bridge under consideration, which is an expected result.

Using the extrapolated probabilities of Table 8.18 and the DBN model of Figure 8.48, we can now forecast the probabilities of the model variables after one period, $t = t_1$; the results are displayed in Figure 8.49. From the histogram of *Ri*, the average risk after one period is forecasted to be \$39.72 M. This increase is mainly due to the rapid increases in different ARs; some of these increases are due to the forecasted increase in traffic volume. All of these appraisal rating increases result in an increased potential of FDC, and hence the increase of risk due to FD.

We assumed so far that *C_O* (consequences that are independent from the bridge conditions such as social/economic consequences) have been constant going from $t = t_0$ to $t = t_1$. Such an assumption is not fairly accurate. In order to estimate the effects of changes of *C_O* on risk forecasts as time marches, we consider a new DBN model as in Figure 8.50. A companion transfer matrix for *C_O* is assumed to be

$$[T]_{C\_O} = \begin{bmatrix} 0.80 & 0.0 \\ 0.20 & 1.0 \end{bmatrix} \tag{8.11}$$

The changes in *C_O* probabilities are now computed as in Table 8.19.

Applying these changes in the DBN up to $t = t_1$, we obtain the marginal probabilities as in Figure 8.51. The average risk in this instant is \$42.11 M, which is higher than when we didn't consider

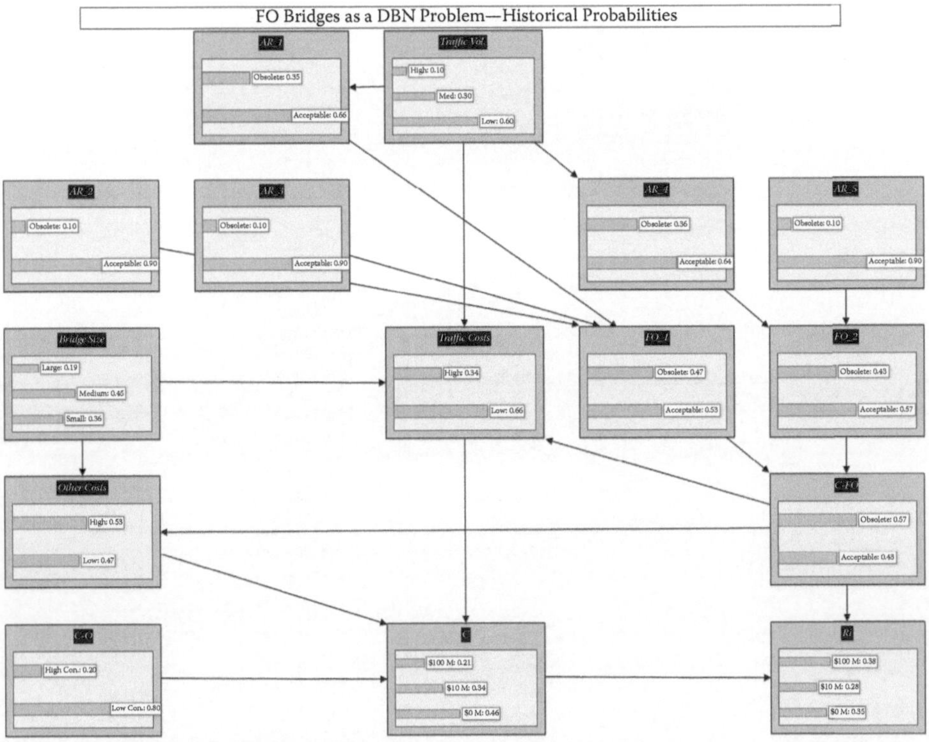

FIGURE 8.46 Historical probabilities of FDC.

TABLE 8.18
Changes of Probabilities for FDC

Bridge Component		**Probabilities at Start of Time Periods** $t = t_0$	$t = t_1$
Traffic_Vol	*Low*	100%	70%
	Medium	0%	20%
	High	0%	10%
AR_2	*Acceptable*	100%	85%
	Obsolete	0%	15%
AR_3	*Acceptable*	100%	90%
	Obsolete	0%	10%
AR_5	*Acceptable*	100%	85%
	Obsolete	0%	15%

bridge-independent consequences changes (Figure 8.49). This indicates the importance of accommodating all possible consequences, including those that are bridge independent, in risk management situation.

8.6.5 Remarks

By following an objective risk-based modeling for functional deficiency conditions, the decision maker can forecast the potential future behavior of the bridge that accommodates not only the physical behavior but also the future behavior of different sources of costs. This allows for basing

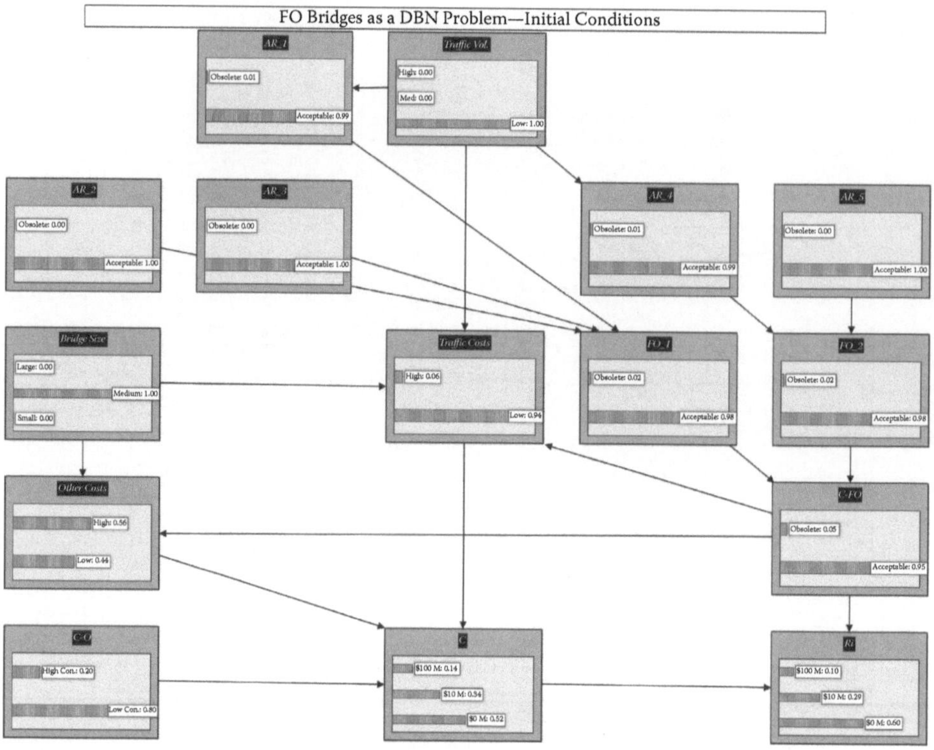

FIGURE 8.47 Marginal probabilities at $t = t_0$ for medium size pristine bridges.

the management decisions on more reasonable and accurate information. Obtaining useful data to conduct such an analysis is still an issue. As big databases are developed, these types of analyses will prove to be useful in the future if not at the present time.

8.7 CASE STUDY 8.7: SUSPENSION BRIDGE SECURITY

8.7.1 Elements of Suspension Bridge Security

8.7.1.1 Overview

The complex nature of suspension bridges coupled with higher security threat and location make mitigating these threats costlier compared to other bridges (Figure 8.52). Alampalli and Ettouney (2015) discussed a three-tier security assessment model for suspension bridges in a risk management framework in such a way that it can be expanded to cover all aspects of risk management associated with the upkeep of suspension bridges. The model can be used as is by bridge owners or can be recalibrated to suit the bridge owner's assessment goals. This section gives a brief overview of this work.

Objective bridge security risk management efforts require a top-down approach. Thus at first, management efforts have to be concentrated on the three risk components *C*, *T*, and *V*; and then build on these efforts. This can be accomplished by subdividing bridge security issues into three tiers to make it easier to objectively manage the security issue in a risk management framework.

Tier I: This tier can compute all risk management components easily and also can interact with reliability (safety) concerns because safety concerns are of interest for estimating vulnerabilities and threats. Tier I ratings are inferred from Tier II ratings (see Table 8.20).

Tier II: This comprises the main components for bridge security. For illustration purposes, 14 main components have been identified and their ratings are inferred from Tier III ratings. For each Tier II consideration/component, from a security viewpoint, evaluation of all relevant Tier III

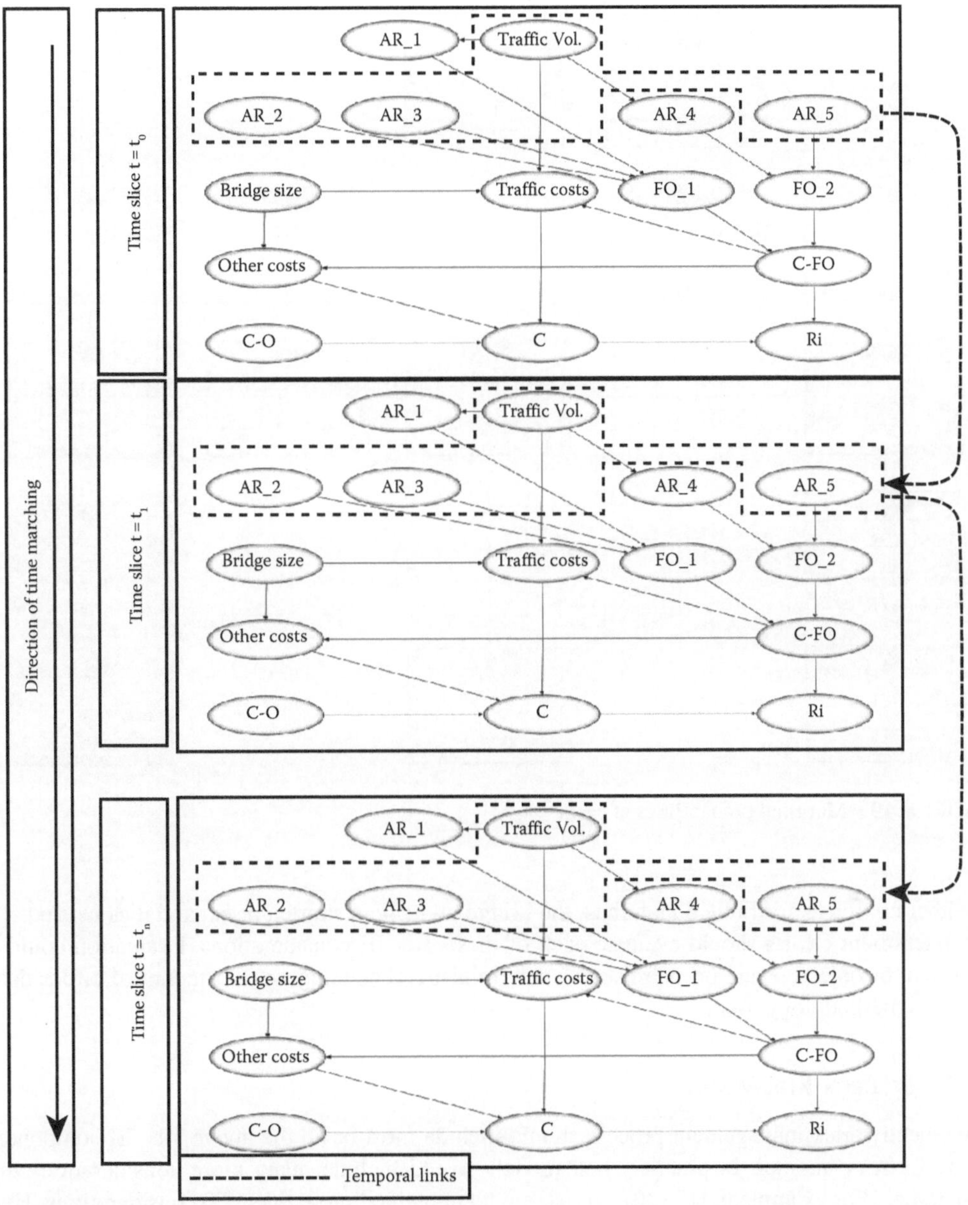

FIGURE 8.48 Time-marching DBN for FDC model.

components under it, vulnerabilities of these components, acceptance/mitigation methods, how to inspect/maintain /rehab, and what to avoid is needed. Care must be exercised not to implement a mitigation effort for one hazard by creating vulnerability of the component to another hazard.

Tier III: This comprises the most detailed components of cable bridge security. Sixty-six such components have been identified in this chapter for illustration purposes. The remainder of this section offers a detailed presentation on assessing these three tiers.

8.7.1.2 Assessment

This section identifies Tier III issues for bridge security risk evaluation. Each Tier II component along with associated Tier III components is presented in Tables 8.21 through 8.34. This information

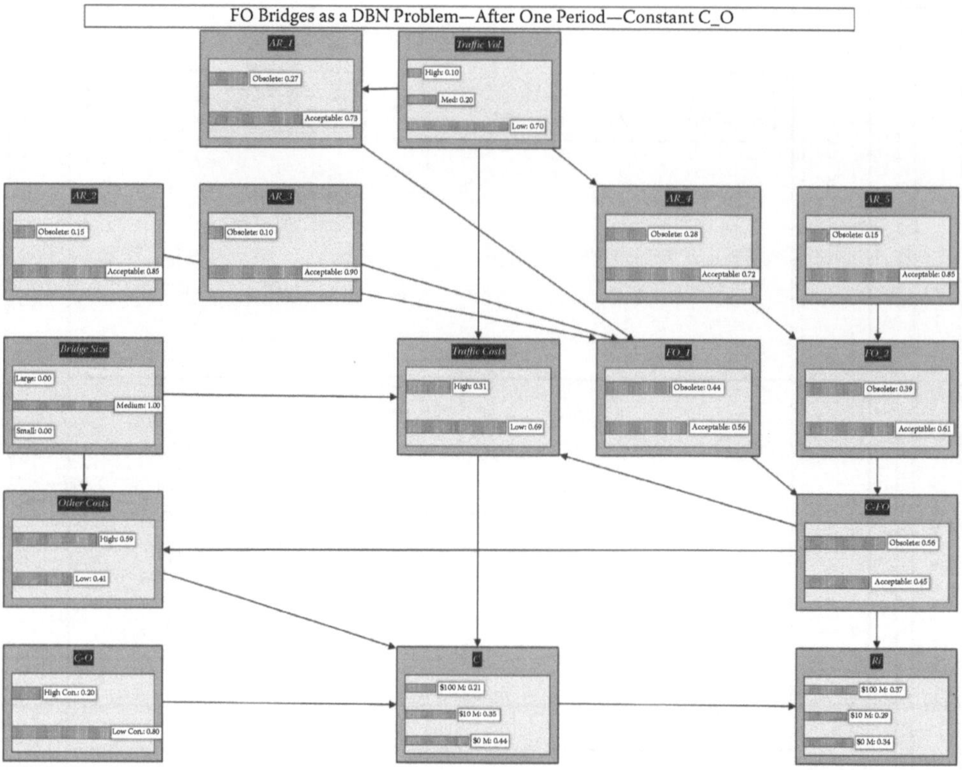

FIGURE 8.49 Marginal probabilities at $t = t_1$.

is neither complete nor limited and, thus, the bridge owner can shorten or expand it as needed. The risk assessment efforts should evaluate each of these Tier III considerations. Evaluation could be subjective or objective and probabilistic or deterministic depending on the amount of effort, detail type, and methodology used.

8.7.2 Security Risk Model

Any objective risk management process should include most or all the above Tier III components in one form or another. In practice, risk models would include many more considerations than these issues. For example, DHS (2011a) risk model contained more than 250 considerations. Here, however, we present a reduced risk model to illustrate most of the basic functional components of a bridge security risk management process with the understanding that for an accurate risk management process, much more resolution and details are still needed beyond what is offered in this section.

8.7.2.1 Model

As discussed above, the complete cable bridge security risk process is demanding and requires such a large effort that we can't exercise it herein. To give a complete practical example of risk-based assessment and mitigation decision-making, we will use a reduced model that will illustrate most major components (sans risk acceptance and multihazards considerations) of bridge security risk management efforts. Figure 8.63 shows the reduced cable bridge security risk model in a BN graph display. Note that the model contains all Tier I risk components (C, T, and V) and most of Tier II components.

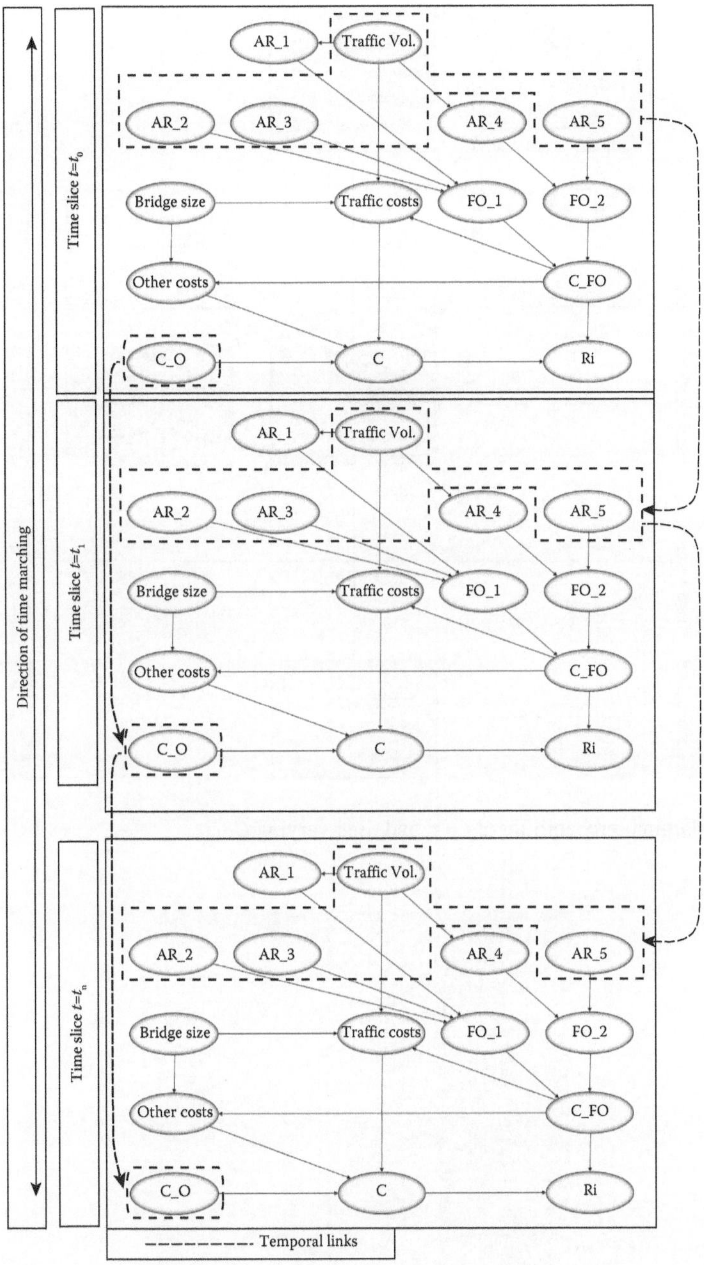

FIGURE 8.50 Time-marching DBN for FDC model with both appraisal ratings and consequences are time-dependent.

TABLE 8.19
Changes of Probabilities of *C_O*

Consequences	Probabilities at Start of Time Periods	
C_O	80%	64%
	20%	36%

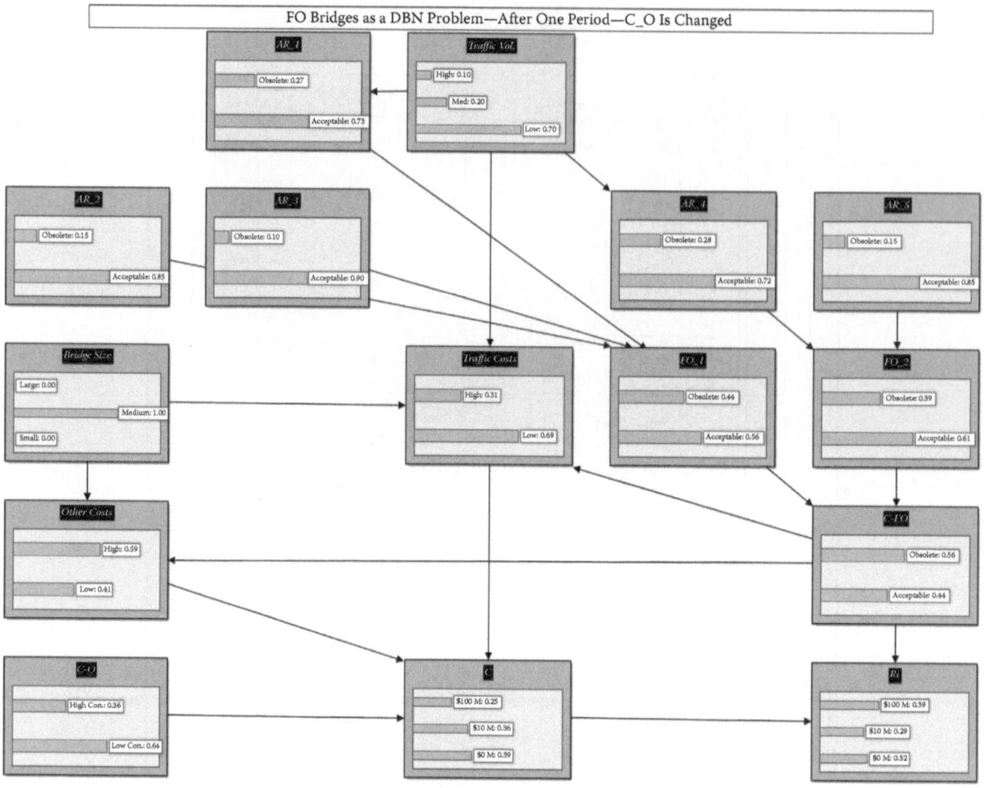

FIGURE 8.51 Marginal probabilities at $t = t_4$ and time-varying C_O.

FIGURE 8.52 Typical suspension bridge.

TABLE 8.20
Cable Bridge Security for Tier I and Tier II Considerations

# Tier II	Tier I C	Tier I T	Tier I V	Tier II Considerations	Is This Suspension Bridge Cable Specific?	Type of Required Analysis
1	Y	Y	Y	Site layout	N	Traffic/WIM/architectural/social
2	Y	Y	Y	Bridge component (suspension cable)	Y	Structural/inspection/SHM/NDT
3	Y	Y	Y	Bridge components (suspenders)	Y	Structural/inspection/SHM/NDT
4	N	Y	Y	Bridge component (other)	N	Structural/inspection/SHM/NDT
5	Y	N	Y	Bridge system	N	Structural/inspection
6	Y	Y	Y	Nonstructural	N	Structural/inspection
7	N	N	Y	Access control	N	Security
8	N	N	Y	Deterrent and other security measures	N	Security
9	Y	Y		Use/functional	N	Functional
10	Y	N	Y	Operational	N	Management
11	Y	N	Y	Special needs	Y	Structural/inspection/SHM/NDT
12	Y	N	N	Economics/social	N	Social/economic
13	N	Y	Y	Other demands	Y	Monitoring
14	N	Y	N	Additional subjective threat factors	N	Social/economic/functional

TABLE 8.21
Tier II and Tier III Issues for Site Layout

Tier II (Inferred)	Number	Tier III (Observable/ Computed)	Comments
1-Site layout	1.01	Average daily traffic (ADT)	From WIM or other sources
	1.02	Locality (urban, suburban, etc.)	See DHS (2011b, 2011c)
	1.03	Adjacent infrastructure (including buildings)	See DHS (2011b, 2011c)
	1.04	Setback rating (for site in general)	See FEMA (2005) and DHS (2011a, 2011b, 2011c)

The reduced risk model includes ten variables (risk network nodes). Table 8.35 shows the variables, their ranks, and the possible states of each variable. A summary of the definitions of nodal variables is shown in Appendix 8.I (Table 8.36).

8.7.2.2 Conditional Probabilities Tables

As usual, the next component in building a network risk model is establishing the CPTs. Appendix 8.VIII shows the CPTs that are used in conjunction with the risk model shown in Figure 8.63. These CP/histograms are generated only for illustrative purposes and needs modification to accommodate individual situations based on relevant experience and judgment.

8.7.2.3 Results

Figure 8.64 shows the results of the prior (historical) marginal probabilities for the particular bridge under consideration. Note that the combined historical results show a probability of high risk of 65%

TABLE 8.22
Tier II and Tier III Issues for Suspension Cable System

Tier II (Inferred)	Number	Tier III (Observable/ Computed)	Comments
2-Bridge component (suspension cable)	2.01	Number of independent suspension ropes and/or cables. See Figure 8.53	Bridge inventory. Might require advanced analysis
	2.02	Number of suspension ropes that can be damaged without causing progressive collapse	Bridge inventory. Might require advanced analysis
	2.03	Condition rating(s)	From inspection reports, see Figures 8.54 through 8.57
	2.04	Control (minimum) setback	Specific to suspension cables. See FEMA (2005) and DHS (2011a, 2011b, 2011c)
	2.05	Multihazards index	Might require analysis. See Ettouney et al. (2005) and Ettouney and Alampalli (2016)
	2.06	Blast retrofit	Blast retrofit can improve vulnerability
	2.07	Accumulated damage	Might require advanced inspection and/or analysis. See NCHRP (2004), Mahmoud (2011) and FHWA (2012)
	2.08	Estimated remaining life	Might require advanced inspection and analysis. See NCHRP (2004), Mahmoud (2011) and FHWA (2012)

FIGURE 8.53 Multiple suspension cables offer a higher degree of redundancy.

FIGURE 8.54 Inspection of suspension cables. (Courtesy of William Moreau.)

FIGURE 8.55 Suspenders and suspension cables. (Courtesy of William Moreau.)

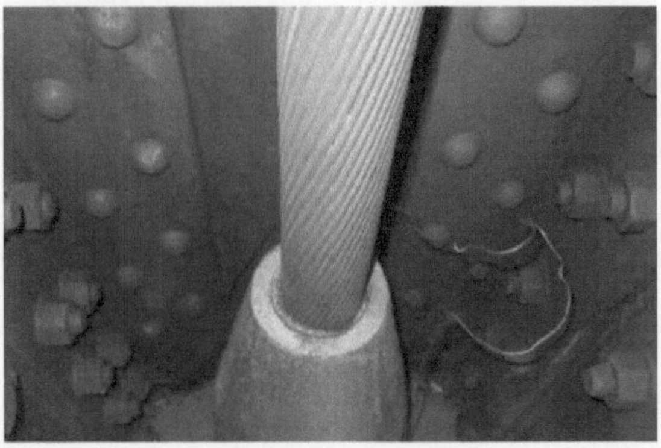

FIGURE 8.56 Connection of a bridge suspender in good condition.

TABLE 8.23
Tier II and Tier III Issues for Suspender System

Tier II (Inferred)	Number	Tier III (Observable/ Computed)	Comments
3-Bridge components (suspenders)	3.01	Number of suspension ropes	Bridge inventory. Might require advanced analysis
	3.02	Number of suspenders that can be broken in a series	Bridge inventory. Might require advanced analysis
	3.03	Condition rating	Inspection reports
	3.04	Control (minimum setback)	Specific to suspenders. See FEMA (2005) and DHS (2011a, 2011b, 2011c)
	3.05	Multihazards index	Might require analysis. See Ettouney et al. (2005) and Ettouney and Alampalli (2016)
	3.06	Blast retrofit	Blast retrofit can improve vulnerability
	3.07	Accumulated damage	Might require advanced inspection and analysis. See NCHRP (2004), Mahmoud (2011) and FHWA (2012)
	3.08	Estimated remaining life	Might require advanced inspection and analysis. See NCHRP (2004), Mahmoud (2011) and FHWA (2012)

TABLE 8.24
Tier II and Tier III Issues for Other Bridge Components

Tier II (Inferred)	Number	Tier III (Observable/Computed)	Comments
4-Bridge component (other)	4.01	Vessel collision (pier type, pier protection, VC under)	Might require advanced analysis
	4.02	Vehicle collision (VC above, distance between suspenders/truss and leftmost or rightmost lane)	Might require advanced analysis
	4.03	Bearing type, setback, and condition	Subjective estimate. Might require advanced analysis

TABLE 8.25
Tier II and Tier III Issues for Bridge System

Tier II (Inferred)	Number	Tier III (Observable/Computed)	Comments
5-Bridge system	5.01	Span type	Subjective estimate
	5.02	Stiffening truss is available or not, see Figure 8.58	Subjective estimate. Might require advanced analysis
	5.03	Fracture critical or not	Subjective estimate. Might require advanced analysis
	5.04	Fire protection	Subjective estimate. Might require advanced analysis in case of fire susceptibility
	5.05	Multihazards analysis performed during design	Require analysis. See Ettouney et al. (2005) and Ettouney and Alampalli (2016)
	5.06	Level of redundancy available	Subjective estimate. Might require advanced analysis to accurately estimate redundancy
	5.07	Progressive collapse considerations	Can be estimated via analysis
	5.08	Deck type (concrete, orthotropic, FRP, etc.)	Subjective estimate. Might require advanced analysis to accurately estimate deck robustness
	5.09	Approach types and characteristics, see Figure 8.57	Subjective estimate. Might require traffic and robustness analysis
	5.10	Load restrictions	Load restrictions indicate deficiency of robustness for live loads. This is an indication of vulnerability to blast loads too. Special care and advanced analysis are needed to accommodate such a condition
	5.11	Height postings	Height-posting might have an effect on blast threat rating. Special analysis is required to estimate such an effect
	5.12	Year built (quality of material and construction are controlled by this)	Older bridges can have degraded performance which might increase vulnerability
	5.13	Year of last retrofit (quality of material and construction are controlled by this)	Retrofits can improve performance
	5.14	Number of bridges separated vertically, see Figure 8.59	Subjective estimate. Might require advanced analysis
	5.15	Ramps underneath	Subjective estimate. Might require advanced analysis

FIGURE 8.57 Although this type of suspension bridge approach is no longer used, its mass offers good security protection.

FIGURE 8.58 Typical stiffening trusses in suspension bridges. (Courtesy of William Moreau.)

FIGURE 8.59 Multilevel suspension bridge.

TABLE 8.26
Tier II and Tier III Issues for Nonstructural Bridge Systems

Tier II (Inferred)	Number	Tier III (Observable/ Computed)	Comments
6-Nonstructural	6.01	Drainage	Might have a multihazards effects during storms/floods
	6.02	Debris accumulation	Might have a multihazards effects during storms/floods
	6.03	Joint/deck condition	Might have multihazards effects
	6.04	Lifelines use (gas, water, and/ or electric conduits). See Figure 8.60	Can have effects on consequences when computing risk ratings

FIGURE 8.60 Lifeline conduits on a suspension bridge cable.

TABLE 8.27
Tier II and Tier III Issues for Access Control

Tier II (Inferred)	Number	Tier III (Observable/Computed)	Comments
7-Access control	7.01	Physical barriers	Can affect overall or specific setbacks
	7.02	Access to towers	Access need to be limited to authorized personnel only
	7.03	Access to climb suspension bridge ropes, see Figure 8.61	Access need to be limited to authorized personnel only
	7.04	Access to anchorage areas, see Figure 8.62	Access need to be limited to authorized personnel only
	7.05	Access for emergency personnel such as fire department	Access need to be limited to authorized personnel only

FIGURE 8.61 Typical tower system in a suspension bridge. (Courtesy of William Moreau.)

FIGURE 8.62 Anchorage area in a cable bridge. (Courtesy of William Moreau.)

TABLE 8.28
Tier II and Tier III Issues for Deterrence and Other Security Measures

Tier II (Inferred)	Number	Tier III (Observable/Computed)	Comments
8-Deterrent and other security measures	8.01	Security cameras	Well-placed security cameras can have deterrence effects
	8.02	Presence of police	A very efficient security measure
	8.03	Other	See FEMA (2005) and DHS (2011a, 2011b, 2011c)

TABLE 8.29
Tier II and Tier III Issues for Use and Functional Considerations

Tier II (Inferred)	Number	Tier III (Observable/Computed)	Comments
9-Use/Functional	9.01	Pedestrians	See FEMA (2005) and DHS (2011a, 2011b, 2011c)
	9.02	Trains	See DHS (2011b, 2011c)
	9.03	Vehicles	See DHS (2011b, 2011c)
	9.04	Combined/shared use	See DHS (2011b, 2011c)
	9.05	Functional use	See NYSDOT vulnerability manuals

TABLE 8.30
Tier II and Tier III Issues for Operational Considerations

Tier II (Inferred)	Number	Tier III (Observable/Computed)	Comments
10-Operational	10.01	Scenario analysis and drills	See DHS (2011b, 2011c)
	10.02	Inspection aspects (SHM, NDT, Visual only)	See Ettouney and Alampalli (2012b)
	10.03	Inspection frequency	Increased inspection frequency can detect degradation of security measures early on

TABLE 8.31
Tier II and Tier III Issues for Special Needs of Suspension Bridges

Tier II (Inferred)	Number	Tier III (Observable/Computed)	Comments
11-Special needs	11.01	In-depth cable inspection	Special inspection can detect degradation of cables early on, see Figure 8.54

TABLE 8.32
Tier II and Tier III Issues for Economic/Social Considerations

Tier II (Inferred)	Number	Tier III (Observable/Computed)	Comments
12-Economics/Social	12.01	Down time after disaster	Effective resilience plans should be prepared to ensure short downtime after disasters
	12.02	Replacement Value	This can have a major effect on consequences in a risk-based management operation. See DHS (2011b, 2011c)
	12.03	Network resiliency	Bridge network need to be accommodated in any effective resilience plans to ensure short downtime after disasters. This will limit undesired consequences after any disaster

TABLE 8.33
Tier II and Tier III Issues for Additional Demands

Tier II (Inferred)	Number	Tier III (Observable/Computed)	Comments
13-Other demands	13.01	Wind strength	Wind can play a major role in the vulnerability rating for cable brides in real time
	13.02	Temperature variations	Extreme temperature swings can play a major role in the vulnerability rating for bridges

while the probability of low risk being only 13%. These results are not surprising because all of the input variables that affect risk tend to be high. For example, the probability of high (demanding/less safe) site condition is 71%, and the probability of high physical vulnerability is 51%. The use of a network to process risks in this manner allows for an account of any desired interaction between variables and produces the results in a form of histograms rather than a single value. Owners can

TABLE 8.34
Tier II and Tier III Issues for Additional Threat/Consequences Factors

Tier II (Inferred)	Number	Tier III (Observable/ Computed)	Comments
14-Additional subjective threat factors	14.01	Significance of bridge	This can have a major effect on threat and consequences in a risk-based management operation. See DHS (2011b, 2011c)
	14.02	Function criticality	This can have a major effect on threat and consequences in a risk-based management operation. See DHS (2011b, 2011c)

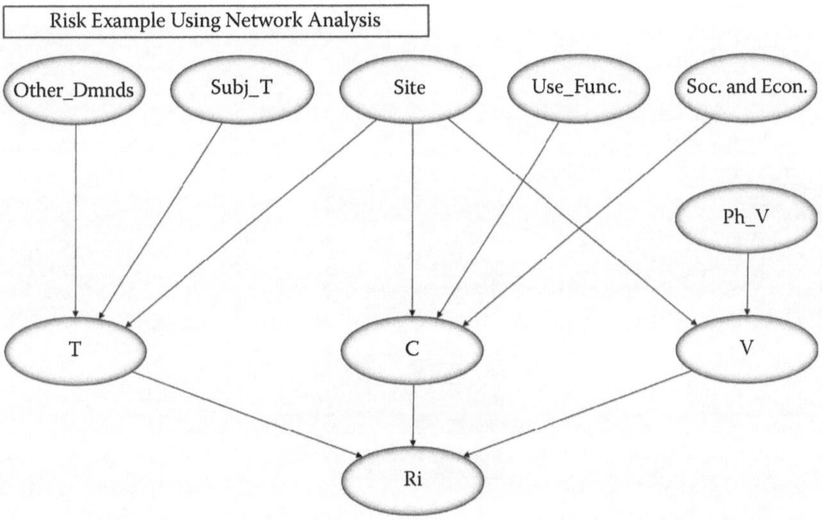

FIGURE 8.63 Simplified security risk model.

TABLE 8.35
States of Nodal Variables

Variable	# of States	Ranks (States)
Other_Demands	2	Low, High
Subj_T	2	Low, High
Site	2	Low, High
Use_Func.	2	Low, High
Soc. & Econ.	3	Low, Med., High
Ph_V	3	Low, Med., High
T	3	Low, Med., High
C	3	Low, Med., High
V	3	Low, Med., High
Ri	3	Low, Med., High

gain more insight into risks and their underlying causes by carefully studying network maps similar to that of Figure 8.64.

If the high risk result for the bridge under consideration is not acceptable, then there is a need to improve the condition. A study of the results of Figure 8.64 shows two potential factors that can be improved upon: site and physical vulnerability. In general, improving other factors is possible, but they require more effort than improving site and physical vulnerability. Hence, let us assume that the owner embarked on a risk mitigation effort for the site and that the bridge's physical vulnerability that resulted in a probability of both low site security and low physical vulnerability of 100%. If we introduce these probabilities as evidence in the risk network, the resulting new risk network is shown in Figure 8.65. Note that the mitigation efforts have now

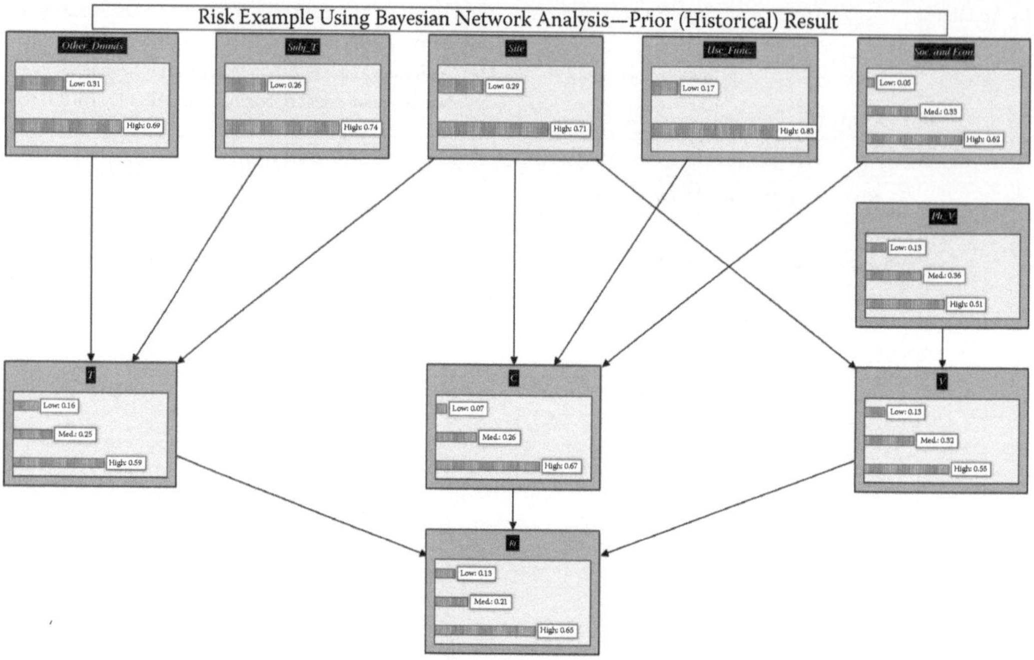

FIGURE 8.64 Historical (prior) risk picture.

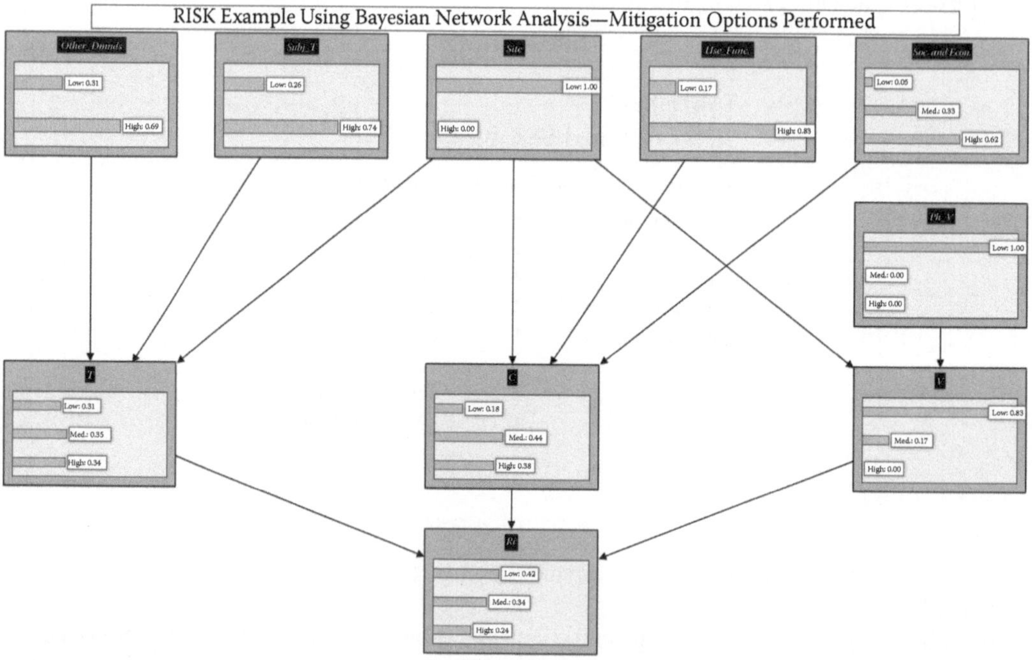

FIGURE 8.65 Risk mitigation results.

improved the probability of low risk to 42% (up from 13%) and lowered the probability of high risk to 24% (down from 64% before the mitigation efforts). It is also worth noting that all underlying risk components (threat, consequences, and vulnerability) have also improved. Thus, this risk network can be used to estimate the effect of various scenarios before taking mitigation action.

Although the above model is fairly simplified, it shows the various steps needed to embark on risk-based management. For risk assessment, the example showed the necessary components that need consideration: uncertainties, historical observations, personal judgment and experiences, intersections and links between variables, and allowances for scalability of the risk network. How the assessment network needs to be employed for developing optimal mitigation decisions is also shown above. Of course, a more detailed risk treatment/risk mitigation process can be followed. The stakeholder may want to follow some of the processes in Chapters 3 through 6, and/or the case study in Section 8.2.

APPENDIX 8.I: DEFINITIONS OF NETWORK NODAL VARIABLES (TABLE 8.36)

TABLE 8.36
Definitions of Nodal Variables

Variable	Used in Case Study #	Definition
Abn.H.	8.2	Estimate of abnormal hazard (earthquake, flood, high wind) that might affect the bridge.
AR_1, AR_2, AR_3, AR_4, AR_5	8.6	Appraisal ratings (AR), which control the functionally deficient condition (FDC). ARs control FDC: *AR_1*: AR for deck geometry; *AR_2*: AR for under clearance: *AR_3*; AR for approach roadway alignment: *AR_4*: AR for structural condition; *AR_5*:AR for waterway adequacy
B1, B2	8.3, 8.4	Objective estimate of operational efficiency of the *i*-th bridge (*Bi*) in the community of the example.
Bridge Size	8.1, 8.5, 8.6	Subjective estimate of bridge size. This variable can be estimated using bridge longest spans, total length, number of spans, number of lanes, etc.
C	8.2, 8.5	Total consequences relating to bridge progressive collapse, including consequences of fracture critical status and other consequences.
C	8.6	Consequences related to traffic costs and other reasons that are beyond the physical bridge system, including social and economic consequences.
C	8.7	Consequences in this example include site, use function, and other social and economic factors.
C.O.	8.2	Estimate of consequences resulting from *Exposure* and *Abn.H.*
C_O	8.5	Same as *C.O.*
C-FO	8.6	Consequences related to functional deficiency/obsolescence of the bridge.
C-O	8.6	Estimate of consequences beyond the system itself, these consequences can be, but are not limited to, cost of replacements, symbolic nature of bridge, bridge locality, and bridge age. In these examples, C.O. is hazard independent.
CR_1, CR_2, CR_3	8.5	Condition ratings (CR), which controls the definition of a severely deteriorated bridge (SDB). Three CRs control SDB: CR_1: CR for superstructure; CR_2: CR for substructure; CR_3: CR for deck.
D1, D2, D3, D4	8.1	Different decisions regarding the critical findings process.
Direct Costs	8.1	Direct costs relating to critical findings (e.g., cost of repairs).
E	8.2	Estimate of exposure of bridge, which is a function of hazards, *Abn.H.*; observations, *S.V.I.*; and fracture critical status, *F.C.B.*

(Continued)

TABLE 8.36 *(Continued)*
Definitions of Nodal Variables

Variable	Used in Case Study #	Definition
F.C.B.	8.2	Variable estimating whether a bridge is a fracture critical bridge.
FO_1	8.6	This variable is a function of the geometry-related ARs (*AR_1*, *AR_2*, and *AR_3*).
FO_2	8.6	This variable is a function of the robustness-related ARs (*AR_4,* and *AR_5*).
Functional Class	8.1	Describes the character of traffic service of the highway under the bridge.
M.T.	8.3, 8.4	Objective estimate of operational efficiency of mass transit station (M.T.) in the community of the example.
Mit.	8.2	Probability of mitigation and its potential adequacy.
OB1, OB2	8.3, 8.4	Objective estimate of operational efficiency of the *i*-th office building (OBi) in the community of the example.
Obey Posting	8.1	Measure of observance of traffic postings by drivers.
Other Costs	8.1	See *C.O.*
Other Costs	8.5	Costs, other than traffic costs, that might be incurred for SDB under consideration.
Other Costs	8.6	Costs, other than traffic costs, that might be incurred for FDC under consideration.
Other_Demands	8.7	Natural (floods, earthquakes, high wind, etc.) and environmental (corrosion, fatigue, wear and tear, etc.) hazards.
Overload Safety	8.1	Measure of safety in relation to overload.
P.S.	8.3, 8.4	Objective estimate of operational efficiency of power station.
Ph_V	8.7	This subjective variable estimates vulnerability of the system itself.
Red_P.S.	8.4	Objective estimate of operational efficiency of power station in the community.
Rel.	8.2	*Rel.* is a metric of the capacity-demands relationship.
Retrofit Extent	8.1	Estimate of retrofit extent that, in this example, is a function of both *V.I.* and *Functional Class.*
Ri	8.2	In this example, risk, *Ri*, is a function of consequences and exposure. Numerous algorithms objectively relate risk to consequences and exposure; see FEMA (2005) and Ettouney and Alampalli (2012a, 2012b).
Ri	8.5	This variable represents subjective risk *SDBC*. It is a function of the consequences, *C*, and the *SDBC* estimator. Note that vulnerability and hazards are implied in the interrelationship between *C* and *SDBC* estimator.
Ri	8.6	This variable represents objective risk in monetary terms. It is a function of the objective consequences, *C*, and the subjective consequences from functional obsolescence status, *C–FO*. Note that vulnerability and hazards are implied in *C–FO*.
Ri	8.7	In this example, risk, *Ri*, is a function of Threat,Vulnerability, and Consequences. Numerous algorithms objectively relate risk to consequences and exposure; see FEMA (2005) and DHS (2011a, 2011b, 2011c) and Ettouney and Alampalli (2012a). Risk in this example is an objective risk that can be expressed by three subjective states.
S.V.I.	8.2	Objective estimate of the result of special visual inspections (*S.V.I*). These types of inspection are usually performed after an abnormal hazard affects the bridge
SDBC	8.5	Estimate of whether this structure has reached a seriously deteriorated condition.
Site	8.7	Objective variable describing site vulnerability and susceptibility to threats.
Soc. & Econ.	8.7	Social and economic factors include, but are not limited to, cost of replacements, cost of traffic delays, symbolic nature, locality, and age.
Subj_T	8.7	Man-made hazards such as sabotage, bomb blast, and other security concerns.
T	8.7	Threat in this example is a function of the site, subjective threat that is usually a man-made threat, as well as other demands that are usually natural hazards as well as environmental hazards.
T1	8.3, 8.4	Objective estimate of operational efficiency of tunnel (*T1*) in the community of the example.

(Continued)

TABLE 8.36 ***(Continued)***
Definitions of Nodal Variables

Variable	Used in Case Study #	Definition
Traffic Costs	8.1	Subjective estimate of traffic costs.
Traffic Costs	8.5	Subjective estimate of traffic costs that are incurred for SDB under consideration. This variable can be estimated subjectively using actual traffic costs.
Traffic Costs	8.6	Subjective estimate of traffic costs that are incurred for SDB under consideration. The CPT of this variable can be estimated subjectively using actual traffic costs.
Traffic Vol.	8.1	Subjective estimate of traffic volume.
Traffic Vol.	8.6	Subjective estimate of traffic volume on this bridge. Traffic volume has a direct effect on *AR_1* because increasing traffic volume requires improved deck geometry.
U	8.1	
Use_Func.	8.7	This variable describes the function of the bridge within the roadway network.
V	8.7	We subdivided vulnerabilities into two categories: physical vulnerability, which pertains to the structural system itself, and vulnerabilities relating to the site. The total vulnerability is *V*.
V.I.	8.1	Results of visual inspection.

APPENDIX 8.II: CPTs FOR CASE STUDY 8.1 (TABLES 8.37–8.47)

TABLE 8.37
CPT of ***Bridge Size***

Bridge size	Small	0.3621
	Medium	0.4525
	Large	0.1854
Histograms for conditional probabilities		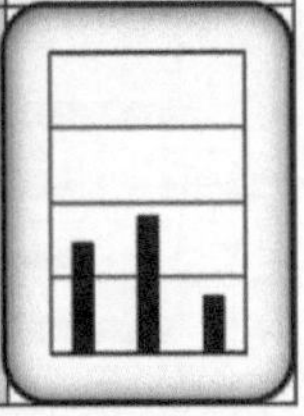

TABLE 8.38
CPT of ***Bridge***

V.I.		Good	Fair	Poor	Service
Bridge	Open	0.9883	0.5783	0.4448	0.1854
	Partial	0.0117	0.4142	0.4967	0.4525
	Close	0.0000	0.0074	0.0585	0.3621
Histograms for conditional probabilities		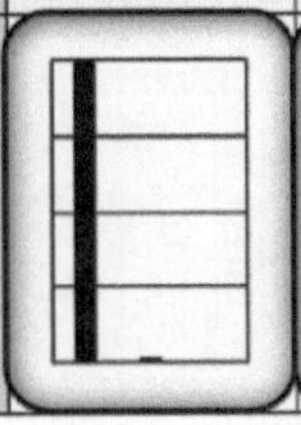	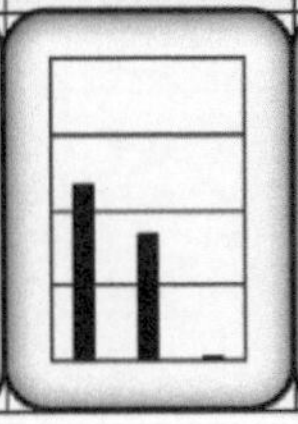	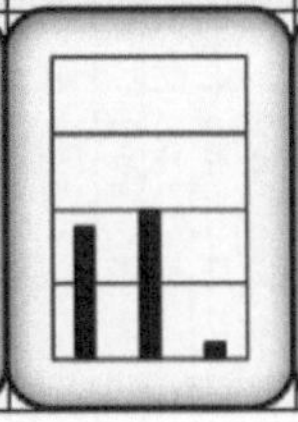	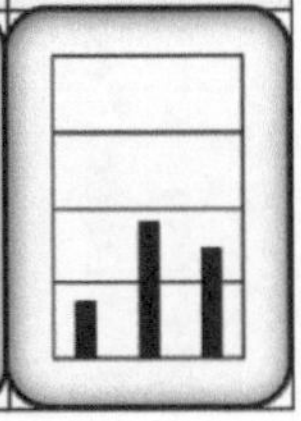

TABLE 8.39
CPT of *Direct Costs*

Bridge size		Small											
Functional class		Local			Collector			Arterial			Interstate		
Retrofit extent		Little	Medium	High	Little	Medium	High	Little	Medium	High	Little	Medium	High
Direct costs	Low	0.9671	0.8547	0.7082	0.8268	0.8537	0.7052	0.8547	0.8018	0.6852	0.7522	0.7644	0.6915
	High	0.0329	0.1453	0.2918	0.1732	0.1463	0.2948	0.1453	0.1982	0.3148	0.2478	0.2356	0.3085

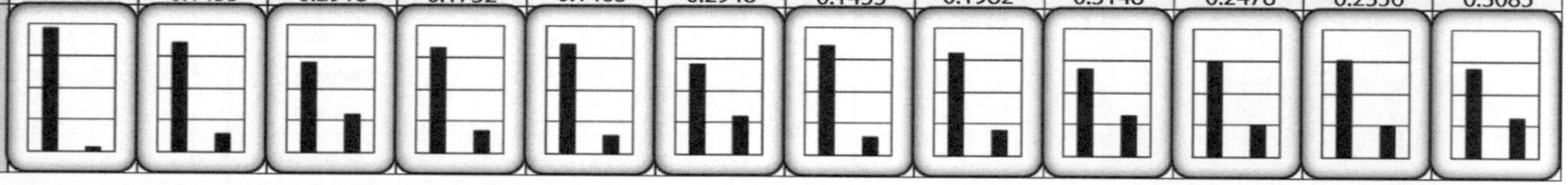

Bridge size		Medium											
Functional class		Local			Collector			Arterial			Interstate		
Retrofit extent		Little	Medium	High	Little	Medium	High	Little	Medium	High	Little	Medium	High
Direct costs	Low	0.7644	0.5718	0.4450	0.6087	0.5000	0.3590	0.5718	0.4149	0.2356	0.5000	0.3148	0.0228
	High	0.2356	0.4282	0.5550	0.3913	0.5000	0.6410	0.4282	0.5851	0.7644	0.5000	0.6852	0.9772

Bridge size		Large											
Functional class		Local			Collector			Arterial			Interstate		
Retrofit extent		Little	Medium	High	Little	Medium	High	Little	Medium	High	Little	Medium	High
Direct costs	Low	0.4045	0.2948	0.2478	0.3913	0.1982	0.1453	0.2948	0.1463	0.0518	0.2918	0.0794	0.0000
	High	0.5955	0.7052	0.7522	0.6087	0.8018	0.8547	0.7052	0.8537	0.9482	0.7082	0.9206	1.0000

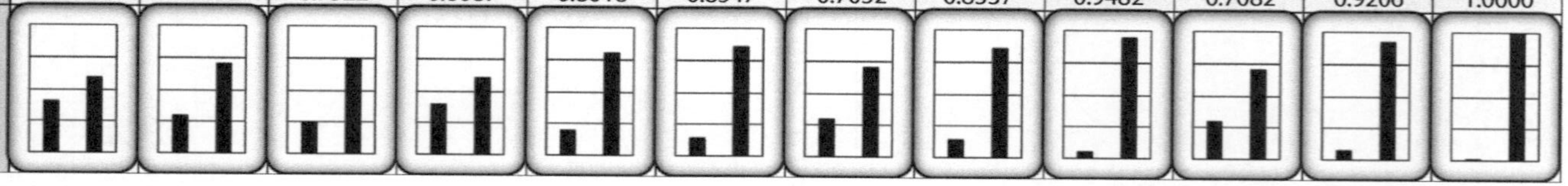

TABLE 8.40
CPT of *Functional Class*

Functional class	Local	0.2453
	Collector	0.3634
	Arterial	0.2796
	Interstate	0.1116
Histograms for conditional probabilities		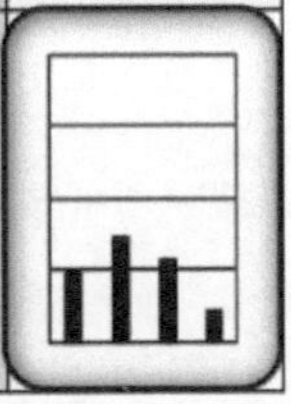

TABLE 8.41
CPT of *Obey Posting*

Bridge		Open	Partial	Close
Obey posting	Yes	1.0000	0.8302	0.9206
	No	0.0000	0.1698	0.0794
Histograms for conditional probabilities		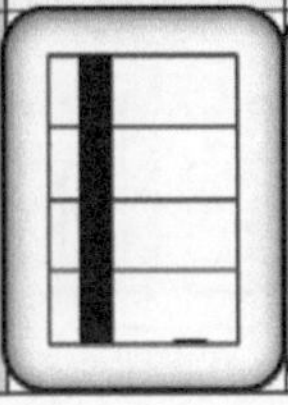	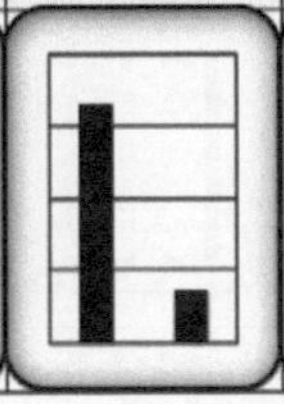	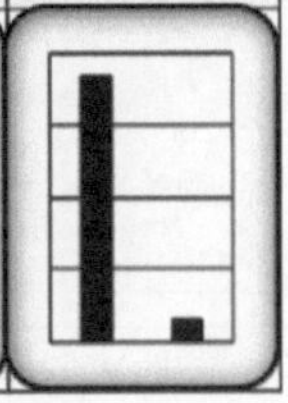

TABLE 8.42
CPT of *Other Costs*

Bridge size		Small	Medium	Large
Other costs	Low	0.9482	0.4450	0.1048
	High	0.0518	0.5550	0.8952
Histograms for conditional probabilities		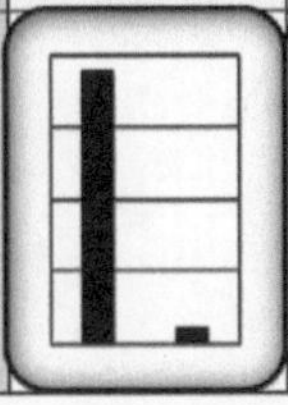	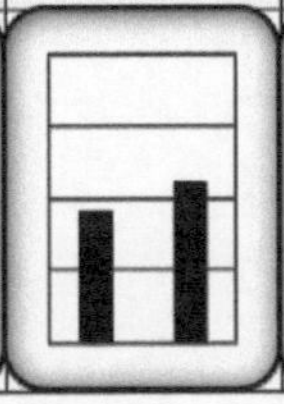	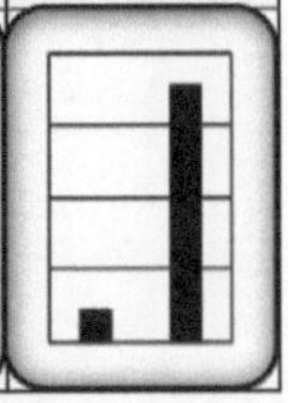

TABLE 8.43
CPT of *Overload Safety*

Obey posting		Yes	No
Overload safety	High	0.9166	0.0329
	Low	0.0834	0.9671
Histograms for conditional probabilities		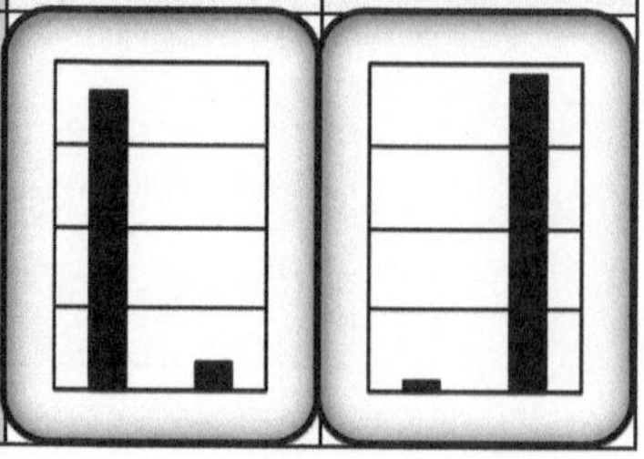	

TABLE 8.44
CPT of *Retrofit Extent*

Functional class		Local				Collector			
V.I.		Good	Fair	Poor	Service	Good	Fair	Poor	Service
Retrofit extent	Little	0.9883	0.6999	0.5353	0.2525	0.9067	0.3553	0.1986	0.0038
	Medium	0.0117	0.2885	0.4292	0.7437	0.0933	0.5527	0.6028	0.7437
	High	0.0000	0.0116	0.0355	0.0038	0.0000	0.0921	0.1986	0.2525
Histograms for conditional probabilities		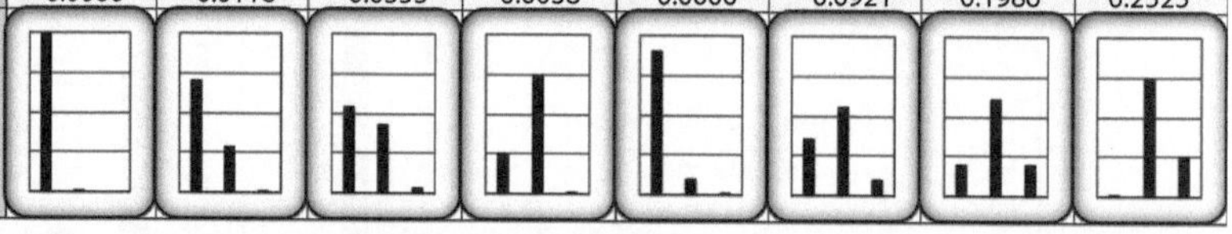							

Functional class		Arterial				Interstate			
V.I.		Good	Fair	Poor	Service	Good	Fair	Poor	Service
Retrofit extent	Little	0.2525	0.0921	0.0355	0.0000	0.0038	0.0355	0.0116	0.0000
	Medium	0.7437	0.5527	0.4292	0.0933	0.7437	0.4292	0.2885	0.0117
	High	0.0038	0.3553	0.5353	0.9067	0.2525	0.5353	0.6999	0.9883
Histograms for conditional probabilities		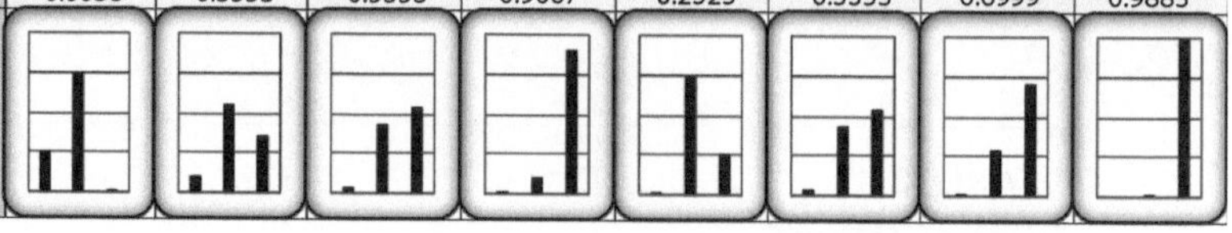							

TABLE 8.45
CPT of *Traffic Costs*

Bridge		Open																	
Overload safety		High									Low								
Traffic vol.		Low			Medium			High			Low			Medium			High		
Bridge size		Small	Medium	Large	Small	Medium	Large	Small	Medium	Large	Small	Medium	Large	Small	Medium	Large	Small	Medium	Large
Traffic costs	Low	1.0000	1.0000	1.0000	0.9856	0.9832	1.9796	0.8990	0.8925	0.8832	0.8302	0.3148	0.0124	0.9067	0.2534	0.0009	0.9772	0.1587	0.0000
	High	0.0000	0.0000	0.0000	0.0144	0.0168	0.0204	0.1010	0.1075	0.1168	0.1698	0.6852	0.9876	0.0933	0.7466	0.9991	0.0228	0.8413	1.0000

Bridge		Partial																	
Overload safety		High									Low								
Traffic vol.		Low			Medium			High			Low			Medium			High		
Bridge size		Small	Medium	Large	Small	Medium	Large	Small	Medium	Large	Small	Medium	Large	Small	Medium	Large	Small	Medium	Large
Traffic costs	Low	0.9832	0.9775	0.9671	0.9832	0.9775	0.9671	0.9832	0.9775	0.9671	0.9482	0.8302	0.5000	0.9883	0.9067	0.5000	0.9998	0.9772	0.5000
	High	0.0168	0.0225	0.0329	0.0168	0.0225	0.0239	0.0168	0.0225	0.0329	0.0518	0.1698	0.5000	0.0117	0.0933	0.5000	0.0002	0.0228	0.5000

Bridge		Close																	
Overload safety		High									Low								
Traffic vol.		Low			Medium			High			Low			Medium			High		
Bridge size		Small	Medium	Large	Small	Medium	Large	Small	Medium	Large	Small	Medium	Large	Small	Medium	Large	Small	Medium	Large
Traffic costs	Low	0.9399	0.9214	0.8891	0.9399	0.9214	0.8891	0.9399	0.9214	0.8891	0.9166	0.8547	0.7252	0.9671	0.9206	0.7879	0.9949	0.9750	0.8650
	High	0.0601	0.0786	0.1109	0.0601	0.0786	0.1109	0.0601	0.0786	0.1109	0.0834	0.1453	0.2748	0.0329	0.0794	0.2121	0.0051	0.0250	0.1350

TABLE 8.46
CPT of *Traffic Vol.*

Traffic vol.	Low	0.3
	Medium	0.5
	High	0.2
Histograms for conditional probabilities		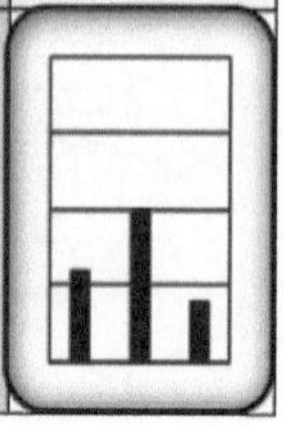

TABLE 8.47
CPT of *V.I.*

V.I.	Good	0.3309
	Fair	0.3772
	Poor	0.2233
	Service	0.0685
Histograms for conditional probabilities		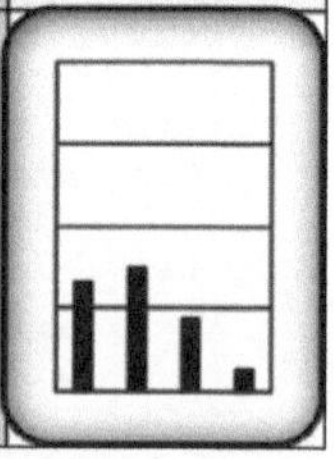

APPENDIX 8.III: CPTs FOR CASE STUDY 8.2 (TABLES 8.48–8.56)

TABLE 8.48
CPT of *Abn.Hz.*

Abn.HZ.	L. Level	0.9772
	H. Level	0.0228
Histograms for conditional probabilities		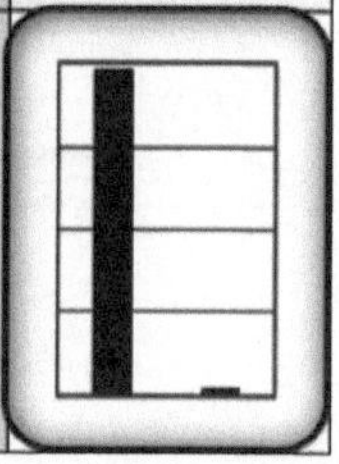

TABLE 8.49
CPT of *C*

Mit.		Adequate		Not Adequate		None (Hazard Occurs Before Mit.)	
C.O.		Low Con.	High Con.	Low Con.	High Con.	Low Con.	High Con.
C	Low C	0.6238	0.0116	0.0378	0.0000	0.0003	0.0000
	Med. C	0.3377	0.2885	0.6364	0.0933	0.1314	0.0117
	High C	0.0385	0.6999	0.3258	0.9067	0.8683	0.9883
Histograms for conditional probabilities		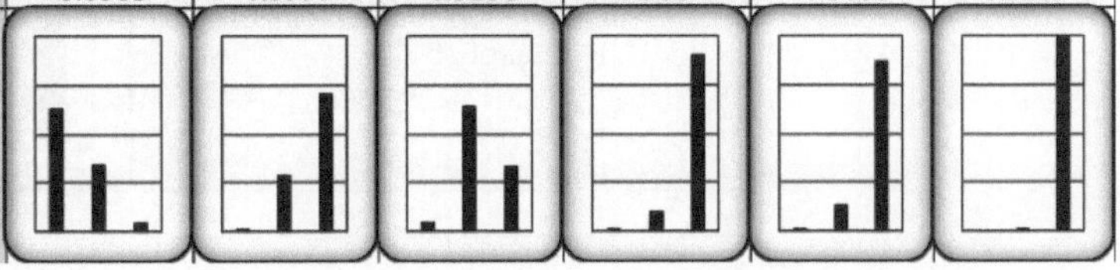					

TABLE 8.50
CPT of *CO*

Abn.Hz.		L. Level			H. Level		
E		L. Level	M. Level	H. Level	L. Level	M. Level	H. Level
C.O.	Low Con.	0.9750	0.5803	0.6087	0.3913	0.1698	0.0014
	High Con.	0.0250	0.4197	0.3913	0.6087	0.8302	0.9986
Histograms for conditional probabilities		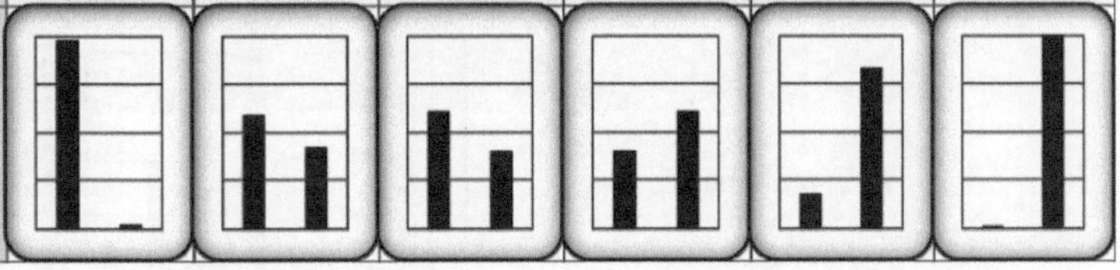					

TABLE 8.51
CPT of *E*

F.C.B.		Yes				No			
S.V.I.		Pristine		Damaged		Pristine		Damaged	
Abn.HZ.		L. Level	H. Level	L. Level	H. Level	L. Level	H. Level	L. Level	H. Level
E	L. Level	0.5143	0.0778	0.0063	0.0000	0.8240	0.1439	0.0745	0.0001
	M. Level	0.3894	0.4258	0.2260	0.0117	0.1726	0.4931	0.3610	0.0800
	H. Level	0.0963	0.4963	0.7677	0.9883	0.0033	0.3629	0.5636	0.9199
Histograms for conditional probabilities		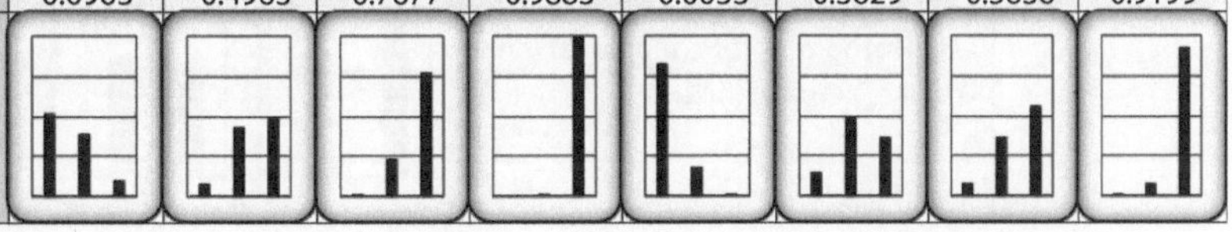							

TABLE 8.52
CPT of *F.C.B.*

F.C.B.	Yes	1.0000
	No	0.0000
Histograms for conditional probabilities		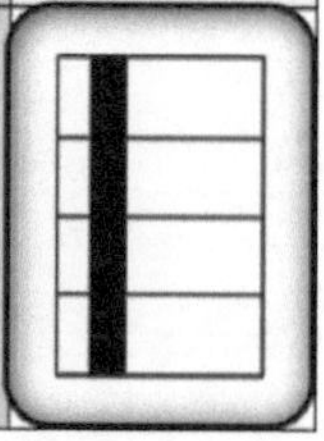

TABLE 8.53
CPT of *Mit.*

F.C.B.		Yes		No	
S.V.I.		Pristine	Damaged	Pristine	Damaged
Mit.	Adequate	0.8601	0.3629	0.9292	0.4963
	Not Adequate	0.1399	0.4931	0.0708	0.4258
	None (Hazard Occurs Before Mit.)	0.0001	0.1439	0.0000	0.0778
Histograms for conditional probabilities		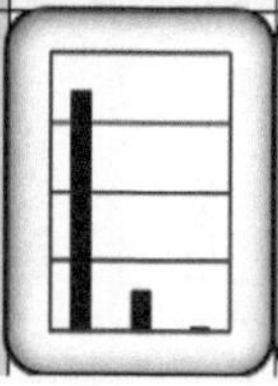	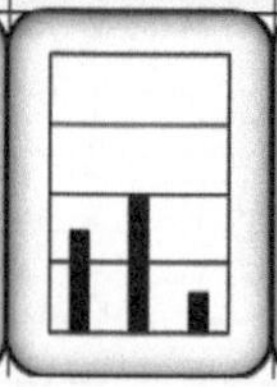	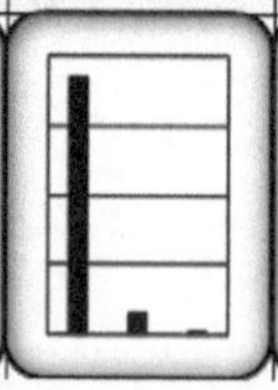	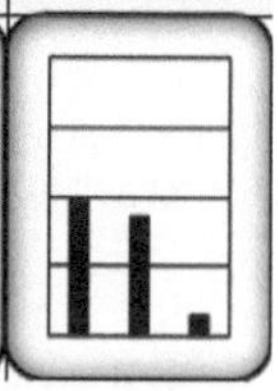

TABLE 8.54
CPT of *Rel.*

E		L. Level	M. Level	H. Level
Rel.	High (Safe)	0.9949	0.4450	0.0834
	Low (Unsafe)	0.0051	0.5550	0.9166
Histograms for conditional probabilities		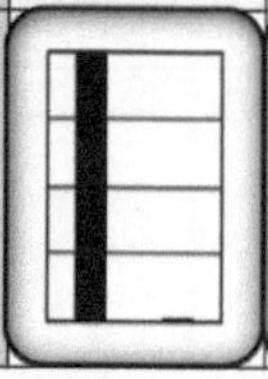	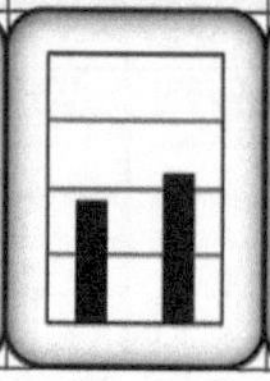	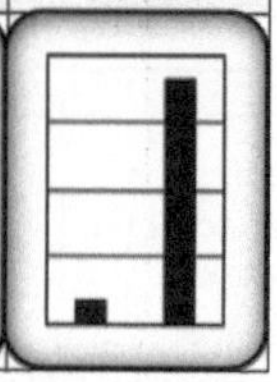

TABLE 8.55
CPT of *Ri*

C		Low C			Med. C			High C		
E		L. Level	M. Level	H. Level	L. Level	M. Level	H. Level	L. Level	M. Level	H. Level
Ri	Low	0.7677	0.4562	0.3621	0.4562	0.1986	0.1077	0.4024	0.1071	0.0063
	Med	0.2260	0.4493	0.4525	0.4493	0.6028	0.4635	0.4399	0.4635	0.2260
	High	0.0063	0.0945	0.1854	0.0945	0.1986	0.4293	0.1576	0.4293	0.7677
Histograms for conditional probabilities										

TABLE 8.56
CPT of *S.V.I.*

Abn.Hz.		L. Level	H. Level
S.V.I.	Pristine	0.8944	0.1587
	Damaged	0.1056	0.8413
Histograms for conditional probabilities		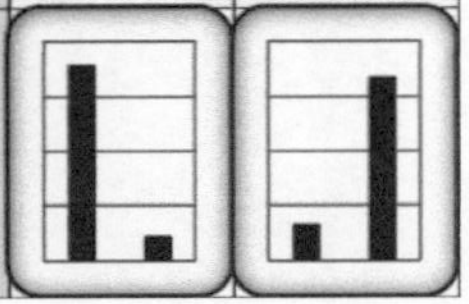	

APPENDIX 8.IV: CPTs AND POTENTIALS FOR CASE STUDY 8.3 (TABLES 8.57–8.66)

TABLE 8.57
Potential ϕ (*OB1, B1*)

OB1		High O.E.	Low O.E.
B1	High O.E.	100	20
	Low O.E.	50	100

TABLE 8.58
Potential ϕ (*MT, B1*)

M.T.		High O.E.	Low O.E.
B1	High O.E.	100	20
	Low O.E.	10	100

TABLE 8.59
Potential ϕ (*MT, T1*)

T1		High O.E.	Low O.E.
M.T.	High O.E.	100	10
	Low O.E.	5	100

TABLE 8.60
Potential ϕ (*B1, T1*)

T1		High O.E.	Low O.E.
B1	High O.E.	100	20
	Low O.E.	20	100

TABLE 8.61
Potential ϕ (*B2, T1*)

B2		High O.E.	Low O.E.
T1	High O.E.	100	20
	Low O.E.	20	100

TABLE 8.62
Potential ϕ (*OB2, B2*)

OB2		High O.E.	Low O.E.
B2	High O.E.	100	50
	Low O.E.	40	100

TABLE 8.63
Potential ϕ (*OB1, OB2*)

OB2		High O.E.	Low O.E.
OB1	High O.E.	100	10
	Low O.E.	70	100

TABLE 8. 64
CPT of *OB2*

P.S.		High O.E.	Low O.E.
OB2	High O.E.	0.75	0.15
	Low O.E.	0.25	0.85
Histograms for conditional probabilities		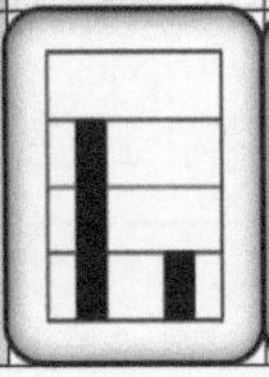	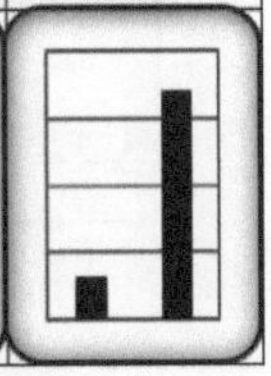

TABLE 8.65
CPT of *P.S.*

P.S.	High O.E.	0.65
	Low O.E.	0.35
Histograms for conditional probabilities		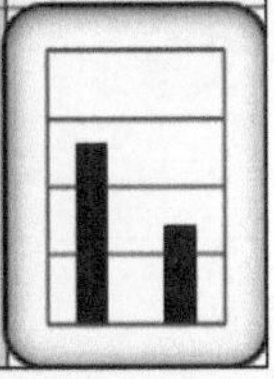

TABLE 8.66
CPT of *T1*

P.S.		High O.E.	Low O.E.
T1	High O.E.	0.8	0.1
	Low O.E.	0.2	0.9
Histograms for conditional probabilities		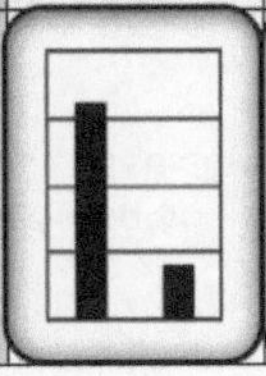	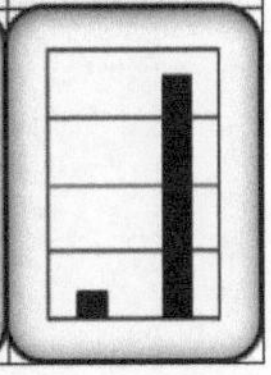

APPENDIX 8.V: CPTs FOR CASE STUDY 8.4 (TABLES 8.67–8.68)

The CPTs of Case Study 8.4 variables *Red_P.S.* and *T1* are shown in Tables 8.67 and 8.68, respectively. The CPTs and potentials of the remainder of the case study are identical to those of Case Study 8.3 (Tables 8.57 through 8.65).

TABLE 8.67
CPT of *Red_P.S.*

Red_P.S.	High O.E.	0.65
	Low O.E.	0.35
Histograms for conditional probabilities		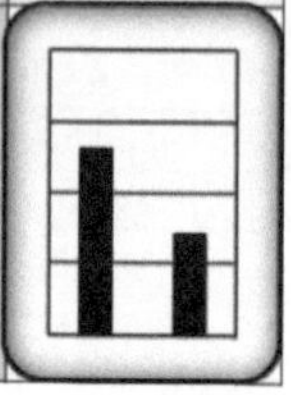

TABLE 8.68
CPT of *T1*

Red_P.S.		High O.E.		Low O.E.	
P.S.		High O.E.	Low O.E.	High O.E.	Low O.E.
T1	High O.E.	0.8	0.8	0.8	0.01
	Low O.E.	0.2	0.2	0.2	0.99
Histograms for conditional probabilities		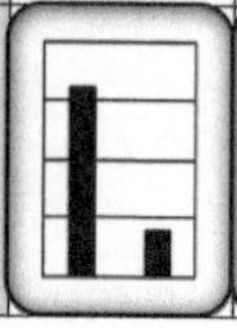	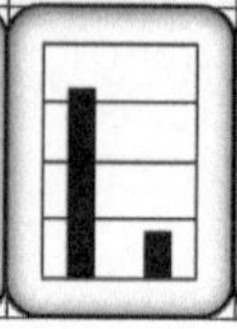	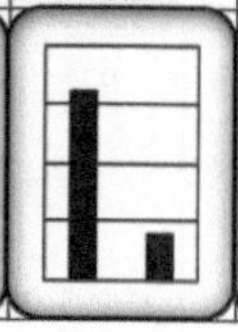	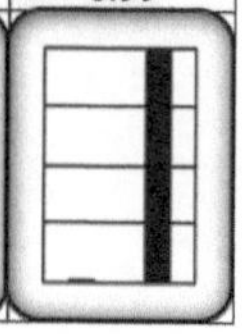

APPENDIX 8.VI: CPTs FOR CASE STUDY 8.5 (TABLES 8.69–8.78)

TABLE 8.69
CPT of *Bridge Size*

Bridge size	Small	0.3621
	Medium	0.4525
	Large	0.1854
Histograms for conditional probabilities		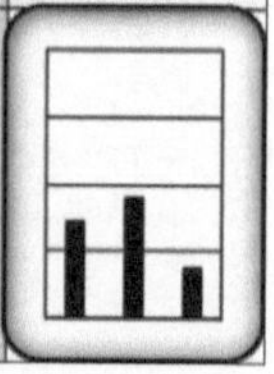

TABLE 8.70
CPT of *C*

Traffic costs		Low				High			
C_O		Low		High		Low		High	
Other costs		Low	High	Low	High	Low	High	Low	High
C	Low	0.8240	0.5353	0.4637	0.0005	0.4379	0.0531	0.1810	0.0001
	Med	0.1726	0.4292	0.4148	0.5262	0.4026	0.4839	0.4129	0.0800
	High	0.0033	0.0355	0.1215	0.4732	0.1595	0.4630	0.4060	0.9199
Histograms for conditional probabilities		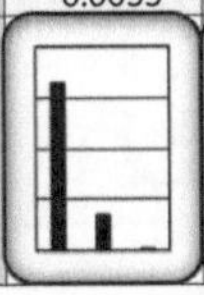	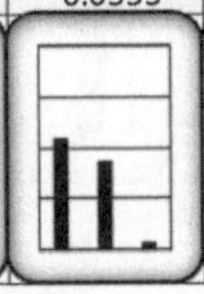	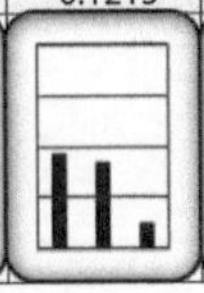	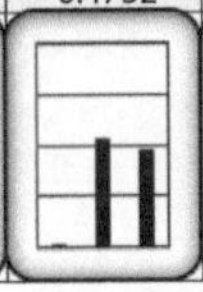	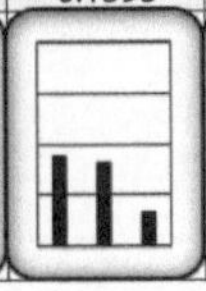	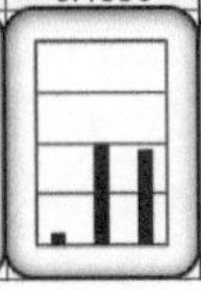	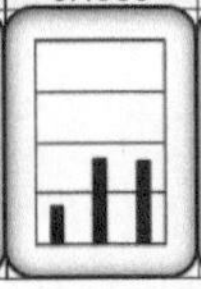	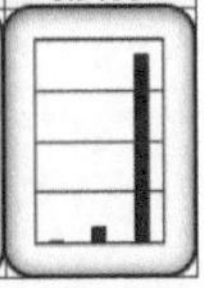

TABLE 8.71
CPT of *C_O*

C_O	Low	0.4
	High	0.6
Histograms for conditional probabilities		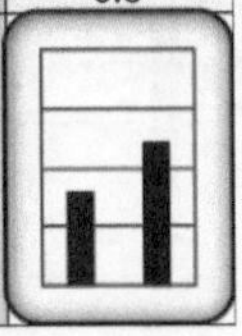

TABLE 8.72
CPT of *CR_1*

CR_1	Pristine	0.9
	Damaged	0.1
Histograms for conditional probabilities		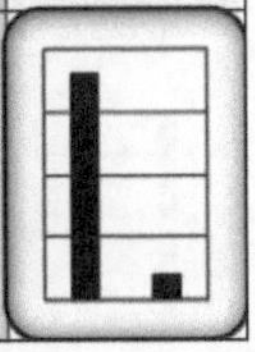

TABLE 8.73
CPT of *CR_2*

CR_2	Pristine	0.9
	Damaged	0.1
Histograms for conditional probabilities		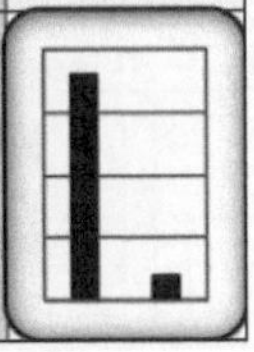

TABLE 8.74
CPT of *CR_3*

CR_3		
CR_3	Pristine	0.9
	Damaged	0.1
Histograms for conditional probabilities		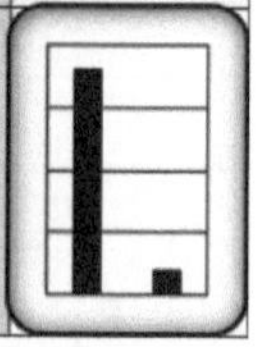

TABLE 8.75
CPT of *Other Costs*

SDBC		No			Yes		
Bridge size		Small	Medium	Large	Small	Medium	Large
Other costs	Low	0.9482	0.4450	0.2710	0.5681	0.3870	0.0002
	High	0.0518	0.5550	0.7290	0.4319	0.6130	0.9998
Histograms for conditional probabilities							

TABLE 8.76
CPT of *Ri*

C		Low		Med		High	
SDBC		No	Yes	No	Yes	No	Yes
Ri	Low	0.9199	0.2397	0.3553	0.0553	0.1854	0.0000
	Med	0.0800	0.4050	0.5527	0.3829	0.4525	0.0117
	High	0.0001	0.3553	0.0921	0.5618	0.3621	0.9883
Histograms for conditional probabilities		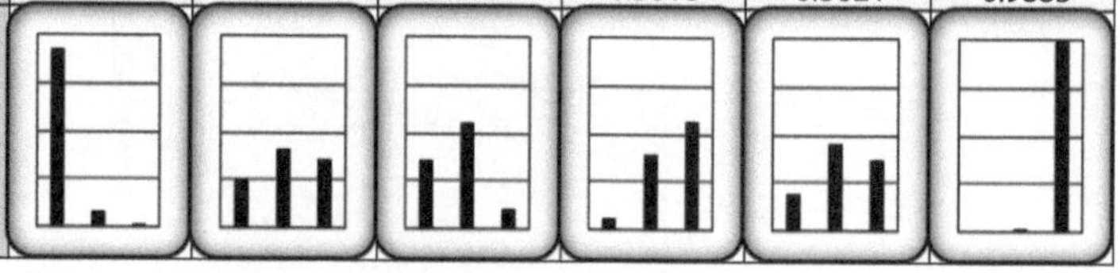					

TABLE 8.77
CPT of *SDBC*

CR_3		Damaged				Damaged			
CR_2		Pristine		Damaged		Pristine		Damaged	
CR_1		Pristine	Damzaged	Pristine	Damaged	Pristine	Damaged	Pristine	Damaged
SDBC	No	0.99	0.01	0.01	0.01	0.01	0.01	0.01	0.01
	Yes	0.01	0.99	0.99	0.99	0.99	0.99	0.99	0.99
Histograms for conditional probabilities		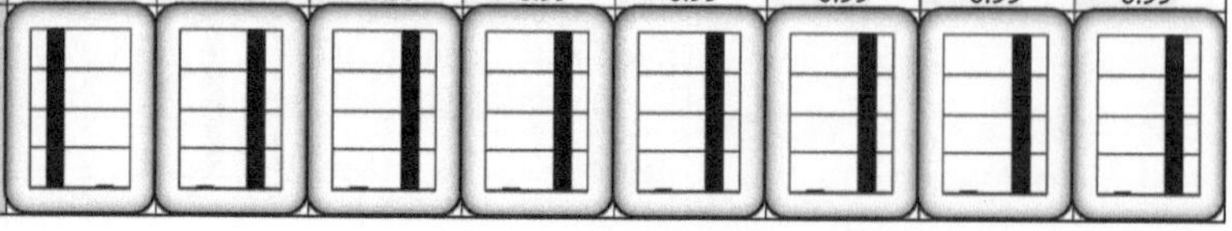							

TABLE 8.78
CPT of *Traffic Costs*

Bridge size		Small		Medium		Large	
SDBC		No	Yes	No	Yes	No	Yes
Traffic costs	Low	0.9671	0.4319	0.7644	0.1698	0.3148	0.0000
	High	0.0329	0.5681	0.2356	0.8302	0.6852	1.0000
Histograms for conditional probabilities		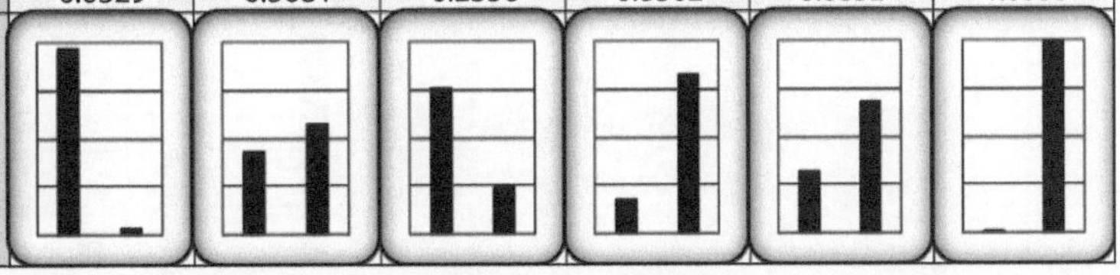					

APPENDIX 8.VII: CPTs FOR CASE STUDY 8.6 (TABLES 8.79–8.93)

TABLE 8.79
CPT of *AR_1*

Traffic Vol.		Low	Med	High
AR_1	Acceptable	0.99	0.2	0.01
	Obsolete	0.01	0.8	0.99
Histograms for conditional probabilities		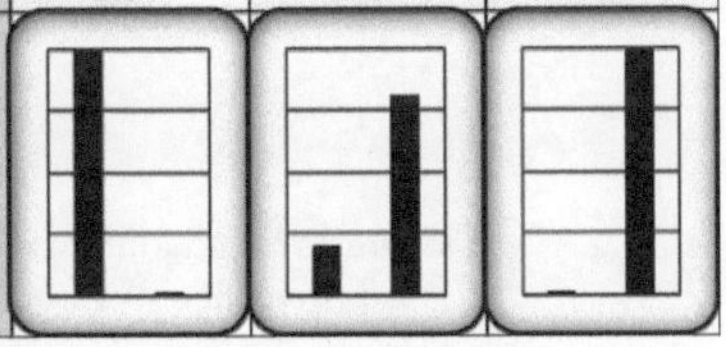		

TABLE 8.80
CPT of *AR_2*

AR_2	Acceptable	0.9
	Obsolete	0.1
Histograms for conditional probabilities		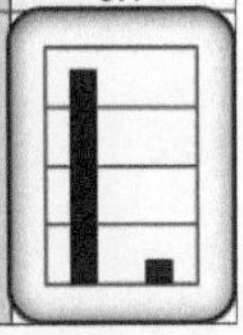

TABLE 8.81
CPT of *AR_3*

AR_3	Acceptable	0.9
	Obsolete	0.1
Histograms for conditional probabilities		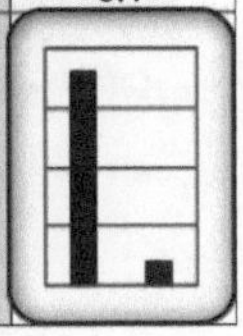

TABLE 8.82
CPT of *AR_4*

Traffic Vol.		Low	Med	High
AR_4	Acceptable	0.99	0.15	0.01
	Obsolete	0.01	0.85	0.99
Histograms for conditional probabilities		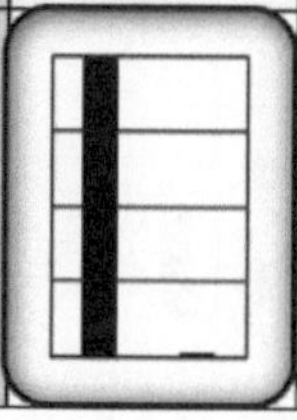	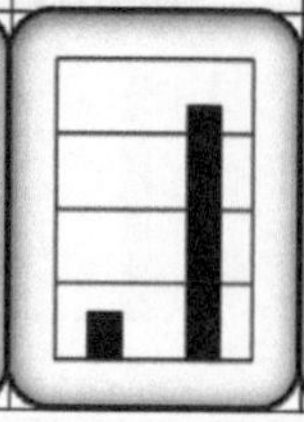	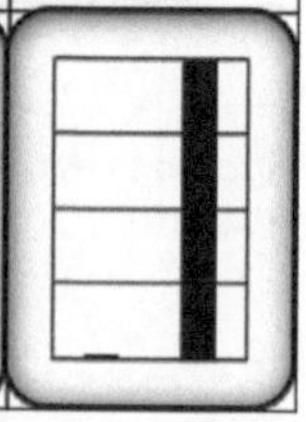

TABLE 8.83
CPT of *AR_5*

AR_5	Acceptable	0.9
	Obsolete	0.1
Histograms for conditional probabilities		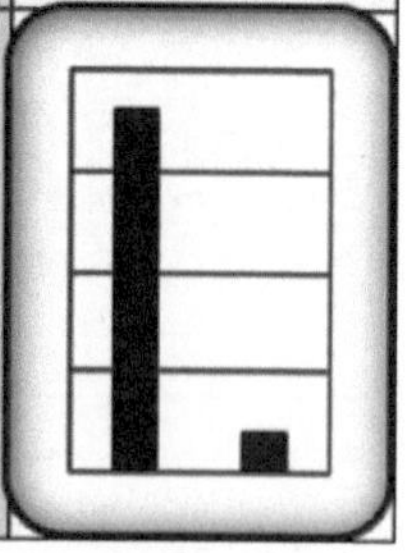

TABLE 8.84
CPT of *Bridge Size*

Bridge size	Small	0.3621
	Medium	0.4525
	Large	0.1854
Histograms for conditional probabilities		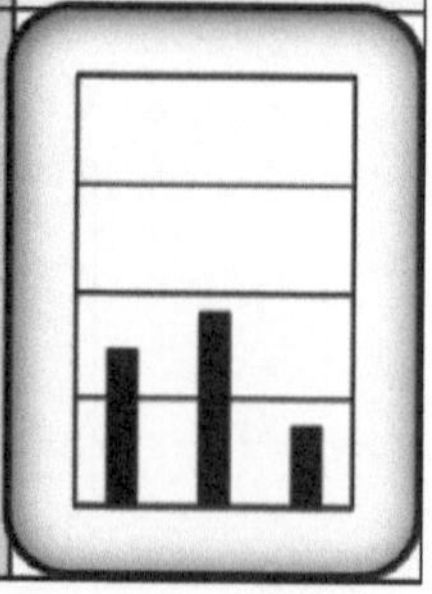

TABLE 8.85
CPT of *C*

Other costs		Low				High			
Traffic costs		Low		High		Low		High	
C_O		Low Con.	High Con.	Low Con.	High Con.	Low Con.	High Con.	Low Con.	High Con.
C	$0 M	0.8240	0.5353	0.4637	0.0005	0.4379	0.0531	0.1810	0.0001
	$10 M	0.1726	0.4292	0.4148	0.5262	0.4026	0.4839	0.4129	0.0800
	$100 M	0.0033	0.0355	0.1215	0.4732	0.1595	0.4630	0.4060	0.9199
Histograms for conditional probabilities		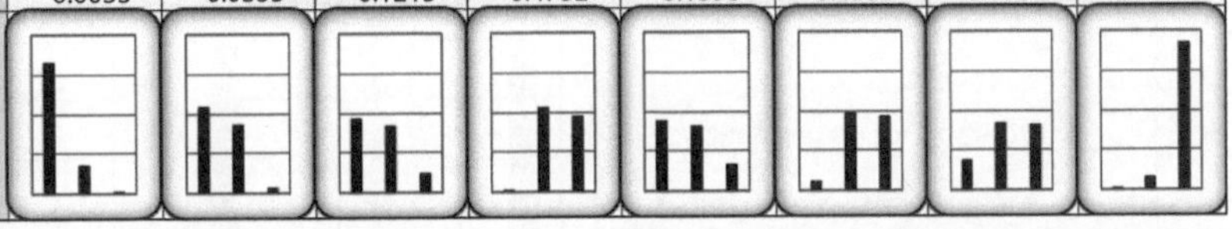							

TABLE 8.86
CPT of *C_FO*

FO_2		Acceptable		Obsolete	
FO_1		Acceptable	Obsolete	Acceptable	Obsolete
C_FO	Acceptable	0.99	0.01	0.01	0.01
	Obsolete	0.01	0.99	0.99	0.99
Histograms for conditional probabilities		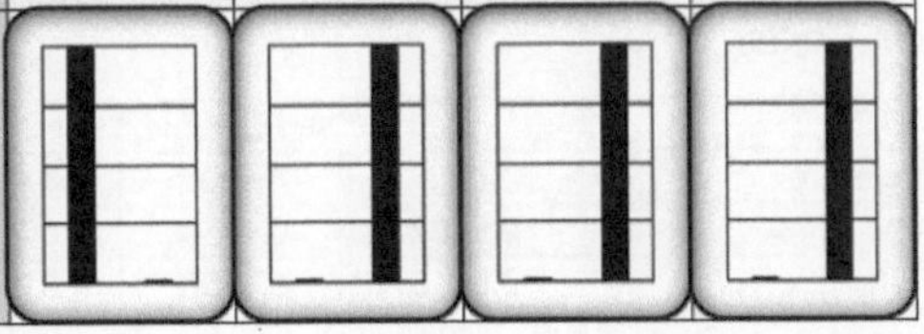			

TABLE 8.87
CPT of *CO*

C_O	Low Con.	0.8
	High Con.	0.2
Histograms for conditional probabilities		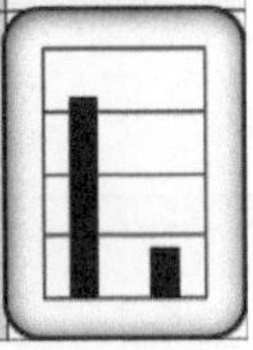

TABLE 8.88
CPT of *FO_1*

AR_3		Acceptable				Obsolete			
AR_2		Acceptable		Obsolete		Acceptable		Obsolete	
AR_1		Acceptable	Obsolete	Acceptable	Obsolete	Acceptable	Obsolete	Acceptable	Obsolete
FO_1	Acceptable	0.99	0.01	0.01	0.01	0.01	0.01	0.01	0.01
	Obsolete	0.01	0.99	0.99	0.99	0.99	0.99	0.99	0.99
Histograms for conditional probabilities		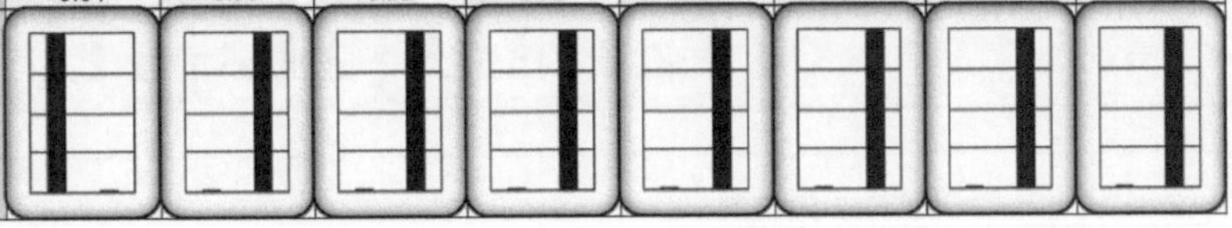							

TABLE 8.89
CPT of *FO_2*

AR_5		Acceptable		Obsolete	
AR_4		Acceptable	Obsolete	Acceptable	Obsolete
FO_2	Acceptable	0.99	0.01	0.01	0.01
	Obsolete	0.01	0.99	0.99	0.99
Histograms for conditional probabilities		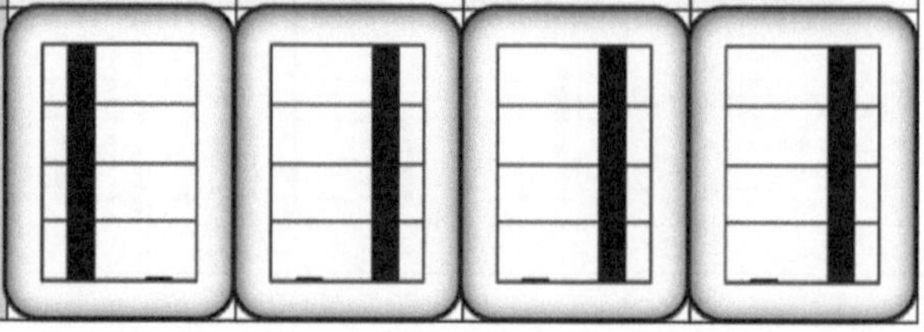			

TABLE 8.90
CPT of *Other Costs*

C-FO		Acceptable			Obsolete		
Bridge size		Small	Medium	Large	Small	Medium	Large
Other costs	Low	0.9482	0.4450	0.2710	0.5681	0.3870	0.0002
	High	0.0518	0.5550	0.7290	0.4319	0.6130	0.9998
Histograms for conditional probabilities							

TABLE 8.91
CPT of *Ri*

C		\$0 M		\$10 M		\$100 M	
C_FO		Acceptable	Obsolete	Acceptable	Obsolete	Acceptable	Obsolete
Ri	\$0 M	0.9199	0.2397	0.3553	0.0553	0.1854	0.0000
	\$10 M	0.0800	0.4050	0.5527	0.3829	0.4525	0.0117
	\$100 M	0.0001	0.3553	0.0921	0.5618	0.3621	0.9883
Histograms for conditional probabilities		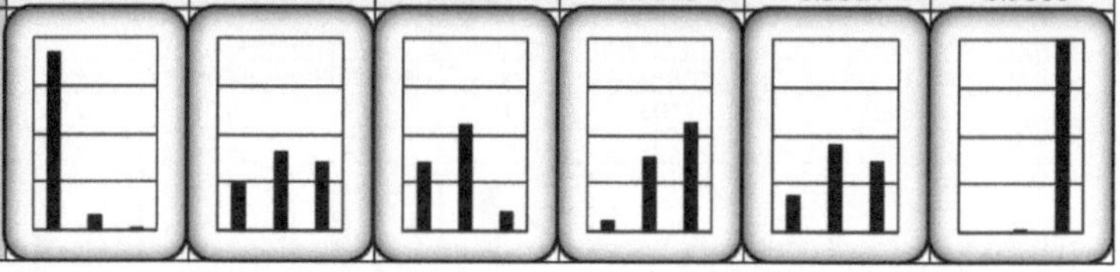					

TABLE 8.92
CPT of *Traffic Vol.*

Traffic Vol.	Low	0.6
	Med	0.3
	High	0.1
Histograms for conditional probabilities		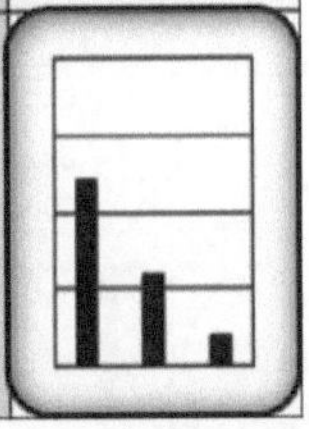

TABLE 8.93
CPT of *Traffic Costs*

Traffic Vol.		Low					
C_FO		Acceptable			Obsolete		
Bridge size		Small	Medium	Large	Small	Medium	Large
Traffic costs	Low	0.9949	0.9482	0.9206	0.8302	0.7644	0.6852
	High	0.0051	0.0581	0.0794	0.1698	0.2356	0.3148
Histograms for conditional probabilities		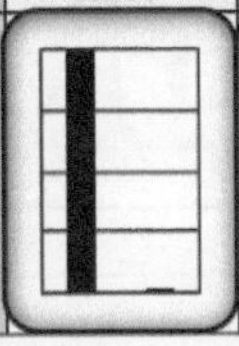	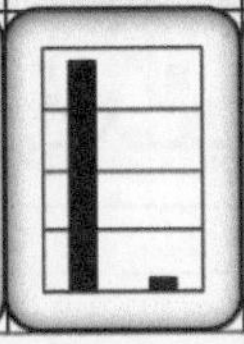	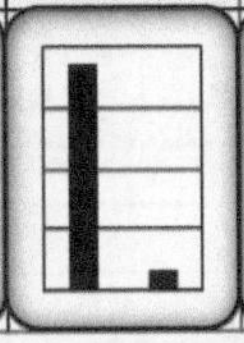	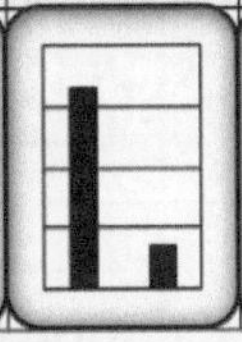	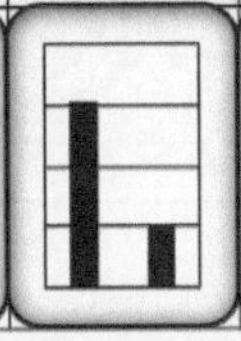	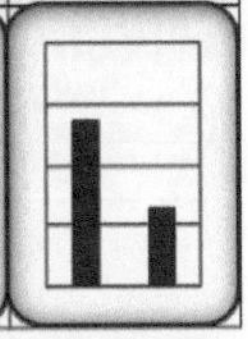

Traffic Vol.		Med					
C-FO		Acceptable			Obsolete		
Bridge size		Small	Medium	Large	Small	Medium	Large
Traffic costs	Low	0.7052	0.6410	0.5718	0.4282	0.3590	0.2948
	High	0.2948	0.3590	0.4282	0.5718	0.6410	0.7052
Histograms for conditional probabilities		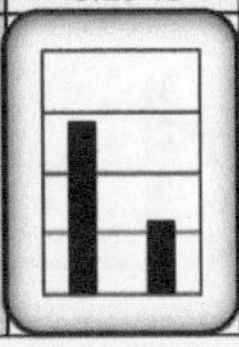	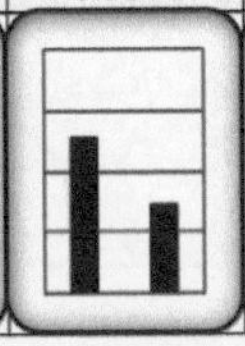	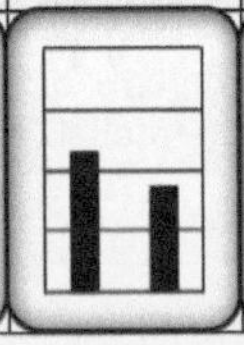	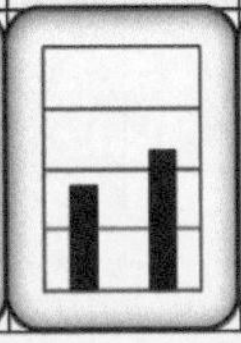	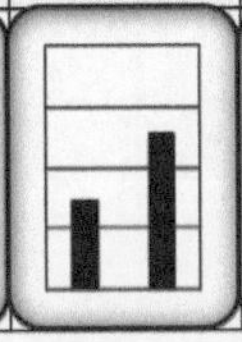	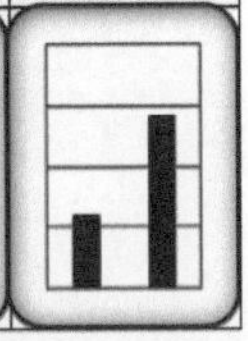

Traffic Vol.		High					
C-FO		Acceptable			Obsolete		
Bridge size		Small	Medium	Large	Small	Medium	Large
Traffic costs	Low	0.3148	0.2356	0.1698	0.0794	0.0518	0.0051
	High	0.6852	0.7644	0.8302	0.9206	0.9482	0.9949
Histograms for conditional probabilities		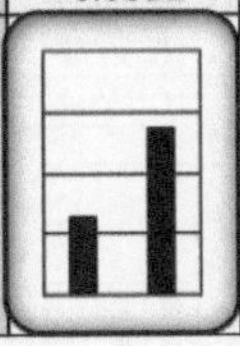	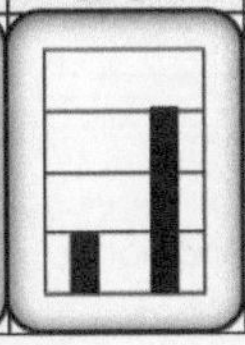	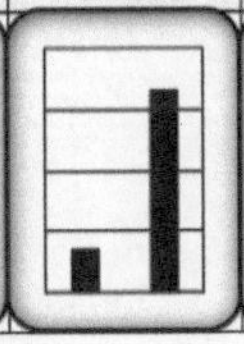	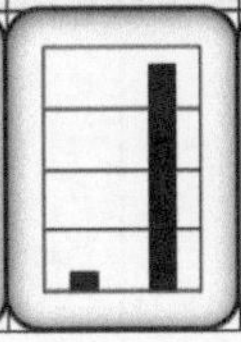	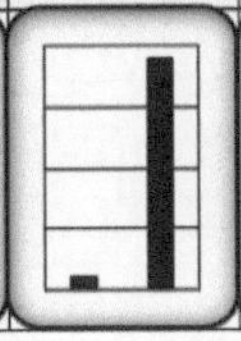	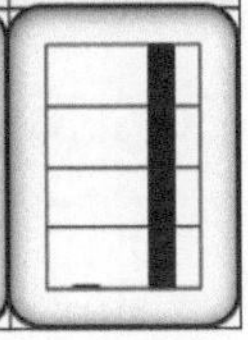

APPENDIX 8.VIII: CPTs FOR CASE STUDY 8.7 (TABLES 8.94–8.103)

TABLE 8.94
CPT of *C*

Site		High					
Soc. and Econ.		High		Med.		Low	
Use_Func.		High	Low	High	Low	High	Low
C	High	0.9675	0.5842	0.6999	0.2242	0.4963	0.1215
	Med.	0.0325	0.3758	0.2885	0.4626	0.4258	0.4148
	Low	0.0000	0.0400	0.0116	0.3133	0.0778	0.4637
Histograms for conditional probabilities		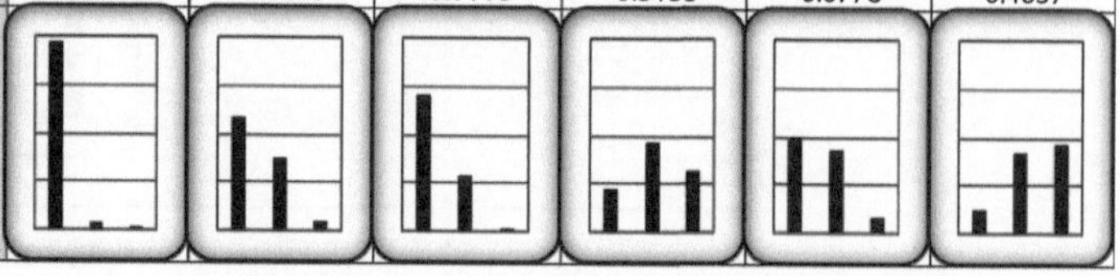					

Site		Low					
Soc. and Econ.		High		Med.		Low	
Use_Func.		High	Low	High	Low	High	Low
C	High	0.5353	0.3243	0.2242	0.0945	0.0385	0.0001
	Med.	0.4292	0.5059	0.4626	0.4493	0.3377	0.0800
	Low	0.0355	0.1698	0.3133	0.4562	0.6238	0.9199
Histograms for conditional probabilities		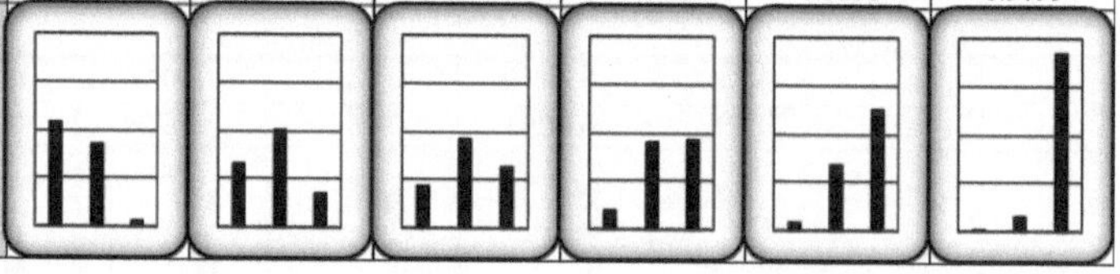					

TABLE 8.95
CPT of *Other Dmnds*

Other Dmnds	High	0.6951
	Low	0.3049
Histograms for conditional probabilities		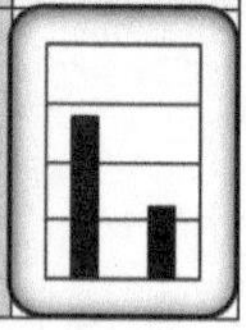

TABLE 8.96
CPT of *Ph_V*

Ph_V	High	0.5129
	Med.	0.3577
	Low	0.1294
Histograms for conditional probabilities		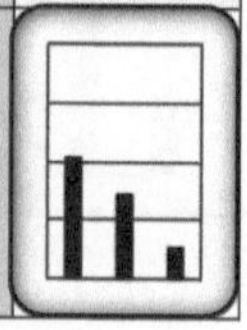

TABLE 8.97
CPT of *Ri*

T		High								
C		High			Med.			Low		
V		High	Med.	Low	High	Med.	Low	High	Med.	Low
Ri	High	0.9883	0.8683	0.6999	0.8683	0.5353	0.3216	0.3588	0.2952	0.2166
	Med.	0.0117	0.1314	0.2885	0.1314	0.4292	0.4253	0.4246	0.4096	0.4246
	Low	0.0000	0.0003	0.0116	0.0003	0.0355	0.2531	0.2166	0.2952	0.3588
Histograms for conditional probabilities										

T		Med.								
C		High			Med.			Low		
V		High	Med.	Low	High	Med.	Low	High	Med.	Low
Ri	High	0.8683	0.5783	0.1986	0.5636	0.2952	0.0754	0.2670	0.0482	0.0003
	Med.	0.1314	0.4142	0.6028	0.3610	0.4096	0.3610	0.4660	0.4707	0.1314
	Low	0.0003	0.0074	0.1986	0.0754	0.2952	0.5636	0.2670	0.4812	0.8683
Histograms for conditional probabilities		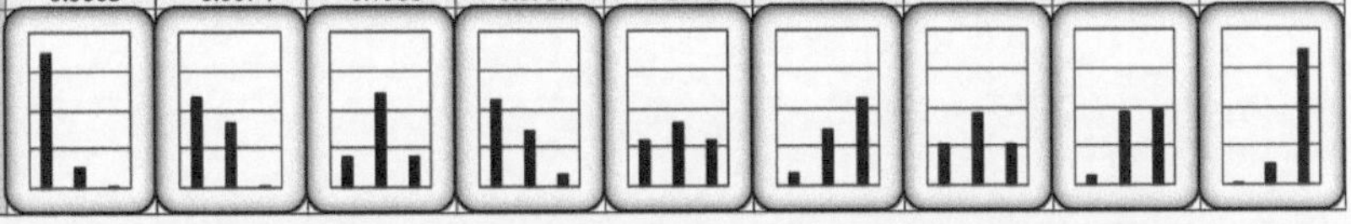								

T		Low								
C		High			Med.			Low		
V		High	Med.	Low	High	Med.	Low	High	Med.	Low
Ri	High	0.0116	0.1215	0.2166	0.2952	0.1215	0.0003	0.2166	0.0207	0.0000
	Med.	0.2885	0.4148	0.4246	0.4096	0.4148	0.1314	0.4246	0.3577	0.0117
	Low	0.6999	0.4637	0.3588	0.2952	0.4637	0.8683	0.3588	0.6216	0.9883
Histograms for conditional probabilities		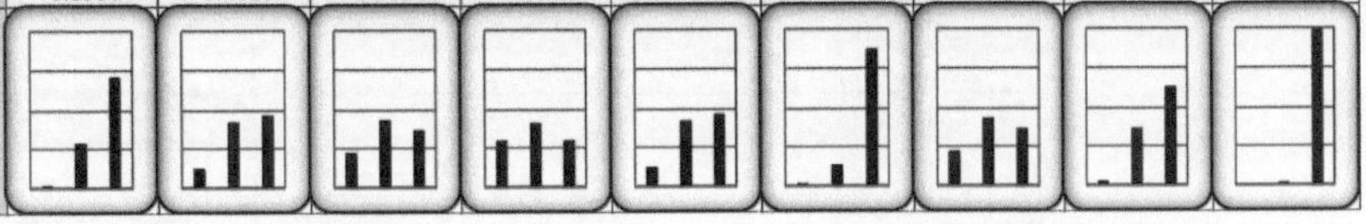								

TABLE 8.98
CPT of *Site*

Site	High	0.7082
	Low	0.2918
Histograms for conditional probabilities		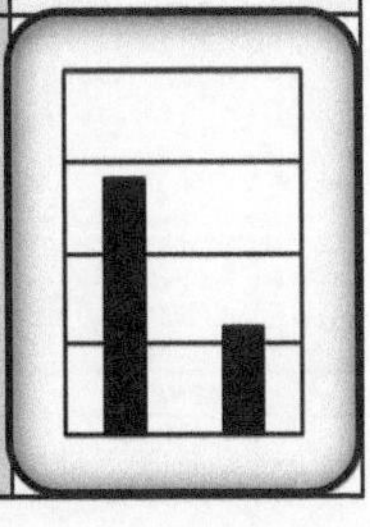

TABLE 8.99
CPT of *Soc.* and *Econ.*

Soc. and Econ.	High	0.6238
	Med.	0.3377
	Low	0.0385
Histograms for conditional probabilities		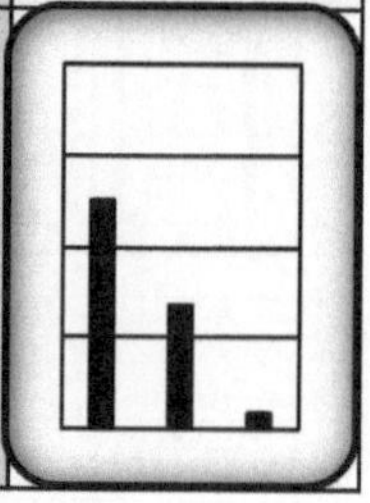

TABLE 8.100
CPT of *Subj_T*

Subj_T	High	0.7382
	Low	0.2618
Histograms for conditional probabilities		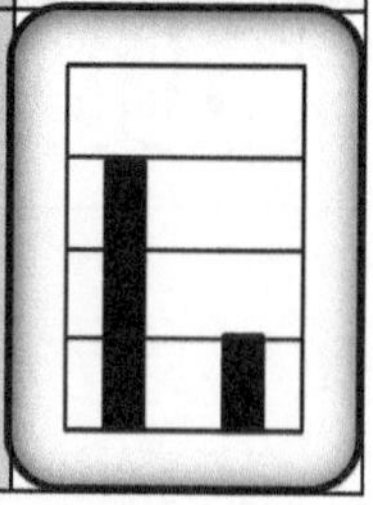

TABLE 8.101
CPT of *T*

Site		High				Low			
Subj_T		High		Low		High		Low	
Other_Dmnds		High	Low	High	Low	High	Low	High	Low
T	High	0.9883	0.5353	0.1215	0.4963	0.5618	0.1683	0.0963	0.0000
	Med.	0.0117	0.4292	0.4148	0.4258	0.3829	0.3819	0.3894	0.0117
	Low	0.0000	0.0355	0.4637	0.0778	0.0553	0.4499	0.5143	0.9883
Histograms for conditional probabilities		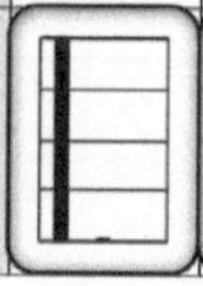	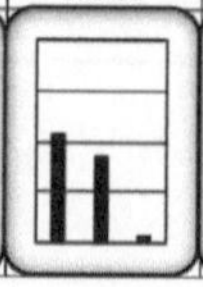	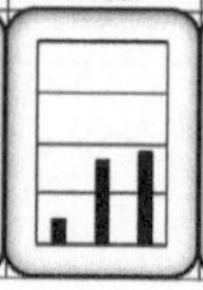	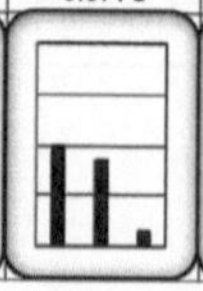	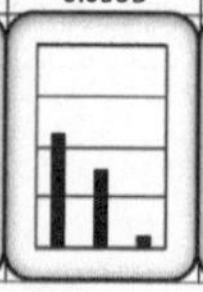	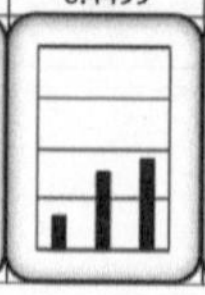	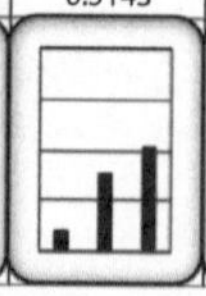	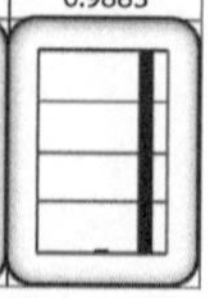

TABLE 8.102
CPT of *Use_Func.*

Use_Func.	High	0.8302
	Low	0.1698
Histograms for conditional probabilities		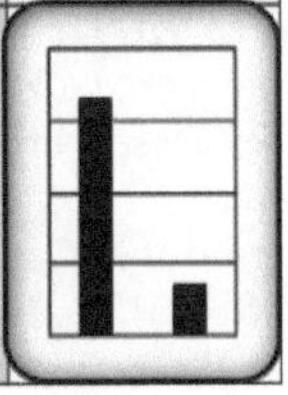

TABLE 8.103
CPT of *V*

Site		High			Low		
Ph_V		High	Med.	Low	High	Med.	Low
V	High	0.9199	0.4448	0.1511	0.5353	0.0921	0.0033
	Med.	0.0800	0.4967	0.4362	0.4292	0.5527	0.1726
	Low	0.0001	0.0585	0.4126	0.0355	0.3553	0.8240
Histograms for conditional probabilities		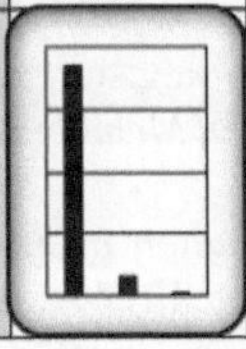	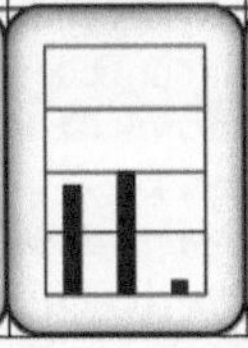	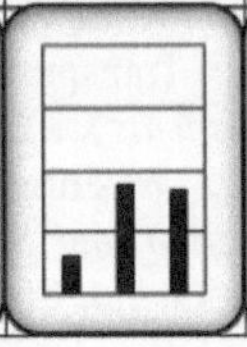	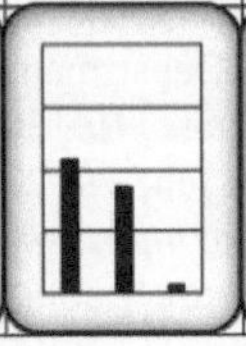	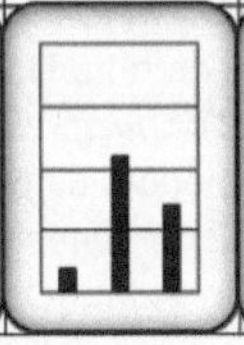	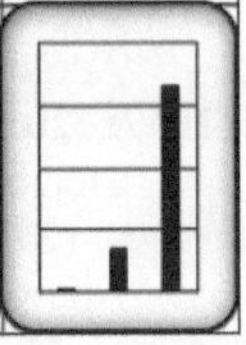

REFERENCES

AASHTO. (1993). *Guidelines for Bridge Management Systems*, American Association of State Highway and Transportation Officials, Washington, DC.

Abudayyeh, O., and Al-Battaineh, H. T. (2003). As-Built Information Model for Bridge Maintenance. *J. Comput. Civ. Eng*. 17(2), 105–112.

Adey, B., Hajdin, R., and Bruhwiler, E. (2003). Risk-Based Approach to the Determination of Optimal Interventions for Bridges Affected by Multiple Hazards. *Eng. Struct*. 25(7), 903–912.

Alampalli, S. (2014). Bridge Maintenance. In: *Bridge Engineering Handbook, Second Edition: Construction and Maintenance*, Chen, W., and Duan, L. (Eds.). Chapter 11, pp. 269–299, CRC Press, Boca Raton, FL.

Alampalli, S., and Ettouney, M. (2013). Multihazards Considerations for Cable-Bridges: A Security Viewpoint. *Proceedings the 8th International Cable Supported Bridge Operator's Conference*, June 3–5, 2013, ICSBOC 2013, Edinburgh, UK.

Alampalli, S., and Ettouney, M. (2015). Security of Suspension Cable Supported Bridges. In: *Suspension Cable Supported Bridges*, Alampalli, S., and Morrow, W. (Eds.). pp. 361–369, CRC Press, Boca Raton, FL.

Chassiakos, A. P., Vagiotas, P., and Theodorakopoulos, D. D. (2005). A Knowledge-Based System for Maintenance Planning of Highway Concrete Bridges. *Adv. Eng. Softw*. 36(11), 740–749.

Cheng, M. Y., and Hoang, N. D. (2012). Risk Score Inference for Bridge Maintenance Project Using Evolutionary Fuzzy Least Squares Support Vector Machine. *J. Comput. Civ. Eng*. 04014003.

DHS. (2008). *Risk Assessment: A How-To Guide to Mitigate Potential Terrorist Attacks*, FEMA 452, United States Department of Homeland Security, Washington, DC.

DHS. (2011a). *Integrated Rapid Visual of Screening of Buildings*, Building and Infrastructure Protection Series. DHS, Washington, DC.

DHS. (2011b). *Integrated Rapid Visual Screening of Mass Transit Stations*, Building and Infrastructure Protection Series, DHS, Washington, DC.

DHS. (2011c). *Integrated Rapid Visual Screening of Tunnels*, Building and Infrastructure Protection Series, DHS, Washington, DC.

DHS. (2011d). *Integrated Rapid Visual Screening of Buildings*, Building and Infrastructure Protection Series. United States Department of Homeland Security, Washington, DC.

Elhag, T. M. S., and Wang, Y.-M. (2007). Risk Assessment for Bridge Maintenance Projects: Neural Networks versus Regression Techniques. *J. Comput. Civ. Eng*. 21(6), 402–409.

Ellingwood, B., Smilowitz, R., Dusenberry, D., Dutinh, D., Lew, H., and Carino, N. (2007). *Best Practices for Reducing the Potential for Progressive Collapse in Buildings*, National Institute of Standards and Technology, U.S. Department of Commerce, Gaithersburg, MD.

Ettouney, M., and Alampalli, S. (2012a). *Infrastructure Health in Civil Engineering: Theory and Components*, CRC Press, Boca Raton, FL.

Ettouney, M., and Alampalli, S. (2012b). *Infrastructure Health in Civil Engineering: Applications and Management*, CRC Press, Boca Raton, FL.

Ettouney, M., and Alampalli, S. (2016). *Multihazard Considerations in Civil Infrastructure*, CRC Press, Boca Raton, FL.

Ettouney, M., Alampalli, S., and Agrawal, A. (2005). *Theory of Multihazards for Bridge Structures*, IO Press, New York.

Evans, K. G., Martin, P., Moliere, D. R., Saynor, M. J., Prendergast, J. B., and Erskine, W. D. (2004). Erosion Risk Assessment of the Jabiluka Mine Site, Northern Territory, Australia. *J. Hydrol. Eng*. 9(6), 512–522.

FEMA. (2005). *Risk Assessment: A How-To Guide to Mitigate Terrorist Attacks*, Risk Management Series, FEMA 452, Federal Emergency Management Agency, Washington FHWA, DC.

FHWA. (1995). *Recording and Coding for the Structure Inventory and Appraisal of the Nation's Bridges*, FHWA report No. FHWA-PD-96-001, FHWA, Washington, DC.

FHWA. (2004). *National Bridge Inspection Standards*, December 14, 2004. 23 CFR Part 650, Federal Highway Administration, US Department of Transportation, Federal Register, Vol. 69, No. 239, Washington, DC.

FHWA. (2012). *Clarification of Requirements for Fracture Critical Members*, Federal Highway Administration Memorandum dated June 20, 2012, Washington, DC.

FHWA. (2012). *Primer for the Inspection and Strength Evaluation of Suspension Bridge Cables*, Federal Highway Administration Publication No. FHWA-IF-11-045, FHWA, Washington, DC.

Gralund, M. S., and Puckett, J. A. (1996). System for Bridge Management in a Rural Environment. *J. Comput. Civ. Eng*. 10(2), 97–105.

Kakadu Board of Management and Parks Australia. (1998). *Kakadu National Park Plan of Management*, Australian Government Publishing Series, Canberra, Australia.

Leavitt, W., and Straight, B. (2013). *Oversized Load may be Cause of I-5 Bridge Collapse in Washington State*, FleetOwner.com. Penton Media [accessed September 23, 2013].

LeRose, C. (2001). The Collapse of the Silver Bridge, *West Virginia Historical Society Quarterly* XV(4).

Lounis, A. (2004). *Risk-Based Maintenance Optimization of Bridge Structures*, 2nd International Colloquium on Advanced Structural Reliability Analysis Network (ASRANet), International Center for Numerical Methods in Engineering, Barcelona, Spain, July 5–7, 2004, pp. 1–9.

Mahmoud, K. M. (2011). BTC Method for Evaluation of Remaining Strength and Service Life of Bridge Cables, Report C-07-11, New York State Department of Transportation, Albany, NY.

NCHRP. (2004). *Guidelines for Inspection and Strength Evaluation of Suspension Bridge Parallel-Wire Cables*, Transportation Research Board, NCHRP Report 534, Transportation Research Board, Washington, DC.

NTSB (National Transportation Safety Board). (1984). *Collapse of a Suspended Span of Route 95 Highway Bridge over the Mianus River*, NTSB, Greenwich, CT.

NTSB (National Transportation Safety Board). (1988). *Collapse of New York Thruway (1–90) Bridge over the Schoharie Creek*, Near Amsterdam, New York. Highway Accident Report: NTSB/HAR-88/02, Washington, DC.

NTSB (National Transportation Safety Board). (2008). *Highway Accident Report*, Collapse of I-35W highway bridge, NTSB, Minneapolis, MN, NTSB/HAR-08/03, PB2008–916203, Notation 7975C.

NYSDOT. (1997). *Bridge Inspection Manual (with updates through 2006)*, New York State Department of Transportation, Albany, NY.

NYSDOT. (2000). *Steel Details Vulnerability Manual*, New York State Department of Transportation, Albany, NY.

NYSDOT. (2008). *Fundamentals of Bridge Maintenance and Inspection*, New York State Department of Transportation, Albany, NY.

NYSDOT. (2010). *Inspection Flagging Procedure for Bridges*. In Engineering Instruction 10-016, New York State Department of Transportation, NYSDOT, Albany, NY.
NYSDOT. (2014). Bridge Inspection Manual, New York State Department of Transportation, Albany, NY.
PONTIS. (1995). *User's manual, Release 3.0*, Cambridge Systematics, Inc., Cambridge, MA.
Saydam, D., Frangopol, D. M., and Dong, Y. (2013). Assessment of Risk Using Bridge Element Condition Ratings. *J. Infrastruct. Syst*. 252–265.
Tehrani, P., and Mitchell, D. (2013). Seismic Risk Assessment of Four-Span Bridges in Montreal Designed Using the Canadian Bridge Design Code. *J. Bridge. Eng*. Paper No. A4014002(9).
Vamvatsikos, D., and Cornell, C. A. (2002). Incremental Dynamic Analysis. *Earthquake Eng. Struct. Dynam*. 31(3), 491–514.
Wang, Y.-M., Liu, J., and Elhag, T. M. S. (2007). A Fuzzy Group Decision Making Approach for Bridge Risk Assessment. *Comput. Ind. Eng*. 53(1), 137–148.
Williamson, E. B., Bayrak, O., Williams, G.D., Davis, C.E., Marchand, K.A., Mckay, A.E., and Wassef, W. (2010). *Blast-Resistant Highway Bridges: Design and Detailing Guidelines*. NCHRP Rep. No. 645, National Cooperative Highway Research Program, Transportation Research Board, Washington, DC.
Yi, Z., Agrawal, A., Ettouney, M., and Alampalli, S. (2014a). Blast Load Effects on Highway Bridges, Part I: Modeling and Blast Load Effects. *J. Bridge Eng*. 19(4).
Yi, Z., Agrawal, A., Ettouney, M., and Alampalli, S. (2014b). *Blast Load Effects on Highway Bridges, Part II: Failure Modes and Multihazard Correlations. J. Bridge Eng*. 19(4).

Appendix A: Truncated Normal Distributions

Normal distribution of a random variable, x, is expressed as

$$f(x) = \frac{1}{\sigma\sqrt{2\pi}} \exp\left(-\frac{(x-\mu)^2}{2\sigma^2}\right) \tag{A.1}$$

The mean and the standard deviation are μ and σ, respectively, and the variance is $V = \sigma^2$. Equation A.1 has infinite tails on both sides of the scale. In many physical applications, it is required to truncate these tails to reasonable physical limits. For a truncated normal distribution, Equation A.1 can be rewritten as (Elishakoff, 1983)

$$f(x) = \frac{A}{\sigma\sqrt{2\pi}} \exp\left(-\frac{(x-\mu)^2}{2\sigma^2}\right) \quad \text{for} \quad a \le x \le b \tag{A.2}$$

$$f(x) = 0 \quad \text{for} \quad a \le x \le b \tag{A.3}$$

$$A = 12erf(k) \tag{A.4}$$

$$erf(k) = \frac{1}{\sqrt{2\pi}} \int_0^k \exp\left(-\frac{y^2}{2}\right) dy \tag{A.5}$$

See Figure A.1 for an illustration of normal and truncated normal distributions. Truncated uniform distributions can be helpful in generating different CPTs required in BN models.

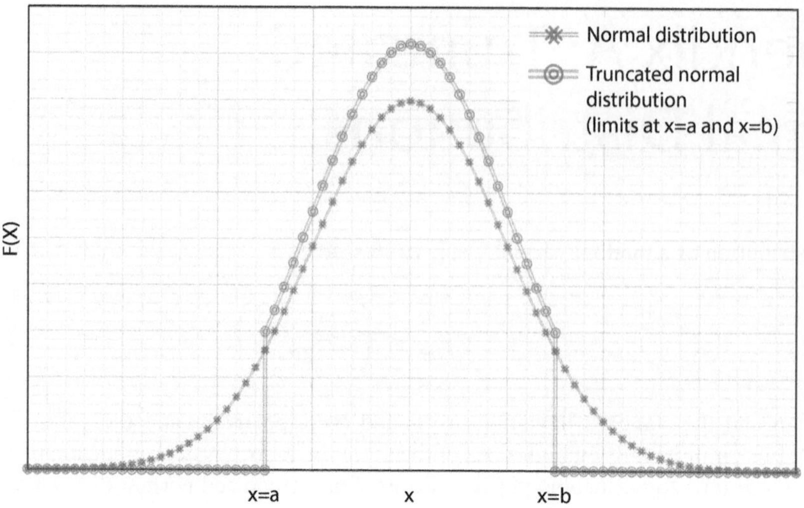

FIGURE A.1 Normal and Truncated Normal PDF.

REFERENCE

Elishakoff, I. (1983). *Probabilistic Methods to the Theory of Structures*. New York: Wiley.

Appendix B: Statistics of Histograms

B.1 OVERVIEW

Since this book uses histograms in many of its models, we offer some basic statistics of histograms which can be helpful.

For a given random variable X, we construct a histogram with a rank size of n by identifying a probability of the i-th rank, $X = x_i$, as $p(X = x_i)$ with $i = 1, 2, \ldots, n$. The values x_i as used throughout this manuscript can be subjective (e.g., *Low*, *Medium*, and *High*) or objective. Of course, $p(X = x_i)$ is always objective. In many of the examples, we use some popular statistics measures of the random variable. For convenience, we offer the formal definitions of these statistic measures below.

B.2 EXPECTED VALUES/AVERAGES

The expected value (average) of X in any given histogram is

$$E(X) = \sum_{i=1}^{i=n} (x_i)\big(p(X - x_i)\big) \tag{B.1}$$

When x_i is objective, its value is obviously the assigned numeric value. When x_i is subjective, we assume that the range of x_i is 0 to 1. We assume that each rank occupies an identical space within that range. Tables B.1–B.4 show the value of x_i for $n = 2$ to $n = 5$, respectively.

TABLE B.1
Values of x_i for $n = 2$

i-th Rank	x_i
1	0.25
2	0.75

TABLE B.2
Values of x_i for $n = 3$

i-th Rank	x_i
1	0.167
2	0.5
3	0.833

TABLE B.3
Values of x_i for $n = 4$

i-th Rank	x_i
1	.125
2	.375
3	.625
4	.875

TABLE B.4
Values of x_i for $n = 5$

i-th Rank	x_i
1	0.1
2	0.3
3	0.5
4	0.7
5	0.9

Source: The justification for those values is self-evident, it can be found in Benjamin and Cornell (1970).

REFERENCE

Benjamin, J and Cornell, A. (1970), *Probability, Statistics, and Decisions for Civil Engineers*, McGraw-Hill, New York.

Index

D

E

F

G

H

I

Q

R

S

stiffening trusses, 459
tower system, 461

T

U

V

W